Principles of Free-electron Lasers

Principles of Free-electron Lasers

H. P. Freund

Science Applications International Corporation
McLean, Virginia

and

T. M. Antonsen, Jr

University of Maryland
College Park, Maryland

CHAPMAN & HALL

London · Weinheim · New York · Tokyo · Melbourne · Madras

Published by Chapman & Hall, 2–6 Boundary Row, London SE1 8HN, UK

Chapman & Hall, 2–6 Boundary Row, London SE1 8HN, UK

Chapman & Hall GmbH, Pappelallee 3, 69469 Weinheim, Germany

Chapman & Hall USA, 115 Fifth Avenue, New York, NY 10003, USA

Chapman & Hall Japan, ITP-Japan, Kyowa Building, 3F, 2-2-1 Hirakawacho, Chiyoda-ku, Tokyo 102, Japan

Chapman & Hall Australia, 102 Dodds Street, South Melbourne, Victoria 3205, Australia

Chapman & Hall India, R. Seshadri, 32 Second Main Road, CIT East, Madras 600 035, India

First edition 1992

Second edition 1996

Typeset in 10/12 pt Times by Thomson Press (India) Ltd, New Delhi

Printed in Great Britain by Hartnolls Ltd, Bodmin, Cornwall

ISBN 0 412 72540 1

A catalogue record for this book is available from the British Library

Library of Congress Catalog Card Number: 95-71227

♾ Printed on permanent acid-free text paper, manufactured in accordance with ANSI/NISO Z39.48-1992 and ANSI/NISO Z39.48-1984 (Permanence of Paper).

To Lena Marion, Anna Jane, Tracy, Thomas Alexander, Margaret Elise and Christine Marie

Contents

Preface

The primary consideration involved in contemplating the utility of a second edition of any book is whether or not the weight of new developments in the field warrants the effort. In the case of free-electron lasers, the field has been growing so rapidly that we judged this to be the case. This rapid growth has occurred both in the number of active free-electron laser experiments and user facilities worldwide, and in the theoretical understanding of the various operating regimes. As such, we felt that the first edition was becoming out-dated, and that a second edition was necessary to give readers a complete description of the current state of the theory of free-electron lasers.

In organizing the material to be included in the second edition, we did not feel it practicable to rewrite the entire volume. If we were to start with a clean slate, then much of the new material would be incorporated directly into the existing chapters. However, in order to minimize the composition costs of the second edition, we chose in most cases to add new chapters rather than rewrite the existing material.

There were, however, compelling reasons for modifying several chapters. Firstly, while our purpose is not that of a review in which a history of the experimental development of free-electron lasers is important, we do provide (1) a brief description of selected experiments in order to illustrate and validate the theory, and (2) an overview of the primary applications of free-electron lasers. This has necessitated some revision in Chapter 1 as well as a substantial updating of the references contained therein. Secondly, the material dealing with spontaneous undulator radiation in Chapter 3 deals exclusively with an infinite and uniform transverse configuration. This is insufficient in the treatment of superradiance in Chapter 15 (new to this edition); hence, we have added some description of the spontaneous radiation in a waveguide in the new chapter. Finally, we have corrected an omission in Chapter 5 dealing with the nonlinear formulation of free-electron laser amplifiers. The first edition was written largely with separate discussions of planar and helical wiggler geometries, and Chapter 5 included discussions of the nonlinear analysis of helical wigglers in both one and three dimensions, but discussed only the three-dimensional analysis of planar wigglers. The lack of a one-dimensional analysis of planar wigglers is not a serious omission, but one which we felt it important to correct. As a result, a new section dealing with this subject has been incorporated into Chapter 5.

With these exceptions, all new material has been organized into four new chapters dealing with (1) wiggler imperfections, (2) the reversed-field configuration, (3) collective effects and (4) amplification of spontaneous emission and superradiance. These are, in our view, the primary fields in which advances have been made since the publication of the first edition.

The issue of the effects of wiggler imperfections has important practical implications in the design of free-electron lasers. A great deal of effort has been expended in the design of wigglers which minimize field imperfections, as well as in the incorporation of external steering magnets to correct for known imperfections. As a result, we felt that the inclusion of a chapter on the effects and importance of wiggler imperfections would be an important addition.

The reversed-field configuration refers to a recent experiment in which a helical wiggler field was used in conjunction with an axial solenoidal field directed anti-parallel to the wiggler. This configuration had not been studied previously since the combined effects of the two fields in this orientation would result in a reduction in the transverse wiggler-induced electron velocity and, in turn, a reduction in the gain. Indeed, this has proven to be the case; however, the reduction in gain occurred along with a relatively high efficiency. The maximum efficiency found in this experiment (which used a uniform helical wiggler) was in the neighbourhood of 27% at a frequency of 35 GHz, which compared favourably with the previous record high efficiency of 35% at the same frequency using a tapered wiggler. As a result, no second edition would be complete without the inclusion of a description of this important experiment.

The treatment of collective effects in free-electron lasers has been addressed in the first edition. It became clear in the three years since the initial publication, however, that there were still misunderstood aspects of both the importance of and subtleties in the theoretical analysis of collective effects. Hence, a chapter discussing these points was felt to be important. This includes both a discussion of the Raman regime in which the beam space-charge wave is important and of the analysis of self-electric and self-magnetic fields due to the DC charge and current densities of the beam. Both cases involve an analysis of several experiments in order to illustrate criteria for evaluation of the importance of collective effects.

The last new chapter deals with superradiance in free-electron lasers. There is some ambiguity in what this term means. In early work on free-electron lasers the term **superradiant amplifier** was used to denote an experiment in which no drive signal was imposed and the radiation grew from noise in a single pass through the wiggler. However, this type of radiation is now often referred to as Self-Amplified Spontaneous Emission (SASE), and the term superradiance is also used to refer to cases in which the radiation pulse breaks up into large-amplitude spikes. The nature of this process was still controversial at the time the first edition was published, and we chose to omit it from the discussion. Since this is no longer the case, we felt it important to include a discussion of these effects in the second edition.

Of course, it is also important to correct the inevitable typographical errors which creep into the text during the typesetting process, and we apologize to any readers of the first edition for the confusion they may have caused. Finally, we refer interested readers to the more recent proceedings of the annual free-electron laser conferences for more complete summaries of the experimental progress in the field [1–3].

Henry P. Freund
T. M. Antonsen, Jr
McLean, Virginia

REFERENCES

1. Free-electron lasers, in *Nuclear Instruments and Methods in Physics Research*, Vol. A318 (ed. Goldstein, J. C. and Newman, B. E.), North-Holland, Amsterdam (1992).
2. Free-electron lasers, in *Nuclear Instruments and Methods in Physics Research*, Vol. A331 (ed. Yamanaka, C. and Mima, K.), North-Holland, Amsterdam (1993).
3. Free-electron lasers, in *Nuclear Instruments and Methods in Physics Research*, Vol. A341 (ed. van Amersfoort, P. W., van der Slot, P. J. M. and Witteman, W. J.), North-Holland, Amsterdam (1994).

Preface to the first edition

At the time that we decided to begin work on this book, several other volumes on the free-electron laser had either been published or were in press. The earliest work of which we were aware was published in 1985 by Dr T. C. Marshall of Columbia University [1]. This book dealt with the full range of research on free-electron lasers, including an overview of the extant experiments. However, the field has matured a great deal since that time and, in our judgement, the time was ripe for a more extensive work which includes the most recent advances in the field.

The fundamental work in this field has largely been approached from two distinct and, unfortunately, separate viewpoints. On the one hand, free-electron lasers at sub-millimetre and longer wavelengths driven by low-energy and high-current electron beams have been pursued by the plasma physics and microwave tube communities. This work has confined itself largely to the high-gain regimes in which collective effects may play an important role. On the other hand, short-wavelength free-electron lasers in the infrared and optical regimes have been pursued by the accelerator and laser physics community. Due to the high-energy and low-current electron beams appropriate to this spectral range, these experiments have operated largely in the low-gain single-particle regimes. The most recent books published on the free-electron laser by Dr C. A. Brau [2] and Drs P. Luchini and H. Motz [3] are excellent descriptions of the free-electron laser in this low-gain single-particle regime. In contrast, it is our intention in this book to present a coherent description of the linear and nonlinear aspects of both the high- and low-gain regimes. In this way, we hope to illustrate the essential unity of the interaction mechanism across the entire spectral range. However, the reader should bear in mind that our own principal research interests derive from the high-gain millimetre and sub-millimetre regime, and the specific examples of experiments we describe are largely confined to this regime. In most cases, however, these cases are adequate to demonstrate the essential physics of the free-electron laser. Indeed, many of the first laboratory demonstrations of the physical principle of the free-electron laser were conducted in the microwave/millimetre-wave regime.

The organization of the book was chosen to appeal to the reader on a variety of levels. It was our intention to write a book which can be approached both by the novice seeking to begin study of the free-electron laser as well as by the expert who wishes to approach the subject at great depth. This has forced us to adopt a multi-level approach to the subject. At the lowest level, the reader unfamiliar with free-electron lasers can find an extensive description of the fundamental physics of the free-electron laser in Chapter 1. This chapter is largely taken from an article on free-electron lasers written by Drs H. P. Freund and R. K. Parker which appeared in the 1991 yearbook of the Academic Press Encyclopedia on Physical Science and Technology [4]. It includes (1) a brief history and summary of the experimental research on the free-electron laser, (2) a description of the essential operating regimes of the free-electron laser, including formulae for the linear gain, nonlinear saturation level, electron beam quality requirements, and efficiency enhancement by means of a tapered wiggler, and (3) a survey of proposed applications. In this regard, an extensive bibliography of the experimental literature is included for the reader who desires to conduct a more complete survey. At the next level, each subsequent chapter includes an introductory section which describes the essential physics to be discussed in that chapter. The highest level is contained in the bulk of each chapter which is devoted to an extensive and in-depth presentation of the appropriate subject.

At the highest level, we have not shrunk from the task of presenting a detailed derivation of each topic of interest, and have included several different methods of derivation of some important quantities. For example, we have employed the Vlasov–Maxwell equations in the study of the linear stability of the free-electron laser in Chapter 4 in the idealized one-dimensional analysis of both the low-gain and high-gain regimes. The purpose of this is twofold. In the first place, it serves to illustrate the relationship between these two operating regimes. In the second place, it allows the effect of a beam thermal spread on the linear gain to be analysed. In contrast, the Vlasov–Maxwell formalism is retained in the three-dimensional stability analysis only in the high-gain regime. The small-signal gain in the three-dimensional analysis is treated by the more conventional approach based upon a phase average of electron motion in the ponderomotive wave formed by the beating of the wiggler and radiation fields. The result obtained by this method, however, is a straightforward extension of that found in the idealized one-dimensional approach, and includes a filling factor which describes the overlap of the electron beam and the radiation field.

The bulk of free-electron laser designs have employed wigglers with either helical or planar symmetry. Heretofore, texts dealing with free-electron lasers typically have concentrated on a discussion of the physics of the interaction for one or the other wiggler geometry, with a brief discussion of the generalization required to obtain the results for the other. However, there are essential differences in the character of the interaction for each of these wiggler designs. Hence, we have chosen to present the linear and nonlinear analyses of each of these configurations. While this approach adds to the length of the presentation, we feel that this is necessary in order to treat the field in adequate depth.

The essential focus of the book is on the theory of free-electron lasers. While the meaning of this term has sometimes been extended to alternate concepts such as cyclotron and Cerenkov masers which also make use of free electrons, we use the term here to refer solely to devices which rely upon a periodic magnetic field to mediate the interaction of the electron beam with the electromagnetic wave. In this regard, we do not include a detailed description of the wide range of experiments conducted in the field. The only experiments which are described in the text are those which illustrate some essential point of the discussion. We have chosen this course both because of constraints on the length of the manuscript and because this aspect of the field is subject to rapid changes due to advances in the technology of electron beams, wiggler designs and optics. Similarly, we have not included examples of the application of the theory of free-electron lasers to speculative designs of coherent ultraviolet and X-ray sources, although the theoretical tools for analysing such designs are included. Such devices are under study at the present time, but are limited by the technology related to the production of high-brightness electron beams and mirrors with high reflectivities at these wavelengths.

More appropriate places for a broad discussion of the experimental base and speculative designs are review papers on the field as well as the proceedings of the annual free-electron laser conferences. We recommend the interested reader to the recent article by C. W. Roberson and P. Sprangle [5] for an extensive summary of the experimental base. In addition, excellent year-by-year summaries of the experimental and speculative literature are to be found in the proceedings of the annual free-electron laser conferences dating back to 1977 [6–17].

Henry P. Freund
T. M. Antonsen, Jr
McLean, Virginia

REFERENCES

1. Marshall, T. C. (1985) *Free-electron Lasers*, Macmillan, New York.
2. Brau, C. A. (1990) *Free-electron Lasers*, Academic Press, Boston.
3. Luchini, P. and Motz, H. (1990) *Undulators and Free-electron Lasers*, Clarendon Press, Oxford.
4. Freund, H. P. and Parker, R. K. (1991) Free-electron Lasers, in *The 1991 Yearbook of the Encyclopedia of Physical Science and Technology*, Academic Press, San Diego, pp. 49–71.
5. Roberson, C. W. and Sprangle, P. (1989) A review of free-electron lasers. *Phys. Fluids*, **B1**, 3.
6. *The Physics of Quantum Electronics: Novel Sources of Coherent Radiation*, Vol. 5 (ed. Jacobs, S. F., Sargent, M. and Scully, M. O.), Addison-Wesley, Reading, Massachusetts (1978).
7. *The Physics of Quantum Electronics: Free-Electron Generators of Coherent Radiation*, Vol. 7 (ed. Jacobs, S. F., Pilloff, H. S., Sargent, M., Scully, M. O. and Spitzer, R.), Addison-Wesley, Reading, Massachusetts (1980).
8. *The Physics of Quantum Electronics: Free-Electron Generators of Coherent Radiation*, Vols. 8 and 9 (ed. Jacobs, S. F., Moore, G. T., Pilloff, H. S., Sargent, M., Scully, M. O. and Spitzer, R.), Addison-Wesley, Reading, Massachusetts (1982).
9. *Bendor Free-electron Laser Conference* (ed. Deacon, D. A. G. and Billardon, M.), *Journal de Physique Colloque*, **C1-44** (1983).

10. *Free-electron Generators of Coherent Radiation* (ed. Brau, C. A., Jacobs, S. F. and Scully, M. O.) Proc. SPIE 453, Bellingham, Washington (1984).
11. Free-electron lasers in *Nuclear Instruments and Methods in Physics Research*, Vol. A237 (ed. Madey, J. M. J. and Renieri, A.) North-Holland, Amsterdam (1985).
12. Free-electron lasers in *Nuclear Instruments and Methods in Physics Research*, Vol. A250 (ed. Scharlemann, E. T. and Prosnitz, D.) North-Holland, Amsterdam (1986).
13. Free-electron lasers in *Nuclear Instruments and Methods in Physics Research*, Vol. A259 (ed. Poole, M. W.) North-Holland, Amsterdam (1987).
14. Free-electron lasers in *Nuclear Instruments and Methods in Physics Research*, Vol. A272 (ed. Sprangle, P., Tang, C. M. and Walsh, J.) North-Holland, Amsterdam (1988).
15. Free-electron lasers in *Nuclear Instruments and Methods in Physics Research*, Vol. A285 (ed. Gover, A. and Granatstein, V. L.) North-Holland, Amsterdam (1989).
16. Free-electron lasers in *Nuclear Instruments and Methods in Physics Research*, Vol. A296 (ed. Elias, L. R. and Kimel, I.) North-Holland, Amsterdam (1990).
17. Free-electron lasers in *Nuclear Instruments and Methods in Physics Research*, Vol. A304 (ed. Buzzi, J. M. and Ortega, J. M.) North-Holland, Amsterdam (1991).

Acknowledgements

The contents of this book are based upon the expertise developed by the authors over several years of research in the field of coherent radiation sources in general, and of free-electron lasers in particular. As such, we would like to express our appreciation to our many collaborators who, in a very real sense, made this book possible. These include Drs Robert H. Jackson, Dean E. Pershing, Robert K. Parker, Phillip Sprangle, Cha-Mei Tang, Victor L. Granatstein, Steven Gold, Charles W. Roberson, Edward Stanford and Raymond Gilbert. Particular gratitude is due to Dr Achintya K. Ganguly for his collaboration in the analysis of free-electron lasers in three dimensions, and Dr Baruch Levush for collaboration in the analysis of free-electron laser oscillators.

Notation

The choice of notation in any discussion of free-electron lasers can present some ambiguities since the symbols have not yet become standardized. For example, different authors tend to use either a_w or K to denote the wiggler parameter. Here, we choose to employ a_w for the wiggler parameter, and to reserve K for the pendulum constant associated with the nonlinear pendulum equation which describes the electron motion in the ponderomotive potential. Another source of ambiguity is our use of N_w to denote either (1) the number of periods in the interaction region, and (2) the number of periods in the tapered-entrance region of the wiggler field. The first use is the more standard of the two; however, the latter is sometimes used in the literature describing the nonlinear simulation of the interaction. Hence, we have chosen to retain both uses of this symbol at the same time being careful to specify the choice as it appears. In practice, the latter use of this symbol is confined to Chapter 5 which discusses the nonlinear analysis of free-electron laser amplifiers. In order to avoid ambiguity as much as possible, we have endeavoured to define symbols whenever they are used, and not just at their first appearance. The following is an abbreviated list of the symbols used in the text.

A_b	Cross-sectional area of the electron beam
A_g	Cross-sectional area of a waveguide
a_w	Wiggler parameter ($\equiv eA_w/m_ec^2$)
A_w	Magnitude of the wiggler vector potential
$\mathbf{A}_w$	Vector potential of the wiggler field
$\alpha_{x,y}$	Ratio of $v_{x,y}/v_\parallel$ for the quasi-steady-state trajectories in planar wigglers
B_0	Magnitude of the axial magnetic field
B_w	Magnitude of the wiggler field
$\mathbf{B}_w$	Wiggler magnetic field
β_w	Coupling coefficient ($\equiv v_w/v_\parallel$)
c	Speed of light *in vacuo*
δa	Normalized vector potential ($\equiv e\delta A/m_ec^2$)
$\delta a_e(t_e)$	Normalized radiation amplitude entering the interaction region of an oscillator at time t_e ($\equiv e\delta A_e(t_e)/m_ec^2$)
δA	Magnitude of the vector potential of the electromagnetic field

$\delta A_e(t_e)$	Radiation amplitude entering the interaction region of an oscillator at time t_e
$\delta \boldsymbol{B}$	The radiation magnetic field
$\delta \boldsymbol{E}$	The radiation electric field
$\delta\phi$	Normalized scalar potential ($\equiv e\delta\phi/m_e c^2$)
$\delta\Phi$	Scalar potential of beam space-charge waves
$\delta\omega_0$	The beam-induced frequency shift for the $m=0$ cavity mode in an oscillator
Δv	Spectral width of a single cavity mode induced by shot noise
e	Magnitude of the electronic charge
$\hat{\boldsymbol{e}}_1, \hat{\boldsymbol{e}}_2, \hat{\boldsymbol{e}}_3$	Basis vectors defining the frame rotating with a helical wiggler (i.e. the wiggler frame)
E_1	Exponential integral function
erf	Error function
ε	Oscillator slippage parameter
$g(\lvert X \rvert)$	Nonlinear gain function for the klystron model
G_m	Linear gain for the mth cavity mode in the oscillator model
$G_{nl}(\lvert X_m \rvert, p_m)$	Nonlinear gain for the mth cavity mode in the oscillator model
γ	Relativistic factor [$\equiv (1-v^2/c^2)^{-1/2}$]
$\gamma_\parallel$	Relativistic factor for the axial velocity [$\equiv (1-v_\parallel^2/c^2)^{-1/2}$]
γ_r	Resonant relativistic factor
η	Efficiency
$I_b(t)$	Normalized beam current at time t in the oscillator model [$\equiv -2\pi e J(t) a_w^2 L^3 \omega_\gamma / m_e v_\parallel^3 v_g \gamma_r$]
I_l	Modified Bessel function of the first kind
I_{st}	Minimum normalized oscillator start current
$J(t)$	Electron beam current density entering the interaction region at time t in the oscillator model
J_l	Regular Bessel function of the first kind
k	Wavenumber
k_{ln}	Wavenumber of the TE_{ln} or TM_{ln} modes
k_w	Wiggler wavenumber [$\equiv 2\pi/\lambda_w$]
K	Pendulum constant
κ_{ln}	Cutoff wavenumber of the TE_{ln} or TM_{ln} modes
L	Length of the interaction region (i.e. wiggler)
L_c	Length of the optical cavity in an oscillator
λ	Wave period
λ_w	Wiggler period
m_e	Electron rest mass
n_b	Ambient electron density
N_T	Total number of electrons within the interaction region
N_w	This can be either (1) the number of wiggler periods in the

	interaction region, or (2) the number of wiggler periods in the entry taper region (Chapter 5)
N_*	Number of electrons transiting the interaction region in time T_d
ω	Angular frequency
ω_b	Electron beam plasma frequency
ω_γ	$(\equiv(\omega/v_\parallel)dv_\parallel/d\gamma_r)$
ω_n	Natural frequency of the nth mode of an oscillator cavity
Ω_0	Relativistic cyclotron frequency corresponding to the axial magnetic field $(\equiv eB_0/\gamma m_e c)$
$\hat{\Omega}_0$	Nonrelativistic cyclotron frequency corresponding to the axial magnetic field $(\equiv eB/m_e c)$
$\hat{\Omega}_w$	Nonrelativistic cyclotron frequency corresponding to the wiggler field $(\equiv eB_w/m_e c)$
Ω_w	Relativistic cyclotron frequency corresponding to the wiggler field $(\equiv eB_w/\gamma m_e c)$
Ω_β	Betatron frequency
ψ	Ponderomotive phase
$p_w, p_\parallel$	Axial and transverse momenta for the steady-state orbits
P_x, P_y, P_z	Canonical momenta
R	Round-trip reflection coefficient for radiation in an oscillator cavity
R_g	Waveguide radius
$t_e(z, t)$	Entrance time of electrons which reach axial position z at time t
T	Round-trip time for radiation in an oscillator cavity $(\equiv L_c/v_g)$
T_a	Arrival time of successive beam micropulses in the oscillator model
T_d	Decay time for radiation in an oscillator cavity
T_δ	Cavity detuning time
$T_m(n)$	Normalized temperature of the mth cavity mode
T_p	Characteristic duration time of micropulses in the oscillator model
$T(\zeta), T_n(\zeta)$	Free-electron laser thermal functions
$\tau(z), \tau(z', z)$	Lagrangian time coordinates
v_1, v_2, v_3	Electron velocity components in the wiggler frame
v_g	Radiation group velocity
v_{ph}	Phase velocity of the ponderomotive wave
$v_\parallel, v_w$	Axial and transverse velocities for the steady-state orbits
$X(t_e)$	Normalized radiation amplitude for oscillator model $(\equiv \omega\omega_\gamma a_w L^2 \delta a_e(t_e)/\gamma_r v_\parallel)$

1 Introduction

In its fundamental concept, the free-electron laser is an extremely adaptable light source which can produce high-power coherent radiation across virtually the entire electromagnetic spectrum. In contrast, gas and solid-state lasers generate light at well-defined wavelengths corresponding to discrete energy transitions within atoms or molecules in the lasing media. Dye lasers are tunable over a narrow spectral range but require a gas laser for optical pumping and operate at relatively low power levels. Further, while conventional lasers typically are characterized by energy conversion efficiencies of only a few per cent, theoretical calculations indicate that the free-electron laser is capable of efficiencies as high as 65% while efficiencies of 40% have been demonstrated in the laboratory.

Applications of free-electron lasers to date range from experiments in solid-state physics to molecular biology, and novel designs are under development for such diverse purposes as communications, radar and ballistic missile defence. At the present time, however, free-electron lasers have been confined largely to the laboratory. Most have been built around available electron accelerators, and although they have the potential to emit light anywhere from the microwave to the ultraviolet, researchers have encountered difficulties in getting them to lase at visible and shorter wavelengths. Only recently have free-electron lasers begun to come into their own as accelerators are designed for their specific needs, and user facilities are set up so that researchers in other disciplines can take advantage of this new source of intense and tunable light.

The free-electron laser was first conceived almost four decades ago and has since operated over a spectrum ranging from microwaves through visible light. There are plans to extend this range to the ultraviolet as well. In a free-electron laser, high-energy electrons emit coherent radiation, as in a conventional laser, but the electrons travel in a beam through a vacuum instead of remaining in bound atomic states within the lasing medium. Because the electrons are free streaming, the radiation wavelength is not constrained by a particular transition between two discrete energy levels. In quantum mechanical terms, the electrons radiate by transitions between energy levels in the continuum and, therefore, radiation is possible over a much larger range of frequencies than is found in a conventional laser. However, the process can be described by classical electromagnetic theory alone.

The radiation is produced by an interaction among three elements: the electron beam, an electromagnetic wave travelling in the same direction as the electrons, and an undulatory magnetic field produced by an assembly of magnets known as a **wiggler** or **undulator**. The distinction in use between these terms is arbitrary; however, 'wiggler' is generally used to describe the periodic magnets in free-electron lasers, while 'undulator' is used for incoherent synchrotron light sources. The wiggler magnetic field acts on the electrons in such a way that they acquire an undulatory motion. The acceleration associated with this curvilinear trajectory is what makes radiation possible. In this process, the electrons lose energy to the electromagnetic wave which is amplified and emitted by the laser. The tunability of the free-electron laser arises because the wavelength of light required for the interaction between these three elements is determined by both the periodicity of the wiggler field and the energy of the electron beam.

Although the basic principle underlying the free-electron laser is relatively simple, the practical application of the concept can be difficult. In 1951 Hans Motz [1–3] of Stanford University first calculated the emission spectrum from an electron beam in an undulatory magnetic field. At the time, coherent optical emission was not expected due to the difficulty of bunching the electron beam at short wavelengths; however, it was recognized that maser (Microwave Amplification through Stimulated Emission of Radiation) operation was possible. Experiments performed by Motz and coworkers shortly thereafter produced both incoherent radiation in the blue-green part of the spectrum, and coherent emission at millimetre wavelengths. The application of undulatory magnetic fields to the maser was invented independently by Robert Phillips [4, 5] in 1957 in search of higher power than was currently available from microwave tubes. The term **ubitron** was used at this time as an acronym for Undulating Beam Interaction. Over the succeeding seven years, Phillips performed an extensive study of the interaction and pioneered many innovative design concepts in use today. Whereas the original microwave experiment at Stanford observed an output power of 1–10 W, Phillips achieved 150 kW at a 5 mm wavelength. However, the full potential of the free-electron laser was unrecognized, and the ubitron programme was terminated in 1964 due to a general shift in interest from vacuum electronics to solid-state physics and quantum electronics.

A resurgence of interest in the concept began in the mid-1970s when the term 'free-electron laser' was coined in 1975 by John Madey [6, 7] to describe an experiment at Stanford University. This experiment [8–10] produced stimulated emission in the infrared spectrum at a wavelength of 10.6 μm using an electron beam from a radio-frequency linear accelerator (r.f. linac). The first optical free-electron laser [11–14] was built using the ACO storage ring at the Université de Paris Sud, and has been tuned over a broad spectrum. More recently, stimulated emission at visible and ultraviolet wavelengths [15] has been reported using the VEPP storage ring at Novosibirsk in Russia. Visible-wavelength free-electron lasers have also been built both at Stanford University [16–18] and by a Boeing Aerospace/Los Alamos National Laboratory collaboration [19–26] based upon r.f. linacs. The r.f. linac has also been the basis for a long-standing infrared free-electron laser experimental programme at Los Alamos [27–48].

In parallel with the work at Stanford, experimenters at several laboratories began work on microwave free-electron lasers, successors to the ubitron. Those projects, at the Naval Research Laboratory [49–64], Columbia University [65–72], the Massachusetts Institute of Technology [73–84], Lawrence Livermore National Laboratory [85–91], TRW [92–4], the École Polytechnique [95, 96] in France, and Hughes [97, 98] differed from the original work by Phillips by using intense relativistic electron beams with currents of the order of a kiloampere and voltages in excess of a megavolt. The principal goal of this effort was the production of high absolute powers, and the results ranged from a peak power of the order of 2 MW at a wavelength of 2.5 mm at Columbia, through 70 MW at a 4 mm wavelength at the Naval Research Laboratory, to a maximum power figure of 1 GW obtained by Livermore at an 8 mm wavelength. This latter result represents an efficiency (defined as the ratio of the output radiation power to the initial electron beam power) of 35%, and was made possible by the use of a nonuniform wiggler field.

At the present time, free-electron lasers have been constructed over the entire

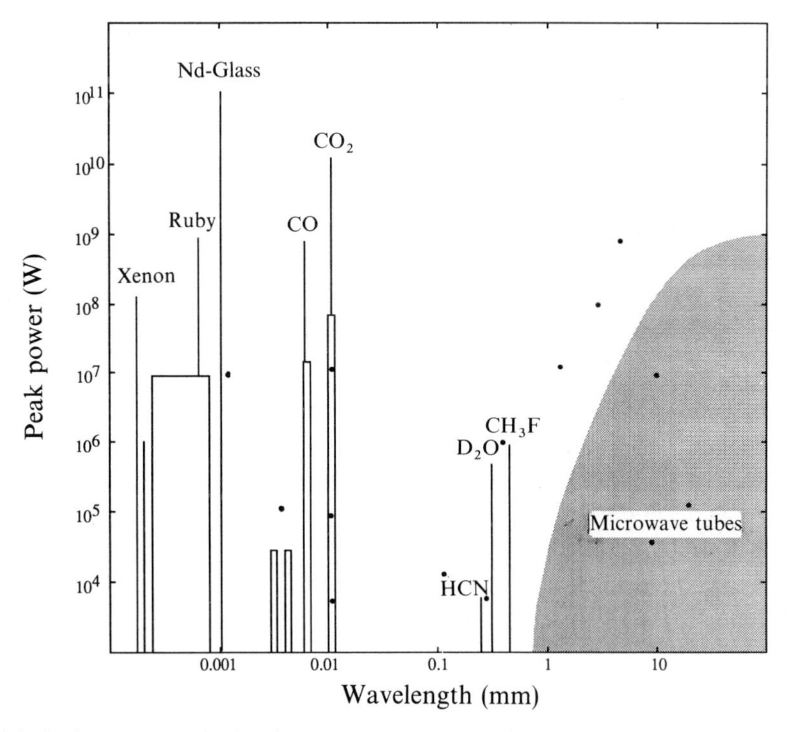

Fig. 1.1 An important criterion in the judgement of radiation sources is the peak power available at specific wavelengths. A comparison of free-electron lasers (represented by dots) with conventional lasers and microwave sources is shown.

electromagnetic spectrum. This spectral range is summarized in Fig. 1.1 in which we plot the peak power of a sample of conventional lasers, microwave tubes and free-electron lasers as a function of wavelength. At wavelengths above 0.1 mm, free-electron lasers already either match or exceed power levels obtainable from conventional technology. At shorter wavelengths, conventional lasers can be found with higher powers than are currently available from free-electron lasers. However, free-electron laser technology is maturing rapidly, and this situation is likely to change in the future.

1.1 PRINCIPLES OF OPERATION

An electron beam which traverses an undulatory magnetic field emits incoherent radiation. Indeed, this is the mechanism employed in synchrotron light sources. In conventional terminology, the periodic magnetic field in synchrotron light sources is referred to as an undulator while that used in free-electron lasers is called a wiggler, although there is no fundamental difference between them. It is necessary for the electron beam to form coherent bunches in order to give rise to the stimulated emission required for a free-electron laser. This can occur when a light wave traverses an undulatory magnetic field such as a wiggler because the spatial variations of the wiggler and the electromagnetic wave combine to produce a beat wave, which is essentially an interference pattern. It is the interaction between the electrons and this beat wave which gives rise to the stimulated emission in free-electron lasers.

This beat wave has the same frequency as the light wave, but its wavenumber is the sum of the wavenumbers of the electromagnetic and wiggler fields. With the same frequency, but a larger wavenumber (and thus a shorter wavelength), the beat wave travels more slowly than the light wave; for this reason it is called a **ponderomotive** wave. Since the ponderomotive wave is the combination of the light wave and the stationary (or magnetostatic) field of the wiggler, it is the effective field experienced by an electron as it passes through the free-electron laser. In addition, since the ponderomotive wave propagates at less than the speed of light *in vacuo* it can be in synchronism with the electrons which are limited by that velocity. Electrons moving in synchronism with the wave are said to be in **resonance** with it and will experience a constant field – that of the portion of the wave with which they are travelling. In such cases, the interaction between the electrons and the ponderomotive wave can be extremely strong.

A good analogy to the interaction between the electrons and the ponderomotive wave is that of a group of surfers and a wave approaching a beach. If the surfers remain stationary in the water the velocity difference between the wave and the surfers is large, and an incoming wave will merely lift them up and down briefly and then return them to their previous level. There is no bulk, or average, translational motion or exchange of energy between the surfers and the wave. But if the surfers 'catch the wave' by paddling so as to match the speed of the wave, then they can gain significant momentum from the wave and be carried inshore. This is the physical basis underlying the resonant interaction in a free-electron laser. However, in a free-electron laser, the electrons amplify the wave, so the situation is more analogous to the surfers 'pushing' on the wave and increasing its amplitude.

The frequency of the electromagnetic wave required for this resonant interaction can be determined by matching the velocities of the ponderomotive wave and the electron beam. This is referred to as the phase-matching condition. The interaction is one in which an electromagnetic wave characterized by an angular frequency ω and wavenumber k and the magnetostatic wiggler with a wavenumber k_w produce a beat wave with the same frequency as the electromagnetic wave but a wavenumber equal to the sum of the wavenumbers of the wiggler and electromagnetic waves (i.e. $k + k_w$). The velocity of the ponderomotive wave is given by the ratio of the frequency of the wave to its wavenumber. As a result, matching this velocity to that of the electron beam gives the resonance condition in a free-electron laser

$$\frac{\omega}{k + k_w} \cong v_z, \tag{1.1}$$

for a beam with a bulk streaming velocity v_z in the z-direction. The z-direction is used throughout to denote both the bulk streaming direction of the electron beam and the symmetry axis of the wiggler field. The dispersion relation between the frequency and wavenumber for waves propagating in free space is $\omega \cong ck$, where c denotes the speed of light *in vacuo*. Combination of the free-space dispersion relation and the free-electron laser resonance condition gives the standard relation for the wavelength as a function of both the electron beam energy and the wiggler period

$$\lambda \cong \frac{\lambda_w}{2\gamma_z^2}, \tag{1.2}$$

where $\gamma_z = (1 - v_z^2/c^2)^{-1/2}$ is the relativistic time-dilation factor which is related to the electron streaming energy, and $\lambda_w = 2\pi/k_w$ is the wiggler wavelength. The wavelength, therefore, is directly proportional to the wiggler period and inversely proportional to the square of the streaming energy. This results in a broad tunability which permits the free-electron laser to operate across virtually the entire electromagnetic spectrum.

How do a magnetostatic wiggler and a forward-propagating electromagnetic wave, both of whose electric and magnetic fields are directed transversely to the direction of propagation, give rise to an axial ponderomotive force which can extract energy from the electron beam? The wiggler is the predominant influence on the electron's motion. In order to understand the dynamical relationships between the electrons and the fields, consider the motion of an electron in a helically symmetric wiggler field. An electron propagating through a magnetic field experiences a force which acts at right angles to both the direction of the field and to its own velocity. The wiggler field is directed transversely to the direction of bulk motion of the electron beam and rotates through 360° in one wiggler period. An electron streaming in the axial direction, therefore, experiences a transverse force and acquires a transverse velocity component upon entry into the wiggler. The resulting trajectory is helical and describes a bulk streaming along the axis of symmetry as well as a transverse circular rotation that lags 180° behind the phase of the wiggler field. The magnitude of the transverse wiggler velocity, denoted by v_w, is proportional to the product of the wiggler amplitude and period. This relationship may be expressed in the form

$$\frac{v_w}{c} \cong 0.934 \frac{B_w \lambda_w}{\gamma_b}, \tag{1.3}$$

where the wiggler period is expressed in units of centimetres, B_w denotes the wiggler amplitude in tesla, and $\gamma_b = 1 + E_b/m_e c^2$ denotes the relativistic time-dilation factor associated with the total kinetic energy E_b of the electron beam (where m_e denotes the rest mass of the electron, and $m_e c^2$ denotes the electron rest energy).

Since the motion is circular, both axial and transverse velocities have a constant magnitude. This is important because the resonant interaction depends upon the axial velocity of the beam. In addition, since the wiggler induces a constant-magnitude transverse velocity, the relation between the total electron energy and the streaming energy can be expressed in terms of the time-dilation factors in the form

$$\gamma_z \cong \frac{\gamma_b}{\sqrt{(1 + 0.872\, B_w^2 \lambda_w^2)}}. \tag{1.4}$$

As a result, the resonant wavelength depends upon the total beam energy, and the wiggler amplitude and period through

$$\lambda \cong (1 + 0.872\, B_w^2 \lambda_w^2) \frac{\lambda_w}{2\gamma_b^2}. \tag{1.5}$$

It is the interaction between the transverse wiggler-induced velocity and the transverse magnetic field of an electromagnetic wave that induces a force normal to both in the axial direction. This is the ponderomotive force. The transverse velocity and the radiation magnetic field are directed at right angles to each other and undergo a simple rotation about the axis of symmetry. A resonant wave must be circularly polarized with a polarization vector that is normal to both the transverse velocity and the wiggler field and which rotates in synchronism with the electrons. This synchronism is illustrated in Fig. 1.2, and is maintained by the aforementioned resonance condition.

In order to understand the energy transfer, we return to the surfer analogy and consider a group of surfers attempting to catch a series of waves. In the attempt to match velocities with the waves, some will catch a wave ahead of the crest and slide forward while others will catch a wave behind the crest and slide backward. As a result, clumps of surfers will collect in the troughs of the waves. Those surfers which slide forward ahead of the wave are accelerated and gain energy at the expense of the wave, while those that slide backward are decelerated and lose energy to the wave. The wave grows if more surfers are decelerated than

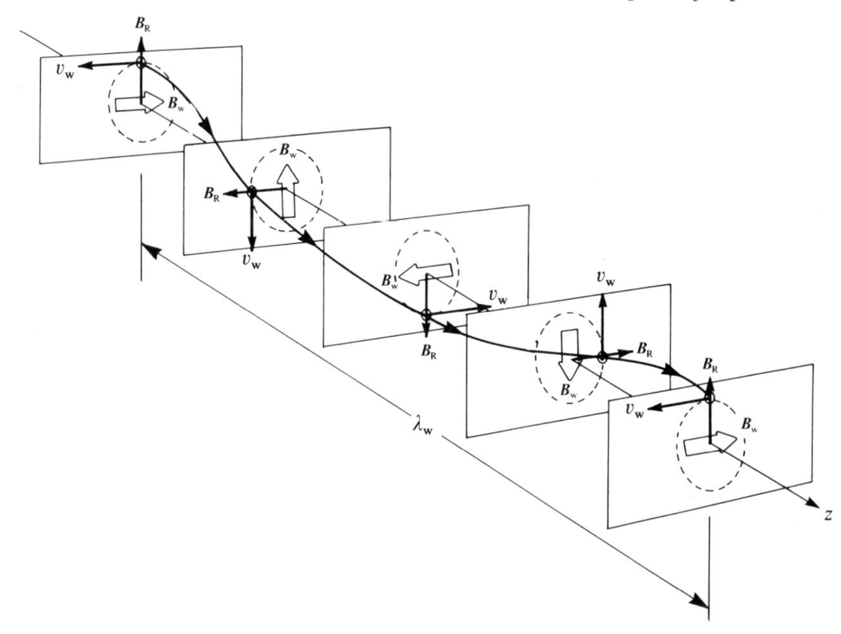

Fig. 1.2 The electron trajectory in a helical wiggler includes bulk streaming parallel to the axis of symmetry as well as a helical gyration. The vector relationships between the wiggler field $\boldsymbol{B}_w$, the transverse velocity $\mathbf{v}_w$, and the radiation field $\boldsymbol{B}_R$ of a resonant wave are shown in the figure projected on to planes transverse to the symmetry axis at intervals of one quarter of a wiggler period. This projection is circular, and the transverse velocity is directed opposite to that of the wiggler. A resonant wave must be circularly polarized with a polarization vector that is normal to both the transverse velocity and the wiggler field and which rotates in synchronism with the electrons. The electrons then experience a slowly varying wave amplitude. The transverse velocity and the radiation field are directed at right angles to each other and undergo a simple rotation. The interaction between the transverse velocity and the radiation field induces a force in the direction normal to both which coincides with the symmetry axis.

accelerated, and there is a net transfer of energy to the wave. The free-electron laser operates by an analogous process. Electrons in near resonance with the ponderomotive wave lose energy to the wave if their velocity is slightly greater than the phase velocity of the wave, and gain energy at the expense of the wave in the opposite case. As a result, wave amplification occurs if the wave lags behind the electron beam.

This process in a free-electron laser is described by a nonlinear pendulum equation. The **ponderomotive phase** $\psi(=(k+k_w)z-\omega t)$ is a measure of the position of an electron in both space and time with respect to the ponderomotive wave. The ponderomotive phase satisfies the circular pendulum equation

$$\frac{d^2}{dz^2}\psi = -K^2 \sin\psi, \tag{1.6}$$

where the pendulum constant is proportional to the square root of the product of the wiggler and radiation fields

$$K \approx 8.29 \frac{\sqrt{(B_w B_R)}}{\gamma_b}. \tag{1.7}$$

Here K is expressed in units of inverse centimetres and the magnetic fields are expressed in tesla. A detailed derivation of the pendulum equation for the free-electron laser is given in the introduction to Chapter 4 (see equation (4.13)). There are two classes of trajectory: trapped and untrapped. The untrapped, or free-streaming, orbits correspond to the case in which the pendulum swings through the full 360° cycle. The electrons pass over the crests of many waves, travelling fastest at the bottom of the troughs and slowest at the crests of the ponderomotive wave. In contrast, the electrons are confined within the trough of a single wave in the trapped orbits. This corresponds to the motion of a pendulum which does not rotate full circle, but is confined to oscillate about the lower equilibrium point. The dynamical process is one in which the pendulum constant evolves during the course of the interaction. Electrons lose energy as the wave is amplified; hence, the electrons decelerate and both the pendulum constant and separatrix grow. Ultimately, the electrons cross the growing separatrix from untrapped to trapped orbits.

In typical operation, electrons entering the free-electron laser are free streaming on untrapped trajectories. The gain in power during this phase of the interaction increases as the cube of the distance z in the case of relatively low-current operation in which the total gain is less than unity. This case is often referred to as the low-gain Compton regime, and the peak power gain in decibels is given by

$$G(dB) \approx 0.59 \frac{\omega_b^2}{\gamma_b c^2 k_w^2} \frac{v_w^2}{c^2} (k_w z)^3, \tag{1.8}$$

where ω_b denotes the **electron plasma frequency**. The detailed derivation of this expression for the gain is given in Chapter 4 (see equation (4.59)). This regime is relevant to operation at short wavelengths in the infrared and optical spectra. These experiments typically employ electron beams generated by radio frequency linear accelerators, microtrons, storage rings and electrostatic accelerators in which the total current is small.

In contrast, experiments operating at microwave and millimetre wavelengths often employ high-current accelerators, and the wave amplification is exponential. A comprehensive derivation of the growth rates in this high-gain regime appears in Chapter 4. In these cases, two distinct regimes are found. The high-gain Compton (sometimes called the **strong-pump**) regime is found when

$$\frac{\omega_b}{\gamma_b^{1/2} c k_w} \ll \frac{1}{16} \gamma_z^3 \left(\frac{v_w}{c}\right)^2. \tag{1.9}$$

In this regime, the maximum gain in the signal power over a wiggler period is given by (see equation (4.94))

$$G(dB) \approx 37.5\left(\frac{v_w^2}{c^2}\frac{\omega_b^2}{\gamma_b c^2 k_w^2}\right)^{1/3} \tag{1.10}$$

and is found at the resonant wavelength.

The opposite limit, referred to as the collective Raman regime, is fundamentally different from either the high- or low-gain Compton regimes. It occurs when the current density of the beam is high enough that the space-charge force exceeds that exerted by the ponderomotive wave. The gain in power over a wiggler period in this regime varies as (see equation (4.97))

$$G(dB) \approx 27.3\frac{v_w}{c}\sqrt{\left(\gamma_z\frac{\omega_b}{\gamma_b c k_w}\right)}. \tag{1.11}$$

In this regime, the space-charge forces result in electrostatic waves which co-propagate with the beam and are characterized by the dispersion relations

$$\omega = k_{sc}v_z \pm \frac{\omega_b}{\gamma_b^{1/2}\gamma_z} \tag{1.12}$$

which describe the relation between the frequency ω and wavenumber k_{sc}. These dispersion relations describe positive- and negative-energy waves corresponding to the '+' and '−' signs, respectively. The interaction results from a stimulated three-wave scattering process. This is visualized from the perspective of the electrons, in which the wiggler field appears to be a backwards propagating electromagnetic wave called a **pump** wave. This pump wave can scatter off the negative-energy electrostatic wave (the **idler**) to produce a forward-propagating electromagnetic wave (the **signal**). The interaction occurs when the wavenumbers of the pump, idler and signal satisfy the condition $k_{sc} = k + k_w$, which causes a shift in the wavelength of the signal to

$$\lambda \approx \frac{\lambda_w}{2\gamma_z^2\left(1 - \dfrac{\omega_b}{\gamma_z\gamma_b^{1/2}ck_w}\right)}. \tag{1.13}$$

Observe that the interaction in the Raman regime is shifted to a somewhat longer wavelength than occurs in the high-gain Compton regime.

Wave amplification can saturate by several processes. The highest efficiency occurs when the electrons are trapped in the ever-deepening ponderomotive wave and undergo oscillations within the troughs. In essence, the electrons are initially free streaming over the crests of the ponderomotive wave. Since they are travelling at a velocity faster than the wave speed, they come upon the wave crests from behind. However, the ponderomotive wave grows together with the radiation field, and the electrons ultimately will come upon a wave which is too high to cross. When this happens, they rebound and become trapped within the trough of the wave. In analogy to the oscillation of a pendulum, the trapped electrons lose energy as they rise and gain energy as they fall toward the bottom of the trough. As a result, the energy transfer between the wave and the electrons is cyclic and the wave amplitude ceases to grow and oscillates with the electron motion in the trough. The ultimate saturation efficiency for this mechanism can be estimated from the requirement that the net change in electron velocity at saturation is equal to twice the velocity difference between the electron beam and the ponderomotive wave. In the case of the low-gain Compton regime, the phase velocity of the ponderomotive wave can be determined from equation (4.62) and the saturation efficiency is estimated to be

$$\eta \approx \frac{1}{2N_w}, \tag{1.14}$$

where N_w denotes the number of periods in the wiggler. In a similar manner, the phase velocity of the ponderomotive wave for the high-gain Compton and collective Raman regimes at the frequency of maximum gain results in a saturation efficiency of

$$\eta \approx \frac{\gamma_b}{2(\gamma_b - 1)}\left(\frac{v_w^2}{2c^2}\frac{\omega_b^2}{\gamma_b c^2 k_w^2}\right)^{1/3} \tag{1.15}$$

in the high-gain Compton regime, and

$$\eta \approx \frac{\gamma_b}{\gamma_b - 1}\frac{\omega_b}{\gamma_z \gamma_b^{1/2} c k_w} \tag{1.16}$$

in the collective Raman regime.

However, this process places stringent requirements on the quality of the electron beam. The preceding formulae apply to the idealized case of a monoenergetic (or cold) beam. This represents a theoretical maximum for the gain and efficiency since each electron has the same axial velocity and interacts with the wave in an identical manner. A monoenergetic beam is physically unrealizable, however, and all beams exhibit a velocity spread which determines a characteristic temperature. Electrons with axial velocities different from the optimal resonant velocity are unable to participate fully in the interaction. If this axial velocity spread is sufficiently large that the entire beam cannot be in simultaneous resonance with the wave, then the fraction of the electron beam which becomes trapped must fall. Ultimately, the trapping fraction falls to the point where the trapping mechanism becomes ineffective, and saturation occurs through the thermalization of the beam. Thus, there are two distinct operating regimes: the **cold**-beam limit characterized by a narrow bandwidth and relatively high efficiencies, and the **thermal** regime characterized by a relatively broader bandwidth and sharply lower efficiencies.

The question of electron beam quality is the most important single issue facing the development of the free-electron laser [99]. In order to operate in the cold-beam regime, the axial velocity spread of the beam must be small. It is convenient to relate the axial velocity spread to an energy spread to obtain an invariant measure of the beam quality suitable for a wide range of electron beams. In the case of the low-gain limit, this constraint on the beam thermal spread is

$$\frac{\Delta E_b}{E_b} \ll \frac{1}{N_w}, \tag{1.17}$$

where ΔE_b represents the beam thermal spread. In the high-gain regimes, the maximum permissible energy spread for saturation by particle trapping is determined by the depth of the ponderomotive or space-charge waves, which is measured by twice the difference between the streaming velocity of the beam and the wave speed. The maximum permissible thermal spread corresponds to this velocity difference, and is one-half the saturation efficiency for either the high-gain Compton or collective Raman regimes, i.e.

$$\frac{\Delta E_b}{E_b} \ll \frac{1}{2}\eta. \tag{1.18}$$

Typically, this energy spread must be approximately 1% or less of the total beam energy for the trapping mechanism to operate at millimetre wavelengths, and decreases approximately with the radiation wavelength. Hence, the requirement on beam quality becomes more restrictive at shorter wavelengths and places greater emphasis on accelerator design.

Two related quantities often used as measures of beam quality are the **emittance** and **brightness**. The emittance measures the collimation of the electron beam, and may be defined in terms of the product of the beam radius and the average pitch angle (i.e. the angle between the velocity and the symmetry axis). It describes a random pitch-angle distribution of the beam which, when the velocities are projected on to the symmetry axis, is equivalent to an axial velocity spread. In general, therefore, even a monoenergetic beam with a nonvanishing emittance displays an axial velocity spread. The electron beam brightness is an analogue of the brightness of optical beams, and is directly proportional to the current and inversely proportional to the square of the emittance. As such, it describes the average current density per unit pitch angle, and measures both the beam intensity and the degree of collimation of the electron trajectories. Since the gain and efficiency increase with increasing beam current for fixed emittance, the brightness is a complementary measure of the beam quality. While it is important to minimize the emittance and maximize the brightness in order to optimize performance, both of these measures relate to the free-electron laser only insofar as they describe the axial velocity spread of the beam.

Typical free-electron laser efficiencies range up to approximately 12%; however, significant enhancements are possible when either the wiggler amplitude or period are systematically tapered. The free-electron laser amplifier at Livermore which achieved a 35% extraction efficiency employed a wiggler with an amplitude that decreased along the axis of symmetry, and contrasts with an observed efficiency of about 6% in the case of a uniform wiggler. The use of a tapered wiggler was pioneered by Phillips in 1960. The technique has received intensive study of late, and tapered wiggler designs have also been shown to be effective at infrared wavelengths in experiments at Los Alamos National Laboratory and using the superconducting r.f. linac at Stanford University [16–18].

The effect of a tapered wiggler is to alter both the transverse and axial velocities. Since the transverse velocity is directly proportional to the product of the amplitude and period, the effect of decreasing either of these quantities gradually is to decrease the transverse velocity and, in turn, increase the axial velocity. The energy extracted during the interaction results in an axial deceleration which drives the beam out of resonance with the wave; hence, efficiency enhancement occurs because the tapered wiggler maintains a relatively constant axial velocity (and phase relationship between the electrons and the wave) over an extended interaction length. The physical basis for this process is discussed in more detail in the introduction to Chapter 5 (see equation (5.12)), which also includes detailed discussions of nonlinear simulations of the tapered wiggler interaction. The enhancement in the tapered wiggler interaction efficiency is proportional to the decrement in the wiggler field of ΔB_w, and satisfies

$$\Delta\eta \approx \frac{0.872 B_w^2 \lambda_w^2}{1 + 0.872 B_w^2 \lambda_w^2} \frac{\Delta B_w}{B_w}. \tag{1.19}$$

In practice, a tapered wiggler is effective only after the bulk of the beam has become trapped in the ponderomotive wave. In single-pass amplifier configurations, therefore, the taper is not begun until the signal has reached saturation in a section of uniform wiggler, and the total extraction efficiency is the sum of the uniform wiggler efficiency and the tapered wiggler increment. Numerical simulations indicate that total efficiencies as high as 65% are possible under the right conditions.

Once particles have been trapped in the ponderomotive wave and begin executing a bounce motion between the troughs of the wave, then the potential exists for exciting secondary emission referred to as sideband waves. These sidebands are caused by the beating of the primary signal with the ponderomotive bounce motion. The bounce period is, typically, much longer than the radiation wavelength and these sidebands are found at wavelengths close to that of the primary signal. The difficulties imposed by the presence of sidebands is that they may compete with and drain energy from the primary signal. This is particularly crucial in long tapered wiggler systems which are designed to trap the beam at an early stage of the wiggler, and then extract a great deal more energy from the beam over an extended interaction length. In these systems, unrestrained sideband growth can be an important limiting factor. As a result, a great deal of effort has been expended on techniques of sideband suppression. One method of sideband suppression was employed in a free-electron laser oscillator at Columbia University [100]. This experiment operated at a 2 mm wavelength in which the dispersion due to the waveguide significantly affected the resonance condition. As a consequence, it was found to be possible by proper choice of the size of the waveguide to shift the sideband frequencies out of resonance with the beam. Experiments on an infrared free-electron laser oscillator at Los Alamos National

Laboratory [41, 42] indicate that it is also possible to suppress sidebands by (1) using a Littrow grating to deflect the sidebands out of the optical cavity or (2) changing the cavity length.

The foregoing description of the principles and theory of the free-electron laser is, necessarily, restricted to the idealized case in which the transverse inhomogeneities of both the electron beam and wiggler field are unimportant. This is sufficient for an exposition of the fundamental physics of the free-electron laser. In practice, however, these gradients can have important consequences on the performance of the free-electron laser. The most important effect is found if the wiggler field varies substantially across the diameter of the electron beam, since the electron response to the wiggler will vary across the beam as well. In practice, this means that an electron at the centre of the beam will experience a different field from that of an electron at the edge of the beam, and the two electrons will follow different trajectories with different velocities. As a result of this, the wave–particle resonance which drives the interaction will be broadened and the gain and efficiency will decline. In essence, therefore, the transverse wiggler inhomogeneity is manifested as an effective beam thermal spread. The bounded nature of the electron beam also affects the interaction since wave growth will occur only in the presence of the beam. Because of this, it is important in amplifier configurations to ensure good overlap between the injected signal and the electron beam. Once such overlap has been accomplished, however, the dielectric response of the electron beam in the presence of the wiggler can act in much the same way as an optical fibre to guide the light through the wiggler refractively.

1.2 QUANTUM-MECHANICAL EFFECTS

The early analyses of free-electron lasers relied upon classical treatments of the interaction physics. The gain in the free-electron laser was reformulated by Madey [6] based upon quantum-mechanical principles, and a great deal of work has since been published on the quantum mechanics of the free-electron laser [101–11]. However, for most applications of practical interest, quantum mechanical effects are negligible and the classical limit of the quantum-mechanical treatments is valid.

In general quantum-mechanical effects can be neglected when the spreading of the electron wave-packet is less than one wave period over the length of the wiggler L. In order to estimate the magnitude of this effect, consider the spreading of a one-dimensional wave-packet. If the electron wave function is represented by a Gaussian wave-packet

$$\psi(z,0) = \frac{1}{(\sqrt{(2\pi)}\Delta z_0)^{1/2}} \exp(ik_0 z) \exp(-z^2/4\Delta z_0^2), \tag{1.20}$$

upon entry to the wiggler with an initial spread or width Δz_0. This may be decomposed into Fourier components

$$\psi(z,0) = \frac{1}{\sqrt{2\pi}} \int_{-\infty}^{\infty} dk\, \psi(k) \exp(ikz), \tag{1.21}$$

where

$$\psi(k) = \left(\frac{2\Delta z_0}{\sqrt{2\pi}}\right)^{1/2} \exp[-(k-k_0)^2 \Delta z_0^2]. \tag{1.22}$$

The time-dependent solution for the wave-function may be constructed using these Fourier amplitudes

$$\psi(z,t) = \frac{1}{\sqrt{2\pi}} \int_{-\infty}^{\infty} dk\, \psi(k) \exp(ikz - i\omega(k)t), \tag{1.23}$$

where

$$\omega(k) = \frac{m_e c^2}{\hbar}\left[\sqrt{\left(1 + \frac{\hbar^2 k^2}{m_e^2 c^2}\right)} - 1\right]. \tag{1.24}$$

For a wave-packet with a narrow spread, this frequency can be expanded about $k = k_0$ as

$$\omega(k) \cong \omega(k_0) + v_0(k-k_0) + \frac{v_0}{\gamma_0^2 k_0}(k-k_0)^2, \tag{1.25}$$

where $v_0 \equiv \hbar k_0/\gamma_0 m_e$ denotes the bulk electron velocity, and $\gamma_0 \equiv (1 + \hbar^2 k_0^2/m_e^2 c^2)^{1/2}$ is the corresponding relativistic factor. Hence, the time-dependent wave-function

$$\psi(z,t) = \frac{1}{(\sqrt{(2\pi)}\Delta z_0)^{1/2}} \frac{\exp(ik_0 z - i\omega(k_0)t)}{\sqrt{(1 + iv_0 t/\gamma_0^2 k_0 \Delta z_0^2)}} \exp\left(-\frac{z^2}{4\Delta z_0^2} \frac{1}{1 + iv_0 t/\gamma_0^2 k_0 \Delta z_0^2}\right). \tag{1.26}$$

Evaluation of $|\psi(z,t)|^2$ shows that the width of the wave-packet increases in time via

$$\Delta z^2(t) = \Delta z_0^2 + \frac{v_0^2 t^2}{\gamma_0^4 k_0^2 \Delta z_0^2}. \tag{1.27}$$

The width of the wave-packet at any given time is minimized by a choice of $\Delta z_0^2 = v_0 t/\gamma_0^2 k_0$. Over the course of the entire length of the wiggler, therefore, we must require that

$$\frac{\lambda_c L}{\gamma_0 \lambda_w} \ll \lambda, \tag{1.28}$$

in order to neglect quantum-mechanical effects in the treatment of the interaction, where $\lambda_c \equiv h/m_e c$ is the Compton wavelength. This is well-satisfied for virtually all cases of practical interest. For example, consider a free-electron laser at a 10.6 μm wavelength which employs a 20 MeV electron beam and has a wiggler which is 100 wiggler periods in length. In this case, $\lambda_c L/\gamma_0 \lambda_w \approx 6 \times 10^{-6}$ μm and the inequality is satisfied by more than six orders of magnitude.

Another requirement for the neglect of quantum mechanical effects is that the electron recoil upon the emission of a photon be small. This criterion may be stated in the form that the downshift in the frequency of the emitted photon due to the electron recoil must be much smaller than the gain linewidth. However, as shown in both Brau [112] and Luchini and Motz [113], this requirement results in a condition which is identical to (1.28).

As shown above, quantum-mechanical effects are expected to be negligible for most free-electron lasers of practical interest. One aspect of free-electron laser operation in which quantum-mechanical effects may play a role, however, is in the oscillator start-up regime. Since free-electron laser oscillators typically start from the spontaneous emission of the electrons propagating through the wiggler, it has been speculated [114] that if the spontaneous emission is weak (i.e. the number of photons emitted is small), then quantum noise may prevent the oscillator from starting up even if the wiggler is long enough that the classical gain formula predicts substantial gain. This is most likely to occur for short-period wigglers in which the wiggler-induced transverse velocity is low, however, and we shall henceforth ignore quantum-mechanical effects and deal with a classical treatment of the interaction in the free-electron laser.

1.3 EXPERIMENTS AND APPLICATIONS

There are three basic experimental configurations for free-electron lasers: amplifiers, oscillators and superradiant amplifiers. In an amplifier, the electron beam is injected into the wiggler in synchronism with the signal to be amplified. The external radiation source which drives the amplifier is referred to as the master oscillator and can be any convenient radiation source such as a conventional laser or microwave tube. As a consequence, this configuration is often referred to as a Master Oscillator Power Amplifier (MOPA). Because amplification occurs during one pass through the wiggler, MOPAs require intense electron beam sources which can operate in the high-gain regime. Oscillators differ from amplifiers in that some degree of reflection is introduced at the ends of the wiggler so that a signal will make multiple passes through the system. The signal is amplified during that part of each pass in which the radiation co-propagates with the electrons, and allows for a large cumulative amplification over many passes even in the event of a low gain per pass. Oscillators are typically constructed to amplify the spontaneous (i.e. shot) noise within the beam, and no outside signal is necessary for their operation. However, a long pulse accelerator is required because a relatively long time may be required to build up to saturation. Superradiant amplifiers are devices in which the shot noise in the beam is amplified over the course of a single pass through the wiggler and, like amplifiers, require high-current accelerators to drive them. Since the shot noise present in the beam is generally broadband, the radiation from a superradiant amplifier is typically characterized by a broader bandwidth than a MOPA.

The optimal configuration and type of accelerator used in a free-electron laser design depends upon the specific application, and issues such as the electron beam quality, energy and current are important considerations in determining

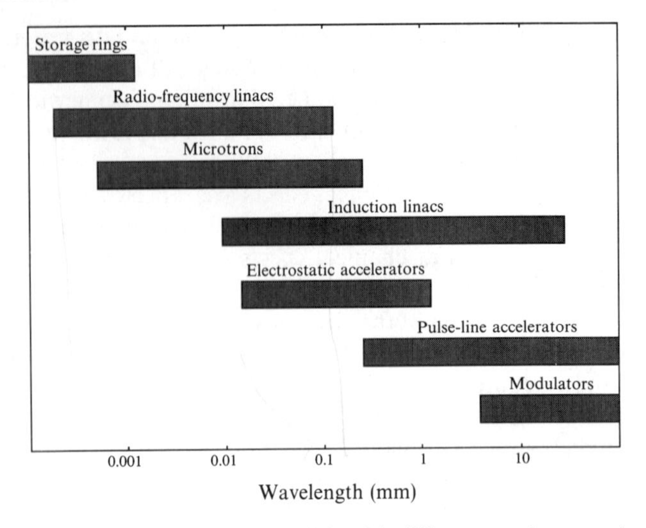

Fig. 1.3 The limits of the wavelengths possible with different accelerators depend both on the electron beam energies and the state of wiggler technology. The approximate wavelength ranges that may be addressed with current accelerator and wiggler technology are indicated.

both the wavelength and power. In general, however, each accelerator type is suited to the production of a limited range of wavelengths, as shown in Fig. 1.3. In addition, the temporal structure of the output light from a free-electron laser corresponds to that of the electron beam. Thus, either a pulsed or continuous electron beam source will give rise to a pulsed or continuous free-electron laser. Free-electron lasers have been constructed using virtually every type of electron source including: storage rings, radio-frequency linear accelerators, microtrons, induction linacs, electrostatic accelerators, pulse-line accelerators and modulators. Since the gain in a free-electron laser increases with current but decreases with energy, accelerators producing low-current/high-energy beams are generally restricted to the low-gain regime. Accelerator types which fall into this category include r.f. linacs, microtrons, storage rings and electrostatic accelerators. In contrast, intense-beam accelerators such as induction linacs, pulse-line accelerators and modulators are suitable electron-beam sources for high-gain systems.

Storage rings are typically characterized by multiple electron pulses continuously circulating through the ring. Each pulse is several nanoseconds in duration, and the output light from a free-electron laser driven by a storage ring is a continuous stream of picosecond bursts. In addition, while storage rings produce high-quality and high-energy beams of low to moderate currents, the electron pulses are recirculated through both the ring and the wiggler and the stability of the ring is disrupted by the extraction of too much energy. Hence, storage rings are feasible for applications that require uniform and continuous short-wavelength radiation sources but do not demand high output powers.

The first successful operation of a storage ring free-electron laser was at the Université de Paris Sud at Orsay [11], and the performance of this system has recently been extended to the VUV by harmonic operation [115]. This experimental configuration was that of an oscillator, and made use of the ACO storage ring which operates at energies and average currents in the range of 160–224 MeV and 16–100 mA. The laser was tuned across a broad band of the visible spectrum, but was first operated at wavelengths in the neighbourhood of approximately 0.65 μm. The peak output power from the oscillator was 60 mW over the 1 nanosecond duration of the micropulses, which corresponds to an intracavity power level of 2 kW. The average power extracted from the system was typically of the order of 75 μW. Higher harmonic emission was also detected in the ultraviolet, however, which posed a problem since radiation at these wavelengths resulted in the ultimate degradation of the optical system. An ultraviolet free-electron laser oscillator has also been achieved using the VEPP storage ring at Novosibirsk [15, 116, 117]. This experiment employed an optical klystron configuration in which the wiggler was composed of two distinct sections separated by a diffusive drift space. In this configuration, the first section operates as a prebuncher for the electron beam which subsequently enhances the gain in the second wiggler section. Operating the storage ring at 350 MeV and a peak current of 6 A, experimenters were able to obtain coherent emission at wavelengths as short as 0.3 μm and average output powers as high as 6 mW.

Radio-frequency linacs employ a series of cavities that contain rapidly varying electromagnetic (r.f.) fields to accelerate streams of electrons. The beams they produce are composed of a sequence of macropulses (typically of microseconds in duration) each of which consists of a train of shorter picosecond pulses. Microtrons produce beams with a temporal structure similar to that of r.f. linacs but, unlike the r.f. linac, are composed of a single accelerating cavity coupled to a magnet that causes the electron beam to recirculate through the cavity many times. The output light from a free-electron laser built with these accelerators, therefore, is similar to that from a storage ring. Recent experiments at Stanford University [16–18] and Boeing Aerospace [24, 118, 119] have also demonstrated the feasibility of the r.f. linac to produce visible light. In addition, r.f. linacs and microtrons are suitable for high-power free-electron lasers since the electrons are not recirculated and the energy extraction is not limited by the disruption of the beam.

Free-electron lasers based upon r.f. linacs have demonstrated operation over a broad spectrum extending from the infrared through the visible. The initial experiments conducted by Madey and coworkers at Stanford University resulted in (1) an amplifier which operated at a wavelength of 10.6 μm with an overall gain of 7%, and (2) a 3.4 μm oscillator which produced peak and average output powers of 7 kW and 0.1 mW, respectively. In collaboration with TRW [120–2], the superconducting r.f. linac (SCA) at Stanford University has been used to drive a free-electron laser oscillator which has demonstrated efficiency enhancement with a tapered wiggler, operation at visible wavelengths, and beam recirculation. The significance of the last point is that the overall wall-plug efficiency can be enhanced by recovery and reuse of the spent electron beam subsequent to its passage through the free-electron laser. The tapered wiggler experiment operated in the infrared at a wavelength of 1.6 μm and peak power levels of 1.3 MW. This yields a peak extraction efficiency of approximately 1.2%, which constitutes an enhancement by a factor of three over the efficiency in the case of an untapered wiggler. Operation at visible wavelengths was also found at 0.52 μm and peak power levels of 21 kW. The superconducting technology embodied in the SCA can enable the r.f. linac to further compete with storage rings by operating in a near steady-state mode. Both energy recovery and enhancement of the extraction efficiency by means of a tapered wiggler were also demonstrated at Los Alamos National Laboratory. Starting in 1981 with a tapered wiggler free-electron laser amplifier which obtained an extraction efficiency of 4% at a wavelength of 10.6 μm, researchers went on to (1) extend that to a 5% extraction efficiency in an oscillator configuration, and (2) demonstrate a 70% energy recovery rate with beam recirculation.

The limitations which storage rings, r.f. linacs and microtrons impose on free-electron laser design stem from restrictions on the peak (or instantaneous) currents which may be obtained, and which limit the peak power from a free-electron laser. High peak powers may be obtained by using induction linacs, pulse-line accelerators or modulators which produce electron beams with currents ranging from several amperes through several thousands of amperes and with pulse times ranging from several tens of nanoseconds through several microseconds. Induction linacs operate by inducing an electromotive force in a cavity through a rapid change in the magnetic field strength. In effect, the electron beam acts as the secondary winding in a transformer. For example, the Advanced Test Accelerator (ATA) at Livermore is an accelerator of this type which achieved energies and currents as high as 50 MeV and 10 kA over a duration of 50 nanoseconds. At lower energies, pulse-line accelerators and conventional microwave-tube modulators are available. Pulse-line accelerators produce beams with energies up to several tens of MeV, currents of several tens of kiloamperes, and pulse times up to 50–100 nanoseconds. As a result, pulse-line accelerators and modulators have been applied exclusively to microwave generation.

Amplifier experiments employing induction linacs have been performed at the Naval Research Laboratory, Lawrence Livermore National Laboratory and the Institute for Laser Engineering at Osaka University in Japan [123–7], and have demonstrated operation from the microwave through the infrared spectra. A superradiant amplifier at the Naval Research Laboratory employed a 650 keV/200 A electron beam and produced 4 MW at a wavelength of one centimetre [55–7]. The Livermore experiments have employed both the Experimental Test Accelerator (ETA) and the ATA. The free-electron laser amplifier experiment at an 8 mm wavelength was conducted with the ETA [85, 86] operating at approximately 3.5 MeV and a current of 10 kA. However, due to beam quality requirements, only about 10% of the beam was found to be usable in the free-electron laser. This ETA-based MOPA has been operated in both uniform and tapered wiggler configurations. In the case of a uniform wiggler, the measured output power was in the neighbourhood of 180 MW, which corresponds to an extraction efficiency of about 6%. A dramatic improvement in the efficiency was achieved, however, using a tapered wiggler. In this case, the total output power rose to 1 GW for an efficiency of 35%. The ATA has been used for a high-power MOPA design at a wavelength of 10.6 μm. Experiments have been conducted at both Livermore and the Institute for Laser Engineering at Osaka on the use of induction linacs in high-power MOPAs at frequencies above 140 GHz for the purposes of radio-frequency heating of magnetically confined plasmas for controlled thermonuclear fusion.

An important consideration in the construction of free-electron lasers with the intense beams generated by these accelerators is that an additional source of focusing is required to confine the electrons against the self-repulsive forces generated by the beam itself. This can be accomplished by the use of additional magnetic fields generated by either solenoid or quadrupole current windings. Quadrupole field windings were employed in the 8 mm amplifier experiment at the Lawrence Livermore National Laboratory; and the interaction mechanism in a free-electron laser is largely, though not entirely, transparent to the effect of the quadrupole field. In contrast, the solenoidal field has a deep and subtle effect on the interaction mechanism. This arises because a solenoidal field results in a precession, called Larmor rotation, about the magnetic field lines which can resonantly enhance the helical motion induced by the wiggler. This enhancement in the transverse velocity associated with the helical trajectory occurs when the Larmor period is comparable to the wiggler period. Since the Larmor period varies with beam energy, this relation can be expressed as

$$B_0 \approx 1.07 \frac{\gamma_b}{\lambda_w}, \tag{1.29}$$

where the solenoidal field B_0 is expressed in tesla and the wiggler period is in centimetres. For fixed beam energies, therefore, this resonant enhancement in the wiggler-induced velocity requires progressively higher solenoidal fields as the wiggler period is reduced.

The effect of a resonant solenoidal magnetic field is to enhance both the gain and saturation efficiency of the interaction. This was demonstrated in a superradiant amplifier experiment using the VEBA pulse line accelerator at the Naval Research Laboratory [51–4]. In this experiment, the output power from the free-electron laser was measured as a function of the solenoidal field as it varied over the range of 0.6–1.6 T. The beam energy and current in the experiment were 1.35 MeV and 1.5 kA, respectively, and the wiggler period was 3.0 cm. It should be remarked that due to the high current in the experiment, the beam was unable to propagate through the wiggler for solenoidal fields below 0.6 T. The magnetic resonance was expected for a solenoidal field in the neighbourhood of 1.3 T, and the experiment showed a dramatic increase in the output power for fields in this range. Other experiments which have demonstrated the effect of a solenoidal field have been performed at the Massachusetts Institute of Technology [73–82], Columbia University [66–70] and the École Polytechnique in France [95, 96].

The maximum enhancement in the output power in the experiment at the Naval Research Laboratory was observed for solenoidal fields slightly above the resonance. In this case, the nature of the interaction mechanism undergoes a fundamental change. In the absence of a solenoidal field (or for fields below the magnetic resonance), the axial velocity of the electrons decreases as energy is lost to the wave while the transverse velocity remains relatively constant. In contrast, the result of the strong solenoidal field is to cause a negative-mass effect in which the electrons accelerate in the axial direction as they lose energy to the wave. The bulk of the energy used to amplify the wave is extracted from the transverse motion of the electrons. Computer simulations of free-electron lasers operating in this strong solenoidal regime indicate that extremely high extraction efficiencies (in the neighbourhood of 50%) are possible without recourse to a tapered wiggler field. However, operation in this regime is precluded below submillimetre wavelengths. The reason for this is that the solenoidal field required to achieve this magnetic resonance varies directly with the beam energy and inversely with the wiggler period. Because of this, impractically high fields are required for wavelengths in the far infrared and below.

The preceding discussion of the effect of a solenoidal magnetic field dealt with solenoidal fields directed parallel to the wiggler field. However, a recent experiment at the Massachusetts Institute of Technology [128, 129] has demonstrated the utility of a solenoidal field directed anti-parallel to the wiggler field. In contrast to the parallel orientation of the wiggler and solenoidal fields, this orientation results in a decrease in the transverse electron velocity and, hence, the linear gain. However, high efficiencies are still possible, and this experiment measured a maximum extraction efficiency of 27% at a frequency of 35 GHz without any tapering of the wiggler field. This compares favourably with the efficiency found in the tapered wiggler experiment using the ETA accelerator at Livermore [85, 86]. A detailed analysis of this experiment is discussed in Chapter 13.

It is important to bear in mind that the high peak versus average power and oscillator versus MOPA distinctions between the different aforementioned accelerator technologies are becoming blurred by advances in the design of both r.f. and induction linacs. On the one hand, the development of laser-driven photocathodes at Los Alamos National Laboratory [43] have increased dramatically both the peak (of the order of 400 A) and average currents achievable with r.f. linacs. As a result, a high-gain MOPA experiment is under design at Los Alamos National Laboratory. It is significant in this regard that a Stanford University collaboration with Rocketdyne, Inc. and TRW [130–33] has already achieved MOPA operation using the Mark III r.f. linac. On the other hand, new induction linacs are under development which may be fired repetitively at a rate of up to several thousand times per second. Successful completion of these development programmes will enable high average power free-electron lasers to be constructed using this technology.

Although their average power is lower than that of linacs, electrostatic accelerators can produce continuous electron beams using charge-recovery techniques. In such a process, the electron beam is recirculated through the wiggler and back into the accelerator in a continuous stream. Using this technology, the electrostatic accelerator holds promise as an electron-beam source for a continuous or long-pulse free-electron laser. However, given practical restrictions on the size of such accelerators, which limit energies and currents, electrostatic accelerators have been restricted to the construction of free-electron laser oscillators which operate from the microwave regime through the infrared spectrum. In particular, a high average power free-electron laser is under development at FOM in The Netherlands for the heating of magnetic fusion reactors which is based upon an electrostatic accelerator with beam recovery [134].

The principal design issue for accelerators at the present time is that of beam quality, since it is crucial to have a low axial velocity spread for efficient operation of the free-electron laser. This process is accomplished in a number of different ways for the various accelerator types. In the low-energy cases relevant to modulators and pulse-line accelerators, this is accomplished by careful design of the cathodes and focusing systems in the electron guns, as well as by attention to the transport system which brings the beam into the wiggler. In r.f. linacs and microtrons, although the ultimate reliability is yet to be determined, the development of laser-driven photocathodes holds promise for the production of exceptionally high brightness electron beams [37, 43, 133, 135, 136].

The breadth of free-electron laser experiments includes many different wiggler configurations, and virtually every type of accelerator in use today. The wiggler has been produced in planar, helical and cylindrical forms by means of permanent magnets, current-carrying coils and hybrid electromagnets with ferrite cores. Helical wiggler fields are produced by a current-carrying bifilar helical coil in which the field increases radially outward from the symmetry axis and provides magnetic focusing to confine the beam against the mutually repulsive forces between the electrons. In a planar wiggler both the transverse and axial components of the velocity oscillate in synchronism with the wiggler. As such, the interaction is determined by the average, or root-mean-square, wiggler field. Because of this, planar wigglers require a stronger field to produce the same effect as a helical wiggler. This is compensated for, however, by the ease of adjustment allowed by a planar design, in which the strengths or positions of the individual magnets can be altered to provide either a uniform or tapered field. In contrast, the only adjustment possible for a bifilar helix is the strength of the field.

One practical constraint on the development of free-electron laser oscillators at the present time is mirror technology for infrared and shorter wavelengths, and relates to both reflectivity and durability. The reflectivity is important since the net gain of an oscillator decreases as the mirror losses increase, and oscillation is possible only if the amplification due to the free-electron laser interaction exceeds the losses at the mirrors. The reflectivity is a measure of this loss rate, and must be kept sufficiently high that the energy losses at the mirrors do not overwhelm the gain. The issue of durability relates to the power level that any specific mirror material can endure without suffering optical damage. In this sense, optical damage refers to a decrease in the reflectivity. Note that the extreme case of the complete burning out of the mirrors might be described as a catastrophic drop in the reflectivity. Problems exist in finding materials with a high enough reflectivity and durability to operate in the infrared and ultraviolet spectra, and even the visible presents problems. For example, the visible free-electron laser oscillator at the Université de Paris Sud experienced mirror degradation due to harmonic emission in the ultraviolet. At extremely high power levels solutions can be found through such techniques as the grazing-incidence mirrors used by Boeing Aerospace in which the optical beam is allowed to expand to the point where the power density on the mirrors is low. In the infrared, oscillator experiments at Los Alamos National Laboratory originally employed a dielectric mirror material with a high reflectivity at low power levels. However, recent observations indicate that nonlinear phenomena occur in this material at high power levels which effectively reduce the reflectivity, and that the use of copper mirrors substantially improves performance. An additional problem occurs at high power levels due to thermal distortion of the optical surface. In order to combat this problem, actively cooled mirrors are under development.

The principal biomedical applications of the free-electron laser are surgery, photocoagulation and cauterization, photodynamic therapy, and the *in vivo* thermal destruction of tissue through a process called photothermolysis. The most common surgical technique is the thermal ablation of tissue which requires a laser producing powers of 10–100 W at a wavelength of approximately 3 μm. This corresponds to a strong absorption resonance of the water molecule characterized by relatively little scattering of the light by the tissues. In contrast, photocoagulation requires a shorter wavelength of approximately 1–1.5 μm which is also strongly absorbed by the water molecule but exhibits a higher degree of scattering throughout the surrounding tissue. It is important to observe that the tunability of the free-electron laser holds the potential for a surgical laser which may be tuned in a single sequential process from 3 μm down to 1 μm to give both clean surgical incisions and cauterization. Another advantage is that the optical pulse can be tailored to meet specific requirements by control of the temporal structure of the electron beam. For example, short pulses are useful in ophthalmic therapy for the surgical disruption of pigmented tissue, while longer pulses are required for retinal photocoagulation. Photodynamic therapies rely on the injection of photosensitive dyes which are concentrated preferentially in malignant tissue. Subsequent irradiation excites a photo-oxidation process which is toxic to the tumorous tissue. The principal dyes are photosensitive at wavelengths between 0.6 μm and 1.7 μm, and are used to treat tumours of the lung, bladder and gastrointestinal tract at early stages in their development as well as for palliative treatment at later stages of growth. At relatively high average powers (up to 100 W), a free-electron laser makes possible the simultaneous treatment of relatively large masses of tissue. However, the high-energy electron beams needed to produce these wavelengths also produce relatively large X-ray fluxes, and the entire facility, including power supply, accelerator, wiggler, optical system and X-ray shielding, is likely to be rather bulky and complex. Hence in consideration of the rapid development of conventional laser sources at these wavelengths, the long-term biomedical applications are envisioned to be (1) in the initial refinement of these therapeutic techniques, (2) as large centralized facilities for tumour treatment, and (3) as experimental research tools.

Applications to research are unimpaired by considerations of the bulk and complexity of a free-electron laser facility, and the first user facility was established by Luis Elias at the University of California at Santa Barbara in 1984 [137–44], and employs a long-pulse 3 MeV electrostatic accelerator. The free-electron laser produces a peak power of as much as 10 kW over a range in wavelengths of 390–1000 μm, and is suitable for a wide range of experiments in the biomedical, solid-state and surface sciences. In the field of photobiology, since the DNA molecule is sensitive to infrared wavelengths, the free-electron laser can study such behaviour as the variation in the DNA mutation rate with wavelength. In addition, experiments have been conceived in the linear and nonlinear excitations of phonons and magnons, ground and excited state Stark splitting, the generation of coherent phonons, phonon amplification by stimulated emission, induced phase transitions, and semiconductor band-gap structure. The latter application may prove relevant to the study of high critical temperature superconductors. This facility is capable of producing kilowatts of tunable radiation in three frequency bands spanning the submillimetre through mid-infrared spectra: 120 GHz–1 THz, 1–5 THz and 5–10 THz.

More recently, two user facilities have been constructed at Stanford University based upon both the Mark III r.f. linac [145–53] and the SCA [16–18, 154]. The Mark III facility, established by John Madey and coworkers, is tunable over the range 0.5–10 μm and produces 60 kW over a pulse time of several microseconds. Due to the electron energies available from the Mark III, however, the fundamental operation occurs at wavelengths in the neighbourhood of 1–3 μm. Shorter wavelengths are achieved through the use of frequency-doubling techniques common to laser engineering. In this case, the tunability, high power and temporal structure offer a unique opportunity to study surgical applications. In particular, experiments have been conducted in the cutting of both bone and soft tissue with encouraging results. Since spot sizes of the order of 100–1000 μm and power densities as high as several megawatts per square centimetre are possible, the cutting mechanism is not thermal ablation but direct plasma formation of the irradiated tissue with extremely clean and localized incisions. In contrast with conventional lasers (which produce lower powers over longer pulse times), the combination of high power and short pulses results in less scar formation and more rapid healing. Indeed, the power densities available with this free-electron laser have raised concern that current optical fibre technology may ultimately prove inadequate to the task of directing the radiation, and research has begun into the development of optical fibres capable of handling higher intensities. In addition, experiments have been conducted to study semiconductor band-gap structure as well as the multi-photon spectroscopy of germanium and poly-acetylene. Recently, the Mark III free-electron laser has been relocated at Duke University.

The SCA free-electron laser facility commonly operates in a continuous mode in which the macropulses are generated at a frequency of 20 Hz (that is, 20 macropulses per second) over a time scale of several hours. A typical macropulse is of 5 ms duration and is composed of a train of 3 ps micropulses separated by 84.6 ns. Average electron beam currents over a macropulse can reach 200 μA with a peak current of 6 A over a micropulse. The peak voltage of the superconducting linac is 75 MeV, but the recirculation system allows this to rise as high as 150 MeV. The free-electron laser has operated over wavelengths ranging from 0.5 μm to 15 μm, and is capable of operation in the ultraviolet. Experiments to date have been concerned with dynamical processes on picosecond time scales. To this end, the picosecond micropulses and the relatively large macropulse separation (which is long enough for conventional optical-pulse selection techniques to be used) produced by the SCA are crucial. For example, the free-electron laser operated at a wavelength of 1.54 μm with a linewidth of 0.08%, and was stable over a time scale of several hours. The output power was approximately 300 kW over a micropulse, 12 W averaged over a macropulse, or 1 W over the longer time scale.

The operation of the SCA is soon to be upgraded by the implementation of a novel electron injector/acceleration scheme. In a conventional r.f. linac, the electron beam is accelerated by a single-frequency r.f. signal which varies sinusoidally over time. Since the amount of electron acceleration depends upon the r.f. power, the electron pulse must be synchronized to the peak of the r.f. signal and its duration kept short. In general, the longer the electron pulse, the larger the variation in r.f. power over the pulse and the greater the energy spread of the electron beam. In order to minimize the beam energy spread, which degrades the performance of the free-electron laser, researchers at the SCA plan to use a composite r.f. signal composed of waves at multiple frequencies. In a manner analogous to the way in which a square-wave signal can be built up from a large composite spectrum of waves, this process will extend the duration of the peak r.f. signal. As a consequence, both the duration and power in the electron beam will increase at little or no cost to the beam quality. It is estimated that the peak and average beam currents will increase to approximately 50 A and 1 mA, respectively, with corresponding increases in the output of the free-electron laser.

Recent user experiments using the SCA free-electron laser have spanned a wide range of fields including solid-state and surface science, molecular chemistry, biophysics and medical science. In solid-state and surface science, the applications have dealt with studies of the picosecond physics of spectroscopy, nonlinear optics, free-carrier lifetimes and dynamics, inter-subband transitions, harmonic generation and hole relaxation. In molecular chemistry, research has been conducted in photon echoes, non-photochemical hole burning, optical dephasing, ultrafast temperature jump spectroscopy, laser ablation, infrared vibrational photon echoes, ultrafast multiphoton up-pumping, ultrafast spectroscopy and vibrational dynamics. Biophysical applications include photon photochemistry, spontaneous vibrational Stark effect spectroscopy and the manipulation of single DNA molecules using optical tweezers. Research in the field of medical science includes microscopy of transient local heating in single living cells, optical tomography and the detection of pathogens in living hosts.

As a measure of the impact of advances in accelerator technology on the free-electron laser, it should be noted that the Mark III linac was the accelerator used by Hans Motz in 1951, and that the present range of experiments was made possible by the continual improvement of the original design. Indeed, a user facility has been built at Vanderbilt University [155, 156] based upon a further improved Mark III linac. This free-electron laser, which began operating in the summer of 1991, operates over a spectral range of 0.2–10 μm, and has produced an average power of 11 W which is the highest average power in an infrared free-electron laser to date. Experiments have been conducted in biology and materials science and include matrix-assisted laser desorption–ionization mass spectroscopy, studies of the electronic structure of semiconductor heterojunctions and laser surgery of hard and soft tissues. In the last case, experiments using radiation at 3–4 μm wavelength have demonstrated the ability to cut bone with previously unrealized precision. Soft-tissue surgery has been conducted using radiation at 6.45 μm wavelength corresponding to a protein band which eliminates tissue by ablation. Modifications to the system are envisioned to provide 10–18 keV X-rays using Compton backscattering of the laser light by the electron beam as well as tunable infrared radiation in the 50–200 μm range using a Smith–Purcell interaction.

In general, the principal advantages of the free-electron laser as a research tool are high intensity (relative to currently available sources), tunability and a temporal structure which is controlled by the characteristics of the accelerator. Tunability permits the selection of specific energy states for study, while an appropriate tailoring of the temporal structure of the pulses allows the time evolution and decay of excited states to be investigated. In addition, it should be remarked that there are no good infrared sources available with wavelengths ranging from 2–5 μm. Important areas of investigation are the bulk and surface properties of semiconductors; in particular, the bandgap structure. Some of the most commercially important semiconductors exhibit bandgaps in the range of 0.25–2.5 μm, and it is important to extend the spectral range of study to within 0.1 μm about this range. A high-intensity, tunable free-electron laser producing pulses of approximately 10–100 ps duration at a repetition rate of several megahertz would permit the study of the dynamic excitation and subsequent decay of electrons into unoccupied energy states. Other applications include laser photochemistry and photophysics which require sources in the visible and ultraviolet spectra. Initial development of an ultraviolet free-electron laser is being conducted at Los Alamos National Laboratory using r.f. linac technology [157]. Since the electron beam is composed of short bursts at high repetition rates, r.f. linacs produce radiation with the desired temporal structure. The principal competition in the ultraviolet comes from incoherent synchrotron light sources which also make use of undulator magnets. The advantage of a free-electron laser over synchrotron sources is that the increased counting rates resulting from a larger photon flux would make practical a large number of currently marginal experiments. A nonexhaustive list of experiments possible with a coherent visible-through-ultraviolet source includes the multi-photon ionization of liquids, the chemistry of combustion and of molecular ions, high-resolution polyatomic and fluorescence spectroscopy, time-resolved resonance Raman spectroscopy, spin-polarized photoemission and photoemission microscopy, magneto-optical studies of rare-earth elements, and studies of optical damage from high-intensity ultraviolet radiation.

Industrial applications are envisioned in materials production and photolithography. The requirements for materials production are sources in the near infrared (2.5–100 μm) and the ultraviolet through soft X-ray spectra, and involve the pyrolytic production of powders for catalysts, the near-stoichiometric production of high-value chemicals and pharmaceuticals, and the pyrolytic and photolytic deposition of thin films on substrates. In photolithography, the surface of a wafer is coated with a layer of a photoresisting substance of which a part is illuminated. The wafer is then processed by the removal of either the exposed or unexposed portions of the resist. Photolithography is primarily done with the optical lines of the mercury arc which occur over the spectral range of 436–357 μm, and the excimer laser is under consideration for this purpose as well. There are three principal lithographic techniques either in use or under consideration. Contact printing is performed by bringing a mask and wafer into contact, while in proximity printing a small space separates these two components. The preferred technique is that of projection printing in which the image of the mask is projected on to the wafer from a greater distance than in contact or proximity printing, which allows the use of larger masks and improves usable mask lifetime. Each of these techniques requires uniform and stable sources of illumination. The sources may be either continuous or pulsed; however, possible difficulties with pulsed sources (i.e. r.f. linac driven free-electron lasers) are (1) that the instantaneous pulse power necessary to give a sufficiently high average power may also be high enough to damage the wafer, and (2) that extremely high uniformity from pulse to pulse is required. In addition, short-wavelength sources may render the non-lithographic direct printing of wafers possible and eliminate the need for photoresists.

Applications also exist for high-power microwaves and submillimetre waves in the fields of communications, radar and plasma heating. We confine the discussion to the heating of a magnetically confined plasma for controlled thermonuclear fusion. The reactor design of greatest interest is the Tokamak, which confines a high-temperature plasma within a toroidal magnetic 'bottle'. A thermonuclear reactor must confine a plasma at high density and temperature for a sufficiently long time to ignite a sustained fusion reaction. In its original conception, the Tokamak was to be heated to ignition by an ohmic heating technique whereby a current is induced by means of a coil threading the torus. As such, the Tokamak acts like the secondary winding in a transformer. However, recent developments indicate the need for auxiliary heating, and the resonant absorption of submillimetre radiation has been proposed for this purpose. The frequencies of interest are the harmonics of the electron cyclotron frequency and range from about 280 GHz through 560 GHz. In the case of the compact ignition torus, which is the proposed design of the next major Tokamak experiment in the United States, the applied magnetic field is 10 T and the resonant wavelengths are in the neighbourhood of 1 millimetre and less. The estimates of the power requirements are, of necessity, relatively crude but indicate the need for an average power of approximately 20 MW over a pulse time of nearly 3 s. At present there are no sources capable of meeting this requirement. However, the free-electron laser has operated in this spectral region at power levels of this order, but over a much shorter pulse time. As such, it represents one of several competing concepts [134, 158–61].

A new user facility has recently become available at FOM in The Netherlands. This is referred to as the Free Electron Laser for Infrared eXperiments (FELIX) facility, and employs a 21 Mev r.f. linac which produces a 3 μs macropulse train of 1–10 ps micropulses with an overall repetition frequency of 5 Hz. Temporal overlap of these pulses is possible and a pulse of approximately 7 ns is achievable. Two spectral bands have been covered thus far: 6–20 μm and 40–60 μm. In addition to medical and materials research, this facility has been used to study many aspects of the eevice physics of free-electron laser oscillators including: phase locking [162], single-mode operation [163], limit cycle oscillations [164], hole coupling and dynamic resonator desynchronization [165, 166], coherent start-up [167] and short pulse effects [168]. Research into medical and materials applications includes: induced two-photon absorption, photon drag effects, inter-subband absorption, sum-frequency generation at surfaces, photoablation of corneal tissue and phosphorescence detection of triplet–triplet transitions.

User facilities also exist in France at the Université de Paris Sud which employ the Super-ACO storage ring [169–75] and an r.f. linac [176–80] system (CLIO). Experiments in time-resolved fluorescence have been conducted using the Super-ACO facility, as well as experiments dealing with solid-state and surface interactions (sum-frequency generation), near-field infrared microscopy and vibrational energy transfer in molecules.

A more controversial application of high-power, long-pulse free-electron lasers was strategic defence against intercontinental ballistic missiles. In this regard, planners envisioned a large-scale ground-based laser which would direct light towards a target by means of both ground-based and orbiting mirrors. Designs based on both amplifier and oscillator free-electron lasers were pursued.

In experiments, a free-electron laser amplifier at Livermore amplified a 14 kW input signal from a carbon dioxide laser at 10.6 μm to a level of approximately 7 MW, a gain of 500 times. Boosting the input beam to 5 MW yielded a saturated power of 50 MW. In initial experiments researchers used a 15 m planar wiggler with a uniform period and amplitude; they have since lengthened the wiggler to 25 m, and use of a tapered version to increase the extraction efficiency further has been considered.

Boeing Aerospace built an experimental free-electron laser oscillator in collaboration with Los Alamos National Laboratory based on a 5-metre-long planar wiggler and an advanced radio-frequency linac. In its initial design, the linac produced electron beams with energies as high as 120 MeV. The oscillator has lased in the red region of the visible spectrum at a wavelength of 0.62 μm and at power levels a billion times that of the normal spontaneous emission within the cavity. The average power over the course of a 100 μs pulse is about 2 kW. The corresponding conversion efficiency is about 1%, but the peak power is a more respectable 40 MW. Even though oscillators typically generate short-wavelength harmonics that can damage the cavity mirrors, no degradation has been observed.

Estimates indicated that pulses of visible or near infrared light at an average power of about 10 to 100 MW over a duration of approximately one second would be required to destroy a missile during its boost phase. This means lengthening the pulses or increasing the peak power levels of existing free-electron lasers by a factor of a million or more. Depending on laser efficiency and target hardness, a collection of ground-based free-electron lasers would require somewhere between 400 MW and 20 GW of power for several minutes during an attack. (For comparison, a large power plant generates about 1 GW.) For these and other reasons, it is not clear whether it will be practical to scale up free-electron lasers to the power levels required. Current and future experiments will help to resolve this question.

1.4 DISCUSSION

This chapter includes a necessarily abridged list of experiments and recently conceived applications of the free-electron laser. The fundamental principles of the free-electron laser are understood at the present time, and the future direction of research is toward evolutionary improvements in electron beam sources (in terms of beam quality and reliability) and wiggler designs. The issues, therefore, are technological rather than physical and the free-electron laser can be expected ultimately to cover the entire spectral range shown. In this regard, it is important to recognize that the bulk of the experiments to date have been performed with accelerators not originally designed for use in a free-electron laser, and issues of the beam quality important to free-electron laser operation were not adequately addressed in the initial designs. As a consequence, the results shown do not represent the full potential of the free-electron laser, although many of the experiments have produced record power levels. Indeed, the only accelerators specifically designed for use with a free-electron laser are the r.f. linac at Boeing Aerospace, an induction linac at Lawrence Livermore National Laboratory, the electrostatic accelerator at the University of California at Santa Barbara and the r.f. linac at FOM (FELIX). In particular, it was found at Los Alamos National Laboratory that wakefields induced by pulses of high peak currents in the accelerator and beam-transport system can result in serious degradation in beam quality, and that the minimization of these effects is an important consideration in the overall design of the beam line in free-electron lasers.

Another important direction for future research is the design of short-period wigglers which permit short wavelength operation with relatively low voltage electron beams. An alternative approach to the production of short wavelength radiation with moderate energy electron beams is the use of higher harmonic interactions. For example, if the third harmonic interaction is employed (i.e. $\omega = (k + 3k_w)v_z$), then the energy requirement for a fixed wavelength output is reduced by a factor of approximately $\sqrt{3}$. Indeed, lasing has been achieved at the third harmonic in experiments at (1) Stanford University using the Mark III linac at a wavelength between 1.4 μm and 1.8 μm [145], and (2) at a wavelength of 4 μm at Los Alamos National Laboratory [47]. The reduction in the voltage requirement has important practical implications in the simplification of accelerator design problems and the reduction in the production of secondary X-rays and neutrons. The combination of improved accelerators and short-period wigglers will reduce both the size and complexity of free-electron laser systems and open doors to a host of new practical applications of the technology. In particular, a recent experiment at Los Alamos National Laboratory achieved lasing in the ultraviolet spectrum at wavelengths of 369–80 nm using a 45 MeV electron beam from an r.f. linac and a pulsed microwiggler with a 1.36 cm period [181]. The laser produced a macropulse (micropulse) output power of 36 ± 5 W (270 ± 70 kW) over 20 ms (6 ps). This represents the first lasing in the ultraviolet with an r.f. linac.

Free-electron laser research and development is international in scope. A good summary of both short [182] and long [183] wavelength free-electron lasers either in operation or under construction is given in the proceedings of the annual free-electron laser conference held at Stanford University in August 1994. The transition from purely device research to user facilities is well underway and many free-electron lasers are either in operation or under construction. In the United States, work continues at Los Alamos National Laboratory on a compact free-electron laser (referred to as the Advanced Free-Electron Laser) which has produced radiation in the 5 μm range using a 20 MeV r.f. linac [184, 185]. A clean-slate design for a photocathode-driven r.f. linac for an infrared free-electron laser oscillator is being built by Northrop–Grumman Corporation at Princeton University [186], and a compact continuous-wave electrostatic-accelerator-based free-electron laser is under construction at the Center for Research in Electro-Optics and Lasers at the University of Central Florida [187]. Further, design work is underway for short-wavelength r.f.-linac-based free-electron lasers at the Continuous Electron Beam Accelerator Facility [188] and at Brookhaven National Laboratory [189]. In addition to the previously mentioned user facilities, free-electron laser research in France includes projects at Bruyères-le-Châtel (ELSA) which has a high-power r.f.-linac-driven infrared free-electron laser [190–4], and at CESTA, at microwave and millimetre wave frequencies [195–7]. Free-electron laser research is also ongoing in the UK [198, 199], Italy [200–9], The Netherlands (a Raman [210, 211] and an infrared experiment (TEUFEL) [212, 213] at the University of Twente), Germany [214–22], Japan [223–39], Sweden [240], China [241–51], South Korea [252, 253], Israel [254] and Russia [255, 256].

REFERENCES

1. Motz, H. (1951) Applications of radiation from fast electron beams. *J. Appl. Phys.*, **22**, 527.
2. Motz, H., Thon, W. and Whitehurst, R. N. (1953) Experiments on radiation by fast electron beams. *J. Appl. Phys.*, **24**, 826.
3. Motz, H. and Nakamura, M. (1959) Radiation of an electron in an infinitely long waveguide. *Ann. Phys.*, **7**, 84.
4. Phillips, R. M. (1960) The ubitron, a high power traveling-wave tube based on a periodic beam interaction in unloaded waveguide. *IRE Trans. Electron Dev.*, **ED-7**, 231.
5. Phillips, R. M. (1988) History of the ubitron. *Nucl. Instr. Meth.*, **A272**, 1.
6. Madey, J. M. J. (1971) Stimulated emission of bremsstrahlung in a periodic magnetic field. *J. Appl. Phys.*, **42**, 1906.
7. Madey, J. M. J., Schwettman, H. A. and Fairbank, W. M. (1973) A free-electron laser. *IEEE Trans. Nucl. Sci.*, **NS-20**, 980.
8. Elias, L. R., Fairbank, W. M., Madey, J. M. J., Schwettman, H. A. and Smith, T. I. (1976) Observation of stimulated emission of radiation by relativistic electrons in a spatially periodic transverse magnetic field. *Phys. Rev. Lett.*, **36**, 717.
9. Deacon, D. A. G., Elias, L. R., Madey, J. M. J., Ramian, G. J., Schwettman, H. A. and Smith, T. I. (1977) First operation of a free-electron laser. *Phys. Rev. Lett.*, **38**, 892.
10. Benson, S. V., Deacon, D. A. G., Eckstein, J. N., Madey, J. M. J., Robinson, K., Smith, T. I. and Taber, R. (1982) Optical autocorrelation of a 3.2 μm free-electron laser. *Phys. Rev. Lett.*, **48**, 235.
11. Billardon, M., Ellaume, P., Ortega, J. M., Bazin, C., Bergher, M., Velghe, M. and Petroff, Y. (1983) First operation of a storage ring free-electron laser. *Phys. Rev. Lett.*, **51**, 1652.
12. Billardon, M., Ellaume, P., Ortega, J. M., Bazin, C., Bergher, M., Velghe, M., Deacon, D. A. G. and Petroff, Y. (1985) Free-electron laser experiment at Orsay: a review. *IEEE J. Quantum Electron.*, **QE-21**, 805.
13. Ortega, J. M., Lapierre, Y., Girard, B., Billardon, M., Ellaume, P., Bazin, C., Bergher, M., Velghe, M. and Petroff, Y. (1985) Ultraviolet coherent generation from an optical klystron. *IEEE J. Quantum Electron.*, **QE-21**, 909.
14. Billardon, M., Ellaume, P., Lapierre, Y., Ortega, J. M., Bazin, C., Bergher, M., Marilleau, J. and Petroff, Y. (1986) The Orsay storage ring free-electron laser: new results. *Nucl. Instr. Meth.*, **A250**, 26.
15. Kuliapanov, G. N., Litvinenko, V. N., Panaev, I. V., Popik, V. M., Skrinsky, A. N., Sokolov, A. S. and Vinokurov, N. A. (1990) The VEPP-3 storage-ring optical klystron: lasing in the visible and ultraviolet regions. *Nucl. Instr. Meth.*, **A296**, 1.
16. Smith, T. I., Schwettman, H. A., Rohatgi, R., Lapierre, Y. and Edighoffer, J. (1987) Development of the SCA free-electron laser for use in biomedical and materials science experiments. *Nucl. Instr. Meth.*, **A259**, 1.
17. Rohatgi, R., Schwettman, H. A., Smith, T. I. and Swent, R. L. (1988) The SCA free-electron laser program: operation in the infrared, visible, and ultraviolet. *Nucl. Instr. Meth.*, **A272**, 32 (1988).
18. Smith, T. I., Frisch, J. C., Rohatgi, R., Schwettman, H. A. and Swent, R. L. (1990) Status of the SCA free-electron laser. *Nucl. Instr. Meth.*, **A296**, 33.
19. Slater, J. M., Adamski, J. L., Quimby, D. C., Churchill, T. L., Nelson, L. Y. and Center, R. E. (1983) Electron spectrum measurements for a tapered-wiggler free-electron laser. *IEEE J. Quantum Electron.*, **QE-19**, 374.
20. Slater, J. M., Churchill, T., Quimby, D. C., Robinson, K. E., Shemwell, D., Valla, A., Vetter, A. A., Adamski, J., Gallagher, W., Kennedy, R., Robinson, B., Shoffstall, D., Tyson, E., Vetter, A. and Yeremian, A. (1986) Visible wavelength free-electron laser oscillator. *Nucl. Instr. Meth.*, **A250**, 228.

21. Robinson, K. E., Quimby, D. C. and Slater, J. M. (1987) The tapered hybrid undulator of the visible free-electron laser oscillator experiment. *IEEE J. Quantum Electron.*, **QE-23**, 1497.
22. Shemwell, D. M., Robinson, K. E., Gellert, R. I., Quimby, D. C., Ross, J. M., Slater, J. M., Vetter, A. A., Trost, D. and Zumdieck, J. (1987) Optical cavities for visible free-electron laser experiments. *IEEE J. Quantum Electron.*, **QE-23**, 1522.
23. Eggleston, J. M. and Slater, J. M. (1987) Baseline conceptual design for high-power free-electron laser ring cavities. *IEEE J. Quantum Electron.*, **QE-23**, 1527.
24. Robinson, K. E., Churchill, T. L., Quimby, D. C., Shemwell, D. M., Slater, J. M., Valla, A. S., Vetter, A. A., Adamski, J., Doering, T., Gallagher, W., Kennedy, R., Robinson, B., Shoffstall, D., Tyson, E., Vetter, A. and Yeremian, A. (1987) Panorama of the visible wavelength free-electron laser oscillator. *Nucl. Instr. Meth.*, **A259**, 49.
25. Lumpkin, A. H., King, N. S. P., Wilke, M. D., Wei, S. P. and Davis, K. J. (1989) Time-resolved spectral measurements for the Boeing free-electron laser experiments. *Nucl. Instr. Meth.*, **A285**, 17.
26. Lumpkin, A. H., Tokar, R. L., Dowell, D. H., Lowrey, A. R., Yeremian, A. D. and Justice, R. E. (1990) Improved performance of the Boeing/LANL free-electron laser experiment: extraction efficiency and cavity-length detuning effects. *Nucl. Instr. Meth.*, **A296**, 169.
27. Warren, R. W., Newnam, B. E., Winston, J. G., Stein, W. E., Young, L. M. and Brau, C. A. (1983) Results of the Los Alamos free-electron laser experiment. *IEEE J. Quantum Electron.*, **QE-19**, 391.
28. Newnam, B. E., Warren, R. W., Sheffield, R. L., Stein, W. E., Lynch, M. T., Fraser, J. S., Goldstein, J. C., Sollid, J. E., Swann, T. A., Watson, J. M. and Brau, C. A. (1985) Optical performance of the Los Alamos free-electron laser. *IEEE J. Quantum Electron.*, **QE-21**, 867.
29. Warren, R. W., Newnam, B. E. and Goldstein, J. C. (1985) Raman spectra and the Los Alamos free-electron laser. *IEEE J. Quantum Electron.*, **QE-21**, 882.
30. Sheffield, R. L., Stein, W. E., Warren, R. W., Fraser, J. S. and Lumpkin, A. H. (1985) Electron beam diagnostics and results for the Los Alamos free-electron laser. *IEEE J. Quantum Electron.*, **QE-21**, 895.
31. Lynch, M. T., Warren, R. W. and Tellerico, P. J. (1985) The effects of linear accelerator noise on the Los Alamos free-electron laser. *IEEE J. Quantum Electron.*, **QE-21**, 904.
32. Watson, J. M. (1986) Status of the Los Alamos free-electron laser. *Nucl. Instr. Meth.*, **A250**, 1.
33. Goldstein, J. C., Newnam, B. E., Warren, R. W. and Sheffield, R. L. (1986) Comparison of the results of theoretical calculations with experimental measurements from the Los Alamos free-electron laser oscillator experiment. *Nucl. Instr. Meth.*, **A250**, 4.
34. Stein, W. E. and Sheffield, R. L. (1986) Electron micropulse diagnostics and results for the Los Alamos free-electron laser. *Nucl. Instr. Meth.*, **A250**, 12.
35. Warren, R. W., Goldstein, J. C. and Newnam, B. E. (1986) Spiking mode operation for a uniform-period wiggler. *Nucl. Instr. Meth.*, **A250**, 19.
36. Fraser, J. S., Sheffield, R. L. and Gray, E. R. (1986) A new high brightness electron injector for free-electron lasers driven by r.f. linacs. *Nucl. Instr. Meth.*, **A250**, 71.
37. Warren, R. W., Feldman, D. W., Newnam, B. E., Bender, S. C., Stein, W. E., Lumpkin, A. H., Lohsen, R. A., Goldstein, J. C., McVey, B. D. and Chan, K. C. D. (1987) Recent results from the Los Alamos free-electron laser. *Nucl. Instr. Meth.*, **A259**, 8.
38. Feldman, D. W., Warren, R. W., Stein, W. E., Fraser, J. S., Spalek, G., Lumpkin, A. H., Watson, J. M., Carlsten, B. F., Takeda, H. and Wang, T. S. (1987) Energy recovery in the Los Alamos free-electron laser. *Nucl. Instr. Meth.*, **A259**, 26.
39. Feldman, D. W., Warren, R. W., Carlsten, B. E., Stein, W. E., Lumpkin, A. H., Bender, S. C., Spalek, G., Watson, J. M., Young, L. M., Fraser, J. S., Goldstein, J. C., Takeda, H., Wang, T. S., Chan, K. C. D., McVey, B. D., Newnam, B. E., Lohsen, R. A.,

Feldman, R. B., Cooper, R. K., Johnson, W. J. and Brau, C. A. (1987) Recent results from the Los Alamos free-electron laser. *IEEE J. Quantum Electron.*, **QE-23**, 1476.
40. Goldstein, J. C., Newnam, B. E. and Warren, R. W. (1988) Sideband suppression by an intracavity optical filter in the Los Alamos free-electron laser oscillator. *Nucl. Instr. Meth.*, **A272**, 150.
41. Warren, R. W. and Goldstein, J. C. (1988) The generation and suppression of synchrotron sidebands. *Nucl. Instr. Meth.*, **A272**, 155.
42. Sheffield, R. L., Gray, E. R. and Fraser, J. S. (1988) The Los Alamos photoinjector program. *Nucl. Instr. Meth.*, **A272**, 222.
43. Carlsten, B. E., Feldman, D. W., Lumpkin, A. H., Sollid, J. E., Stein, W. E. and Warren, R. W. (1988) Emittance studies at the Los Alamos National Laboratory free-electron laser. *Nucl. Instr. Meth.*, **A272**, 247.
44. Warren, R. W., Sollid, J. E., Feldman, D. W., Stein, W. E., Johnson, W. J., Lumpkin, A. H. and Goldstein, J. C. (1989) Near-ideal lasing with a uniform wiggler. *Nucl. Instr. Meth.*, **A285**, 1.
45. Feldman, D. W., Takeda, H., Warren, R. W., Sollid, J. E., Stein, W. E., Johnson, W. J., Lumpkin, A. H. and Feldman, R. B. (1989) High extraction efficiency experiments with the Los Alamos free-electron laser. *Nucl. Instr. Meth.*, **A285**, 11.
46. Warren, R. W., Haynes, L. C., Feldman, D. W., Stein W. E. and Gitomer, S. J. (1990) Lasing on the third harmonic. *Nucl. Instr. Meth.*, **A296**, 84.
47. Lumpkin, A. H., Feldman, D. W., Sollid, J. E., Warren, R. W., Stein, W. E., Johnson, W. J., Watson, J. M., Newnam, B. E. and Goldstein, J. C. (1990) First direct observation of free-electron lasing from $\lambda = 20$ to 45 μm. *Nucl. Instr. Meth.*, **A296**, 181.
48. Feldman, D. W., Bender, S. C., Carlsten, B. E., Early, J., Feldman, R. B., Johnson, W. J. D., Lumpkin, A. H., O'Shea, P. G., Stein, W. E., Sheffield, R. L. and McKenna, K. (1991) Performance of the Los Alamos HIBAF accelerator at 17 MeV. *Nucl. Instr. Meth.*, **A304**, 224.
49. Granatstein, V. L., Schlesinger, S. P., Herndon, M., Parker, R. K. and Pasour, J. A. (1977) Production of megawatt submillimeter pulses by stimulated magneto-Raman scattering. *Appl. Phys. Lett.*, **30**, 384.
50. McDermott, D. B., Marshall, T. C., Schlesinger, S. P., Parker, R. K. and Granatstein, V. L. (1978) High-power free-electron laser based on stimulated Raman backscattering. *Phys. Rev. Lett.*, **41**, 1368.
51. Parker, R. K., Jackson, R. H., Gold, S. H., Freund, H. P., Granatstein, V. L., Efthimion, P. C., Herndon, M. and Kinkead, A. K. (1982) Axial magnetic field effects in a collective-interaction free-electron laser at millimeter wavelengths. *Phys. Rev. Lett.*, **48**, 238.
52. Jackson, R. H., Gold, S. H., Parker, R. K., Freund, H. P., Efthimion, P. C., Granatstein, V. L., Herndon, M., Kinkead, A. K., Kosakowski, J. E. and Kwan, T. J. T. (1983) Design and operation of a collective millimeter-wave free-electron laser. *IEEE J. Quantum Electron.*, **QE-19**, 346.
53. Gold, S. H., Black, W. M., Freund, H. P., Granatstein, V. L., Jackson, R. H., Efthimion, P. C. and Kinkead, A. K. (1983) Study of gain, bandwidth, and tunability of a millimeter-wave free-electron laser operating in the collective regime. *Phys. Fluids*, **26**, 2683.
54. Gold, S. H., Black, W. M., Freund, H. P., Granatstein, V. L. and Kinkead, A. K. (1984) Radiation growth in a millimeter-wave free-electron laser operating in the collective regime. *Phys. Fluids*, **27**, 746.
55. Pasour, J. A., Lucey, R. F. and Kapetanakos, C. A. (1984) Long-pulse, high-power free-electron laser with no external beam focusing. *Phys. Rev. Lett.*, **53**, 1728.
56. Gold, S. H., Hardesty, D. L., Kinkead, A. K., Barnett, L. R. and Granatstein, V. L. (1984) High-gain 35 GHz free-electron laser amplifier experiment. *Phys. Rev. Lett.*, **52**, 1218.

57. Pasour, J. A., Lucey, R. F. and Roberson, C. W. (1984) Long pulse free-electron laser driven by a linear induction accelerator, in *Free-Electron Generators of Coherent Radiation* (ed. Brau, C. A., Jacobs, S. F. and Scully, M. O.), Proc. SPIE 453, p. 328.
58. Pasour, J. A. and Gold, S. H. (1985) Free-electron laser experiments with and without a guide magnetic field: a review of the millimeter-wave free-electron laser research at the Naval Research Laboratory. *IEEE J. Quantum Electron.*, **QE-21**, 845.
59. Gold, S. H., Ganguly, A. K., Freund, H. P., Fliflet, A. W., Granatstein, V. L., Hardesty, D. L. and Kinkead, A. K. (1986) Parametric behavior of a high-gain 35 GHz free-electron laser amplifier with guide magnetic field. *Nucl. Instr. Meth.*, **A250**, 366.
60. Mathew, J. and Pasour, J. A. (1986) High-gain, long-pulse free-electron laser oscillator. *Phys. Rev. Lett.*, **56**, 1805.
61. Pasour, J. A., Mathew, J. and Kapetanakos, C. (1987) Recent results from the Naval Research Laboratory experimental free-electron laser program. *Nucl. Instr. Meth.*, **A259**, 94.
62. Jackson, R. H., Pershing, D. E. and Wood, F. (1987) Naval Research Laboratory ubitron amplifier cold test results. *Nucl. Instr. Meth.*, **A259**, 99.
63. Pershing, D. E., Jackson, R. H., Freund, H. P. and Bluem, H. (1989) The K_u-band ubitron experiment at the Naval Research Laboratory. *Nucl. Instr. Meth.*, **A285**, 56.
64. Pershing, D. E., Jackson, R. H., Bluem, H. and Freund, H. P. (1990) Fundamental mode amplifier performance of the Naval Research Laboratory ubitron. *Nucl. Instr. Meth.*, **A296**, 199.
65. Birkett, D. S., Marshall, T. C., Schlesinger, S. P. and McDermott, D. B. (1981) A submillimeter free-electron laser experiment. *IEEE J. Quantum Electron.*, **QE-17**, 1348.
66. Masud, J., Marshall, T. C., Schlesinger, S. P. and Yee, F. G. (1986) Gain measurements from start-up and spectrum of a Raman free-electron laser oscillator. *Phys. Rev. Lett.*, **56**, 1567.
67. Masud, J., Marshall, T. C., Schlesinger, S. P., Yee, F. G., Fawley, W. M., Scharlemann, E. T., Yu, S. S., Sessler, A. M. and Sternbach, E. J. (1987) Sideband control in a millimeter-wave free-electron laser. *Phys. Rev. Lett.*, **58**, 763.
68. Masud, J., Marshall, T. C., Schlesinger, S. P. and Yee, F. G. (1987) Regenerative gain in a Raman free-electron laser oscillator. *IEEE J. Quantum Electron.*, **QE-23**, 1594.
69. Yee, F. G., Masud, J., Marshall, T. C. and Schlesinger, S. P. (1987) Power and sideband studies of a Raman free-electron laser. *Nucl. Instr. Meth.*, **A259**, 104.
70. Yee, F. G., Marshall, T. C. and Schlesinger, S. P. (1988) Efficiency and sideband observations of a Raman free-electron laser oscillator with a tapered undulator. *IEEE Trans. Plasma Sci.*, **PS-16**, 162.
71. Cai, S. Y., Chang, S. P., Dodd, J. W., Marshall, T. C. and Tang, H. (1988) Optical guiding in a Raman free-electron laser: computation and experiment. *Nucl. Instr. Meth.*, **A272**, 136.
72. Dodd, J. W. and Marshall, T. C. (1990) Spiking radiation in the Columbia free-electron laser. *Nucl. Instr. Meth.*, **A296**, 4.
73. Fajans, J., Bekefi, G., Yin, Y. Z. and Lax, B. (1984) Spectral measurements from a tunable, Raman free-electron laser. *Phys. Rev. Lett.*, **53**, 246.
74. Fajans, J., Bekefi, G., Yin, Y. Z. and Lax, B. (1985) Microwave studies of a tunable free-electron laser in combined axial and wiggler magnetic fields. *Phys. Fluids*, **28**, 1995.
75. Fajans, J., Wurtele, J. S., Bekefi, G., Knowles, D. S. and Xu, K. (1986) Nonlinear power saturation and phase in a collective free-electron laser amplifier. *Phys. Rev. Lett.*, **57**, 579.
76. Fajans, J. and Bekefi, G. (1986) Measurements of amplification and phase shift in a free-electron laser. *Phys. Fluids*, **29**, 3461.

77. Fajans, J. and Bekefi, G. (1987) Effect of electron beam temperature on the gain of a collective free-electron laser. *IEEE J. Quantum Electron.*, **QE-23**, 1617.
78. Hartemann, F., Xu, K., Bekefi, G., Wurtele, J. S. and Fajans, J. (1987) Wave-profile modification induced by the free-electron laser interaction. *Phys. Rev. Lett.*, **59**, 1177.
79. Hartemann, F., Xu, K. and Bekefi, G. (1988) Pulse compression in a free-electron laser amplifier. *Nucl. Instr. Meth.*, **A272**, 125.
80. Hartemann, F., Xu, K. and Bekefi, G. (1988) Generation of short pulses of coherent electromagnetic radiation in a free-electron laser amplifier. *IEEE J. Quantum Electron.*, **QE-24**, 105.
81. Leibovitch, C., Xu, K. and Bekefi, G. (1988) Effects of electron prebunching on the radiation growth rate of a collective free-electron laser amplifier. *IEEE J. Quantum Electron.*, **QE-24**, 1825.
82. Xu, K., Bekefi, G. and Leibovitch, C. (1989) Observations of field profile modifications in a Raman free-electron laser amplifier. *Phys. Fluids B*, **1**, 2066.
83. Kirkpatrick, D. A., Bekefi, G., DiRienzo, A. C., Freund, H. P. and Ganguly, A. K. (1989) A millimeter and submillimeter wavelength free-electron laser. *Phys. Fluids B*, **1**, 1511.
84. Xu, K. and Bekefi, G. (1990) Experimental study of multiple frequency effects in a free-electron laser amplifier. *Phys. Fluids B*, **2**, 678.
85. Orzechowski, T. J., Anderson, B. R., Fawley, W. M., Prosnitz, D., Scharlemann, E. T., Yarema, S. M., Hopkins, D. B., Paul, A. C., Sessler, A. M. and Wurtele, J. S. (1985) Microwave radiation from a high-gain free-electron laser amplifier. *Phys. Rev. Lett.*, **54**, 889.
86. Orzechowski, T. J., Anderson, B. R., Clark, J. C., Fawley, W. M., Paul, A. C., Prosnitz, D., Scharlemann, E. T., Yarema, S. M., Hopkins, D. B., Sessler, A. M. and Wurtele, J. S. (1986) High-efficiency extraction of microwave radiation from a tapered-wiggler free-electron laser. *Phys. Rev. Lett.*, **57**, 2172.
87. Orzechowski, T. J., Scharlemann, E. T. and Hopkins, D. B. (1987) Measurement of the phase of the electromagnetic wave in a free-electron laser amplifier. *Phys. Rev. A*, **35**, 2184.
88. Jong, R. A. and Scharlemann, E. T. (1987) High gain free-electron laser for heating and current drive in the ALCATOR-C Tokamak. *Nucl. Instr. Meth.*, **A259**, 254.
89. Throop, A. L., Fawley, W. M., Jong, R. A., Orzechowski, T. J., Prosnitz, D., Scharlemann, E. T., Stever, R. D., Westenskow, G. A., Hopkins, D. B., Sessler, A. M., Evangelides, S. G. and Kreischer, K. E. (1988) Experimental results of a high gain microwave free-electron laser operating at 140 GHz. *Nucl. Instr. Meth.*, **A272**, 15.
90. Jong, R. A. and Stone, R. R. (1989) Induction-linac based free-electron laser amplifier for fusion applications. *Nucl. Instr. Meth.*, **A285**, 387.
91. Throop, A. L., Jong, R. A., Allen, S. L., Atkinson, D. P., Clark, J. C., Felker, B., Ferguson, S. W., Makowski, M. A., Nexsen, W. E., Rice, B. W., Stallard, B. W. and Turner, W. C. (1990) 140 GHz microwave free-electron laser experiments using ELF-II. *Nucl. Instr. Meth.*, **A296**, 41.
92. Boehmer, H., Caponi, M. Z., Edighoffer, J., Fornaca, S., Munch, J., Neil, G., Saur, B. and Shih, C. C. (1981) An experiment on free-electron laser efficiency enhancement with a variable wiggler. *IEEE Trans. Nucl. Sci.*, **NS-28**, 3156.
93. Boehmer, H., Caponi, M. Z., Edighoffer, J., Fornaca, S., Munch, J., Neil, G. R., Saur, B. and Shih, C. C. (1982) Variable wiggler free-electron laser experiment. *Phys. Rev. Lett.*, **48**, 141.
94. Arnush, D., Boehmer, H., Caponi, M. Z. and Shih, C. C. (1982) Design of a high power CW free-electron maser. *Int. J. Electron.*, **53**, 605.
95. Felch, K. L., Vallier, L., Buzzi, J. M., Drossart, P., Boehmer, H., Doucet, H. J., Etlicher, B., Lamain, H. and Rouillé, C. (1981) Collective free-electron laser studies. *IEEE J. Quantum Electron.*, **QE-17**, 1354.
96. Buzzi, J. M. (1983) Collective free-electron lasers, in *Fifth International Conference on*

High-Power Particle Beams (ed. Briggs, R. and Toepfer, A. J.), LLNL, Livermore, California, p. 546.

97. Dolezal, F. A., Harvey, R. J., Palmer, A. J. and Schumacher, R. W. (1984) Mode structure of a low-gain low-voltage free-electron laser, in *Free-Electron Generators of Coherent Radiation* (ed. Brau, C. A., Jacobs, S. F. and Scully, M. O.), Proc. SPIE 453, p. 356.
98. Harvey, R. J. and Dolezal, F. A. (1986) The Hughes low-voltage free-electron laser program. *Nucl. Instr. Meth.*, **A250**, 274.
99. Roberson, C. W. (1985) Free-electron laser beam quality. *IEEE J. Quantum Electron.*, **QE-21**, 860.
100. Masud, J., Marshall, T. C., Schlesinger, S. P., Yee, F. G., Fawley, W. M., Scharlemann, E. T., Yu, S. S., Sessler, A. M. and Sternbach, E. J. (1987) Sideband control in a millimeter-wave free-electron laser. *Phys. Rev. Lett.*, **58**, 763.
101. Colson, W. B. (1976) Theory of a free-electron laser. *Phys. Lett.*, **A59**, 187.
102. Becker, W. (1980) Multiphoton analysis of the free-electron laser. *Opt. Commun.*, **33**, 69.
103. Becker, W., Scully, M. O. and Zubairy, M. S. (1982) Generation of squeezed coherent states via a free-electron laser. *Phys. Rev. Lett.*, **48**, 475.
104. Becker, W. and Zubairy, M. S. (1982) Photon statistics of a free-electron laser. *Phys. Rev. A*, **25**, 2200.
105. Becker, W. and McIver, J. K. (1983) Fully quantized many-particle theory of a free-electron laser, *Phys. Rev. A*, **27**, 1030.
106. Dattoli, G. and Renieri, A. (1983) Free-electron laser quantum aspects. *J. Physique*, **44**, C1–125.
107. Bosco, P., Colson, W. B. and Freedman, R. A. (1983) Quantum/classical mode evolution in free-electron laser oscillators. *IEEE J. Quantum Electron.*, **QE-19**, 272.
108. Dattoli, G. and Richetta, M. (1984) Free-electron laser quantum theory: comments on Glauber coherence, antibunching and squeezing. *Opt. Commun.*, **50**, 165.
109. Ciocci, F., Dattoli, G. and Richetta, M. (1984) Analysis of the harmonic Raman-Nath equation. *J. Phys. A*, **17**, 1333.
110. Dattoli, G., Gallardo, J., Renieri, A. and Richetta, M. (1985) Quantum statistical properties of a free-electron laser amplifier. *IEEE J. Quantum Electron.*, **QE-21**, 1069.
111. Gea-Banacloche, J., Moore, G. T., Schlicher, R. R., Scully, M. O. and Walther, H. (1987) Soft X-ray free-electron laser with a laser undulator. *IEEE J. Quantum Electron.*, **QE-23**, 1558.
112. Brau, C. A. (1990) *Free-Electron Lasers*, Academic Press, Boston.
113. Luchini, P. and Motz, H. (1990) *Undulators and Free-Electron Lasers*, Clarendon Press, Oxford.
114. Colson, W. B. (1990) A limitation on the start-up of compact free-electron lasers. *Nucl. Instr. Meth.*, **A296**, 348.
115. Prazeres, R., Guyot-Sionnest, P., Ortega, J. M., Jaroszynski, D., Billardon, M., Couprie, M. E., Velghe, M. and Petroff, Y. (1991) Coherent harmonic generation in VUV with the optical klystron on the storage ring Super-ACO. *Nucl. Instr. Meth.*, **A304**, 72.
116. Couprie, M. E., Gavrilov, N. G., Kulipanov, G. N., Litvinenko, V. N., Panaev, I. V., Popik, V. M., Skrinsky, A. N. and Vinokurov, N. A. (1991) The results of lasing linewidth narrowing on the VEPP-3 storage ring optical klystron. *Nucl. Instr. Meth.*, **A304**, 47.
117. Litvinenko, V. N. and Vinokurov, N. A. (1991) Lasing spectrum and temporal structure in storage ring free-electron lasers: theory and experiment. *Nucl. Instr. Meth.*, **A304**, 66.
118. Dowell, D. H., Laucks, M. L., Lowrey, A. R., Bemes, M., Currie, A., Johnson, P., McCrary, K., Milliman, L., Lancaster, C., Adamski, J., Pistoresi, D., Shoffstall, D. R., Bentz, M., Burns, R., Guha, J., Hudyma, R., Sun, K., Tomita, W., Mower, W., Bender,

S., Goldstein, J. C., Lumpkin, A. H., McVey, B. D., Tokar, R. L. and Shemwell, D. (1991) First operation of a free-electron laser using a ring resonator. *Nucl. Instr. Meth.*, **A304**, 1.

119. Laucks, M. L., Dowell, D. H., Lowrey, A. R., Bemes, M., Currie, A., Johnson, P., McCrary, K., Adamski, J., Pistoresi, D., Shoffstall, D. R., Bentz, M., Burns, R., Hudyma, R., Sun, K., Mower, W., Bender, S., Goldstein, J. C., Lumpkin, A. H., McVey, B. D., Tokar, R. L. and Shemwell, D. (1991) Optical measurements on the Boeing free-electron laser ring resonator experiment. *Nucl. Instr. Meth.*, **A304**, 25.
120. Frisch, J. C. and Edighoffer, J. A. (1990) Time-dependent measurements on the SCA free-electron laser. *Nucl. Instr. Meth.*, **A296**, 9.
121. Edighoffer, J. A., Boehmer, H., Caponi, M. Z., Fornaca, S., Munch, J., Neil, G. R., Saur, B. and Shih, C. C. (1983) Free-electron laser small signal gain measurement at 10.6 μm. *IEEE J. Quantum Electron.*, **QE-19**, 316.
122. Edighoffer, J. A., Neil, G. R., Hess, C. E., Smith, T. I., Fornaca, S. W. and Schwettman, H. A. (1984) Variable-wiggler free-electron laser oscillation. *Phys. Rev. Lett.*, **52**, 344.
123. Ohigashi, N., Mima, K., Miyamoto, S., Imasaki, K., Kitagawa, Y., Fujita, M., Kuruma, S. I., Nakai, S. and Yamanaka, C. (1987) Free-electron laser experiments with the pulse power beam at the Institute for Laser Engineering at Osaka. *Nucl. Instr. Meth.*, **A259**, 111.
124. Akiba, T., Fukuda, M., Noma, M., Inoue, N., Tateishi, K., Imasaki, K., Miyamoto, S., Ohigashi, N., Mima, K., Nakai, S. and Yamanaka, C. (1987) Development of an inductive voltage accumulating system for a free-electron laser. *Nucl. Instr. Meth.*, **A259**, 115.
125. Mima, K., Kitagawa, Y., Akiba, T., Imasaki, K., Kuruma, S., Ohigashi, N., Miyamoto, S., Fujita, M., Nakayama, S., Tsunawaki, Y., Motz, H., Taguchi, T., Nakai, S. and Yamanaka, C. (1988) Experiment and theory on CO_2 laser powered wiggler and induction linac free-electron laser. *Nucl. Instr. Meth.*, **A272**, 106.
126. Mima, K., Imasaki, K., Kuruma, S., Akiba, T., Ohigashi, N., Tsunawaki, Y., Tanaka, K., Yamanaka, C. and Nakai, S. (1989) Theory and experiment for the induction linac free-electron laser. *Nucl. Instr. Meth.*, **A285**, 47.
127. Hiramatsu, S., Ebihara, K., Kimura, Y., Kishiro, J., Kumada, M., Kurino, H., Mizumaki, Y., Ozaki, T. and Takayama, K. (1989) Proposal for an X-band single-stage free-electron laser. *Nucl. Instr. Meth.*, **A285**, 83.
128. Conde, M. E. and Bekefi, G. (1991) Experimental study of a 33.3 GHz free-electron laser amplifier with a reversed axial guide magnetic field. *Phys. Rev. Lett.*, **67**, 3082.
129. Conde, M. E. and Bekefi, G. (1992) Amplification and superradiant emission from a 33.3 GHz free-electron laser with a reversed axial guide magnetic field. *IEEE Trans. Plasma Sci.*, **20**, 240.
130. Bhowmik, A., Curtin, M. S., McMullin, W. A., Benson, S. V., Madey, J. M. J., Richman, B. A. and Vintro, L. (1988) First operation of the Rocketdyne/Stanford free-electron laser. *Nucl. Instr. Meth.*, **A272**, 10.
131. Curtin, M., Bennett, G., Burke, R., Bhowmik, A., Metty, P., Benson, S. V. and Madey, J. M. J. (1990) First demonstration of a free-electron laser driven by electrons from a laser-irradiated photocathode. *Nucl. Instr. Meth.*, **A296**, 127.
132. Bhowmik, A., Curtin, M. S., McMullin, W. A., Benson, S. V., Madey, J. M. J., Richman, B. A. and Vintro, L. (1990) Initial results from the free-electron laser master oscillator power amplifier experiment. *Nucl. Instr. Meth.*, **A296**, 20.
133. Lazar, N. H., Boehmer, H., Fornaca, S., Hauss, B., Quon, B., Roy, N., Talmadge, S., Thompson, H. Jr and Neil, G. (1991) High brightness injector for a high power superconducting rf free-electron laser. *Nucl. Instr. Meth.*, **A304**, 243.
134. Urbanus, W. H., van der Geer, C. A. J., Bongers, W. A., van Dijk, G., Elzendoorn, B. S. Q., van Ieperen, J. P., van der Linden, A., Manintveld, P., Sterk, A. B., Tulupov, A. V., Verhoeven, A. G. A., van der Wiel, M. J., Varfolomeev, A. A., Ivanchenkov,

S. N., Khelbnikov, A. S., Bratman, V. L. and Denisov, G. G. (1995) A free-electron maser for fusion. *Nucl. Instr. Meth.*, **A358**, 155.

135. Balleyguier, P., Blésés, J. P., Bloquet, A., Bonetti, C., De Brion, J. P., Brisset, J., Dei-Cas, R., Di Crescenzo, J., Fourdin, P., Haouat, G., Joly, R., Joly, S., Laspalles, C., Laget, J. P., Leboutet, H., Schumann, F., Striby, S. and Vouillarmet, J. (1991) Production of high-brightness electron beams for the ELSA free-electron laser. *Nucl. Instr. Meth.*, **A304**, 308.
136. Takeda, H., Davis, K. and Delo, L. (1991) Beam quality simulation of the Boeing photoinjector accelerator for the MCTD project. *Nucl. Instr. Meth.*, **A304**, 396.
137. Elias, L. R., Ramian, G., Hu, R. J. and Amir, A. (1986) Observation of single-mode operation in a free-electron laser. *Phys. Rev. Lett.*, **57**, 424.
138. Amir, A., Elias, L. R., Gregoire, D. J., Kotthaus, J., Ramian, G. J. and Stern, A. (1986) Spectral characteristics of the University of California at Santa Barbara free-electron laser 400 μm experiment. *Nucl. Instr. Meth.*, **A250**, 35.
139. Elias, L. R. (1987) Free-electron laser research at the University of California at Santa Barbara. *IEEE J. Quantum Electron.*, **QE-23**, 1470.
140. Ramian, G. and Elias, L. R. (1988) The new University of California at Santa Barbara far-infrared free-electron laser. *Nucl. Instr. Meth.*, **A272**, 81.
141. Amir, A., Hu, R. J., Kielmann, F., Mertz, J. and Elias, L. R. (1988) Injection locking experiment at the University of California at Santa Barbara free-electron laser. *Nucl. Instr. Meth.*, **A272**, 174.
142. Amir, A., Knox-Seith, J. F. and Warden, M. (1991) Narrow-bandwidth operation of a free-electron laser enforced by seeding. *Phys. Rev. Lett.*, **66**, 29.
143. Elias, L. R. and Kimmel, I. (1990) Confirmation of single-mode free-electron laser operation. *Nucl. Instr. Meth.*, **A296**, 144.
144. Kaminski, J. (1990) Current applications of the University of California at Santa Barbara free-electron laser. *Nucl. Instr. Meth.*, **A296**, 784.
145. Benson, S. V. and Madey, J. M. J. (1989) Demonstration of harmonic lasing in a free-electron laser. *Phys. Rev. A*, **39**, 1579.
146. Benson, S. V., Madey, J. M. J., Schultz, J., Marc, M., Wadensweiler, W., Westenskow, G. A. and Velghe, M. (1986) The Stanford Mark III infrared free-electron laser. *Nucl. Instr. Meth.*, **A250**, 39.
147. Benson, S. V., Schultz, J., Hooper, B. A., Crane, R. and Madey, J. M. J. (1988) Status report on the Stanford Mark III infrared free-electron laser. *Nucl. Instr. Meth.*, **A272**, 22.
148. Carlson, K., Fann, W. and Madey, J. M. J. (1988) Spatial distribution of visible coherent harmonics generated by the Mark III free-electron laser. *Nucl. Instr. Meth.*, **A272**, 92.
149. Hooper, B. A., Benson, S. V., Cutolo, A. and Madey, J. M. J. (1988) Experimental results of two stage harmonic generation with picosecond pulses on the Stanford Mark III free-electron laser. *Nucl. Instr. Meth.*, **A272**, 96.
150. La Sala, J. E., Deacon, D. A. G. and Madey, J. M. J. (1988) Optical guiding measurements on the Mark III free-electron laser oscillator. *Nucl. Instr. Meth.*, **A272**, 141.
151. Bamford, D. J. and Deacon, D. A. G. (1989) Measurement of the coherent harmonics emitted in the Mark III free-electron laser. *Nucl. Instr. Meth.*, **A285**, 23.
152. Bamford, D. J. and Deacon, D. A. G. (1990) Harmonic generation experiments on the Mark III free-electron laser. *Nucl. Instr. Meth.*, **A296**, 89.
153. Benson, S. V., Fann, W. S., Hooper, B. A., Madey, J. M. J., Szarmes, E. B., Richman, B. and Vintro, L. (1990) A review of the Stanford Mark III infrared free-electron laser program. *Nucl. Instr. Meth.*, **A296**, 110.
154. Schwettman, H. A., Smith, T. I. and Swent, R. L. (1994) The Stanford Free-Electron Laser Center. *Nucl. Instr. Meth.*, **A341**, ABS 19.

155. Edwards, G. S. and Tolk, N. H. (1988) Vanderbilt University Free-Electron Laser Research Center for biomedical and materials research. *Nucl. Instr. Meth.*, **A272**, 37.
156. Brau, C. A. and Mendenhall, M. H. (1994) Medical and materials research at the Vanderbilt University free-electron laser center. *Nucl. Instr. Meth.*, **A341**, ABS 21.
157. Goldstein, J. C., McVey, B. D. and Newnam, B. E. (1990) Optical design and performance of an XUV free-electron laser oscillator. *Nucl. Instr. Meth.*, **A296**, 288.
158. Booske, J. H., Radack, D. J., Antonsen, T. M., Bidwell, S. W., Carmel, Y., Destler, W. W., Freund, H. P., Granatstein, V. L., Latham, P. E., Levush, B., Mayergoyz, I. D. and Serbeto, A. (1990) Design of high-average-power near-millimeter free-electron laser oscillators using short-period wigglers and sheet electron beams. *IEEE Trans. Plasma Sci.*, **PS-18**, 399.
159. Bidwell, S. W., Radack, D. J., Antonsen, T. M. Jr, Booske, J. H., Carmel, Y., Destler, W. W., Granatstein, V. L., Levush, B., Latham, P. E., Mayergoyz, I. D. and Zhang, Z. X. (1991) A high-average-power tapered free-electron laser amplifier at submillimeter frequencies using sheet electron beams and short-period wigglers. *Nucl. Instr. Meth.*, **A304**, 187.
160. van Amersfoort, P. W., Urbanus, W. H., Verhoeven, A. G. A., Verheul, A., Sterk, A. B., van Ingen, A. M. and van der Wiel, M. J. (1991) An electrostatic free-electron maser for fusion: design considerations. *Nucl. Instr. Meth.*, **A304**, 168.
161. Zhang, Z. X., Granatstein, V. L., Destler, W. W., Levush, B., Antonsen, T. M. Jr, Rodgers, J. and Cheng, S. (1994) First operation of a 94 GHz sheet beam free-electron laser amplifier. *Nucl. Instr. Meth.*, **A341**, 76.
162. Oepts, D., Bakker, R. J., Jaroszynski, D. A., van der Meer, A. F. G. and van Amersfoort, P. W. (1992) Induced and spontaneous interpulse phase locking in a free-electron laser. *Phys. Rev. Lett.*, **68**, 3543.
163. Oepts, D., van der Meer, A. F. G., Bakker, R. J. and van Amersfoort, P. W. (1993) Selection of single-mode radiation from a short pulse free-electron laser. *Phys. Rev. Lett.*, **70**, 3255.
164. Jaroszynski, D. A., Bakker, R. J., van der Meer, A. F. G., Oepts, D. and van Amersfoort, P. W. (1993) Experimental observation of limit-cycle oscillations in a short pulse free-electron laser. *Phys. Rev. Lett.*, **70**, 3412.
165. Faatz, B., Best, R. W. B., Oepts, D. and van Amersfoort, P. W. (1993) Hole coupling in free-electron lasers. *IEEE J. Quantum Electron.*, **QE-29**, 2229.
166. Bakker, R. J., Knippels, G. M. H., van der Meer, A. F. G., Oepts, D., Jaroszynski, D. A. and van Amersfoort, P. W. (1993) Dynamic desynchronization of a free-electron laser resonator. *Phys. Rev. E*, **48**, R3256.
167. Jaroszynski, D. A., Bakker, R. J., van der Meer, A. F. G., Oepts, D. and van Amersfoort, P. W. (1993) Coherent start-up of an infrared free-electron laser. *Phys. Rev. Lett.*, **71**, 3798.
168. Bakker, R. J., Jaroszynski, D. A., van der Meer, A. F. G., Oepts, D. and van Amersfoort, P. W. (1994) Short-pulse effects in a free-electron laser. *IEEE J. Quantum Electron.*, **QE-30**, 1635.
169. Robinson, K. E., Deacon, D. A. G., Velghe, M. F. and Madey, J. M. J (1983) Laser induced bunch lengthening on the ACO storage ring free-electron laser. *IEEE J. Quantum Electron.*, **QE-19**, 365.
170. Ortega, J. M. (1986) Harmonic generation in the VUV on a storage ring and prospects for Super-ACO at Orsay. *Nucl. Instr. Meth.*, **A250**, 203.
171. Billardon, M., Ellaume, P., Ortega, J. M., Bazin, C., Bergher, M., Couprie, M. E., Lapierre, Y., Petroff, Y., Prazeres, R. and Velghe, M. (1987) Status of the Orsay free-electron laser experiment. *Nucl. Instr. Meth.*, **A259**, 72.
172. Prazérès, R., Ortega, J. M., Bazin, C., Bergher, M., Billardon, M., Couprie, M. E., Velghe, M. and Petroff, Y. (1988) Coherent harmonic generation in the vacuum ultraviolet spectral range on the storage ring ACO. *Nucl. Instr. Meth.*, **A272**, 68.
173. Couprie, M. E., Billardon, M., Velghe, M., Bazin, C., Bergher, M., Fang, H., Ortega,

J. M., Petroff, Y. and Prazérès, R. (1988) Optical properties of multilayer mirrors exposed to synchrotron radiation. *Nucl. Instr. Meth.*, **A272**, 166.

174. Couprie, M. E., Bazin, C., Billardon, M. and Velghe, M. (1989) Spontaneous emission of the Super-ACO free-electron laser optical klystron DOMINO. *Nucl. Instr. Meth.*, **A285**, 31.
175. Couprie, M. E., Billardon, M., Velghe, M., Bazin, C., Ortega, J. M., Prazérès, R. and Petroff, Y. (1990) Free-electron laser oscillation on the Super-ACO storage ring at Orsay. *Nucl. Instr. Meth.*, **A296**, 13.
176. Garzella, D., Couprie, M. E., Delboulbé, A., Hara, T. and Billardon, M. (1994) Temporal behavior and longitudinal instabilities on the Super-ACO free-electron laser. *Nucl. Instr. Meth.*, **A341**, 24.
177. Ortega, J. M., Bergher, M., Chaput, R., Dael, A., Velghe, M., Petroff, Y., Bourdon, J. C., Belbeoch, R., Brunet, P., Dabin, Y., Mouton, B., Perrine, J. P., Plouvier, E., Pointal, R., Renard, M., Roch, M., Rodier, J., Roudier, P., Thiery, Y., Bourgeois, P., Carlos, P., Hezard, C., Fagot, J., Fallou, J. L., Garganne, P., Malglaive, J. C. and Tran, D. T. (1989) CLIO: collaboration for an infrared laser at Orsay. *Nucl. Instr. Meth.*, **A285**, 97.
178. Glotin, F., Berset, J. M., Chaput, R., Jaroszynski, D. A., Ortega, J. M. and Prazérès, R. (1994) Bunch length measurements on CLIO. *Nucl. Instr. Meth.*, **A341**, 49.
179. Prazérès, R., Glotin, F. and Ortega, J. M. (1994) Optical mode analysis on the CLIO infrared free-electron laser. *Nucl. Instr. Meth.*, **A341**, 54.
180. Ortega, J. M. (1994) Operation of the CLIO infrared user facility. *Nucl. Instr. Meth.*, **A341**, 138.
181. O'Shea, P. G., Bender, S. C., Boyd, D. A., Early, J. W., Feldman, D. W., Fortgang, C. M., Goldstein, J. C., Newnam, B. E., Sheffield, R. L., Warren, R. W. and Zhang, T. J. (1994) Demonstration of ultraviolet lasing with a low energy electron beam. *Nucl. Instr. Meth.*, **A341**, 7.
182. Colson, W. B. (1995) Status of short wavelength free-electron lasers. *Nucl. Instr. Meth.*, **A358**, 555.
183. Freund, H. P. and Granatstein, V. L. (1995) Status of long wavelength free-electron lasers. *Nucl. Instr. Meth.*, **A358**, 551.
184. Nguyen, D. C., Austen, R. H., Chan, K. C. D., Fortgang, C., Johmson, W. J. D., Goldstein, J., Gierman, S. M., Kinross-Wright, J., Kong, S. H., Meier, K. L., Plato, J. G., Russell, S. J., Sheffield, R. L., Sherwood, B. A., Timmer, C. A., Warren, R. W. and Webber, M. E. (1994) Initial performance of the Los Alamos Advanced Free-Electron Laser. *Nucl. Instr. Meth.*, **A341**, 29.
185. Sheffield, R. L., Austen, R. H., Chan, K. C. D., Gierman, S. M., Kinross-Wright, J. M., Kong, S. H., Nguyen, D. C., Russell, S. J. and Timmer, C. A. (1994) Performance of the high brightness linac for the Advanced Free-Electron Laser initiative at Los Alamos. *Nucl. Instr. Meth.*, **A341**, 371.
186. Lehrman, I. S., Krishnaswamy, J., Hartley, R. A., Efthimion, P. C. and Austen, R. H. (1995) The Grumann compact infrared free-electron laser (CIRFEL). *Nucl. Instr. Meth.*, **A358**, ABS 5.
187. Tecimer, M. and Elias, L. R. (1994) Hybrid microundulator for the CREOL compact cw-free-electron laser. *Nucl. Instr. Meth.*, **A341**, ABS 126.
188. Neil, G. R., Benson, S. V., Bisognano, J. J., Chao, Y., Douglas, D., Dylla, H. F., Harwood, L., Leemann, C. W., Liu, H., Liger, P., Machie, D., Neuffer, D. V., Rode, C., Simrock, S. N., Sinclair, C. K., VanZeijts, J. and Yunn, B. (1995) Status report of the CEBAF infrared free-electron laser. *Nucl. Instr. Meth.*, **A341**, ABS 39.
189. Ben-Zvi, I. (1995) Milestone experiments for single pass UV/X-ray free-electron lasers. *Nucl. Instr. Meth.*, **A358**, 52.
190. Dei-Cas, R., Balleyguier, P., Bardy, J., Bertin, A., Bonetti, Cl., Cocu, F., De Brion, J. P., Frehaut, J., Haouat, G., Girardeau-Montaut, J. P., Herscovici, A., Iracane, D., Joly, R., Laget, J. P., Leboutet, H., Marmouget, J. G., Pranal, Y., Sigaud, J., Striby,

S., Vouillarmet, J. and Yvon, P. with the collaboration of Thomsen-CSF (1989) Photo-injector, accelerator chain and wiggler development programs for a high peak power of free-electron laser. *Nucl. Instr. Meth.*, **A285**, 320.

191. Dei-Cas, R., Balleyguier, P., Bertin, A., Beuve, M. A., Binet, A., Bloquet, A., Bois, R., Bonetti, C., De Brion, J. P., Cocu, F., Di Crescenzo, J., Fourdin, P., Fréhaut, J., Guilloud, M., Haouat, G., Herscovici, A., Iracane, D., Joly, R., Jouys, J. C., Laget, J. P., Laspalles, C., Leboutet, H., Marmouget, J. G., Masseron, D., De Penquer, Y., Pranal, Y., Sigaud, J., Striby, S., Véron, D., Vouillarmet, J., Aucouturier, J., Bensussan, A., Simon, M., Dubrovin, A., Le Meur, G., Adam, J. C. and Héron, A. (1990) Status report on the low-frequency photo-injector and on the infrared free-electron laser experiment (ELSA). *Nucl. Instr. Meth.*, **A296**, 209.
192. Dei-Cas, R., Balleyguier, P., Beuve, M. A., Binet, A., Biésès, J. P., Bloquet, A., Bois, R., Bonetti, Cl., Brisset, J., De Brion, J. P., Di Crescenzo, J., Dolique, J. M., Faujour, R., Fourdin, P., Fréhaut, J., Guilloud, M., Haouat, G., Iracane, D., Joly, R., Joly, S., Laget, J. P., Laspalles, Cl., Leboutet, H., Marmouget, J. G., Masseron, D., Michaud, P., De Penquer, Y., Pranal, Y., Seguin, S., Striby, S., Schumann, F. and Vouillarmet, J. (1991) Overview of the free-electron laser activities at Bruyères-le-Châtel. *Nucl. Instr. Meth.*, **A304**, 215.
193. Guimbal, Ph., Joly, S., Iracane, D., Fréhaut, J., Balleyguier, P., Binet, A., Biésès, J. P., DiCrescenzo, J., Fontenay, V., Fourdin, P., Haouat, G., Marmouget, J. G., Pranal, Y., Sabary, F., Striby, S. and Touati, D. (1994) First results in the saturation regime with the ELSA free-electron laser. *Nucl. Instr. Meth.*, **A341**, 43.
194. Joly, S., Dowell, D. H., Balleyguier, P., de Brion, J. P., Couillard, Ch., Coacolo, J. L., Di Crescenzo, J., Fourdin, P., Haouat, G., Laget, J. P., Loulergue, A., Raimbourg, J., Sabary, F., Schumann, F. and Striby, S. (1994) Magnetic bunching and brightness measurements at ELSA. *Nucl. Instr. Meth.*, **A341**, 386.
195. Bottollier-Curtet, H., Gardelle, J., Bardy, J., Bonnafond, C., Devin, A., Germain, G., Labrouche, J., Launspach, J., Le Taillandier, P. and de Mascureau, J. (1991) First free-electron laser experiment in the millimeter range of C.E.S.T.A.: the ONDINE experiment. *Nucl. Instr. Meth.*, **A304**, 197.
196. Launspach, J., Angles, J. M., Angles, M., Anthouard, P., Bardy, J., Bonnafond, C., Bottollier-Curtet, H., Devin, A., Ehyarts, P., Eyl, P., Gardelle, J., Germain, G., Grua, P., Labrouche, J., de Mascureau, J., Le Taillandier, P., Stadnikoff, W. and Thevenot, M. (1991) A high power induction linac free-electron laser application at C.E.S.T.A. *Nucl. Instr. Meth.*, **A304**, 368.
197. Rullier, J. L., Gardelle, J., Labrouche, J. and Le Taillandier, P. (1995) Strong coupling operation of a free-electron laser amplifier with an axial magnetic field. *Nucl. Instr. Meth.*, **A358**, 118.
198. Kong, G., Lucas, J. and Stuart, R. A. (1991) Low-voltage ubitrons for industrial purposes. *Nucl. Instr. Meth.*, **A304**, 238.
199. Dearden, G., Quirk, E. G., Al-Shamma'a, A. I., Stuart, R. A. and Lucas, J. (1994) Results from the Liverpool prototype industrial free-electron laser. *Nucl. Instr. Meth.*, **A341**, 80.
200. Ambrosio, M., Barbarino, G. C., Castellano, M., Cavallo, N., Cevenini, F., Masullo, M. R., Patteri, P., Preger M. and Cutolo, A. (1986) Progress report on the LELA experiment. *Nucl. Instr. Meth.*, **A250**, 239.
201. Bizzarri, U., Ciocci, F., Dattoli, G., De Angelis, A., Gallerano, G. P., Giabbai, I., Giordano, G., Letardi, T., Messina, G., Mola, A., Picardi, L., Renieri, A., Sabia, E., Vignati, A., Fiorentino, E. and Marino, A. (1986) Above threshold operation of the ENEA free-electron laser. *Nucl. Instr. Meth.*, **A250**, 254.
202. Patteri, P., Preger, M. A., Ambrosio, M., Barbarino, G., Castellano, M., Cavallo, N., Cevenini, F. and Masullo, M. R. (1987) Optical cavity alignment and mirror damage in the LELA free-electron laser experiment. *Nucl. Instr. Meth.*, **A259**, 88.

203. Cavallo, N., Cevenini, F., Masullo, M. R., Castellano, M., Guiducci, S., Patteri, P., Preger, M. A. and Serio, M. (1988) Final report on the LELA experiment. *Nucl. Instr. Meth.*, **A272**, 64.
204. Ciocci, F., Dattoli, G., De Angelis, A., Dipace, A., Doria, A., Giannessi, L., Gallerano, G. P., Renieri, A., Sabia, E., Torre, A. and Jaroszynski, D. A. (1990) Recent results of the ENEA-Frascati undulator free-electron laser experiment. *Nucl. Instr. Meth.*, **A296**, 75.
205. Castellano, M., Ghigo, A., Patteri, P., Sanelli, C., Serio, M., Tazzioli, F., Cavallo, N., Cevenini, F., Ciocci, F., Dattoli, G., Dipace, A., Gallerano, G. P., Renieri, A., Sabia, E. and Torre, A. (1990) The free-electron laser project in the Frascati INFN laboratories with the SC accelerator LISA. *Nucl. Instr. Meth.*, **A296**, 159.
206. Bonifacio, R., De Salvo, L., Pierini, P. and Young, L. (1990) A modified version of the ELFA layout. *Nucl. Instr. Meth.*, **A296**, 205.
207. Boscolo, I., Pellicoro, M., Stagno, V., Variale, V., Calbrese, R. and Tecchio, L. (1990) SEAFEL experiment: design and analysis. *Nucl. Instr. Meth.*, **A296**, 297.
208. Castellano, M., Ghigo, A., Patteri, P., Sanelli, C., Serio, M., Tazzioli, F., Cavallo, N., Cevenini, F., Ciocci, F., Dattoli, G., Dipace, A., Gallerano, G. P., Renieri, A., Sabia, E. and Torre, A. (1991) Status report of the IR free-electron laser project on the superconducting linac LISA at LNF-Frascati. *Nucl. Instr. Meth.*, **A304**, 204.
209. Castellano, M., Ferrario, M., Minestrini, M., Patteri, P., Tazzioli, F., Cevenini, F., Ciocci, F., Dattoli, G., DiPace, A., Gallerano, G. P., Renieri, A., Sabia, E., Catani, L. and Tazzari, S. (1994) Status of the commissioning of the SC linac LISA for the "SURF" free-electron laser. *Nucl. Instr. Meth.*, **A341**, ABS 43.
210. van der Slot, P. J. M., Penman, C. and Witteman, W. J. (1991) Status report of the Twente University Raman free-electron laser. *Nucl. Instr. Meth.*, **A304**, 268.
211. Zambon, P., Witteman, W. J. and van der Slot, P. J. M. (1994) Comparison between a free-electron laser amplifier and oscillator. *Nucl. Instr. Meth.*, **A341**, 88.
212. Ernst, G. J., Witteman, W. J., Haselhoff, E. H., Botman, J. I. M., Delhez, J. L. and Hagedoorn, H. L. (1990) Status of the Dutch TEUFEL project. *Nucl. Instr. Meth.*, **A296**, 304.
213. Botman, J. I. M., Delhez, J. L., Hagedoorn, H. L. Kleeven, W. J. G. M., Knoben, M. H. M., Timmermans, C. J., Webers, G. A., Ernst, G. J., Vershuur, J. W. J. and Witteman, W. J. (1994) Developments of the TEUFEL injector racetrack microtron. *Nucl. Instr. Meth.*, **A341**, 402.
214. Aab, V., Alrutz-Ziemssen, K., Gräf, H.-D., Richter, A. and Gaupp, A. (1988) The Darmstadt near-infrared free-electron laser project. *Nucl. Instr. Meth.*, **A272**, 53.
215. Wille, K. (1988) The free-electron laser storage ring project DELTA. *Nucl. Instr. Meth.*, **A272**, 59.
216. Aab, V., Alrutz-Ziemssen, K., Genz, H., Gräf, H.-D., Richter, A., Weise, H. and Gaupp, A. (1989) Status of the Darmstadt near-infrared free-electron laser project. *Nucl. Instr. Meth.*, **A285**, 76.
217. Nölle, D., Brinker, F., Negrazus, M., Schirmer, D. and Wille, K. (1990) DELTA, a new storage ring free-electron laser facility at the University of Dortmund. *Nucl. Instr. Meth.*, **A296**, 263.
218. Alrutz-Ziemssen, K. J., Auerhammer, J., Genz, H., Gräf, H.-D., Richter, A., Töpper, J. and Weise, H. (1991) Status of the Darmstadt near-infrared free-electron laser project. *Nucl. Instr. Meth.*, **A304**, 159.
219. Auerhammer, J., Genz, H., Gräf, H.-D., Grill, W., Hahn, R., Richter, A., Schlott, V., Thomas, F., Töpper, J., Wiese, H., Wesp,T. and Wiencken, M. (1994). First observation of amplification of spontaneous emission achieved with the Darmstadt infrared free-electron laser. *Nucl. Instr. Meth.*, **A341**, 63.
220. Nölle, D., Geisler, A., Ridder, M., Schmidt, T. and Wille, K. (1994) Progress on the FELICITA I experiment at DELTA. *Nucl. Instr. Meth.*, **A341**, ABS 5.

221. Ridder, M. (1994) FELICITA II a possible high gain free-electron laser at the storage ring DELTA. *Nucl. Instr. Meth.*, **A341**, ABS 11.
222. Renz, G. and Spindler G. (1995) Status of the Stuttgart Raman free-electron laser project. *Nucl. Instr. Meth.*, **A358**, ABS 13.
223. Kamamura, Y., Toyoda, K. and Kawai, M. (1987) Self mode-locked oscillation in a free-electron laser. *Nucl. Instr. Meth.*, **A259**, 107.
224. Kawarasaki, Y., Ohkubo, M., Mashiko, K., Sugimoto, M., Sawamura, M., Yoshikawa, H., Takabe, M., and Shikazono, N. (1989) A linac for a free-electron laser oscillator. *Nucl. Instr. Meth.*, **A285**, 338.
225. Ohkubo, M., Sugimoto, M., Sawamura, M., Mashiko, K., Minehara, E., Takabe, M., Sasabe, J. and Kawarasaki, Y. (1990) Status of the JAERI free-electron laser system. *Nucl. Instr. Meth.*, **A296**, 270.
226. Yamada, K., Yamazaki, T., Sugiyama, S., Tomimasu, T., Mikado, T., Chiwaki, M., Suzuki, R. and Ohgaki, H. (1991) Free-electron laser experiment at ETL. *Nucl. Instr. Meth.*, **A304**, 86.
227. Mima, K., Akiba, T., Imasaki, K., Ohigashi, N., Tsunawaki, T., Taguchi, T., Kuruma, S. I., Nakai, S. and Yamanaka, C. (1991) Distributed feedback and gas-loaded free-electron lasers driven by induction linac SHVS. *Nucl. Instr. Meth.*, **A304**, 93.
228. Kawamura, Y., Lee, B. C., Kawai, M. and Toyoda, K. (1991) Experimental studies on various characteristics of a waveguide mode free-electron laser using cold relativistic electron beams. *Nucl. Instr. Meth.*, **A304**, 111.
229. Shiho, M., Sakamoto, K., Maebara, S., Watanabi, A., Kishimoto, Y., Oda, H., Kawasaki, S., Nagashima, T. and Maeda, H. (1991) JAERI millimeter wave free-electron laser experiments with a focusing wiggler. *Nucl. Instr. Meth.*, **A304**, 141.
230. Hajima, R., Ohashi, H., Kondo, S. and Akiyama, M. (1991) Status of the free-electron laser experiment at UT/NERL. *Nucl. Instr. Meth.*, **A304**, 230.
231. Yamanaka, C. (1991) New Japanese free-electron laser program at Osaka. *Nucl. Instr. Meth.*, **A304**, 276.
232. Nishimura, E., Saeki, K., Abe, S., Kobayashi, A., Morii, Y., Keishi, T., Tomimasu, T., Hajima, R., Hara, T., Ohashi, H., Akiyama, M., Kondo, S., Yoshida, Y., Ueda, T., Kobayashi, T., Uesaka, M. and Miya, K. (1994) Optical performance of the UT-FEL at first lasing. *Nucl. Instr. Meth.*, **A341**, 39.
233. Okuda, S., Ohkuma, J., Suemine, S., Ishida, S., Yamamoto, T., Okada, T. and Takeda, S. (1994) Amplification of spontaneous emission with two high-brightness electron bunches of the ISIR linac. *Nucl. Instr. Meth.*, **A341**, 59.
234. Asakawa, M., Sakamoto, N., Inoue, N., Yamamoto, T., Mima, K., Nakai, S., Chen, J., Fujita, M., Imasaki, K., Yamanaka, C., Agari, T., Asakuma, T., Ohigashi, N. and Tsunawaki, Y. (1994) A millimeter-range free-electron laser experiment using coherent synchrotron radiation emitted from electron bunches. *Nucl. Instr. Meth.*, **A341**, 72.
235. Mizuno, T., Ohtuki, T., Ohsima, T. and Saito, H. (1994) Spectrum analysis of a circular free-electron laser. *Nucl. Instr. Meth.*, **A341**, 98.
236. Takayama, K., Kishiro, J., Ozaki, T., Ebihara, K., Hiramatsu, S. and Katoh, H. (1994) Microwave transport in the ion-channel guided free-electron laser. *Nucl. Instr. Meth.*, **A341**, 109.
237. Shiho, M., Kawasaki, S., Sakamoto, K., Maeda, H., Ishizuka, H., Watanabe, Y., Tokuchi, A., Yamashita, Y. and Nakajima, S. (1994) Design and construction of an induction linac for a mm wave free-electron laser for fusion research. *Nucl. Instr. Meth.*, **A341**, 412.
238. Yamazaki, T., Yamada, K., Sugiyama, S., Ohgaki, H., Sei, N., Mikado, T., Noguchi, T., Chiwaki, M., Suzuki, R., Kawai, M., Yokoyama, M. and Hamada, S. (1994) Present status of the NIJI-IV free-electron lasers. *Nucl. Instr. Meth.*, **A341**, ABS 3.
239. Sugimoto, M., Takao, M., Sawamura, M., Nagai, R., Kato, R., Kikuzawa, N. Minehara, E. J., Ohkubo, M., Kawarasaki, Y. and Suzuki Y. (1994) Progress of the infrared free-electron laser development at JAERI. *Nucl. Instr. Meth.*, **A341**, ABS 41.

240. Werin, S., Eriksson, M., Larsson, J., Persson, A. and Svanberg, S. (1991) Harmonic generation at the MAX-lab undulator: report of the first results. *Nucl. Instr. Meth.*, **A304**, 81.
241. Du, X. W., Ding, W., Su, Y., Chu, C. and Fu, E. S. (1988) The present status of free-electron laser research work in China. *Nucl. Instr. Meth.*, **A272**, 29.
242. Jialin, X., Jiejia, Z., Youzhi, W. and Shicai, Z. (1988) Design considerations of the Beijing free-electron laser project. *Nucl. Instr. Meth.*, **A272**, 40.
243. Liu, W., Wu, T., Yang, T., Weng, Z. and Ma, Y. (1988) A free-electron laser project at the Institute of Atomic Energy. *Nucl. Instr. Meth.*, **A272**, 50.
244. Su, Y., Huang, S., Chen, Y., Shen, Y., Fu, S., Liu, X., Li, Z. and Hu, K. (1988) A microwave free-electron laser experiment. *Nucl. Instr. Meth.*, **A272**, 147.
245. Shenggang, L. (1989) A survey of free-electron laser activity in China. *Nucl. Instr. Meth.*, **A285**, 63.
246. Xie, J., Zhuang, J., Wang, Y., Zhong, S., Ying, R. and Mao, C. (1990) Progress report on the Beijing free-electron laser project. *Nucl. Instr. Meth.*, **A296**, 244.
247. Wang, M. C., Wang, Z., Chen, J., Lu, Z. and Zhang, L. (1991) Experiments of a Raman free-electron laser with distributed feedback cavity. *Nucl. Instr. Meth.*, **A304**, 116.
248. Yang, T. L., Zhai, X. L., Zhou, W. Z., Song, Z. H., Cao, D. Z., Pan, L. H., Bu, S. F., Shi, X. Z. and Liu, C. (1991) Rf-driven free-electron laser activity at CIAE. *Nucl. Instr. Meth.*, **A304**, 280.
249. Xie, J., Zhuang, J., Huang, Y., Li, Y., Lin, S., Mao, C., Ying, R., Zhong, Y., Zhang, L., Wu, G., Zhang, Y., Li, L., Fu, E. and Liu W. (1994) First lasing of the Beijing free-electron laser. *Nucl. Instr. Meth.*, **A341**, 34.
250. Hui, Z., Zhou, C., Wu, R., Deng, J., Chen, Y., Ding, B., Tang, L., Zhang, J., Meng, F., Tao, Z., Yang, Z., Tian, S., Dong, Z. and Wu, S. (1994) Microwave free-electron laser amplifier experiments on the SG-1 device. *Nucl. Instr. Meth.*, **A341**, 85.
251. Zhou, W. Z., Yang, T. L., Zhai, X. L., Shi, X. Z., Jin, Z. M., Pu, D. X., Sun, Q., Lu, Y. Z. and Chen, W. K. (1994) Progress report of the free-electron laser injector at CIAE. *Nucl. Instr. Meth.*, **A341**, ABS 37.
252. Lee, K. C. and Chung, K. H. (1994) Construction of a microwave free-electron laser and studies of its characteristics. *Nucl. Instr. Meth.*, **A341**, ABS 51.
253. Cho, S. O., Lee, B. C., Kim, S. K., Jeong, Y. U., Choi, B. H., Lee, J. and Chung, H. (1994) Development of an electrostatic accelerator for a millimeter-wave free-electron laser. *Nucl. Instr. Meth.*, **A341**, ABS 55.
254. Cohen, M., Draiznin, M., Goldring, A., Gover, A., Pinhasi, Y., Wachtell, J., Yakover, Y., Sokolowski, J., Mandelbaum, B., Rosenberg, S., Shiloh, Y., Hazak, G., Levine, L. M. and Shahal, O. (1994) The Israeli tandem electrostatic accelerator free-electron laser – status report. *Nucl. Instr. Meth.*, **A341**, ABS 57.
255. Kaminsky, A. A., Kaminsky, A. K., Sarantsev, V. P., Sedykh, S. N., Sergeev, A. P. and Silivra, A. A. (1994) Investigation of a microwave free-electron laser with a reversed guide field. *Nucl. Instr. Meth.*, **A341**, 105.
256. Agafonov, A., Gevorgyan, G., Krastelev, E., Kustov, A., Lebedev, A., Martynchuk, N., Mikhalev, P., Fedotov, V., Zakharov, S. and Yablokov, B. (1994) Linac driver for the free-electron laser project at the P. N. Lebedev Institute. *Nucl. Instr. Meth.*, **A341**, 375.

2

The wiggler field and electron dynamics

The electron trajectories in the external magnetostatic fields in free-electron lasers are fundamental to any understanding of the operational principles, and have been the object of study for a considerable time [1–9]. The basic concept relies upon a spatially periodic magnetic field to induce an oscillatory motion in the electron beam, and the emission of radiation is derived from the corresponding acceleration. The specific character of the wiggler field can take on a variety of forms exhibiting both helical and planar symmetries. The most common wiggler configurations which have been employed to date include helically symmetric fields generated by bifilar current windings and linearly symmetric fields generated by alternating stacks of permanent magnets. However, wiggler fields generated by rotating quadrupole fields (helical symmetry) and pinched solenoidal fields (cylindrical symmetry) have also been considered.

The helical wiggler field generated by a bifilar helical current winding results in electron trajectories which exhibit circular motion in the plane transverse to the axis of symmetry of the wiggler. As a result, the parallel component of the velocity is relatively constant. This is an important consideration from the standpoint of the interaction between the beam and the radiation fields. In cylindrical coordinates, the field generated by a bifilar helix can be expressed in the form

$$\mathbf{B}_w(\mathbf{x}) = 2B_w\left[I_1'(\lambda)\hat{\mathbf{e}}_r \cos\chi - \frac{1}{\lambda} I_1(\lambda)\hat{\mathbf{e}}_\theta \sin\chi + I_1(\lambda)\hat{\mathbf{e}}_z \sin\chi \right], \tag{2.1}$$

where B_w denotes the wiggler amplitude, $\lambda = k_w r$, $\chi = \theta - k_w z$, $k_w(=2\pi/\lambda_w$, where λ_w is the wiggler period) denotes the wiggler wavenumber, and I_n and I_n' denote the modified Bessel function of the first kind of order n and its derivative, respectively. This field exhibits a local minimum on-axis (i.e. at $r = 0$) which acts to focus and to confine the beam against the effects of self-generated electric and magnetic fields. However, a uniform axial solenoidal field $\mathbf{B}_0(=B_0\hat{\mathbf{e}}_z)$ is often employed in conjunction with a helical wiggler in order to provide enhanced focusing for extremely intense electron beams. In such cases, the combined influence of the wiggler and axial magnetic fields has important consequences on the character of both the electron trajectories and the interaction.

Linearly symmetric wiggler fields induce an oscillatory electron motion in the plane transverse to that of the wiggler field itself. As a result, all components of the velocity oscillate about some mean value. While this is not an advantageous quality in comparison with the helical wiggler from the standpoint of the resonant interaction in the free-electron laser, the planar wiggler has the advantage of greater ease of adjustment than is possible with a helical wiggler. Planar wigglers may be constructed of alternating stacks of either permanent magnets or electromagnets. In its simplest form, in which the pole faces of the individual magnets are flat, the wiggler field may be expressed as

$$\boldsymbol{B}_w(\boldsymbol{x}) = B_w(\hat{\boldsymbol{e}}_y \cosh k_w y \sin k_w z + \hat{\boldsymbol{e}}_z \sinh k_w y \cos k_w z). \tag{2.2}$$

In this case, the transverse component of the wiggler field is directed along the y-axis and the electron motion is predominantly in the x–z plane. This form of planar wiggler has a local minimum at the symmetry plane for $y = 0$ and, hence, provides a focusing force for the electron beam in the direction normal to the wiggler-induced motion of the electron beam. If additional focusing is required, then the pole faces may be tapered to provide for an inhomogeneity in the plane of the wiggler-induced motion [10, 11]. In the case in which the pole faces are parabolically shaped, the wiggler field is of the form

$$\begin{aligned}\boldsymbol{B}_w(\boldsymbol{x}) = B_w\Bigg\{\cos k_w z\left[\hat{\boldsymbol{e}}_x \sinh\left(\frac{k_w x}{\sqrt{2}}\right)\sinh\left(\frac{k_w y}{\sqrt{2}}\right) + \hat{\boldsymbol{e}}_y \cosh\left(\frac{k_w x}{\sqrt{2}}\right)\cosh\left(\frac{k_w y}{\sqrt{2}}\right)\right]\\ - \sqrt{2}\hat{\boldsymbol{e}}_z \cosh\left(\frac{k_w x}{\sqrt{2}}\right)\sinh\left(\frac{k_w y}{\sqrt{2}}\right)\sin k_w z\Bigg\}.\end{aligned} \tag{2.3}$$

An axial solenoidal field is not employed to provide enhanced focusing in conjunction with a planar wiggler because the combined fields result in a $\boldsymbol{B}_0 \times \nabla \boldsymbol{B}_w$ drift which causes the beam to diverge rapidly [5].

While alternate wiggler concepts have been studied analytically, there has been little effort to perfect them in the laboratory. In the case of a rotating quadrupole configuration [12–15], the wiggler field vanishes along the axis of symmetry. For this reason, the radial profile of the electron beam employed with such a wiggler must be hollow. This requirement presents difficulties in generating beams of sufficiently low emittance and energy spread to ensure high-efficiency operation. Configurations employing pinched solenoidal fields are often referred to as **lowbitrons** in the literature [16, 17]. These devices operate by a different principle from the typical free-electron laser since the magnetic field itself does not impart a coherent transverse velocity to the beam, and the beam must be created with a large transverse velocity component. For this reason, lowbitrons can be more properly considered as a form of axially modulated cyclotron laser than as a free-electron laser. For these reasons, we shall focus our attention on both planar and bifilar helical wiggler configurations.

2.1 HELICAL WIGGLER CONFIGURATIONS

For the sake of generality, we shall consider the electron dynamics in a configuration which consists of both a helical wiggler and a uniform axial solenoidal field. The orbits in such a case are determined by the Lorentz force equation

$$\frac{d}{dt}\boldsymbol{p} = -\frac{e}{c}\mathbf{v} \times (\boldsymbol{B}_0\hat{\boldsymbol{e}}_z + \boldsymbol{B}_w), \tag{2.4}$$

where e denotes the charge of the electron, and $\mathbf{v}$ and $\boldsymbol{p}$ denote the velocity and momentum, respectively. This set of equations displays a wide variation of properties which depends upon the amplitudes of the wiggler and solenoidal fields, the wiggler period and the electron energy. For the sake of clarity of presentation it is often useful first to consider the idealized one-dimensional approximation in which the electron displacement from the axis of symmetry remains much less than a wiggler period (i.e. $r \ll \lambda_w$). In this limit, the helical wiggler field assumes the particularly simple form

$$\boldsymbol{B}_w = B_w(\hat{\boldsymbol{e}}_x \cos k_w z + \hat{\boldsymbol{e}}_y \sin k_w z), \tag{2.5}$$

which exhibits no transverse inhomogeneities to within terms of order $(k_w r)^2$. This wiggler field model describes an idealized picture of the interaction in a free-electron laser, since the lack of transverse inhomogeneities also implies that there will be no variation in either the electron trajectories or the interaction with the electromagnetic field across the beam profile.

2.1.1 Idealized one-dimensional trajectories

The trajectories in the idealized one-dimensional limit may be obtained very simply in the absence of an axial solenoidal field by noting that the vector potential of the wiggler is

$$\boldsymbol{A}_w = -\frac{B_w}{k_w}(\hat{\boldsymbol{e}}_x \cos k_w z + \hat{\boldsymbol{e}}_y \sin k_w z). \tag{2.6}$$

Therefore, both x and y are ignorable coordinates and the corresponding canonical momenta, which we denote by P_x and P_y, are constants of the motion. Since the total energy is conserved, the relativistic factor $\gamma(\equiv (1 - v^2/c^2)^{-1/2})$ is also a constant of the motion. As a result, the trajectories are given by

$$v_x = \frac{P_x}{\gamma m_e} + v_w \cos k_w z, \tag{2.7}$$

$$v_y = \frac{P_y}{\gamma m_e} + v_w \sin k_w z, \tag{2.8}$$

$$\frac{v_z}{c} = \sqrt{\left[1 - \frac{1}{\gamma^2} - \frac{v_w^2 + V_\perp^2}{c^2} + 2\frac{v_w V_\perp}{c^2}\cos(k_w z - \varphi)\right]}, \tag{2.9}$$

where m_e is the electron mass, c is the speed of light *in vacuo*, $v_w \equiv -\Omega_w/k_w$ is the wiggler-induced velocity, $V_\perp \equiv (P_x^2 + P_y^2)^{1/2}/\gamma m_e$, and $\varphi \equiv \tan^{-1}(P_y/P_x)$. Observe that the magnitude of both the transverse and axial components of the velocity are constant in the limit in which both P_x and P_y vanish, and the orbit describes a helix which is in phase with the wiggler field.

In the presence of an axial solenoidal field the orbits are more complicated, since x and y are no longer ignorable coordinates and the associated canonical momenta are not constants of the motion. For convenience, we shall work in the coordinate frame which rotates with the wiggler field and is defined by the basis vectors: $\hat{e}_1 = \hat{e}_x \cos k_w z + \hat{e}_y \sin k_w z$, $\hat{e}_2 = -\hat{e}_x \sin k_w z + \hat{e}_y \cos k_w z$, and $\hat{e}_3 = \hat{e}_z$. We shall henceforth refer to this as the **wiggler frame**. In this coordinate frame, substitution of the idealized wiggler model into the Lorentz force equation yields

$$\frac{d}{dt} v_1 = -(\Omega_0 - k_w v_3) v_2, \tag{2.10}$$

$$\frac{d}{dt} v_2 = (\Omega_0 - k_w v_3) v_1 - \Omega_w v_3, \tag{2.11}$$

$$\frac{d}{dt} v_3 = \Omega_w v_2, \tag{2.12}$$

where $\Omega_0 \equiv eB_0/\gamma m_e c$ and $\Omega_w \equiv eB_w/\gamma m_e c$. Two constants of the motion are readily obtained. The first, as in the absence of the axial field, is the total energy or γ. The second may be obtained by elimination of v_2 from equations (2.10) and (2.12), which yields

$$\frac{d}{dt} v_1 = -\frac{1}{\Omega_w} (\Omega_0 - k_w v_3) \frac{d}{dt} v_3. \tag{2.13}$$

As a result, a second constant of the motion is

$$V_1 = v_1 - \frac{(\Omega_0 - k_w v_3)^2}{2 k_w \Omega_w}. \tag{2.14}$$

The solutions to the orbit equations (2.10–12) may be reduced to quadrature by means of these constants.

Steady-state trajectories

Before we obtain closed-form solutions for the orbits, it is instructive to consider the steady-state solution [2, 6] for which v_1, v_2 and v_3 are constants. If the derivatives in equations (2.10–12) vanish, then the solutions are $v_1 = v_w$, $v_2 = 0$ and $v_3 = v_\parallel$, where

$$v_w \equiv \frac{\Omega_w v_\parallel}{\Omega_0 - k_w v_\parallel} \tag{2.15}$$

describes the wiggler-induced transverse velocity. In contrast to the orbits in the absence of an axial magnetic field, the presence of the axial field establishes a preferred direction of propagation through the wiggler in which propagation parallel to the axial field results in an enhancement in the transverse velocity. The constants which determine the transverse and axial velocities $(v_w, v_\parallel)$ are related through the requirement that the energy is conserved; hence,

$$v_w^2 + v_\parallel^2 = \left(1 - \frac{1}{\gamma^2}\right)c^2. \tag{2.16}$$

One effect of the axial solenoidal field is to increase the magnitude of the transverse wiggler-induced velocity resonantly when the Larmor period associated with the axial field is comparable to the wiggler period (i.e. $\Omega_0 \approx k_w v_\parallel$).

Stability of the steady-state trajectories

The stability of these steady-state solutions was first examined by Friedland [2] by means of an expansion $v_1 = v_m + \delta v_1$, $v_2 = \delta v_2$ and $v_3 = v_\parallel + \delta v_3$, where the velocity perturbations are assumed to be small. Expansion of the orbit equations (2.10–12) to first order in the perturbed velocity yields

$$\frac{d}{dt}\delta v_1 = -(\Omega_0 - k_w v_\parallel)\delta v_2, \tag{2.17}$$

$$\frac{d}{dt}\delta v_2 = (\Omega_0 - k_w v_\parallel)\delta v_1 - \Omega_0 \beta_w \delta v_3, \tag{2.18}$$

$$\frac{d}{dt}\delta v_3 = \Omega_w \delta v_2, \tag{2.19}$$

where $\beta_w \equiv v_w/v_\parallel$. Differentiation of equation (2.18) and substitution of the derivatives of δv_1 and δv_3 from equations (2.17) and (2.18) gives

$$\left(\frac{d^2}{dt^2} + \Omega_r^2\right)\delta v_2 = 0, \tag{2.20}$$

where the natural response frequency of the perturbation Ω_r is defined by

$$\Omega_r^2 \equiv (\Omega_0 - k_w v_\parallel)[(1 + \beta_w^2)\Omega_0 - k_w v_\parallel]. \tag{2.21}$$

As a result, the steady-state trajectories are unstable whenever $\Omega_r^2 < 0$. It should be remarked that in the absence of the wiggler $\Omega_r^2 = (\Omega_0 - k_w v_\parallel)^2$. The character of the perturbations in this limit, therefore, describes stable Larmor rotation in the axial magnetic field.

Substitution of (2.15) into (2.16) yields a quartic polynomial in the axial velocity which may be solved for $v_\parallel$ as a function of B_0, B_w, k_w and γ. There are in general four solutions for $v_\parallel$ for any given sets of the parameters. Of these solutions, one corresponds to backwards propagation (i.e. $v_\parallel < 0$), which will be ignored. The remaining steady-state solutions may be divided into two classes corresponding to the cases in which: $\Omega_0 < k_w v_\parallel$ referred to as Group I, and $\Omega_0 > k_w v_\parallel$ referred to as Group II. A representative solution for the axial velocity is shown in Fig. 2.1

as a function of the axial magnetic field for a beam energy of 1 MeV ($\gamma = 2.957$) and $\Omega_w/ck_w = 0.05$. The question of the stability of these orbits is illustrated in Fig. 2.2 in which Ω_r^2 is plotted as a function of the axial magnetic field. The Group I trajectories are, in general, multivalued functions of the axial field, and unstable

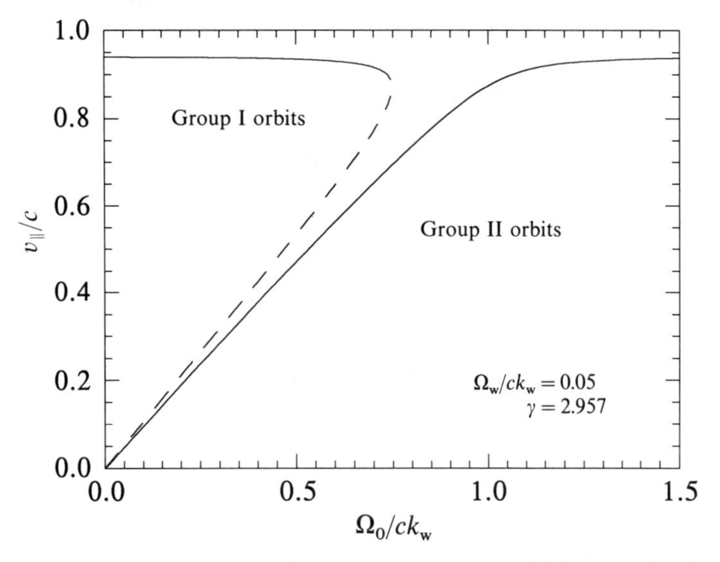

Fig. 2.1 Graph of the axial velocity of the steady-state orbits as a function of the solenoidal magnetic field showing the Group I and Group II orbits. The dashed line indicates unstable Group I trajectories.

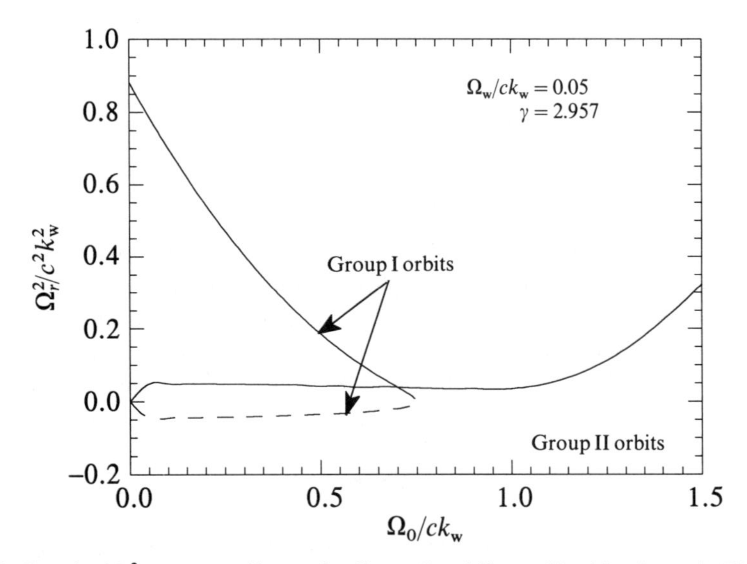

Fig. 2.2 Graph of Ω_r^2 corresponding to the Group I and Group II orbits shown in Fig. 2.1.

orbits occur whenever

$$v_{\parallel} < \frac{\Omega_0}{k_w}\left[1 + \left(\frac{\Omega_w}{\Omega_0}\right)^{2/3}\right]. \tag{2.22}$$

The unstable Group I trajectories are indicated by a dashed line in the figure. This limits the Group I trajectories to axial magnetic fields below some critical value (given by the criterion $\Omega_0/ck_w < 0.75$ for the present choice of parameters). The stable trajectories decrease monotonically with increasing B_0 up to this point corresponding to the increase in the transverse velocity as the resonance (at $\Omega_0 \approx k_w v_{\parallel}$) is approached. In contrast, Group II orbits are always stable, and the axial velocity increases monotonically with the axial magnetic field.

The steady-state trajectories describe helical orbits in the combined wiggler and axial magnetic fields. Observe that for the Group I trajectories, $v_w = -\Omega_w/k_w$ when the axial magnetic field vanishes and that this corresponds with the results given in equations (2.7) and (2.8) in the case for which $P_x = P_y = 0$. It should be remarked that the transverse velocity given in (2.15) undergoes a change in sign between the Group I and Group II orbits. It will be shown in the succeeding section that this does not indicate a change in the direction of rotation; rather, it indicates a phase shift of the trajectory in which the phases of the Group II orbits are shifted by 180° relative to both the wiggler and the Group I orbits.

Negative-mass trajectories

One further characteristic of the steady-state orbits merits discussion; specifically, the variation in the axial velocity with the electron energy. Implicit differentiation of equations (2.15) and (2.16) indicates that

$$\frac{d}{d\gamma}v_{\parallel} = \frac{c^2}{\gamma\gamma_{\parallel}^2 v_{\parallel}}\Phi, \tag{2.23}$$

where $\gamma_{\parallel} = (1 - v_{\parallel}^2/c^2)^{-1/2}$, and

$$\Phi \equiv 1 - \frac{\gamma_{\parallel}^2\beta_w^2\Omega_0}{(1+\beta_w^2)\Omega_0 - k_w v_{\parallel}}. \tag{2.24}$$

The function Φ is shown in Fig. 2.3 as a function of the axial magnetic field for both the stable Group I and Group II orbits. Observe that for Group I orbits, Φ increases monotonically from unity (at $B_0 = 0$) and exhibits a singularity at the transition to orbital instability (at which point both $\Omega_r = 0$ and $(1+\beta_w^2)\Omega_0 = k_w v_{\parallel}$). The behaviour of Φ for the Group II orbits is more interesting, however, since it is negative whenever

$$\frac{\gamma_{\parallel}^2}{\gamma^2}\Omega_0 < k_w v_{\parallel}. \tag{2.25}$$

This implies the existence of a **negative-mass regime** in which the axial velocity will increase with decreasing energy.

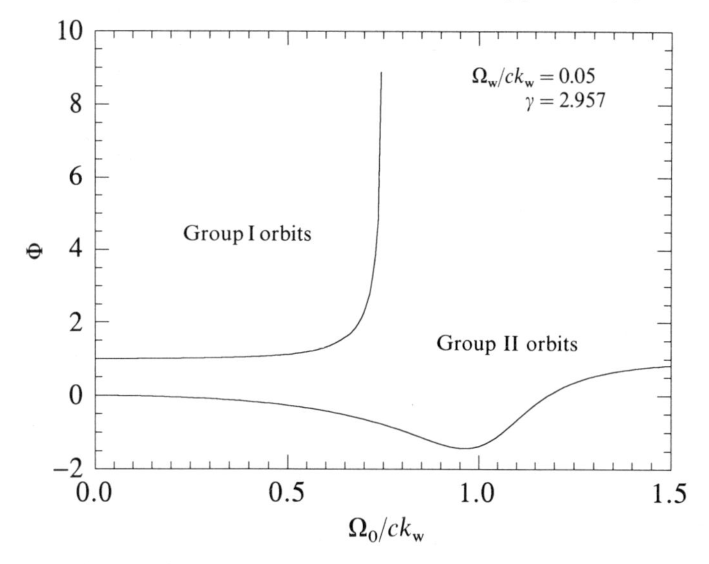

Fig. 2.3 Graph of Φ as a function of the axial magnetic field corresponding to the Group I (stable) and Group II orbits shown in Fig. 2.1.

General integration of the orbit equations

The orbit equations (2.10–12) can be reduced to a single integrable differential equation [6] using the constants of the motion. Since the total energy is a conserved quantity $v_2^2 = v_0^2 - v_1^2 - v_3^2$, where $v_0(\equiv(1-1/\gamma^2)^{1/2}c)$ denotes the magnitude of the velocity. Elimination of v_1 using equation (2.14) and substitution of the result into equation (2.12) yields an equation for $u_3(\equiv v_3/c - \Omega_0/ck_w)$

$$\left(\frac{d}{dt}u_3\right)^2 + c^2k_w^2U(u_3) = 0, \tag{2.26}$$

where the pseudopotential $U(u_3)$ is a quartic polynomial in u_3 defined by

$$U(u_3) = \tfrac{1}{4}u_3^4 + \frac{\Omega_w}{ck_w}\left(\frac{V_1}{c} + \frac{\Omega_w}{ck_w}\right)u_3^2 + 2\frac{\Omega_0}{ck_w}\left(\frac{\Omega_w}{ck_w}\right)^2 u_3 + \left(\frac{\Omega_w}{ck_w}\right)^2\left(\frac{V_1^2}{c^2} + \frac{\Omega_0^2}{c^2k_w^2} - \frac{v_0^2}{c^2}\right). \tag{2.27}$$

The form of the pseudopotential determines the character of the solutions to equation (2.26). We observe that real solutions are possible only when $U(u_3) \leqslant 0$ and, as a result, the roots of the pseudopotential define the bounds on the oscillations in u_3. An example is shown in Fig. 2.4 in which a family of curves of the pseudopotential is plotted versus u_3 for $\Omega_0/ck_w = 0.10$, $\Omega_w/ck_w = 0.01$, an electron energy of 1 MeV, and $V_1/c = -0.50$, -0.94 and -1.50. For the case in which $V_1/c = -1.50$ the pseudopotential has four real roots and the orbit is characterized by oscillations in u_3 over the approximate ranges of ± 0.21 through ± 0.11.

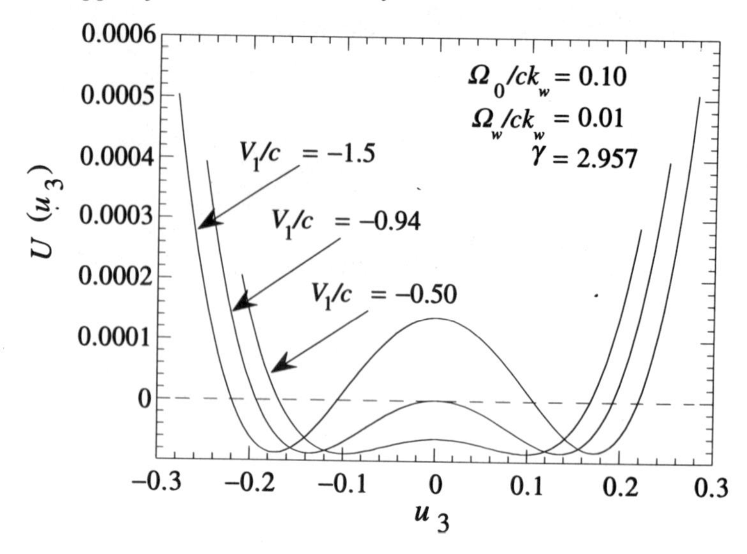

Fig. 2.4 Plots of the pseudopotential for characteristic values of $\Omega_0/ck_w = 0.10$, $\Omega_w/ck_w = 0.01$, an electron energy of 1 MeV and $V_1/c = -0.50$, -0.94 and -1.50.

Two of these roots tend to coalesce as V_1 increases until, at $V_1/c = -0.94$, they become degenerate and the pseudopotential at the central maximum vanishes. An orbit exists at this point with a uniform axial velocity, but it corresponds to one of the unstable steady-state trajectories. As V_1 increases further, the degenerate roots become a complex-conjugate pair, and there are only two real roots for the pseudopotential which define a large-amplitude oscillation which for $V_1/c = -0.50$ extends over the approximate range of from -0.16 through 0.16.

An example of the form of the pseudopotential in the vicinity of a stable steady-state orbit for $u_3 \approx -1.94$ is shown in Fig. 2.5 for $\Omega_0/ck_w = 1.0$, $\Omega_w/ck_w = 0.1$, $V_1/c = -18.8$ and an energy of 1 MeV. For this choice of parameters, the root of the pseudopotential corresponds to electron streaming anti-parallel to the solenoidal magnetic field with a velocity of $v_\parallel/c \approx -0.94$. For $V_1/c > -18.8$ there are no real roots to the quartic and no trajectories are possible. In the opposite case in which $V_1/c < -18.8$, the orbits are oscillatory.

Equation (2.26) can be integrated in closed form in terms of the Jacobi elliptic functions, which describe anharmonic oscillations. We first consider the case in which only two real roots exist and the pseudopotential is of the form $U(u_3) = (u_3 - \alpha_+)(u_3 - \alpha_-)[(u_3 - \rho)^2 + \zeta^2]$, where it is assumed that α_+, α_-, ρ and ζ are real and that $\alpha_+ > \alpha_-$. The oscillation is confined to within the range $\alpha_- \leqslant u_3 \leqslant \alpha_+$ and formal integration of equation (2.26) gives

$$\int_{\alpha_-}^{u_3(t)} \frac{dx}{\sqrt{\{(x-\alpha_-)(\alpha_+ - x)[(x-\rho)^2 + \zeta^2]\}}} = \pm ck_w\tau, \tag{2.28}$$

where $\tau = t - t_0$ and t_0 is the time at which the trajectory passes through $u_3 = \alpha_-$.

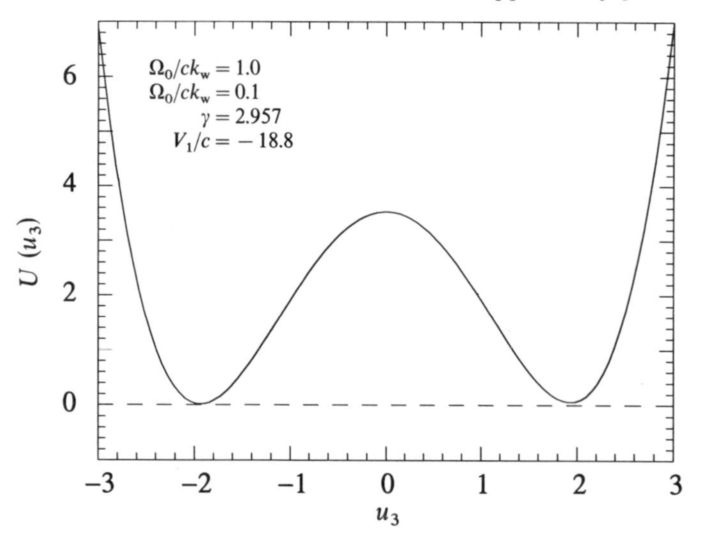

Fig. 2.5 Plot of the pseudopotential corresponding to a stable steady-state trajectory for $\Omega_0/ck_w = 1.0$, $\Omega_w/ck_w = 0.1$, $V_1/c = -18.8$, and an electron energy of 1 MeV.

Integration of equation (2.28) yields

$$\pm ck_w\tau = \frac{1}{\sqrt{pq}} F\left(2\cot^{-1}\sqrt{\left[\frac{q(\alpha_+ - u_3)}{p(u_3 - \alpha_-)}\right]}, k\right), \tag{2.29}$$

where $F(x, k)$ is the incomplete elliptic integral of argument x and modulus k, $p^2 \equiv (\rho - \alpha_+)^2 + \zeta^2$, $q^2 \equiv (\rho - \alpha_-)^2 + \zeta^2$, and

$$k = \frac{1}{2}\sqrt{\left[\frac{(\alpha_+ - \alpha_-)^2 - (p - q)^2}{pq}\right]}. \tag{2.30}$$

Inversion of equation (2.29) yields an expression for the axial velocity in terms of the Jacobi elliptic functions

$$\frac{v_\parallel}{c} = \frac{\Omega_0}{ck_w} + \frac{\alpha_+ q\,\mathrm{sn}^2(ck_w\sqrt{pq}\,\tau, k) + \alpha_- p[1 \pm \mathrm{cn}(ck_w\sqrt{pq}\,\tau, k)]^2}{q\,\mathrm{sn}^2(ck_w\sqrt{pq}\,\tau, k) + p[1 \pm \mathrm{cn}(ck_w\sqrt{pq}\,\tau, k)]^2}, \tag{2.31}$$

where sn and cn denote the Jacobi elliptic sine and cosine functions, respectively. Observe that the steady-state trajectories are recovered for the case in which the roots are degenerate (i.e. $\alpha = \alpha_+ = \alpha_-$), since $v_\parallel/c = \Omega_0/ck_w + \alpha$ in this limit.

In the case in which there are four real roots, the pseudopotential is of the form $U(u_3) = (u_3 - \alpha_+)(u_3 - \alpha_-)(u_3 - \beta_+)(u_3 - \beta_-)$, where $\alpha_- < \alpha_+ < \beta_- < \beta_+$. Oscillatory solutions occur within the ranges $\alpha_- \leqslant u_3 \leqslant \alpha_+$ and $\beta_- \leqslant u_3 \leqslant \beta_+$. Therefore we formally integrate

$$\int_{\alpha_-}^{u_3(t)} \frac{dx}{\sqrt{[(x - \alpha_-)(\alpha_+ - x)(x - \beta_+)(x - \beta_-)]}} = \pm ck_w\tau \tag{2.32}$$

for orbits within the lower range, and

$$\int_{\beta_-}^{u_3(t)} \frac{dx}{\sqrt{[(x-\alpha_-)(\alpha_+-x)(\beta_+-x)(x-\beta_-)]}} = \pm ck_w\tau, \tag{2.33}$$

for orbits within the upper range. Integration yields, respectively,

$$F(x,\kappa) = \pm\mu ck_w\tau, \tag{2.34}$$

where $2\mu \equiv [(\beta_+-\alpha_+)(\beta_--\alpha_-)]^{1/2}$, the modulus

$$\kappa \equiv \sqrt{\left[\frac{(\beta_+-\beta_-)(\alpha_+-\alpha_-)}{(\beta_+-\alpha_+)(\beta_--\alpha_-)}\right]}, \tag{2.35}$$

and

$$x = \begin{cases} \sin^{-1}\sqrt{\left[\dfrac{(\beta_+-\alpha_+)(u_3-\alpha_-)}{(\alpha_+-\alpha_-)(\beta_+-u_3)}\right]}; & \alpha_- \leqslant u_3 \leqslant \alpha_+ \\ \sin^{-1}\sqrt{\left[\dfrac{(\beta_+-\alpha_+)(u_3-\beta_-)}{(\beta_+-\beta_-)(u_3-\alpha_+)}\right]}; & \beta_- \leqslant u_3 \leqslant \beta_+ \end{cases} \tag{2.36}$$

The axial velocity can be found by inversion of equation (2.34)

$$\frac{v_\parallel}{c} = \frac{\Omega_0}{ck_w} + \frac{\alpha_-(\beta_+-\alpha_+) + \beta_+(\alpha_+-\alpha_-)\mathrm{sn}^2(\mu ck_w\tau,\kappa)}{(\beta_+-\alpha_+) + (\alpha_+-\alpha_-)\mathrm{sn}^2(\mu ck_w\tau,\kappa)}, \tag{2.37}$$

in the lower region, and

$$\frac{v_\parallel}{c} = \frac{\Omega_0}{ck_w} + \frac{\beta_-(\beta_+-\alpha_+) - \alpha_+(\beta_+-\beta_-)\mathrm{sn}^2(\mu ck_w\tau,\kappa)}{(\beta_+-\alpha_+) - (\beta_+-\beta_-)\mathrm{sn}^2(\mu ck_w\tau,\kappa)}, \tag{2.38}$$

in the upper region. As in the case in which there are only two real roots, these orbits reduce to the steady-state trajectories in the limit in which the roots become degenerate.

The general solutions for the trajectories can be expanded to obtain orbits which are arbitrarily close to the steady-state trajectories; however, it is more instructive to expand the orbit equations directly. In order to simplify the comparison with the trajectories in the absence of the axial magnetic field, we treat the orbit equations in Cartesian coordinates and separate the x- and y-components to obtain

$$\left(\frac{d^2}{dt^2} + \Omega_0^2\right)\left(v_x + \frac{\Omega_w}{k_w}\cos k_w z\right) = \frac{\Omega_w\Omega_0}{k_w}(\Omega_0 + k_w v_z)\cos k_w z, \tag{2.39}$$

$$\left(\frac{d^2}{dt^2} + \Omega_0^2\right)\left(v_y + \frac{\Omega_w}{k_w}\sin k_w z\right) = \frac{\Omega_w\Omega_0}{k_w}(\Omega_0 + k_w v_z)\sin k_w z, \tag{2.40}$$

$$\frac{d}{dt}v_z = -\Omega_w(v_x \sin k_w z - v_y \cos k_w z). \tag{2.41}$$

Under the assumption that $v_z = v_\parallel + \delta v_z(t)$ where $|\delta v_z(t)| \ll v_\parallel$, we find that the transverse velocity is given by

$$v_x \cong v_w \cos k_w z + V_\perp \cos(\Omega_0 t + \varphi), \tag{2.42}$$

$$v_y \cong v_w \sin k_w z + V_\perp \sin(\Omega_0 t + \varphi), \tag{2.43}$$

where $V_\perp$ and φ are constants and we require that $V_\perp^2 \ll v_w^2$. Substitution of (2.42) and (2.43) into (2.41) shows that the axial velocity satisfies

$$v_z \cong v_\parallel - \frac{v_w V_\perp}{v_\parallel} \cos(k_w z - \Omega_0 t - \varphi), \tag{2.44}$$

where the average axial velocity satisfies the equation

$$v_\parallel \equiv c\sqrt{\left(1 - \frac{1}{\gamma^2} - \frac{v_w^2 + V_\perp^2}{c^2}\right)}. \tag{2.45}$$

The solution given by equations (2.42–44) describe the combined effect of the wiggler-induced oscillation and Larmor gyromotion due to the axial solenoidal field, and are valid as long as the magnitude of the Larmor motion is small. The solution also reduces to that shown in equations (2.7–9) in the absence of the axial field.

2.1.2 Trajectories in a realizable helical wiggler

The trajectories in the one-dimensional idealized wiggler are valid under the assumption that the electron displacement from the axis of symmetry is much less than a wiggler period (i.e. $k_w r \ll 1$). Since the radius of curvature of an orbit is proportional to the transverse velocity, this implies that the idealized analysis is valid as long as $v_w/v_\parallel \ll 1$. As a consequence, a three-dimensional analysis [3, 7, 8] of the trajectories is required for axial magnetic fields in the neighbourhood of the resonance at $\Omega_0 \approx k_w v_\parallel$.

The fundamental equations governing the electron trajectories in a realizable helical wiggler (2.1) and an axial magnetic field are

$$\frac{d}{dt}v_1 = -(\Omega_0 - k_w v_\parallel + 2\Omega_w I_1(\lambda)\sin\chi)v_2 + \Omega_w v_3 I_2(\lambda)\sin 2\chi, \tag{2.46}$$

$$\frac{d}{dt}v_2 = (\Omega_0 - k_w v_\parallel + 2\Omega_w I_1(\lambda)\sin\chi)v_1 - \Omega_w v_3[I_1(\lambda) + I_2(\lambda)\cos 2\chi], \tag{2.47}$$

$$\frac{d}{dt}v_3 = \Omega_w v_2[I_1(\lambda) + I_2(\lambda)\cos 2\chi] - \Omega_w v_1 I_2(\lambda)\sin 2\chi, \tag{2.48}$$

$$\frac{d}{dt}\lambda = k_w(v_1\cos\chi + v_2\sin\chi), \tag{2.49}$$

and

$$\frac{d}{dt}\chi = \frac{k_w}{\lambda}(v_2\cos\chi - v_1\sin\chi - \lambda v_3), \tag{2.50}$$

in the wiggler frame, where $\lambda \equiv k_w r$, and $\chi \equiv \theta - k_w z$.

In addition to the total energy, a second constant of the motion can be obtained from the symmetry properties of the vector potential which, in cylindrical coordinates, can be expressed in the form

$$\boldsymbol{A}(\boldsymbol{x}) = 2\frac{B_w}{k_w}\frac{1}{\lambda}I_1(\lambda)\hat{\boldsymbol{e}}_r \cos\chi - \left(\frac{1}{2}B_0 - 2\frac{B_w}{k_w}\sin\chi\right)\hat{\boldsymbol{e}}_\theta. \tag{2.51}$$

The Lagrangian associated with this vector potential has the functional form $\mathscr{L} = \mathscr{L}(r, \chi, \dot{r}, \dot{\chi}, \dot{z})$ which is independent of z. As a consequence, the z-component of the canonical momentum [7]

$$P_z = \frac{\gamma m_e}{k_w}\left[\lambda^2\left(\frac{d}{dt}\theta - \frac{1}{2}\Omega_0\right) + k_w v_3 - 2\Omega_w \lambda I_1(\lambda)\sin\chi\right], \tag{2.52}$$

is also a conserved quantity. This is sometimes referred to as the **helical invariant**.

Steady-state trajectories

The helical orbits were first obtained by Diament [3] by the requirement of steady-state solutions in which v_1, v_2, v_3, λ and χ are constant. This implies that $v_1 = v_w$, $v_2 = 0$, $v_3 = v_\parallel$, $\chi = \pm\pi/2$ and $\lambda = \lambda_0$, where $v_\parallel$ is constant, $\lambda_0 = \mp v_w/v_\parallel$ and

$$v_w = \frac{2\Omega_w v_\parallel I_1(\lambda_0)/\lambda_0}{\Omega_0 - k_w v_\parallel \pm 2\Omega_w I_1(\lambda_0)}. \tag{2.53}$$

Observe that this result for the transverse wiggler-induced velocity (2.53) reduces to that found for the idealized wiggler (2.15) in the limit as $\lambda_0 \rightarrow 0$. The complete determination of the orbit requires knowledge of v_w, $v_\parallel$ or λ_0. Specification of any one of these is sufficient to determine the other two. This is accomplished by means of the energy-conservation requirement which can be expressed in the form

$$\sqrt{\left(\frac{\gamma^2 - 1}{1 + \lambda_0^2}\right)} = \frac{\Omega_0}{ck_w} \pm 2\frac{\Omega_w}{ck_w}\left(\frac{1 + \lambda_0^2}{\lambda_0^2}\right)I_1(\lambda_0). \tag{2.54}$$

Solution of these equations produces the general form of the Group I and Group II orbits discussed for the case of the idealized wiggler corresponding to $\chi = \pm\pi/2$, respectively. It should be remarked that while v_w changes sign between the Group I and II orbits, the direction of rotation of these steady-state helical trajectories does not change. The sign change indicates, rather, a relative phase shift of 180° in the orbits.

Stability of the steady-state trajectories

The stability of these orbits is determined by perturbation of the orbit equations about the steady-state trajectories: $v_1 = v_w + \delta v_1$, $v_2 = \delta v_2$, $v_3 = v_\parallel + \delta v_3$,

$\chi = \pm\pi/2 + \delta\chi$ and $\lambda = \lambda_0 + \delta\lambda$. The orbit equations, therefore, can be written as

$$\frac{d}{dt}\delta v_1 = -\frac{(\Omega_0 - k_w v_\parallel)}{(1+\lambda_0^2)}\delta v_2 - 2\Omega_w v_\parallel I_2(\lambda_0)\delta\chi, \tag{2.55}$$

$$\frac{d}{dt}\delta v_2 = \frac{(\Omega_0 - k_w v_\parallel)}{(1+\lambda_0^2)}\delta v_1 - \frac{v_w}{v_\parallel}\frac{(\Omega_0 + \lambda_0^2 k_w v_\parallel)}{(1+\lambda_0^2)}\delta v_3 - 2\Omega_w v_\parallel\left[\left(\frac{1+\lambda_0^2}{\lambda_0}\right)I_2(\lambda_0) + I_1(\lambda_0)\right]\delta\lambda, \tag{2.56}$$

$$\frac{d}{dt}\delta v_3 = 2\Omega_w\left[v_w I_2(\lambda_0)\delta\chi + \frac{1}{\lambda_0}I_1(\lambda_0)\delta v_2\right], \tag{2.57}$$

$$\frac{d}{dt}\delta\chi = -k_w\left(\delta v_3 \pm \frac{1}{\lambda_0}\delta v_1 + \frac{1}{\lambda_0}v_\parallel\delta\lambda\right), \tag{2.58}$$

$$\frac{d}{dt}\delta\lambda = \pm k_w(\delta v_2 - v_w\delta\chi), \tag{2.59}$$

to lowest order in the perturbations.

The perturbed quantities can be isolated from this system of coupled first-order differential equations and the result can be written as a set of homogeneous higher order differential equations. Thus [7],

$$\left(\frac{d^2}{dt^2} + \Omega_+^2\right)\left(\frac{d^2}{dt^2} + \Omega_-^2\right)\begin{bmatrix}\delta v_2\\ \delta\chi\end{bmatrix} = 0, \tag{2.60}$$

and

$$\frac{d}{dt}\left(\frac{d^2}{dt^2} + \Omega_+^2\right)\left(\frac{d^2}{dt^2} + \Omega_-^2\right)\begin{bmatrix}\delta v_1\\ \delta v_3\\ \delta\lambda\end{bmatrix} = 0, \tag{2.61}$$

where

$$\Omega_\pm^2 \equiv \tfrac{1}{2}(\omega_1^2 + \omega_2^2) \pm \tfrac{1}{2}\sqrt{[(\omega_1^2 - \omega_2^2)^2 + k_w v_\parallel \Omega_w \omega_3^2]}, \tag{2.62}$$

$$\omega_1^2 \equiv k_w^2 v_\parallel^2 + 2\Omega_w k_w v_w\left(\frac{1+\lambda_0^2}{\lambda_0^2}\right)I_2(\lambda_0), \tag{2.63}$$

$$\omega_2^2 \equiv \frac{(\Omega_0 - k_w v_\parallel)^2}{1+\lambda_0^2} - 2\Omega_w k_w v_w\left(\frac{1+\lambda_0^2}{\lambda_0^2}\right)I_2(\lambda_0), \tag{2.64}$$

and

$$\omega_3^2 \equiv 8\frac{v_\parallel}{v_w}(\Omega_0 - 2k_w v_\parallel)[\Omega_0 I_2(\lambda_0) + \lambda_0 k_w v_\parallel(I_1(\lambda_0) + 2\lambda_0 I_2(\lambda_0))]. \tag{2.65}$$

It is evident that $\Omega_+^2 \geqslant 0$, so that orbital instability occurs only when $\Omega_-^2 < 0$. In order to obtain an orbital-instability criterion, therefore, it is more convenient to deal with the product $\Omega_-^2\Omega_+^2$. This is negative, and orbital instability occurs whenever

$$[(1+\lambda_0^2)\Omega_0 - k_w v_\parallel]Z(\lambda_0) - \lambda_0^2 k_w v_\parallel W(\lambda_0) < 0, \tag{2.66}$$

where

$$Z(\lambda_0) \equiv (1+\lambda_0^2)I_1'(\lambda_0) - \frac{2}{\lambda_0}I_1(\lambda_0), \tag{2.67}$$

and

$$W(\lambda_0) \equiv (1+\lambda_0^2)I_1'(\lambda_0) - \frac{1-\lambda_0^2}{\lambda_0}I_1(\lambda_0). \tag{2.68}$$

Observe that this instability criterion (2.66) reduces to that found in the idealized wiggler limit (2.21) when $\lambda_0 \ll 1$.

Solutions for these steady-state trajectories as a function of the axial magnetic field which correspond to those found in the idealized limit are shown in Fig. 2.6, in which the dashed lines indicate unstable orbits. These orbits correspond to the idealized steady-state trajectories shown in Fig. 2.1, which are reproduced here for comparison. The steady-state orbits are close to those obtained in the idealized limit whenever $\lambda_0 \ll 1$ which, in general, requires that $\Omega_w/ck_w \ll 1$. However, λ_0 can approach and even exceed unity near the resonance at $\Omega_0 \approx k_w v_\parallel$ due to the enhancement in both the transverse velocity and λ_0. It is in this regime, as shown in the figure, that the effects of a realizable wiggler are most pronounced. The discrepancy with the idealized trajectories is most in evidence for the Group II trajectories, which exhibit an orbital instability not found in the idealized wiggler limit.

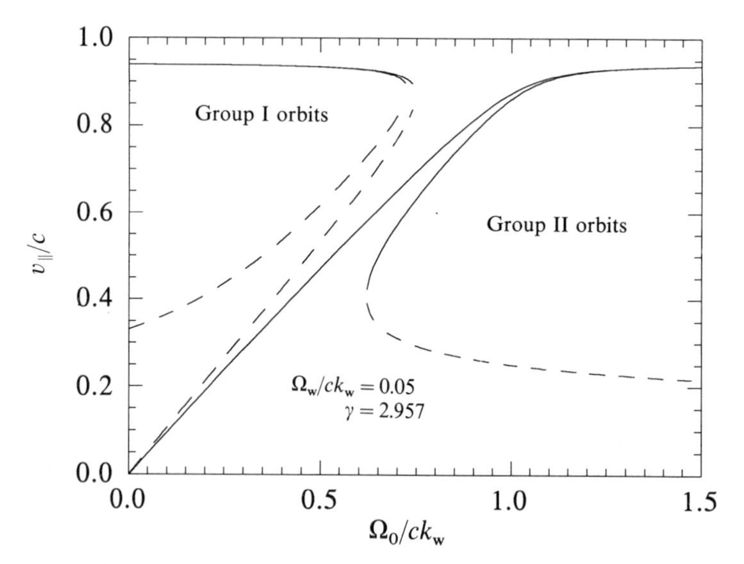

Fig. 2.6 Graph of the axial velocity of the steady-state orbits as a function of the axial magnetic field for both the idealized and realizable wiggler models. The dashed lines indicate unstable trajectories.

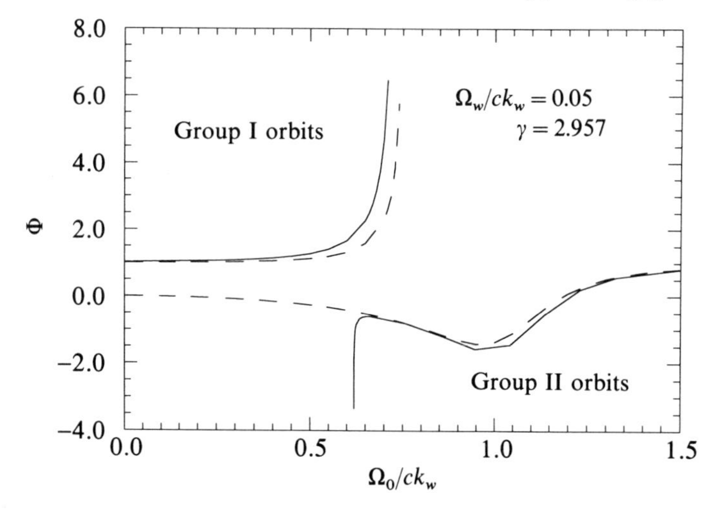

Fig. 2.7 Graphs of Φ as a function of the axial magnetic field for both the idealized (dashed) and realizable (solid) wiggler representations.

Negative-mass trajectories

The negative-mass regime found for Group II orbits in the idealized limit also occurs for the realizable wiggler. In particular [7],

$$\frac{d}{d\gamma} v_\parallel = \frac{c^2}{\gamma\gamma_\parallel^2 v_\parallel} \Phi(\lambda_0), \tag{2.69}$$

where

$$\Phi(\lambda_0) \equiv 1 - \frac{\gamma_\parallel^2 \lambda_0^2 [\Omega_0 Z(\lambda_0) - k_w v_\parallel W(\lambda_0)]}{[(1+\lambda_0^2)\Omega_0 - k_w v_\parallel] Z(\lambda_0) - \lambda_0^2 k_w v_\parallel W(\lambda_0)}. \tag{2.70}$$

This is similar to equation (2.23) found in the idealized wiggler limit, and we note that the generalized $\Phi(\lambda_0)$ reduces to equation (2.24) in the idealized wiggler limit in which $\lambda_0 \ll 1$. The form of $\Phi(\lambda_0)$ as a function of the axial field is shown in Fig. 2.7 for parameters which correspond to the steady-state orbits shown in Fig. 2.6. Observe that, as in the idealized limit, $\Phi(\lambda_0)$ exhibits singularities at the transitions to orbital instability which now occur for both Group I and Group II orbits.

Generalized trajectories: Larmor and betatron oscillations

The homogeneous equations for the perturbed quantities (equations (2.60) and (2.61)) describe orbits that, while close to the steady-state orbits, describe non-helical, axis-encircling trajectories. The solutions to equation (2.60) are of the form [7]

$$\delta v_2 = V_+ \sin(\Omega_+ t - \phi_+) + V_- \sin(\Omega_- t - \phi_-), \tag{2.71}$$

and

$$\delta\chi = X_+ \sin(\Omega_+ t - \theta_+) + X_- \sin(\Omega_- t - \theta_-), \tag{2.72}$$

where $V_\pm, X_\pm, \phi_\pm$ and $\theta_\pm$ are the integration constants. Substitution of these solutions into the first-order equations indicates that δv_2 and $\delta\chi$ are related by $\theta_\pm = \phi_\pm$ and

$$X_\pm = \frac{V_\pm}{v_w}\frac{k_w v_\parallel(\Omega_0 - 2k_w v_\parallel)}{\Omega_\pm^2 - \omega_1^2}. \tag{2.73}$$

As a consequence,

$$\delta v_1 = \Delta_+ V_+ \cos(\Omega_+ t - \phi_+) + \Delta_- V_- \cos(\Omega_- t - \phi_-), \tag{2.74}$$

$$\delta v_3 = \pm\lambda_0 \delta v_1, \tag{2.75}$$

$$\frac{\delta\lambda}{\lambda_0} = \rho_+ \frac{V_+}{v_w}\frac{k_w v_\parallel}{\Omega_+}\cos(\Omega_+ t - \phi_+) + \rho_- \frac{V_-}{v_w}\frac{k_w v_\parallel}{\Omega_-}\cos(\Omega_- t - \phi_-), \tag{2.76}$$

where

$$\Delta_\pm \equiv 2\frac{\Omega_w}{\Omega_\pm}\frac{v_\parallel}{v_w}\left[\frac{1}{\lambda_0}I_1(\lambda_0) + \frac{k_w v_\parallel(\Omega_0 - 2k_w v_\parallel)}{\Omega_\pm^2 - \omega_1^2}I_2(\lambda_0)\right], \tag{2.77}$$

$$\rho_\pm \equiv 1 - \frac{k_w v_\parallel(\Omega_0 - 2k_w v_\parallel)}{\Omega_\pm^2 - \omega_1^2}. \tag{2.78}$$

These non-helical trajectories exhibit oscillations at the frequencies $\Omega_\pm$ in addition to the large-scale, wiggler-induced oscillation. In order to determine the physical basis for these oscillations, we consider the characteristics of these frequencies in more detail. A plot of the variation of Ω_+ and Ω_- versus the axial magnetic field is shown in Fig. 2.8 for parameters corresponding to the stable steady-state orbits shown in Fig. 2.6. Expansion of $\Omega_\pm$ in powers of λ_0 in the absence of an axial magnetic field indicates that to lowest nontrivial order

$$\Omega_\pm \approx k_w v_\parallel \pm \Omega_\beta, \tag{2.79}$$

where $\Omega_\beta \equiv \Omega_w/\sqrt{2}$ defines the betatron frequency. As a result, transformation from the wiggler frame to Cartesian coordinates indicates that these oscillations are degenerate in the absence of an axial magnetic field and correspond to betatron oscillations due to the inhomogeneous magnetic field of the wiggler. Betatron oscillations arise because the wiggler field is a minimum on axis and, therefore, exerts a restoring force on the electrons. In the absence of an axial magnetic field, this restoring force gives rise to uncoupled oscillations in x and y. It is evident from Fig. 2.8 that $\Omega_+ \approx k_w v_\parallel$ for all values of the axial field except in the vicinity of the resonance at $\Omega_0 \approx k_w v_\parallel$. In contrast, Ω_- varies widely as a function of the

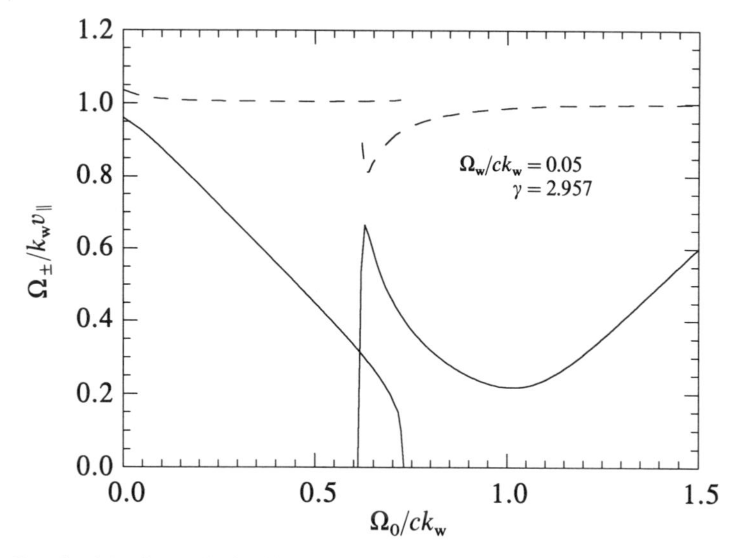

Fig. 2.8 Graph of $\Omega_+/k_w v_\parallel$ (dashed line) and $\Omega_-/k_w v_\parallel$ (solid line) versus the axial magnetic field for parameters corresponding to the stable steady-state orbits shown in Fig. 2.6.

axial field. Indeed, for sufficiently strong axial fields that $2\lambda_0\Omega_w\Omega_0 < \Omega_0^2 < 4k_w^2 v_\parallel^2$

$$\Omega_+ \approx k_w v_\parallel - \frac{\Omega_\beta^2}{\Omega_0} \frac{k_w v_\parallel}{\Omega_0 - k_w v_\parallel}, \tag{2.80}$$

and

$$\Omega_- \approx |k_w v_\parallel - \Omega_0|, \tag{2.81}$$

correct to within terms of order λ_0^2. Here Ω_- describes the rapid Larmor oscillations, while Ω_+ describes a modified betatron oscillation. The presence of the axial magnetic field modifies both the frequency and character of the betatron oscillation. The modification of the frequency is evident by comparison of equations (2.79) and (2.80). The character of the oscillation is altered because the presence of an axial field results in a coupling of the x and y motion which results in an elliptic oscillation in the transverse plane.

The aforementioned orbits refer to motion which departs only marginally from the steady-state helical trajectories and is axis encircling. Indeed, the steady-state trajectories are axis-centred. However, not all motion is axis encircling. The opposite limit in which the electron displacement from the symmetry axis is large in comparison with all rapid oscillatory motion (i.e. either wiggler or Larmor) is referred to as the **guiding-centre** approximation. In order to treat this limit [7], we decompose the electron position and velocity into $\boldsymbol{x} = \boldsymbol{x}_0 + \boldsymbol{x}_c$ and $\mathbf{v} = \mathbf{v}_0 + \mathbf{v}_c$, where $(\boldsymbol{x}_0, \mathbf{v}_0)$ describes the rapid oscillatory motion about the slowly moving guiding-centre $(\boldsymbol{x}_c, \mathbf{v}_c)$ and it is assumed that $|\boldsymbol{r}_0| \ll |\boldsymbol{r}_c|$ (where $\boldsymbol{r}_c$ and $\boldsymbol{r}_0$ denote the transverse components of $\boldsymbol{x}_0$ and $\boldsymbol{x}_c$) while $|\mathbf{v}_c| \ll |\mathbf{v}_0|$. A schematic illustration of this decomposition is illustrated in Fig. 2.9. In this regime, the electron reacts to

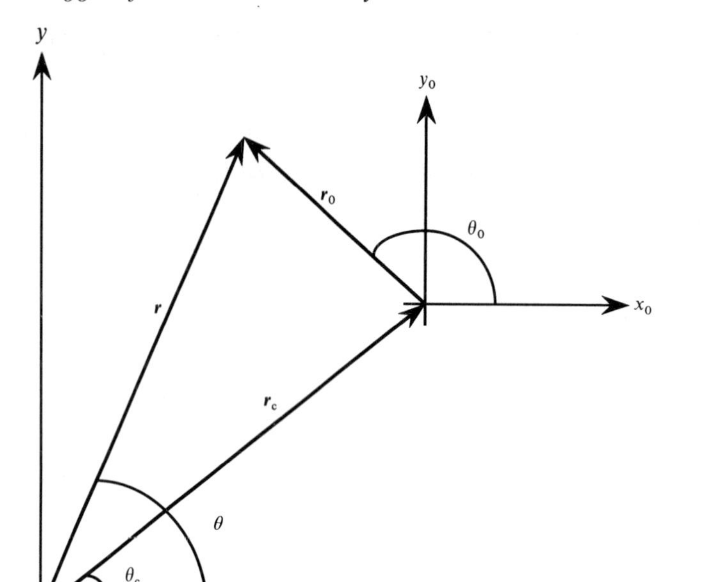

Fig. 2.9 Schematic representation of the vector relationships between the electron position $\boldsymbol{r}$, the guiding-centre location $\boldsymbol{r}_c$ and the wiggler-induced component of the electron trajectory $\boldsymbol{r}_0$.

the local magnetic field at the guiding-centre, and we may expand

$$\boldsymbol{B}(\boldsymbol{x}) \cong \boldsymbol{B}(\boldsymbol{x}_c) + \boldsymbol{x}_0 \cdot \nabla_c \boldsymbol{B}(\boldsymbol{x}_c), \tag{2.82}$$

where ∇_c denotes the divergence with respect to the guiding-centre coordinates. The motion of the guiding-centre is obtained by means of an average of the orbit's equations over a wiggler period, which yields

$$\frac{d}{dt}\mathbf{v}_c \cong -\Omega_0 \mathbf{v}_c \times \hat{\boldsymbol{e}}_z - \frac{e}{\gamma m_e c}\langle \mathbf{v}_0 \times [\boldsymbol{B}(\boldsymbol{x}_c) + \boldsymbol{x}_0 \cdot \nabla_c \boldsymbol{B}(\boldsymbol{x}_c)]\rangle$$

$$-\frac{e}{\gamma m_e c}\mathbf{v}_c \times \langle \boldsymbol{x}_0 \cdot \nabla_c \boldsymbol{B}(\boldsymbol{x}_c)\rangle, \tag{2.83}$$

where $\langle \boldsymbol{B}(\boldsymbol{x}_c)\rangle = B_0 \hat{\boldsymbol{e}}_z$. In addition, it is assumed that $\langle \boldsymbol{r}_0 \rangle = 0$ and $\langle \mathbf{v}_0 \rangle = v_\parallel \hat{\boldsymbol{e}}_z$. The equation governing the rapid oscillatory motion

$$\frac{d}{dt}\mathbf{v}_0 \cong -\frac{e}{\gamma m_e c}\mathbf{v}_0 \times [\boldsymbol{B}(\boldsymbol{x}_c) + \boldsymbol{x}_0 \cdot \nabla_c \boldsymbol{B}(\boldsymbol{x}_c)] + \frac{e}{\gamma m_e c}\mathbf{v}_0 \times \langle \boldsymbol{B}(\boldsymbol{x}_c) + \boldsymbol{x}_0 \cdot \nabla_c \boldsymbol{B}(\boldsymbol{x}_c)\rangle$$

$$-\frac{e}{\gamma m_e c}\mathbf{v}_c \times [\boldsymbol{B}_w(\boldsymbol{x}_c) + \boldsymbol{x}_0 \cdot \nabla_c \boldsymbol{B}(\boldsymbol{x}_c) - \langle \boldsymbol{x}_0 \cdot \nabla_c \boldsymbol{B}(\boldsymbol{x}_c)\rangle], \tag{2.84}$$

is obtained by subtraction of the guiding-centre motion from the complete orbit equation.

If we solve the equation for the rapid oscillatory motion to lowest order in $k_w|\boldsymbol{r}_0|$, then only the first term survives and we write

$$\frac{d}{dt}\mathbf{v}_0 \cong -\frac{e}{\gamma m_e c}\mathbf{v}_0 \times \boldsymbol{B}(\boldsymbol{x}_c). \tag{2.85}$$

This equation essentially describes the effect of the local magnetic field at the electron guiding-centre upon the trajectory. The orbit equations in the wiggler frame, therefore, take the form

$$\frac{d}{dt}v_{01} \cong -[\Omega_0 - k_w v_{03} + 2\Omega_w I_1(\lambda_c)\sin\chi_c]v_{02} + \Omega_w v_{03} I_2(\lambda_c)\sin 2\chi_c, \tag{2.86}$$

$$\begin{aligned}\frac{d}{dt}v_{02} \cong {} & [\Omega_0 - k_w v_{03} + 2\Omega_w I_1(\lambda_c)\sin\chi_c]v_{01} \\ & - \Omega_w v_{03}[I_0(\lambda_c) + I_2(\lambda_c)\cos 2\chi_c],\end{aligned} \tag{2.87}$$

$$\frac{d}{dt}v_{03} \cong \Omega_w v_{02}[I_0(\lambda_c) + I_2(\lambda_c)\cos 2\chi_c] - \Omega_w v_{01} I_2(\lambda_c)\sin 2\chi_c, \tag{2.88}$$

where $\lambda_c \equiv k_w r_c$ and $\chi_c \equiv \theta_c - k_w z$. These equations are formally identical to those found for the steady-state trajectories (2.55–7) under the substitution of the guiding-centre position. Since the guiding-centre position is assumed to be fixed on the rapid time scale, a quasi-steady-state solution $v_{01} = v_{0w}$, $v_{02} = 0$ and $v_{03} = v_{0\parallel}$ is obtained where

$$v_{0w} = \frac{\Omega_w v_{0\parallel}[I_0(\lambda_c) + I_2(\lambda_c)\cos 2\chi_c]}{[\Omega_0 - k_w v_{0\parallel} + 2\Omega_w I_1(\lambda_c)\sin\chi_c]}. \tag{2.89}$$

This solution is similar to the helical steady-state orbits discussed previously. However, while the values of v_{0w} and $v_{0\parallel}$ are constant on the rapid wiggler time scale, they vary slowly with the position of the guiding-centre.

The slow-time-scale motion of the guiding-centre can be determined by substitution of the lowest-order solution for $(\boldsymbol{x}_0, \mathbf{v}_0)$ into equation (2.81). Since $\langle \mathbf{v}_0 \times \boldsymbol{x}_0 \cdot \nabla_c \boldsymbol{B}(\boldsymbol{x}_c)\rangle - B_w(v_{0w}/v_{0\parallel})\hat{e}_z$ and $\langle \mathbf{v}_0 \times \boldsymbol{B}(\boldsymbol{x}_c)\rangle = -B_w k_w v_{0w} r_c/2$, the equation governing the guiding-centre motion can be expressed as

$$\frac{d}{dt}\mathbf{v}_c = \omega_c \hat{\boldsymbol{e}}_z \times \mathbf{v}_c + \frac{1}{2}\Omega_w k_w v_{0w}\boldsymbol{r}_c, \tag{2.90}$$

where

$$\omega_c \equiv \Omega_0 - \Omega_w \frac{v_{0w}}{v_{0\parallel}}. \tag{2.91}$$

In general, the guiding-centre motion is a coupled oscillation in (x_c, y_c) which can be separated into the following two fourth-order differential equations

$$\left(\frac{d^2}{dt^2}+\omega_+^2\right)\left(\frac{d^2}{dt^2}+\omega_-^2\right)\begin{bmatrix} x_c \\ y_c \end{bmatrix}=0, \tag{2.92}$$

where

$$\omega_\pm^2 \equiv \frac{1}{2}(\omega_c^2-\Omega_w k_w v_{0w}) \pm \frac{\omega_c}{2}\sqrt{(\omega_c^2-\Omega_w k_w v_{0w})} \tag{2.93}$$

subject to the requirement that $|\omega_\pm^2| \ll k_w^2 v_{0\parallel}^2$.

In the limit in which the axial magnetic field disappears, the frequencies $\omega_\pm \cong \Omega_\beta$ and the (x_c, y_c) motion of the guiding-centre reduces to well-known uncoupled betatron oscillations in the inhomogeneous wiggler

$$\begin{bmatrix} x_c(t) \\ \dot{x}_c(t) \end{bmatrix} = \begin{bmatrix} \cos\Omega_\beta t & \dfrac{1}{\Omega_\beta}\sin\Omega_\beta t \\ -\Omega_\beta \sin\Omega_\beta t & \cos\Omega_\beta t \end{bmatrix} \begin{bmatrix} x_c(0) \\ \dot{x}_c(0) \end{bmatrix}, \tag{2.94}$$

and

$$\begin{bmatrix} y_c(t) \\ \dot{y}_c(t) \end{bmatrix} = \begin{bmatrix} \cos\Omega_\beta t & \dfrac{1}{\Omega_\beta}\sin\Omega_\beta t \\ -\Omega_\beta \sin\Omega_\beta t & \cos\Omega_\beta t \end{bmatrix} \begin{bmatrix} y_c(0) \\ \dot{y}_c(0) \end{bmatrix}. \tag{2.95}$$

This motion describes an oscillation in which the electron trajectories pass through the symmetry axis at $x_c = y_c = 0$. It arises because the wiggler field strength increases with the displacement from the symmetry axis and, therefore, exerts a restoring force on the electron trajectories.

The presence of the axial magnetic field couples the (x_c, y_c) motion and results in a precession of the guiding-centre around the symmetry axis. If the axial magnetic field is sufficiently strong that $\Omega_0^2 \gg |\Omega_w k_w v_{0w}|$, then $\omega_c \cong \Omega_0$, $\omega_+^2 \cong \Omega_0^2 - \Omega_w k_w v_{0w}$, and

$$\omega_- \cong \frac{\Omega_\beta^2}{\Omega_0}\frac{k_w v_{0\parallel}}{\Omega_0 - k_w v_{0\parallel}}. \tag{2.96}$$

Observe that this is precisely the modified betatron frequency found previously by perturbation about the steady-state trajectories (equation (2.80)). As a result, $\omega_\pm$ characterize modified Larmor and betatron motion, respectively, which now describe ellipses. In particular, electrons in the combined magnetic fields execute betatron motion characterized by

$$\begin{bmatrix} x_c(t) \\ y_c(t) \end{bmatrix} = \begin{bmatrix} \cos\omega_- t & -\alpha\sin\omega_- t \\ \dfrac{1}{\alpha}\sin\omega_- t & \cos\omega_- t \end{bmatrix} \begin{bmatrix} x_c(0) \\ y_c(0) \end{bmatrix}, \tag{2.97}$$

where

$$\alpha \equiv \frac{2\Omega_0\omega_-}{2\Omega_0^2 + \Omega_w k_w v_{0w}}. \tag{2.98}$$

A similar expression can be found for the modified Larmor motion.

2.2 PLANAR WIGGLER CONFIGURATIONS

Because the wiggler magnitude is itself oscillatory in the case of planar wiggler fields, both the transverse and axial components of the electron trajectories are periodic as well. This is in marked contrast to a helical wiggler, for which steady-state orbits characterized by constant-magnitude transverse and axial velocities are possible. The absence of steady-state solutions for the electron trajectories in planar wiggler geometries implies that the wave–particle resonance condition can be satisfied only in an average sense, and that electrons will drift into and out of resonance with the wave over the course of a wiggler period. In practice, this means that the interaction will be governed by the root-mean-square value of the wiggler field and that the effective wiggler field is reduced. In comparison with the effect of a helical wiggler, therefore, a planar wiggler field must be approximately 71% higher in order to have a comparable effect. In view of this disadvantage, the widespread use of planar wigglers is attributed to the ease with which they may be adjusted. In particular, the ability to adjust the field strength of the wiggler in the direction of the axis of symmetry is important for efficiency-enhancement schemes which employ nonuniform (i.e. tapered) wiggler fields. The effect of axial wiggler tapers on the electron trajectories will be discussed in the next section.

2.2.1 Idealized one-dimensional trajectories

In the limit in which the electron displacements from the plane of symmetry are small in comparison with a wiggler period (i.e. $k_w|x| \ll 1$ and $k_w|y| \ll 1$), both planar wiggler configurations given by equation (2.2) may be represented in the form

$$\boldsymbol{B}_w = B_w \hat{\boldsymbol{e}}_y \sin k_w z, \tag{2.99}$$

with the corresponding vector potential

$$\boldsymbol{A}_w = -\frac{B_w}{k_w} \hat{\boldsymbol{e}}_x \cos k_w z. \tag{2.100}$$

As in the case of the idealized one-dimensional limit of the helical wiggler, both x and y are ignorable coordinates, and the transverse canonical momenta are constants of the motion. The third constant of the motion is the total energy. As a consequence, v_y is also a constant of the motion and the x- and z-components of the velocity are given by

$$v_x = \frac{P_x}{\gamma m_e} + v_w \cos k_w z, \tag{2.101}$$

$$\frac{v_z}{c} = \sqrt{\left(1 - \frac{1}{\gamma^2} - \frac{V_\perp^2}{c^2} - \frac{v_w^2}{c^2}\cos^2 k_w z - 2\frac{v_w V_x}{c^2}\cos k_w z\right)}, \tag{2.102}$$

where $P_{x,y}$ denote the canonical momenta, $v_w \equiv -\Omega_w/k_w$ is the wiggler-induced transverse velocity, $V_\perp \equiv (P_x^2 + P_y^2)^{1/2}/\gamma m_e$, and $V_{x,y} \equiv P_{x,y}/\gamma m_e$. Observe that the assumption of small displacements from the symmetry plane is equivalent to the requirement that $|v_w/v_\parallel| \ll 1$.

In contrast with the helical wiggler, where the orbits describe steady-state helices in the limit in which the canonical momenta vanish, there are no steady-state orbits for a planar wiggler. In this case, the transverse and axial velocity components can be written in the form

$$v_x = v_w \cos k_w z, \tag{2.103}$$

which describes the wiggler-induced oscillation in the transverse direction and

$$v_z = v_\parallel \sqrt{\left(1 - \frac{v_w^2}{2v_\parallel^2} \cos 2k_w z\right)}, \tag{2.104}$$

where

$$\frac{v_\parallel}{c} \equiv \sqrt{\left(1 - \frac{1}{\gamma^2} - \frac{v_w^2}{2c^2}\right)}, \tag{2.105}$$

denotes the bulk axial velocity. Observe that this bulk axial velocity corresponds to an average transverse velocity which corresponds to the root-mean-square value of the wiggler field. In addition, it should be remarked that the oscillatory component of the axial velocity is of second order in the wiggler magnitude and is, therefore, negligible under most circumstances.

Quasi-steady-state trajectories

In the presence of an axial magnetic field, the canonical momenta are no longer conserved and the orbits take on an even more complex nature. Using the Lorentz force equation (2.4), we write the orbit equations in the form

$$\frac{d}{dt} v_x = -\Omega_0 v_y + v_z \Omega_w \sin k_w z, \tag{2.106}$$

$$\frac{d}{dt} v_y = \Omega_0 v_x, \tag{2.107}$$

$$\frac{d}{dt} v_z = -v_x \Omega_w \sin k_w z. \tag{2.108}$$

In order to obtain quasi-steady-state solutions we assume that the orbits are characterized by a large bulk axial motion with a small-amplitude oscillation superimposed upon it. As in the case shown in equations (2.103–6), this oscillation in the axial velocity is of second order in the wiggler amplitude. Therefore, $v_z = v_\parallel$ (constant) to within first order in the wiggler amplitude, and we obtain

$$\left(\frac{d^2}{dt} + \Omega_0^2\right) v_x = k_w v_\parallel^2 \Omega_w \cos k_w z, \tag{2.109}$$

and

$$\left(\frac{d^2}{dt^2} + \Omega_0^2\right) v_y = \Omega_0 \Omega_w v_\parallel \sin k_w z. \tag{2.110}$$

The homogeneous solutions to these equations describe the Larmor rotation due to the solenoidal magnetic field. This Larmor motion is dependent upon the initial conditions of the orbit and can, in principle, be made arbitrarily small. In contrast, the wiggler-induced motion corresponds to the particular solution and describes an ellipse in the x–y-plane characterized by

$$\mathbf{v}_\perp = \alpha_x \hat{e}_x \cos k_w z + \alpha_y \hat{e}_y \sin k_w z, \tag{2.111}$$

where

$$\alpha_x \equiv \frac{\Omega_w k_w v_\parallel^2}{\Omega_0^2 - k_w^2 v_\parallel^2}, \tag{2.112}$$

and

$$\alpha_y = \frac{\Omega_w \Omega_0 v_\parallel}{\Omega_0^2 - k_w^2 v_\parallel^2}. \tag{2.113}$$

Observe that in the limit in which the solenoidal magnetic field vanishes $\alpha_x = -\Omega_w/k_w, \alpha_y = 0$ and the result given in equation (2.103) is recovered. These orbits show a resonant enhancement in the magnitude of the transverse velocity when $\Omega_0 \approx k_w v_\parallel$, which is similar to that found for helical wigglers. The average axial velocity over a wiggler period for this trajectory is obtained from the conservation of energy using the root-mean-square values of the transverse velocity components, and we find that

$$\frac{v_\parallel^2}{c^2}\left[1 + \frac{\Omega_w^2(\Omega_0^2 + k_w^2 v_\parallel^2)}{2(\Omega_0^2 - k_w^2 v_\parallel^2)^2}\right] = 1 - \frac{1}{\gamma^2}. \tag{2.114}$$

It should be noted that this equation is symmetric in $v_\parallel$ and yields trajectories which are independent of the direction of propagation. This contrasts with the helical wiggler which establishes a preferred direction of propagation with respect to the orientation of the axial magnetic field. The second-order oscillation in the axial velocity is found by substitution of equation (2.111) into equation (2.108) to be

$$v_z = v_\parallel + \frac{1}{4}\frac{\Omega_w^2 v_\parallel}{\Omega_0^2 - k_w^2 v_\parallel^2} \cos 2k_w z. \tag{2.115}$$

The bulk axial velocity can be obtained by solution of equation (2.114) as a function of the wiggler field strength and period, the axial magnetic field and the electron energy. The characteristic solutions are very similar to those found in the case of a helical wiggler field and can be divided into Group I ($\Omega_0 < k_w v_\parallel$) and Group II ($\Omega_0 > k_w v_\parallel$) orbits. These solutions are illustrated in Fig. 2.10 as a function of the axial magnetic field for parameters consistent with the steady-state solutions shown for the helical wiggler in Fig. 2.1. Observe that only those solutions for $v_\parallel > 0$ are shown as the curves are symmetric about the direction of propagation.

Negative-mass trajectories

These quasi-steady-state trajectories exhibit similar positive and negative mass regimes as found in the case of a helical wiggler. This is shown by the implicit

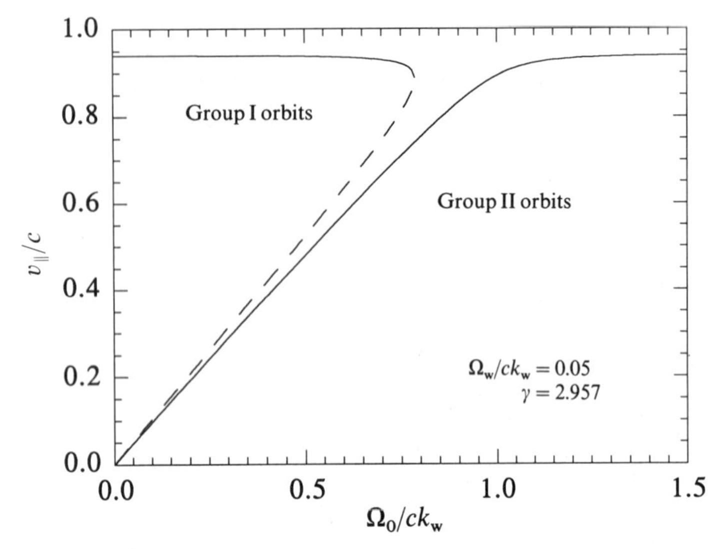

Fig. 2.10 Graph of the axial velocity of the quasi-steady-state orbits as a function of the solenoidal magnetic field showing the Group I and Group II orbits.

differentiation of equation (2.114) with respect to γ, which gives

$$\frac{d}{d\gamma} v_{\parallel} = \frac{c^2}{\gamma \gamma_{\parallel}^2 v_{\parallel}} \Phi_p, \tag{2.116}$$

where

$$\Phi_p \equiv 1 - \frac{\gamma_{\parallel}^2 \Omega_w^2 \Omega_0^2 (\Omega_0^2 + 3k_w^2 v_{\parallel}^2)}{2(\Omega_0^2 - k_w^2 v_{\parallel}^2)^3 + \Omega_w^2 \Omega_0^2 (\Omega_0^2 + 3k_w^2 v_{\parallel}^2)} \tag{2.117}$$

is the analogue of the function Φ (equation (2.24)) for a planar wiggler. A plot of the function Φ_p is given in Fig. 2.11 corresponding to the quasi-steady-state trajectories shown in Fig. 2.10. As in the case of helical wiggler fields, these quasi-steady-state orbits exhibit a negative mass regime for Group II orbits close to the resonance at $\Omega_0 \approx k_w v_{\parallel}$.

The general character of the quasi-steady-state trajectories in a planar wiggler is similar to that of the steady-state trajectories in a helical wiggler. The major distinctions, however, are that (1) there is no preferred direction of propagation with respect to the axial solenoidal field, (2) the projection of the orbits in the x–y-plane in a planar wiggler is elliptic rather than circular, and (3) the magnitudes of the transverse components of the velocity are determined by the root-mean-square magnitude of the wiggler. This latter distinction implies that a planar wiggler must be approximately 41% larger in magnitude than a helical wiggler in order to have a comparable effect.

There are two effects which we have ignored in the idealized treatment of the quasi-steady-state trajectories. The first is the oscillations in the axial velocity. However, these oscillations are of second order in the wiggler magnitude and

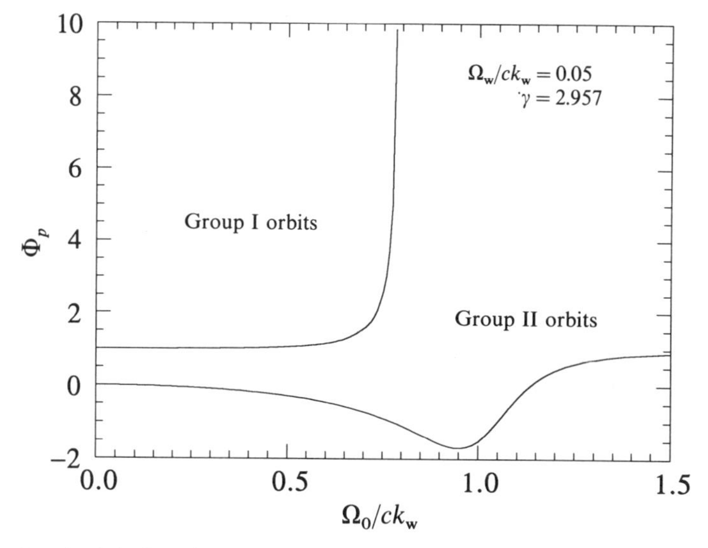

Fig. 2.11 Graph of the function Φ_p to the quasi-steady-state trajectories shown in Fig. 2.10.

become important for large-amplitude wigglers and for the generation of harmonic radiation. The second effect arises within the context of a three-dimensional analysis which employs a self-consistent wiggler field. In this case, a $\boldsymbol{B}_0 \times \nabla \boldsymbol{B}_w$ drift appears which drives the electron beam away from the symmetry plane and will ultimately result in the loss of the beam to the drift tube wall. It is for this reason that alternate methods of beam focusing are usually employed in conjunction with planar wigglers.

2.2.2 Trajectories in realizable planar wigglers

When the electron trajectories diverge substantially from the plane of symmetry (i.e. $k_w|x| \approx 1$ and $k_w|y| \approx 1$), then the wiggler inhomogeneities become important and the idealized wiggler model breaks down. We first treat the case of the wiggler model given in equation (2.2) for flat pole faces, which can be described by a vector potential of the form

$$\boldsymbol{A}_w(y, z) = -\frac{B_w}{k_w} \hat{\boldsymbol{e}}_x \cosh k_w y \cos k_w z. \tag{2.118}$$

Since the vector potential is independent of x, this is an ignorable coordinate and the canonical momentum in the x-direction is a conserved quantity. This allows us to eliminate one of the equations of motion. However, before we treat this case it is instructive to consider the effect of an axial solenoidal magnetic field upon the trajectories, particularly in regard to the aforementioned $\boldsymbol{B}_0 \times \nabla \boldsymbol{B}_w$ drift.

The orbit equations in a realizable planar wiggler with flat pole faces take the form

$$\frac{d}{dt}v_x = -\Omega_0 v_y - \Omega_w(v_y \sinh k_w y \cos k_w z - v_z \cosh k_w y \sin k_w z), \tag{2.119}$$

$$\frac{d}{dt}v_y = \Omega_0 v_x + \Omega_w v_x \sinh k_w y \cos k_w z, \tag{2.120}$$

$$\frac{d}{dt}v_z = -\Omega_w v_x \cosh k_w y \sin k_w z, \tag{2.121}$$

where we observe that the total energy is a conserved quantity. A second constant of the motion corresponds to the x-component of the canonical momentum. Equation (2.118) can be integrated to obtain v_x as a function of the canonical momentum P_x and the (y, z) coordinates, and we obtain

$$v_x = \frac{P_x}{\gamma m_e} - \Omega_0 y + v_w \cosh k_w y \cos k_w z, \tag{2.122}$$

where $v_w = -\Omega_w/k_w$.

Gradient drifts due to an axial magnetic field

Under the assumption that the orbit does not diverge significantly from the symmetry plane, we may expand in powers of $k_w y$ about the quasi-steady-state solutions obtained previously (equations (2.111–13)) and write that

$$\mathbf{v}_\perp = \alpha_x \hat{e}_x \cos k_w z + \alpha_y \hat{e}_y \sin k_w z + \delta \mathbf{v}_\perp, \tag{2.123}$$

and

$$v_z = v_\parallel + \delta v_z, \tag{2.124}$$

where we have neglected contributions of second order in the wiggler amplitude to the quasi-steady-state trajectories. We now expand equations (2.119–21) to first order in $k_w y$ and average the resulting equations over a wiggler period. As a result, the perturbations in the transverse velocities are governed by

$$\left(\frac{d^2}{dt^2} + \Omega_0^2 + \Omega_\beta^2\right)\delta v_x \cong -y\frac{\Omega_0 \Omega_\beta^2 k_w^2 v_\parallel^2}{\Omega_0^2 - k_w^2 v_\parallel^2}, \tag{2.125}$$

and

$$\left(\frac{d^2}{dt^2} + \Omega_0^2 - \frac{\Omega_\beta^2 k_w^2 v_\parallel^2}{\Omega_0^2 - k_w^2 v_\parallel^2}\right)\delta v_y \cong 0, \tag{2.126}$$

where Ω_β denotes the betatron frequency. The latter equation (2.126) describes a modified betatron oscillation in the y-direction which arises from the wiggler inhomogeneity. Observe that the frequency of this oscillation reduces to Ω_β in the limit in which the solenoidal magnetic field vanishes. The homogeneous solution to equation (2.125) represents a modified Larmor oscillation in the combined wiggler and solenoidal fields. However, the particular solution describes the $\boldsymbol{B}_0 \times \nabla \boldsymbol{B}_w$ drift due to the combined fields. This bulk drift in the x-direction varies as

$$\delta v_x \cong -y \frac{B_w^2}{2B_0^2 + B_w^2} \frac{\Omega_0 k_w^2 v_\parallel^2}{\Omega_0^2 - k_w^2 v_\parallel^2}, \tag{2.127}$$

so that

$$x(y, z) \cong -yz \frac{B_w^2}{2B_0^2 + B_w^2} \frac{\Omega_0 k_w^2 v_\parallel}{\Omega_0^2 - k_w^2 v_\parallel^2}. \tag{2.128}$$

We observe that the magnitude of this drift is (1) proportional to the displacement from the plane of symmetry in the y-direction, and (2) resonant at $\Omega_0 \approx k_w v_\parallel$.

As a consequence of this drift, the electron beam can readily be driven to the wall of the drift tube. In order to minimize particle losses, this imposes stringent conditions on the parameters of free-electron lasers which employ this configuration. Since the drift is in the x-direction and is proportional to the y-displacement from the plane of symmetry, we must require that the maximum thickness of the beam in the y-direction, Δy_b, must be sufficiently thin that

$$\Delta y_b < \frac{L_x}{k_w L_z} \frac{(\Omega_0^2 - k_w^2 v_\parallel^2)}{\Omega_0 k_w v_\parallel} \frac{2B_0^2 + B_w^2}{B_w^2}, \tag{2.129}$$

where L_x defines the x-dimension of the drift tube, and L_z is the axial length of the wiggler. The maximum permissible thickness of the beam in the y-direction, therefore, decreases as the wiggler length increases and as the resonance at $\Omega_0 \approx k_w v_\parallel$ is approached. As a result, few attempts have been made to use a planar wiggler in conjunction with an axial magnetic field, and alternative means of enhancing beam confinement in planar wiggler configurations have been employed.

Betatron oscillations

Few experiments have employed both a planar wiggler and a solenoidal magnetic field due to the $\boldsymbol{B}_0 \times \nabla \boldsymbol{B}_w$ drift. In the absence of the solenoidal magnetic field, the x-component of the transverse velocity is determined by

$$v_x = \frac{P_x}{\gamma m_e} + v_w \cosh k_w y \cos k_w z. \tag{2.130}$$

For simplicity, we consider the limit in which $P_x = 0$. Substitution of equation (2.130) into equation (2.119) for v_y yields

$$\frac{d}{dt} v_y = -\frac{\Omega_w^2}{k_w} \sinh k_w y \cosh k_w y \cos^2 k_w z. \tag{2.131}$$

This equation describes both rapid oscillations with a period half that of the wiggler, and slow betatron oscillations. If we average this equation over a wiggler period to remove the rapid oscillations, then we find that slow betatron oscillations are described by

$$\frac{d^2}{dt^2} y \cong -\frac{\Omega_w^2}{2k_w} \sinh k_w y \cosh k_w y. \tag{2.132}$$

This nonlinear second-order differential equation reduces to simple harmonic oscillations when the electron displacement from the symmetry axis is small. In this limit, $k_w y \ll 1$ and to lowest order we obtain

$$\left(\frac{d^2}{dt^2} + \Omega_\beta^2\right) y \cong 0. \tag{2.133}$$

Solution of this equation results in the typical form for betatron oscillations given in equations (2.95) and (2.126). It should be remarked that the magnitude of the betatron oscillations is independent of the wiggler amplitude and period, and is dependent on the initial conditions on the electrons; specifically, on the magnitudes of the initial velocity and displacement from the plane of symmetry.

In the case of large-amplitude excursions from the plane of symmetry, equation (2.132) describes a nonlinear oscillation which can be described in terms of the elliptic functions. Equation (2.132) can be integrated once to obtain

$$\left(\frac{du}{d\tau}\right)^2 \cong K^2 - \sinh^2 u, \tag{2.134}$$

where K is an integration constant, $u = k_w y$ and $\tau = \Omega_\beta t$. This describes oscillatory motion in which $\sinh^2 u < K^2$. Equation (2.134) takes on the usual form of an elliptic differential equation under the transformation $w = \sinh u$, for which we obtain

$$\left(\frac{dw}{d\tau}\right)^2 \cong (1 + w^2)(K^2 - w^2). \tag{2.135}$$

This has the solution

$$w = \frac{\kappa K \,\mathrm{sn}[\sqrt{(1+K^2)}\tau, \kappa]}{\sqrt{\{1 - \kappa^2 \,\mathrm{sn}^2[\sqrt{(1+K^2)}\tau, \kappa]\}}}, \tag{2.136}$$

subject to the initial condition $w(\tau = 0) = 0$, where sn denotes the Jacobi elliptic sine function, and

$$\kappa \equiv \frac{K}{\sqrt{(1+K^2)}}. \tag{2.137}$$

Hence, the large-amplitude betatron oscillation is described by

$$y(t) = \frac{1}{k_w} \sinh^{-1}\left[\frac{\kappa K \,\mathrm{sn}[\Omega_\beta\sqrt{(1+K^2)}t, \kappa]}{\sqrt{\{1 - \kappa^2 \,\mathrm{sn}^2[\Omega_\beta\sqrt{(1+K^2)}t, \kappa]\}}}\right], \tag{2.138}$$

subject to the initial condition $y(t = 0) = 0$. Equation (2.138) reduces to the small-amplitude result in the limit in which $K^2 \ll 1$.

The effect of parabolic pole faces

Betatron oscillations describe the electron response to the inhomogeneity of the wiggler which provides a restoring force that confines the electron beam in the neighbourhood of the symmetry plane. However, the inhomogeneity is manifested only in the y-direction in the case of a planar wiggler with flat pole faces (equation (2.2)), and there is no restoring force to enhance the confinement of the electron beam in the x-direction. Both solenoidal and quadrupole magnetic fields have been employed to confine the electron beam against spreading in the x-direction. However, these techniques are not entirely satisfactory since orbital resonances between these fields and the wiggler can affect the interaction strength deleteriously. An alternative technique is to use a wiggler with tapered pole faces such as that given in equation (2.3), which describes the effect of parabolic pole faces on the wiggler field. This technique for improving the confinement of the electron beam has the advantage that the resulting betatron oscillations occur at low frequencies and couple only weakly to the electromagnetic waves. Using equation (2.3), we find that the detailed equations which describe the electron trajectories in this wiggler are

$$\frac{d}{dt}v_x = \Omega_w \cosh\left(\frac{k_w x}{\sqrt{2}}\right)\left[\sqrt{2}\,v_y \sinh\left(\frac{k_w y}{\sqrt{2}}\right)\sin k_w z + v_z \cosh\left(\frac{k_w y}{\sqrt{2}}\right)\cos k_w z\right], \tag{2.139}$$

$$\frac{d}{dt}v_y = -\Omega_w \sinh\left(\frac{k_w y}{\sqrt{2}}\right)\left[\sqrt{2}v_x \cosh\left(\frac{k_w x}{\sqrt{2}}\right)\sin k_w z + v_z \sinh\left(\frac{k_w x}{\sqrt{2}}\right)\cos k_w z\right], \tag{2.140}$$

and

$$\frac{d}{dt}v_z = -\Omega_w \cos k_w z\left[v_x \cosh\left(\frac{k_w x}{\sqrt{2}}\right)\cosh\left(\frac{k_w y}{\sqrt{2}}\right) - v_y \sinh\left(\frac{k_w x}{\sqrt{2}}\right)\sinh\left(\frac{k_w y}{\sqrt{2}}\right)\right]. \tag{2.141}$$

Observe that there are no ignorable coordinates for this form of the wiggler, and the only constant of the motion is the total energy.

In order to obtain solutions for the equations of motion, we assume that displacements from the symmetry plane are much less than the wiggler period and expand in powers of $k_w x$ and $k_w y$. The principal component of the wiggler-induced oscillation remains in the x-direction for the planar wiggler with parabolic pole faces (equation (2.3)). As a consequence, there will be two components of the motion in the x-direction: a rapid oscillation at the wiggler period, and slow variation due to the wiggler inhomogeneity. For simplicity, we shall assume that both components of the motion are small, and expand the orbit equations about the idealized trajectories (equations (2.103) and (2.104))

$$\mathbf{v}^{(0)} \cong -v_w \hat{e}_x \sin k_w z + v_\parallel \hat{e}_z\left(1 + \frac{v_w^2}{4v_\parallel^2}\cos 2k_w z\right), \tag{2.142}$$

which is correct to within terms of order $(v_w/v_\parallel)^2$. Observe that the form we employ for the planar wiggler with parabolic pole faces introduces a phase shift in the idealized trajectories relative to the planar wiggler with flat pole faces (equation (2.2)). Expanding about this solution, we write $x = (v_w/k_w v_\parallel)\cos k_w z + \delta x$, $y = \delta y$, $z = v_\parallel t + \delta z$, $v_x = -v_w \sin k_w z + \delta v_x$, $v_y = \delta v_y$ and $v_z = v_\parallel + \delta v_z$. To lowest order in the perturbed quantities, therefore, the orbit equations governing the transverse velocities are

$$\frac{d}{dt}\delta v_x \cong -\frac{1}{2}\Omega_\beta^2 \delta x, \tag{2.143}$$

and

$$\frac{d}{dt}\delta v_y \cong -\frac{1}{2}\Omega_\beta^2 \delta y, \tag{2.144}$$

where we have performed an average over the wiggler period. The inhomogeneity introduced by the parabolic pole faces results in a modified form of the betatron oscillation at a frequency of $\Omega_\beta/\sqrt{2}$. The detailed solution for this modified betatron oscillation in the x-direction is

$$\begin{bmatrix} \delta x(t) \\ \delta v_x(t) \end{bmatrix} = \begin{bmatrix} \cos \dfrac{\Omega_\beta}{\sqrt{2}} t & \dfrac{\sqrt{2}}{\Omega_\beta} \sin \dfrac{\Omega_\beta}{\sqrt{2}} t \\ -\dfrac{\Omega_\beta}{\sqrt{2}} \sin \dfrac{\Omega_\beta}{\sqrt{2}} t & \cos \dfrac{\Omega_\beta}{\sqrt{2}} t \end{bmatrix} \begin{bmatrix} \delta x(0) \\ \delta v_x(0) \end{bmatrix}. \tag{2.145}$$

A similar form is found for the betatron oscillation in the y-direction.

The effects of the betatron oscillations in the x- and y-directions on the axial component of the velocity are not identical. Expansion of equation (2.141) to first order in the perturbed quantities shows that the axial velocity is dependent upon δv_x in the following manner

$$\frac{d}{dt}\delta v_z \cong -\Omega_w \delta v_x \cos k_w v_\parallel t. \tag{2.146}$$

As a result, substitution of the form for the betatron oscillation from equation (2.145) gives a perturbed axial velocity which varies as

$$\delta v_z \cong \frac{\Omega_w}{2}\left[\frac{1}{\Omega_+}\left(\frac{\Omega_\beta}{\sqrt{2}}\delta x(0)\cos\Omega_+ t + \delta v_x(0)\sin\Omega_+ t\right) + \frac{1}{\Omega_-}\left(\frac{\Omega_\beta}{\sqrt{2}}\delta x(0)\cos\Omega_- t - \delta v_x(0)\sin\Omega_- t\right)\right], \tag{2.147}$$

where $\Omega_\pm \equiv k_w v_\parallel \pm \Omega_\beta/\sqrt{2}$. The result of a planar wiggler with parabolic pole faces, therefore, is to introduce sideband oscillations in the axial velocity about the wiggler period. In the event that the displacement from the plane of symmetry in the x-direction is large (i.e. when $|\delta x| \approx \Omega_w/k_w^2 v_\parallel$, the sideband perturbations to the axial velocity are comparable in magnitude to the oscillation with a wavenumber of $2k_w$), then these sideband oscillations can have a significant impact on the operation of the free-electron laser.

2.3 TAPERED WIGGLER CONFIGURATIONS

Axially nonuniform (i.e. tapered) magnetostatic fields are important for the purpose of enhancing the extraction efficiency in free-electron lasers. The electron beam loses energy to the wave during the course of the interaction and decelerates in the axial direction. The interaction saturates when the electron beam loses sufficient energy to drop out of resonance with the wave. The purpose of a tapered magnetic field is to transfer energy from the transverse motion of the beam to accelerate the beam and maintain the wave–particle resonance over an extended interaction region.

This reacceleration of the electron beam can be accomplished by means of a wiggler field in which either the amplitude or period varies in the axial direction. In addition, an axially tapered solenoidal magnetic field can also be used for this purpose. This process can be most easily described under the assumption of a slow variation in the magnetostatic fields in which the scale length of the field inhomogeneity is much longer than the wiggler period; specifically,

$$\frac{1}{B_{w,0}}\left|\frac{d}{dz}B_{w,0}\right| \ll k_w, \tag{2.148}$$

and

$$\left|\frac{d}{dz}k_w\right| \ll k_w^2. \tag{2.149}$$

The tapered fields, therefore, can be treated as a small perturbation, and the orbits can be obtained by expansion about the trajectories obtained in the uniform field limit.

2.3.1 The idealized one-dimensional limit

In the case of a configuration which consists of a helical wiggler field and a solenoidal magnetic field, we expand about the steady-state trajectories. The magnitude of the axial velocity is dependent upon the magnitudes of the wiggler and solenoidal magnetic fields and upon the period of the wiggler field, and varies with these parameters as

$$\frac{d}{dz}v_\parallel = \frac{\partial v_\parallel}{\partial B_w}\frac{d}{dz}B_w + \frac{\partial v_\parallel}{\partial B_0}\frac{d}{dz}B_0 + \frac{\partial v_\parallel}{\partial k_w}\frac{d}{dz}k_w. \tag{2.150}$$

It is assumed implicitly within the context of this formulation that the trajectory varies continuously from one steady-state orbit to another. Because the total energy is conserved, the derivatives of $v_\parallel$ can be obtained by implicit differentiation of $v_w^2 + v_\parallel^2 = (1-\gamma^{-2})c^2$, where v_w is given by equations (2.15) and (2.53) for the idealized and realistic wigglers, respectively. Therefore, the axial acceleration occurs at the expense of a transverse deceleration. In the idealized limit, we find that

$$\frac{\partial v_\parallel}{\partial B_w} = \frac{-1}{B_w}\frac{\Omega_w v_w}{(1+\beta_w^2)\Omega_0 - k_w v_\parallel}, \tag{2.151}$$

$$\frac{\partial v_\parallel}{\partial k_w} = -\frac{v_w^2}{(1+\beta_w^2)\Omega_0 - k_w v_\parallel}, \tag{2.152}$$

and

$$\frac{\partial v_\parallel}{\partial B_0} = \frac{1}{B_0}\frac{v_\parallel \beta_w^2 \Omega_0}{(1+\beta_w^2)\Omega_0 - k_w v_\parallel}, \tag{2.153}$$

Substitution of these derivatives into equation (2.150) gives the rate of change of the axial velocity

$$\frac{d}{dz}v_\parallel = -\frac{\beta_w^2 v_\parallel}{(1+\beta_w^2)\Omega_0 - k_w v_\parallel}\left(\frac{\Omega_0 - k_w v_\parallel}{B_w}\frac{d}{dz}B_w - \frac{k_w v_\parallel}{\lambda_w}\frac{d}{dz}\lambda_w - \frac{\Omega_0}{B_0}\frac{d}{dz}B_0\right). \tag{2.154}$$

It is important to observe that in order to accelerate the electron beam, both the wiggler amplitude and period must be tapered downward (i.e. $dB_w/dz < 0$ and $d\lambda_w/dz < 0$). However, the situation is more complex in the case of the solenoidal field. Observe that $(1+\beta_w^2)\Omega_0 < k_w v_\parallel$ for the stable Group I orbits; hence, the solenoidal field amplitude must be tapered downward to achieve an axial acceleration of the electrons. In contrast, $\Omega_0 > k_w v_\parallel$ for Group II trajectories and the solenoidal field must be tapered upward (i.e. increased) in order to accelerate the beam. Observe that in the limit in which the solenoidal magnetic field vanishes, the variation in the axial velocity is given by

$$\frac{d}{dz}v_\parallel = -\frac{\Omega_w^2}{k_w^2 v_\parallel}\left(\frac{1}{B_w}\frac{d}{dz}B_w + \frac{1}{\lambda_w}\frac{d}{dz}\lambda_w\right). \tag{2.155}$$

2.3.2 The realizable three-dimensional formulation

These results may be generalized to treat the realistic helical wiggler given by equation (2.1), for which

$$\frac{\partial v_\parallel}{\partial B_w} = \frac{1}{B_w}\frac{2\Omega_w v_w (1+\lambda_0^2)^2 I_1^2(\lambda_0)/\lambda_0^2}{[(1+\lambda_0^2)\Omega_0 - k_w v_\parallel]Z(\lambda_0) - \lambda_0^2 k_w v_\parallel\, W(\lambda_0)}, \tag{2.156}$$

$$\frac{\partial v_\parallel}{\partial k_w} = \frac{\lambda_0 v_\parallel^2 (1+\lambda_0^2) I_1(\lambda_0)}{[(1+\lambda_0^2)\Omega_0 - k_w v_\parallel]Z(\lambda_0) - \lambda_0^2 k_w v_\parallel\, W(\lambda_0)}, \tag{2.157}$$

and

$$\frac{\partial v_\parallel}{\partial B_0} = -\frac{1}{B_0}\frac{\lambda_0 v_\parallel \Omega_0 (1+\lambda_0^2) I_1(\lambda_0)}{[(1+\lambda_0^2)\Omega_0 - k_w v_\parallel]Z(\lambda_0) - \lambda_0^2 k_w v_\parallel\, W(\lambda_0)}. \tag{2.158}$$

As a consequence, the axial velocity varies as

$$\frac{d}{dz}v_\parallel = \frac{\lambda_0 v_\parallel (1+\lambda_0^2) I_1(\lambda_0)}{[(1+\lambda_0^2)\Omega_0 - k_w v_\parallel]Z(\lambda_0) - \lambda_0^2 k_w v_\parallel\, W(\lambda_0)} \times\left(\frac{\Omega_0 - k_w v_\parallel}{B_w}\frac{d}{dz}B_w - \frac{k_w v_\parallel}{\lambda_w}\frac{d}{dz}\lambda_w - \frac{\Omega_0}{B_0}\frac{d}{dz}B_0\right). \tag{2.159}$$

This differs from the result obtained in the idealized limit only in terms of an overall factor which describes the effect of large displacements from the axis of symmetry, and reduces to that result in the limit as $\lambda_0 \to 0$.

2.3.3 Planar wiggler geometries

The corresponding axial acceleration due to a tapered planar wiggler magnetic field can be determined in an analogous manner by perturbation about the quasi-steady-state orbits. In this case, however, we neglect the solenoidal field. This is because the presence of an axial magnetic field results in a transverse drift which, over the course of the tapered wiggler, will cause a loss of the beam to the walls of the drift tube. As a result, the acceleration of the bulk axial velocity (equation (2.105)) due to gradients in the amplitude and period is given by

$$\frac{d}{dz}v_\parallel = -\frac{\Omega_w^2}{2k_w^2 v_\parallel}\left(\frac{1}{B_w}\frac{d}{dz}B_w + \frac{1}{\lambda_w}\frac{d}{dz}\lambda_w\right). \tag{2.160}$$

Observe that this is identical to the result obtained for the helical wiggler in the absence of the solenoidal magnetic field under the substitution of the root-mean-square value of the wiggler field amplitude.

REFERENCES

1. Blewett, J. P. and Chasman, R. (1977) Orbits and fields in the helical wiggler. *J. Appl. Phys.*, **48**, 2692.
2. Friedland, L. (1980) Electron beam dynamics in combined guide and pump magnetic fields for free-electron laser applications. *Phys. Fluids*, **23**, 2376.
3. Diament, P. (1981) Electron orbits and stability in realizable and unrealizable wigglers of free-electron lasers. *Phys. Rev. A.*, **23**, 2537.
4. Jones, R. D. (1981) Constants of the motion in a helical magnetic field. *Phys. Fluids*, **24**, 564.
5. Pasour, J. A., Mako, F. and Roberson, C. W. (1982) Electron drift in a linear magnetic wiggler with an axial guide field. *J. Appl. Phys.*, **53**, 7174.
6. Freund, H. P. and Drobot, A. T. (1982) Relativistic electron trajectories in free-electron lasers with an axial guide field. *Phys. Fluids*, **25**, 736.
7. Freund, H. P. and Ganguly, A. K. (1985) Electron orbits in free-electron lasers with helical wiggler and axial guide magnetic fields. *IEEE J. Quantum Electron.*, **QE-21**, 1073.
8. Fajans, J., Kirkpatrick, D. A. and Bekefi, G. (1985) Off-axis electron orbits in realistic helical wigglers for free-electron laser applications. *Phys. Rev. A*, **32**, 3448.
9. Littlejohn, R. G. and Kaufman, A. N. (1987) Hamiltonian structure of particle motion in an ideal helical wiggler with guide field. *Phys. Lett. A*, **120**, 291.
10. Phillips, R. M. (1988) History of the ubitron. *Nucl. Instr. Meth.*, **A272**, 1.
11. Scharlemann, E. T. (1985) Wiggler plane focussing in linear wigglers. *J. Appl. Phys.*, **58**, 2154.
12. Levush, B., Antonsen, T. M. Jr, Manheimer, W. M. and Sprangle, P. (1985) A free-electron laser with a rotating quadrupole wiggler. *Phys. Fluids*, **28**, 2273.
13. Levush, B., Antonsen, T. M. Jr and Manheimer, W. M. (1986) Spontaneous radiation of an electron beam in a free-electron laser with a quadrupole wiggler. *J. Appl. Phys.*, **60**, 1584.

14. Antonsen, T. M. Jr and Levush, B. (1987) Nonlinear theory of a quadrupole free-electron laser. *IEEE J. Quantum Electron.*, **QE-23**, 1621.
15. Chang, S. F., Eldridge, O. C. and Sharer, J. E. (1988) Analysis and nonlinear simulation of a quadrupole wiggler free-electron laser at millimeter wavelengths. *IEEE J. Quantum Electron.*, **QE-24**, 2309.
16. Shefer, R. E. and Bekefi, G. (1981) Cyclotron emission from intense relativistic electron beams in uniform and rippled magnetic fields. *Int. J. Electronics*, **51**, 569.
17. McMullin, W. A. and Bekefi, G. (1982) Stimulated emission from relativistic electrons passing through a spatially periodic longitudinal magnetic field. *Phys. Rev. A*, **25**, 1826.

3

Incoherent undulator radiation

The spontaneous synchrotron radiation produced by individual electrons executing undulatory trajectories in a magnetostatic field is incoherent and is the radiation mechanism used in synchrotron light sources. The magnetostatic fields in these devices are formally identical to those employed in free-electron lasers but are commonly referred to as **undulators** rather than wigglers. The reason for this is that electron synchrotrons produce high-energy electron beams which permit the use of extremely long-period undulations. The use of long-period undulations makes possible the production of relatively large-amplitude magnetostatic fields which are required to ensure the production of a relatively high radiation intensity from this incoherent mechanism. In contrast, since the free-electron laser relies on a coherent emission process, the wiggler magnets employed can be of shorter periods and lower amplitudes. However, incoherent synchrotron radiation is produced in free-electron lasers as well [1–5].

3.1 TEST PARTICLE FORMULATION

The time-averaged power radiated by this incoherent process can be calculated from Poynting's theorem by means of the equation

$$P = -\lim_{T\to\infty} \frac{1}{T}\int_{-T/2}^{T/2} dt \int d^3x\, \boldsymbol{E}(\boldsymbol{x},t)\cdot\boldsymbol{J}(\boldsymbol{x},t), \tag{3.1}$$

where $\boldsymbol{E}(\boldsymbol{x},t)$ denotes the microscopic radiation field, and

$$\boldsymbol{J}(\boldsymbol{x},t) = -e\sum_{j=1}^{N_b} \mathbf{v}_j(t)\delta[\boldsymbol{x}-\boldsymbol{x}_j(t)] \tag{3.2}$$

is the microscopic source current consisting of the sum of all the N_b electrons in the beam during the time $-T/2 < t < T/2$. The radiated power can be expressed in terms of the Fourier amplitudes of the microscopic fields and source currents by noting that for frequency ω and wavenumber $\boldsymbol{k}$

$$P = -2(2\pi)^4 \lim_{T\to\infty}\frac{1}{T}\int_0^\infty d\omega \int d^3k\, \mathrm{Re}[\boldsymbol{E}_{k,\omega}\cdot\boldsymbol{J}^*_{k,\omega}], \tag{3.3}$$

where the asterisk (*) denotes the complex conjugate, and the Fourier transforms are defined as

$$f_{k,\omega} = \int_{-\infty}^{\infty} dt \int d^3x \exp(i\omega t - i\boldsymbol{k}\cdot\boldsymbol{x}) f(\boldsymbol{x}, t). \tag{3.4}$$

A self-consistent relation between the fields and the source current depends upon the dielectric properties of the beam, and can be expressed in the form

$$\Lambda_{k,\omega}\cdot \boldsymbol{E}_{k,\omega} = \frac{4\pi i}{\omega}\boldsymbol{J}_{k,\omega}, \tag{3.5}$$

where $\Lambda_{k,\omega}$ denotes the dispersion tensor

$$\Lambda_{k,\omega} = \frac{c^2}{\omega^2}\left(\boldsymbol{kk} - k^2\boldsymbol{I}\right) + \varepsilon, \tag{3.6}$$

$\boldsymbol{I}$ denotes the unit dyadic, and $\boldsymbol{\varepsilon}$ is the dielectric tensor of the beam. For the case of a diffuse beam in which the wave frequency greatly exceeds the electron plasma frequency, $\boldsymbol{\varepsilon} \cong \boldsymbol{I}$ and equation (3.5) can be inverted to give

$$\boldsymbol{E}_{k,\omega} = -\frac{4\pi i\omega}{(\omega^2 - c^2k^2)}\left(\boldsymbol{J}_{k,\omega} - \frac{c^2}{\omega^2}\boldsymbol{kk}\cdot\boldsymbol{J}_{k,\omega}\right). \tag{3.7}$$

The power radiated per unit frequency, volume V, and solid angle subtended by $\boldsymbol{k}$ is referred to as the **emissivity**, and can be expressed as

$$\eta(\omega, \Omega_k) \equiv \frac{1}{V}\frac{dP}{d\omega d\Omega_k} = (2\pi)^6\frac{\omega^2}{Vc^3}\lim_{T\to\infty}\frac{1}{T}\left(|\boldsymbol{J}_{k,\omega}|^2 - \frac{1}{k^2}|\boldsymbol{k}\cdot\boldsymbol{J}_{k,\omega}|^2\right)_{k=\omega/c}, \tag{3.8}$$

where V is the total volume of the interaction region. The radiation spectrum, therefore, can be determined from (3.8) based upon a knowledge of the single-particle orbits, which is necessary to specify the source current.

The Fourier transform of the source current is

$$\boldsymbol{J}_{k,\omega} = -\frac{e}{(2\pi)^4}\sum_{j=1}^{N_b}\lim_{T\to\infty}\int_{-T/2}^{T/2} dt\mathbf{v}_j(t)\exp[i\omega t - i\boldsymbol{k}\cdot\boldsymbol{x}_j(t)], \tag{3.9}$$

and the electron trajectories must be specified as well. For the case of a helical wiggler configuration, the electron trajectories are given by the steady-state orbits; hence

$$\mathbf{v} = v_w(\hat{\boldsymbol{e}}_x \cos k_w z + \hat{\boldsymbol{e}}_y \sin k_w z) + v_\parallel \hat{\boldsymbol{e}}_z, \tag{3.10}$$

and

$$\boldsymbol{x} = \boldsymbol{x}_0 + \frac{v_w}{v_\parallel k_w}[\hat{\boldsymbol{e}}_x(\sin k_w z - \sin k_w z_0) - \hat{\boldsymbol{e}}_y(\cos k_w z - \cos k_w z_0)] + v_\parallel t\hat{\boldsymbol{e}}_z, \tag{3.11}$$

where v_w and $v_\parallel$ are the transverse and axial velocities, and $\boldsymbol{x}_0$ denotes the initial position. In view of the Bessel function identity

$$\exp(ib\sin\theta) = \sum_{n=-\infty}^{\infty} J_n(b)\exp(in\theta), \tag{3.12}$$

where J_n is the regular Bessel function of the first kind, it is apparent that

$$\begin{aligned}\mathbf{v}_j(t)\exp(i\omega t - i\boldsymbol{k}\cdot\boldsymbol{x}_j) = \exp(-i\boldsymbol{k}\cdot\boldsymbol{x}_{0j}) \sum_{n=-\infty}^{\infty} \boldsymbol{V}_n(b_j)\exp[i\omega t - i(k_\parallel + nk_w)v_{j\parallel}t] \\ \times \exp[ib_j\sin(k_w z_0 - \theta) - in(k_w z_0 - \theta)],\end{aligned} \tag{3.13}$$

where $b_j \equiv k_\perp v_{jw}/k_w v_{j\parallel}$, the wavenumber is $\boldsymbol{k} = k_\perp(\hat{\boldsymbol{e}}_x\cos\varphi + \hat{\boldsymbol{e}}_y\sin\varphi) + k_\parallel\hat{\boldsymbol{e}}_z$, and

$$\boldsymbol{V}_n(b) \equiv \frac{nk_w v_\parallel}{k_\perp} J_n(b)\hat{\boldsymbol{e}}_\perp + iv_w J_n'(b)\hat{\boldsymbol{e}}_\varphi + v_\parallel J_n(b)\hat{\boldsymbol{e}}_z \tag{3.14}$$

in the basis $\hat{\boldsymbol{e}}_\perp = \hat{\boldsymbol{e}}_x\cos\varphi + \hat{\boldsymbol{e}}_y\sin\varphi$ and $\hat{\boldsymbol{e}}_\varphi = -\hat{\boldsymbol{e}}_x\sin\varphi + \hat{\boldsymbol{e}}_y\cos\varphi$. As a result, the source current is

$$\begin{aligned}\boldsymbol{J}_{k,\omega} = -\frac{e}{(2\pi)^4}\lim_{T\to\infty}\sum_{j=1}^{N_b}\exp(-i\boldsymbol{k}\cdot\boldsymbol{x}_{0j})\sum_{n=-\infty}^{\infty}\boldsymbol{V}_n(b_j)\frac{\sin(\Delta\omega_{jn}T/2)}{\Delta\omega_{jn}/2} \\ \times \exp[ib_j\sin(k_w z_0 - \theta) - in(k_w z_0 - \theta)],\end{aligned} \tag{3.15}$$

where $\Delta\omega_{jn} \equiv \omega - (k_\parallel + nk_w)v_{j\parallel}$ defines the frequency mismatch parameter.

In the computation of the quadratic forms which appear in the emissivity (3.8), we impose a random phase approximation and obtain

$$\begin{aligned}\eta(\omega,\Omega_k) = \frac{e^2\omega^2}{2\pi^2 Vc^3}\lim_{T\to\infty}\sum_{j=1}^{N_b}\sum_{n=-\infty}^{\infty}\left(|\boldsymbol{V}_n(b_j)|^2 - \frac{1}{k^2}|\boldsymbol{k}\cdot\boldsymbol{V}_n(b_j)|^2\right) \\ \times \frac{\sin^2(\Delta\omega_{jn}T/2)}{\Delta\omega_{jn}^2 T/2}\bigg|_{k=\omega k}.\end{aligned} \tag{3.16}$$

Observe that this expression describes emission at each harmonic of the wiggler resonance. We now convert the discrete sum over individual electrons into a continuous integral over the beam distribution function over the parallel component of the momentum $F_b(p_\parallel)$ by making the replacement

$$\frac{1}{V}\sum_{j=1}^{N_b} \to n_b\int_0^\infty dp_\parallel F_b(p_\parallel), \tag{3.17}$$

subject to the normalization condition that

$$\int_0^\infty dp_\parallel F_b(p_\parallel) = 1. \tag{3.18}$$

Observe that only one degree of freedom remains under the assumption of the steady-state trajectories. For a specific choice of the magnetic fields, the individual electron trajectories may be specified by means of either the total momentum (which is equivalent to the total energy) or the axial component of the momentum, where the relation between these quantities is specified in equation (2.16) in the one-dimensional idealized limit and equation (2.54) for the three-dimensional realizable wiggler representation. The emissivity assumes the form

$$\eta(\omega,\theta)=\frac{e^2n_b\omega^2}{2\pi^2c^3}\lim_{T\to\infty}\sum_{n=-\infty}^{\infty}\int_0^\infty dp_\parallel v_w^2F_b(p_\parallel)\frac{\sin^2(\Delta\omega_nT/2)}{\Delta\omega_n^2T/2}\times\left[\left(\cos\theta-\frac{k}{nk_w}\sin^2\theta\right)^2\frac{n^2}{b^2}J_n^2(b)+J_n'^2(b)\right]\Bigg|_{k=\omega/c}, \tag{3.19}$$

upon substitution for $V_n(b)$. This describes the total emissivity due to contributions from all harmonics. Observe that the emissivity is independent of the angle $\varphi[=\tan^{-1}(k_y/k_x)]$, and the emission is azimuthally symmetric. It should be remarked that in the conversion of the discrete sum over electrons to the continuous integral over the beam distribution we have made use of the fact that the electron trajectories have been constrained to follow the steady-state orbits. In this case, apart from the field information, only one parameter (the total momentum or energy) is required to specify v_w and $v_\parallel$ for the orbit. In addition, we have not restricted steady-state trajectories to the idealized one-dimensional form. Equations (3.10) and (3.11) are valid for either the one-dimensional or three-dimensional solution, subject to the inclusion of the guiding-centre motion in $\mathbf{x}_0$ in the realizable case. This inclusion does not affect the emissivity (3.18) since the guiding-centre oscillations occur over a much longer period than does the emitted radiation.

The situation is somewhat different for the case of a planar wiggler field since the azimuthal symmetry present for the helical wiggler geometry does not exist. Using the quasi-steady-state orbits described in section 2.2.1 to treat the trajectories in the presence of both a planar wiggler and a solenoidal magnetic field, we find that the emissivity is of the form

$$\begin{aligned}\eta(\omega,\theta,\varphi)=\frac{e^2n_b\omega^2}{2\pi^2c^3}&\lim_{T\to\infty}\sum_{l,n=-\infty}^{\infty}\int_0^\infty dp_\parallel\alpha_x^2F_b(p_\parallel)J_l^2(b_y)\frac{\sin^2(\Delta\omega_{n-l}T/2)}{\Delta\omega_{n-l}^2T/2}\\&\times\left\{\left[\sin^2\varphi+\cos^2\varphi\left(\cos\theta-\frac{k}{nk_w}\sin\theta\right)^2\right]\frac{n^2}{b_x^2}J_n^2(b_x)\right.\\&\left.+[\sin^2\varphi\cos^2\theta+\cos^2\varphi]\frac{\alpha_y^2}{\alpha_x^2}J_n'^2(b_x)\right\}\Bigg|_{k=\omega/c},\end{aligned} \tag{3.20}$$

where $b_x\equiv k_x\alpha_x/k_wv_\parallel$, $b_y\equiv k_y\alpha_y/k_wv_\parallel$, and we have employed the quasi-steady-state trajectories given in equation (2.111)

$$\mathbf{v}=\alpha_x\hat{e}_x\cos k_wz+\alpha_y\hat{e}_y\sin k_wz+v_\parallel\hat{e}_z. \tag{3.21}$$

It should be remarked that the oscillatory component of the axial velocity has been neglected in the derivation of equation (3.19). This is valid under the assumption of the quasi-steady-state trajectories subject to the requirements that $\alpha_x^2\ll1$ and $\alpha_y^2\ll1$. In this case, the emissivity in the presence of an axial solenoidal magnetic field does depend upon the angle φ. This means that the spontaneous emission from an electron beam propagating through a planar wiggler is not azimuthally symmetric as in the case of a helical wiggler. This polarization in the direction of propagation reaches its most extreme limit when the solenoidal magnetic field vanishes, at which point $\alpha_y = 0$ and only the $l = 0$ term in the summation survives.

3.2 THE COLD BEAM REGIME

The cold (i.e. monoenergetic) beam limit is obtained in the regime in which the width of the frequency spectrum is determined primarily by the bounded nature of the system (i.e. the axial length $L = v_\parallel T$ of the interaction region) rather than the thermal spread of the beam. This condition can be expressed as

$$\frac{\Delta v_\parallel}{v_\parallel} < \frac{\lambda}{L}, \tag{3.22}$$

where $\Delta v_\parallel$ denotes the axial velocity spread. In this regime, the distribution can be written as $F_b(p) = \delta(p - p_0)$ where p_0 is the bulk momentum of the beam, and the emissivity in the case of a helical wiggler becomes

$$\eta(\omega, \theta) = \frac{e^2 n_b \omega^2 v_w^2 L}{4\pi^2 c^3 v_\parallel} \sum_{n=-\infty}^{\infty} \left[\left(\cos\theta - \frac{k}{nk_w} \sin^2\theta \right)^2 \frac{n^2}{b^2} J_n^2(b) + J_n'^2(b) \right] \frac{\sin^2 \Theta_n}{\Theta_n^2}, \tag{3.23}$$

where θ $(\equiv \tan^{-1}(k_\perp / k_\parallel))$ denotes the polar angle between the wavevector and the symmetry axis, and $\Theta_n \equiv \Delta\omega_n L/2v_\parallel$. Observe that equation (3.23) is to be evaluated for the steady-state solutions of v_w and $v_\parallel$ corresponding to a total energy given by $\gamma_0 = (1 + p_0^2/m_e^2 c^2)^{1/2}$.

The emission band at each harmonic is determined by the spectral function $\sin^2\Theta_n/\Theta_n^2$, which is plotted in Fig. 3.1. The spontaneous emission spectrum peaks in the limit of a vanishing frequency mismatch for which $\omega \equiv (k_\parallel + nk_w)v_\parallel$. The bandwidth is given by the FWHM points of the spectral function which occur at $\Theta \cong \pm 1.4$; hence

$$\frac{\Delta\omega}{\omega} < \frac{\lambda_w}{L}, \tag{3.24}$$

where the centre frequency

$$\omega = \frac{nk_w v_\parallel}{1 - \beta_\parallel \cos\theta}, \tag{3.25}$$

and $\beta_\parallel \equiv v_\parallel / c$. This reduces to the more familiar form $\omega \cong 2\gamma_\parallel^2 k_w v_\parallel$ for the fundamental resonance frequency for parallel propagation in the highly relativistic limit (i.e. $v_\parallel \cong c$).

The emission peaks for propagation in the direction of the symmetry axis; hence, $k_\perp \ll k$ and we focus on the limit in which $b \ll 1$. Retaining only the lowest

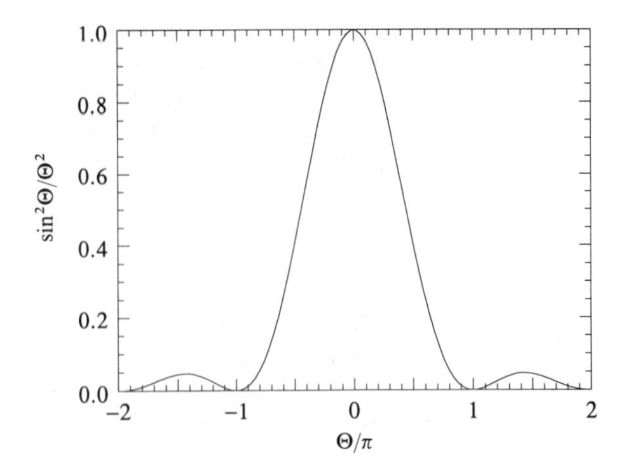

Fig. 3.1 Graph of the spectral function for the spontaneous radiation from a cold beam.

order terms in the power series expansion of the Bessel functions,

$$\frac{n}{b} J_n(b) \cong J'_n(b) \cong \frac{b^{n-1}}{2^n(n-1)!}, \tag{3.26}$$

we obtain an expression for the total emissivity due to the sum of each harmonic contribution

$$\eta(\omega, \theta) = \frac{e^2 n_b \omega^2 v_w^2 L}{4\pi^2 c^3 v_\parallel} \sum_{n=-\infty}^{\infty} \left[1 + \left(\cos\theta - \frac{k}{nk_w} \sin^2\theta \right)^2 \right] \frac{b^{2n-2}}{4^n[(n-1)!]^2} \frac{\sin^2 \Theta_n}{\Theta_n^2}. \tag{3.27}$$

Observe that the emissivity vanishes for forward-directed propagation (i.e. $k_\perp = 0$) at the higher harmonics, and the emission is concentrated in the forward direction only for frequencies in the neighbourhood of the fundamental. In particular, the emissivity at frequencies in the neighbourhood of the fundamental resonance is

$$\eta(\omega, \theta) = \frac{e^2 n_b \omega^2 v_w^2 L}{16\pi^2 c^3 v_\parallel} \left[1 + \left(\cos\theta - \frac{k}{nk_w} \sin^2\theta \right)^2 \right] \frac{\sin^2 \Theta_1}{\Theta_1^2}, \tag{3.28}$$

which, for propagation parallel to the symmetry axis, reduces to

$$\eta(\omega, \theta = 0) = \frac{e^2 n_b \omega^2 v_w^2 L}{8\pi^2 c^3 v_\parallel} \frac{\sin^2 \Theta_1}{\Theta_1^2}. \tag{3.29}$$

The emission at the fundamental resonance, therefore, is directly proportional to the square of the transverse wiggler-induced velocity.

In order to illustrate that the emission at the fundamental is sharply peaked about the forward direction, we consider radiation at the *fixed* frequency

$$\omega = \frac{k_w v_\parallel}{1 - \beta_\parallel}, \tag{3.30}$$

which defines the centre frequency in the limit of forward propagation. As a consequence,

$$\Theta_1 = \frac{k_w L}{2}\left[\frac{2\beta_\parallel^2 \sin^2(\theta/2)}{1-\beta_\parallel} - (1-\beta_\parallel)\right], \tag{3.31}$$

and the spectral function decreases rapidly with increases in θ. Indeed, the polar angle at the FWHM point ($\theta = \theta_{1/2}$) is

$$\theta_{1/2} \cong 1.3\sqrt{\left(\frac{\lambda}{L}\right)}, \tag{3.32}$$

where λ denotes the radiation wavelength. Since the wavelength is, typically, much smaller than the length of the system $\theta_{1/2} \ll 1$ and is comparable to the frequency bandwidth.

3.3 THE TEMPERATURE-DOMINATED REGIME

The spectral band of the incoherent radiation is dominated by the thermal spread of the beam in the limit in which

$$\frac{\Delta v_\parallel}{v_\parallel} > \frac{\lambda}{L}. \tag{3.33}$$

In this regime, we observe that

$$\lim_{T\to\infty} \frac{\sin^2 xT/2}{x^2 T/2} = \pi\delta(x), \tag{3.34}$$

and the emissivity for a helical wiggler configuration is of the form

$$\eta(\omega,\theta) = \frac{e^2 n_b \omega^2}{2\pi c^3} \sum_{n=-\infty}^{\infty} \int_0^\infty dp_\parallel\, v_w^2 F_b(p_\parallel)\delta[\omega - (k_\parallel + nk_w)v_\parallel]$$

$$\times\left[\left(\cos\theta - \frac{k}{nk_w}\sin^2\theta\right)^2 \frac{n^2}{b^2} J_n^2(b) + J_n'^2(b)\right]\Bigg|_{k=\omega/c}. \tag{3.35}$$

As in the cold beam limit, only the fundamental contribution ($n = 1$) survives for forward-directed emission and the emissivity reduces to

$$\eta(\omega,\theta=0) = \frac{e^2 n_b \omega^2}{4\pi c^3} \int_0^\infty dp_\parallel v_w^2 F_b(p_\parallel)\delta\left[\omega - \left(\frac{\omega}{c} + k_w\right)v_\parallel\right], \tag{3.36}$$

which varies as the square of the wiggler-induced transverse velocity.

The spectral bandwidth of the emission is determined by the thermal spread of the electron beam. For illustrative purposes, we restrict the analysis to the case of forward propagation and assume an electron beam which has a Gaussian temperature spread

$$F_b(p_\parallel) = \frac{2\exp[-(p_\parallel - p_0)^2/\Delta p_\parallel^2]}{\sqrt{\pi}\Delta p_\parallel[1 + \mathrm{erf}(p_0/\Delta p_\parallel)]}, \tag{3.37}$$

where p_0 and $\Delta p_\parallel$ describe the bulk streaming momentum and the momentum spread respectively, and

$$\mathrm{erf}(x) = \frac{2}{\sqrt{\pi}} \int_0^x dt \exp(-t^2), \tag{3.38}$$

denotes the error function. The emission occurs at a resonant axial phase velocity

$$\frac{v_{ph}}{c} = \frac{\omega}{\omega + ck_w}, \tag{3.39}$$

which is associated with a transverse velocity

$$v_w(v_{ph}) = \frac{\hat{\Omega}_w v_{ph}}{\hat{\Omega}_0 - \gamma k_w v_{ph}}, \tag{3.40}$$

where $\hat{\Omega}_{0,w} \equiv eB_{0,w}/m_e c$. This variation in the axial and transverse velocities with frequency corresponds to a variation in the energy which satisfies the equation

$$v_{ph}^2 + v_w^2(v_{ph}) = \left(1 - \frac{1}{\gamma^2}\right) c^2. \tag{3.41}$$

This defines a quartic polynomial equation for γ as a function of frequency which specifies the Group I and Group II trajectories. As a consequence, the emissivity for this choice of distribution is

$$\eta(\omega, \theta = 0) = \frac{e^2 n_b \omega}{2\pi^{3/2} c} \frac{p_{ph}}{\Delta p_\parallel} \frac{v_w^2(v_{ph})}{c^2} \frac{[\beta_{ph}^2 + (1 - \beta_{ph}^2)\Phi(v_{ph})]}{(1 - \beta_{ph}^2)\Phi(v_{ph})} \times \frac{\exp[-(p_{ph} - p_0)^2/\Delta p_\parallel^2]}{[1 + \mathrm{erf}(p_0/\Delta p_\parallel)]}, \tag{3.42}$$

where $\beta_{ph} \equiv v_{ph}/c$, $p_{ph} \equiv \gamma m_e v_{ph}$ is the associated resonant momentum, and Φ is defined in equation (2.24).

This expression for the emissivity simplifies considerably in the limit in which the solenoidal magnetic field vanishes. In this case, $\Phi = 1$,

$$v_w = -\frac{\hat{\Omega}_w}{\gamma k_w}, \tag{3.43}$$

and the energy is given by

$$\gamma = \gamma_{ph} \sqrt{\left(1 + \frac{\hat{\Omega}_w^2}{c^2 k_w^2}\right)}, \tag{3.44}$$

where $\gamma_{ph} \equiv (1 - v_{ph}^2/c^2)^{-1/2}$. The emissivity, therefore, becomes

$$\eta(\omega, \theta = 0) = \frac{e^2 n_b \omega}{2\pi^{3/2} c} \frac{p_{ph}}{\Delta p_\parallel} \frac{\hat{\Omega}_w^2/c^2 k_w^2}{1 + \hat{\Omega}_w^2/c^2 k_w^2} \frac{\exp[-(p_{ph} - p_0)^2/\Delta p_\parallel^2]}{[1 + \mathrm{erf}(p_0/\Delta p_\parallel)]}. \tag{3.45}$$

The frequency spectrum is governed by the Gaussian, and peak emission occurs for $p_{ph} = p_0$ which corresponds to the frequency

$$\omega_0 = \frac{k_w v_0}{1 - v_0/c}, \tag{3.46}$$

where the streaming velocity is $v_0 \equiv p_0/\gamma_0 m_e$, and

$$\gamma_0 \equiv \sqrt{\left(1 + \frac{\hat{\Omega}_w^2}{c^2 k_w^2} + \frac{p_0^2}{m_e^2 c^2}\right)}. \tag{3.47}$$

The width of the frequency spectrum, under the assumption that $\Delta p_{\parallel} \ll p_0$,

$$\frac{\Delta\omega}{\omega_0} \cong 2\gamma_{\parallel}^2 \frac{\Delta v_{\parallel}}{v_0}, \tag{3.48}$$

where $\gamma_{\parallel} \equiv (1 - v_0^2/c^2)^{-1/2}$, and $\Delta v_{\parallel} \equiv \gamma_0 \gamma_{\parallel}^2 m_e \Delta p_{\parallel}$. Observe that the criterion which defines the cold beam versus temperature-dominated regimes is obtained by comparison of the respective frequency bandwidths.

REFERENCES

1. Motz, H. and Nakamura, M. (1959) Radiation of an electron in an infinitely long waveguide. *Ann. Phys.*, **7**, 84.
2. Kroll, N. M. (1980) Relativistic synchrotron radiation in a medium and its implications for stimulated electromagnetic shock radiation, in *Physics of Quantum Electronics: Free-Electron Generators of Coherent Radiation*, Vol. 7 (ed. Jacobs, S. F., Pilloff, H. S., Sargent, M., Scully, M. O. and Spitzer, R.), Addison-Wesley, Reading, Massachusetts, p. 335.
3. Madey, J. M. J. (1979) Relationship between mean radiated energy, mean squared radiated energy and spontaneous power spectrum in a power series expansion of the equations of motion in a free-electron laser. *Nuovo Cimento*, **50B**, 64.
4. Freund, H. P., Sprangle, P., Dillenburg, D., da Jornada, E. H., Liberman, B. and Schneider, R. S. (1981) Coherent and incoherent radiation from free-electron lasers with an axial guide field. *Phys. Rev. A*, **24**, 1965.
5. Kim, K. J. (1989) Characteristics of synchrotron radiation, in *The Physics of Particle Accelerators* (ed. Month, M. and Dienes, M.), AIP Conf. Proceedings No. 184, American Institute of Physics, New York, p. 565.

4

Coherent emission: linear theory

In order to give rise to stimulated emission, it is necessary for the electron beam to respond in a collective manner to the radiation field and to form coherent bunches. This can occur when a light wave traverses an undulatory magnetic field such as a wiggler because the spatial variations of the wiggler and the electromagnetic wave combine to produce a beat wave, which is essentially an interference pattern. It is the interaction between the electrons and this beat wave which gives rise to the stimulated emission in free-electron lasers. In the case of a magnetostatic wiggler, this beat wave has the same frequency as the light wave, but its wavenumber is the sum of the wavenumbers of the electromagnetic and wiggler fields. As a result, the phase velocity of the beat wave is less than that of the electromagnetic wave and it is called a ponderomotive wave. Since the ponderomotive wave propagates at less than the speed of light *in vacuo* it can be in synchronism with electrons which are limited by that velocity.

Our purpose in this chapter is to give a detailed discussion of the free-electron laser as a linear gain medium as well as to provide a comprehensive derivation of the relevant formulae for the gain in various configurations in both the idealized one-dimensional and the realistic three-dimensional limits. To this end, we derive the expressions for the gain in both the low- and high-gain regimes. The low-gain regime is relevant to short-wavelength free-electron laser oscillators driven by high-energy but low-current electron beams. In contrast, the results in the high (exponential) gain regime are usually described in terms of a dispersion equation and are appropriate to free-electron laser amplifiers driven by intense relativistic electron beams. The linear theory of the interaction has been discussed extensively in the literature. Early work concentrated upon the idealized one-dimensional regime [1–19], and dealt predominantly with helical wiggler configurations. Additional work included planar wiggler configurations as well and dealt with the effects of a beam thermal spread [20–2], and an axial solenoidal field [23–31], as well as realistic multi-dimensional models of the interaction [32–44].

The interaction between the electron beam and the ponderomotive wave is governed by a nonlinear pendulum equation. In order to derive the dynamical equations, we consider the Lorentz force equations of an electron subject to an idealized helical wiggler field (i.e. equation (2.5)), a uniform axial solenoidal field ($\equiv B_0\hat{e}_z$), and to a circularly polarized plane wave of frequency ω and wavenumber

k given by the vector potential

$$\delta A(z,t) = \delta A[\hat{e}_x \cos(kz - \omega t) - \hat{e}_y \sin(kz - \omega t)], \tag{4.1}$$

where δA denotes the amplitude. As a result, the orbit equations in the wiggler frame are of the form

$$\frac{d}{dt}v_1 = -\left(\frac{\hat{\Omega}_0}{\gamma} - k_w v_3\right)v_2 + \frac{e\delta A}{\gamma m_e c}\left[\left(\omega - \omega\frac{v_1^2}{c^2} - kv_3\right)\sin\psi - \omega\frac{v_1 v_2}{c^2}\cos\psi\right], \tag{4.2}$$

$$\frac{d}{dt}v_2 = \left(\frac{\hat{\Omega}_0}{\gamma} - k_w v_3\right)v_1 - \frac{\hat{\Omega}_w}{\gamma}v_3 + \frac{e\delta A}{\gamma m_e c}\left[\left(\omega - \omega\frac{v_2^2}{c^2} - kv_3\right)\cos\psi - \omega\frac{v_1 v_2}{c^2}\sin\psi\right], \tag{4.3}$$

$$\frac{d}{dt}v_3 = \frac{\hat{\Omega}_w}{\gamma}v_2 + \frac{e\delta A}{\gamma m_e c}\left(k - \omega\frac{v_3}{c^2}\right)(v_1\sin\psi + v_2\cos\psi), \tag{4.4}$$

and

$$\frac{d}{dt}\gamma = \frac{e\delta A}{m_e c^2}\frac{\omega}{c}(v_1\sin\psi + v_2\cos\psi), \tag{4.5}$$

where $\hat{\Omega}_{0,w} \equiv eB_{0,w}/m_e c, \gamma$ is the relativistic factor, and $\psi \equiv (k + k_w)z - \omega t$ is the ponderomotive phase.

Equations (4.2–5) are solved by perturbation about the steady-state trajectories to first order in the electromagnetic field δA. To this end, we expand $v_1 = v_w + \delta v_1$, $v_2 = \delta v_2$, $v_3 = v_\parallel + \delta v_3$, $\gamma = \gamma_0 + \delta\gamma$, where $(\gamma_0, v_w, v_\parallel)$ are the constants which characterize the steady-state trajectory. The first-order equations, therefore, are

$$\frac{d}{dt}\delta v_1 = -(\Omega_0 - k_w v_\parallel)\delta v_2 + \frac{e\delta A}{\gamma_0 m_e c}\left(\omega - \omega\frac{v_w^2}{c^2} - kv_\parallel\right)\sin\psi \tag{4.6}$$

$$\frac{d}{dt}\delta v_2 = (\Omega_0 - k_w v_\parallel)\delta v_1 - \Omega_0\frac{v_w}{v_\parallel}\delta v_3 - \frac{1}{\gamma_0}k_w v_w v_\parallel \delta\gamma + \frac{e\delta A}{\gamma_0 m_e c}(\omega - kv_\parallel)\cos\psi, \tag{4.7}$$

$$\frac{d}{dt}\delta v_3 = \Omega_w \delta v_2 + \frac{e\delta A}{\gamma_0 m_e c}\left(k - \omega\frac{v_\parallel}{c^2}\right)v_w \sin\psi, \tag{4.8}$$

and

$$\frac{d}{dt}\delta\gamma = \frac{e\delta A}{m_e c^2}\frac{\omega}{c}v_w\sin\psi, \tag{4.9}$$

where $\Omega_{0,w} \equiv eB_{0,w}/\gamma_0 m_e c$. Differentiation of equation (4.7) allows us to eliminate δv_2, δv_3 and $\delta\gamma$; hence, we obtain the following second-order differential equation for δv_2

$$\left(\frac{d^2}{dt^2}+\Omega_r^2\right)\delta v_2=\frac{e\delta A}{\gamma_0 m_e c}\sin\psi$$
$$\times\{\Omega_0[\omega-(1+\beta_w^2)kv_\parallel]+(\omega-kv_\parallel)[\omega-(k+2k_w)v_\parallel]\}, \quad (4.10)$$

where Ω_r^2 is defined in equation (2.21). Observe that this procedure is formally identical to the orbital stability analysis described in Chapter 2, and that the homogeneous form of equation (4.10) is identical to equation (2.20). Substitution of the particular solution to equation (4.10) into equation (4.8) under the assumption that the electrons are in near resonance with the wave (i.e. $\omega\approx(k+k_w)v_\parallel$) yields

$$\frac{d}{dt}\delta v_3=\frac{e\delta A}{\gamma_0\gamma_\parallel^2 m_e c}(k+k_w)v_w\Phi\sin\psi, \quad (4.11)$$

where $\gamma_\parallel^2\equiv(1-v_\parallel^2/c^2)^{-1}$, and Φ is defined in equation (2.24).

The pendulum equation which governs the evolution of the ponderomotive phase follows readily from (4.11) since

$$\frac{d}{dt}\delta v_3\cong\frac{v_\parallel^2}{k+k_w}\frac{d^2}{dz^2}\psi. \quad (4.12)$$

As a result, we find that

$$\frac{d^2}{dz^2}\psi=-K^2\sin\psi, \quad (4.13)$$

where the pendulum constant is given by

$$K^2\equiv-\frac{e\delta A}{\gamma_0\gamma_\parallel^2 m_e c v_\parallel^2}(k+k_w)^2 v_w\Phi. \quad (4.14)$$

Observe that the pendulum constant is proportional to the product of the wiggler and electromagnetic field amplitudes. In the absence of the axial solenoidal field the pendulum constant reduces to

$$K^2=\frac{e\delta A}{\gamma_0^4 m_e v_\parallel^2}(k+k_w)^2\frac{\hat{\Omega}_w}{k_w c}\left(1+\frac{\hat{\Omega}_w^2}{k_w^2c^2}\right). \quad (4.15)$$

The presence of the solenoidal field can enhance the magnitude of the pendulum constant. For the Group I trajectories (which include the limit in which the solenoidal field vanishes), v_w is less than zero and increases in magnitude with B_0 as the magnetic resonance at $\Omega_0\approx k_w v_\parallel$ is approached. In addition, Φ is greater than unity and exhibits a singularity at the transition to orbital instability (i.e. when $\Omega_r=0$). As a consequence, the pendulum constant increases rapidly with the solenoidal field which gives rise to significant enhancements in the axial bunching of the electrons. The situation is similar for Group II orbits in the vicinity of the magnetic resonance. In this case, the phase of the bunching process is preserved since v_w is greater than zero while Φ is less than zero. However, as the axial solenoidal field continues to increase Φ vanishes when

$$\gamma_{\parallel}^2 \Omega_0 = \gamma_0^2 k_w v_{\parallel}, \tag{4.16}$$

and the bunching process increases. At this point, the coherent interaction disappears. At still higher values of the axial solenoidal field, both v_w and Φ are positive, and the ponderomotive phase which controls the bunching process is shifted 180° in phase.

The pendulum equation can be reduced to

$$\frac{1}{2}\left(\frac{d\psi}{dz}\right)^2 + U(\psi) = H, \tag{4.17}$$

where H has the form of the Hamiltonian or total energy of the system, and

$$U(\psi) = -K^2 \cos\psi \tag{4.18}$$

is the ponderomotive potential. The electron trajectories through the wiggler, therefore, may be expressed as

$$\frac{d\psi}{dz} = \pm\sqrt{(2H + 2K^2 \cos\psi)}. \tag{4.19}$$

Observe that the first derivative of the phase (i.e. $d\psi/dz = (k + k_w) - \omega/v_z$) is a measure of the electron streaming velocity; hence, this equation effectively describes the electron velocity as a function of the phase of the ponderomotive wave.

There are two classes of trajectory: trapped and untrapped. The **separatrix** describes the transition between trapped and untrapped orbits, and occurs when $H = K^2$. Hence, the separatrix is defined by a pair of curves in the phase space for $(d\psi/dz, \psi)$

$$\frac{d\psi}{dz} = \pm\begin{cases} 2K\cos(\psi/2); & K^2 > 0 \\ 2|K|\sin(\psi/2); & K^2 < 0 \end{cases}. \tag{4.20}$$

Free-streaming, untrapped orbits are characterized by $H > |K^2|$ and occupy that region of the phase space outside of the separatrix,

$$\left|\frac{d\psi}{dz}\right| > \begin{cases} 2K\cos(\psi/2); & K^2 > 0 \\ 2|K|\sin(\psi/2); & K^2 < 0 \end{cases}. \tag{4.21}$$

The trapped orbits are those for which $H < |K^2|$ within the bounds of the separatrix. The free-streaming orbits correspond to the case in which the pendulum swings through the full 360° cycle. The electrons pass over the crests of many waves, travelling fastest at the bottom of the troughs and slowest at the crests of the ponderomotive wave. In contrast, the electrons are confined within the trough of a single wave in the trapped orbits. This corresponds to the motion of a pendulum which does not rotate full circle, but is confined to oscillate about the lower equilibrium point.

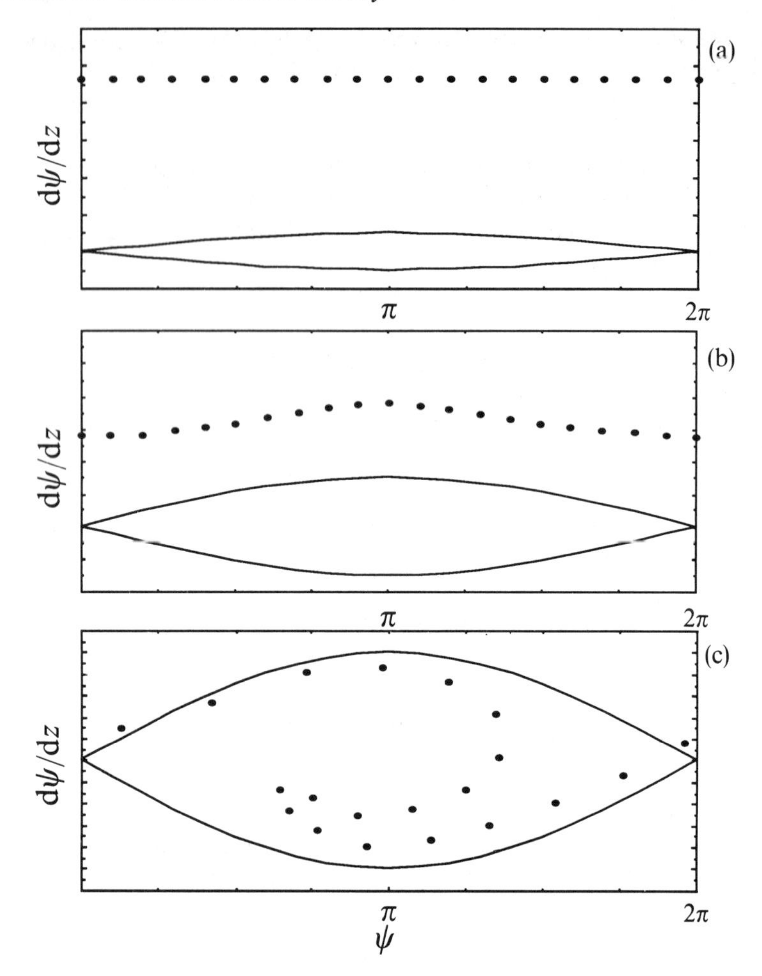

Fig. 4.1 Schematic of the electron phase-space evolution in free-electron lasers. The electron beam is initially (a) monoenergetic. During the linear phase of the interaction (b), the wave grows in amplitude and the separatrix expands. The bulk of the beam executes trapped orbits at saturation (c). The phase-space distribution shown represents an electron beam in which the electrons which are still losing energy to the wave are balanced by those which are gaining energy from the wave.

The dynamical evolution of the electron phase space during the coherent emission process is illustrated in Fig. 4.1, and is one in which the pendulum constant evolves during the course of the interaction. We assume that the electrons are initially characterized by the same axial velocity and describe a horizontal line in phase space. Since the amplification process has only just begun, the wave is of small amplitude and the separatrix encloses a small area of phase space. The electrons lose energy as the wave is amplified; hence, the electrons decelerate and both the pendulum constant and separatrix grow. During the linear phase of the interaction, as illustrated in Fig. 4.1(b), the electrons have only begun to form bunches and remain on untrapped trajectories outside of the separatrix. Ultimately, the electrons cross the growing separatrix from untrapped to trapped orbits. The interaction saturates after the electrons have executed approximately half of an oscillation in the ponderomotive well. At this point, the electrons which are still losing energy to the wave are balanced by those electrons which are gaining energy at the expense of the wave. In this chapter, we deal with the amplification mechanism during the linear phase of the interaction.

The bulk of this chapter is devoted to a derivation and discussion of the linear gain for both helical and planar wiggler configurations. The analysis of the linear gain in the idealized one-dimensional limit is presented in section 4.1 based upon a solution of the Vlasov–Maxwell equations. This is not the only technique used in the literature to obtain the linear gain; however, it has the advantage of providing a unified formalism which can treat both configurations of interest in the low- and the high-gain regimes, as well as the inclusion of the effects of an electron beam thermal spread.

Within the context of this analysis, section 4.1 begins with a general statement of the approach valid for both helical and planar wigglers. The first configuration to be dealt with in depth is that of a helical wiggler. In this case, we include the presence of an axial solenoidal magnetic field in the interest of generality. The axial magnetic field is not universally employed for this wiggler geometry; however, it does provide for the enhanced focusing of the electron beam against the effects of beam spreading due to self-electric and magnetic fields, and can result in enhancements in both the gain and saturation efficiency of the interaction. The expression for the gain in the low-gain regime is given in equation (4.59), and implicitly includes the effect of the axial field. However, the limit in which the axial magnetic field vanishes can be taken in a straightforward manner. The general dispersion equation (4.85) which describes the high-gain regime also includes the effect of an axial magnetic field, and the solutions of this equation are discussed in a variety of operating regimes. The dispersion equation in the limiting case in which the axial field vanishes is given explicitly in equation (4.87), and its solutions are given in both the high-gain Compton (4.94) and collective Raman (4.97) regimes. The effect of the axial magnetic field on the high-gain Compton (4.104) and the collective Raman regimes (4.106) is discussed as well for the case in which $\Phi > 0$ (for a definition of Φ see equation (2.24)). The presence of an axial magnetic field has a fundamental impact on the interacting wave modes, and the beam space-charge mode can itself be driven unstable when $\Phi < 0$. The growth rate of the unstable beam space-charge wave is given in equation (4.82), and its impact upon the growth rate of the free-electron laser is illustrated in more detail in equation (4.108) and in Fig. 4.9.

Thermal effects of the electron beam on the linear gain are also discussed, and a general dispersion equation is derived subject to the assumption that the beam is monoenergetic but exhibits a pitch-angle spread. Since the free-electron laser operates by means of an axial bunching mechanism, the axial energy spread of the beam is the most important performance criterion, and this pitch-angle spread translates into an axial energy and momentum spread. The specific distribution we employ is given in equations (4.113) and (4.114), and the general dispersion equation which results (in the absence of an axial magnetic field) appears in equation (4.128). The thermal effect is described in terms of a general thermal function (4.125), which is also shown to be appropriate for planar wiggler configurations.

The analysis of the linear gain in planar wiggler configurations in the idealized one-dimensional limit is structured in a similar way to that of the helical wiggler analysis. The formulation is a unified Vlasov–Maxwell treatment in which the low-gain limit is obtained first, and is given in equation (4.159). This analysis is sufficiently general to include the effects of the oscillations in the axial velocity which have been described for a planar wiggler (see Chapter 2), as well as the gain at the fundamental and odd harmonics. A general dispersion equation is found in the exponential gain regime which describes both the high-gain Compton and collective Raman regimes (4.170) at the fundamental and odd harmonics in the limit of a vanishing beam thermal spread. This analysis is generalized to include thermal effects using the same beam distribution function as in the case of a helical wiggler. The general dispersion equation which is obtained (4.180) includes the fundamental as well as the harmonic gain. The thermal effects are described by a generalized thermal function (4.182) which recovers that found for helical wigglers for the case of the gain at the fundamental resonance. A general comparison of the differences between the planar and helical wiggler configurations is discussed.

The formulation of the linear gain for realizable three-dimensional configurations is restricted to the case of helical wigglers. The linear gain regime in planar wigglers is discussed as a subset of the material in Chapter 5 on the nonlinear analysis of planar wiggler configurations. In analysing the low-gain limit for realizable configurations, we have adopted an alternative approach to the Vlasov–Maxwell formalism used in the one-dimensional analyses. In this case, we have used a single-particle phase-average technique to obtain the gain expressed in equation (4.235). The result described in that equation is similar to that found in equation (4.59) for the one-dimensional regime, and differs in two primary respects. In the first place, the orbit parameters which appear in the gain expression now relate to the three-dimensional counterparts of the one-dimensional helical steady-state trajectories (see Chapter 2). In the second place, a filling-factor has been derived which describes the overlap of the beam and the radiation field which is set to unity for the idealized one-dimensional limit. It should be noted that the analysis presented in this chapter deals with the mode polarization specific to propagation in a waveguide, and the filling-factor is appropriate to the TE and TM modes of a loss-free cylindrical waveguide. It should be observed that harmonic emission is possible in the context of a three-dimensional analysis of the helical wiggler configuration, and that equation (4.235) describes the gain for harmonics as well as for the fundamental. We return to a Vlasov–Maxwell formulation in a loss-free cylindrical waveguide for the linear analysis of the high-gain regime, and obtain a general dispersion equation in terms of the determinant of the matrix which couples the waveguide and beam space-charge modes (4.279). This dispersion equation is then solved numerically and compared with several specific experiments.

4.1 LINEAR STABILITY IN THE IDEALIZED LIMIT

In this section, we shall derive the linearized dispersion equation for the free-electron laser in the idealized one-dimensional representation within the context of a linearized Vlasov–Maxwell formalism. The first step in this process is the development of a general formalism which is applicable to both the helical and planar configurations. The Vlasov equation in the combined wiggler and axial magnetostatic fields, and the electromagnetic fields is

$$\left[\frac{\partial}{\partial t}+\mathbf{v}\cdot\nabla - e\left\{\delta\boldsymbol{E}(z,t)+\frac{1}{c}\mathbf{v}\times[B_0\hat{\boldsymbol{e}}_z+\boldsymbol{B}_w(z)+\delta\boldsymbol{B}(z,t)]\right\}\cdot\nabla_p\right]f_b(z,\boldsymbol{p},t)=0, \tag{4.22}$$

where $f_b(z,\boldsymbol{p},t)$ is the distribution function of the electron beam, $\delta\boldsymbol{E}(z,t)$ and $\delta\boldsymbol{B}(z,t)$ denote the fluctuating electric and magnetic fields of the wave, and

$$\nabla_p \equiv \hat{\boldsymbol{e}}_x\frac{\partial}{\partial p_x}+\hat{\boldsymbol{e}}_y\frac{\partial}{\partial p_y}+\hat{\boldsymbol{e}}_z\frac{\partial}{\partial p_z}. \tag{4.23}$$

The Vlasov equation is linearized by expanding the distribution in powers of the fluctuating fields. To this end we write $f_b(z,\boldsymbol{p},t)=F_b(z,\boldsymbol{p})+\delta f_b(z,\boldsymbol{p},t)$ where F_b and δf_b are the equilibrium and perturbed components of the distribution, and it is assumed that the perturbed distribution is of the order of the fluctuating fields and $|\delta f_b|\ll|F_b|$. The equilibrium distribution must satisfy the lowest-order Vlasov equation

$$\left\{\frac{\partial}{\partial t}+\mathbf{v}\cdot\nabla-\frac{e}{c}\mathbf{v}\times[B_0\hat{\boldsymbol{e}}_z+\boldsymbol{B}_w(z)]\cdot\nabla_p\right\}F_b(z,\boldsymbol{p})=0. \tag{4.24}$$

This is satisfied for any equilibrium distribution which is a function of the constants of the motion. As discussed in Chapter 2, these constants are the total energy (or momentum) as well as the canonical momenta for both helical and planar wigglers in the one-dimensional representation; hence, we may express the equilibrium distribution in the form $F_b(z,\boldsymbol{p})=F_b(P_x,P_y,p)$. It should be remarked, however, that the canonical momenta are only approximate constants of the motion in the presence of an axial solenoidal magnetic field. We shall discuss this in more detail later. Correct to first order in the fluctuation fields, the perturbed distribution satisfies

$$\begin{aligned}\left\{\frac{\partial}{\partial t}+\mathbf{v}\cdot\nabla-\frac{e}{c}\mathbf{v}\times[B_0\hat{\boldsymbol{e}}_z+\boldsymbol{B}_w(z)]\cdot\nabla_p\right\}&\delta f_b(z,\boldsymbol{p},t)\\ &=e\left[\delta\boldsymbol{E}(z,t)+\frac{1}{c}\mathbf{v}\times\delta\boldsymbol{B}(z,t)\right]\cdot\nabla_p F_b.\end{aligned} \tag{4.25}$$

The perturbed Vlasov equation may be solved by the method of characteristics in which we integrate

$$\delta f_b[z, \boldsymbol{p}, \tau(z)] = e\int_0^z \frac{dz'}{v_z(z')}\left\{\delta \boldsymbol{E}[z, \tau(z')] + \frac{1}{c}\mathbf{v}(z') \times \delta \boldsymbol{B}[z, \tau(z')]\right\}\cdot\boldsymbol{\nabla}_p F_b \quad (4.26)$$

over the unperturbed trajectories under the assumption that the perturbations are negligibly small at time $t = 0$. Observe that we treat the case of spatial growth and have adopted Lagrangian coordinates in which $\mathbf{v}(z)$ denotes the unperturbed velocity of an electron as a function of the axial position, and

$$\tau(z) \equiv t_0 + \int_0^z \frac{dz'}{v_z(z', t_0)}, \quad (4.27)$$

represents the time it takes an electron to reach a particular axial position after crossing the $z = 0$ plane at time t_0.

The solution to the perturbed Vlasov equation is solved in conjunction with Maxwell's equations. We choose to deal with the scalar $\delta\varphi(z, t)$ and vector potentials $\delta \boldsymbol{A}_\perp(z, t)$ in the Coulomb gauge. Note that since we treat a one-dimensional model, the scalar and vector potentials describe plane waves. Hence, the vector potential represents a purely transverse electromagnetic wave. In terms of this representation, Maxwell's equations are

$$\left(\boldsymbol{\nabla}^2 - \frac{1}{c^2}\frac{\partial^2}{\partial t^2}\right)\delta \boldsymbol{A}_\perp = -\frac{4\pi}{c}\delta \boldsymbol{J}_\perp, \quad (4.28)$$

and

$$\frac{\partial^2}{\partial t \partial z}\delta\varphi = 4\pi\delta J_z. \quad (4.29)$$

Observe that the scalar potential is described in terms of the z-component of Ampère's law rather than with Poisson's equation. The perturbed source current is given in terms of the perturbed distribution function as follows

$$\delta \boldsymbol{J}(z, t) = -\frac{e}{m_e}\int d^3p \frac{1}{\gamma}\boldsymbol{p}\delta f_b(z, \boldsymbol{p}, t). \quad (4.30)$$

The dispersion equation governing the growth and/or damping of the electromagnetic field is obtained by the simultaneous solution of equations (4.26), (4.28) and (4.29).

4.1.1 Helical wiggler configurations

In treating a helical wiggler configuration, we shall include the effect of the axial solenoidal magnetic field in the interest of generality. However, the limit in which the axial field vanishes is evident, and will be discussed explicitly. In addition, the discussion of the effect of beam thermal effects on the gain is restricted to the limit in which the axial field vanishes. The formulation itself follows that described in references [9, 18]. As discussed in Chapter 2, the effect of the axial field is to enhance both the transverse wiggler-induced velocity and the gain of the free-electron laser when the solenoidal field is strong enough that $\Omega_0 \approx k_w v_\parallel$. However, it should be noted that the canonical momenta are only approximate constants of the motion in the presence of the axial field. Hence, we used equations (2.42–4) to describe the unperturbed trajectories and make the implicit assumption that $V_\perp \ll v_w[V_\perp^2 \equiv (P_x^2 + P_y^2)/\gamma^2 m_e^2]$. Within the context of this orbit representation, the canonical momenta describe the Larmor rotation of the electrons in the solenoidal field.

In treating the helical wiggler geometry, we begin with the application of the Vlasov–Maxwell formalism to this configuration under the assumption of a plane-wave representation for the vector and scalar potentials. This general analysis provides the framework necessary to treat both the low- and exponential-gain regimes. The gain in the low-gain regime is presented in equation (4.59), in which we have neglected the space-charge fields. This is usually appropriate since the low-gain regime is generally applicable to free-electron laser oscillators which are driven by high-energy but low-current accelerators such as r.f. linacs, microtrons and storage rings.

We subsequently treat the high-gain regime which applies to free-electron lasers driven by intense-beam accelerators, such as pulse-line accelerators, and induction linacs. A derivation of the general dispersion equation including the effect of an axial magnetic field is given in the ideal beam limit in which thermal effects can be neglected (see equations (4.76) and (4.83)). This includes both the high-gain Compton and collective Raman regimes. The effect of the axial magnetic field on the stability of the beam space-charge wave is discussed, as is the general effect of the axial magnetic field on the structure of the interacting electromagnetic modes (see Figs 4.4 and 4.5). The transition between the high-gain Compton and collective Raman regimes as a function of the beam density is illustrated in Fig. 4.6, and the expressions for the gain in these two regimes are given in equations (4.94) and (4.97), respectively, for the case in which the orbit parameter $\Phi > 0$ (see equation (2.24)). The case in which $\Phi < 0$ represents a negative-mass regime in which the axial velocity accelerates as the electrons lose energy. This regime is also discussed, and it is shown that the bandwidth of the interaction is broadened substantially in this regime.

The effect of an electron-beam thermal spread is treated for a configuration in which the axial field is absent in the interest of simplicity. The analysis is presented under the assumption that the electron beam is monoenergetic but exhibits a spread in pitch angle. This is adequate to treat the effect of an axial energy (or momentum) spread. A distribution function which describes such a beam is given in equations (4.113) and (4.114), and a general dispersion equation is derived. The general dispersion equation (4.128) describes the effect of the axial energy spread in terms of a generalized thermal function T (see equation (4.125)). It is shown that this function approaches unity in the limit in which the axial energy spread vanishes and the ideal beam limit is recovered.

The source currents

Under the assumption of plane wave solutions, the vector and scalar potentials for a wave with angular frequency ω are of the form

$$\delta \boldsymbol{A}_{\perp}(z,t) = \frac{1}{2}\delta \hat{\boldsymbol{A}}_{\perp}(z)\exp(-i\omega t) + \text{c.c.}, \tag{4.31}$$

and

$$\delta\varphi(z,t) = \frac{1}{2}\delta\hat{\varphi}(z)\exp(-i\omega t) + \text{c.c.} \tag{4.32}$$

After transformation to the basis

$$\hat{\boldsymbol{e}}_{\pm} \equiv \frac{1}{2}(\hat{\boldsymbol{e}}_x \pm i\hat{\boldsymbol{e}}_y), \tag{4.33}$$

which is convenient for the description of left- and right-hand circularly polarized electromagnetic waves, the perturbed distribution function can be written as

$$\delta f_b[z,\boldsymbol{p},\tau(z)] \equiv \delta\hat{f}_b(z,\boldsymbol{p})\exp[-i\omega\tau(z)] + \text{c.c.}, \tag{4.34}$$

where

$$\delta\hat{f}_b(z,\boldsymbol{p}) = \frac{e}{2c}\left[D_+\left(\frac{\partial}{\partial P_x} + i\frac{\partial}{\partial P_y}\right) + D_-\left(\frac{\partial}{\partial P_x} - i\frac{\partial}{\partial P_y}\right) + D_z\frac{\partial}{\partial p}\right]F_b(P_x, P_y, p). \tag{4.35}$$

The orbit integrals in equation (4.35) are defined as

$$D_{\pm} \equiv \int_0^z dz' \frac{\exp[i\omega\tau(z,z')]}{v_z(z')}\exp[\pm i\Omega_0\tau(z')]\left[i\omega - v_z(z')\frac{\partial}{\partial z'}\right]\delta\hat{A}_{\pm}(z'), \tag{4.36}$$

and

$$D_z \equiv \frac{1}{p}\int_0^z dz' \frac{\exp[i\omega\tau(z,z')]}{v_z(z')} \times\{-cp_z(z')\delta\hat{\varphi}(z') + i\omega[p_-(z')\delta\hat{A}_+(z') + p_+(z')\delta\hat{A}_-(z')]\}, \tag{4.37}$$

where $p_{\pm} \equiv p_x \mp ip_y$, $\tau(z,z') \equiv \tau(z) - \tau(z')$,

$$\delta\hat{A}_{\pm} \equiv \frac{1}{2}(\delta\hat{A}_x \mp i\delta\hat{A}_y), \tag{4.38}$$

denotes the amplitudes of the circularly polarized electromagnetic waves.

The source current

$$\delta \boldsymbol{J}(z,t) = [\delta\hat{J}_+(z)\hat{\boldsymbol{e}}_+ + \delta\hat{J}_-(z)\hat{\boldsymbol{e}}_- + \delta\hat{J}_z(z)\hat{\boldsymbol{e}}_z]\exp(-i\omega t) + \text{c.c.} \tag{4.39}$$

is determined by means of the perturbed distribution as follows

$$\begin{pmatrix}\delta\hat{J}_{\pm}(z)\\ \delta\hat{J}_z(z)\end{pmatrix} = -\frac{e}{m_e}\int dP_x dP_y dp\frac{p}{\gamma p_z}\delta\hat{f}_b(z,\boldsymbol{p})\begin{pmatrix}p_{\pm}\\ p_z\end{pmatrix}. \tag{4.40}$$

Substitution of the solution for the perturbed distribution from equation (4.35), we obtain

$$\delta\hat{J}_{\pm}(z)=\frac{e^2}{2m_ec}\int dP_xdP_ydp\frac{p}{\gamma}\left[\left(\frac{\partial}{\partial P_x}+i\frac{\partial}{\partial P_y}\right)\left(\frac{p_\pm}{p_z}D_+\right)+\left(\frac{\partial}{\partial P_x}-i\frac{\partial}{\partial P_y}\right)\right.$$
$$\left.\times\left(\frac{p_\pm}{p_z}D_-\right)-\frac{p_\pm}{p_z}D_z\frac{\partial}{\partial p}\right]F_b(P_x,P_y,p), \tag{4.41}$$

and

$$\delta\hat{J}_z(z)=\frac{e^2}{2m_ec}\int dP_xdP_ydp\frac{p}{\gamma}\left[\left(\frac{\partial}{\partial P_x}+i\frac{\partial}{\partial P_y}\right)D_+\right.$$
$$\left.+\left(\frac{\partial^2}{\partial P_x}-i\frac{\partial}{\partial P_y}\right)D_--D_z\frac{\partial}{\partial p}\right]F_b(P_x,P_y,p). \tag{4.42}$$

The dispersion equation is obtained by substitution of the source current into the wave equations

$$\left(\frac{\partial^2}{\partial z^2}+\frac{\omega^2}{c^2}\right)\delta\hat{A}_{\pm}=-\frac{4\pi}{c}\delta\hat{J}_{\pm}, \tag{4.43}$$

and

$$\frac{\partial}{\partial z}\delta\hat{\varphi}(z)=\frac{4\pi i}{\omega}\delta\hat{J}_z. \tag{4.44}$$

An ideal electron beam is both monoenergetic and has a vanishing pitch-angle spread (i.e. $P_x=P_y=0$). We first impose the condition that the pitch-angle spread of the beam vanishes, and assume an equilibrium distribution of the form

$$F_b(P_x,P_y,p)=n_b\delta(P_x)\delta(P_y)G_b(p), \tag{4.45}$$

where n_b denotes the ambient density of the electron beam, and the distribution in thc total momentum satisfies the normalization condition

$$\int_0^\infty dp\frac{p}{p_z}G_b(p)=1. \tag{4.46}$$

As a consequence, the source currents take the form

$$\delta\hat{J}_{\pm}(z)=\frac{\omega_b^2}{8\pi c}\int_0^\infty dp\frac{p}{p_z\gamma}\left\{\exp[\mp i\Omega_0 t(z)]\left(2+\frac{p_+p_-}{p_z^2}\right)D_\pm\right.$$
$$+\exp[\pm i\Omega_0 t(z)]\frac{p_\pm^2}{p_z^2}D_\mp+p_\pm\left(\frac{\partial}{\partial P_x}+i\frac{\partial}{\partial P_y}\right)D_+$$
$$\left.+p_\pm\left(\frac{\partial}{\partial P_x}-i\frac{\partial}{\partial P_y}\right)D_--p_\pm D_z\frac{\partial}{\partial p}\right\}_{P_x=P_y=0}G_b(p), \tag{4.47}$$

and

$$\delta\hat{J}_z(z)=\frac{\omega_b^2}{8\pi c}\int_0^\infty dp\frac{p}{\gamma}\left[\left(\frac{\partial}{\partial P_x}+i\frac{\partial}{\partial P_y}\right)D_+\right.$$
$$\left.+\left(\frac{\partial}{\partial P_x}-i\frac{\partial}{\partial P_y}\right)D_--D_z\frac{\partial}{\partial p}\right]_{P_x=P_y=0}G_b(p), \tag{4.48}$$

where $\omega_b^2 \equiv 4\pi e^2 n_b/m_e$ is the square of the electron-beam plasma frequency. If we require that $P_x = P_y = 0$, then $p_\pm = p_w \exp(\mp ik_w z)$, $\tau(z,z') = (z - z')/v_\parallel$,

$$D_\pm = -\exp[\pm i\Omega_0(z)]\left\{\delta\hat{A}_\pm(z) - \delta\hat{A}_\pm(0)\exp[i(\omega \mp \Omega_0)z/v_\parallel]\right.$$

$$\left.\mp i\frac{\Omega_0}{v_\parallel}\int_0^z dz'\,\delta\hat{A}_\pm(z')\exp[i(\omega \pm \Omega_0)\tau(z,z')]\right\}, \tag{4.49}$$

$$D_z = \frac{\gamma m_e c}{p}\int_0^z dz' \exp[i\omega\,\tau(z,z')]\left\{-\frac{\partial}{\partial z'}\delta\hat{\varphi}(z') + \frac{i\omega}{c}\left[\frac{p_-}{p_\parallel}\delta\hat{A}_+(z') + \frac{p_+}{p_\parallel}\delta\hat{A}_-(z')\right]\right\}, \tag{4.50}$$

and $p_{w,\parallel} \equiv \gamma m_e v_{w,\parallel}$ is determined by the steady-state trajectories. In addition, we observe that

$$\left(\frac{\partial}{\partial P_x} \pm i\frac{\partial}{\partial P_y}\right)D_\pm\bigg|_{P_x = P_y = 0} = -\frac{p_w}{p_\parallel^2}\frac{\{\exp[\pm i(k_w - \Omega_0/v_\parallel)z] - 1\}}{k_w v_\parallel - \Omega_0}$$

$$\times [\Omega_0 \delta\hat{A}_\pm(z)\exp(\pm i\Omega_0 z/v_\parallel) \mp \omega\delta\hat{A}_\pm(0)\exp(\pm i\omega z/v_\parallel)]$$

$$\pm \frac{i\Omega_0}{v_\parallel}\frac{p_w}{p_\parallel^2}\exp(i\omega z/v_\parallel)\int_0^z dz'\,\delta\hat{A}_\pm(z')\exp[-i(\omega/v_\parallel - k_w)z']$$

$$\times\left\{1 \pm \frac{\omega}{k_w v_\parallel - \Omega_0}[\exp(\pm i(k_w - \Omega_0/v_\parallel)(z - z')) - 1]\right.$$

$$\left. - \frac{\Omega_0}{k_w v_\parallel - \Omega_0}[\exp(\mp i(k_w - \Omega_0/v_\parallel)z') - 1]\right\}. \tag{4.51}$$

The calculation of the linear dispersion requires the evaluation of the source currents in equations (4.47) and (4.48). In the remainder of this section we develop the solutions in both the low- and high-gain regimes.

The low-gain regime

The low-gain, tenuous-beam limit is relevant to free-electron laser configurations in which the electron beam current is low and the gain of the signal in a single pass through the wiggler is less than unity. In this regime, the beam plasma frequency $\omega_b \ll \omega$ and collective effects due to beam space-charge waves are negligible. In addition, the coupling between the left- and right-hand circularly polarized electromagnetic modes scales with $(v_w/v_\parallel)^2$, which is assumed to be small in the low-gain limit. Hence, we can focus attention on either of these modes. We choose to consider the right-hand circularly polarized mode which can be represented in the low-gain regime as [9]

$$\delta\hat{A}_+(z) = \delta\hat{A}_+(0)\exp\left[i\int_0^z dz'\,k_+(z')\right], \tag{4.52}$$

where we assume that $|\mathrm{Im}\,k_+(z)| \ll |\mathrm{Re}\,k_+(z)|$. Under the additional assumption that $k_+(z) = k + \delta k_+(z)$, where k is independent of axial position and $|\delta k_+(z)| \ll k$, the dispersion equation for the real part of the wavenumber is

$$\frac{\omega^2}{c^2} - k^2 \cong 0, \tag{4.53}$$

and for the growth rate we obtain

$$\operatorname{Im} \delta k_+(z) \cong -\frac{\omega_b^2}{4c} \int_0^\infty dp\, \beta_w^2 \frac{\sin \Delta k z}{\Delta k} \left\{ m_e \frac{\partial}{\partial p} G_b(p) + \frac{p}{\gamma p_\parallel} \frac{\Delta k}{k_w v_\parallel - \Omega_0} \left[1 + \frac{\Omega_0 (k_w v_\parallel - \Omega_0 + c\Delta k)}{c\Delta k(\omega - \Omega_0 - k v_\parallel)} \right] G_b(p) \right\}, \tag{4.54}$$

where

$$\Delta k \equiv \frac{\omega}{v_\parallel} - k - k_w, \tag{4.55}$$

is the detuning parameter. Observe that $\Delta k = 0$ frequencies in the vicinity of the free-electron laser resonance $\omega = (k + k_w) v_\parallel$.

The gain in power over an interaction length L is defined as

$$G(L) \equiv -2 \int_0^L dz \operatorname{Im} \delta k_+(z). \tag{4.56}$$

It should be remarked that it has been implicitly assumed in the derivation that the gain is weak and that $G(L) < 1$. Integration of the growth rate gives

$$G(L) \cong -\frac{\omega_b^2 L}{2c} \int_0^\infty dp G_b(p) \left\{ \frac{m_e L}{2} \frac{\partial}{\partial p} \left(\beta_w^2 \frac{\sin^2 \Theta}{\Theta^2} \right) - \beta_w^2 \frac{p}{\gamma p_\parallel} \frac{\sin^2 \Theta}{\Theta} \frac{1}{k_w v_\parallel - \Omega_0} \left[1 + \frac{\Omega_0 (k_w v_\parallel - \Omega_0 + c\Delta k)}{c\Delta k(\omega - \Omega_0 - k v_\parallel)} \right] \right\}, \tag{4.57}$$

where $\Theta \equiv \Delta k L/2$.

Under the assumption that the electron beam is cold (i.e. monoenergetic), the distribution function can be expressed in the form

$$G_b(p) = \frac{p_\parallel}{p} \delta(p - p_0), \tag{4.58}$$

where p_0 denotes the bulk momentum of the beam. The gain in this limit can be expressed as [9]

$$G(L) \cong \frac{\omega_b^2 L^3 k}{8 \gamma_0 \gamma_\parallel^2 v_\parallel^2} \beta_w^2 \Phi F(\Theta), \tag{4.59}$$

where $\gamma_0 \equiv (1 + p_0^2/m_e^2 c^2)^{1/2}$, $\beta_w \equiv v_w / v_\parallel$, $(v_w, v_\parallel)$ are the transverse and axial velocities of the steady-state orbits, $\gamma_\parallel^2 \equiv (1 - v_\parallel^2/c^2)^{-1}$, Φ is defined in equation (2.24), and

$$F(\Theta) = \frac{d}{d\Theta} \left(\frac{\sin \Theta}{\Theta} \right)^2 \tag{4.60}$$

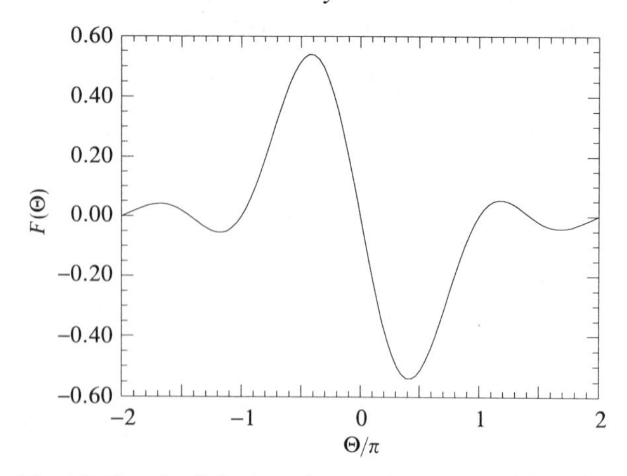

Fig. 4.2 Graph of the free-electron laser spectral function.

defines the spectral function. The spectral function is shown in Fig. 4.2, and exhibits the extrema $F(\pm 1.3) \approx \mp 0.54$. Observe that $\Phi = 1$ and $\beta_w = -\Omega_w/k_w v_\parallel$ in the limit in which the axial solenoidal magnetic field vanishes. Observe that the spectral variation of the emission in the low-gain regime as shown in Fig. 4.2 is proportional to the derivative of the spectral dependence of the spontaneous emission (see Chapter 3). This property of the gain is referred to as Madey's theorem [4].

The maximum permissible thermal spread under which the expression (4.59) for the gain is valid can be estimated by comparison of the bandwidth characterized by the spectral function with the bandwidth induced by a thermal spread. The frequency spread characterized by the spectral function is similar to that found for the spontaneous emission from a cold beam (3.24), i.e. $\Delta\omega/\omega \approx \lambda_w/L$. The effect of a thermal spread in the axial velocity of the beam is to cause a shift in the resonant frequency of the order of $\Delta\omega/\omega \approx 2\gamma_\parallel^2 \Delta_\parallel v_\parallel / v_\parallel$. As a consequence, the cold beam approximation is valid whenever

$$\frac{\Delta v_\parallel}{v_\parallel} \ll \frac{\lambda}{L}, \tag{4.61}$$

which is identical to the condition found to obtain for the spontaneous emission for a cold beam (3.22).

In the case of either (1) weak axial magnetic fields corresponding to the Group I trajectories (which applies in the limit in which the axial field vanishes), or (2) strong magnetic fields corresponding to Group II orbits with $\Phi > 0$ the peak gain occurs for $\Theta \approx -1.3$ which corresponds to a resonant frequency

$$\omega \cong \frac{k_w v_\parallel}{1 - \beta_\parallel}\left(1 - \frac{2.6}{k_w L}\right), \tag{4.62}$$

where $\beta_\parallel \equiv v_\parallel / c$. However, for solenoidal fields corresponding to Group II steady-state trajectories for which $\Phi < 0$, there is a phase shift in the interaction due to the negative mass effect. In this case, peak gain occurs for $\Theta \approx 1.3$ which defines a resonant frequency of

$$\omega \cong \frac{k_w v_\parallel}{1-\beta_\parallel}\left(1+\frac{2.6}{k_w L}\right). \tag{4.63}$$

The fundamental physics contained in the result (4.59) for the low-gain regime is explained in the introductory material for this chapter, and is summarized in equation (1.8) in units of decibels. The result is identical to that found by a single-particle analysis either from quantum-mechanical principles [4] or from classical considerations [8, 15], and describes the effect of electron bunching in the ponderomotive potential formed by the beating of the wiggler and radiation fields. The gain in the low-gain regime will be rederived later in this chapter in the analysis of three-dimensional effects. The derivation in that case will rely upon a phase average over the untrapped electron trajectories which makes plain the essential physics of the interaction.

The high-gain regime

The high-gain regime is applicable to intense-beam free-electron lasers which are usually operated in an amplifier mode. The fields in this regime exhibit exponential growth of the fluctuation fields, and we assume that

$$\delta\hat{A}_\pm(z) = \delta\hat{A}_\pm(0)\exp(ik_\pm z), \tag{4.64}$$

and

$$\delta\hat{\varphi}(z) = \delta\hat{\varphi}(0)\exp(ikz). \tag{4.65}$$

As a result, the source currents can be expressed as

$$\delta\hat{J}_\pm(z) = -\frac{\omega_b^2}{8\pi c}\delta\hat{A}_\pm(z)\left[2\left\langle\frac{p}{\gamma p_\parallel}\frac{\omega - k_\pm v_\parallel}{\omega - \Omega_0 - k_\pm v_\parallel}\right\rangle + a_w^2 S_\pm(k_\pm,\omega)\right]$$
$$- a_w\frac{ck^2}{8\pi}\delta\hat{\varphi}(z)\exp(\mp ik_w z)\,\chi_a(k,\omega), \tag{4.66}$$

and

$$\delta\hat{J}_z(z) = -\frac{\omega k}{8\pi}\delta\hat{\varphi}(z)\,\chi(k,\omega) - \frac{\omega k_w}{8\pi}a_w[\delta\hat{A}_+(z)\exp(ik_w z)\sigma_+(k_+,\omega)$$
$$+ \delta\hat{A}_-(z)\exp(-ik_w z)\sigma_-(k_-,\omega)], \tag{4.67}$$

where $a_w \equiv eA_w/m_e c^2$, the average is defined as

$$\langle(\ldots)\rangle \equiv \int_0^\infty dp\, G_b(p)(\ldots), \tag{4.68}$$

the dielectric function of the uncoupled beam-plasma waves is

$$\chi(k,\omega) \equiv \frac{\omega_b^2}{k}\int_0^\infty dp\frac{m_e}{\omega - kv_\parallel}\frac{\partial}{\partial p}G_b(p), \tag{4.69}$$

the second-order dielectric function of the electromagnetic waves is

$$S_{\pm}(k_{\pm},\omega) \equiv \int_0^{\infty} dp \frac{p}{\gamma^3 p_{\parallel}} G_b(p) \frac{k_w^2 c^2}{(k_w v_{\parallel} - \Omega_0)^2} \left[1 + \frac{\Omega_0}{k_w v_{\parallel} - \Omega_0} \right.$$
$$\left. \times \left(\frac{\omega \mp \Omega_0 + k_w v_{\parallel}}{\omega \mp \Omega_0 - k_{\pm} v_{\parallel}} - \frac{\omega \mp k_w v_{\parallel}}{\omega \mp k_w v_{\parallel} - k_{\pm} v_{\parallel}} \right) \right]$$
$$- \omega c^2 \int_0^{\infty} dp \frac{k_w^2 v_{\parallel}}{\gamma^2 (k_w v_{\parallel} - \Omega_0)^2} \frac{m_e}{\omega \mp k_w v_{\parallel} - k_{\pm} v_{\parallel}} \frac{\partial}{\partial p} G_b(p), \tag{4.70}$$

and the dielectric functions which describe the mode–mode couplings are

$$\chi_a(k,\omega) \equiv \frac{\omega_b^2}{k} \int_0^{\infty} dp \frac{k_w v_{\parallel}}{k_w v_{\parallel} - \Omega_0} \frac{m_e}{\gamma(\omega - k v_{\parallel})} \frac{\partial}{\partial p} G_b(p), \tag{4.71}$$

and

$$\sigma_{\pm}(k_{\pm},\omega) \equiv \frac{\omega_b^2}{\omega} \int_0^{\infty} dp \frac{p}{\gamma^2 p_{\parallel}} G_b(p) \frac{\Omega_0}{(k_w v_{\parallel} - \Omega_0)^2}$$
$$\times \left(\frac{\omega \mp k_w v_{\parallel}}{\omega \mp v_w v_{\parallel} - k_{\pm} v_{\parallel}} - \frac{\omega}{\omega \mp \Omega_0 - k_{\pm} v_{\parallel}} \right)$$
$$+ \frac{\omega_b^2}{\omega} \int_0^{\infty} dp \frac{\omega}{\gamma (k_w v_{\parallel} - \Omega_0)} \frac{m_e v_{\parallel}}{\omega \mp k_w v_{\parallel} - k_{\pm} v_{\parallel}} \frac{\partial}{\partial p} G_b(p). \tag{4.72}$$

Substitution of these expressions for the source current into Maxwell's equations (4.43) and (4.44) yields three coupled equations for the amplitudes of the vector and scalar potentials

$$\Lambda_{\pm}(k \mp k_w, \omega) \delta \hat{A}_{\pm}(0) + \frac{a_w}{2} \frac{c^2 k^2}{\omega^2} \chi_a(k,\omega) \delta \hat{\varphi}(0) = 0, \tag{4.73}$$

$$[1 + \chi(k,\omega)] \delta \hat{\varphi}(0) + a_w \frac{k_w}{k} [\sigma_+(k - k_w, \omega) \delta \hat{A}_+(0) + \sigma_-(k + k_w, \omega) \delta \hat{A}_-(0)] = 0, \tag{4.74}$$

where

$$\Lambda_{\pm}(k \mp k_w, \omega) \equiv 1 - \frac{c^2 k_{\pm}^2}{\omega^2} - \frac{\omega_b^2}{\omega^2} \left\langle \frac{p}{\gamma p_{\parallel}} \frac{\omega - k_{\pm} v_{\parallel}}{\omega \mp \Omega_0 - k_{\pm} v_{\parallel}} \right\rangle - \frac{\omega_b^2}{2\omega^2} a_w^2 S_{\pm}(k_{\pm}, \omega) \tag{4.75}$$

describes the dispersion of the pure electromagnetic modes in the presence of the wiggler. Observe that the wavelength matching conditions between the scalar potential and the right- and left-hand circularly polarized vector potentials imply that $k_{\pm} = k \mp k_w$.

The general dispersion equation If we now impose the cold beam distribution function (4.58) in the evaluation of the coupled dispersion equations and take the determinant of the matrix of coefficients, then we obtain a dispersion equation of the form

$$(\omega - kv_\parallel)^2 - \kappa^2 v_\parallel^2 = -\frac{\beta_w^2}{2}\frac{\omega_b^2}{\gamma_0}\left[\frac{\alpha_+(k-k_w,\omega)}{\varepsilon_+(k-k_w,\omega)} + \frac{\alpha_-(k+k_w,\omega)}{\varepsilon_-(k+k_w,\omega)}\right], \tag{4.76}$$

to second order in the wiggler amplitude, where all orbit quantities denote the steady-state trajectories computed using a total energy given by $\gamma_0 = (1 + p_0^2/m_e^2c^2)^{1/2}$,

$$\kappa^2 v_\parallel^2 \equiv \frac{\omega_b^2}{\gamma_0\gamma_\parallel^2}\Phi, \tag{4.77}$$

$$\varepsilon_\pm(k_\pm,\omega) \equiv 1 - \frac{\omega_b^2(\omega - k_\pm v_\parallel)}{\gamma_0(\omega^2 - k_\pm^2c^2)(\omega \mp \Omega_0 - k_\pm v_\parallel)}, \tag{4.78}$$

$$\begin{aligned}
\alpha_\pm(k_\pm,\omega) \equiv{}& \frac{1}{\omega^2 - k_\pm^2c^2}\left\{\beta_\parallel^2\left(\omega^2 - k^2c^2 - \frac{\omega_b^2}{\gamma_0}\right) + \beta_\parallel^2\frac{\omega_b^2}{\gamma_0(\omega - kv_\parallel)^2}\right.\\
&\times\left[(1-\Psi^2)\left(kc - \omega\frac{v_\parallel}{c}\right)^2 + \frac{(1-\Phi)}{\gamma_\parallel^2}(\omega^2 - c^2k^2)\right]\\
&+\frac{\omega_b^2}{\gamma_0\omega}\beta_\parallel\Psi\frac{\Omega_0}{k_wv_\parallel - \Omega_0}\left(\frac{\omega \mp k_wv_\parallel}{\omega - kv_\parallel} - \frac{\omega}{\omega \mp \Omega_0 - k_\pm v_\parallel}\right)\\
&+\frac{\Omega_0}{k_wv_\parallel - \Omega_0}[(\omega - kv_\parallel)^2 - \kappa^2v_\parallel^2]\left(\frac{3\omega \mp k_wv_\parallel}{\omega - kv_\parallel} - \frac{\omega \mp \Omega_0 \pm k_wv_\parallel}{\omega \mp \Omega_0 - k_\pm v_\parallel}\right)\\
&\left.-\frac{1-\Phi}{\gamma_\parallel^2}\frac{\omega[(\omega - kv_\parallel)^2 - \kappa^2v_\parallel^2]}{\omega - kv_\parallel}\left(\frac{\omega \mp 2kv_\parallel}{\omega - kv_\parallel} + \frac{2\Omega_0}{k_wv_\parallel - \Omega_0}\right)\right\},
\end{aligned} \tag{4.79}$$

and we have defined

$$\Psi \equiv 1 - \frac{\Omega_0}{\beta_\parallel(kc - \omega v_\parallel/c)}\frac{(1+\beta_w^2)k_wv_\parallel - \omega}{(1+\beta_w^2)\Omega_0 - k_wv_\parallel}. \tag{4.80}$$

It should be observed that the left-hand side of equation (4.76) describes the uncoupled dispersion equation for the beam-plasma waves in the presence of the wiggler, while $\varepsilon_\pm(k_\pm,\omega)$ describes the dispersion functions for the circularly polarized electromagnetic modes in the absence of the wiggler. Hence, this equation describes the coupling of the electromagnetic modes to the electrostatic beam-plasma waves due to the presence of the wiggler field.

The stability of beam-plasma modes Before proceeding further, we observe that the electrostatic beam-plasma modes are themselves modified in the presence of the wiggler field. If we neglect the coupling to the electromagnetic waves, then the beam-plasma modes satisfy the dispersion equation [28]

$$(\omega - kv_\parallel)^2 = \frac{\omega_b^2}{\gamma_0\gamma_\parallel^2}\Phi. \tag{4.81}$$

This implies that the effect of the wiggler is to modify the beam-plasma frequency based upon the response of the beam to the axial electric field perturbations (see equation (2.23)). As a consequence, the effective plasma frequency is increased by the combined presence of the wiggler and solenoidal fields for the Group I trajectories since $\Phi > 1$ for these orbits. In contrast, $|\Phi| < 1$ for the Group II trajectories, and the beam-plasma frequency decreases relative to that found in the absence of either magnetostatic field. Indeed, the effective plasma frequency is imaginary for those Group II orbits for which $\Phi < 0$, which corresponds to an unstable beam-plasma wave in which the growth rate is

$$\operatorname{Im} k = \frac{\omega_b}{\gamma_0^{1/2}\gamma_\parallel}|\Phi|^{1/2}. \tag{4.82}$$

This is a particularly interesting regime for the free-electron laser because the coupling of the electromagnetic wave is to an unstable beam-plasma mode. This has the effect of increasing the free energy available to the interaction and broadening the bandwidth of the instability. This regime will be discussed in more detail in regard to both the linear and nonlinear analyses of the interaction.

A reduced form of the dispersion equation The wiggler field provides for the coupling between the beam space-charge wave and either polarization state of the electromagnetic field, as well as a growth mechanism for the electromagnetic waves in the absence of the beam-plasma waves. In the former case, coherent amplification occurs by three-wave coherent Raman scattering in which the magnetostatic wiggler represents the pump wave, the beam-plasma mode represents the idler, and the output signal is the daughter wave. The latter case is coherent Compton scattering in which the wiggler (that appears as a backwards-propagating electromagnetic wave in the beam frame) scatters off the electron beam to produce the output signal. We choose, without loss of generality, to focus on the right-hand mode. As a consequence, we assume that $|\varepsilon_+(k_+,\omega)| \ll |\varepsilon_-(k_-,\omega)|$ and neglect the term in $[\varepsilon_-(k_-,\omega)]^{-1}$ in equation (4.76). If we assume in addition that $\omega_b^2/\gamma_0\omega^2 \ll 1$, then the dispersion equation can be cast into the substantially simpler form [18]

$$\begin{aligned}
&[(\omega - kv_\parallel)^2 - \kappa^2 v_\parallel^2]\left[\omega^2 - k_+^2 c^2 - \frac{\omega_b^2(\omega - k_+ v_\parallel)}{\gamma_0(\omega - \Omega_0 - k_+ v_\parallel)}\right] \\
&\quad \cong -\frac{\beta_w^2}{2}\frac{\omega_b^2}{\gamma_0}\left[\beta_\parallel^2\left(\omega^2 - k^2c^2 - \frac{\omega_b^2(\omega - k_+ v_\parallel)}{\gamma_0(\omega - \Omega_0 + k_+ v_\parallel)}\right) + \omega^2\frac{1-\Phi}{\gamma_\parallel^2}\right],
\end{aligned} \tag{4.83}$$

where we have imposed the wavenumber matching condition $k_+ = k - k_w$. This equation describes the coupling of the electromagnetic modes with the positive- and negative-energy beam-plasma waves (denoted by $\omega = (k_+ + k_w)v_\parallel \pm \kappa v_\parallel$, respectively when $\kappa > 0$). The peak gain occurs near the intersections of the uncoupled dispersion curves of the beam-plasma and electromagnetic modes; hence, we may make the replacements

$$\omega^2 - k^2c^2 - \frac{\omega_b^2(\omega - k_+ v_\parallel)}{\gamma_0(\omega - \Omega_0 - k_+ v_\parallel)} \cong -k_w(2k_+ + k_w)c^2 \tag{4.84}$$

and $\omega \approx (k_+ + k_w)v_\parallel$ on the right-hand side of equation (4.83). As a result, the dispersion equation becomes

$$[(\omega - kv_\parallel)^2 - \kappa^2 v_\parallel^2]\left[\omega^2 - k_+^2 c^2 - \frac{\omega_b^2(\omega - k_+ v_\parallel)}{\gamma_0(\omega - \Omega_0 - k_+ v_\parallel)}\right] \cong \frac{\beta_w^2}{2}\frac{\omega_b^2}{\gamma_0 \gamma_\parallel^2}\Phi\omega^2. \quad (4.85)$$

It is important to observe that growth is found near the intersections of the electrostatic beam-plasma mode and the circularly polarized electromagnetic mode, in particular in the neighbourhood of the intersection between the electromagnetic wave and the negative-energy beam-plasma mode.

Dispersive effects on the interaction In the absence of a solenoidal magnetic field both polarization states of the electromagnetic modes have the same dispersion relation (i.e. $\omega^2 = k^2c^2 + \omega_b^2/\gamma_0$), which is shown schematically in Fig. 4.3 in the first quadrant of the (ω, k) plane. The situation is more complex in the presence of the solenoidal magnetic field in which both escape and cyclotron branches exist (see Fig. 4.4), and the escape branch is characterized by a cutoff at

$$\omega_{co} = \frac{1}{2}\Omega_0\left[1 + \sqrt{\left(1 + \frac{4\omega_b^2}{\gamma_0 \Omega_0^2}\right)}\right]. \quad (4.86)$$

Since the slope of the negative-energy beam-plasma mode is identical to that of the cyclotron resonance line (i.e. $\omega \approx \Omega_0 + k_+ v_\parallel$), the particular branch of the electromagnetic dispersion curve which is resonant depends upon the magnitude of the axial velocity. When $\Omega_0 < (k_w - \kappa)v_\parallel$ the intersection is possible only with the escape branch of the electromagnetic waves for which $\omega > \Omega_0 + k_+ v_\parallel$. This case includes the limit in which the axial solenoidal field vanishes. Observe that this condition is satisfied for Group I trajectories because for most cases of practical interest $\kappa \ll k_w$. In contrast, when $\Omega_0 > (k_w - \kappa)v_\parallel$ corresponding to the

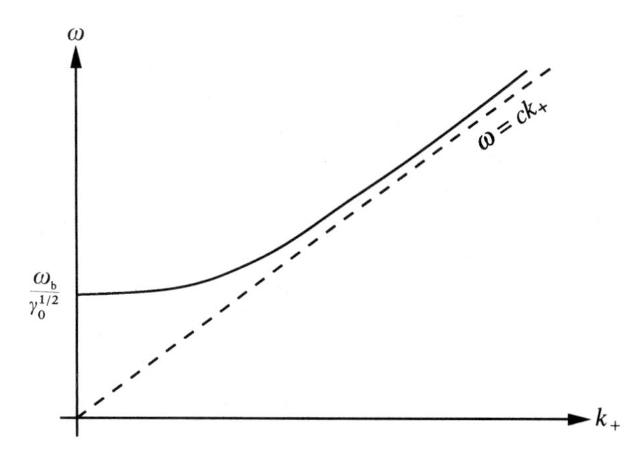

Fig. 4.3 Schematic drawing of the circularly polarized electromagnetic modes in the absence of a solenoidal magnetic field.

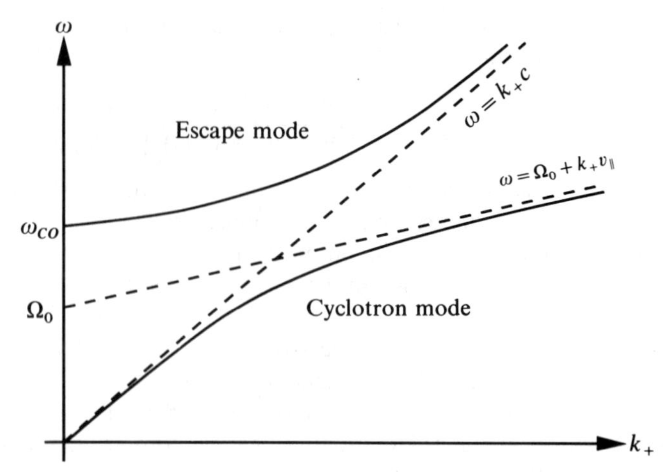

Fig. 4.4 Schematic of the dispersion relation of the right-hand circularly polarized electromagnetic mode in the presence of a solenoidal magnetic field showing both the escape and cyclotron branches.

Group II trajectories, the intersection is possible only with the electromagnetic cyclotron mode.

The Compton and Raman regimes The terminology of a weak axial solenoidal magnetic field refers to the regime in which $\Omega_0 \leqslant k_w v_\parallel$. This corresponds to electrons executing Group I trajectories in which $\Phi \geqslant 1$, and includes the limit in which the axial solenoidal field vanishes. As a consequence, the beam space-charge mode is stable, and the only effect of the solenoidal magnetic field is to enhance both the transverse wiggler-induced velocity and the effective plasma frequency.

The simplest case is that in which the solenoidal field vanishes, and the dispersion equation reduces to

$$[(\omega - kv_\parallel)^2 - \kappa^2 v_\parallel^2]\left(\omega^2 - k_+^2 c^2 - \frac{\omega_b^2}{\gamma_0}\right) \cong \frac{\beta_w^2}{2}\frac{\omega_b^2}{\gamma_0 \gamma_\parallel^2}\omega^2, \tag{4.87}$$

which is a quartic polynomial equation for k. This describes a dispersion equation in which the electromagnetic mode is coupled to the positive- and negative-energy beam-plasma modes. The form of the solutions for the coupled modes is shown schematically in Fig. 4.5. Growth is found in the vicinity of the intersection of the negative-energy beam-plasma wave and the forward-propagating electromagnetic mode; hence, the resonant frequency of the interaction occurs in the vicinity of two intersection points

$$\omega \cong \gamma_\parallel^2 (k_w - \kappa) v_\parallel \left\{1 \pm \beta_\parallel \sqrt{\left[1 - \frac{\kappa^2}{(k_w - \kappa)^2}\right]}\right\}, \tag{4.88}$$

which reduces to the well-known free-electron laser resonance condition ($\omega = 2\gamma_\parallel^2 k_w v_\parallel$) in the limit in which $\kappa \ll k_w$. In addition, since we are interested in

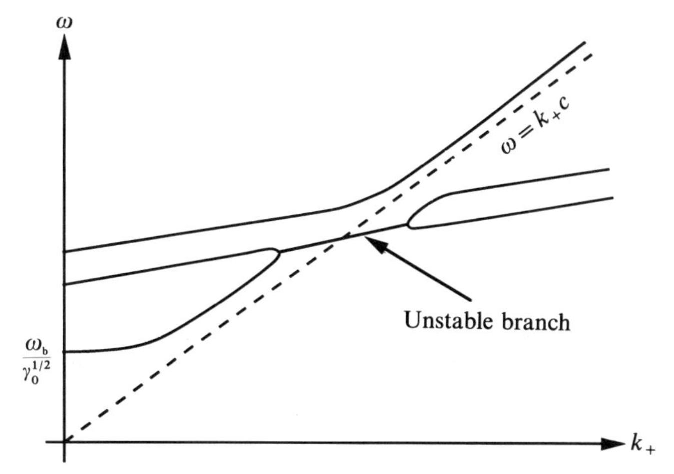

Fig. 4.5 Schematic illustration of the coupled dispersion relation for the electromagnetic mode and the positive- and negative-energy beam-plasma modes. As shown, it is the coupling with the negative-energy beam-plasma mode which leads to instability.

the propagation of waves in the first quadrant (i.e. $\omega > 0, k_+ > 0$) we may make the approximation

$$\omega^2 - k_+^2 c^2 - \frac{\omega_b^2}{\gamma_0} \cong -2c^2 k_+ (k_+ - K), \tag{4.89}$$

on the right-hand side of equation (4.87), where $K^2 \equiv (\omega^2 - \omega_b^2/\gamma_0)^{1/2}/c$. Therefore under the transformation

$$k = \frac{\omega}{v_\parallel} + \kappa + \delta k, \tag{4.90}$$

the dispersion equation reduces to a cubic in δk

$$\delta k(\delta k + 2\kappa)(\delta k - \Delta k) \cong -\frac{\beta_w^2}{2} \frac{\omega_b^2}{\gamma_0 c^2} \frac{k_w}{\beta_\parallel}, \tag{4.91}$$

where

$$\Delta k \equiv k_w + K - \frac{\omega}{v_\parallel} - \kappa \tag{4.92}$$

defines the wavenumber mismatch parameter. There are two principal regimes of interest in the solution of the dispersion equation (4.91) corresponding to the low- and high-density regimes.

The high-gain Compton regime (sometimes called the **strong-pump regime**) is found when the effect of the ponderomotive potential is greater than that of the space-charge potential of the beam-plasma waves. In this limit $|\delta k| \gg 2\kappa$, and the dispersion equation reduces to the cubic equation

$$\delta k^2(\delta k - \Delta k) \cong -\frac{\beta_w^2}{2} \frac{\omega_b^2}{\gamma_0 c^2} \frac{k_w}{\beta_\parallel}. \tag{4.93}$$

The maximum growth is found to correspond to zero detuning, and we find that the complex roots are (see equation (1.8))

$$\frac{\delta k_{\max}}{k_w} \cong \frac{1}{2}(1 \pm i\sqrt{3})\left(\frac{\beta_w^2}{2\beta_\parallel}\frac{\omega_b^2}{\gamma_0 c^2 k_w^2}\right)^{1/3}. \tag{4.94}$$

In order to neglect the effect of the space-charge fields, therefore, we must require that

$$\frac{\omega_b}{\gamma_0^{1/2} c k_w} \ll \frac{1}{16}\frac{v_w^2}{c^2}\gamma_\parallel^3. \tag{4.95}$$

The opposite case in which space-charge fields dominate over the ponderomotive potential is referred to as the Raman (or collective) regime. In this limit, $|\delta k| \ll 2\kappa$ and the dispersion equation reduces to a quadratic equation for δk

$$\delta k(\delta k - \Delta k) \cong -\frac{\beta_w^2}{4}\frac{\omega_b}{\gamma_0^{1/2} c} k_w, \tag{4.96}$$

where peak growth is again found for zero detuning. This peak growth rate is given by (see equation (1.11))

$$\frac{\delta k_{\max}}{k_w} \cong i\frac{\beta_w}{2}\sqrt{\left(\frac{\omega_b}{\gamma_0^{1/2} c k_w}\right)}, \tag{4.97}$$

which is valid for sufficiently high-beam densities that

$$\frac{\omega_b}{\gamma_0^{1/2} c k_w} \gg \frac{1}{16}\frac{v_w^2}{c^2}\gamma_\parallel^3. \tag{4.98}$$

The transition between the Compton and Raman regimes The detailed effect of increasing beam density upon the growth rate of the instability is shown in Fig. 4.6 in which we plot the magnitude of the growth rate versus the beam density for $\Omega_w/ck_w = 0.05$ and $\gamma_0 = 2.957$ (corresponding to a beam energy of 1 MeV). The dashed line represents the growth rate as calculated for the high-gain Compton regime (4.94) which scales as $\omega_b^{2/3}$. For this choice of parameters, $\beta_w^2 \approx 0.00282$, $\gamma_\parallel^2 \approx 8.72$, and the transition between the high-gain Compton and the collective Raman regimes occurs for $\omega_b/ck_w \approx 0.004$. It is evident from the figure that the growth rate tracks the solution for the high-gain Compton up to this point, but diverges from it as the beam density continues to increase. For beam densities such that $\omega_b/\gamma_0^{1/2}ck_w > 0.01$, the solution follows the Raman solution (4.97) in which the growth rate scales as $\omega_b^{1/2}$. The discrepancy between the actual solution for the growth rate and the prediction for the high-gain Compton regime in the dashed line indicates that collective effects involved in the stimulated Raman scattering process tend to reduce the linear growth rate.

The effect of an axial magnetic field

The dispersion equation (4.85) is, in general, a quintic polynomial in k when an axial solenoidal field is present. If we make the restriction that $k_+ > 0$, then the

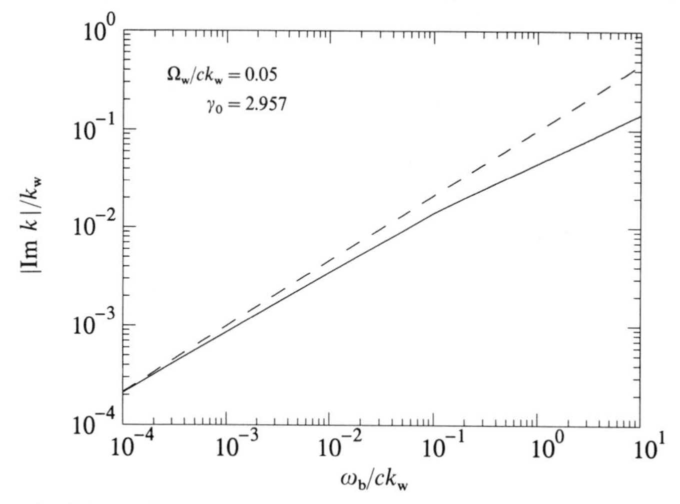

Fig. 4.6 Graph of the peak growth rate for zero detuning (i.e. $\Delta k = 0$) is shown in the solid line as a function of the beam-plasma frequency. The dashed line shows an increase as the 2/3 power of the beam-plasma frequency as found in the high-gain Compton regime. The onset of the collective Raman regime is evident from the decrease in the slope of the curve.

dispersion equation can be reduced to a quartic polynomial

$$[(\omega - kv_\parallel)^2 - \kappa^2 v_\parallel^2](k_+ - K_+)(k_+ - K_-) \cong -\beta_w^2 \frac{\omega_b^2}{\gamma_0 c^2} \frac{\omega^2}{4\gamma_\parallel^2 K} \Phi\left(k_+ - \frac{\omega - \Omega_0}{v_\parallel}\right), \tag{4.99}$$

where

$$K_\pm \equiv \frac{1}{2}\left(K + \frac{\omega - \Omega_0}{v_\parallel}\right) \pm \frac{1}{2}\sqrt{\left(\Delta K^2 + 2\frac{\omega_b^2}{\gamma_0 c^2}\frac{\Omega_0}{Kv_\parallel}\right)}, \tag{4.100}$$

and $\Delta K \equiv K - (\omega - \Omega_0)/v_\parallel$. It is clear that the nature of the instability depends upon the sign of Φ, which determines the effective plasma frequency as well as mediating the ponderomotive potential.

The case of a weak magnetic field In the weak magnetic field regime we deal with the Group I orbits for which $\Phi \geqslant 1$ and the intersection between the beam-plasma mode and the electromagnetic wave occurs on the escape branch. Substantial simplification of the dispersion equation (4.99) is found in the limit in which the cyclotron resonance effects on the dispersion of the electromagnetic wave can be neglected. This implies that the wave frequency $\omega \gg \omega_{co}$. In this regime, the dispersion equation can be approximated by

$$\delta k(\delta k + 2\kappa)(\delta k - \Delta k) \cong -\frac{\beta_w^2}{2}\frac{\omega_b^2}{\gamma_0 c^2}\frac{k_w}{\beta_\parallel}\Phi, \tag{4.101}$$

in which the effective plasma frequency (i.e. $\omega_b^2\Phi$) is used in κ. In addition, the resonant frequency is given approximately by equation (4.88) subject to the substitution of the effective plasma frequency. As a consequence, the dispersion equation in the high-gain Compton regime is

$$\delta k^2(\delta k - \Delta k) \cong -\frac{\beta_w^2}{2}\frac{\omega_b^2}{\gamma_0 c^2}\frac{k_w}{\beta_\parallel}\Phi. \tag{4.102}$$

The criterion for the high-gain Compton regime is now

$$\frac{\omega_b}{\gamma_0^{1/2} c k_w}\Phi^{1/2} \ll \frac{1}{16}\frac{v_w^2}{c^2}\gamma_\parallel^3, \tag{4.103}$$

and at maximum growth

$$\frac{\delta k_{\max}}{k_w} \cong \frac{1}{2}(1 + i\sqrt{3})\left(\frac{\beta_w^2}{2\beta_\parallel}\frac{\omega_b^2}{\gamma_0 c^2 k_w^2}\Phi\right)^{1/3}. \tag{4.104}$$

The Raman regime is found in the opposite limit to satisfy

$$\delta k(\delta k - \Delta k) \cong -\frac{\beta_w^2}{4}\frac{\omega_b}{\gamma_0^{1/2} c}\Phi^{1/2} k_w, \tag{4.105}$$

which has the solution at maximum growth of

$$\frac{\delta k_{\max}}{k_w} \cong i\frac{\beta_w}{2}\sqrt{\left(\frac{\omega_b}{\gamma_0^{1/2} c k_w}\Phi^{1/2}\right)}. \tag{4.106}$$

The effect of the axial solenoidal field in this regime is to enhance the growth rates due to enhancements in both the transverse wiggler-induced velocity v_w and the effective plasma frequency through Φ. However, since this enhancement occurs at the expense of a reduction in $v_\parallel$, the interaction frequency at fixed energy must decrease. This is illustrated in Fig. 4.7 in which we plot the magnitude of the maximum growth rate and the corresponding frequency as a function of the solenoidal magnetic field for $\Omega_w/ck_w = 0.05$, $\gamma_0 = 2.957$ (for a beam energy of 1 MeV), and $\omega_b/ck_w = 0.1$. For these parameters, which correspond to the steady-state orbit obtained in Chapter 2, the Group I orbits exhibit the transition to orbital instability at $\Omega_0/ck_w \approx 0.750$ and this defines the maximum value of the solenoidal field for which an interaction is possible. Observe that the growth-rate increases to a singularity at this transition (which is due to the singularity in Φ and the effective plasma frequency), but that this singularity is an artefact of the approximations implicit in the linearization. It is also evident that the frequency corresponding to peak growth decreases with increasing Ω_0 over this range. The increase in the growth rate and the decrease in the frequency of the interaction is accompanied by an increase in the bandwidth over which growth is found. This is illustrated in Fig. 4.8 in which we plot the magnitude of the growth rate versus frequency for $\Omega_0/ck_w = 0$ and 0.5.

The case of a strong magnetic field The terminology of the strong axial magnetic field refers to the case in which $\Omega_0 \geq k_w v_\parallel$ and the electrons are on Group II

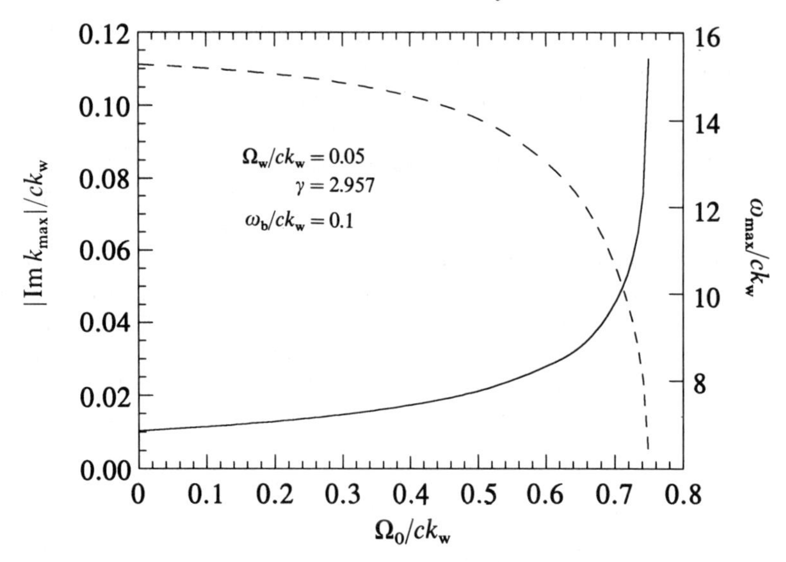

Fig. 4.7 Graph of the magnitude of the maximum growth rate (solid line) and the frequency corresponding to maximum growth (dashed line) as a function of the axial solenoidal field for Group I trajectories.

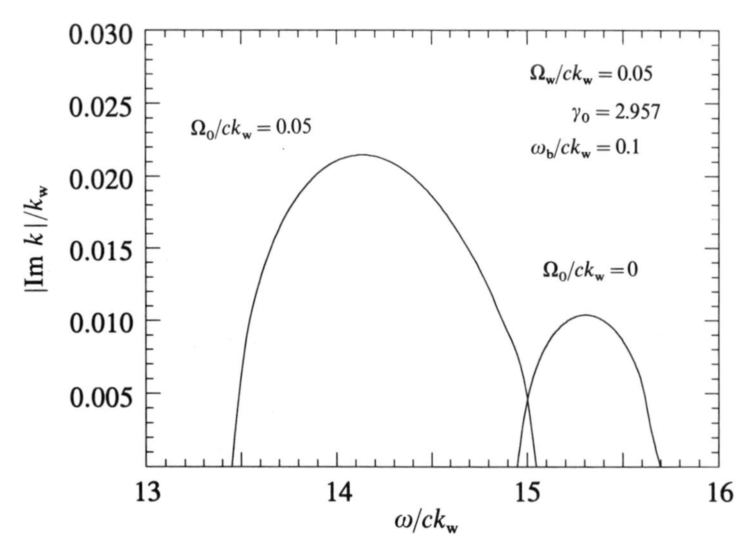

Fig. 4.8 Graph of the magnitude of the growth rate versus frequency for $\Omega_w/ck_w = 0.05$, $\gamma_0 = 2.957$ and $\Omega_0/ck_w = 0$ and 0.05.

orbits. There are two distinct regimes for these orbits which relate to (1) the negative-mass limit in the vicinity of the magnetic resonance at $\Omega_0 \approx k_w v_\parallel$ in which the axial velocity of the beam increases with decreases in the electron energy, and (2) a more usual positive-mass case found at still higher axial magnetic fields. The dispersion equation in the positive-mass limit is identical to that found for the weak axial magnetic field limit. However, since $\lim_{B_0 \to \infty} v_w = 0$ and $\lim_{B_0 \to \infty} \Phi = 1$, the growth rate in this limit is small. As a consequence, we shall focus our attention on the negative-mass regime.

The character of the beam-plasma mode in the negative-mass regime has already been shown to be unstable. As a consequence, the circularly polarized electromagnetic mode couples with an unstable beam-plasma wave, which greatly enhances the bandwidth of the instability. In general, the growth rate must be found by solution of the full quartic dispersion equation (4.99) by numerical methods. This solution of the growth rate as a function of frequency is shown in Fig. 4.9 for $\Omega_w/ck_w = 0.05$, $\gamma_0 = 2.957$, $\omega_b/ck_w = 0.1$ and $\Omega_0/ck_w = 1.0$. These parameters correspond to both the idealized steady-state trajectories discussed in Chapter 2, and we find that $\beta_\parallel = 0.8744$ and $\Phi = -1.3685$. Note that at the point at which $\Phi = 0$ (which occurs for $\Omega_0/ck_w \approx 1.2$), the coupling coefficient on the right-hand side of the dispersion equation (4.85) vanishes along with the interaction. It is evident from the figure that there are two unstable bands. For the negative-mass regime, the frequency of the beam-plasma mode varies as $\omega = (k + k_w)v_\parallel$ and there is no contribution due to the effective plasma frequency. Hence, the intersection point between the electromagnetic mode and the beam-plasma mode occurs at

$$\omega \cong \gamma_\parallel^2 k_w v_\parallel \left[1 + \beta_\parallel \sqrt{\left(1 - \frac{\omega_b^2}{\gamma_0 \gamma_\parallel^2 k_w^2 v_\parallel^2} \right)} \right]. \tag{4.107}$$

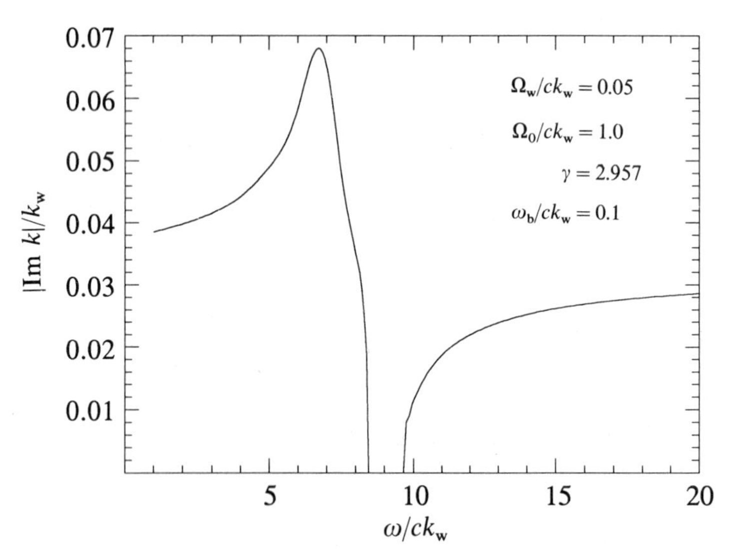

Fig. 4.9 Graph of the magnitude of the growth rate as a function of frequency for a strong axial magnetic field corresponding to the negative-mass regime of the Group II electron orbits.

For the present choice of parameters, this frequency occurs at $\omega/ck_w \approx 6.94$ which corresponds to the point of maximum gain shown in the figure.

The additional unstable band that is found for $\omega/ck_w > 9.6$ arises from the unstable nature of the beam-plasma wave. For this frequency range in which $|\Delta k| \gg |\delta k|$, the dispersion equation may be approximated by

$$k^2 - 2\frac{\omega}{v_\parallel}k + \frac{\omega^2}{v_\parallel^2} + |\kappa^2|\left[1 - \frac{\beta_w^2}{4}\beta_\parallel\gamma_\parallel^2(1+\beta_\parallel)\right] \cong 0, \tag{4.108}$$

which has the approximate solution

$$k \cong \frac{\omega}{v_\parallel} \pm i|\kappa|\sqrt{\left(1 - \frac{\beta_w^2}{4}\beta_\parallel\gamma_\parallel^2(1+\beta_\parallel\right)}. \tag{4.109}$$

This mode is clearly a modified beam-plasma wave, where it should be observed that the wiggler acts as a stabilizing influence which reduces the effective growth rate of the instability. Indeed, this instability vanishes whenever

$$\frac{\beta_w^2}{4}\beta_\parallel\gamma_\parallel^2(1+\beta_\parallel) > 1. \tag{4.110}$$

Thermal effects on the instability

In the treatment of thermal effects upon the growth rate of the free-electron laser we consider the regime in which the axial solenoidal field is absent. In general, however, the resonance condition $\omega = (k + k_w)v_\parallel$ implies that thermal effects become important when the ratio of the axial velocity spread $\Delta v_\parallel$ to the bulk axial velocity of the beam is $\Delta v_\parallel/v_\parallel \approx \mathrm{Im}\,k/(\mathrm{Re}\,k + k_w)$. In the high-gain Compton regime, this condition implies that thermal effects cannot be neglected when

$$\frac{\Delta v_\parallel}{v_\parallel} \approx \frac{\sqrt{3}}{4\gamma_\parallel^2}\left(\frac{\beta_w^2}{2\beta_\parallel^2}\frac{\omega_b^2}{\gamma_0 k_w^2 v_\parallel^2}\right)^{1/3}. \tag{4.111}$$

Similarly, thermal effects are important in the collective Raman regime when

$$\frac{\Delta v_\parallel}{v_\parallel} \approx \frac{\beta_w}{4\gamma_\parallel^2\beta_\parallel}\sqrt{\left(\frac{\omega_b}{\gamma_0^{1/2}ck_w}\right)}. \tag{4.112}$$

The aforementioned formulations of the growth rate in the Raman and high-gain Compton regimes break down when these conditions are satisfied, and a more general formulation must be developed.

Since the free-electron laser operates by means of an axial bunching mechanism, it is the axial velocity spread which is most important. As a consequence, in the treatment of thermal effects on the linear stability properties we shall impose the simplification that the electron beam is monoenergetic but exhibits a pitch-angle spread [23]. The effect of the pitch-angle spread is to include velocity spreads in both the axial and transverse directions, and may be described by a distribution function of the form

$$F_b(P_x, P_y, p) = n_b G_\perp(P_x, P_y)\frac{p_z}{p}\delta(p - p_0), \tag{4.113}$$

where $G_\perp(P_x, P_y)$ represents the transverse distribution. For convenience we shall assume that this transverse distribution takes the form of a Gaussian

$$G_\perp(P_x, P_y) = \frac{1}{\pi\Delta P^2}\exp(-P_\perp^2/\Delta P^2), \tag{4.114}$$

where $P_\perp^2 \equiv P_x^2 + P_y^2$, and ΔP represents the thermal spread.

Under the assumption that $\Delta P \ll p_0$, the electron orbits can be represented in terms of the perturbed steady-state trajectories given in equations (2.42–4). In the absence of an axial solenoidal field, these orbits are of the form

$$p_\pm = (P_x \mp iP_y) + p_w \exp(\mp ik_w z), \tag{4.115}$$

and

$$p_z = \sqrt{[P_\parallel^2 - 2p_w(P_x \cos k_w z + P_y \sin k_w z)]}, \tag{4.116}$$

where $\gamma_0 \equiv (1 + p_0^2/m_e^2c^2)^{1/2}$, $p_w \equiv \gamma_0 m_e v_w$, $P_\parallel^2 \equiv p_0^2 - p_w^2 - P_x^2 - P_y^2$, and $v_w \equiv -\Omega_w/k_w$.

For simplicity, we neglect the left-hand circular-polarization state and treat the high-gain regime in which the vector and scalar potentials may be expressed as in equations (4.64) and (4.65). As a consequence, the source currents can be written as

$$\begin{aligned}\delta\hat{J}_+(z) = \frac{e^2}{2m_e c}\int dP_x dP_y dp \frac{p}{\gamma p_z}\bigg[\left(2 + \frac{p_+ p_-}{p_z^2}\right)D_+ \\ + p_+\left(\frac{\partial}{\partial P_x} + i\frac{\partial}{\partial P_y}\right)D_+ - p_+ D_z \frac{\partial}{\partial p}\bigg]F_b(P_x, P_y, p),\end{aligned} \tag{4.117}$$

and

$$\delta\hat{J}_z(z) = \frac{e^2}{2m_e c}\int dP_x dP_y dp \frac{p}{\gamma}\bigg[\left(\frac{\partial}{\partial P_x} + i\frac{\partial}{\partial P_y}\right)D_+ - D_z\frac{\partial}{\partial p}\bigg]F_b(P_x, P_y, p), \tag{4.118}$$

where the orbit integrals are

$$D_\pm = \delta A_\pm(0)\exp[i\omega t(z, 0)] - \delta A_\pm(z), \tag{4.119}$$

and

$$D_z = i\frac{\gamma m_e c}{p}\int_z^z dz' \exp[i\omega t(z, z')]\bigg[-k\delta\hat{\varphi}(z') + \frac{\omega}{c}\frac{p_-}{p_z}\delta\hat{A}_+(z')\bigg]. \tag{4.120}$$

The orbit integrals may be evaluated using the perturbed steady-state trajectories (4.115) and (4.116). We retain only the lowest-order contributions due to the perturbed steady-state trajectories in P_x and P_y which appear in the resonance condition. In this limit, the source currents become

$$\delta\hat{J}_+(z) \cong -\frac{e^2}{m_e c}\delta\hat{A}_+(z)\int dP_x dP_y dp\left[\frac{p}{\gamma p_\parallel}\left(1+\frac{\beta_w^2}{2}\right)-\frac{\beta_w^2}{2}\frac{\omega m_e v_\parallel}{\omega - kV_\parallel}\frac{\partial}{\partial p}\right]F_b(P_x, P_y, p)$$
$$-\frac{e^2}{2m_e}p_w\delta\hat{\varphi}(z)\exp(-ik_w z)\int dP_x dP_y dp\frac{k}{\gamma(\omega - kV_\parallel)}\frac{\partial}{\partial p}F_b(P_x, P_y, p), \tag{4.121}$$

and

$$\delta\hat{J}_z(z) \cong \frac{e^2}{2m_e c}p_w\delta\hat{A}_+(z)\exp(ik_w z)\int dP_x dP_y dp\frac{\omega}{\gamma(\omega - kV_\parallel)}\frac{\partial}{\partial p}F_b(P_x, P_y, p)$$
$$-\frac{e^2}{2m_e}\delta\hat{\varphi}(z)\int dP_x dP_y dp\frac{m_e k v_\parallel}{\omega - kV_\parallel}\frac{\partial}{\partial p}F_b(P_x, P_y, p), \tag{4.122}$$

where $V_\parallel \equiv P_\parallel/\gamma_0 m_e$ is the axial velocity corresponding to the generalized steady-state trajectory, $v_\parallel \equiv (p_0^2 - p_w^2)/\gamma_0 m_e$, and $\beta_w^2 \equiv v_w^2/v_\parallel^2$. Note that $v_\parallel$ and v_w denote the axial and transverse velocities for the steady-state trajectory corresponding to γ_0.

The derivatives of the distribution (4.114) with respect to p which appear in the above expressions for the source currents may be integrated by parts and the results substituted into Maxwell's equations to give

$$\left[\omega^2 - k_+^2 c^2 - \frac{\omega_b^2}{\gamma_0}\left(1 - \frac{a_w^2}{\gamma_0^2}\frac{\omega^2 - k^2 c^2}{(\omega - kv_\parallel)^2}T(\zeta)\right)\right]\delta\hat{A}_+(0)$$
$$= \frac{\omega_b^2}{2\gamma_0}\frac{a_w}{\gamma_0}\frac{ck(ck - \omega\beta_\parallel)}{(\omega - kv_\parallel)^2}T(\zeta)\delta\hat{\varphi}(0), \tag{4.123}$$

and

$$\left[(\omega - kv_\parallel)^2 - \frac{\omega_b^2}{\gamma_0\gamma_\parallel^2}T(\zeta)\right]\delta\hat{\varphi}(0) = \frac{\omega_b^2}{\gamma_0}\frac{a_w}{\gamma_0}\left(1 - \frac{\omega}{ck}\beta_\parallel\right)T(\zeta)\delta\hat{A}_+(0), \tag{4.124}$$

where we identify $k = k_+ + k_w$ from the wavenumber matching condition, $\beta_\parallel \equiv v_\parallel/c$, $\gamma_\parallel^2 \equiv (1 - v_\parallel^2/c^2)^{-1}$, and $T(\zeta)$ defines the thermal function. The thermal function which describes the effect of the pitch-angle spread on the instability is defined as [23]

$$T(\zeta) \equiv \zeta[1 - \zeta\exp(\zeta)E_1(\zeta)], \tag{4.125}$$

for argument

$$\zeta \equiv \frac{\gamma_0^2 m_e^2}{\Delta P^2}\left(\frac{\omega^2}{k^2} - v_\parallel^2\right), \tag{4.126}$$

where

$$E_1(\zeta) \equiv \int_\zeta^\infty dt\frac{\exp(-t)}{t}, \tag{4.127}$$

denotes the exponential integral function defined over the domain $|\arg\zeta| < \pi$.

The dispersion equation which results from this formulation is

$$\left[(\omega - kv_{\parallel})^2 - \frac{\omega_b^2}{\gamma_0 \gamma_{\parallel}^2} T(\zeta)\right]\left[\omega^2 - k_+^2 c^2 - \frac{\omega_b^2}{\gamma_0}\right]$$
$$= -\frac{\omega_b^2 a_w^2}{2\gamma_0 \gamma_0^2} T(\zeta)\left[\omega^2 - k^2 c^2 - \frac{\omega_b^2}{\gamma_0} T(\zeta)\right], \quad (4.128)$$

correct to lowest nontrivial order in a_w. In order to verify that this dispersion equation reproduces that found in the idealized beam limit (4.87), we observe that $\lim_{\Delta P \to 0} |\zeta| = \infty$. Expanding the exponential integral function in the asymptotic limit, therefore, we find that $\lim_{\Delta P \to 0} T(\zeta) = 1$ and that the ideal beam-dispersion equation (4.87) is recovered. Thermal effects become dominant wherever $\mathrm{Im}\, k/(\mathrm{Re}\, k + k_w) = \Delta v_{\parallel}/v_{\parallel}$, where the wavenumber is to be evaluated at the peak growth rate in the ideal beam limit. On the basis of the perturbed trajectories, it is clear that $\Delta v_{\parallel}/v_{\parallel} \approx \Delta P^2/2p_0^2$; hence, thermal effects are important when

$$\frac{\Delta P^2}{p_{\parallel}^2} \approx \frac{\sqrt{3}}{2\gamma_{\parallel}^2}\left(\frac{\beta_w^2}{2\beta_{\parallel}^2} \frac{\omega_b^2}{\gamma_0 k_w^2 v_{\parallel}^2}\right)^{1/3}, \quad (4.129)$$

in the high-gain Compton regime, and

$$\frac{\Delta P^2}{p_{\parallel}^2} \approx \frac{\beta_w}{2\gamma_{\parallel}^2 \beta_{\parallel}} \sqrt{\left(\frac{\omega_b}{\gamma_0^{1/2} ck_w}\right)}, \quad (4.130)$$

in the collective Raman regime.

The effect of the thermal spread on the linear growth rate is threefold. In the first place, the wider range of axial velocities introduced thereby results in a broader resonance condition in which the unstable frequency band increases. In the second place, the fact that the bulk axial velocity decreases means that the centre frequency of the gain band also decreases. In the third place, the peak growth rate decreases with increasing ΔP. Each of these properties is illustrated in Fig. 4.10 in which we solve equation (4.128) numerically for the growth rate, and plot the magnitude of the growth rate as a function of the frequency for $\Omega_w/ck_w = 0.05$, $\gamma_0 = 2.957$ and $\omega_b/ck_w = 0.1$. Observe that the growth rate peaks for $|\mathrm{Im}\, k|/k_w \approx 0.011$ in the absence of the thermal spread, and decreases by over 100% as the thermal spread increases to $\Delta P/p_0 = 5\%$. Comparison with Fig. 4.6 for this value of the plasma frequency indicates that this example is in the intermediate range between the high-gain Compton and the collective Raman regimes.

The detailed variation in the peak growth rate and the frequency corresponding to peak growth as a function of ΔP is illustrated in Fig. 4.11. As shown in the figure, the peak growth rate remains relatively constant for $\Delta P/p_0 < 2\%$ and decreases rapidly thereafter. As a consequence, thermal effects become dominant for $\Delta P/p_0 \geqslant 3\%$. The reason for the decline in the growth rate is that progressively more electrons are driven out of resonance with the wave as the thermal spread in the axial energy (i.e. velocity) increases. The effect can be quantified by

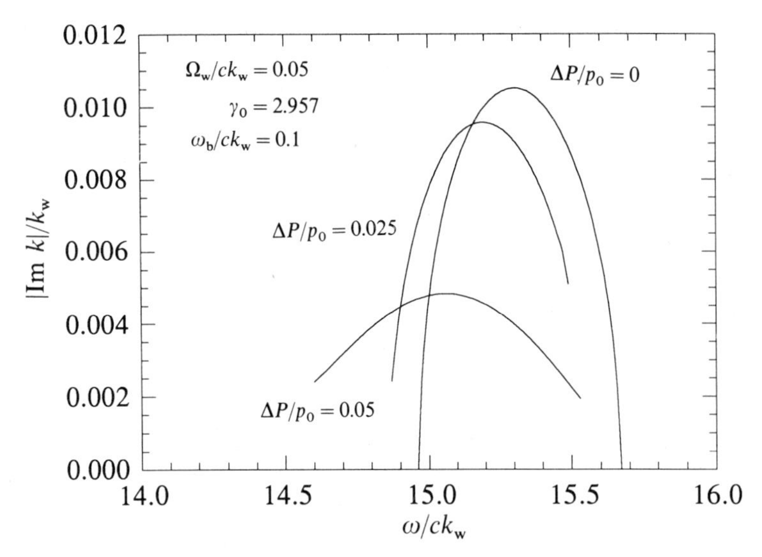

Fig. 4.10 Graph of the magnitude of the growth rate versus frequency for $\Delta P/p_0 = 0$, 0.025 and 0.05, and $\Omega_w/ck_w = 0.05$, $\gamma_0 = 2.957$ and $\omega_b/ck_w = 0.1$.

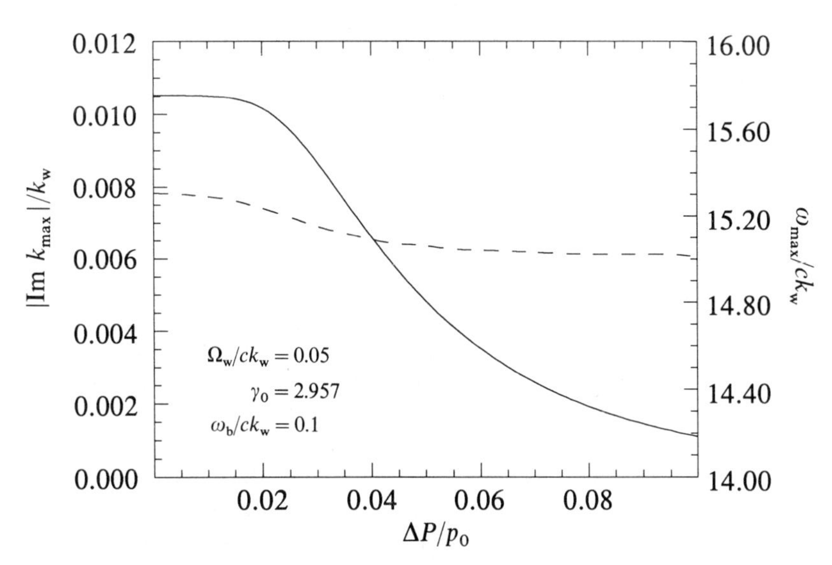

Fig. 4.11 Graph of the magnitude of the maximum growth rate (solid line) and the corresponding frequency (dashed line) as a function of $\Delta P/p_0$ for $\Omega_w/ck_w = 0.05$, $\gamma_0 = 2.957$ and $\omega_b/ck_w = 0.1$.

consideration of the resonance condition $\omega \approx (k + k_w)v_z$, from which it is apparent that thermal effects dominate whenever $(\mathrm{Im}\, k)v_{\parallel} \approx (\mathrm{Re}\, k + k_w)\Delta v_{\parallel}$. Since this example is neither in the Raman nor in the high-gain Compton regimes, we expect that for thermal effects to be important $\Delta v_{\parallel}/v_{\parallel} \approx \mathrm{Im}\, k/(1 + \mathrm{Re}\, k) \approx 0.067\%$ (where

$\mathrm{Im}\, k/k_w \approx 0.011$ and $\mathrm{Re}\, k/k_w \approx 15.4$ at peak growth). This yields an estimate of $\Delta P/p_0 \approx 3.7\%$, which is in reasonable agreement with the numerical solution for the growth rate shown in the figure. The frequency at which peak growth is found drops approximately 2% over this range.

4.1.2 Planar wiggler configurations

The physics of the free-electron laser interaction in planar wiggler configurations is similar but not identical to that for helical wiggler configurations. The principal difference arises due to the oscillations in the axial velocity which were discussed in Chapter 2 (see equation (2.105)), which have two major effects on the interaction. The first effect is that the bulk-resonance condition and the bulk gain are determined by the average value of the axial and transverse velocities. Hence, the effective wiggler field is reduced from the peak on-axis field by a factor of $\sqrt{2}$ (i.e. the root-mean-square value of the wiggler amplitude). The second effect is that the oscillation in the axial velocity results in an interaction at the even harmonics (i.e. $\omega \approx (k + lk_w)v_\parallel$, where l is an even integer). The oscillation in the axial velocity also results in an oscillation in the gain about the bulk value. This effect is beyond the scope of this chapter, however, and will be discussed in some detail in Chapter 5 in which a nonlinear analysis and simulation of the interaction in the presence of a planar wiggler is presented.

The analysis of planar wiggler configurations is presented in a similar manner to that for the helical wiggler. The low-gain regime is discussed after a basic discussion of the application of the Vlasov–Maxwell formalism to this wiggler geometry. The expression for the gain in this regime (4.159) includes the effects of the oscillation in the axial velocity both to the bulk gain and to the excitation of harmonics.

A derivation of the general dispersion equation in the high-gain regime is also presented for the case of an ideal beam (see equation (4.170)), and the similarities to that found for the helical wiggler are discussed. A derivation of the dispersion equation subject to the inclusion of beam thermal effects follows. The beam distribution used is identical to that employed for the helical wiggler (see equations (4.113) and (4.114)). The dispersion equation so derived includes a generalized thermal function which reduces to that found for the helical wiggler at the fundamental resonance, but which also describes the effect of the axial energy spread on the harmonic interaction in the high-gain regime. In addition, the effect of an axial magnetic field on the interaction is also discussed, and a dispersion equation in the ideal beam limit is derived (4.205).

Within the context of the interaction in a planar wiggler configuration, we employ the idealized wiggler representation given in equation (2.99) and perform a perturbation analysis about the associated single-particle trajectories. The y-component of the velocity is a constant of the motion in this geometry, $v_y = P_y/\gamma m_e$ where P_y is the y-component of the canonical momentum. The x-component of the velocity is given in equation (2.101), and we may approximate the axial velocity as

$$v_z = V_\parallel \sqrt{\left(1 - \frac{v_w^2}{2V_\parallel^2}\cos 2k_w z - 2\frac{P_x}{P_\parallel}\frac{v_w}{V_\parallel}\cos k_w z\right)}, \tag{4.131}$$

where P_x denotes the canonical momentum in the x-direction, $v_w \equiv -\Omega_w/k_w$ is the wiggler-induced velocity,

$$V_\parallel \equiv c\sqrt{\left(1 - \frac{1}{\gamma^2} - \frac{v_w^2}{2c^2} - \frac{P_\perp^2}{\gamma^2 m_e^2 c^2}\right)}, \tag{4.132}$$

$P_\perp^2 \equiv P_x^2 + P_y^2$, and we define $P_\parallel \equiv \gamma m_e V_\parallel$. Note that this expression reduces to the bulk axial velocity (2.106) found in the limit in which the canonical momentum vanishes.

The source currents

The interaction occurs principally for plane waves polarized in the direction of the wiggler-induced oscillation, and the vector and scalar potentials for a wave with angular frequency ω are of the form

$$\delta A(z,t) = \frac{1}{2}\delta\hat{A}(z)\hat{e}_x \exp(-i\omega t) + \text{c.c.}, \tag{4.133}$$

and

$$\delta\varphi(z,t) = \frac{1}{2}\delta\hat{\varphi}(z)\exp(-i\omega t) + \text{c.c.} \tag{4.134}$$

As in the case of the helical wiggler analysis, the perturbed distribution function takes the form

$$\delta\hat{f}_b(z,\boldsymbol{p}) = \frac{e}{2c}\left(D_x\frac{\partial}{\partial P_x} + D_z\frac{\partial}{\partial p}\right)F_b(P_x, P_y, p), \tag{4.135}$$

where the orbit integrals are defined as

$$D_x \equiv \int_0^z dz' \frac{\exp[-i\omega\tau(z,z')]}{v_z(z')}\left[i\omega - v_z(z')\frac{\partial}{\partial z'}\right]\delta\hat{A}(z'), \tag{4.136}$$

and

$$D_z \equiv \frac{1}{p}\int_0^z dz' \frac{\exp[i\omega\tau(z,z')]}{v_z(z')}\left[-cp_z(z')\frac{\partial}{\partial z'}\delta\hat{\varphi}(z') + i\omega p_x(z')\delta\hat{A}(z')\right], \tag{4.137}$$

$\tau(z,z') \equiv \tau(z) - \tau(z')$, and $\tau(z)$ is the Lagrangian time-variable defined in equation (4.27). Observe that D_x may be integrated directly to give

$$D_x = -\delta\hat{A}(z) + \delta\hat{A}(0)\exp[i\omega\tau(z,0)]. \tag{4.138}$$

The source current

$$\delta\boldsymbol{J}(z,t) = [\delta\hat{J}_x(z)\hat{e}_x + \delta\hat{J}_z(z)\hat{e}_z]\exp(-i\omega t) + \text{c.c.}, \tag{4.139}$$

is determined by integration over the perturbed distribution

$$\delta\hat{\mathbf{J}}(z) = -\frac{e}{m_e}\int dP_x dP_y dp \frac{p}{\gamma}\delta\hat{f}_b\left(\frac{p_x}{p_z}\hat{e}_x + \hat{e}_z\right). \tag{4.140}$$

Substitution of the perturbed distribution (4.135) yields, after integration by parts over P_x,

$$\delta\hat{J}_x(z) = \frac{e}{m_e}\int dP_x dP_y dp \frac{p}{\gamma}\left[\frac{\partial}{\partial P_x}\left(\frac{p_x}{p_z}D_x\right) - \frac{p_x}{p_z}D_z\frac{\partial}{\partial p}\right]F_b(P_x, P_y, p), \tag{4.141}$$

and

$$\delta\hat{J}_z(z) = \frac{e}{m_e}\int dP_x dP_y dp \frac{p}{\gamma}\left[\frac{\partial}{\partial P_x}D_x - D_z\frac{\partial}{\partial p}\right]F_b(P_x, P_y, p). \tag{4.142}$$

The determination of the linear-dispersion equation requires the evaluation of the source currents for representative distribution functions. The orbit integrals D_x and D_z which appear in the source currents represent an integration over the unperturbed electron trajectories in the planar wiggler. The characteristic trajectories in a planar wiggler differ from those in a helical wiggler in that the magnitudes of the axial and transverse velocity components are not constant but, rather, oscillate at harmonics of the wiggler period. This, in turn, introduces harmonic components into the dynamics of the interaction. In particular, we observe that in a planar wiggler the Lagrangian time-variable characteristic of the electron trajectories is of the approximate form

$$\tau(z) \cong t_0 + \frac{z}{V_\parallel} + \frac{P_x}{P_\parallel}\frac{v_w}{V_\parallel}\frac{1}{k_w V_\parallel}\sin k_w z + \frac{v_w^2}{8V_\parallel^2}\frac{1}{k_w V_\parallel}\sin 2k_w z, \tag{4.143}$$

where it is assumed that both $v_w < V_\parallel$ and $P_x < P_\parallel$. Observe that existence of a nonvanishing canonical momentum introduces an oscillation at the wiggler period into the trajectory.

The low-gain regime

The analysis of the low-gain regime for planar-wiggler configurations is analogous to that for the helical-wiggler system. We assume that the beam is sufficiently diffuse that space-charge forces are negligible, and focus attention on the linearly polarized electromagnetic wave which can be represented in the form

$$\delta\hat{A}(z) = \sum_{n=-\infty}^{\infty}\delta\hat{A}_n(0)\exp\left(i\int_0^z dz' k_n(z')\right), \tag{4.144}$$

where $k_n(z) \equiv k(z) + nk_w$, and we assume that $|\text{Im}\, k(z)| \ll |\text{Re}\, k(z)|$. We impose the additional assumption that $k(z) = k_0 + \delta k(z)$, where k_0 is real and independent of z and $|\delta k(z)| \ll k_0$. In the tenuous beam limit the dispersion of the wave is given approximately by the free-space value

$$\frac{\omega^2}{c^2} - k_{0_n}^2 \cong 0, \tag{4.145}$$

where $k_{0n} \equiv k_0 + nk_w$, and the gain is determined by the imaginary part of $\delta k(z)$. Gain is driven by the x-component of the source current, and from Maxwell's equations we obtain to lowest order in $\delta k(z)$ that

$$\sum_{n=-\infty}^{\infty} k_{0n}\delta k(z)\delta \hat{A}_n(z) \cong \frac{4\pi}{c}\delta \hat{J}_x(z). \tag{4.146}$$

The imaginary part of the x-component of the source current is determined primarily by the term in D_z, which can be written approximately as

$$D_z \cong \frac{ie}{2cp}\omega\gamma m_e \sum_{n=-\infty}^{\infty} \delta \hat{A}_n(z)\exp[-ik_{0n}z' + i\omega\tau(z)]$$

$$\times \int_0^z dz' \frac{v_x(z')}{v_z(z')}\exp[ik_{0n}z' - i\omega\tau(z')]. \tag{4.147}$$

Under the assumption that P_x and $P_y < |\gamma m_e v_w|$, the orbit integral in equation (4.147) can be reduced to

$$\int_0^z dz' \frac{v_x(z')}{v_z(z')}\exp[ik_{0n}z' - i\omega\tau(z')] \cong i\frac{v_w}{V_\parallel}\exp(-i\omega t_0)$$

$$\times \sum_{l,m=-\infty}^{\infty} (-1)^{l+m}\frac{l}{b_1}J_l(b_1)J_m(b_2)\frac{\exp(-i\Delta k_{n+l+2m}z) - 1}{\Delta k_{n+l+2m}}, \tag{4.148}$$

where

$$\Delta k_n \equiv \frac{\omega}{V_\parallel} - k_{0n}, \tag{4.149}$$

$$b_1 \equiv \frac{\omega}{k_w}\frac{v_w}{V_\parallel}\frac{P_x}{V_\parallel P_\parallel}, \tag{4.150}$$

$$b_2 \equiv \frac{\omega}{k_w V_\parallel}\frac{v_w^2}{8V_\parallel^2}, \tag{4.151}$$

and we have made use of the Bessel function identity

$$\exp(ib\sin\theta) = \sum_{n=-\infty}^{\infty} J_n(b)\exp(in\theta). \tag{4.152}$$

As a consequence,

$$\frac{p_x}{p_\parallel}D_z \cong -\frac{e\omega}{2cp}\gamma m_e \frac{v_w^2}{4V_\parallel^2}\sum_{n,l,m=-\infty}^{\infty} \delta \hat{A}_n(z)J_l^2(b_1)K_m^{(\pm)}(b_2)$$

$$\times \frac{1 - \exp(i\Delta k_{n+l+2m\pm 1}z)}{\Delta k_{n+l+2m\pm 1}}, \tag{4.153}$$

where

$$K_m^{(\pm)}(b_2) \equiv [J_m(b_2) - J_{m\pm1}(b_2)]^2, \tag{4.154}$$

and we have made use of the Bessel function recursion relation

$$\frac{2n}{x} J_n(x) \equiv J_{n-1}(x) + J_{n+1}(x). \tag{4.155}$$

Observe that this expression for D_z contains harmonic contributions to the interaction. The presence of a nonvanishing canonical momentum is related to a beam pitch-angle spread, and results in resonant interactions at all harmonics (i.e. $\omega \approx (k_0 + lk_w) V_\parallel$, for integer l). Harmonics arise in the absence of a pitch-angle spread due to the oscillation of the axial velocity due to the planar geometry. In this case, odd harmonics are excited in which $\omega \approx [k_0 + (1 + 2m)k_w] V_\parallel$.

The gain in power over an interaction length L is given in equation (4.56) which, using equation (4.152), can be expressed as

$$G(L) \cong \frac{\omega_b^2 L^2}{16c} \sum_{l,m=-\infty}^{\infty} \iint dP_x dP_y G_\perp(P_x, P_y) \int_0^\infty dp \frac{p}{\gamma p_z} G_b(p)$$
$$\times \frac{v_w^2}{\gamma_\parallel^2 V_\parallel^4} J_l^2(b_1) K_m^{(\pm)}(b_2) F(\Theta_{l+2m\pm1}), \tag{4.156}$$

where we have assumed a distribution function of the form

$$F_b(P_x, P_y, p) = n_b G_\perp(P_x, P_y) G_b(p), \tag{4.157}$$

for an ambient density n_b, $\omega_b^2 \equiv 4\pi e^2 n_b/m_e$, $\gamma_\parallel \equiv (1 - V_\parallel^2/c^2)^{-1/2}$, and $F(\Theta_{l+2m\pm1})$ is the spectral function defined in equation (4.60) for

$$\Theta_n \equiv \frac{L}{2}\left(\frac{\omega}{V_\parallel} - k_0 - nk_w\right). \tag{4.158}$$

Equation (4.156) describes the small-signal gain corresponding to two Doppler-shifted frequencies, i.e. $\omega \approx [k_0 + (l + 2m + 1)k_w] V_\parallel$ and $\omega \approx [k_0 + (l + 2m - 1)k_w] V_\parallel$. The former case corresponds to the frequency upshift associated with the slowly varying ponderomotive potential, while the latter case describes a frequency downshift. In order to understand the underlying physical mechanism for these two resonance conditions, recall that the free-electron laser operates by means of a beating between the wiggler and radiation fields. The upper beat wave is associated with the slowly varying ponderomotive potential and a frequency upshift. In contrast, the lower beat wave is associated with a frequency downshift. In particular, for a wave satisfying the upshifted resonance condition, the lower beat wave describes a rapid oscillation with a period half that of the wiggler. Since this rapid oscillation is nonresonant it will produce no net gain of the signal and will be neglected, henceforth, in the linear formulation. However, we shall return to this issue within the context of the nonlinear formulation.

Restricting the discussion to the resonant frequency upshift, we consider the small-signal gain in the cold beam limit for which $G_\perp(P_x, P_y) = \delta(P_x)\delta(P_y)$, and

$G_b(p)$ is given in equation (4.58). In this regime $b_1 = 0$ and only the $l = 0$ term survives. Therefore, the gain for the fundamental and each even harmonic contribution in the cold regime is

$$G(L) \cong \frac{\omega_b^2 L^2 k_0}{16\gamma_0 \gamma_\parallel^2 v_\parallel^2} \beta_w^2 K_m^{(+)}(b_2) F(\Theta_{2m+1}), \tag{4.159}$$

where $v_\parallel$ is the bulk axial velocity defined in equation (2.105), $\gamma_0 \equiv (1 + p_0^2/m_e^2c^2)^{1/2}$, $\beta_w \equiv |v_w/v_\parallel|$, and

$$b_2 \cong \frac{a_w^2(1+2m)}{2(2+a_w^2)}, \tag{4.160}$$

for $a_w \equiv eA_w/m_ec^2$. The small-signal gain for the planar wiggler configuration, therefore, is similar to that found for a helical wiggler (4.59). There are, however, three distinctions which merit discussion. In the first place, the magnitudes of the transverse and axial components of the velocity are oscillatory for planar wigglers; hence, the average value of the transverse velocity is reduced by a factor of $\sqrt{2}$ with respect to that found for a helical wiggler of comparable magnitude. This implies, in turn, that for a fixed magnitude of the wiggler field, the gain is reduced by half with respect to that of a helical wiggler. In the second place, since the axial velocity is also oscillatory, the electrons will maintain no fixed resonance with the wave. Hence, the gain will be modulated, and the average gain will be reduced. This effect is described by the Bessel function factor K_m. In the third and last place, the oscillatory nature of the axial velocity permits resonant interactions to occur at the odd harmonics. In contrast, harmonic interactions do not appear within the context of the idealized one-dimensional formulation of the interaction for a helical wiggler since the axial velocity is constant in the limit in which the canonical momenta vanish. This last distinction, however, is somewhat arbitrary since harmonic radiation does appear for a helical wiggler in the three-dimensional formulation, and will be discussed for both linear and nonlinear formulations.

The high-gain regime

The analysis for the high-gain regime follows that discussed in reference [23]. We express the vector and scalar potentials by application of Floquet's theorem for periodic systems in the form

$$\delta\hat{A}(z) = \sum_{n=-\infty}^{\infty} \delta\hat{A}_n \exp(ik_n z), \tag{4.161}$$

and

$$\delta\hat{\varphi}(z) = \sum_{n=-\infty}^{\infty} \delta\hat{\varphi}_n \exp(ik_n z), \tag{4.162}$$

where $k_n = k + nk_w$. We first impose the condition that the pitch-angle spread of the beam vanishes, and assume a distribution of the form given in equation (4.45). Since the gain is exponential in this regime, we may neglect the initial value contributions to the orbit integrals. As a consequence, the source current can be expressed as

$$\delta\hat{J}_x(z) = -\frac{\omega_b^2}{8\pi c}\sum_{n=-\infty}^{\infty}\delta\hat{A}_n\exp(ik_nz)\int_0^\infty dp\frac{p}{\gamma p_z}G_b(p)\left(1+\frac{p_x^2}{p_z^2}\right)$$
$$+\frac{\omega_b^2}{32\pi c}\sum_{n,m=-\infty}^{\infty}\delta\hat{A}_n\exp(ik_nz)\int_0^\infty dp K_m^{(\pm)}(b_2)\frac{\omega m_e\beta_w^2 v_\parallel}{\omega-k_{n+2m\pm1}v_\parallel}\frac{\partial}{\partial p}G_b(p)$$
$$-\frac{\omega_b^2}{16\pi}\sum_{n,m=-\infty}^{\infty}\delta\hat{\varphi}_n\exp(ik_{n\mp1}z)\int_0^\infty dp J_m(b_2)[J_m(b_2)-J_{m\pm1}(b_2)]$$
$$\times\frac{m_e k_n v_w}{\omega-k_{n+2m}v_\parallel}\frac{\partial}{\partial p}G_b(p), \tag{4.163}$$

and

$$\delta\hat{J}_z(z) = -\frac{\omega_b^2}{8\pi}\sum_{n,m=-\infty}^{\infty}\delta\hat{\varphi}_n\exp(ik_nz)\int_0^\infty dp\, J_m^2(b_2)\frac{m_e k_n v_\parallel}{\omega-k_{n+2m}v_\parallel}\frac{\partial}{\partial p}G_b(p)$$
$$+\frac{\omega_b^2}{16\pi c}\sum_{n,m=-\infty}^{\infty}\delta\hat{A}_{n\mp1}\exp(ik_nz)\int_0^\infty dp J_m(b_2)[J_m(b_2)-J_{m\pm1}(b_2)]$$
$$\times\frac{m_e\omega v_w}{\omega-k_{n+2m}v_\parallel}\frac{\partial}{\partial p}G_b(p), \tag{4.164}$$

where $K_m^{(\pm)}$ and b_2 are defined in equations (4.151) and (4.154).

We now assume a cold beam distribution as in equation (4.58) and integrate by parts in p. Retaining only the terms which correspond to the Doppler upshift in frequency, we find that the source currents may be approximated by

$$\delta\hat{J}_x(z) \cong -\frac{\omega_b^2}{8\pi\gamma_0 c}\sum_{n=-\infty}^{\infty}\delta\hat{A}_n\exp(ik_nz)\left[1-\frac{v_w^2}{4c^2}\sum_{m=-\infty}^{\infty}K_m^{(+)}(b_2)\frac{\omega^2-c^2k_{n+2m+1}^2}{(\omega-k_{n+2m+1}v_\parallel)^2}\right]$$
$$+\frac{\omega_b^2}{16\pi\gamma_0}\sum_{n,m=-\infty}^{\infty}k_n\delta\hat{\varphi}_n\exp(ik_{n-1}z)J_m(b_2)[J_m(b_2)-J_{m+1}(b_2)]$$
$$\times\frac{v_w(k_{n+2m}-\omega v_\parallel/c^2)}{(\omega-k_{n+2m}v_\parallel)^2}, \tag{4.165}$$

and

$$\delta\hat{J}_z \cong \frac{\omega_b^2}{8\pi\gamma_0\gamma_\parallel^2}\sum_{n,m=-\infty}^{\infty}k_n\delta\hat{\varphi}_n\exp(ik_nz)J_m^2(b_2)\frac{\omega}{(\omega-k_{n+2m}v_\parallel)^2}$$
$$-\frac{\omega_b^2}{16\pi\gamma_0}\frac{\omega}{c}\sum_{n,m=-\infty}^{\infty}\delta\hat{A}_{n-1}\exp(ik_nz)J_m(b_2)[J_m(b_2)$$
$$-J_{m+1}(b_2)]\frac{v_w(k_{n+2m}-\omega v_\parallel/c^2)}{(\omega-k_{n+2m}v_\parallel)^2}. \tag{4.166}$$

Note that in the derivation of equation (4.164) we have made use of the identity

$$\sum_{m=-\infty}^{\infty} [J_m(x) - J_{m\pm 1}(x)]^2 = 2. \tag{4.167}$$

Substitution of these forms for the source currents into Maxwell's equations yields the following set of coupled equations for the vector and scalar potentials

$$\left[(\omega - k_{n+2m}v_\parallel)^2 - \frac{\omega_b^2}{\gamma_0\gamma_\parallel^2}\right]\delta\hat{\varphi}_n J_m(b_2)$$
$$\cong -\frac{\omega_b^2}{\gamma_0 k_n}\frac{v_w}{2c}\delta\hat{A}_{n-1}\left(k_{n+2m} - \omega\frac{v_\parallel}{c^2}\right)[J_m(b_2) - J_{m+1}(b_2)], \tag{4.168}$$

and

$$\left[\omega^2 - c^2k_{n-1}^2 - \frac{\omega_b^2}{\gamma_0}\left(1 - \frac{v_w^2}{4c^2}K_m^{(+)}(b_2)\frac{\omega^2 - c^2k_{n+2m}^2}{(\omega - k_{n+2m}v_\parallel)^2}\right)\right]\delta\hat{A}_{n-1}$$
$$\cong -\frac{\omega_b^2}{2\gamma_0}\frac{k_n v_w}{(\omega - k_{n+2m}v_\parallel)^2}\delta\hat{\varphi}_n\left(k_{n+2m} - \omega\frac{v_\parallel}{c^2}\right)J_m(b_2)[J_m(b_2) - J_{m+1}(b_2)], \tag{4.169}$$

where we have focused on a single harmonic contribution. The dispersion equation is obtained by the requirement that the determinant of the coefficients vanishes, which gives

$$\left[(\omega - k_{n+2m}v_\parallel)^2 - \frac{\omega_b^2}{\gamma_0\gamma_\parallel^2}\right]\left(\omega^2 - c^2k_{n-1}^2 - \frac{\omega_b^2}{\gamma_0}\right)$$
$$\cong -\frac{v_w^2}{4c^2}\frac{\omega_b^2}{\gamma_0}K_m^{(+)}(b_2)\left(\omega^2 - c^2k_{n+2m}^2 - \frac{\omega_b^2}{\gamma_0}\right). \tag{4.170}$$

This dispersion equation is similar to that found for the helical wiggler configuration in equation (4.83). The differences associated with the planar wiggler are the same as those found for the low-gain regime; specifically, (1) the wiggler field is replaced by the root-mean-square amplitude (i.e. β_w is replaced by $\beta_w/\sqrt{2}$), (2) the $K_m^\pm$ factor describes the effect of the oscillatory axial velocity, and (3) the amplification of signals at harmonic frequencies is present in the idealized one-dimensional formulation.

Thermal effects on the instability

In the analysis of the thermal regime in planar wiggler configurations, we employ the same distribution (4.113) used in the treatment of thermal effects in helical wiggler geometries, which describes a monoenergetic beam with a pitch-angle spread. The dominant contribution of the axial thermal spread occurs within the resonance condition; hence, if we restrict the analysis to the resonance associated with the Doppler upshift in frequency then the source currents can be expressed in the form

$$\delta\hat{J}_x(z) \cong -\frac{\omega_b^2}{8\pi\gamma_0 c}\sum_{n=-\infty}^{\infty}\delta\hat{A}_n\exp(ik_n z)\left[1-\frac{v_w^2}{4c^2}\sum_{l,m=-\infty}^{\infty}K_m^{(+)}(b_2)\right.$$
$$\left.\times\iint dP_x dP_y G_\perp(P_x,P_y)J_l^2(b_1)\frac{\omega^2-c^2k_{n+l+2m+1}^2}{(\omega-k_{n+l+2m+1}V_\parallel)^2}\right]$$
$$+\frac{\omega_b^2}{16\pi\gamma_0}\sum_{n,l,m=-\infty}^{\infty}k_n\delta\hat{\varphi}_n\exp(ik_{n-1}z)J_m(b_2)[J_m(b_2)-J_{m+1}(b_2)]$$
$$\times\iint dP_x dP_y G_\perp(P_x,P_y)J_l^2(b_1)\frac{v_w\left(k_{n+l+2m}-\omega\dfrac{v_\parallel}{c^2}\right)}{(\omega-k_{n+l+2m}V_\parallel)^2}, \tag{4.171}$$

and

$$\delta\hat{J}_z \cong \frac{\omega_b^2}{8\pi\gamma_0\gamma_\parallel^2}\sum_{n,m=-\infty}^{\infty}k_n\delta\hat{\varphi}_n\exp(ik_n z)J_m^2(b_2)$$
$$\times\iint dP_x dP_y G_\perp(P_x,P_y)J_l^2(b_1)\frac{\omega}{(\omega-k_{n+l+2m}V_\parallel)^2}$$
$$-\frac{\omega_b^2}{16\pi\gamma_0}\frac{\omega}{c}\sum_{n,l,m=-\infty}^{\infty}\delta\hat{A}_{n-1}\exp(ik_n z)J_m(b_2)[J_m(b_2)-J_{m+1}(b_2)]$$
$$\times\iint dP_x dP_y G_\perp(P_x,P_y)J_l^2(b_1)\frac{v_w\left(k_{n+l+2m}-\omega\dfrac{v_\parallel}{c^2}\right)}{(\omega-k_{n+l+2m}V_\parallel)^2}, \tag{4.172}$$

where $V_\parallel$ and $v_\parallel$ are the bulk axial velocities with (4.132) and without (2.105), the pitch-angle spread, respectively. The integrals over the canonical momenta in equations (4.171) and (4.172) may be evaluated using the transverse distribution function (4.114), and we find that in the limit in which $P_\perp \ll p_\parallel$

$$\iint dP_x dP_y G_\perp(P_x,P_y)\frac{J_l^2(b_1)}{(\omega-kV_\parallel)^2} \cong \frac{T_l(\zeta)}{(\omega-kv_\parallel)^2}, \tag{4.173}$$

where

$$T_l(\zeta) \equiv \frac{\zeta^2}{2\pi}\int_0^{2\pi}d\phi\int_0^{\infty}dz\exp(-z)\frac{J_l^2(b_1)}{(z+\zeta)^2}, \tag{4.174}$$

$$\zeta \equiv \frac{\gamma_0^2 m_e^2}{\Delta P^2}\left(\frac{\omega^2}{k^2}-v_\parallel^2\right), \tag{4.175}$$

and we may write that

$$b_1 = \frac{\omega}{k_w v_\parallel}\frac{v_w}{v_\parallel}\frac{\Delta P}{p_\parallel}z^{1/2}\cos\phi. \tag{4.176}$$

If we now select a specific harmonic (i.e. for fixed l and m), then the coupled mode equations may be written as

$$\left[(\omega - k_{n+l+2m}v_{\|})^2 - \frac{\omega_b^2}{\gamma_0\gamma_{\|}^2} T_l(\zeta_{n+l+2m})\right]\delta\hat{\varphi}_n J_m(b_2)$$
$$\cong -\frac{\omega_b^2}{\gamma_0 k_n}\frac{v_w}{2c}\delta\hat{A}_{n-1}\, T_l(\zeta_{n+l+2m})\left(k_{n+l+2m} - \omega\frac{v_{\|}}{c^2}\right)[J_m(b_2) - J_{m+1}(b_2)], \quad (4.177)$$

and

$$\left\{\omega^2 - c^2k_{n-1}^2 - \frac{\omega_b^2}{\gamma_0}\left[1 - \frac{v_w^2}{4c^2}K_m^{(+)}(b_2)\frac{\omega^2 - k_{n+l+2m}^2c^2}{(\omega - k_{n+l+2m}v_{\|})^2}T_l(\zeta_{n+l+2m})\right]\right\}\delta\hat{A}_{n-1}$$
$$\cong -\frac{\omega_b^2}{2\gamma_0(\omega - k_{n+l+2m}v_{\|})^2}\frac{k_n v_w}{}\delta\hat{\varphi}_n\, T_l(\zeta_{n+l+2m})$$
$$\times\left(k_{n+l+2m} - \omega\frac{v_{\|}}{c^2}\right)J_m(b_2)[J_m(b_2) - J_{m+1}(b_2)], \quad (4.178)$$

where

$$\zeta_{n+l+2m} \equiv \frac{\gamma_0^2 m_e^2}{\Delta P^2}\left(\frac{\omega^2}{k_{n+l+2m}^2} - v_{\|}^2\right). \quad (4.179)$$

The dispersion equation is found by requiring that the determinant of the coefficients vanishes. Therefore, for a specific choice of harmonic interaction, the dispersion equation which results is expressed as a straightforward generalization of that found in the cold beam limit (4.170); specifically,

$$\left[(\omega - k_{n+l+2m}v_{\|})^2 - \frac{\omega_b^2}{\gamma_0\gamma_{\|}^2}T_l(\zeta_{n+l+2m})\right]\left(\omega^2 - c^2k_{n-1}^2 - \frac{\omega_b^2}{\gamma_0}\right)$$
$$\cong -\frac{v_w^2}{4c^2}\frac{\omega_b^2}{\gamma_0}K_m^{(+)}(b_2)\,T_l(\zeta_{n+l+2m})\left(\omega^2 - c^2k_{n+l+2m}^2 - \frac{\omega_b^2}{\gamma_0}T_l(\zeta_{n+l+2m})\right). \quad (4.180)$$

This dispersion equation, which includes the effect of an axial energy spread for a planar configuration, is similar to that found for the corresponding case for a helical wiggler geometry (4.128). The differences are as stated previously in that (1) the wiggler amplitude is replaced by the root-mean-square value, (2) the oscillation in the axial velocity introduces modifications in $K_m^{(\pm)}(b_2)$ and $J_l^2(b_1)$, and (3) harmonic amplification is found in the one-dimensional formulation. The effect of the pitch-angle spread on the axial velocity is the source of the lth harmonic contribution, which has the effect of modifying the thermal function T_l. In order to describe this effect in more detail, we assume that $b_1 \ll 1$, which is valid as long as $P_\perp \ll p_{\|}$. As a result, we expand

$$J_l(b_1) \cong \frac{1}{(l!)^2}\left(\frac{v_w}{2v_{\|}}\right)^{2l}\left(\frac{\omega}{k_w v_{\|}}\right)^{2l}\left(\frac{\Delta P}{p_{\|}}\right)^{2l} z^l \cos^{2l}\phi. \quad (4.181)$$

As a consequence,

$$T_l(\zeta) \cong \frac{(2l)!}{(l!)^4}\left(\frac{v_w}{2v_\parallel}\right)^{2l}\left(\frac{\omega}{k_w v_\parallel}\right)^{2l}\left(\frac{\Delta P}{p_\parallel}\right)^{2l}\zeta^2 \exp(\zeta)\int_\zeta^\infty dt\frac{\exp(-t)}{t^2}(t-\zeta)^l. \quad (4.182)$$

As in the case of the helical wiggler, the thermal function may be expressed in terms of the exponential integral function (4.127). To this end we observe that

$$\zeta^2 \exp(\zeta)\int_\zeta^\infty dt\frac{\exp(-t)}{t^2}(t-\zeta)^l = (-1)^l\zeta^{l+1}[1-(\zeta+l)\exp(\zeta)E_1(\zeta)] + U_l(\zeta), \quad (4.183)$$

where

$$U_l(\zeta) \equiv \begin{cases} 0; & l<2 \\ \displaystyle\sum_{k=2}^{l}\sum_{n=0}^{k-2}\frac{(-1)^{k-n}l!\,n!\,\zeta^{l-n}}{k!(l-k)!(k-2-n)!}; & l \geqslant 2 \end{cases}. \quad (4.184)$$

As a consequence,

$$T_l(\zeta) \cong \frac{(2l)!}{(l!)^4}\left(\frac{v_w}{2v_\parallel}\right)^{2l}\left(\frac{\omega}{k_w v_\parallel}\right)^{2l}\left(\frac{\Delta P}{p_\parallel}\right)^{2l}\{(-1)^l\zeta^{l+1} \times [1-(\zeta+l)\exp(\zeta)E_1(\zeta)] + U_l(\zeta)\}. \quad (4.185)$$

We observe that for $l=0$ the planar wiggler thermal function reduces to that found for the helical wiggler (4.125), i.e.

$$T_0(\zeta) \cong \zeta[1-\zeta\exp(\zeta)E_1(\zeta)]. \quad (4.186)$$

This will reproduce a thermal response for the interaction at the fundamental which is similar to that found for the helical geometry. The interaction at harmonics will be discussed in Chapter 7, in which it will be shown that the harmonic response is more sensitive to thermal effects.

The effect of an axial magnetic field

It has been shown in Chapter 2 that the single-particle orbits in a combined planar wiggler and axial guide field configuration exhibit a drift motion which can result in the loss of the electron beam to the walls of the drift tube. As a consequence, great care must be exercised in the design of such a system to ensure that substantial beam loss does not occur over the interaction length of the device. Under the assumption, therefore, that substantial beam loss does not occur, the presence of the axial magnetic field can substantially modify many of the characteristics of the interaction [31]. In the first place, the electrons execute elliptic trajectories in the transverse plane (as opposed to the circular trajectories in helical wiggler geometries) that result in the excitation of elliptically polarized electromagnetic waves. In the second place, the enhancements in the gain which are present in the combined helical wiggler and axial guide field configurations when the Larmor period and wiggler periods are comparable are also present for the planar wiggler systems.

In order to treat the essential features of this configuration, it is instructive to simplify the analysis to the cold beam regime in which $v_w \ll v_\parallel$ and to neglect the oscillatory components of the axial velocity. The source currents and charge density can be obtained from a fluid analysis of the interaction. The fundamental equations describing the beam are those of continuity

$$\frac{\partial}{\partial t} n + \nabla \cdot (n\mathbf{v}) = 0. \tag{4.187}$$

momentum transfer

$$\frac{d}{dt}\mathbf{v} = -\frac{e}{\gamma m_e}\left[\left(\boldsymbol{I} - \frac{1}{c^2}\mathbf{v}\mathbf{v}\right)\cdot\delta\boldsymbol{E} + \frac{1}{c}\mathbf{v}\times(\boldsymbol{B}_0 + \boldsymbol{B}_w + \delta\boldsymbol{B})\right], \tag{4.188}$$

and energy balance

$$\frac{d}{dt}\gamma = -\frac{e}{m_e c^2}\mathbf{v}\cdot\delta\boldsymbol{E}, \tag{4.189}$$

where n and $\mathbf{v}$ are the macroscopic density and velocity of the beam, γ is the relativistic factor corresponding to the bulk energy of the beam, $\boldsymbol{I}$ is the unit dyadic, $\boldsymbol{B}_0$ is the axial solenoidal field, $\boldsymbol{B}_w$ is the idealized planar wiggler representation (2.99), and $\delta\boldsymbol{E}$ and $\delta\boldsymbol{B}$ are the fluctuating electric and magnetic fields.

These equations are solved by means of a perturbation expansion to first order in the electromagnetic fields. To this end, we write $n = n_b + \delta n$, $\gamma = \gamma_0 + \delta\gamma$, and

$$\mathbf{v} = \alpha_x \hat{\boldsymbol{e}}_x \cos k_w z + \alpha_y \hat{\boldsymbol{e}}_y \sin k_w z + v_\parallel \hat{\boldsymbol{e}}_z + \delta\mathbf{v}, \tag{4.190}$$

where n_b describes the ambient beam density, $\alpha_{x,y}$ are defined in equations (2.112) and (2.113), and $v_\parallel$ denotes the unperturbed bulk axial velocity of the beam. Observe that the unperturbed velocity is described by the single-particle electron trajectory in the combined magnetostatic field. We choose to express the electromagnetic field in terms of the vector and scalar potentials (in the Coulomb gauge), and employ Floquet's theorem for periodic systems to write

$$\delta\hat{\boldsymbol{A}}(z) = \sum_{n=-\infty}^{\infty} \delta\hat{\boldsymbol{A}}_n \exp(ik_n z), \tag{4.191}$$

for the vector potential, where $k_n = k + nk_w$. The scalar potential is written as in equation (4.162). Observe that in the Coulomb gauge the z-component of the vector potential must vanish.

Solution of the perturbed fluid equations shows that the density perturbation is given by

$$\delta\hat{n}_m = n_b \frac{k_m}{\Delta\omega_m}\delta\hat{v}_{z,m}, \tag{4.192}$$

where $\Delta\omega_m \equiv \omega - k_m v_\parallel$. The perturbed velocity, in turn, is given approximately by

$$(\Delta\omega_m^2 - \Omega_0^2)\delta\hat{v}_{x,m\pm1} \cong \frac{e}{2\gamma_0\gamma_\parallel^2 m_e}\frac{k_m}{v_\parallel \Delta\omega_m}\delta\hat{\varphi}_m$$
$$\times [\alpha_x\beta_\parallel^2\gamma_\parallel^2\Delta\omega_{m\pm1}^2 \mp \alpha_y\Omega_0(\Phi_p\Delta\omega_{m\pm2} + \beta_\parallel^2\gamma_\parallel^2\Delta\omega_{m\pm1})]$$
$$+\frac{e}{\gamma_0 m_e c}\delta\hat{A}_{x,m\pm1}\left\{\Delta\omega_{m\pm1}^2 - \frac{\alpha_x^2}{2c^2}\frac{\omega\Delta\omega_{m\pm1}(\Delta\omega_{m\pm1}^2 + \Omega_0^2)}{\Delta\omega_m\Delta\omega_{m\pm2}}\right.$$
$$\left.\pm\frac{\alpha_y}{4c}\Omega_0\frac{\Delta\omega_{m\pm2}}{\Delta\omega_m}\left[\alpha_x\left(k_{m\pm1} - \omega\frac{v_\parallel}{c^2}\right) - \Omega_w\Delta_m U_m^{(\pm)}\right]\right\}$$
$$+\frac{ie}{\gamma_0 m_e c}\Omega_0\delta\hat{A}_{y,m\pm1}\left\{\Delta\omega_{m\pm1}\left(1 - \frac{\alpha_x^2 + \alpha_y^2}{2c^2}\frac{\omega\,\Delta\omega_{m\pm1}}{\Delta\omega_m\Delta\omega_{m\pm2}}\right)\right.$$
$$\left.-\frac{\alpha_y}{4c}\frac{\Delta\omega_{m\pm2}}{\Delta\omega_m}\left[\alpha_y\left(k_{m\pm1} - \omega\frac{v_\parallel}{c^2}\right) \mp \Omega_w\Delta_m V_m^{(\pm)}\right]\right\}, \tag{4.193}$$

$$(\Delta\omega_m^2 - \Omega_0^2)\delta\hat{v}_{y,m\pm1} \cong \mp\frac{ie}{2\gamma_0\gamma_\parallel^2 m_e}\frac{k_m}{v_\parallel\Delta\omega_m}\delta\hat{\varphi}_m$$
$$\times [\alpha_y\beta_\parallel^2\gamma_\parallel^2\Delta\omega_{m\pm1}\Delta\omega_{m\pm2} \mp \Phi_p\Omega_0(\alpha_x\Delta\omega_{m\pm1} \mp \alpha_y\Omega_0)]$$
$$+\frac{ie}{\gamma_0 m_e c}\Omega_0\delta\hat{A}_{x,m\pm1}\left\{\Delta\omega_{m\pm1}\left(1 - \frac{\alpha_x^2}{c^2}\frac{\omega\Delta\omega_{m\pm1}}{\Delta\omega_m\Delta\omega_{m\pm2}}\right)\right.$$
$$\left.-\frac{(\alpha_x\Delta\omega_{m\pm1} \mp \alpha_y\Omega_0)}{4v_\parallel\Delta\omega_m}\left[\alpha_x\left(k_{m\pm1} - \omega\frac{v_\parallel}{c^2}\right) - \Omega_w\Delta_m U_m^{(\pm)}\right]\right\}$$
$$+\frac{e}{\gamma_0 m_e c}\delta\hat{A}_{y,m\pm1}\left\{\Delta\omega_{m\pm1}^2 - \frac{\alpha_y^2}{2c^2}\frac{\omega\Delta\omega_{m\pm1}(\Delta\omega_{m\pm1}^2 + k_w^2 v_\parallel^2)}{\Delta\omega_m\Delta\omega_{m\pm2}}\right.$$
$$\left.\pm\frac{\Omega_0(\alpha_x\Delta\omega_{m\pm1} \mp \alpha_y\Omega_0)}{4v_\parallel\Delta\omega_m}\left[\alpha_y\left(k_{m\pm1} - \omega\frac{v_\parallel}{c^2}\right) \mp \Omega_w\Delta_m V_m^{(\pm)}\right]\right\}, \tag{4.194}$$

and

$$\Delta\omega_m\delta\hat{v}_{z,m} \cong -\frac{e}{\gamma_0\gamma_\parallel^2 m_e}k_m\Phi_p\delta\hat{\varphi}_m + \frac{e}{2\gamma_0 m_e c}\delta\hat{A}_{x,m+1}[\alpha_x(ck_{m+1} - \omega\beta_\parallel) - \Omega_w\Delta_m U_m^{(+)}]$$
$$+\frac{e}{2\gamma_0 m_e c}\delta\hat{A}_{x,m-1}[\alpha_x(ck_{m-1} - \omega\beta_\parallel) - \Omega_w\Delta_m U_m^{(-)}]$$
$$+\frac{ie}{2\gamma_0 m_e c}\delta\hat{A}_{y,m+1}[\alpha_y(ck_{m+1} - \omega\beta_\parallel) - \Omega_w\Delta_m V_m^{(+)}]$$
$$-\frac{ie}{2\gamma_0 m_e c}\delta\hat{A}_{y,m-1}[\alpha_y(ck_{m-1} - \omega\beta_\parallel) - \Omega_w\Delta_m V_m^{(-)}], \tag{4.195}$$

where Φ_p is given in equation (2.117), and $\Omega_{0,w} \equiv |eB_{0,w}/\gamma_0 m_e c|$. The coefficients which appear in these expressions for the perturbed velocity are defined as

$$U_m^{(\pm)} \equiv \mp \frac{\Delta\omega_{m\pm 1}}{\Delta\omega_{m\pm 1}^2 - \Omega_0^2}\left[\Delta\omega_{m\pm 1} - \frac{\alpha_x^2}{2c^2}\frac{\omega(\Delta\omega_{m\pm 1}^2 + \Omega_0^2)}{\Delta\omega_m \Delta\omega_{m\pm 2}}\right]$$
$$\mp \frac{\alpha_x^2}{4c^2}\left(\Delta\omega_{m\mp 1} \pm \frac{\alpha_y}{\alpha_x}\Omega_0\right)\frac{\omega\Delta\omega_{m\mp 1}}{\Delta\omega_m(\Delta\omega_{m\mp 1}^2 - \Omega_0^2)}$$
$$+ \frac{\alpha_x^2}{2c^2}\left(k_{m\pm 1} - \omega\frac{v_\parallel}{c^2}\right)\frac{\beta_\parallel \Omega_0^2(\Omega_0^2 + 3k_w^2 v_\parallel^2)}{(\Delta\omega_{m+1}^2 - \Omega_0^2)(\Omega\omega_{m-1}^2 - \Omega_0^2)}, \tag{4.196}$$

$$V_m^{(\pm)} \equiv \pm \frac{\Omega_0 \Delta\omega_{m\pm 1}}{\Delta\omega_{m\pm 1}^2 - \Omega_0^2}\left(1 - \frac{\alpha_x^2 + \alpha_y^2}{2c^2}\frac{\omega\Delta\omega_{m\pm 1}}{\Delta\omega_m \Delta\omega_{m\pm 2}}\right)$$
$$- \frac{\alpha_x \alpha_y}{4c^2}\left(\Delta\omega_{m\mp 1} \pm \frac{\alpha_y}{\alpha_x}\Omega_0\right)\frac{\omega\Delta\omega_{m\mp 1}}{\Delta\omega_m(\Delta\omega_{m\mp 1}^2 - \Omega_0^2)}$$
$$\pm \frac{\alpha_y^2}{2c^2}\left(k_{m\pm 1} - \omega\frac{v_\parallel}{c^2}\right)\frac{\beta_\parallel \Omega_0(\Omega_0^2 + 3k_w^2 v_\parallel^2)}{(\Delta\omega_{m+1}^2 - \Omega_0^2)(\Delta\omega_{m-1}^2 - \Omega_0^2)}, \tag{4.197}$$

and

$$\Delta_m \equiv \frac{2v_\parallel(\Delta\omega_{m+1}^2 - \Omega_0^2)(\Delta\omega_{m-1}^2 - \Omega_0^2)}{2v_\parallel(\Delta\omega_{m+1}^2 - \Omega_0^2)(\Delta\omega_{m-1}^2 - \Omega_0^2) + \alpha_y \Omega_0 \Omega_w(\Omega_0^2 + 3k_w^2 v_\parallel^2)}. \tag{4.198}$$

Observe that the y-component of the perturbed velocity vanishes in the limit in which the axial magnetic field vanishes.

The source current may be expressed in terms of the perturbed density and velocity in the following manner

$$\delta\hat{\boldsymbol{J}}_{\perp,m\pm 1} = -en_b\left[\delta\hat{\mathbf{v}}_{\perp,m\pm 1} + \frac{1}{2}(\alpha_x \hat{\boldsymbol{e}}_x + \alpha_y \hat{\boldsymbol{e}}_y)\frac{k_m}{\Delta\omega_m}\delta v_{z,m}\right]. \tag{4.199}$$

The perturbed source current and charge density are substituted into the Maxwell–Poisson equations

$$\left(\frac{\omega^2}{c^2} - k_{m\pm 1}^2\right)\delta\hat{\boldsymbol{A}}_{\perp,m\pm 1} = -\frac{4\pi}{c}\delta\hat{\boldsymbol{J}}_{\perp,m\pm 1}, \tag{4.200}$$

and

$$k_m^2 \delta\hat{\varphi}_m = 4\pi e\,\delta\hat{n}_m, \tag{4.201}$$

to obtain the dispersion equation. Since the interaction excites elliptically polarized waves, it is convenient to transform to the basis which describes right- and left-hand circularly polarized waves and we define

$$\delta\hat{A}_m^{(\pm)} \equiv \delta\hat{A}_{x,m} \mp i\delta\hat{A}_{y,m}. \tag{4.202}$$

As a consequence, the dispersion equations describing these modes corresponding to the Doppler upshifted resonance are given by

$$\left(\Delta\omega_m^2 - \frac{\omega_b^2}{\gamma_0 \gamma_\parallel^2}\Phi_p\right)\left(\omega^2 - c^2 k_{m-1}^2 - \frac{\omega_b^2 \Delta\omega_{m-1}}{\gamma_0(\Delta\omega_{m-1} \mp \Omega_0)}\right) \cong \frac{\omega_b^2 \Omega_w \Psi}{\gamma_0(\Omega_0 \mp k_w v_\parallel)^2} k_{m-1} k_w v_\parallel^2, \tag{4.203}$$

where the '$\mp$' refers to the right- and left-hand modes respectively, and

$$\Psi \equiv 1 - \frac{\alpha_y^2(\Omega_0^2 + 3k_w^2 v_\parallel^2)}{2v_\parallel^2(\Omega_0^2 - k_w^2 v_\parallel^2) + \alpha_y^2(\Omega_0^2 + 3k_w^2 v_\parallel^2)}, \tag{4.204}$$

which reduces to unity in the limit in which the axial solenoidal field vanishes. Observe that the interaction differs from that found for a helical wiggler, in which the transverse projection of the single-particle orbits is circular. In that case, the amplified wave has a right-hand circular polarization only. In contrast, the amplified wave has an elliptic polarization for a planar wiggler, and the dispersion equations describe the amplification of both right- and left-hand modes.

In the limit in which the axial solenoidal field vanishes, we expect to recover the previously derived results (4.170) for the dispersion equation at the fundamental resonance for a planar wiggler geometry. Indeed, it is clear that the dispersion equations (4.203) are identical for the right- and left-hand modes. This implies that these modes have identical growth rates and that, as expected, the aggregate wave is linearly polarized in the x-direction. The dispersion equation for this mode reduces to

$$\left(\Delta\omega_m^2 - \frac{\omega_b^2}{\gamma_0\gamma_\parallel^2}\right)\left(\omega^2 - c^2k_{m-1}^2 - \frac{\omega_b^2}{\gamma_0}\right) \cong \frac{v_w^2}{2c^2}\frac{\omega_b^2}{2\gamma_0}k_{m-1}k_w, \tag{4.205}$$

which corresponds to equation (4.170) for the fundamental resonance in the limit in which the canonical momentum vanishes and $v_w \ll c$.

4.2 LINEAR STABILITY IN THREE DIMENSIONS

In the formulation of the linear-stability theory of free-electron lasers in realizable (i.e. three-dimensional) configurations, we shall restrict the analysis to the treatment of helical wiggler/axial solenoidal magnetic field configurations as illustrated in equation (2.1), in which the electron beam propagates through a cylindrical drift tube. The analysis [34, 35] involves a perturbation expansion of the Lorentz force equations about the single-particle trajectories to first order in the electromagnetic field. To this end, we employ the nonhelical, axis-encircling trajectories described in equations (2.71–8). These orbits represent perturbations about the helical steady-state trajectories which include oscillatory components at both the betatron and Larmor periods. For the sake of simplicity, if we neglect the effect of the betatron oscillations on the trajectory, then these orbits can be expressed in rectangular coordinates in the form

$$p_x = p_w \cos k_w z + \alpha_+[P_x \cos(k_w z - \Omega_- t) - P_y \sin(k_w z - \Omega_- t)] \\ + \alpha_-[P_x \cos(k_w z + \Omega_- t) + P_y \sin(k_w z + \Omega_- t)], \tag{4.206}$$

$$p_y = p_w \sin k_w z + \alpha_+[P_x \sin(k_w z - \Omega_- t) + P_y \cos(k_w z - \Omega_- t)] \\ + \alpha_-[P_x \sin(k_w z + \Omega_- t) - P_y \cos(k_w z + \Omega_- t)], \tag{4.207}$$

and

$$p_z = p_\| - \frac{p_w}{p_\| \sqrt{(1+\lambda_0^2)}}(P_x \cos\Omega_- t + P_y \sin\Omega_- t), \tag{4.208}$$

where $(p_w, p_\|)$ denote the transverse and axial momenta corresponding to the steady-state trajectory, $\lambda_0 \equiv |p_w/p_\||$, P_x and P_y represent the canonical momenta, and

$$\alpha_\pm \equiv \frac{1 \pm \sqrt{(1+\lambda_0^2)}}{2\sqrt{(1+\lambda_0^2)}}. \tag{4.209}$$

Observe that $\lim_{\lambda_0 \to 0}\Omega_- = k_w v_\| - \Omega_0$, $\lim_{\lambda_0 \to 0}\alpha_+ = 1$, and $\lim_{\lambda_0 \to 0}\alpha_- = 0$. As a consequence, these orbits reduce to equations (2.42) and (2.43) in the idealized one-dimensional limit and describe the effect of the combined wiggler and Larmor oscillations.

4.2.1 The low-gain regime

The analysis of the low-gain regime for a three-dimensional realizable configuration differs from that used for the idealized one-dimensional treatments of both the helical and planar wiggler geometries. In those cases, we employed a unified Vlasov–Maxwell treatment for both the low- and high-gain regimes. However, we employ the more usual phase-average technique to treat the more general three-dimensional case. This technique results in an identical expression for the gain, at considerable reduction in mathematical complexity. The general expression obtained in this case (4.235) reduces to the previously discussed one-dimensional result (4.59) in the limit in which the three-dimensional effects are negligible. The principal differences between the one- and three-dimensional results are twofold. In the first place, the orbital parameters contained in these expressions are those given by the one- and three-dimensional steady-state trajectories discussed in Chapter 2. In the second place, the three-dimensional result contains a **filling factor** which describes the effect of the overlap between the electron beam and the radiation field. In the one-dimensional limit, this filling factor is unity.

It should be observed that we employ a waveguide-mode decomposition for the electromagnetic field rather that a Gaussian-mode decomposition which is often more appropriate to short-wavelength free-electron laser oscillators. However, the generalization to the Gaussian-mode case can be effected merely by the modification of the filling factor. We have chosen to employ the waveguide-mode decomposition in the interest of providing a unified formalism between the low- and high-gain regimes. The high-gain regime is more appropriate to the treatment of longer-wavelength free-electron laser amplifiers (submillimetre wavelengths and longer) driven by intense electron-beam sources. As a consequence, the waveguide is the dominant dispersive effect in these configurations.

Since it is our intention to treat the effects of transverse inhomogeneities in a self-consistent manner [34], these effects must also be included in the electromagnetic field. In this regard, we observe that the drift tube acts as a waveguide and that the electromagnetic field must satisfy the boundary conditions on the walls. In order to ensure that these boundary conditions are satisfied, we represent the electromagnetic field in terms of an ensemble of the TE and TM modes in the vacuum waveguide. The vector potential of the electromagnetic field can, therefore, be expressed in cylindrical coordinates as

$$\delta\mathbf{A}(\boldsymbol{x},t)=\sum_{l=0,n=1}^{\infty}\delta A_{ln}(z)\left[\frac{l}{\kappa_{ln}r}J_l(\kappa_{ln}r)\hat{e}_r\sin\alpha_l+J_l'(\kappa_{ln}r)\hat{e}_\theta\cos\alpha_l\right],\qquad(4.210)$$

for the TE modes, and

$$\delta A(\boldsymbol{x},t)=\sum_{l=0,n=1}^{\infty}\delta A_{ln}(z)\left[J_l'(\kappa_{ln}r)\hat{e}_r\cos\alpha_i-\frac{l}{\kappa_{ln}r}J_l(\kappa_{ln}r)\hat{e}_\theta\cos\alpha_i\right.$$
$$\left.+\frac{\kappa_{ln}}{k}J_l(\kappa_{ln}r)\hat{e}_z\sin\alpha_l\right],\qquad(4.211)$$

for the TM modes, where for frequency ω and wavenumber k the phase is defined as

$$\alpha_l\equiv\int_0^z dz'\,k(z')+l\theta-\omega t,\qquad(4.212)$$

J_l and J_l' denote the regular Bessel function of the first kind and its derivative, and κ_{ln} defines the cutoff wavenumber. For convenience we have suppressed the mode indices on the wavenumber k which, for fixed frequency, will differ for each mode. In this representation, the cutoff wavenumbers for the TE_{ln} mode is given by $\kappa_{ln}\equiv x_{ln}'/R_g$, where $J_l'(x_{ln}')=0$, and R_g denotes the radius of the waveguide. The cutoff wavenumber for the TM_{ln} mode is given by $\kappa_{ln}\equiv x_{ln}/R_g$, where $J_l(x_{ln})=0$. The coherent amplification process is included under the assumption that the amplitude and the wavenumber vary slowly with respect to the wavelength of the radiation.

In order to determine the gain, we make use of Maxwell's equations under the assumption that the beam density is sufficiently diffuse that the space-charge effects can be ignored. Observe that the divergence of the vector potential vanishes in the vacuum waveguide, as well as for the TE modes under the assumption of a slowly varying amplitude and wavenumber. However, the divergence of the vector potential does not vanish within the context of this formulation for the TM modes due to the presence of an axial component of the electric field. In this case,

$$\nabla\cdot\delta A=\sum_{l=0,n=1}^{\infty}\kappa_{ln}\frac{d}{dz}\left(\frac{1}{k}\delta A_{ln}\right)J_l(\kappa_{ln}r)\sin\alpha_l.\qquad(4.213)$$

The effect of the divergence of the vector potential on the evolution of the TM modes will be discussed in detail in the context of the nonlinear analysis of the interaction in Chapter 5. For the present discussion, we shall assume that the wave is far above cutoff (i.e. $\kappa_{ln}\ll k$) and neglect this contribution. The vector potential, therefore, satisfies Maxwell's equations

$$\left(\nabla^2 - \frac{1}{c^2}\frac{\partial^2}{\partial t^2}\right)\delta \boldsymbol{A}(\boldsymbol{x}, t) = -\frac{4\pi}{c}\delta \boldsymbol{J}(\boldsymbol{x}, t), \tag{4.214}$$

where $\delta \boldsymbol{J}$ represents the source current.

Observe that these assumptions are equivalent to the condition that $kR_g \ll 1$, and the principal characteristic of the waveguide structure included in the analysis is the radial localization of the modes. Since we are concerned with the low-gain regime, we make the further assumption that the beam is sufficiently tenuous that it only provides for a gain medium and that the dispersion relation for both the TE and TM modes is given approximately by the free-space limit

$$\omega \equiv ck. \tag{4.215}$$

This is valid as long as the frequency is much greater than both the beam plasma frequency and the waveguide cutoff frequencies. The growth of each TE or TM mode is determined by the diagonalization of Maxwell's equations in view of the azimuthal and radial mode structure of the waveguide modes. Using the orthogonality properties of the Bessel functions, it can be shown that the evolution of the slowly-varying amplitude is governed by

$$\frac{d}{dz}\delta A_{ln} = -\frac{1}{ck}\frac{x_{ln}'^2}{(x_{ln}'^2 - l^2)J_l^2(x_{ln}')}\frac{\omega}{\pi R_g^2}\int_{-\pi/\omega}^{\pi/\omega} dt \int_0^{2\pi} d\theta \int_0^{R_g} drr[C_l\delta J_1 - S_l\delta J_2], \tag{4.216}$$

for the TE_{ln} mode, and

$$\frac{d}{dz}\delta A_{ln} = \frac{1}{ck}\frac{1}{J_l'^2(x_{ln})}\frac{\omega}{\pi R_g^2}\int_{-\pi/\omega}^{\pi/\omega} dt \int_0^{2\pi} d\theta \int_0^{R_g} drr(S_l\delta J_1 + C_l\delta J_2), \tag{4.217}$$

for the TM_{ln} mode, where we have neglected second-order derivatives of the amplitude and phase (i.e. terms of order dk/dz), $(\delta J_1, \delta J_2)$ denote the transverse components of the source current in the wiggler frame, $\chi = \theta - k_w z$,

$$S_l \equiv J_{l-1}(\kappa_{ln}r)\sin[\psi_l + (l-1)\chi] - J_{l+1}(\kappa_{ln}r)\sin[\psi_l + (l+1)\chi], \tag{4.218}$$

$$C_l \equiv J_{l-1}(\kappa_{ln}r)\cos[\psi_l + (l-1)\chi] + J_{l+1}(\kappa_{ln}r)\cos[\psi_l + (l+1)\chi], \tag{4.219}$$

and

$$\psi_l = -\omega t_0 + \int_0^z dz'\left[k + lk_w - \frac{\omega}{v_z(z')}\right], \tag{4.220}$$

defines the ponderomotive phase. Note that t_0 denotes the entry time at which the electron crosses the $z = 0$ plane.

We now assume the existence of an idealized electron beam (i.e. monoenergetic with a vanishing pitch-angle spread) which is executing axi-centred, steady-state trajectories. Hence, $v_1 = v_w, v_2 = 0, v_3 = v_\parallel, r = r_0 \equiv \mp v_w/k_w v_\parallel$, and $\chi = \pm\pi/2$ for the Group I and Group II orbits, respectively. The beam configuration is that of a thin axi-centred helix in which the beam density is uniform within the range

$r_0 - \Delta R \leqslant r \leqslant r_0$ for $\Delta R \ll r_0$, and the azimuthal position rotates with $\theta = k_w z \pm \pi/2$. The source current which corresponds to this case is characterized by $\delta J_2 = 0$ and

$$\delta J_1 = -2\pi e n_b v_w \Delta R \delta(r - r_0)\delta(\chi \mp \pi/2) \int_{-\infty}^{\infty} dt_0 \delta(t - t_0 - z/v_\parallel), \quad (4.221)$$

where n_b is the ambient density of the beam, and we have assumed that the beam is continuous (i.e. uniform in entry time). Substitution of this source current into equations (4.216) and (4.217) indicates that the amplitude evolves as

$$\frac{d}{dz}\delta a_{ln} = \pm \frac{\omega_b^2}{c^3 k} \frac{x_{ln}'^2}{(x_{ln}'^2 - l^2) J_l^2(x'_{ln})} \frac{r_0 \Delta R}{R_g^2} v_w J_l'(\kappa_{ln} r_0) \left\langle \sin\left(\psi_l \pm \frac{l\pi}{2}\right)\right\rangle, \quad (4.222)$$

for the TE$_{ln}$ mode, and

$$\frac{d}{dz}\delta a_{ln} = \pm \frac{\omega_b^2}{c^3 k} \frac{1}{J_l'^2(x_{ln})} \frac{r_0 \Delta R}{R_g^2} v_w \frac{l}{\kappa_{ln} r_0} J_l(\kappa_{ln} r_0) \left\langle \cos\left(\psi_l + \frac{l\pi}{2}\right)\right\rangle, \quad (4.223)$$

for the TM$_{ln}$ mode. Here $\delta a_{ln} \equiv e\delta A_{ln}/m_e c^2$, $\omega_b^2 \equiv 4\pi e^2 n_b/m_e$, and the averaging operator is defined over the initial phase $\psi_{l0} = -\omega t_0$ as

$$\langle(\cdots)\rangle \equiv \frac{1}{2\pi}\int_0^{2\pi} d\psi_{l0}\sigma_\parallel(\psi_{l0})(\cdots), \quad (4.224)$$

where $\sigma_\parallel$ defines the electron distribution in initial phase.

In order to determine the gain for each waveguide mode, we must first obtain solutions for the ponderomotive phase to first order in the mode amplitudes. Generalization of the derivation of the pendulum equation to three dimensions by perturbation about the axi-centred, steady-state trajectories to first order in each waveguide mode yields

$$\frac{d^2}{dz^2}\psi_l = \mp K_l^2 J_l'(\kappa_{ln} r_0) \sin\left(\psi_l \pm \frac{l\pi}{2}\right), \quad (4.225)$$

for the TE modes, and

$$\frac{d^2}{dz^2}\psi_l = \mp K_l^2 \frac{l}{\kappa_{ln} r_0} J_l(\kappa_{ln} r_0) \cos\left(\psi_l \pm \frac{l\pi}{2}\right), \quad (4.226)$$

for the TM modes, where

$$K_l^2 \equiv \frac{c v_w (k + l k_w)^2}{\gamma_0 \gamma_\parallel^2 v_\parallel^2} \Phi(k_w r_0)\delta a_{ln}, \quad (4.227)$$

and Φ is defined in equation (2.70).

In the linear-gain regime, we seek to solve the pendulum equations for untrapped trajectories. These solutions are of the form $\psi_l = \psi_{l0} + \Delta k_l z + \delta\psi_l$, where $\Delta k_l \equiv (k + l k_w) - \omega/v_\parallel$ and we assume that $|\delta\psi_l/\Delta k_l z| \ll 1$. Expansion of the pendulum equations to first order in the electromagnetic field and $\delta\psi_l$ yields

$$\frac{d^2}{dz^2}\delta\psi_l \cong \mp K_l^2 J_l'(\kappa_{ln}r_0)\sin\left(\psi_{l0} + \Delta k_l z \pm \frac{l\pi}{2}\right), \tag{4.228}$$

for the TE_{ln} mode, and

$$\frac{d^2}{dz^2}\delta\psi_l \cong \mp K_l^2 \frac{l}{\kappa_{ln}r_0} J_l(\kappa_{ln}r_0)\cos\left(\psi_{l0} + \Delta k_l z \pm \frac{l\pi}{2}\right), \tag{4.229}$$

for the TM_{ln} mode. Hence, the solutions are

$$\delta\psi_l \cong \pm\frac{K_l^2}{\Delta k_l^2} J_l'(\kappa_{ln}r_0)\left[\sin\left(\psi_{l0} + \Delta k_l z \pm \frac{l\pi}{2}\right) - \sin\left(\psi_{l0} \pm \frac{l\pi}{2}\right)\right.$$
$$\left. - \Delta k_l z\cos\left(\psi_{l0} \pm \frac{l\pi}{2}\right)\right], \tag{4.230}$$

for the TE_{ln} mode, and

$$\delta\psi_l \cong \pm\frac{K_l^2}{\Delta k_l^2}\frac{l}{\kappa_{ln}r_0} J_l(\kappa_{ln}r_0)\left[\cos\left(\psi_{l0} + \Delta k_l z \pm \frac{l\pi}{2}\right) - \cos\left(\psi_{l0} \pm \frac{l\pi}{2}\right)\right.$$
$$\left. + \Delta k_l z\sin\left(\psi_{l0} \pm \frac{l\pi}{2}\right)\right], \tag{4.231}$$

for the TM_{ln} mode, subject to the initial conditions $\delta\psi_l(0) = 0$ and $d\psi_l(0)/dz = 0$. As a consequence of these untrapped solutions, the phase averages appearing in equations (4.222) and (4.223) are

$$\left\langle \sin\left(\psi_l \pm \frac{l\pi}{2}\right)\right\rangle \cong \pm\frac{K_l^2}{2\Delta k_l^2} J_l'(\kappa_{ln}r_0)(\sin\Delta k_l z - \Delta k_l z\cos\Delta k_l z), \tag{4.232}$$

for the TE_{ln} mode, and

$$\left\langle \cos\left(\psi_l \pm \frac{l\pi}{2}\right)\right\rangle \cong \pm\frac{K_l^2}{2\Delta k_l^2}\frac{l}{\kappa_{ln}r_0} J_l(\kappa_{ln}r_0)(\sin\Delta k_l z - \Delta k_l z\cos\Delta k_l z), \tag{4.233}$$

for the TM_{ln} mode in the random phase limit (i.e. $\sigma_\parallel(\psi_{l0}) = 1$).

The linear power gain over an axial distance L is obtained by integration of the derivatives of the field amplitudes

$$G(L) \equiv \frac{2}{\delta a_{ln}(0)}\int_0^L dz\frac{d}{dz}\delta a_{ln}(z), \tag{4.234}$$

under the assumption that $G(L)$ is much less than unity. Substitution of the expressions for the phase averages for the untrapped trajectories into equations (4.222) and (4.223), therefore, yields

$$G(L) \cong \frac{\omega_b^2 L^3 k}{8\gamma_0\gamma_\parallel^2 v_\parallel^2}\beta_w^2\Phi(k_w r_0)\, T_{ln}F(\Theta_l), \tag{4.235}$$

where F is the spectral function defined in equation (4.60), $\Theta_l \equiv \Delta k_l L/2$,

$\beta_w^2 \equiv v_w^2/v_\parallel^2$, and

$$T_{ln} \equiv \begin{cases} \dfrac{2r_0 \Delta R}{R_g^2} \dfrac{x_{ln}'^2}{x_{ln}'^2 - l^2} \dfrac{J_l'^2(\kappa_{ln} r_0)}{J_l^2(x_{ln}')}; & \text{TE}_{ln} \text{ mode} \\ \dfrac{2r_0 \Delta R}{R_g^2} \dfrac{l^2}{k_{ln}^2 r_0^2} \dfrac{J_l^2(\kappa_{ln} r_0)}{J_l'^2(x_{ln})}; & \text{TM}_{ln} \text{ mode} \end{cases} \tag{4.236}$$

is a mode- and beam-dependent geometric factor which describes the overlap of the electron beam and the radiation field and is sometimes referred to as the **filling factor**. Observe this expression for the gain (4.234) is similar to that found in the idealized one-dimensional limit (4.59) when the electron trajectories may be approximated by the idealized steady-state orbits and the filling factor is unity. The spectral features of the linear gain (i.e. the bandwidth, and the frequency corresponding to the peak gain) are similar to those found in the idealized description.

One feature of the three-dimensional analysis which does not appear in one dimension, however, is the appearance of gain at harmonics of the fundamental resonance frequency. Typically, peak gain is found for $\Theta_l \approx \mp 1.3$ corresponding to Group I and Group II orbits, respectively. This implies that growth is found at the fundamental and all harmonics, and that the peak gain occurs for

$$\omega \cong \frac{lk_w v_\parallel}{1 - \beta_\parallel} \left(1 \mp \frac{2.6}{lk_w L} \right). \tag{4.237}$$

Indeed, this implies the existence of a **selection rule** whereby a TE or TM mode with the azimuthal mode number l is resonant at the lth harmonic and will support gain. The physical basis for the appearance of harmonics in the three-dimensional analysis of free-electron lasers based upon helical wiggler fields is the azimuthal variation of the electromagnetic field. Since the electrons execute a helical trajectory, they excite circularly polarized waves. Resonant modes are those in which the polarization vector rotates in synchronism with the electrons. If the polarization vector rotates once per wiggler period, then the interaction is resonant at the fundamental. An lth harmonic interaction corresponds to waves in which the polarization vector rotates l times per wiggler period. The phase of the TE and TM modes of a cylindrical waveguide varies as $\exp(ikz + il\theta - i\omega t)$. Since the azimuthal trajectory of the electrons in a helical wiggler varies as $\theta \approx k_w z$, this implies that a TE_{ln} or TM_{ln} mode will be resonant (i.e. the electrons will experience a near-constant phase) when $\omega \approx (k + lk_w)v_\parallel$. The growth of harmonics for a helical wiggler configuration, therefore, arises from a different mechanism than is found for a planar wiggler, and does not require harmonic oscillations in the electron trajectories. The question of harmonic radiation will be discussed in some depth in Chapter 7 for both planar and helical wigglers.

4.2.2 The high-gain regime

The high-gain regime is distinguished from the previously described low-gain regime in that the dielectric effects of the electron beam are sufficiently strong that the mode structure of the electromagnetic wave can differ significantly from that of the vacuum waveguide. In this regime, a self-consistent model of the interaction must be used to describe both the resonant amplification of the signal and the beam- and wiggler-driven modifications to the vacuum dispersion [35]. The configuration we consider for this purpose is that of a relativistic electron beam propagating through a cylindrical waveguide in the presence of a helical wiggler (2.1) and an axial guide magnetic field. Within this context, we shall assume that the electrons are executing the helical, steady-state trajectories described in equations (2.53), (2.54) and (4.206–8) and consider the first-order perturbations to these trajectories due to the electromagnetic field.

We analyse the spatial amplification of a signal for which the source current and charge density are obtained from the moments of the perturbed distribution function

$$\delta f_b[\boldsymbol{r}(z), z, \boldsymbol{p}, \tau(z)] = e\int_0^z \frac{dz'}{v_z[\boldsymbol{r}(z'), z']}\{\delta \boldsymbol{E}[\boldsymbol{r}(z'), z', \tau(z')] + \frac{1}{c}\mathbf{v}[\boldsymbol{r}(z'), z'] \times \delta \boldsymbol{B}[\boldsymbol{r}(z'), z', \tau(z')]\}\cdot\nabla_p F_b(P_x, P_y, p), \tag{4.238}$$

where the equilibrium distribution F_b is a function of the constants of the motion (P_x, P_y, p), $\delta\boldsymbol{E}$ and $\delta\boldsymbol{B}$ are the fluctuating electric and magnetic fields, $\boldsymbol{r}(z)$ is the position of the electron in the x–y-plane at the axial position z, $\mathbf{v}[\boldsymbol{r}(z), z]$ is the electron velocity, and

$$\tau[\boldsymbol{r}(z), z] \equiv t_0 + \int_0^z \frac{dz'}{\mathbf{v}_z[\mathbf{r}(z'), z']}, \tag{4.239}$$

is the sum of the entry time t_0 and the time required for an electron to travel from $[\boldsymbol{r}(z=0), z=0]$ at the start of the interaction region to $[\boldsymbol{r}(z'), z']$. In addition, we work with the vector and scalar potentials which are written as

$$\delta\boldsymbol{A}(\boldsymbol{x}, t) = \frac{1}{2}\delta\hat{\boldsymbol{A}}(\boldsymbol{x})\exp(-i\omega t) + \text{c.c.}, \tag{4.240}$$

and

$$\delta\varphi(\boldsymbol{x}, t) = \frac{1}{2}\delta\hat{\varphi}(\boldsymbol{x})\exp(-i\omega t) + \text{c.c.} \tag{4.241}$$

Integration of equation (4.238) with respect to the vector basis which describes circularly polarized electromagnetic waves (4.33) yields a perturbed distribution of the form

$$\delta\hat{f}_b[\boldsymbol{r}(z), z] = \frac{e}{2c}\left[D_+\left(\frac{\partial}{\partial P_x} + i\frac{\partial}{\partial P_y}\right) + D_-\left(\frac{\partial}{\partial P_x} - i\frac{\partial}{\partial P_y}\right) + D_z\frac{\partial}{\partial p}\right]F_b(P_x, P_y, p), \tag{4.242}$$

which is formally identical to equation (4.35) derived for the idealized one-dimensional limit. In the three-dimensional regime, however,

$$D_z \equiv \frac{1}{2p}\int_0^z dz' \frac{\exp[i\omega\tau(z, z')]}{v_z[\boldsymbol{r}(z'), z']}[-c(2p_z\nabla_{z'} + p_-\nabla_+ + p_+\nabla_-)\delta\hat{\varphi} + 2i\omega(p_z\delta\hat{A}_z + p_-\delta\hat{A}_+ + p_+\delta\hat{A}_-)], \tag{4.243}$$

and

$$D_{\pm} \equiv \int_0^z dz' \frac{\exp[i\omega t(z,z')]}{v_z[\mathbf{r}(z'),z']} \left\{ \alpha_+ \exp(\pm i\theta_-) \left[-\frac{c}{2}\nabla_{\pm}\delta\hat{\varphi} + \left(i\omega - v_z \frac{\partial}{\partial z'} \right) \delta\hat{A}_{\pm} \right. \right.$$

$$\left. + \frac{v_z}{2}\nabla_{\pm}\delta\hat{A}_z \mp \frac{v_{\pm}}{2}(\nabla_-\delta\hat{A}_+ - \nabla_+\delta\hat{A}_-) \right]$$

$$- \alpha_- \exp(\mp i\theta_+) \left[-\frac{c}{2}\nabla_{\mp}\delta\hat{\varphi} + \left(i\omega - v_z \frac{\partial}{\partial z'} \right) \delta\hat{A}_{\mp} \right.$$

$$\left. \left. + \frac{v_z}{2}\nabla_{\mp}\delta\hat{A}_z \pm \frac{v_{\mp}}{2}(\nabla_-\delta\hat{A}_+ - \nabla_+\delta\hat{A}_-) \right] \right\}, \tag{4.244}$$

where $\tau(z,z') \equiv \tau(z) - \tau(z'), \theta_{\pm} \equiv k_w z \pm \Omega_- \tau(z)$, and

$$\nabla_{\pm} \equiv \frac{\partial}{\partial x} \mp i \frac{\partial}{\partial y}. \tag{4.245}$$

The source currents

The source current and charge density are found by computation of the appropriate moments of the perturbed distribution as

$$\delta\hat{J}_{\pm} = -\frac{e}{m_e}\int dP_x dP_y dp \frac{pp_{\pm}}{\gamma p_z \sqrt{(1+\lambda_0^2)}} \delta\hat{f}_b, \tag{4.246}$$

$$\delta\hat{J}_z = -\frac{e}{m_e}\int dP_x dP_y dp \frac{p}{\gamma \sqrt{(1+\lambda_0^2)}} \delta\hat{f}_b, \tag{4.247}$$

for the components of the current, and

$$\delta\hat{\rho} = -e\int dP_x dP_y dp \frac{p}{p_z \sqrt{(1+\lambda_0^2)}} \delta\hat{f}_b, \tag{4.248}$$

for the charge density. By application of Floquet's theorem for periodic systems, the axial and azimuthal structure of the fields and sources are expressed in the form

$$f(r,\theta,z) = \sum_{l,n=-\infty}^{\infty} f_{l,n}(r) \exp[i(k+nk_w)z + il\theta], \tag{4.249}$$

for some arbitrary function $f(r,\theta,z)$. Substitution of the perturbed distribution function into the expressions for the source currents and charge densities results in harmonic contributions to the source current $\delta\hat{J}_{l,n}$ and charge density $\delta\hat{\rho}_{l,n}$, each of which depends upon many harmonics of $\delta\hat{A}_{l,n}$ and $\delta\hat{\varphi}_{l,n}$. However, in the limit in which $\omega \gg \Omega_-$ and the vector and scalar potentials vary slowly in r with respect to the wavelength of the radiation, then the sources assume the comparatively simple forms

$$\delta\hat{J}^{(\pm)}_{l,n} \cong -\frac{i\omega_b^2}{8\pi c}\int_0^{\infty} dp \frac{1}{\gamma} \left\{ [p_w(\hat{H}^{(\pm)}_{l,m} + \hat{H}^{(\mp)}_{l\mp 2,n\pm 2}) - ip_{\parallel}(ck_{n\pm 1}\delta\hat{\varphi}_{l\mp 1,n\pm 1} - \omega\delta\hat{A}_{l\mp 1,n\pm 1,z})] \right.$$

$$\times \frac{p_w}{\omega - k_{l+n}v_{\parallel}} \frac{1}{p}\frac{\partial}{\partial p} - \frac{2\alpha_+^2}{\omega \pm \Omega_- - k_{l+n}v_{\parallel}} \hat{L}^{(\pm)}_{l,n}$$

$$\mp \frac{\lambda_0^2}{\sqrt{(1+\lambda_0^2)}}\frac{\omega}{\Omega_-}(R_{l,n}^{(\pm)}\hat{L}_{l,n}^{(\pm)} - R_{l,n}^{(\mp)}\hat{L}_{l+2,n\pm 2}^{(\mp)})\Bigg\} G_b(p), \tag{4.250}$$

and

$$\delta\hat{\rho}_{l,n} \cong -\frac{i\omega_b^2}{8\pi c}\int_0^\infty dp \frac{1}{\gamma v_w}\Bigg\{[p_w(\hat{H}_{l-1,n+1}^{(-)} + \hat{H}_{l+1,n-1}^{(+)}) - ip_\parallel(ck_n\hat{\varphi}_{l,n} - \omega\delta\hat{A}_{l,n,z})]$$

$$\times \frac{p_w}{\omega - k_{l+n}v_\parallel}\frac{1}{p}\frac{\partial}{\partial p} + \frac{\lambda_0^2}{\sqrt{(1+\lambda_0^2)}}\frac{\omega}{\Omega_-}$$

$$\times\left(R_{l,n}^{(-)}\hat{L}_{l-1,n+1}^{(-)} - R_{l,n}^{(+)}\hat{L}_{l+1,n-1}^{(+)} + \frac{v_w}{2}\hat{K}_{l,n}S_{l,n}\right)\Bigg\} G_b(p), \tag{4.251}$$

where we have assumed a distribution function of the form

$$F_b(P_x, P_y, p) = n_b\delta(P_x)\delta(P_y)G_b(p), \tag{4.252}$$

for an ambient beam density n_b, $\omega_b^2 \equiv 4\pi e^2 n_b/m_e$, $\delta J_{l,n}^{(\pm)} \equiv (\delta J_{l,n})_r \mp i(\delta J_{l,n})_\theta$, and

$$\delta\hat{A}_{l,n}^{(\pm)} \equiv \frac{1}{2}[(\delta\hat{A}_{l,n})_r \mp i(\delta\hat{A}_{l,n})_\theta]. \tag{4.253}$$

In addition, we have defined

$$\hat{L}_{l,n}^{(\pm)} \equiv i(\omega - k_n v_\parallel)\delta\hat{A}_{l,n}^{(\pm)} - \frac{1}{2}\nabla_l^{(+)}(c\delta\hat{\varphi}_{l,n} - v_\parallel\delta\hat{A}_{l,n,z}), \tag{4.254}$$

$$\hat{H}_{l,n}^{(\pm)} \equiv i\omega\delta\hat{A}_{l,n}^{(\pm)} - \frac{c}{2}\nabla_l^{(\pm)}\delta\hat{\varphi}_{l,n}, \tag{4.255}$$

$$\hat{K}_{l,n} \equiv \nabla_{l-1}^{(-)}\delta\hat{A}_{l,n}^{(+)} - \nabla_{l+1}^{(+)}\delta\hat{A}_{l,n}^{(-)}, \tag{4.256}$$

$$R_{l,n}^{(\pm)} \equiv \frac{\alpha_+}{\omega \pm \Omega_- - k_{l+n}v_\parallel} - \frac{\alpha_-}{\omega \mp \Omega_- - k_{l+n}v_\parallel} - \frac{1}{\omega - k_{l+n}v_\parallel}, \tag{4.257}$$

and

$$S_{l,n} \equiv \frac{1}{\omega - \Omega_- - k_{l+n}v_\parallel} - \frac{1}{\omega + \Omega_- + k_{l+n}v_\parallel} - \frac{2}{\omega - k_{l+n}v_\parallel}. \tag{4.258}$$

where $\nabla_l^{(\pm)} \equiv \partial/\partial r \pm l/r$. Observe that $\delta\hat{J}_{l,n,z}$ has been omitted because the specification of a gauge condition allows us to eliminate one of the components of the four-vector formed by the vector and scalar potentials, and we choose to deal with $\delta\hat{A}^{(\pm)}$ and $\delta\hat{\varphi}$.

The dispersion equation

The dispersion equation is obtained by substitution of the sources into the Maxwell–Poisson equations

$$\left[\frac{1}{r}\frac{d}{dr}r\frac{d}{dr} + p_n^2 - \frac{(l \mp 1)^2}{r^2}\right]\delta\hat{A}_{ln}^{(\pm)} = -\frac{4\pi}{c}\delta\hat{J}_{ln}^{(\pm)}, \tag{4.259}$$

$$\left[\frac{1}{r}\frac{d}{dr}r\frac{d}{dr} + p_n^2 - \frac{l^2}{r^2}\right]\delta\hat{\varphi}_{l,n} = -8\pi\delta\hat{\rho}_{l,n}, \tag{4.260}$$

as well as the Lorentz gauge condition

$$k_n \delta \hat{A}_{l,n,z} - \frac{\omega}{c} \delta \hat{\varphi}_{l,n} = i(\nabla^{(-)}_{l-1} \delta \hat{A}^{(+)}_{l,n} + \nabla^{(+)}_{l+1} \delta \hat{A}^{(-)}_{l,n}), \tag{4.261}$$

where $p_n^2 \equiv \omega^2/c^2 - k_n^2$. In order to proceed, we must specify the distribution function $G_b(p)$. In this regard, we observe that there is a one-to-one correspondence between the total energy and the radius of the trajectory for the steady-state orbits. Hence, a small energy spread in the distribution of steady-state orbits is equivalent to a narrow radial profile. For simplicity, therefore, we shall assume a monoenergetic distribution with a sharp radial profile and choose $G_b(p) = N(p)\delta[p - p(r)]$, where

$$p(r) = m_e c \sqrt{(1 + \lambda_0^2)} \left[\frac{\Omega_0}{ck_w} \pm 2 \frac{\Omega_w}{ck_w} \frac{1 + \lambda_0^2}{\lambda_0^2} I_1(\lambda_0) \right], \tag{4.262}$$

defines the relationship between the momentum and the orbit radius, and $N(p)$ models the density profile. In general, the Maxwell–Poisson equations define a system of equations in which each harmonic contribution is coupled to many others. However, a substantial simplification is found by retaining only the dominant coupling terms, which is valid as long as $\omega_b/\gamma^{1/2} ck_w \ll 1$. In this case, we find that

$$\delta \hat{J}^{(\pm)}_{l,n} \cong \frac{c}{4\pi} \left[\Lambda^{(\pm)}_{l,n} \delta \hat{A}^{(\pm)}_{l,n} + T_{l\mp 1, n\pm 1} \delta \hat{\varphi}_{l\mp 1, n\pm 1} + V_{l\mp 2, n\pm 2} \delta \hat{A}^{(\mp)}_{l\mp 2, n\pm 2} \right], \tag{4.263}$$

and

$$\delta \hat{\rho}_{l,n} \cong \frac{1}{4\pi} \left[\chi_{l,n} \delta \hat{\varphi}_{l,n} + W_{l+1,n-1} \delta \hat{A}^{(+)}_{l+1,n-1} + W_{l-1,n+1} \delta \hat{A}^{(-)}_{l-1,n+1} \right], \tag{4.264}$$

where we note that $N(p)$ is an implicit function of r,

$$\Lambda^{(\pm)}_{l,n} \equiv -\frac{\omega_b^2}{\gamma c^2} N(r) \left[\omega - k_n v_\parallel \left(1 + \frac{l^2}{2k_n^2 r^2} \right) \right] \times \left[\frac{\alpha_+^2}{\omega \pm \Omega_- - k_{l+n} v_\parallel} + \frac{\alpha_-^2}{\omega \mp \Omega_- - k_{l+n} v_\parallel} \right] + V_{l\pm 1, n\mp 1}, \tag{4.265}$$

$$\chi_{l,n} \equiv -\frac{\omega_b^2}{\gamma c^2} \Phi(\lambda_0) N(r) \frac{k_{l+n}}{k_n} \frac{\omega^2 - c^2 k_n^2}{\gamma^2 (1 + \lambda_0^2)(\omega - k_{l+n} v_\parallel)^2}, \tag{4.266}$$

$$T^{(\pm)}_{l,n} \equiv -\frac{\omega_b^2}{2\gamma c^2} \frac{p_w}{p_\parallel} \Phi(\lambda_0) N(r) \frac{\omega}{ck_n} \frac{\omega^2 - c^2 k_n^2}{\gamma^2 (1 + \lambda_0^2)(\omega - k_{l+n} v_\parallel)^2}, \tag{4.267}$$

$$W_{l\pm 1, n\mp 1} \equiv -\frac{\omega_b^2}{\gamma c^2} \frac{p_w}{p_\parallel} \Phi(\lambda_0) N(r) \frac{\omega k_{l+n} c}{\gamma^2 (1 + \lambda_0^2)(\omega - k_{l+n} v_\parallel)^2}, \tag{4.268}$$

and

$$V_{l\pm 1, n\mp 1} \equiv -\frac{\omega_b^2}{2\gamma c^2} \Phi(\lambda_0) N(r) \frac{\lambda_0^2 \omega^2}{\gamma^2 (1 + \lambda_0^2)(\omega - k_{l+n} v_\parallel)^2}, \tag{4.269}$$

where $\Phi(\lambda_0)$ is given by equation (2.70). As a consequence, we obtain the following set of coupled differential equations

$$\left(\frac{1}{r}\frac{d}{dr}r\frac{d}{dr}+p_{n\mp1}^2-\frac{l^2}{r^2}\right)\delta\hat{A}_{l\pm1,n\mp1}^{(\pm)}$$

$$=-\Lambda_{l\pm1,n\mp1}^{(\pm)}\delta\hat{A}_{l\pm1,n\mp1}^{(\pm)}-T_{l,n}^{(\pm)}\delta\hat{\varphi}_{l,n}-V_{l\mp1,n\pm1}\delta\hat{A}_{l\mp1,n\pm1}^{(\mp)}, \tag{4.270}$$

and

$$\left(\frac{1}{r}\frac{d}{dr}r\frac{d}{dr}+p_n^2-\frac{l^2}{r^2}\right)\delta\hat{\varphi}_{l,n}=-\chi_{l,n}\delta\hat{\varphi}_{l,n}-W_{l+1,n-1}\delta\hat{A}_{l+1,n-1}^{(+)}$$

$$-W_{l-1,n+1}\delta\hat{A}_{l-1,n+1}^{(-)}. \tag{4.271}$$

In order to solve these coupled second-order differential equations, we must specify the boundary conditions appropriate for a cylindrical waveguide of radius R_g. We assume the walls to be grounded at zero potential, and we may express the boundary conditions in the form

$$\delta\hat{A}_{l,n}^{(+)}(R_g)=\delta\hat{A}_{l,n}^{(-)}(R_g), \tag{4.272}$$

and

$$\delta\hat{\varphi}_{l,n}(R)=\frac{d}{dr}\left[r\left(\delta\hat{A}_{l,n}^{(+)}+\delta\hat{A}_{l,n}^{(-)}\right)\right]\bigg|_{r=R}=0. \tag{4.273}$$

In the thin-beam limit we assume, as in the previous section, that the electron beam density is within the range $r_0-\Delta R\leqslant r\leqslant r_0$ (where $\Delta R\ll r_0$) and we write $N(r)=\Delta R\,\delta(r-r_0)$ where r_0 denotes the radius of the steady-state orbit. Hence, we shall also assume that the vector and scalar potentials are continuous at $r=r_0$, and use the Maxwell–Poisson equations to obtain jump conditions for the derivatives of the potentials across the beam. This thin-beam condition is equivalent to a small energy spread on the beam, which is of the order of

$$\frac{\Delta\gamma}{\gamma}=\frac{\gamma^2-1}{\Phi(\lambda_0)-1}\,\frac{\lambda_0^2}{1+\lambda_0^2}\,\frac{\Delta R}{r_0}, \tag{4.274}$$

and is assumed to be small.

Equations (4.270) and (4.271) describe the coupling between five harmonic components: $\delta\hat{\varphi}_{l,n}$, $\delta\hat{A}_{l+1,n-1}^{(\pm)}$ and $\delta\hat{A}_{l-1,n+1}^{(\pm)}$. The solutions are given in terms of the Bessel and Neumann functions, and are of the form

$$\delta\hat{\varphi}_{l,n}=A_{l,n}J_l(p_nr), \tag{4.275}$$

$$\delta\hat{A}_{l,n}^{(\pm)}=A_{l,n}^{(\pm)}J_{l\mp1}(p_nr), \tag{4.276}$$

for $0\leqslant r<r_0$, and

$$\delta\hat{\varphi}_{l,n}=B_{l,n}J_l(p_nr)+C_{l,n}\,Y_l(p_nr), \tag{4.277}$$

$$\delta\hat{A}_{l,n}^{(\pm)}=B_{l,n}^{(\pm)}J_{l\mp1}(p_nr)+C_{l,n}^{(\pm)}\,Y_{l\mp1}(p_nr), \tag{4.278}$$

for $r > r_0$, where J_l and Y_l denote the Bessel and Neumann functions. Thus, each harmonic component requires three coefficients to characterize the solution throughout the waveguide, for a total of 15 coefficients in all. Two of these coefficients for each component are determined from the boundary and continuity conditions at the waveguide wall and the electron beam. The third coefficient is found from the jump condition at the beam, which is obtained by multiplying the Maxwell–Poisson equations by r and integrating over $r_0 - \varepsilon \leqslant r \leqslant r_0 + \varepsilon$ in the limit as $\varepsilon \to 0^+$. The jump condition results in a 5×5 matrix equation for the coefficients: $A_{l,n}, A^{(+)}_{l+1,n-1}, A^{(-)}_{l-1,n+1}, A^{(+)}_{l-1,n+1}$ and $A^{(-)}_{l+1,n-1}$.

Observe that the coupling of the field components through their coefficients $A^{(\pm)}_{l\mp1,n\pm1}$ occur through the boundary conditions and not through the source terms. By eliminating these coefficients, therefore, we may reduce the problem to a 3×3 matrix equation

$$\begin{bmatrix} \varepsilon_{l,n} & -\frac{\pi}{2} r_0 \Delta R \bar{W}_{l+1,n-1} & -\frac{\pi}{2} r_0 \Delta R \bar{W}_{l-1,n+1} \\ \frac{\pi}{2} r_0 \Delta R \bar{T}_{l,n} & \varepsilon^{(+)}_{l+1,n-1} & \frac{\pi}{2} r_0 \Delta R \bar{V}_{l-1,n+1} \\ \frac{\pi}{2} r_0 \Delta R \bar{T}_{l,n} & \frac{\pi}{2} r_0 \Delta R \bar{V}_{l+1,n-1} & \varepsilon^{(-)}_{l-1,n+1} \end{bmatrix} \begin{bmatrix} A_{l,n} \\ A^{(+)}_{l+1,n-1} \\ A^{(-)}_{l-1,n+1} \end{bmatrix} = 0, \tag{4.279}$$

where

$$\varepsilon_{l,n} \equiv D_{l,n} - \frac{\pi}{2} r_0 \Delta R \bar{\chi}_{l,n}, \tag{4.280}$$

$$\varepsilon^{(\pm)}_{l\pm1,n\mp1} \equiv D^{(\pm)}_{l\pm1,n\mp1} + \frac{\pi}{2} r_0 \Delta R (\bar{\Lambda}^{(\pm)}_{l\pm1,n\mp1} + \bar{\Lambda}^{(\mp)}_{l\pm1,n\mp1} M^{(l\pm2)}_{l\mp1,n\pm1}), \tag{4.281}$$

and $\bar{W}_{l\pm1,n\mp1}$, $\bar{V}_{l\pm1,n\mp1}$, $\bar{\Lambda}^{(\pm)}_{l\pm1,n\mp1}$, $\bar{T}_{l,n}$ and $\bar{\chi}_{l,n}$ denote those quantities specified in equations (4.243–7) in which the factor $N(r)$ has been removed. In addition,

$$D_{l,n} \equiv \frac{J_l(\zeta_n)}{J_l(\xi_n)[J_l(\xi_n)\, Y_l(\zeta_n) - J_l(\zeta_n)\, Y_l(\xi_n)]}, \tag{4.282}$$

$$D^{(\pm)}_{l\pm1,n\mp1} \equiv \frac{2J_{l\pm1}(\zeta_{n\mp1}) J'_{l\pm1}(\zeta_{n\mp1})}{J_l(\zeta_{n\mp1}) S^{(l\pm2)}_{l\pm1,n\mp1}}, \tag{4.283}$$

$$M^{(\pm)}_{l\mp1,n\pm1} \equiv \frac{J_{l\pm2}(\zeta_{n\mp1}) \Gamma^{(l\pm2)}_{l\pm1}(\zeta_{n\mp1})}{J_l(\zeta_{n\mp1}) S^{(l\pm2)}_{l\pm1,n\mp1}}, \tag{4.284}$$

where $\zeta_n \equiv p_n R_g$ and $\xi_n \equiv p_n r_0$. In addition,

$$\Gamma^{(l)}_{k,m} \equiv Y_l(\xi_m) \frac{d}{d\zeta_m} J^2_k(\zeta_m) - J_l(\xi_m) \frac{d}{d\zeta_m} [J_k(\zeta_m)\, Y_k(\zeta_m)], \tag{4.285}$$

$$S^{(l\pm2)}_{k,m} \equiv \Gamma^{(l)}_{k,m} + \frac{\pi}{2} r_0 \Delta R \bar{\Lambda}^{(\mp)}_{k,m} J_{l\pm2}(\xi_m) \Psi^{(l\pm2)}_{k,m}, \tag{4.286}$$

$$\Psi_{k,m}^{(l\pm 2)} \equiv Y_{l\pm 2}(\xi_m)\Gamma_{k,m}^{(l)} - J_{l\pm 2}(\xi_m)\left[Y_l(\xi_m)\frac{d}{d\zeta_m}[J_k(\zeta_m)\, Y_k(\zeta_m)] - J_l(\xi_m)\frac{d}{d\zeta_m}\, Y_k^2(\zeta_m)\right]. \tag{4.287}$$

The dispersion equation is found by setting the determinant of this interaction matrix to zero.

Substantial simplification occurs in the resonant approximation for which $|\omega - k_{l+n}v_\parallel| \ll \omega, |k_{l+n}v_\parallel|$ and

$$\bar{\chi}_{l,n} \cong -\frac{\omega_b^2}{\gamma c^2}\Phi(\lambda_0)\frac{\omega}{k_n v_\parallel}\frac{\omega^2 - c^2 k_n^2}{\gamma^2(1+\lambda_0^2)(\omega - k_{l+n}v_\parallel)^2}, \tag{4.288}$$

$$\bar{W}_{l\pm 1,n\mp 1} \cong -\frac{\omega_b^2}{\gamma c^2}\frac{p_w}{p_\parallel}\Phi(\lambda_0)\frac{\omega^2}{\gamma^2\beta_\parallel(1+\lambda_0^2)(\omega - k_{l+n}v_\parallel)^2}, \tag{4.289}$$

$$\begin{aligned}\bar{\Lambda}_{l\pm 1,n\mp 1}^{(\pm)} \cong &-\frac{\omega_b^2}{\gamma c^2}\left[\omega - k_n v_\parallel\left(1 + \frac{l^2}{2k_n^2 r_0^2}\right)\right] \\ &\times\left[\frac{\alpha_+^2}{\omega \pm \Omega_- - k_{l+n}v_\parallel} + \frac{\alpha_-^2}{\omega \mp \Omega_- - k_{l+n}v_\parallel}\right] \\ &-\lambda_0^2\frac{\omega_b^2}{2\gamma c^2}\Phi(\lambda_0)\frac{\omega^2}{\gamma^2(1+\lambda_0^2)(\omega - k_{l+n}v_\parallel)^2},\end{aligned} \tag{4.290}$$

$$\bar{T}_{l,n} \cong \frac{1}{2}\frac{v_w}{c}\bar{\chi}_{l,n}, \tag{4.291}$$

$$\bar{V}_{l\pm 1,n\mp 1} \cong \frac{1}{2}\frac{v_w}{c}\bar{W}_{l\pm 1,n\mp 1}. \tag{4.292}$$

For all cases of practical interest $v_w \ll c$ and $|\bar{V}_{l\pm 1,n\mp 1}| \ll |\bar{W}_{l\pm 1,n\mp 1}|$. As a consequence, the terms in $\bar{V}_{l\pm 1,n\mp 1}$ can be ignored. This is equivalent to the neglect of any direct coupling between the electromagnetic modes given by $\delta\hat{A}_{l\pm 1,n\mp 1}^{(\pm)}$. In addition, we can neglect the coupling to the $\delta\hat{A}_{l\mp 1,n\pm 1}^{(\pm)}$, so that

$$\varepsilon_{l\pm 1,n\mp 1}^{(\pm)} \cong \frac{2J_{l\pm 1}(\zeta_{n\mp 1})J'_{l\pm 1}(\zeta_{n\mp 1})}{J_l(\zeta_{n\mp 1})S_{l\pm 1,n\mp 1}^{(l\pm 2)}} + \frac{\pi}{2}r_0\Delta R\bar{\Lambda}_{l\pm 1,n\mp 1}^{(\pm)}. \tag{4.293}$$

Within the context of this approximation, the dispersion equation is of the form

$$\varepsilon_{l,n} \cong \frac{\lambda_0^2}{1+\lambda_0^2}\frac{\omega_b^2}{2\gamma c^2}\left(\frac{\pi}{2}r_0\Delta R\right)^2\frac{\omega^2\Phi(\lambda_0)}{\gamma^2(\omega - k_{l+n}v_\parallel)^2}\bar{\chi}_{l,n}\left[\frac{1}{\varepsilon_{l+1,n-1}^{(+)}} + \frac{1}{\varepsilon_{l-1,n+1}^{(-)}}\right]. \tag{4.294}$$

Finally, if the solution is restricted to the first quadrant in the (ω, k_n) plane, $|\varepsilon_{l-1,n+1}^{(-)}| \gg |\varepsilon_{l+1,n-1}^{(+)}|$, and the dispersion equation can be approximated by

$$\varepsilon_{l,n}\varepsilon_{l+1,n-1}^{(+)} \cong \frac{\lambda_0^2}{1+\lambda_0^2}\frac{\omega_b^2}{2\gamma c^2}\left(\frac{\pi}{2}r_0\Delta R\right)^2\Phi(\lambda_0)\frac{\omega^2}{\gamma^2(\omega - k_{l+n}v_\parallel)^2}\bar{\chi}_{l,n}. \tag{4.295}$$

This equation describes the coupling between the beam space-charge modes and a circularly polarized electromagnetic mode (either TE or TM modes of the waveguide).

Numerical solution of the dispersion equation

Equation (4.295) describes the coupling between the beam space-charge mode and a right-hand circularly polarized waveguide mode. In order to illustrate the nature of the gain in the free-electron laser, the complete dispersion equation (4.294) is solved for a representative set of parameters: $\gamma = 3.5$, $\omega_b/\gamma^{1/2}ck_w = 0.1$, $\Omega_w/ck_w = 0.05$, $\Delta R/r_0 = 0.1$, $k_w R_g = 1.5$, and a wide range of axial guide fields corresponding to both Group I and Group II orbits. For this choice of parameters, the transitions to orbital instability (and the corresponding singularities in Φ) occur for $\Omega_0/ck_w \approx 0.75$ for the Group I and $\Omega_0/ck_w \approx 0.62$ for the Group II trajectories. Hence, there is a substantial overlap in the range of axial fields for which these trajectories occur. While the growth rates found at these orbital stability transitions are also singular, it should be recognized that the linear-stability analysis breaks down in this regime. In addition, both Φ and the growth rate vanish for Group II orbits when $\Omega_0/ck_w \approx 1.25$.

The normalized magnitude of the growth rate $|\mathrm{Im}\, k_n|/k_w$ of the TE_{11} mode is plotted in Fig. 4.12 for these parameters with $\Omega_0/ck_w = 0.0$ and 0.5. This corresponds to Group I orbits. The waveguide cutoff occurs at a frequency $\omega_{co}/ck_w \approx 1.23$ for this mode and the two peaks correspond to the upper and lower intersections between the space-charge and waveguide modes. An approximate expression for the location of these intersection points can be obtained on the basis of the dispersion equation for the vacuum waveguide modes (i.e. in the absence of the electron beam) and the negative-energy space-charge wave. In this limit, it can be shown that the intersections occur at

$$\frac{\omega}{ck_w} \cong \gamma_\parallel^2 \beta_\parallel \left\{ \left(1 - \frac{\omega_b}{\gamma^{1/2}\gamma_\parallel k_w v_\parallel}\right) \pm \beta_\parallel \sqrt{\left[\left(1 - \frac{\omega_b}{\gamma^{1/2}\gamma_\parallel k_w v_\parallel}\right)^2 - \frac{\omega_{co}^2}{\gamma_\parallel^2 k_w^2 v_\parallel^2}\right]}\right\}, \tag{4.296}$$

where $\beta_\parallel \equiv v_\parallel/c$. It should be emphasized that this is only an approximate expression for the intersection points because (1) the dielectric effect of the beam can modify the dispersion relation of the waveguide mode significantly, and (2) the waveguide geometry reduces the effective plasma frequency. As shown in the figure, the peak growth rate increases rapidly with the axial magnetic field corresponding to the increase (decrease) in the transverse (axial) beam velocity and the increase in Φ. In addition, the resonant frequency is sensitive to the axial velocity and the upper intersection point decreases rapidly with the change in the axial velocity associated with the increasing axial magnetic field.

The peak growth rates for the Group I trajectories, as well as the frequencies corresponding to peak growth, are plotted in Fig. 4.13 as a function of the axial solenoidal field. The singularity at the orbital instability transition ($\Omega_0/ck_w \approx 0.75$) is evident from this figure. It is clear that the frequency of the upper (lower)

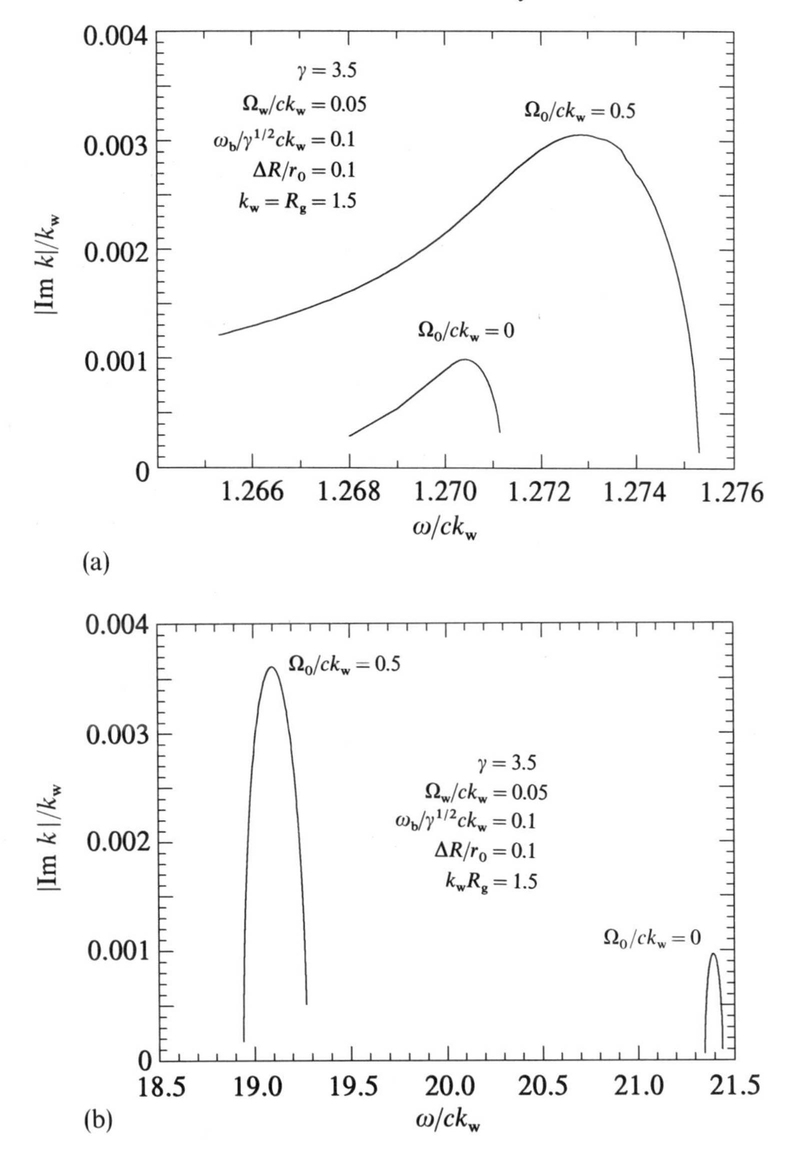

Fig. 4.12 Plot of the normalized growth rate at the lower (a) and upper (b) intersections as a function of frequency for the TE_{11} mode and Group I orbits.

intersection point decreases (increases) as the axial velocity decreases with increasing axial field strength. As a result, the intersections tend to coalesce with decreasing $v_\parallel$; however, the TE_{11} mode cutoff for the chosen parameters is sufficiently low that coalescence does not occur for the Group I orbits, and the two lines remain well separated. While the growth rate corresponding to the upper intersection exceeds that of the lower intersection over the entire range of

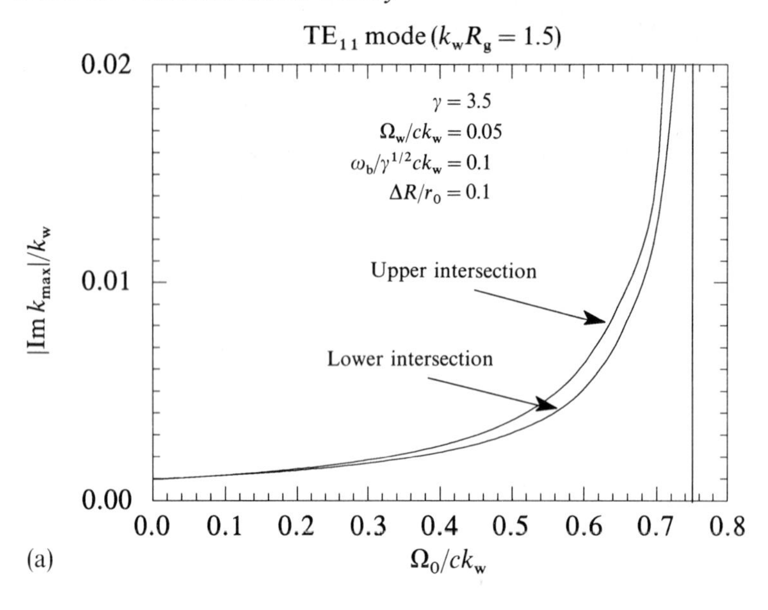

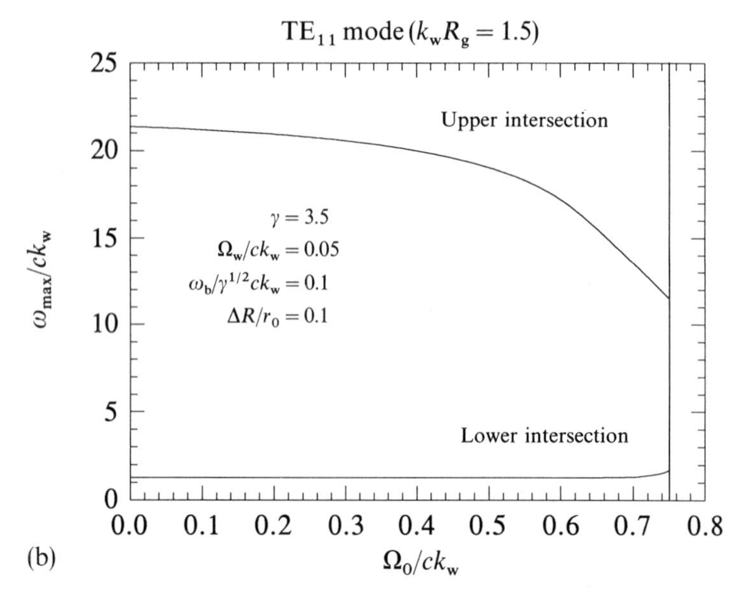

Fig. 4.13 The maximum growth rate (a) and corresponding frequency (b) of the TE$_{11}$ mode as a function of the axial magnetic field for the Group I orbits.

the Group I trajectories for the parameters illustrated, it should not be construed that this holds in general.

The maximum growth rates and corresponding frequencies are plotted in Fig. 4.14 as a function of the axial magnetic field for Group II trajectories. The growth rate for the Group II trajectories exhibits characteristics similar to those

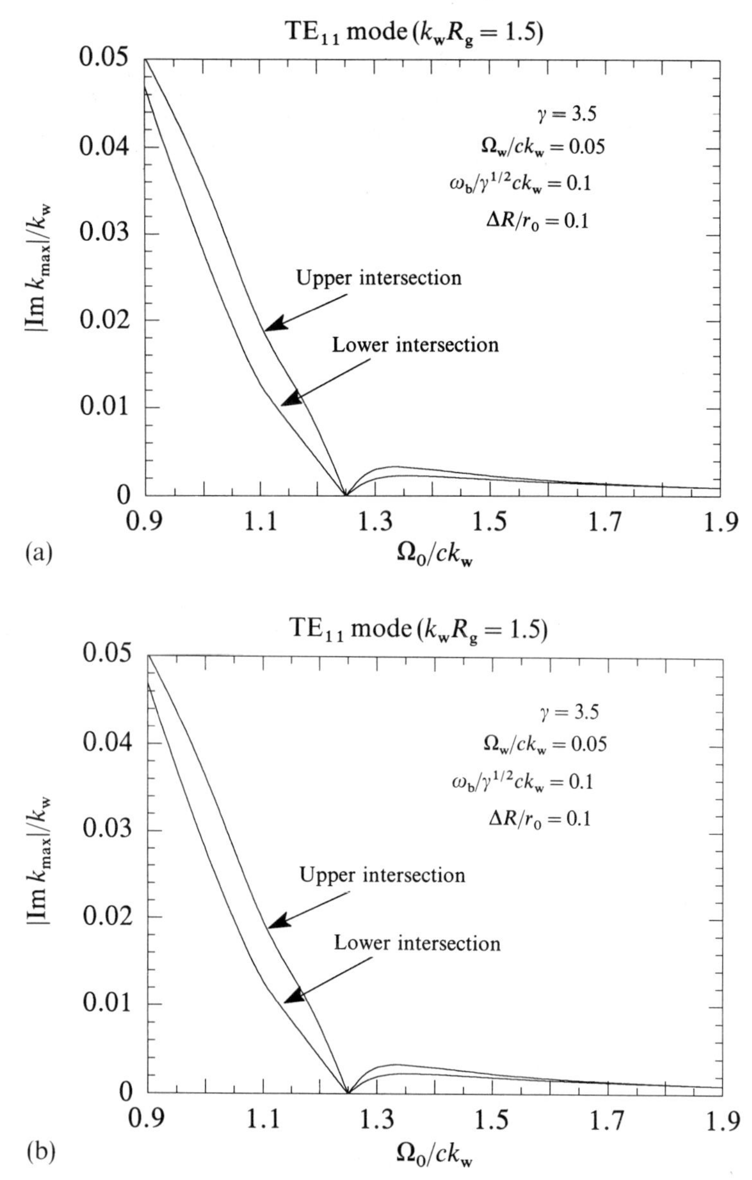

Fig. 4.14 The maximum growth rate (a) and corresponding frequency (b) of the TE$_{11}$ mode as a function of the axial magnetic field for the Group II orbits.

found for Group I trajectories. Specifically, (1) that the growth rate tends to increase in the vicinity of the singularity in Φ at $\Omega_0/ck_w \approx 0.62$, and (2) that the upper and lower intersections tend to coalesce as the axial velocity decreases in the neighbourhood of the magnetic resonance. Differences between the case of Group I and Group II trajectories, however, arise from several effects. In the first place, the resonance at $\Omega_0 \approx k_w v_\parallel$ can be more closely approached for the Group II trajectories, and the growth rate is correspondingly larger than that found for the Group I case. In the second place, Φ vanishes at the transition between the negative- and positive-mass regimes ($\Omega_0/ck_w \approx 1.25$ for the chosen parameters), and the growth rate vanishes at this point. In the third place, the space-charge wave is itself unstable in the negative-mass regime (i.e. $\Phi < 0$). In this case, the interaction is found at the approximate intersections between the vacuum waveguide modes and the beam resonance line ($\omega = (k + k_w)v_\parallel$), which yields

$$\frac{\omega}{ck_w} \cong \gamma_\parallel^2 \beta_\parallel \left[1 \pm \beta_\parallel \sqrt{\left(1 - \frac{\omega_{co}^2}{\gamma_\parallel^2 k_w^2 v_\parallel^2} \right)} \right]. \tag{4.297}$$

Finally, an interaction is possible for the TM_{11} mode as well for the chosen parameters. Indeed, the growth rates are comparable to those found for the TE_{11} mode, and the character of the interaction differs only insofar as the cutoff frequency is higher.

4.2.3 Comparison with experiment

The aforementioned dispersion equation has been applied to the analysis of collective Raman free-electron laser experiments conducted at Columbia

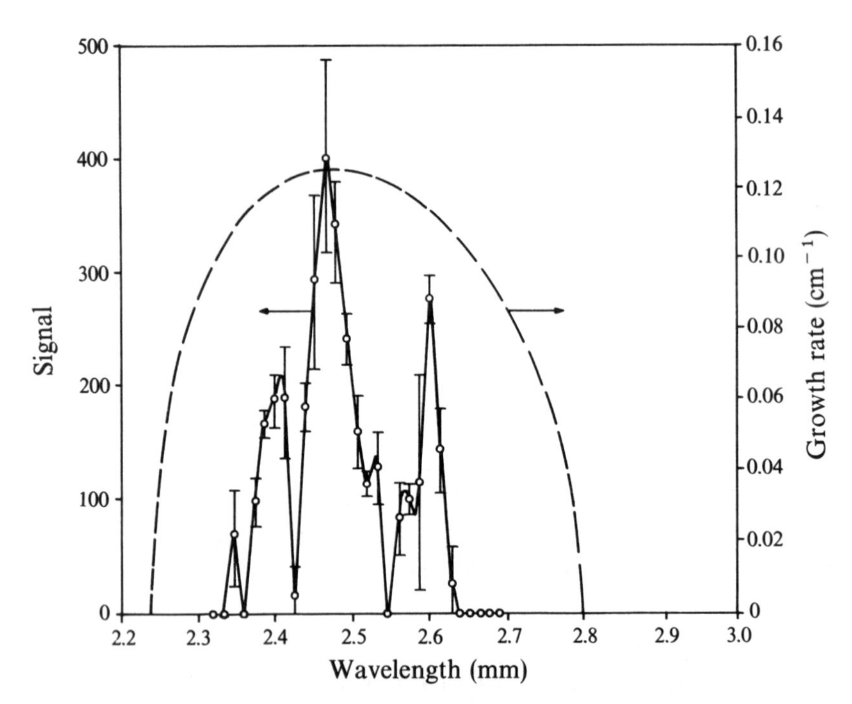

Fig. 4.15 Comparison of the observed spectrum of the Columbia University Raman free-electron laser experiment [45] and the theoretical prediction of the growth rate as a function of wavelength.

University [45] and the Massachusetts Institute of Technology [46]. Both of these experiments employed helical wigglers and axial solenoidal magnetic fields generated by current-carrying coils and, hence, were variable over a wide range of parameters. In addition, the principal resonance in each case was with the TE_{11} mode. The principal distinctions between these experiments involved the differences in the parameter ranges which could be accessed. As a result, these experiments together covered an extremely wide range of operating parameters.

The free-electron laser at Columbia University employed a beam with an energy of 700 keV and a current of 200 A. The wiggler was characterized by a period of 1.45 cm. A specific comparison with the three-dimensional analysis dealt with the case of a wiggler field of 760 G and a solenoidal field of 9.5 kG. As a consequence, the experiment operated with Group I trajectories for which $v_\parallel/c \approx 0.900$ and $\lambda_0 \approx 0.124$. Since the waveguide radius was 0.3 cm, the high-frequency intersection occurred at a frequency of approximately 124 GHz at a wavelength of approximately 2.5 mm. Agreement between the observed spectrum (both as regards the frequency and growth rate) and the three-dimensional linear theory was good. A comparison of the observed spectrum and the calculated growth rates are given in Fig. 4.15 as a function of wavelength. As shown in the figure, the peak in the spectrum is found at a wavelength of approximately 2.47 mm in both the experiment and the theory. The observed exponential growth rate was

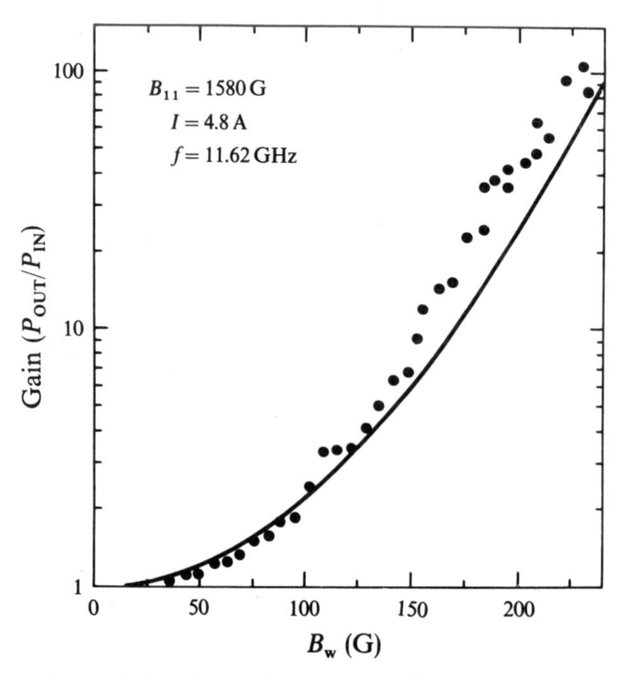

Fig. 4.16 Comparison of the observed spectrum of the Raman free-electron laser experiment at MIT and the theoretical prediction of the growth rate as a function of wiggler amplitude [46].

approximately $0.1\,\mathrm{cm}^{-1}$, which is also in good agreement with the theoretical result of $0.12\,\mathrm{cm}^{-1}$.

The experiment at the Massachusetts Institute of Technology was conducted at much lower voltages and currents, although the current density was sufficiently high for this experiment to be in the collective Raman regime as well. The specific parameters of this experiment involved a beam voltage and current of 160 keV and 4.8 A, respectively, with a beam radius of approximately 2.6 mm. The wiggler field amplitude was variable up to approximately 1.5 kG with a period of 3.3 cm, and the axial solenoidal field could achieve a maximum field strength of approximately 7 kG. As in the case of the experiment at Columbia University, the interaction was resonant with the TE_{11} mode, which had a cutoff frequency of approximately 6.92 GHz for the choice of a drift tube with a 2.54 cm inner diameter. An example of the comparison of the theoretically predicted growth rates and those measured in the experiments are shown in Fig. 4.16. The specific parameters used in this figure involved a beam current of 4.8 A and an axial guide field of 1.58 kG. The free-electron laser was run in an amplifier mode with a drive signal at a frequency of 11.62 GHz. The power gain was then measured as a function of the wiggler field amplitude over a range of up to 300 G. As is evident from the figure, the comparison between the experiment and the theoretical prediction is quite good over the entire range of wiggler-field amplitudes studied.

REFERENCES

1. Sukhatme, V. P. and Wolff, P. A. (1973) Stimulated Compton scattering as a radiation source – theoretical limitations. *J. Appl. Phys.*, **44**, 2331.
2. Kwan, T. J. T., Dawson, J. M. and Lin, A. T. (1977) Free-electron laser. *Phys. Fluids*, **20**, 581.
3. Kroll, N. M. and McMullin, W. A. (1978) Stimulated emission from relativistic electrons passing through a spatially periodic transverse magnetic field. *Phys. Rev. A*, **17**, 300.
4. Madey, J. M. J. (1979) Relationship between mean radiated energy, mean squared radiated energy and spontaneous power spectrum in a power series expansion of the equations of motion in a free-electron laser. *Nuovo Cimento*, **50B**, 64.
5. Kwan, T. J. T. and Dawson, J. M. (1979) Investigation of the free-electron laser with a guide magnetic field. *Phys. Fluids*, **22**, 1089.
6. Bernstein, I. B. and Hirshfield, J. L. (1979) Amplification on a relativistic electron beam in a spatially periodic transverse magnetic field. *Phys. Rev. A*, **20**, 1661.
7. Sprangle, P., Smith, R. A. and Granatstein, V. L. (1979) Free-electron lasers and stimulated scattering from relativistic electron beams, in *Infrared and Millimeter Waves*, Vol. 1 (ed. K. J. Button), Academic Press, New York, p. 279.
8. Friedland, L. and Hirshfield, J. L. (1980) Free-electron laser with a strong axial magnetic field. *Phys. Rev. Lett.*, **44**, 1456.
9. Freund, H. P., Sprangle, P., Dillenburg, D., da Jornada, E. H., Liberman, B. and Schneider, R. S. (1981) Coherent and incoherent radiation from free-electron lasers with an axial guide field. *Phys. Rev. A*, **24**, 1965.
10. Bernstein, B. and Friedland, L. (1981) Theory of free-electron laser in combined helical pump and axial guide magnetic field. *Phys. Rev. A*, **23**, 816.
11. Gover, A. and Sprangle, P. (1981) A generalized formulation of free-electron lasers in the low-gain regime including transverse velocity spread and wiggler incoherence. *J. Appl. Phys.*, **52**, 599.
12. Stenholm, S. T. and Bambini, A. (1981) Single-particle theory of the free-electron laser in a moving frame. *IEEE J. Quantum Electron.*, **QE-17**, 1363.

13. Shih, C. C. and Yariv, A. (1981) Inclusion of space-charge effects with Maxwell's equations in the single-particle analysis of free-electron lasers. *IEEE J. Quantum Electron.*, **QE-17**, 1387.
14. Coisson, R. (1981) Energy loss calculation of gain in a plane sinusoidal free-electron laser. *IEEE J. Quantum Electron.*, **QE-17**, 1409.
15. Colson, W. B. (1981) The nonlinear wave equation for higher harmonics in free-electron lasers. *IEEE J. Quantum Electron.*, **QE-17**, 1417.
16. Kroll, N. M., Morton, P. L. and Rosenbluth, M. N. (1981) Free-electron lasers with variable parameter wigglers. *IEEE J. Quantum Electron.*, **QE-17**, 1436.
17. Uhm, H. S. and Davidson, R. C. (1981) Free-electron laser instability for a relativistic annular electron beam in a helical wiggler field. *Phys. Fluids*, **24**, 2348.
18. Freund, H. P., Sprangle, P., Dillenburg, D., da Jornada, E. H., Schneider, R. S. and Liberman, B. (1982) Collective effects on the operation of free-electron lasers with an axial guide field. *Phys. Rev. A*, **26**, 2004.
19. Davies, J. A., Davidson, R. C. and Johnston, G. L. (1985) Compton and Raman free-electron laser stability properties for a cold electron beam propagating through a helical magnetic field. *J. Plasma Phys.*, **33**, 387.
20. Grover, L. K. and Pantell, R. H. (1985) Simplified analysis of free-electron lasers using Madey's theorem. *IEEE J. Quantum Electron.*, **QE-21**, 944.
21. Ibanez, L. F. and Johnston, S. (1983) Finite-temperature effects in free-electron lasers. *IEEE J. Quantum Electron.*, **QE-19**, 339.
22. Jerby, E. and Gover, A. (1985) Investigation of the gain regimes and gain parameters of the free-electron laser dispersion equations. *IEEE J. Quantum Electron.*, **QE-21**, 1041.
23. Freund, H. P., Davidson, R. C. and Kirkpatrick, D. A. (1991) Thermal effects on the linear gain in free-electron lasers. *IEEE J. Quantum Electron.*, **QE-27**, 2550.
24. Friedland, L. and Fruchtman, A. (1982) Amplification on relativistic electron beams in combined helical and axial magnetic fields. *Phys. Rev. A*, **25**, 2693.
25. Uhm, H. S. and Davidson, R. C. (1982) Helically distorted relativistic beam equilibria for free-electron laser applications. *J. Appl. Phys.*, **53**, 2910.
26. McMullin, W. A. and Davidson, R. C. (1982) Low-gain free-electron laser near cyclotron resonance. *Phys. Rev. A*, **25**, 3130.
27. Uhm, H. S. and Davidson, R. C. (1983) Free-electron laser instability for a relativistic solid electron beam in a helical wiggler field. *Phys. Fluids*, **26**, 288.
28. Freund, H. P. and Sprangle, P. (1983) Unstable electrostatic beam modes in free-electron laser systems. *Phys. Rev. A*, **28**, 1835.
29. Grebogi, C. and Uhm, H. S. (1986) Vlasov susceptibility of relativistic magnetized plasma and application to free-electron lasers. *Phys. Fluids*, **29**, 1748.
30. Ginzburg, N. S. (1987) Diamagnetic and paramagnetic effects in free-electron lasers. *IEEE Trans. Plasma Sci.*, **PS-15**, 411.
31. Freund, H. P., Davidson, R. C. and Johnston, G. L. (1990) Linear theory of the collective Raman interaction in a free-electron laser with a planar wiggler and an axial guide field. *Phys. Fluids B*, **2**, 427.
32. Cary, J. R. and Kwan, T. J. T. (1981) Theory of off-axis mode production by free-electron lasers. *Phys. Fluids*, **24**, 729.
33. Kwan, T. J. T. and Cary, J. R. (1981) Absolute and convective instabilities in two-dimensional free-electron lasers. *Phys. Fluids*, **24**, 899.
34. Freund, H. P., Johnston, S. and Sprangle, P. (1983) Three-dimensional theory of free-electron lasers with an axial guide field. *IEEE J. Quantum Electron.*, **QE-19**, 322.
35. Freund, H. P. and Ganguly, A. K. (1983) Three-dimensional theory of the free-electron laser in the collective regime. *Phys. Rev. A*, **28**, 3438.
36. Luchini, P. and Solimeno, S. (1985) Gain and mode-coupling in a three-dimensional free-electron laser: a generalization of Madey's theorem. *IEEE J. Quantum Electron.*, **QE-21**, 952.
37. Rosenbluth, M. N. (1985) Two-dimensional effects in free-electron lasers. *IEEE J. Quantum Electron.*, **QE-21**, 966.

38. Steinberg, B. Z., Gover, A. and Ruschin, S. (1987) Three-dimensional theory of free-electron lasers in the collective regime. *Phys. Rev. A*, **36**, 147.
39. Elliot, C. J. and Schmitt, M. J. (1987) Small-signal gain for a planar free-electron laser with a period magnetic field. *IEEE Trans. Plasma Sci.*, **PS-15**, 319.
40. Fruchtman, A. (1988) High-density thick beam free-electron laser. *Phys. Rev. A*, **37**, 4259.
41. Antonsen, T. M. and Latham, P. E. (1988) Linear theory of a sheet beam free-electron laser. *Phys. Fluids*, **31**, 3379.
42. Tripathi, V. K. and Liu, C. S. (1989) A slow wave free-electron laser. *IEEE Trans. Plasma Sci.*, **PS-17**, 583.
43. Fruchtman, A. and Weitzner, H. (1989) Raman free-electron laser with transverse density gradients. *Phys. Rev. A*, **39**, 658.
44. Yu, L. H., Krinsky, S. and Gluckstern, R. L. (1990) Calculation of universal scaling function for free-electron laser gain. *Phys. Rev. Lett.*, **64**, 3011.
45. Masud, J., Marshall, T. C., Schlesinger, S. P., Yee, F. G., Fawley, W. M., Scharlemann, E. T., Yu, S. S., Sessler, A. M and Sternbach, E. J. (1987) Sideband control in a millimeter-wave free-electron laser. *Phys. Rev. Lett.*, **58**, 763.
46. Fajans, J. and Bekefi, G. (1986) Measurement of amplification and phase-shift in a free-electron laser. *Phys. Fluids*, **29**, 3461.

5

Coherent emission: nonlinear theory

The self-consistent nonlinear theory of the free-electron laser describes the interaction through the linear regime and includes the saturation of the growth mechanism. Saturation can occur through a variety of mechanisms. For an ideal beam which is both monoenergetic and has a vanishing pitch-angle spread, saturation occurs by means of electron trapping in the ponderomotive potential. In the thermal regime, saturation occurs by a different process. In this case, the axial energy spread of the beam (which can arise due to either a distribution in the total energy of the beam electrons or pitch-angle spread) gives rise to a broad-band emission spectrum. As a result, a **quasilinear** saturation mechanism is operative in which the beam undergoes turbulent diffusion in momentum space. Since the growth rate in this regime is proportional to the slope of the distribution function, the turbulent diffusion acts to form a plateau in momentum space which *flattens out* the distribution of the beam. As a result, the axial energy spread of the beam increases, and the instability is quenched when the slope of the distribution falls to zero. As might be expected, however, the saturation efficiency in the thermal regime is much reduced relative to that found for a sufficiently cold beam in which saturation occurs through the particle-trapping mechanism. As a consequence, we shall focus attention on the latter case in this chapter.

The saturation mechanism in the particle-trapping regime is illustrated in Fig. 4.1. As shown in the figure, the separatrix grows during the linear phase of the interaction up to the point at which the beam electrons cross the separatrix. This point occurs approximately when the separatrix encloses the initial phase space position (i.e. in $d\psi/dz$) of the beam. The trapped electrons subsequently undergo a rotation in the ponderomotive well during which they give up energy to the wave during alternate half cycles. As a result, the growth of the wave continues until the beam electrons reach the bottom of the trough. At this point, the leading electrons begin to absorb energy from the wave, and saturation occurs when this energy absorption is balanced by the energy still being lost by the trailing electrons.

An *ad hoc* estimate of the saturation efficiency obtained by this process is found under the assumption that the individual electrons in the beam lose an amount of energy comparable to twice the difference in the bulk axial velocity of the beam, v_b, and the phase velocity of the trapping potential, v_{ph}, which is formed by the aggregate of the ponderomotive and space-charge potentials. The loss in beam kinetic energy due to this deceleration $\Delta v_b \approx -2(v_b - v_{ph})$ is given by

$$\Delta T \approx 2\gamma\gamma_\parallel^2 m_e v_b(v_{ph} - v_b), \tag{5.1}$$

where $\gamma_\parallel \equiv (1 - v_b^2/c^2)^{-1/2}$. The interaction efficiency, η, is defined by the ratio of the energy lost by the beam to the initial beam kinetic energy; specifically,

$$\eta \approx \frac{2\gamma\gamma_\parallel^2 v_b}{(\gamma - 1)c}\frac{(v_b - v_{ph})}{c}. \tag{5.2}$$

The phase velocity is given by $v_{ph} \approx \omega/(\mathrm{Re}\,k + k_w)$, where the real part of the wavenumber differs in the collective Raman and high-gain Compton regimes. Typically,

$$\mathrm{Re}\,k \approx \frac{\omega}{v_b} + \frac{\omega_b}{\gamma^{1/2}\gamma_\parallel v_b} + \mathrm{Re}\,\delta k, \tag{5.3}$$

where δk is given by solutions to equations (4.93) and (4.96) for the high-gain Compton and collective Raman regimes, respectively, in the idealized one-dimensional regime. In the high-gain Compton regime the ponderomotive potential is dominant and

$$\mathrm{Re}\,k \approx \frac{\omega}{v_b} + \frac{1}{2}k_w\left(\frac{v_w^2}{2c^2}\frac{\omega_b^2}{\gamma c^2 k_w^2}\right)^{1/3}, \tag{5.4}$$

at the frequency corresponding to maximum growth. Equation (1.14) for the efficiency in the high-gain Compton regime is recovered using this expression to calculate the phase velocity of the trapping ponderomotive potential. In contrast, the space-charge potential dominates in the collective Raman regime, and from equation (4.97)

$$\mathrm{Re}\,k \approx \frac{\omega}{v_b} + \frac{\omega_b}{\gamma^{1/2}\gamma_\parallel v_b}, \tag{5.5}$$

at the frequency of maximum growth. In this case, the efficiency is determined by equation (1.15). While these estimates of the saturation efficiency were made on the basis of the idealized one-dimensional model of the linear stability analysis, the *ad hoc* phase-trapping argument may be generalized to the three-dimensional as well by the simple expedient of using the results of the theory for the real part of the wavenumber.

The estimates of the saturation efficiency obtained from the phase-trapping arguments are applicable to configurations employing uniform wigglers (i.e. wigglers with constant amplitudes and periods). However, as first demonstrated experimentally by Phillips [1] and subsequently by others [2–5], it is possible to enhance the saturation efficiency by using a nonuniform wiggler in which either the amplitude or period is tapered in the direction of the symmetry axis [6–13]. In addition, for configurations which include an axial solenoidal magnetic field, the efficiency may be enhanced by tapering this field as well [14]. The effect of tapering either the wiggler or solenoidal magnetic fields is to alter both the transverse and axial velocities of the beam electrons. The fundamental concept underlying this mechanism is based upon the fact that as the beam gives up energy to the wave, it decelerates in the axial direction until ultimately it drops out of resonance. The fields must be tapered, therefore, in such a way as to accelerate the beam and maintain the phase relationship necessary for the resonant interaction. It should be remarked that in the negative-mass regime appropriate to strong axial solenoidal fields and Group II orbits, the effect of the loss of energy by the beam to the wave is to accelerate the beam in the direction of the symmetry axis. In this regime, therefore, the fields must be tapered in the opposite sense so as to decelerate the beam in order to achieve an enhancement in the saturation efficiency. In either case, however, the process is effective only if the beam has crossed the separatrix and has become trapped in the ponderomotive potential. For this reason, the field taper must begin at the point at which the beam crosses the separatrix, and the total efficiency is the sum of the efficiency in the uniform wiggler region and that determined by the tapered fields.

In order to treat the effect of the field taper analytically, we perturb the orbit equations about the steady-state trajectories to first order in both the radiation field and the gradient in the wiggler and solenoidal magnetic fields. For the sake of simplicity, we consider the case of a helical wiggler field in the idealized one-dimensional limit, and the amplification of a circularly polarized electromagnetic wave as given by equation (4.1). The combined effect of the electromagnetic wave and the gradients in the magnetostatic fields is obtained by the combination of equations (2.154) and (4.11), and the nonlinear pendulum equation may be expressed as

$$\frac{d^2}{dz^2}\psi = -K^2(\sin\psi - \sin\psi_{res}), \tag{5.6}$$

where K^2 is given by equation (4.14), and the resonant phase ψ_{res} is

$$\sin\psi_{res} = -\frac{k_w}{K^2\lambda_w}\frac{d}{dz}\lambda_w - \frac{k+k_w}{K^2}\frac{\beta_w^2}{(1+\beta_w^2)\Omega_0 - k_w v_\parallel} \times\left(\frac{\Omega_0 - k_w v_\parallel}{B_w}\frac{d}{dz}B_w - \frac{k_w v_\parallel}{\lambda_w}\frac{d}{dz}\lambda_w - \frac{\Omega_0}{B_0}\frac{d}{dz}B_0\right). \tag{5.7}$$

Integration of this equation results in an equation similar to that found for a uniform magnetostatic field (4.17), with a ponderomotive potential

$$U(\psi) = -K^2(\cos\psi + \psi\sin\psi_{res}). \tag{5.8}$$

The tapered field, therefore, results in either a linear increase or decrease in the ponderomotive potential. The difference between the ponderomotive potential for a uniform and tapered wiggler is illustrated in Fig. 5.1 for a case in which $\sin\psi_{res} < 0$. As a result, the motion is similar to that of a ball rolling down a bumpy hill and accelerating as it falls. For Group I orbits in which $\Omega_0 < k_w v_\parallel$,

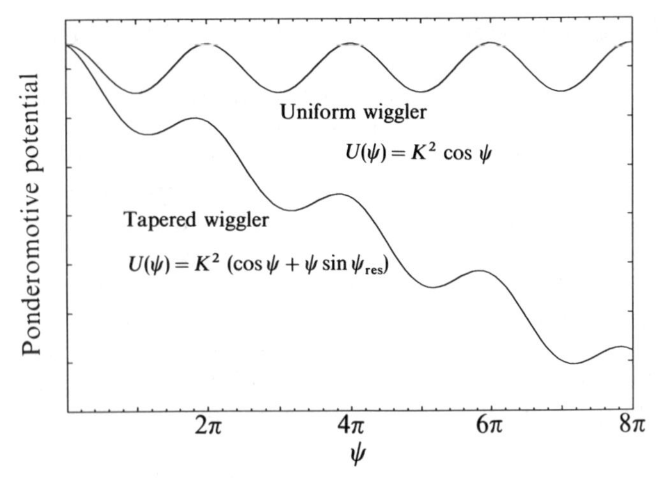

Fig. 5.1 A comparison of the ponderomotive potential for uniform and tapered magnetostatic fields.

this acceleration implies that the wiggler field amplitude and period and the axial solenoidal field must be tapered downward for efficiency enhancement to occur. The situation is more complicated for Group II orbits in which $\Omega_0 > k_w v_\parallel$. In this case, the wiggler period and axial solenoidal field must be tapered downward when $\Phi < 0$ while the wiggler field amplitude must be tapered upward in order to achieve acceleration. As a consequence, efficiency enhancement in the negative-mass regime requires either an upward taper in the wiggler amplitude or a downward taper in the wiggler period and the axial solenoidal field. In contrast, when $\Phi > 0$ acceleration is found when either the wiggler period is tapered downward or the axial field and wiggler period are tapered upward.

The magnitude of the efficiency enhancement can be calculated by noting that

$$\frac{d\gamma}{dz} = \frac{\omega}{v_\parallel^2}\left(\frac{d}{d\gamma}v_\parallel\right)^{-1}\frac{d^2}{dz^2}\psi. \tag{5.9}$$

Substitution of equations (2.23) and (5.6), therefore, gives

$$\frac{d\gamma}{dz} = \beta_w \frac{e\delta A}{m_e c^2}\frac{\omega}{c}(\sin\psi - \sin\psi_{res}). \tag{5.10}$$

Under the assumption of a random phase approximation (which is appropriate for a fully filled ponderomotive well), the average rate of energy loss is determined by the resonant phase (i.e. $\langle \sin\psi \rangle = 0$); hence,

$$\left\langle \frac{d\gamma}{dz} \right\rangle = -\frac{\gamma_0 \beta_\parallel^2}{2\Phi}\left[\frac{1}{\lambda_w}\frac{d}{dz}\lambda_w + \frac{2\gamma_\parallel^2\beta_w^2}{(1+\beta_w^2)\Omega_0 - k_w v_\parallel} \times \left(\frac{\Omega_0 - k_w v_\parallel}{B_w}\frac{d}{dz}B_w - \frac{k_w v_\parallel}{\lambda_w}\frac{d}{dz}\lambda_w - \frac{\Omega_0}{B_0}\frac{d}{dz}B_0\right)\right]. \tag{5.11}$$

As a consequence, the efficiency enhancement, $\Delta\eta$, corresponding to changes in the field amplitudes ΔB_w and ΔB_0 and the wiggler period $\Delta\lambda_w$ over some axial length Δz is

$$\Delta\eta = -\frac{\gamma_0\beta_\parallel^2}{2(\gamma_0 - 1)\Phi}\left[\frac{\Delta\lambda_w}{\lambda_w} + \frac{2\gamma_\parallel^2\beta_w^2}{(1+\beta_w^2)\Omega_0 - k_w v_\parallel} \times \left((\Omega_0 - k_w v_\parallel)\frac{\Delta B_w}{B_w} - k_w v_\parallel \frac{\Delta\lambda_w}{\lambda_w} - \Omega_0 \frac{\Delta B_0}{B_0}\right)\right]. \tag{5.12}$$

In the limit in which the axial solenoidal field vanishes and the wiggler period is fixed, this result reduces to that given in equation (1.18).

The above discussion of the nonlinear interaction mechanism is primarily concerned with the interaction of a wave(s) of a single frequency with the electron beam. In this case, a nonlinear formalism can be derived based upon an average of Maxwell's equations over a wave period to remove the fast time-scale oscillation from the nonlinear analysis, which greatly simplifies the numerical procedures. This chapter will be primarily concerned with the development of such slow time-scale formulations, as well as the application of the analyses to the description of the fundamental physics of the nonlinear saturation mechanism and to the analysis of specific experiments. This formulation of the nonlinear interaction in free-electron lasers is applicable primarily to single-pass amplifier configurations, although it can be applied to the analysis of the saturation efficiency in narrow-band oscillators which operate in a single-frequency regime. However, a detailed treatment of oscillator configurations is presented in Chapter 9.

5.1 ONE-DIMENSIONAL ANALYSIS: HELICAL WIGGLERS

There is an extensive literature dealing with idealized one-dimensional models of the interaction in free-electron lasers [6–27] including both helical and planar wiggler configurations, both amplifiers [6, 14–17, 22–4, 26, 27] and oscillators [13, 18–21, 25], and with slow time-scale formulations [6, 14, 17, 24] as well as full-scale one-dimensional particle-in-cell simulations [15, 26, 27]. Slow time-scale formulations are suitable for the description of free-electron laser amplifiers in which it is reasonable to assume that the wave phase and amplitude vary slowly with respect to the wavelength and frequency; hence, Maxwell's equations may be averaged over a wave period. This results in an enormous increase in the computational efficiency since the integration of the Maxwell–Lorentz equations need not resolve the fast time scale corresponding to the wave frequency. In contrast, the Maxwell–Lorentz equations must be integrated on a fast time scale corresponding to the wave period in particle-in-cell techniques. In this section, we shall concentrate on the slow time-scale formulation of free-electron laser amplifiers.

5.1.1 The dynamical equations

The physical configuration used to develop a self-consistent, one-dimensional nonlinear formulation of the free-electron laser interaction consists of a magnetostatic field which is composed of both a solenoidal guide magnetic field and a helical wiggler field; specifically,

$$\boldsymbol{B}(z) = B_0\hat{e}_z + B_w(z)(\hat{e}_x\cos k_w z + \hat{e}_y \sin k_w z), \tag{5.13}$$

where B_0 and $B_w(z)$ denote the amplitudes of the solenoidal and wiggler fields, and k_w is the wiggler wavenumber. The wiggler amplitude has been assumed to be a slowly varying function of axial position in order to treat either the injection of the beam into the wiggler or efficiency enhancement by means of a tapered wiggler. The problem of the injection of the beam into the wiggler is of considerable practical importance, because it is necessary to minimize the fluctuations of the beam about the steady-state trajectories in order to minimize a uniformly resonant interaction. For this purpose, therefore, the wiggler field is assumed to increase adiabatically from zero to a constant value over N_w wiggler periods as follows [17]

$$B_w(z) = \begin{cases} B_w \sin^2\left(\dfrac{k_w z}{4N_w}\right); & 0 \leqslant z \leqslant N_w\lambda_w \\ B_w; & N_w\lambda_w < z. \end{cases} \tag{5.14}$$

Observe that in order to neglect the fringing fields associated with the wiggler gradient, we assume implicitly that the wiggler amplitude varies slowly over a wiggler (i.e. $N_w \gg 1$). Efficiency enhancement by means of a tapered magnetostatic field will be treated within the context of the three-dimensional analysis.

The field equations

The fluctuating electromagnetic and electrostatic fields are treated using the vector and scalar potentials in the Coulomb gauge, and we assume that these fields are of the form

$$\delta\boldsymbol{A}(z,t) = \delta A(z)[\hat{e}_x \cos\alpha_+(z,t) - \hat{e}_y \sin\alpha_+(z,t)], \tag{5.15}$$

and

$$\delta\Phi(z,t) = \delta\Phi(z)\cos\alpha(z,t), \tag{5.16}$$

where the phase of an electromagnetic wave of frequency ω and wavenumber k_+ is given by

$$\alpha_+(z,t) = \int_0^z dz' k_+(z') - \omega t, \tag{5.17}$$

and

$$\alpha(z,t) = \int_0^z dz' k(z') - \omega t, \tag{5.18}$$

is the phase of the space-charge wave with the same frequency but wavenumber k. This is equivalent to a WKB formulation in which it is assumed implicitly that the amplitudes and wavenumbers vary slowly over a wavelength.

In the Coulomb gauge, the Maxwell–Poisson equations are of the form

$$\left(\frac{\partial^2}{\partial z^2} - \frac{1}{c^2}\frac{\partial^2}{\partial t^2}\right)\delta \boldsymbol{A}(z,t) = -\frac{4\pi}{c}\,\delta \boldsymbol{J}_\perp(z,t), \tag{5.19}$$

$$\frac{\partial^2}{\partial z\,\partial t}\,\delta\Phi(z,t) = 4\pi\delta J_z(z,t), \tag{5.20}$$

and

$$\frac{\partial^2}{\partial z^2}\,\delta\Phi(z,t) = -4\pi\delta\rho(z,t), \tag{5.21}$$

where $\delta\boldsymbol{J}(z,t)$ and $\delta\rho(z,t)$ are the nonlinear current and charge densities. The charge and current densities can be written [17] as averages over the entry time t_0 (defined as the time at which an electron crosses the $z=0$ plane)

$$\delta\boldsymbol{J}(z,t) = -en_b v_{z0}\int_{-\infty}^{\infty} dt_0\,\sigma(t_0)\mathbf{v}(t,t_0)\frac{\delta[t-\tau(z,t_0)]}{|v_z(t,t_0)|}, \tag{5.22}$$

and

$$\delta\rho(z,t) = -en_b v_{z0}\int_{-\infty}^{\infty} dt_0\,\sigma(t_0)\frac{\delta[t-\tau(z,t_0)]}{|v_z(t,t_0)|}, \tag{5.23}$$

where n_b is the ambient electron density, v_{z0} is the initial axial velocity, $\mathbf{v}(t,t_0)$ is the velocity of an electron at time t which crossed the entry plane at time $t_0, \sigma(t_0)$ is the distribution in entry times, and

$$\tau(z,t_0) = t_0 + \int_0^z \frac{dz'}{v_z(z',t_0)}, \tag{5.24}$$

is a Lagrangian time coordinate which measures the time at which an electron reaches the axial position z, and $v_z(z,t_0)$ denotes the velocity of an electron at axial position z which crossed the entry plane at time t_0. It should be remarked that it is assumed implicitly in the charge and current densities that the electron beam is monoenergetic and with a vanishing pitch-angle spread. This is not essential to the formulation, and a more general momentum space distribution will be employed in the context of the three-dimensional formulation. In addition, it should be noted that the average over entry times is equivalent to a discrete numbering system for the electrons.

A set of coupled nonlinear differential equations for the slowly varying amplitudes and phases is obtained by the straightforward substitution of the representations for the fields in equations (5.15) and (5.16) into the Maxwell–Poisson equations. Substitution of the representation of the vector potential into Maxwell's equations, for example, yields

$$\left(\frac{d^2}{dz^2} + \frac{\omega^2}{c^2} - k_+^2\right)\delta A\cos\alpha_+ - 2k_+^{1/2}\frac{d}{dz}(k_+^{1/2}\delta A)\sin\alpha_+ = -\frac{4\pi}{c}\delta J_x, \tag{5.25}$$

and

$$\left(\frac{d^2}{dz^2}+\frac{\omega^2}{c^2}-k_+^2\right)\delta A\sin\alpha_+ + 2k_+^{1/2}\frac{d}{dz}(k_+^{1/2}\delta A)\cos\alpha_+ = \frac{4\pi}{c}\delta J_y. \quad (5.26)$$

We now multiply equation (5.25) by $\cos\alpha_+$, equation (5.26) by $\sin\alpha_+$, average over a wave period, and add the resulting equations to obtain

$$\left(\frac{d^2}{dz^2}+\frac{\omega^2}{c^2}-k_+^2\right)\delta a = \frac{\omega_b^2}{c^2}\beta_{z0}\frac{\omega}{2\pi}\int_0^{2\pi/\omega}dt\int_{-\infty}^{\infty}dt_0\sigma(t_0)\frac{\delta[t-\tau(z,t_0)]}{|v_z(z,t_0)|}$$
$$\times(v_x\sin\alpha_+ - v_y\sin\alpha_+), \quad (5.27)$$

after substitution of the source currents. By an analogous procedure, we obtain the complementary equation

$$2k_+^{1/2}\frac{d}{dz}(k_+^{1/2}\delta a) = -\frac{\omega_b^2}{c^2}\beta_{z0}\frac{\omega}{2\pi}\int_0^{2\pi/\omega}dt\int_{-\infty}^{\infty}dt_0\sigma(t_0)\frac{\delta[t-\tau(z,t_0)]}{|v_z(z,t_0)|}$$
$$\times(v_x\sin\alpha_+ + v_y\cos\alpha_+), \quad (5.28)$$

where $\beta_{z0}\equiv v_{z0}/c, \omega_b$ is the plasma frequency denoted by the ambient beam density, and $\delta a\equiv e\delta A(z)/m_ec^2$ is the normalized amplitude of the vector potential. Similarly, Poisson's equation yields the following equations for the evolution of the amplitude and wavenumber of the space-charge potential

$$\left(\frac{d^2}{dz^2}-k^2\right)\delta\varphi = 2\frac{\omega_b^2}{c^2}v_{z0}\frac{\omega}{2\pi}\int_0^{2\pi/\omega}dt\int_{-\infty}^{\infty}dt_0\sigma(t_0)\frac{\delta[t-\tau(z,t_0)]}{|v_z(z,t_0)|}\cos\alpha, \quad (5.29)$$

and

$$2k^{1/2}\frac{d}{dz}(k^{1/2}\delta\varphi) = -2\frac{\omega_b^2}{c^2}v_{z0}\frac{\omega}{2\pi}\int_0^{2\pi/\omega}dt\int_{-\infty}^{\infty}dt_0\sigma(t_0)\frac{\delta[t-\tau(z,t_0)]}{|v_z(z,t_0)|}\sin\alpha, \quad (5.30)$$

where $\delta\varphi\equiv e\delta\Phi(z)/m_ec^2$ is the normalized amplitude of the space-charge potential.

The double integrals which appear in equations (5.27–30) are reducible to a single integral over the entry time of a **beamlet** of electrons which cross the entry plane within one wave period under a **quasistatic** assumption. In order to understand this concept, observe that all the beam electrons execute identical trajectories in the absence of the fluctuating fields. For a steady-state amplifier model in which the fluctuating fields vary in axial position, but not in time, two electrons which enter the interaction region separated in time by integral multiples of the wave period will execute identical trajectories. As a consequence, we assume that

$$\mathbf{v}(t+2\pi N/\omega, t_0+2\pi N/\omega) = \mathbf{v}(t,t_0), \quad (5.31)$$

and

$$\tau(z, t_0+2\pi N/\omega) = \tau(z,t_0), \quad (5.32)$$

for integer N. The integrands in equations (5.27–30) are of the generic form $G(z,t,t_0)\delta[t-\tau(z,t_0)]$, where

$$G(z,t+2\pi N/\omega,t_0+2\pi N/\omega)=G(z,t,t_0). \tag{5.33}$$

On the basis of this symmetry condition, the above-mentioned double integrals reduce to

$$\begin{aligned}\int_0^{2\pi/\omega} dt \int_{-\infty}^{\infty} dt_0\, G(z,t,t_0)\delta[t-\tau(z,t_0)]\\ = \int_0^{2\pi/\omega} dt_0 \int_{-\infty}^{\infty} dt G(z,t,t_0)\delta[t-\tau(z,t_0)]\\ = \int_0^{2\pi/\omega} dt_0\, G[z,\tau(z,t_0),t_0].\end{aligned} \tag{5.34}$$

Hence, the problem is reduced to the consideration of only one beamlet of electrons which enters the interaction region during a wave period. This has the effect of greatly reducing the computational requirements since there is no necessity of filling the entire interaction region with electrons.

The equations governing the evolution of the slowly varying amplitudes and phases are simplified considerably using this symmetry property of the integrands (5.34). Transforming to the wiggler frame in which $v_x=v_1\cos k_w z-v_2\sin k_w z$, $v_y=v_1\sin k_w z+v_2\cos k_w z$, and $v_3=v_z$, we find that [17]

$$\left(\frac{d^2}{dz^2}+\frac{\omega^2}{c^2}-k_+^2\right)\delta a=\frac{\omega_b^2}{c^2}\beta_{z0}\left\langle\frac{v_1\cos\psi-v_2\sin\psi}{|v_3|}\right\rangle, \tag{5.35}$$

$$2k_+^{1/2}\frac{d}{dz}(k_+^{1/2}\delta a)=-\frac{\omega_b^2}{c^2}\beta_{z0}\left\langle\frac{v_1\sin\psi-v_2\cos\psi}{|v_3|}\right\rangle, \tag{5.36}$$

$$\left(\frac{d^2}{dz^2}-k^2\right)\delta\varphi=2\frac{\omega_b^2}{c^2}v_{z0}\left\langle\frac{\cos\psi_{sc}}{|v_3|}\right\rangle, \tag{5.37}$$

and

$$2k^{1/2}\frac{d}{dz}(k^{1/2}\delta\varphi)=-2\frac{\omega_b^2}{c^2}v_{z0}\left\langle\frac{\sin\psi_{sc}}{|v_3|}\right\rangle, \tag{5.38}$$

where the ponderomotive phase is defined in Lagrangian coordinates as

$$\psi=\psi_0+\int_0^z dz'\left(k_++k_w-\frac{\omega}{v_3}\right), \tag{5.39}$$

and the phase of the space-charge wave is

$$\psi_{sc}=\psi_0+\int_0^z dz'\left(k-\frac{\omega}{v_3}\right). \tag{5.40}$$

Observe that in the Raman regime, the wavenumbers of the space-charge mode and the electromagnetic mode differ by the wiggler wavenumber (i.e. $k\approx k_++k_w$).

As a consequence, $\psi \approx \psi_{sc}$ during the course of the interaction. The averaging operator is defined over the initial phase $\psi_0 = -\omega t_0$ as

$$\langle(\cdots)\rangle \equiv \frac{1}{2\pi}\int_{-\pi}^{\pi} d\psi_0 \sigma(\psi_0)(\cdots). \tag{5.41}$$

The electron orbit equations

In order to complete the formulation, the electron orbit equations in the presence of the static and fluctuating fields must be specified. Since we treat the case of a steady-state amplifier configuration, we integrate in z and write the Lorentz force equations in the form [17]

$$\frac{d}{dz}u_1 = -(\hat{\Omega}_0 - k_w u_3)\frac{u_2}{u_3} + c\left[\delta a\left(\frac{\omega}{v_3} - k_+\right)\sin\psi + \cos\psi\frac{d}{dz}\delta a\right], \tag{5.42}$$

$$\frac{d}{dz}u_2 = (\hat{\Omega}_0 - k_w u_3)\frac{u_1}{u_3} - \hat{\Omega}_w + c\left[\delta a\left(\frac{\omega}{v_3} - k_+\right)\cos\psi - \sin\psi\frac{d}{dz}\delta a\right], \tag{5.43}$$

and

$$\frac{d}{dz}u_3 = \hat{\Omega}_w\frac{u_2}{u_3} + ck_+\delta a\frac{u_1\sin\psi + u_2\cos\psi}{u_3} - c\frac{d}{dz}\delta a\frac{u_1\cos\psi - u_2\sin\psi}{u_3}$$
$$-\frac{c^2}{v_3}\left(k\delta\varphi\sin\psi_{sc} - \frac{d}{dz}\delta\varphi\cos\psi_{sc}\right), \tag{5.44}$$

where $\boldsymbol{u} \equiv \boldsymbol{p}/m_e$ denotes a momentum-like variable, $\hat{\Omega}_{0,w} \equiv eB_{0,w}/m_e c$, and B_w is given by equation (5.14). In addition, we integrate a time-like coordinate for each electron

$$\frac{d}{dz}\psi = k_+ + k_w - \frac{\omega}{v_3}. \tag{5.45}$$

It is important to note that there is no necessity to obtain a reduced form of the Lorentz force equations by averaging over a wiggler period. It has been assumed implicitly that the amplitude and phase are slowly varying functions of position, and this occurs only in the vicinity of the wave–particle resonance at $\omega \approx (k_+ + k_w)v_3$. When this is satisfied, the ponderomotive phase is also a slowly varying quantity (i.e. $d\psi/dz \approx 0$), and the Lorentz force equations describe trajectories which vary slowly with respect to the wave period as well. Hence, the orbit equations describe implicitly a slowly varying interaction near resonance, and no wiggler average is required.

The numerical procedure

Together, equations (5.35–8) for the fields and (5.42–4) for an ensemble of electrons represent a complete and self-consistent formulation of the interaction in the free-electron laser. The field equations constitute a set of four second-order differential equations for the amplitudes and phases of the fluctuating fields. This may be reduced to a set of eight first-order differential equations for $\delta a, \Gamma_+, k_+, \Delta\phi_+, \delta\varphi, \Gamma, k$ and $\Delta\phi$, where $\Delta\phi_+$ and $\Delta\phi$ denote the relative phases defined by

$$\Delta\phi_+ \equiv \int_0^z dz' [k_+(z') - k_+(0)], \tag{5.46}$$

$$\Delta\phi \equiv \int_0^z dz' [k(z') - k(0)], \tag{5.47}$$

and Γ_+ and Γ define the growth rates (i.e. the logarithmic derivatives) of the field vector and scalar potentials. These equations may be expressed as

$$\frac{d}{dz}\delta a = \Gamma_+ \delta a, \tag{5.48}$$

$$\frac{d}{dz}\Gamma_+ = -\left(\frac{\omega^2}{c^2} + \Gamma_+^2 - k_+^2\right) + \frac{\omega_b^2}{c^2}\frac{\beta_{z0}}{\delta a}\left\langle \frac{v_1 \cos\psi - v_2 \sin\psi}{|v_3|}\right\rangle, \tag{5.49}$$

$$\frac{d}{dz}\Delta\phi_+ = k_+, \tag{5.50}$$

$$\frac{d}{dz}k_+ = -2k_+\Gamma_+ - \frac{\omega_b^2}{c^2}\frac{\beta_{z0}}{\delta a}\left\langle \frac{v_1 \sin\psi + v_2 \cos\psi}{|v_3|}\right\rangle, \tag{5.51}$$

$$\frac{d}{dz}\delta\varphi = \Gamma\delta\varphi, \tag{5.52}$$

$$\frac{d}{dz}\Gamma = k^2 - \Gamma^2 + 2\frac{\omega_b^2}{c^2}\frac{v_{z0}}{\delta\varphi}\left\langle \frac{\cos\psi_{sc}}{|v_3|}\right\rangle, \tag{5.53}$$

$$\frac{d}{dz}\Delta\phi = k, \tag{5.54}$$

$$\frac{d}{dz}k = -2k\Gamma - 2\frac{\omega_b^2}{c^2}\frac{v_{z0}}{\delta\varphi}\left\langle \frac{\sin\psi_{sc}}{|v_3|}\right\rangle. \tag{5.55}$$

The problem, therefore, consists in the simultaneous solution of $4N + 8$ first-order differential equations, where N denotes the total number of electrons, subject to the initial conditions. The numerical procedure involves the evaluation of the particle averages for the source current and charge density at each step in the integration, and the subsequent incrementation of the field quantities and the electron trajectories. This may be accomplished by a variety of integration schemes including the Adams–Moulton predictor/corrector as well as Runge–Kutta algorithms. The quantities that are averaged in the dynamical equations are functions of the instantaneous values of the momenta and phases, which are *implicit* functions of the *initial* phases of the electrons. It is the *initial* phase which appears in the averaging operator and represents a convenient method of discretizing (i.e. counting) the electrons based upon their initial phase relative to the wave. The most efficient method of discretization for a free-electron laser is to use an Nth-order Gaussian quadrature technique to distribute the electrons in their initial phases. Within this context, efficient refers to the convergence of a discrete sum over particles to the continuous integral with the fewest number of particles in the simulation. In comparison, Simpson's rule requires at least twice as many electrons to achieve the same degree of accuracy.

The initial conditions

The specific initial conditions appropriate to the analysis are those of a continuous and uniform electron beam, and we choose $\sigma(\psi_0)=1$ for $-\pi<\psi_0<\pi$. This describes an electron beam with an initial random phase distribution. The beam is also assumed to be monoenergetic and propagating paraxially (i.e. with a zero emittance). As a consequence, $v_{z0}=(1-\gamma_0^{-2})^{1/2}c$, where γ_0 is relativistic factor corresponding to the total beam energy. It is important to recognize, however, that the subsequent evolution of the beam is integrated self-consistently, and the beam may bunch in axial phase as well as develop both energy and pitch-angle spreads due to the nature of the interaction. The initial conditions of the wave fields are chosen as follows. Since the wiggler field increases adiabatically from zero at the entry plane, the growth rate of the vector potential is initially zero as well. The wavenumbers of the vector and scalar potentials are chosen to satisfy the uncoupled dispersion equations for a circularly polarized electromagnetic wave

$$\omega^2-k_+^2c^2-\frac{\omega_b^2(\omega-k_+v_{z0})}{\gamma_0(\omega-\Omega_0-k_+v_{z0})}=0, \tag{5.56}$$

and the beam-plasma mode

$$(\omega-kv_{z0})^2-\kappa^2v_{z0}^2=0, \tag{5.57}$$

of angular frequency ω. Since we treat the case of a steady-state amplifier model, the initial amplitude of the vector potential can be selected arbitrarily to represent the amplitude of the injected signal. However, the scalar potential must be chosen to represent the initial level of fluctuations (i.e. noise) in the beam. This is conveniently determined by means of the z-component of Maxwell's equations (5.21). Following a procedure analogous to that used for Poisson's equation, we find that the scalar potential and its derivative must also satisfy

$$\delta\varphi=-2\frac{\omega_b^2}{\omega kc^2}v_{z0}\langle\cos\psi_{sc}\rangle, \tag{5.58}$$

and

$$\frac{d}{dz}\delta\varphi=-2\frac{\omega_b^2}{\omega k_wc^2}v_{z0}\langle\sin\psi_{sc}\rangle. \tag{5.59}$$

For a random initial phase distribution modelled by a discrete and finite number of electrons the phase averages in these equations are not identically zero, and they may be used to define a convenient initial fluctuation level for the space-charge modes. Typically, the initial average over the $\langle \sin \psi_{sc} \rangle \approx 0$, and we may choose $\Gamma(z=0)=0$. However, the average over $\cos \psi_{sc}$ can be relatively large due to the discretization, and we choose

$$\delta\varphi(z=0) = -2\frac{\omega_b^2}{\omega k c^2} v_{z0} \langle \cos \psi_0 \rangle. \tag{5.60}$$

5.1.2 Electron beam injection

Before proceeding to the complete solution of the dynamical equations, we turn to the question of the injection of the electron beam into the wiggler. This is an issue of practical importance since the beam is not created in the wiggler field, and the electron optics required to transport the beam from the accelerator to the wiggler can often be a complicated system in itself. The specific question we examine here is the effect of an adiabatic increase in the wiggler amplitude such as that described in equation (5.14) on the trajectories of electrons which are propagating paraxially upon entry to the wiggler.

The appropriate equations of motion are a subset of the complete orbit equations given in equations (5.42–4), and are given as

$$\frac{d}{dz} v_1 = -(\Omega_0 - k_w v_3)\frac{v_2}{v_3}, \tag{5.61}$$

$$\frac{d}{dz} v_2 = (\Omega_0 - k_w v_3)\frac{v_1}{v_3} - \Omega_w, \tag{5.62}$$

and

$$\frac{d}{dz} v_3 = \Omega_w \frac{v_2}{v_3}, \tag{5.63}$$

where $\Omega_{0,w} \equiv |eB_{0,w}/\gamma m_e c|$, and γ is a constant of the motion. These equations are formally identical to equations (2.10–12) except that we choose to integrate in the axial position rather than time and that the wiggler amplitude is not constant but is given in equation (5.14).

The desired result is to inject the beam on to the steady-state trajectories in which the axial velocity is a constant. This ensures that the wave–particle resonance will be maintained over an extended interaction length, and maximizes the gain and efficiency of the interaction. In order to determine the efficacy of the adiabatic entry taper, the first example we consider is one in which the axial solenoidal field is absent, and the steady-state trajectories in the wiggler correspond to the Group I orbits. The specific parameters of this example are a beam energy corresponding to $\gamma = 3.5$, and a wiggler field in which $\Omega_w/ck_w = 0.05$ and the entry taper is effected over 10 wiggler periods (i.e. $N_w = 10$). The results of the integration of equations (5.61–3) for these parameters are shown in Fig. 5.2 in which we plot v_1 and v_2 versus axial distance. In this case, the injection process works well and the electron trajectory exhibits small oscillations about the steady-state orbit for

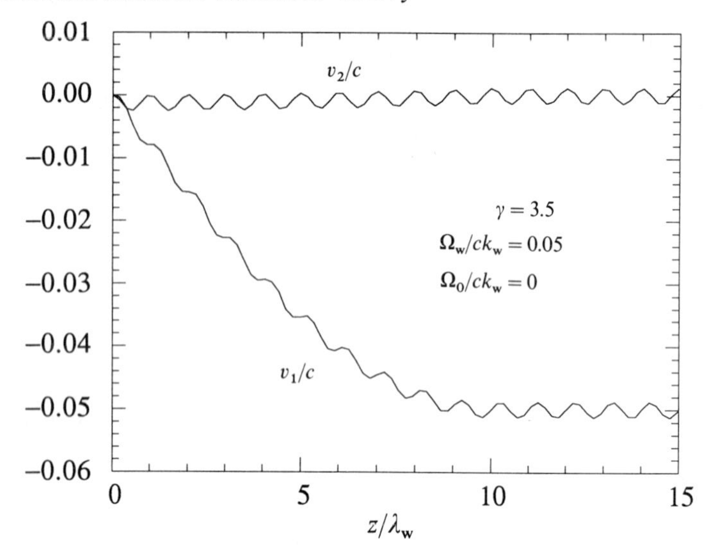

Fig. 5.2 Plot of the single-particle trajectories versus axial distance for $\Omega_0/ck_w = 0$, $\Omega_w/ck_w = 0.05$, $\gamma = 3.5$ and $N_w = 10$.

which $v_1/c \approx -0.05$, $v_2 \approx 0$ and $v_3/c \approx 0.957$. As is evident from the figure, the bulk value for the magnitude of v_1 increases with the adiabatic rise in B_w, after which small oscillations about the mean value corresponding to the steady-state trajectory are found in the constant-B_w region. The value of v_2 remains small over the entire course of the trajectory. The behaviour of v_3, while not shown explicitly, also exhibits small oscillations (of the order of 1%) about the value for the steady-state orbit.

Thus, we conclude that the adiabatic wiggler entry taper is capable of injecting electrons on near-steady-state trajectories. However, it should be observed that the magnitude of the fluctuation about the steady-state orbit increases as the axial solenoidal field approaches values near the resonance for $\Omega_0 \approx k_w v_\parallel$. Indeed, it becomes increasingly difficult to inject electrons on to these near-steady-state trajectories as the orbital instability point is approached ($\Omega_0/ck_w \approx 0.76$ for the present choice of parameters). At this point the orbits differ widely from the steady-state trajectories and exhibit large fluctuations in velocity. As a result, it becomes increasingly difficult either to inject or to propagate a coherent beam through the system near the gyroresonance. An example of injection for parameters corresponding to Group II orbits is shown in Fig. 5.3 in which we plot the individual components of the velocity as a function of axial position for $\Omega_0/ck_w = 1.0$, $\Omega_w/ck_w = 0.05$, $\gamma = 3.5$ and $N_w = 10$. These parameters correspond to Group II orbits in the negative-mass regime close to the gyroresonance. Although the orbital instability is not found for Group II orbits in the idealized one-dimensional limit, the axial velocity of a steady-state orbit near the gyroresonance is significantly less than the initial velocity of paraxial flow, and the

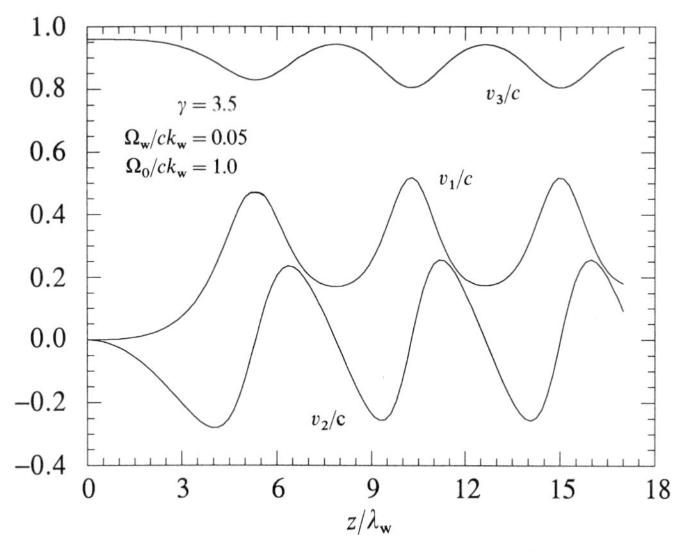

Fig. 5.3 Plot of the single-particle trajectories versus axial distance for $\Omega_0/ck_w = 1.0$, $\Omega_w/ck_w = 0.05$, $\gamma = 3.5$ and $N_w = 10$.

orbits which result exhibit large fluctuations about mean values corresponding to the steady-state orbit.

In view of these results regarding the injection of an electron beam into a combined helical wiggler and axial solenoidal field, we conclude that large-scale fluctuations in the electron velocity accompany injection whenever the axial field is close to gyroresonance. Indeed, these fluctuations act to counter the enhancements found in the linear gain of the interaction demonstrated in Chapter 4 for free-electron laser operation near the gyroresonance.

5.1.3 Numerical solution of the dynamical equations

The dynamical equations (5.48–55) for the fields and (5.42–5) for each electron have been solved numerically. In the case of the idealized one-dimensional model, the ensemble of electrons is discretized using Simpson's rule with a total of 61 electrons per wave period. The use of larger numbers of electrons was found to result in discrepancies of considerably less than 1%. As in the case of the single-particle trajectories, the wiggler-field amplitude increases adiabatically to a constant level, and the effect of the injection mechanism on the gain and saturation of the interaction can be studied directly. Particular interest is centred on the degradation in the gain and saturation efficiency due to large-amplitude fluctuations in the electron trajectories in the vicinity of the gyroresonance.

The first example we consider corresponds to field parameters shown in Fig. 5.2, and we choose $\Omega_0/ck_w = 0$, $\Omega_w/ck_w = 0.05$, $N_w = 10$, $\gamma = 3.5$ and $\omega_b/\gamma^{1/2}ck_w = 0.1$. These parameters correspond to the boundary between the high-gain Compton and the collective Raman regimes. We initialize the vector potential arbitrarily

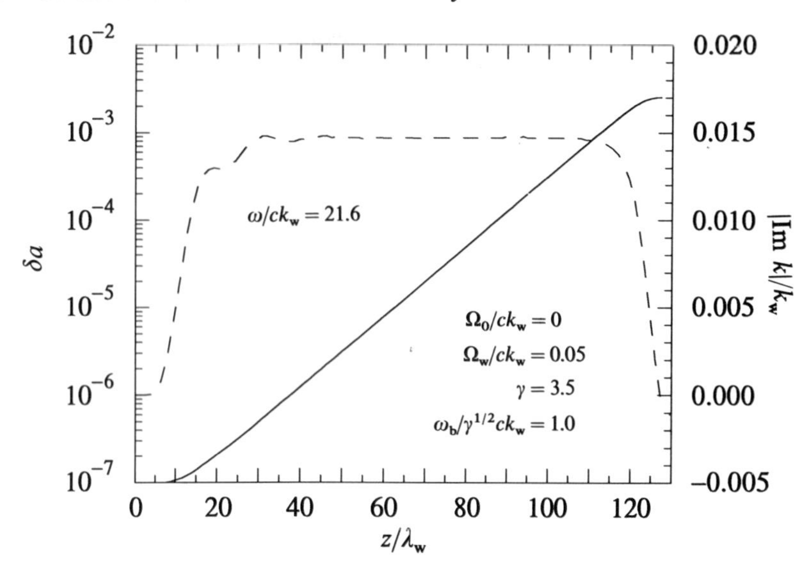

Fig. 5.4 Graphs of (a) the amplitude of the radiation field (solid line) and (b) the magnitude of the growth rate (dashed line) versus axial position for $\Omega_0/ck_w = 0$, $\Omega_w/ck_w = 0.05$, $\gamma = 3.5$, $N_w = 10$ and $\omega_b/\gamma^{1/2}ck_w = 0.1$.

in this case at $\delta a(z = 0) = 1.0 \times 10^{-7}$, and find from equation (5.60) that the initial value of the scalar potential is $\delta\varphi(z = 0) = 1.09 \times 10^{-6}$. Since we are interested primarily in the electromagnetic radiation, we confine our attention henceforth to the vector potential. Since the trajectories within the uniform wiggler obtained for these parameters are close to the steady-state trajectories, we expect to find that the growth rate in the linear phase of the interaction is close to that found from the linear dispersion analysis in Chapter 4. The results of the simulation for these parameters are shown in Fig. 5.4 in which we plot both the amplitude δa and the growth rate Γ_+ versus axial position for a frequency $\omega/ck_w = 21.6$ which corresponds to maximum growth. Observe that $k/k_w \approx 21.6$ for this case as well. It is evident from the figure that an extended region of exponential growth occurs after an initial transient phase (for $z/\lambda_w \leqslant 29$). The growth rate during this phase of the interaction as computed from the simulation is $\Gamma_+/k_w \approx 0.0146$, which is in good agreement with the growth rate of $\Gamma_{lin}/k_w \approx 0.0145$ computed on the basis of the linear dispersion equation (4.83).

Fluctuations in the growth rate are of the order of $\Delta\Gamma_+/k_w \approx \pm 0.0002$ which is to be expected based upon the orbit calculation. Saturation begins to occur at $z/\lambda_w \approx 114.1$, after which the growth rate rapidly decreases to zero at $z/\lambda_w \approx 127.3$. At saturation, the radiation field amplitude is $\delta a_{sat} \approx 2.56 \times 10^{-3}$. The saturation efficiency is given in terms of these results by

$$\eta \cong \frac{1}{(\gamma - 1)\beta_\|} \frac{ck}{\omega} \left(\frac{\omega}{\omega_b} \delta a_{sat} \right)^2, \tag{5.64}$$

which implies that $\eta \approx 3.65\%$ for this case. Observe that for these parameters, $v_\parallel/c \approx 0.957$, $v_{ph}/c \approx 0.956$ and the estimate of the saturation efficiency made on the basis of the *ad hoc* phase-trapping arguments in equation (5.2) is $\eta_{est} \approx 3.2\%$, which is in good agreement with the simulation results.

Increases in the axial solenoidal field result in increasing levels of fluctuations in the electron velocities about the steady-state trajectories, and this effect will compete with the increases in the growth rate and efficiency which correspond to increases in the transverse wiggler-induced velocity due to operation near the gyroresonance. In order to illustrate this effect we consider the case for which $\Omega_0/ck_w = 0.5$, $\Omega_w/ck_w = 0.05$, $N_w = 10$, $\gamma = 3.5$ and $\omega_b/\gamma^{1/2}ck_w = 0.1$. These parameters correspond to Group I trajectories with $v_\parallel/c \approx 0.953$ and $v_w/c \approx -0.105$. In this case, both the growth rate and the level of oscillations in the growth rate have increased relative to those shown in Fig. 5.4. The results of the simulation yield an average growth rate of $\Gamma_+/k_w \approx 0.030$ with a fluctuation of $\Delta\Gamma_+/k_w \approx \pm 0.003$ at the frequency of peak growth $\omega/ck_w = 19.4$. This is also in substantial agreement with the linear dispersion analysis which yields a growth rate of $\Gamma_{lin}/k_w \approx 0.029$. The increased growth rate results in a shorter distance to saturation of $z/\lambda_w \approx 67.5$ at a field level of $\delta a_{sat} \approx 3.30 \times 10^{-3}$. This corresponds to an efficiency of 4.92%.

Further increases in the axial magnetic field above this level (but still corresponding to Group I trajectories) lead to still larger fluctuations in both the orbits and the linear growth rate, until a point of diminishing returns is reached at which the orbital fluctuations result in a degradation in both the growth rate and the efficiency. A transitional case is illustrated in Fig. 5.5, in which we plot

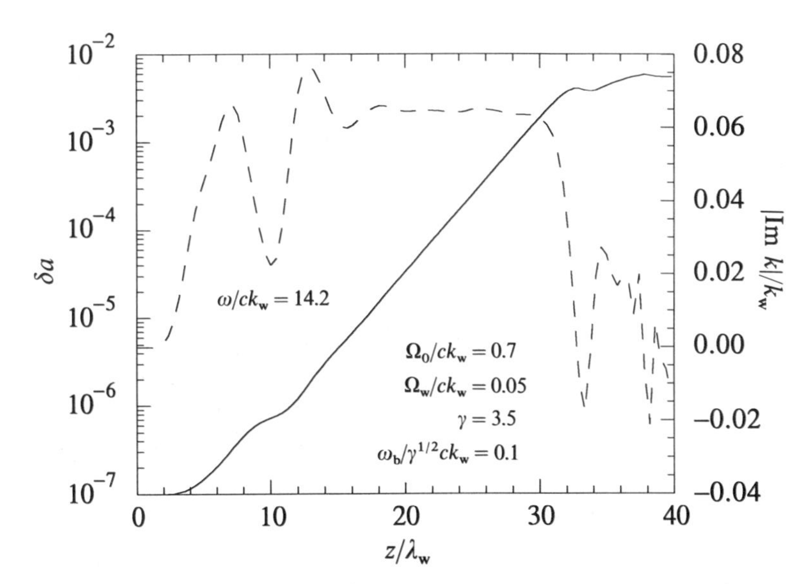

Fig. 5.5 Graphs of (a) the amplitude of the radiation field (solid line) and (b) the magnitude of the growth rate (dashed line) versus axial position for $\Omega_0/ck_w = 0.7$, $\Omega_w/ck_w = 0.05$, $\gamma = 3.5$, $N_w = 10$ and $\omega_b/\gamma^{1/2}ck_w = 0.1$.

the amplitude and the growth rate of the vector potential for $\Omega_0/ck_w = 0.7$, $\Omega_w/ck_w = 0.05$, $N_w = 10$, $\gamma = 3.5$ and $\omega_b/\gamma^{1/2}ck_w = 0.1$ at the frequency of maximum growth of $\omega/ck_w = 14.2$. Observe that this is close to, but still below, the orbital instability transition at $\Omega_0/ck_w \approx 0.76$. The increased level of fluctuations in the growth rate is apparent from the figure although there is an extended range for $20 < z/\lambda_w < 30$ over which the growth rate is relatively constant. The growth rate found in simulation for this example, $\Gamma_+/k_w \approx 0.063$ (which is close to the result of the linear theory of $\Gamma_{lin}/k_w \approx 0.060$), is still increased relative to that found for $\Omega_0/ck_w = 0.5$, but the saturation efficiency has decreased to $\eta \approx 4.02\%$ corresponding to a field level of $\delta a_{sat} \approx 4.09 \times 10^{-3}$.

For higher levels of the axial solenoidal field such that $\Omega_0/ck_w > 0.76$, the orbits correspond to perturbations about the Group II trajectories. However, as seen in Fig. 5.3, large oscillations about the steady-state orbits are characteristic of the low bulk axial velocities found near gyroresonance in this regime. The implications of such orbital behaviour are that (1) the growth rate must also exhibit large-scale oscillations, and (2) the resonant frequency is relatively low. An example of this regime is given in Fig. 5.6 in which we plot the amplitude and growth rate of the vector potential for $\Omega_0/ck_w = 1.1$, $\Omega_w/ck_w = 0.05$, $N_w = 10$, $\gamma = 3.5$ and $\omega_b/\gamma^{1/2}ck_w = 0.1$ at the frequency of maximum growth of $\omega/ck_w = 11.4$. The system exhibits the expected large-scale fluctuation in the growth rate of $\Delta\Gamma_+/k_w \approx \pm 0.036$ about a mean value of $\Gamma_+/k_w \approx 0.072$ after the transients corresponding to launching losses have decayed away. Note that the linear theory predicts a growth rate of $\Gamma_{lin}/k_w \approx 0.056$ which is within the range of fluctuation found in simulation.

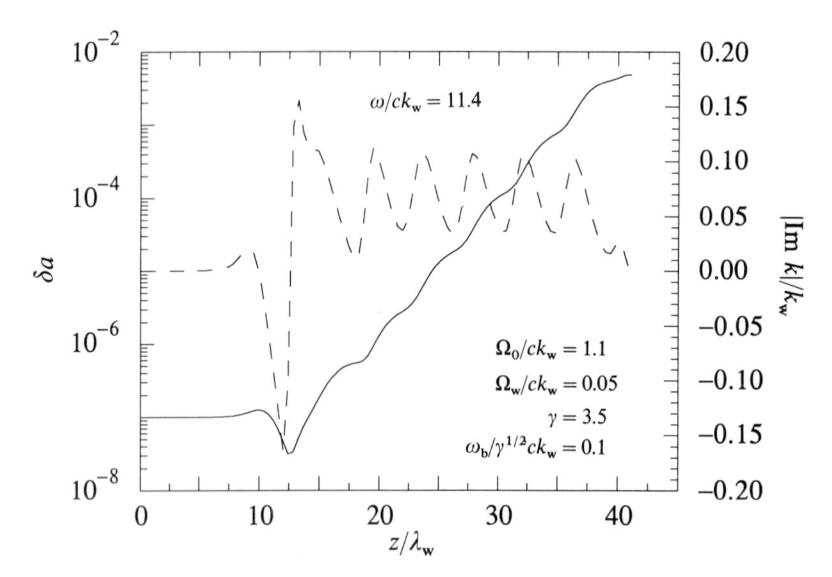

Fig. 5.6 Graphs of (a) the amplitude of the radiation field (solid line) and (b) the magnitude of the growth rate (dashed line) versus axial position for $\Omega_0/ck_w = 1.1$, $\Omega_w/ck_w = 0.05$, $\gamma = 3.5$, $N_w = 10$ and $\omega_b/\gamma^{1/2}ck_w = 0.1$.

Saturation occurs at $z/\lambda_w \approx 41.1$ for $\delta a_{sat} \approx 4.91 \times 10^{-3}$ corresponding to an efficiency of 3.88%.

Further increases in the axial solenoidal field correspond to increases in the resonant frequency and decreases in the oscillations of the orbits about the steady-state trajectories. As a consequence, the evolution of the radiation field becomes more regular and saturation efficiency rises with the improved wave–particle

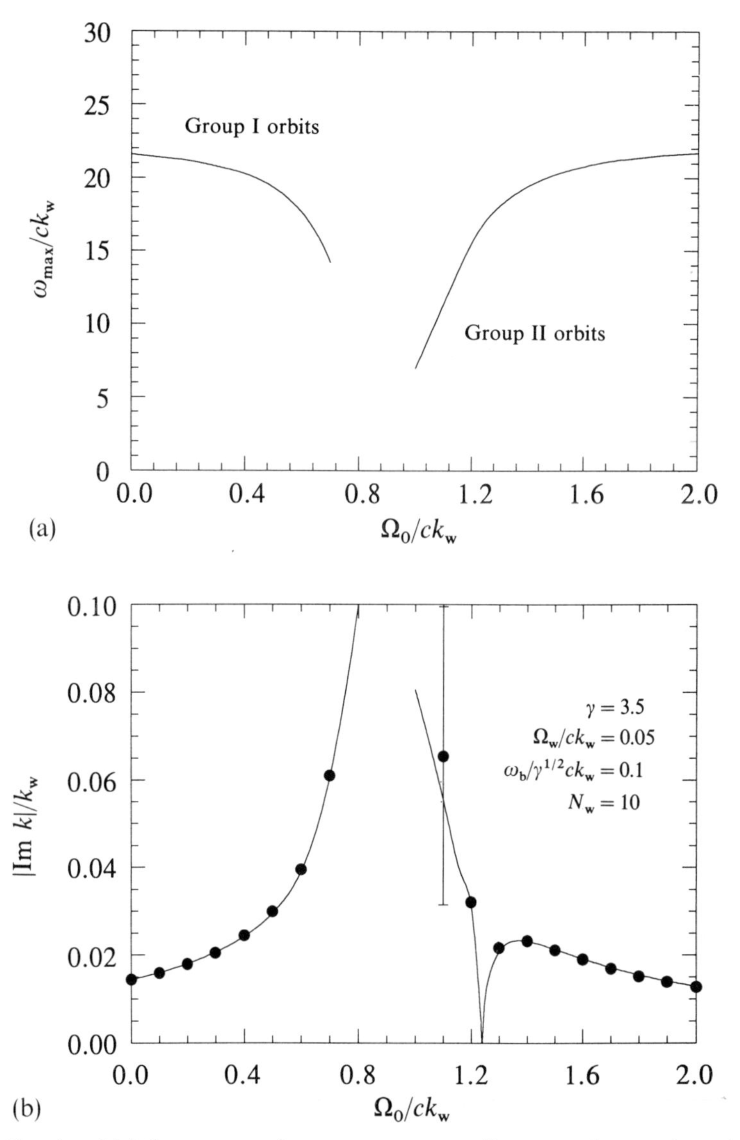

Fig. 5.7 Graphs of (a) the resonant frequency corresponding to peak growth, and (b) the magnitude of the peak growth rate versus Ω_0/ck_w for $\Omega_w/ck_w = 0.05$, $\gamma = 3.5$, $N_w = 10$ and $\omega_b/\gamma^{1/2}ck_w = 0.1$ [17].

resonance. For example, when $\Omega_0/ck_w = 1.5$ the frequency of peak growth occurs for $\omega/ck_w = 20.3$, and a growth rate of $\Gamma_+/k_w \approx 0.021$ is found in simulation with a fluctuation of only approximately 1% about this value. It should be remarked that a growth rate of $\Gamma_{lin}/k_w \approx 0.021$ is also recovered from the linear theory. The saturation efficiency found for this case has risen to a level of 5.02%, which is comfortably higher than the peak efficiency observed in simulation for parameters corresponding to Group I orbits.

An overview of the effect of the axial magnetic field on the interaction frequency and the linear growth rate is shown in Fig. 5.7 in which ω/ck_w and Γ_+/k_w are plotted as functions of the axial field strength. The curves for the frequency represent the variation in the resonant frequency at maximum growth found from the intersection of the uncoupled dispersion curves for the appropriate value of the axial velocity of the steady-state trajectory. These values represent values used in the simulations. The solid lines in the plot of the growth rate represent the calculated value from the linear theory for the appropriate steady-state orbit, while the dots are used to denote the results found in the simulation. As seen in the figure, the agreement between the linear and nonlinear analyses is excellent. For all cases shown, only the values of the growth rate for $\Omega_0/ck_w = 1.1$ (note that the bar indicated for this point denotes an oscillation range) for the linear and nonlinear analyses differ by more than 2%. This is due to the relatively large oscillation in the trajectory about the steady-state orbit corresponding to this case. Note also that the zero for $\Omega_0/ck_w \approx 1.25$ corresponds to the upper bound of the Group II orbit negative-mass regime at which Φ (2.24) and the ponderomotive potential vanish.

The variation in the saturation efficiency with the axial magnetic field is shown in Fig. 5.8 in which we plot the distance to saturation and the saturation efficiency versus Ω_0/ck_w. It is clear from the figure that substantial enhancements in the efficiency are possible over that found in the absence of an axial magnetic field. However, a point of diminishing returns is found as the axial magnetic field approaches the gyroresonance, and the interaction efficiency declines due to the oscillations in the electron trajectories for axial fields closer to the resonance. For parameters corresponding to Group I orbits, the maximum efficiency is approximately 5% for $\Omega_0/ck_w = 0.5$ for the chosen parameters which constitutes a 37% enhancement over the efficiency found when no axial field is present. However, the greatest enhancements in the efficiency correspond to axial magnetic fields associated with Group II trajectories in the negative-mass regime for which the maximum efficiency approaches 8.1%. It is important to bear in mind, however, that these enhancements in the efficiency occur at the expense of a reduced interaction frequency.

5.1.4 The phase-space evolution of the electron beam

The saturation mechanism is particle trapping in the combined ponderomotive and space-charge fields. An example of this is shown in Figs. 5.9 and 5.10 in which the positions of the electrons in the phase space $(\psi, d\psi/dz)$ are plotted for the

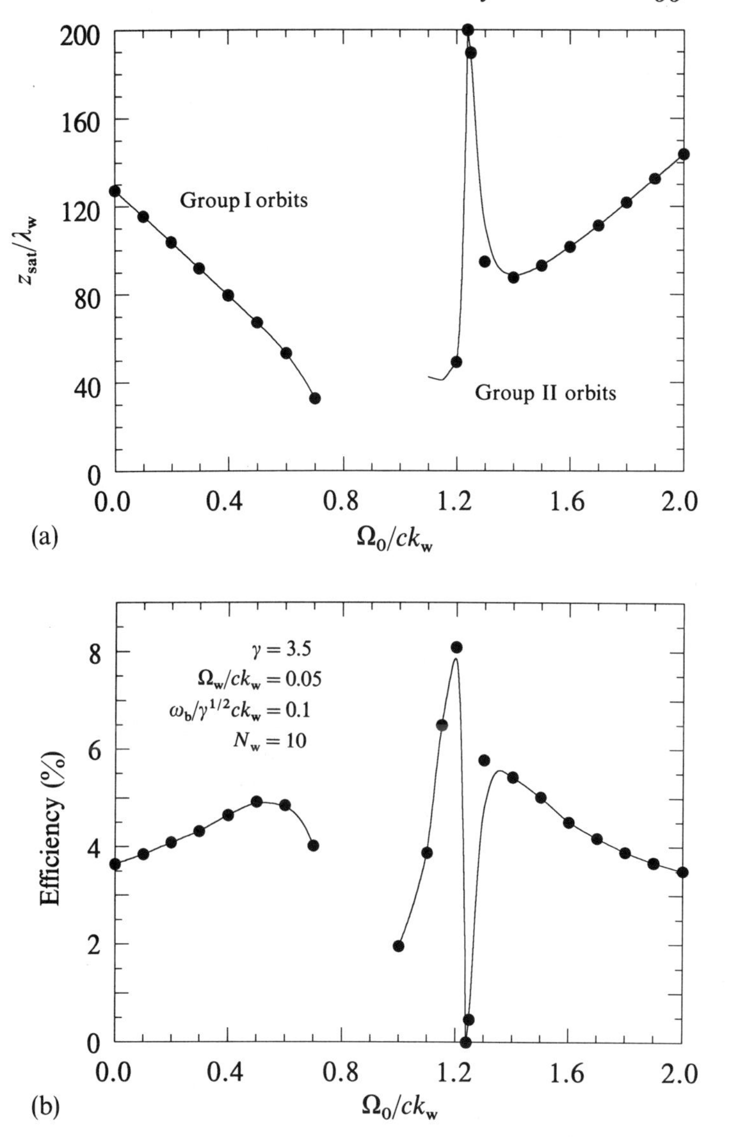

Fig. 5.8 Graphs of (a) the distance to saturation, and (b) the saturation efficiency versus Ω_0/ck_w for $\Omega_w/ck_w = 0.05$, $\gamma = 3.5$, $N_w = 10$ and $\omega_b/\gamma^{1/2}ck_w = 0.1$.

case in which $\Omega_0/ck_w = 0.5$, which corresponds to Group I trajectories. The initial phase-space distribution reflects a uniform phase-space distribution in which the electrons are propagating paraxially with identical axial velocities; hence, the initial phase-space distribution describes a horizontal line for which $d\psi/dz = k_+ + k_w - \omega/v_{z0}$. For the parameters appropriate to this example, the initial value of $d\psi/dz \approx 0.149$. Figure 5.9 represents the phase space during the linear phase of the interaction after the electrons have lost energy but still prior to saturation.

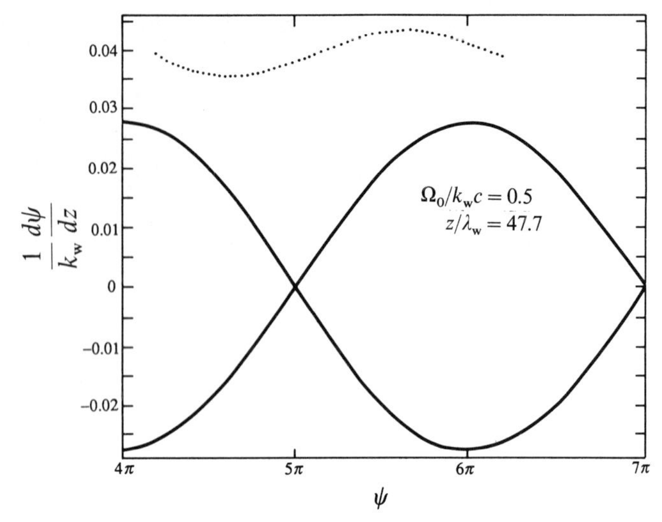

Fig. 5.9 Plot of the electron phase space during the linear exponential growth phase of the interaction for $\Omega_0/ck_w = 0.5$.

The solid lines in the figure represent the approximate separatrix as determined from equation (4.20), which is included in the figure for illustrative purposes.

The dynamics of the nonlinear pendulum motion is included implicitly in the simulation by means of the integration of the Lorentz force equations. It is evident from the figure that as the electrons lose energy they decelerate and fall in phase space (i.e. $d\psi/dz$ decreases). The early phases of the process of axial bunching in the ponderomotive potential are shown in the figure, but the electrons have not

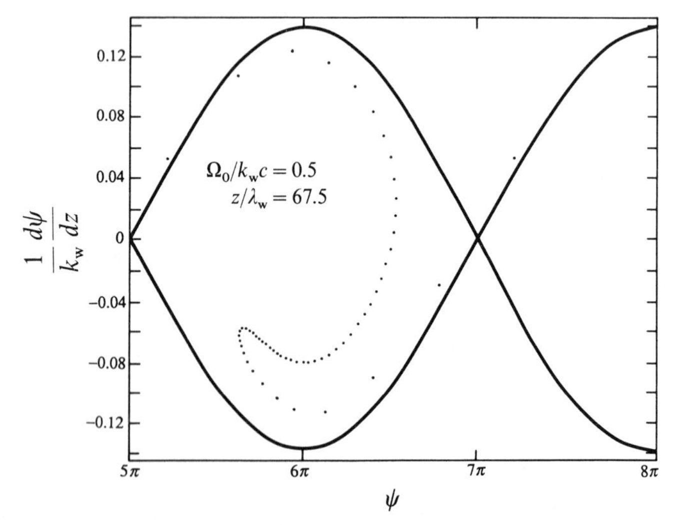

Fig. 5.10 Plot of the electron phase space at saturation of the interaction for $\Omega_0/ck_w = 0.5$.

yet crossed the separatrix on to trapped orbits. The phase-space distribution at saturation is shown in Fig. 5.10, which shows a snapshot of the electrons as they execute a clockwise rotation within the ponderomotive well. At this point the energy transfer of the electrons to the wave in the first half cycle of the oscillation is exactly balanced by the energy transfer from the wave to the electrons in the second half cycle. Note that the entire beam has not been trapped by this process.

An example of the phase-space evolution for an axial magnetic field corresponding to Group II trajectories is shown in Figs 5.11 and 5.12, where the solid line again

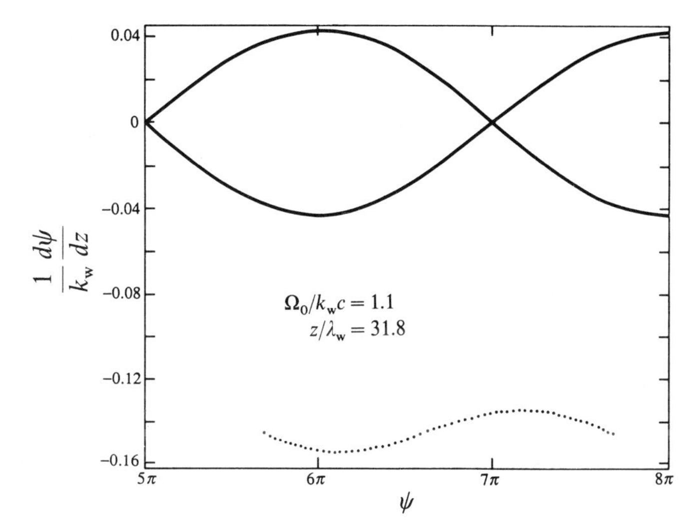

Fig. 5.11 Plot of the electron phase space during the linear exponential growth phase of the interaction for $\Omega_0/ck_w = 1.1$.

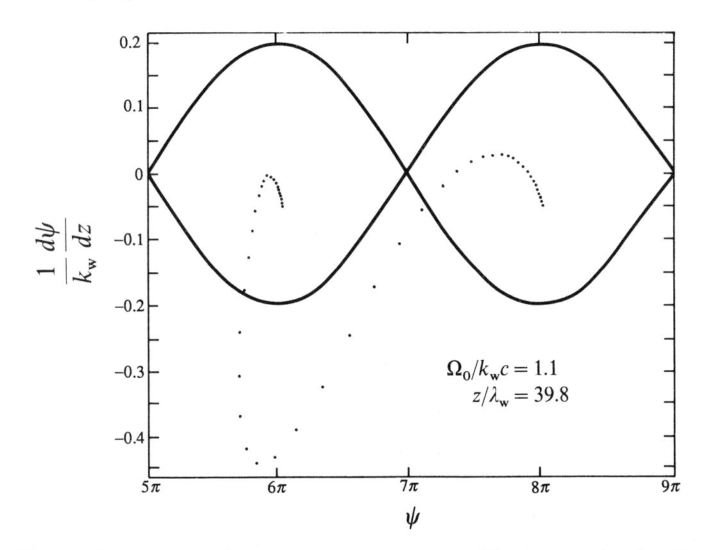

Fig. 5.12. Plot of the electron phase space at saturation of the interaction for $\Omega_0/ck_w = 1.1$.

indicates the analytic approximation for the separatrix and not a result from the simulation. This case corresponds to the choice of $\Omega_0/ck_w = 1.1$ discussed in regard to Fig. 5.6. Note that while the initial phase-space distribution corresponds to $d\psi/dz \approx 0.149$ as in the case of $\Omega_0/ck_w = 0.5$, this example describes the negative-mass regime in which the axial velocity in the uniform-wiggler region is much smaller than the paraxial velocity of the beam on entry to the wiggler. For this reason, the bulk axial velocity in the uniform-wiggler region is $v_\parallel/c \approx 0.922$, corresponding to $d\psi/dz \approx -0.063$. Hence, the electron phase-space distribution describes a line in the lower half-plane at the start of the uniform-wiggler region and, since this is in the negative-mass regime, the electrons accelerate in the axial direction (i.e. $d\psi/dz$ increases) as energy is lost to the wave. This process is clearly indicated in Fig. 5.11, which illustrates the electron phase-space distribution at $z/\lambda_w = 31.8$ which is a late stage of the linear phase of the interaction. The electron phase space at saturation is shown in Fig. 5.12.

5.1.5 Comparison with experiment

Experimental observations of the variation of the saturation efficiency with the axial magnetic field have been obtained at the Naval Research Laboratory [28]. This experiment employed an intense relativistic electron beam with a voltage of 1.35 MeV and a current of 1.5 kA, and a helical wiggler field with a 3.0 cm period and an amplitude variable up to approximately 4.0 kG. The axial solenoidal field was variable up to approximately 20 kG. This experiment was operated as

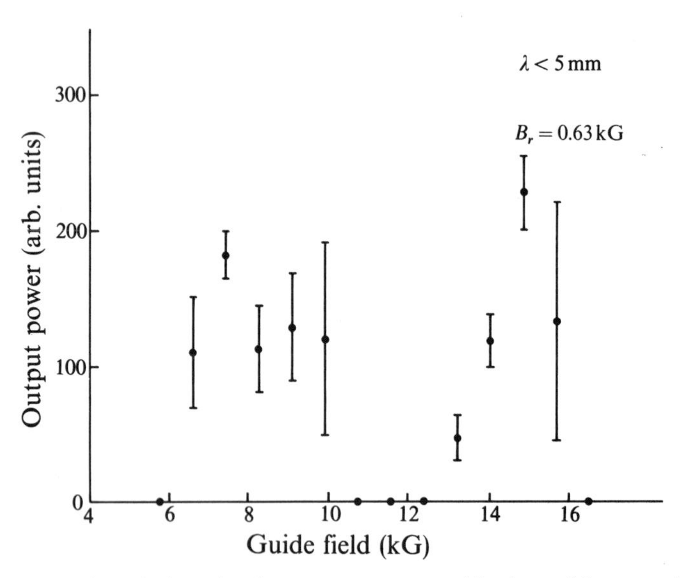

Fig. 5.13 Measured variations in the output power with the axial magnetic field for experiments performed at the Naval Research Laboratory [28] with a 1.35 MeV/1.5 kA electron beam, and a wiggler field with a period of 3.0 cm and an amplitude of 630 G.

a superradiant amplifier in which the signal grew from noise on the beam, and reached saturated efficiencies of the order of 2–3%. An example of the output power as a function of the axial field strength is shown in Fig. 5.13 for a wiggler field strength of 630 G, and it is evident that the peak emission was found for trong axial fields corresponding to Group II orbit parameters. As seen in the igure, there is no emission in the range of 10–13 kG which corresponds to the vicinity of the gyroresonance in which the orbits executed extremely large-amplitude oscillations. The lack of emission for low axial magnetic fields was due to the fact that it was impossible to propagate such an intense beam without a sufficiently high axial magnetic field to confine the beam against the effects of self-electric and magnetic fields. In addition, the cutoff in the emission for strong axial magnetic fields corresponds to the transition between the negative-mass and positive-mass regimes of the Group II trajectories.

5.2 ONE-DIMENSIONAL ANALYSIS: PLANAR WIGGLERS

We now turn to a slow time-scale nonlinear analysis of planar wiggler configurations. In many ways the analysis is similar to that described for the helical wiggler; however, there are interesting differences in the interaction dynamics for planar wigglers. In the first place, we do not include an axial guide field in the analysis due to the existence, as described in Section 2.2.2, of a $\boldsymbol{B}_0 \times \nabla \boldsymbol{B}_w$ drift which causes the beam to diverge. In the second place, in contrast to the case of a helical wiggler, there are no orbits in a planar wiggler for which the axial electron velocity is constant. This (1) leads to the presence of harmonic interactions within the context of a one-dimensional analysis, and (2) results in instantaneous oscillations in the power and phase of the wave due to the lower beat wave. Harmonic interactions are possible with both helical and planar wigglers, but do not appear in the one-dimensional treatment of the interaction in a helical wiggler. The nature of harmonic emission in both helical and planar wigglers will be discussed in some detail in Chapter 7. The effect of the lower beat wave is suppressed by the symmetry properties of a helical wiggler, but results in an oscillation with a period of $\lambda_w/2$ in a planar wiggler in both the power and phase of the wave.

5.2.1 The dynamical equations

The physical configuration employed in the treatment of a self-consistent model of the interaction in a planar wiggler-based free-electron laser is founded upon an idealized representation of the planar magnetostatic field given in equation (2.2). Under the assumption that the electron displacement from the plane of symmetry is small so that $|k_w y| \ll 1$, this field can be approximated as

$$\boldsymbol{B}_w(\boldsymbol{x}) = B_w(z)\hat{\boldsymbol{e}}_y \sin k_w z, \tag{5.65}$$

where the wiggler amplitude is assumed to increase adiabatically as in equation

(5.14). As in the case of the helical wiggler, we shall model the injection of the beam into the wiggler. Observe that the wiggler-induced transverse velocity is in the x-direction; hence, the beam will couple to an electromagnetic wave polarized in the x-direction as well.

The formulas governing a planar wiggler configuration are closely similar to those governing the helical wiggler geometry, and the derivation of the dynamical equations closely follows that described in Section 5.1.

The field equations

The vector and scalar potentials of the fluctuating electromagnetic and electrostatic waves are given in terms of a plane-wave approximation in the form

$$\delta \mathbf{A}(z,t) = \delta A(z)\hat{\mathbf{e}}_x \cos\alpha_+(z,t) \tag{5.66}$$

and

$$\delta\Phi(z,t) = \delta\Phi(z)\cos\alpha(z,t), \tag{5.67}$$

where the amplitudes and phases are assumed to vary slowly in z, and the phases are defined as in equations (5.17) and (5.18). Following the procedure described in section 5.1, substitution of these fields into the Maxwell–Poisson equations yields

$$\left(\frac{d^2}{dz^2} + \frac{\omega^2}{c^2} - k_+^2\right)\delta A\cos\alpha_+ - 2k_+^{1/2}\frac{d}{dz}(k_+^{1/2}\delta A)\sin\alpha_+ = -\frac{4\pi}{c}\delta J_x \tag{5.68}$$

and

$$\left(\frac{d^2}{dz^2} - k^2\right)\delta\Phi\cos\alpha - 2k^{1/2}\frac{d}{dz}(k^{1/2}\delta\Phi)\sin\alpha = -4\pi\delta\rho, \tag{5.69}$$

where the source current and charge density are defined in equations (5.22) and (5.23).

Multiplying equation (5.68) by $\cos\alpha_+$ and $\sin\alpha_+$ and averaging over a wave period yields

$$\left(\frac{d^2}{dz^2} + \frac{\omega^2}{c^2} - k_+^2\right)\delta A = -\frac{4\omega}{c}\int_0^{2\pi/\omega} dt\,\delta J_x\cos\alpha_+ \tag{5.70}$$

and

$$2k_+^{1/2}\frac{d}{dz}(k_+^{1/2}\delta A) = \frac{4\omega}{c}\int_0^{2\pi/\omega} dt\,\delta J_x\sin\alpha_+. \tag{5.71}$$

Substitution of the x-component of the source current into these equations yields the dynamical equations for the vector potential

$$\left(\frac{d^2}{dz^2} + \frac{\omega^2}{c^2} - k_+^2\right)\delta a = 2\frac{\omega_b^2}{c^2}\beta_{z0}\left\langle\frac{v_x}{|v_z|}\cos\alpha_+\right\rangle \tag{5.72}$$

and

$$2k_+^{1/2}\frac{d}{dz}(k_+^{1/2}\delta a) = -2\frac{\omega_b^2}{c^2}\beta_{z0}\left\langle\frac{v_x}{|v_z|}\sin\alpha_+\right\rangle, \tag{5.73}$$

where $\delta a \equiv e\delta A/m_ec^2$, ω_b denotes the beam plasma frequency, $\beta_{z0} \equiv v_{z0}/c$ (v_{z0} denotes the initial electron velocity), and the averaging operator is defined in equation (5.41). A similar procedure using the $\cos\alpha$ and $\sin\alpha$ on Poisson's equation yields the dynamical equations for the scalar potential

$$\left(\frac{d^2}{dz^2} - k^2\right)\delta\varphi = 2\frac{\omega_b^2}{c^2}v_{z0}\left\langle\frac{\cos\alpha}{|v_z|}\right\rangle, \tag{5.74}$$

and

$$2k^{1/2}\frac{d}{dz}(k^{1/2}\delta\varphi) = -2\frac{\omega_b^2}{c^2}v_{z0}\left\langle\frac{\sin\alpha}{|v_z|}\right\rangle, \tag{5.75}$$

where $\delta\varphi \equiv e\delta\Phi/m_ec^2$.

The electron orbit equations

Equations (5.72–74) describe the evolution of the vector and scalar potentials, and are the analogue of equations (5.35–38) for the helical configuration. In order to complete the formulation, we must also specify the Lorentz force equations governing the trajectories of an ensemble of electrons. The Lorentz force equations can be written in the form

$$\frac{1}{c}\frac{d}{dz}u_x = \frac{\hat{\Omega}_w}{c}\sin k_wz + \left(\frac{\omega}{v_z} - k_+\right)\delta a\sin\alpha_+ + \cos\alpha_+\frac{d}{dz}\delta a, \tag{5.76}$$

$$\frac{d}{dz}u_y = 0, \tag{5.77}$$

$$\begin{aligned}\frac{1}{c}\frac{d}{dz}u_z = &-\frac{\hat{\Omega}_w}{c}\frac{v_x}{v_z}\sin k_wz + \frac{v_x}{v_z}\left(k_+\delta a\sin\alpha_+ - \cos\alpha_+\frac{d}{dz}\delta a\right)\\ &-\frac{c}{v_z}\left(k\delta\varphi\sin\alpha - \cos\alpha\frac{d}{dz}\delta\varphi\right),\end{aligned} \tag{5.78}$$

where $\hat{\Omega}_w \equiv eB_w/m_ec$, $\boldsymbol{u} = \boldsymbol{p}/m_e$ denotes a normalized momentum and

$$\frac{d}{dz}\psi = k_+ + k_w - \frac{\omega}{v_z} \tag{5.79}$$

describes the evolution of the ponderomotive phase.

The numerical procedure and initial conditions

The numerical procedure and the initial conditions associated with the integration of these dynamical equations follow analogously that described for the helical configuration in section 5.1. That is, we rewrite the two sets of second-order differential equations for the amplitude and phase of the scalar and vector potentials in terms of eight first-order differential equations which are integrated simultaneously with the orbit equations for an ensemble of electrons. The initial conditions of the particles and fields are also chosen as in the case of the helical wiggler; thus, we also use the noise level associated with the discrete initial electron distribution to determine the initial phase averages of the beam and, hence, the initial value of the scalar potential.

One difference with the numerical procedure used for the helical wiggler geometry is that Gaussian quadrature rather than Simpson's rule is used for the particle averages. The reason for this is that the Gaussian quadrature is more computationally efficient and the same accuracy is obtained using 20 electrons with Gaussian quadrature as Simpson's rule produces with 61 electrons. For this reason, Gaussian quadrature is employed in all subsequent numerical discussions.

5.2.2 Numerical solution of the dynamical equations

We consider a case analogous to that studied for the helical wiggler and shown in Fig. 5.4. Specifically, we choose $\Omega_0/ck_w = 0$, $N_w = 10$, $\gamma = 3.5$ and $\omega_b/\gamma^{1/2}ck_w = 0.1$. However, the wiggler field value must be increased relative to that used for the helical wiggler due to the oscillation in the wiggler-induced transverse velocity associated with the planar symmetry. Noting that it is the rms wiggler field amplitude which is significant, we choose $\Omega_w/ck_w = 0.071$, which is equivalent to the value of $\Omega_w/ck_w = 0.05$ used for the helical wiggler in the preceding section. The initial level of the vector potential is, as for the helical wiggler, chosen to be $\delta a(z=0) = 10^{-7}$. The bulk resonance condition for these parameters is $\omega/ck_w \approx 21.6$.

The results of the simulation are shown in Fig. 5.14. The initial value of the scalar potential, however, is $\delta\varphi(z=0) \approx 2.04 \times 10^{-12}$ which is much smaller than the value obtained in the helical wiggler example (for which $\delta\varphi(z=0) \approx 1.04 \times 10^{-6}$). The reason for this is that the initial value for the scalar potential is dependent upon the accuracy of the initial phase average, and the Gaussian quadrature technique being used for the planar wiggler is more accurate than Simpson's rule which was used for the helical wiggler. Be that as it may, the results for the planar wiggler are in substantial agreement with those found for the helical wiggler. The evolution of the vector potential versus axial position is shown in Fig. 5.14 in which it is evident that the bulk growth of the vector potential is exponential and the amplitude grows from its initial value to a saturation value of $\delta a_{sat} \approx 3.6 \times 10^{-3}$ over a length of 131 wiggler periods. The efficiency for planar polarization is given in terms of the normalized amplitude by one half the value shown in equation (5.64). This yields

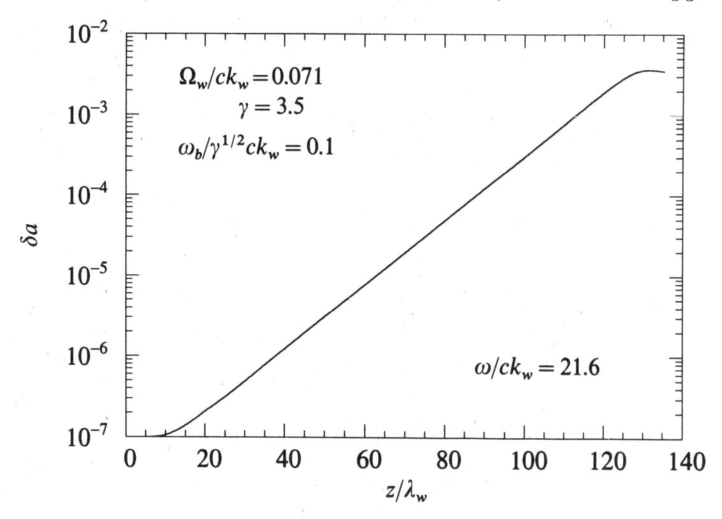

Fig. 5.14 Evolution of the amplitude of the vector potential versus axial position.

a maximum extraction efficiency of 3.68%. These results are comparable to that found for the helical wiggler where saturation occurred at $z_{sat}/\lambda_w \approx 127$ for $\delta a_{sat} \approx 2.6 \times 10^{-3}$ corresponding to an efficiency of 3.65%. Indeed, the bulk growth rate obtained from the figure in the exponential growth region of $30 \leqslant z/\lambda_w \leqslant 100$ is $\Gamma_{avg}/k_w \approx 0.0147$, which is very close to the value of 0.0146 found for the helical wigger. However, a close examination of the data indicates that a fast oscillation is present with a period of $\lambda_w/2$, which corresponds to the lower beat wave. This effect will be described in more detail in regard to three-dimensional simulations of planar wiggler geometries described in section 5.4 and in Chapter 8.

5.3 THREE-DIMENSIONAL ANALYSIS: HELICAL WIGGLERS

The three-dimensional analyses of free-electron laser amplifiers have also received considerable attention [5, 29–50] from a variety of formulations. One technique for dealing with the transverse gradients associated with the wiggler and radiation fields is to use a numerical field-solver for the transverse variation of the electromagnetic radiation [5, 29, 31, 32, 39]. An alternative technique is to expand the field in terms of the normal modes determined by the specific geometry of the free-electron laser. This can involve the normal modes supported by a waveguide [34–8, 40–50] or the Gaussian resonator modes supported by confocal mirrors [30, 33]. The electron trajectories have been treated both by means of a **wiggler-averaged** approximation in which the Lorentz force equations are reduced to a simplified form by means of an average over a wiggler period [5, 29–33, 39, 50] as well as by integration over the complete Lorentz force equations [34–8, 40–9]. Due to the expanded computational requirements, the number of particle-in-cell analyses of free-electron laser amplifiers in three dimensions is limited [35]. In the remainder of this chapter, we shall rely on a slow time-scale formulation in which the electromagnetic wave is treated by a normal-mode analysis based upon a waveguide-mode expansion in conjunction with a non-wiggler-averaged orbit analysis. The formalism we adopt is described in references [34, 36–8, 40].

5.3.1 The general formulation

The single-frequency analysis described in the preceding section is readily generalizable to three dimensions. The wiggler field model we employ is the general form for a helical wiggler given in equation (2.1)

$$\boldsymbol{B}_w(\boldsymbol{x}) = 2B_w(z)\left[I_1'(\lambda)\hat{\boldsymbol{e}}_r \cos\chi - \frac{1}{\lambda} I_1(\lambda)\hat{\boldsymbol{e}}_\theta \sin\chi + I_1(\lambda)\hat{\boldsymbol{e}}_z \sin\chi \right], \tag{5.80}$$

where the wiggler amplitude is assumed to vary adiabatically to model both the injection of the beam and the enhancement of the interaction efficiency by means of a tapered wiggler amplitude. Hence, as in equation (5.14), we choose

$$B_w(z) = \begin{cases} B_w \sin^2\left(\dfrac{k_w z}{4N_w}\right); & 0 \leqslant z \leqslant N_w \lambda_w \\ B_w; \quad N_w \lambda_w < z \leqslant z_0 \\ B_w[1 + k_w \varepsilon_w (z - z_0)]; \quad z_0 < z. \end{cases} \tag{5.81}$$

In addition, we include the presence of an axial solenoidal magnetic field which can also be tapered to achieve an enhancement of the efficiency. A self-consistent model of a uniformly tapered solenoidal field which is both curl and divergence free is

$$\boldsymbol{B}_0(z) = \begin{cases} B_0 \hat{\boldsymbol{e}}_z; \quad z \leqslant z_0 \\ B_0[1 + k_w \varepsilon_0 (z - z_0)]\hat{\boldsymbol{e}}_z - \dfrac{1}{2} B_0 \lambda \varepsilon_0 \hat{\boldsymbol{e}}_r; \quad z_0 \leqslant z. \end{cases} \tag{5.82}$$

In equations (5.81) and (5.82), the scale lengths for variation of the field amplitudes are assumed to be constant (i.e. linear variation) and defined as

$$\varepsilon_{0,w} \equiv \frac{1}{k_w} \frac{d}{dz} \ln B_{0,w}. \tag{5.83}$$

As a consequence, it is possible to vary the degree of taper as well as the point at which the taper begins. It should be remarked that the wiggler field model is valid (i.e. curl and divergence free) only as long as $\varepsilon_w \ll 1$; however, the tapered axial solenoidal field model is valid for arbitrary ε_0.

The radiation field can be expressed as a superposition of any convenient orthogonal basis such as waveguide or Gaussian resonator modes. We observe that the drift tube through which the electron beam passes in a free-electron laser constitutes a waveguide which specifies the boundary conditions which the electromagnetic field must satisfy. For the purposes of the present analysis, therefore, we employ waveguide modes.

The field equations

In the treatment of a helical wiggler configuration, we assume the drift tube is cylindrical and expand the electromagnetic field in terms of a superposition of the TE and TM modes of a loss-free cylindrical waveguide in the absence of the electron beam. Thus, we write the vector potential of the radiation field in cylindrical coordinates in the form

$$\delta\boldsymbol{A}(\boldsymbol{x},t)=\sum_{\substack{l=0\\n=1}}^{\infty}\delta A_{ln}(z)\left[\frac{l}{\kappa_{ln}r}J_l(\kappa_{ln}r)\hat{\boldsymbol{e}}_r\sin\alpha_{ln}+J_l'(\kappa_{ln}r)\hat{\boldsymbol{e}}_\theta\cos\alpha_{ln}\right],\tag{5.84}$$

for the TE modes, and

$$\delta\boldsymbol{A}(\boldsymbol{x},t)=\sum_{\substack{l=0\\n=1}}^{\infty}\delta A_{ln}(z)\left[J_l'(\kappa_{ln}r)\hat{\boldsymbol{e}}_r\cos\alpha_{ln}-\frac{l}{\kappa_{ln}r}J_l(\kappa_{ln}r)\hat{\boldsymbol{e}}_\theta\sin\alpha_{ln}+\frac{\kappa_{ln}}{k_{ln}}J_l(\kappa_{ln}r)\hat{\boldsymbol{e}}_z\sin\alpha_{ln}\right],\tag{5.85}$$

for the TM modes, where for frequency ω and wavenumber $k_{ln}(z)$

$$\alpha_{ln}\equiv\int_0^z dz'\,k_{ln}(z')+l\theta-\omega t.\tag{5.86}$$

In equations (5.84) and (5.85), J_l and J_l' denote the regular Bessel function of the first kind and its derivative, and κ_{ln} describes the cutoff wavenumber of each mode. In the case of the TE modes, $\kappa_{ln}\equiv x_{ln}'/R_g$, where $J_l'(x_{ln}')=0$ and R_g is the waveguide radius. For the TM modes, $\kappa_{ln}\equiv x_{ln}/R_g$, where $J_l(x_{ln})=0$. It is implicitly assumed that the mode amplitudes $\delta A_{ln}(z)$ and wavenumbers $k_{ln}(z)$ vary slowly over a wavelength. Space-charge fields are neglected at the present stage of the analysis.

The microscopic source current can be written in terms of the initial conditions of the beam as in the idealized one-dimensional analysis. A three-dimensional representation of the source current is of the form

$$\begin{aligned}\delta\boldsymbol{J}(\boldsymbol{x},t)=&-en_b\iiint d^3p_0\,v_{z0}F_b(\boldsymbol{p}_0)\iint_{A_g}dx_0\,dy_0\,\sigma_\perp(x_0,y_0)\\&\times\int_{-T/2}^{T/2}dt_0\sigma_\parallel(t_0)\mathbf{v}(z;x_0,y_0,t_0,\boldsymbol{p}_0)\delta[x-x(z;x_0,y_0,t_0,\boldsymbol{p}_0)]\\&\times\delta[y-y(z;x_0,y_0,t_0,\boldsymbol{p}_0)]\frac{\delta[t-\tau(z;x_0,y_0,t_0,\boldsymbol{p}_0)]}{|v_z(z;x_0,y_0,t_0,\boldsymbol{p}_0)|},\end{aligned}\tag{5.87}$$

where n_b is the ambient beam density, $A_g(\equiv \pi R_g^2)$ denotes the cross-sectional area of the waveguide, T denotes the duration of the electron beam pulse, v_{z0} is the initial axial velocity of the electron, (x_0, y_0) describes the cross-sectional position of the electron as it crosses the $z=0$ plane at time t_0, $\sigma_\perp(x_0, y_0)$ is the cross-sectional distribution of the beam, $\sigma_\parallel(t_0)$ is the initial distribution in entry times, $\boldsymbol{p}_0$ is the initial momentum of the electron, $F_b(\boldsymbol{p}_0)$ is the momentum space distribution of the beam, and the Lagrangian time coordinate is defined as

$$\tau(z; x_0, y_0, t_0, \boldsymbol{p}_0) \equiv t_0 + \int_0^z \frac{dz'}{v_z(z'; x_0, y_0, t_0, \boldsymbol{p}_0)}. \tag{5.88}$$

The initial distributions are assumed to be subject to the normalization conditions

$$\iint_{A_g} dx_0 dy_0 \sigma_\perp(x_0, y_0) = A_g, \tag{5.89}$$

$$\int_{-T/2}^{T/2} dt_0 \sigma_\parallel(t_0) = T, \tag{5.90}$$

and

$$\iiint d^3 p_0 F_b(\boldsymbol{p}_0) = 1. \tag{5.91}$$

Finally, we note that the generalized form of the quasi-static assumption is also imposed; specifically, that $\mathbf{v}(z; x_0, y_0, \boldsymbol{p}_0, t_0) = \mathbf{v}(z; x_0, y_0, \boldsymbol{p}_0, t_0 + 2\pi N/\omega)$ and $\boldsymbol{x}_\perp(z; x_0, y_0, \boldsymbol{p}_0, t_0) = \boldsymbol{x}_\perp(z; x_0, y_0, \boldsymbol{p}_0, t_0 + 2\pi N/\omega)$ for integer N.

The dynamical equations for the evolution of the slowly varying amplitude and wavenumber of each mode are obtained by direct substitution of the representations of the vector potentials into Maxwell's equations

$$\left(\nabla^2 - \frac{1}{c^2}\frac{\partial^2}{\partial t^2}\right)\delta \boldsymbol{A} - \nabla(\nabla \cdot \delta \boldsymbol{A}) = -\frac{4\pi}{c}\delta \boldsymbol{J}. \tag{5.92}$$

Observe that the divergence of the vector potential vanishes for the vacuum state (i.e. in the absence of the electron beam). In the case of the TE modes the divergence of the vector potential vanishes even in the presence of the electron beam, and substitution of the waveguide mode representation (5.84) into Maxwell's equations gives

$$\begin{aligned}\sum_{l,n} \frac{l}{\kappa_{ln} r} J_l(\kappa_{ln} r)\Bigg[\left(\frac{d^2}{dz^2} + \frac{\omega^2}{c^2} - k_{ln}^2 - \kappa_{ln}^2\right)\delta A_{ln} \sin \alpha_{ln} \\ + 2k_{ln}^{1/2}\frac{d}{dz}(k_{ln}^{1/2}\delta A_{ln})\cos \alpha_{ln}\Bigg] = -\frac{4\pi}{c}\delta J_r,\end{aligned} \tag{5.93}$$

and

$$\begin{aligned}\sum_{l,n} J_l'(\kappa_{ln} r)\Bigg[\left(\frac{d^2}{dz^2} + \frac{\omega^2}{c^2} - k_{ln}^2 - \kappa_{ln}^2\right)\delta A_{ln} \cos \alpha_{ln} \\ - 2k_{ln}^{1/2}\frac{d}{dz}(k_{ln}^{1/2}\delta A_{ln})\sin \alpha_{ln}\Bigg] = -\frac{4\pi}{c}\delta J_\theta.\end{aligned} \tag{5.94}$$

Equations (5.93) and (5.94) are reduced further by averaging them over a wave period, as in the case of the idealized one-dimensional formulation, and orthogonalization over both the azimuthal angle and the radial position. For the TE modes, the relevant Bessel function orthogonality relation is of the form

$$\int_0^{R_g} dr\, r\left[J_l'(\kappa_{ln}r)J_l'(\kappa_{lm}r)+\frac{l}{\kappa_{ln}r}\frac{l}{\kappa_{lm}r}J_l(\kappa_{ln}r)J_l(\kappa_{lm}r)\right]=\frac{x_{ln}'^2-l^2}{2\kappa_{ln}^2}\delta_{n,m}J_l^2(x_{ln}'). \tag{5.95}$$

As a consequence, the dynamical equations take the form in cylindrical coordinates

$$\left(\frac{d^2}{dz^2}+\frac{\omega^2}{c^2}-k_{ln}^2-\kappa_{ln}^2\right)\delta A_{ln}=-\frac{4\kappa_{ln}^2}{c(x_{ln}'^2-l^2)J_l^2(x_{ln}')}\frac{\omega}{\pi}\int_{-\pi/\omega}^{\pi/\omega}dt\int_0^{2\pi}d\theta\int_0^{R_g}dr\, r$$
$$\times\left[\frac{l}{\kappa_{ln}r}J_l(\kappa_{ln}r)\delta J_r\sin\alpha_{ln}+J_l'(\kappa_{ln}r)\delta J_\theta\cos\alpha_{ln}\right], \tag{5.96}$$

and

$$2k_{ln}^{1/2}\frac{d}{dz}(k_{ln}^{1/2}\delta A_{ln})=-\frac{4\kappa_{ln}^2}{c(x_{ln}'^2-l^2)J_l^2(x_{ln}')}\frac{\omega}{\pi}\int_{-\pi/\omega}^{\pi/\omega}dt\int_0^{2\pi}d\theta\int_0^{R_g}dr\, r$$
$$\times\left[\frac{l}{\kappa_{ln}r}J_l(\kappa_{ln}r)\delta J_r\cos\alpha_{ln}-J_l'(\kappa_{ln}r)\delta J_\theta\sin\alpha_{ln}\right], \tag{5.97}$$

for each TE_{ln} mode. Observe also that in the absence of the electron beam, the source currents vanish and these equations reduce to the vacuum solutions; specifically, $\omega^2/c^2=k_{ln}^2+\kappa_{ln}^2$ and $d(\delta A_{ln})/dz=0$.

The analysis is more complicated for the TM modes because the assumption of a slowly varying amplitude implies that the divergence of the vector potential no longer vanishes. This arises due to the z-component of the vector potential, and we find that

$$(\nabla\cdot\delta\boldsymbol{A})=\sum_{l,n}\kappa_{ln}J_l(\kappa_{ln}r)\frac{d}{dz}\left(\frac{1}{k_{ln}}\delta A_{ln}\right)\sin\alpha_{ln}. \tag{5.98}$$

As a consequence, substitution of the TM mode representation into Maxwell's equations yields equations of the form

$$\sum_{l,n}J_n'(\kappa_{ln}r)\left[\left(\frac{d^2}{dz^2}+\frac{\omega^2}{c^2}-k_{ln}^2-\kappa_{ln}^2\right)\delta A_{ln}\cos\alpha_{ln}\right.$$
$$\left.-\left\{2k_{ln}^{1/2}\frac{d}{dz}(k_{ln}^{1/2}\delta A_{ln})+\kappa_{ln}^2\frac{d}{dz}\left(\frac{1}{k_{ln}}\delta A_{ln}\right)\right\}\sin\alpha_{ln}\right]=-\frac{4\pi}{c}\delta J_r, \tag{5.99}$$

$$\sum_{l,n}\frac{l}{\kappa_{ln}r}J_l(\kappa_{ln}r)\left[\left(\frac{d^2}{dz^2}+\frac{\omega^2}{c^2}-k_{ln}^2-\kappa_{ln}^2\right)\delta A_{ln}\sin\alpha_{ln}\right.$$
$$\left.+\left\{2k_{ln}^{1/2}\frac{d}{dz}(k_{ln}^{1/2}\delta A_{ln})+\kappa_{ln}^2\frac{d}{dz}\left(\frac{1}{k_{ln}}\delta A_{ln}\right)\right\}\cos\alpha_{ln}\right]=\frac{4\pi}{c}\delta J_\theta, \tag{5.100}$$

and

$$\sum_{l,n} \frac{\kappa_{ln}}{k_{ln}} J_l(\kappa_{ln} r)\left[\left(\frac{\omega^2}{c^2} - k_{ln}^2 - \kappa_{ln}^2\right)\delta A_{ln} \sin\alpha_{ln} + k_{ln}\frac{d}{dz}\delta A_{ln}\cos\alpha_{ln}\right] = -\frac{4\pi}{c}\delta J_z. \tag{5.101}$$

Note that the z-component of Maxwell's equations cannot be ignored in the derivation of the dynamical equations. The transverse components represented in equations (5.99) and (5.100) are averaged over a wave period and orthogonalized in r and θ in a manner analogous to that used for the TE modes. In the case of the TM modes, however, the Bessel function orthogonality relation is of the form

$$\int_0^{R_g} dr\, r\left[J_l'(\kappa_{ln} r)J_l'(\kappa_{lm} r) + \frac{l}{\kappa_{ln} r}\frac{l}{\kappa_{lm} r} J_l(\kappa_{ln} r) J_l(\kappa_{lm} r)\right] = \frac{x_{ln}^2}{2\kappa_{ln}^2}\delta_{n,m} J_l'^2(x_{ln}). \tag{5.102}$$

As a consequence, the transverse components of Maxwell's equations reduce to equations of the form

$$\left(\frac{d^2}{dz^2} + \frac{\omega^2}{c^2} - k_{ln}^2 - \kappa_{ln}^2\right)\delta A_{ln} = -\frac{4}{cR_g^2 J_l'^2(x_{ln}')}\frac{\omega}{\pi}\int_{-\pi/\omega}^{\pi/\omega} dt \int_0^{2\pi} d\theta \int_0^{R_g} dr\, r \times\left[J_l'(\kappa_{ln} r)\delta J_r \cos\alpha_{ln} - \frac{l}{\kappa_{ln} r} J_l(\kappa_{ln} r)\delta J_\theta \sin\alpha_{ln}\right], \tag{5.103}$$

and

$$2\left(k_{ln} + \frac{\kappa_{ln}^2}{k_{ln}}\right)^{1/2} \frac{d}{dz}\left[\left(k_{ln} + \frac{\kappa_{ln}^2}{k_{ln}}\right)^{1/2}\delta A_{ln}\right] - \frac{\kappa_{ln}^2}{k_{ln}}\frac{d}{dz}\delta A_{ln} = \frac{4}{cR_g^2 J_l'^2(x_{ln}')}\frac{\omega}{\pi}\int_{-\pi/\omega}^{\pi/\omega} dt \int_0^{2\pi} d\theta \int_0^{R_g} dr\, r \times\left[J_l'(\kappa_{ln} r)\delta J_r \sin\alpha_{ln} + \frac{l}{\kappa_{ln} r} J_l(\kappa_{ln} r)\delta J_\theta \cos\alpha_{ln}\right]. \tag{5.104}$$

The z-component of Maxwell's equations as represented by equation (5.101) must also be averaged over a wave period and orthogonalized in r and θ. In this case, the Bessel function orthogonality relation is of the form

$$\int_0^{R_g} dr\, r\, J_l(\kappa_{ln} r) J_l(\kappa_{lm} r) = \frac{R_g^2}{2}\delta_{n,m} J_l'^2(x_{ln}), \tag{5.105}$$

and we obtain the equations

$$\left(\frac{\omega^2}{c^2} - k_{ln}^2 - \kappa_{ln}^2\right)\delta A_{ln} = -\frac{4k_{ln}}{c\kappa_{ln} R_g^2 J_l'^2(x_{ln}')}\frac{\omega}{\pi}\int_{-\pi/\omega}^{\pi/\omega} dt \times \int_0^{2\pi} d\theta \int_0^{R_g} dr\, r\, J_l(\kappa_{ln} r)\delta J_z \sin\alpha_{ln}, \tag{5.106}$$

and

$$k_{ln}\frac{d}{dz}\delta A_{ln} = -\frac{4k_{ln}}{c\kappa_{ln}R_g^2 J_l'^2(x_{ln}')}\frac{\omega}{\pi}\int_{-\pi/\omega}^{\pi/\omega} dt \int_0^{2\pi} d\theta \int_0^{R_g} dr\, r J_l(\kappa_{ln}r)\delta J_z \cos\alpha_{ln}. \tag{5.107}$$

Multiplying equations (5.106) and (5.107) by κ_{ln}^2/k_{ln}^2 and adding the result to equations (5.103) and (5.104), we obtain

$$\begin{aligned}
&\frac{d^2}{dz^2} + \left(1+\frac{\kappa_{ln}^2}{k_{ln}^2}\right)\left(\frac{\omega^2}{c^2} - k_{ln}^2 - \kappa_{ln}^2\right)\delta A_{ln} \\
&\quad = -\frac{4}{cR_g^2 J_l'^2(x_{ln}')}\frac{\omega}{\pi}\int_{-\pi/\omega}^{\pi/\omega} dt \int_0^{2\pi} d\theta \int_0^{R_g} dr\, r \\
&\quad \times \left[J_l'(\kappa_{ln}r)\delta J_r \cos\alpha_{ln} - \left(\frac{l}{\kappa_{ln}r}\delta J_\theta - 2\frac{\kappa_{ln}}{k_{ln}}\delta J_z\right)J_l(\kappa_{ln}r)\sin\alpha_{ln}\right],
\end{aligned} \tag{5.108}$$

and

$$\begin{aligned}
&2\left(k_{ln}+\frac{\kappa_{ln}^2}{k_{ln}}\right)^{1/2}\frac{d}{dz}\left[\left(k_{ln}+\frac{\kappa_{ln}^2}{k_{ln}}\right)^{1/2}\delta A_{ln}\right] = \frac{4}{cR_g^2 J_l'^2(x_{ln}')}\frac{\omega}{\pi}\int_{-\pi/\omega}^{\pi/\omega} dt \int_0^{2\pi} d\theta \int_0^{R_g} dr\, r \\
&\quad \times \left[J_l'(\kappa_{ln}r)\delta J_r \sin\alpha_{ln} + \left(\frac{l}{\kappa_{ln}r}\delta J_\theta - 2\frac{\kappa_{ln}}{k_{ln}}\delta J_z\right)J_l(\kappa_{ln}r)\cos\alpha_{ln}\right].
\end{aligned} \tag{5.109}$$

Equations (5.108) and (5.109) are the dynamical equations for each TM_{ln} mode, and are analogous to equations (5.96) and (5.97) derived for the TE_{ln} modes.

The final form for the dynamical equations is obtained by substitution of the source current. It should be remarked at this point that the δ-functions in the source current are used to eliminate the integrals over r, θ and t, and that the symmetry property illustrated in equations (5.31–33) also applied in three dimensions. As a result, the dynamical equations for the TE_{ln} mode take the form

$$\left[\frac{d^2}{dz^2} + \left(\frac{\omega^2}{c^2} - k_{ln}^2 - \kappa_{ln}^2\right)\right]\delta a_{ln} = \frac{\omega_b^2}{c^2} H_{ln}\left\langle \frac{v_1 T_{ln}^{(+)} + v_2 W_{ln}^{(+)}}{|v_3|}\right\rangle, \tag{5.110}$$

and

$$2k_{ln}^{1/2}\frac{d}{dz}(k_{ln}^{1/2}\delta a_{ln}) = \frac{\omega_b^2}{c^2} H_{ln}\left\langle \frac{v_1 W_{ln}^{(-)} - v_2 T_{ln}^{(-)}}{|v_3|}\right\rangle, \tag{5.111}$$

where $\delta a_{ln} \equiv e\delta A_{ln}/m_e c^2$, $\omega_b^2 \equiv 4\pi e^2 n_b/m_e$, (v_1, v_2, v_3) are the components of the electron velocity in the frame rotating with the wiggler. The analogous equations for the TM_{ln} mode take the form

$$\begin{aligned}
&\left[\frac{d^2}{dz^2} + \left(1+\frac{\kappa_{ln}^2}{k_{ln}^2}\right)\left(\frac{\omega^2}{c^2} - k_{ln}^2 - \kappa_{ln}^2\right)\right]\delta a_{ln} \\
&\quad = \frac{\omega_b^2}{c^2} H_{ln}\left\langle \frac{v_1 T_{ln}^{(+)} + v_2 W_{ln}^{(+)}}{|v_3|} + 2\frac{\kappa_{ln}}{k_{ln}} J_l(\kappa_{ln}r)\sin\alpha_{ln}\right\rangle,
\end{aligned} \tag{5.112}$$

and

$$2\left(k_{ln}+\frac{\kappa_{ln}^2}{k_{ln}}\right)^{1/2}\frac{d}{dz}\left[\left(k_{ln}+\frac{\kappa_{ln}^2}{k_{ln}}\right)^{1/2}\delta a_{ln}\right]$$
$$=\frac{\omega_b^2}{c^2}H_{ln}\left\langle\frac{v_1 W_{ln}^{(-)}-v_2 T_{ln}^{(-)}}{|v_3|}+2\frac{\kappa_{ln}}{k_{ln}}J_l(\kappa_{ln}r)\cos\alpha_{ln}\right\rangle. \tag{5.113}$$

In the preceding equations H_{ln}, $T_{ln}^{(\pm)}$ and $W_{ln}^{(\pm)}$ are mode-dependent quantities defined as

$$H_{ln}\equiv\begin{cases}\dfrac{x_{ln}'^2}{(x_{ln}'^2-l^2)J_l^2(x_{ln}')}; \text{TE}_{ln}\text{ mode}\\ \dfrac{1}{J_l'^2(x_{ln})}; \text{TM}_{ln}\text{ mode,}\end{cases} \tag{5.114}$$

and

$$T_{ln}^{(\pm)}\equiv\begin{cases}F_{ln}^{(\pm)}\sin\psi_{ln}+G_{ln}^{(\pm)}\cos\psi_{ln};\ \text{TE}_{ln}\text{ mode}\\ F_{ln}^{(\mp)}\cos\psi_{ln}-G_{ln}^{(\mp)}\sin\psi_{ln};\ \text{TM}_{ln}\text{ mode,}\end{cases} \tag{5.115}$$

$$W_{ln}^{(\pm)}\equiv\begin{cases}F_{ln}^{(\mp)}\cos\psi_{ln}-G_{ln}^{(\mp)}\sin\psi_{ln};\text{TE}_{ln}\text{ mode}\\ -(F_{ln}^{(\pm)}\sin\psi_{ln}+G_{ln}^{(\pm)}\cos\psi_{ln});\text{TE}_{ln}\text{ mode,}\end{cases} \tag{5.116}$$

where

$$\psi_{ln}\equiv\psi_0+\int_0^z dz'\left(k_{ln}+lk_w-\frac{\omega}{v_z}\right) \tag{5.117}$$

denotes the ponderomotive phase, $\psi_0\equiv-\omega t_0$ is the initial phase,

$$F_{ln}^{(\pm)}\equiv J_{l-1}(\kappa_{ln}r)\cos[(l-1)\chi]\pm J_{l+1}(\kappa_{ln}r)\cos[(l+1)\chi], \tag{5.118}$$

and

$$G_{ln}^{(\pm)}\equiv J_{l-1}(\kappa_{ln}r)\sin[(l-1)\chi]\pm J_{l+1}(\kappa_{ln}r)\sin[(l+1)\chi]. \tag{5.119}$$

The particle average defined in equations (5.110–13) is defined over the *initial* conditions of the beam. Observe that the formulation includes the effect of an arbitrary number of TE and TM modes as long as they all have identical frequencies.

In order to specify the particle average we first choose an initial momentum-space distribution which describes a monoenergetic beam with a pitch-angle spread. This distribution corresponds to the physical model of beam thermal effects employed in the linearized theory described in Chapter 4. The difference between the present model and the linear theory is that we specify only the *initial* momentum space distribution here, which describes the beam *prior* to injection into the wiggler. The complete nonlinear formulation includes the description of the evolution of the beam in its injection, propagation through the wiggler, and interaction with the electromagnetic field. Hence, the subsequent evolution of the beam, including the possible growth of the thermal spread, is included self-consistently in the formulation. The specific distribution we choose is of the form

$$F_b(\mathbf{p}_0) = A\exp[-(p_{z0} - p_0)^2/\Delta p_\parallel^2]\delta(p_0^2 - p_\perp^2 - p_{z0}^2)H(p_{z0}), \tag{5.120}$$

where $H(p_{z0})$ is the Heaviside function, the normalization constant is given by

$$A = \left\{\pi\int_0^{p0} dp_{z0}\exp[-(p_{z0} - p_0)^2/\Delta p_\parallel^2]\right\}^{-1}, \tag{5.121}$$

p_0 describes the total momentum of the beam, and $\Delta p_\parallel$ describes the axial momentum spread. This axial momentum spread can be related to the axial energy spread by the relation

$$\frac{\Delta\gamma_z}{\gamma_0} = 1 - \frac{1}{\sqrt{\left[1 + 2(\gamma_0^2 - 1)\dfrac{\Delta p_\parallel}{p_0}\right]}}, \tag{5.122}$$

where $\gamma_0 = (1 + p_0^2/m^2c^2)^{1/2}$ is the relativistic factor corresponding to the total energy. The averaging operator is defined in terms of this momentum space distribution as

$$\begin{aligned}\langle(\cdots)\rangle = &\frac{A}{4\pi A_g}\int_0^{2\pi} d\phi_0 \int_0^{p_0} dp_{z0}\beta_{z0}\exp[-(p_{z0} - p_0)^2/\Delta p_\parallel^2] \\ &\times \iint_{A_g} dx_0 dy_0 \sigma_\perp(x_0, y_0)\int_0^{2\pi} d\psi_0 \sigma_\parallel(\psi_0)(\cdots),\end{aligned} \tag{5.123}$$

where $\beta_{z0} \equiv v_{z0}/c$, $\phi_0 \equiv \tan^{-1}(p_{y0}/p_{x0})$, and p_{z0} defines the initial axial momentum of the electron. It is important to recognize that this average implicitly includes the effect of the overlap of the electron beam with the transverse mode structure of the radiation field (often included in one-dimensional analyses in an *ad hoc* manner by the inclusion of a filling factor) in a self-consistent manner.

It is important to emphasize here that this distribution describes the effect of an *initial* momentum spread on the beam. The effective momentum spread which develops during the course of the interaction due to (1) the injection of the beam into the wiggler, (2) wiggler inhomogeneities (i.e. Larmor and betatron oscillations), and (3) the interaction of the electrons with the electromagnetic field is included implicitly in the formulation. The implicit inclusion of these effects is one of the advantages of this formulation in which the complete set of three-dimensional Lorentz force equations is integrated for the ensemble of electrons, and is carried over in the discussions of planar wiggler configurations in Section 5.4 and the inclusion of collective effects in Section 5.5.

The electron orbit equations

The formulation is inherently multi-mode in the sense that an arbitrary number of TE and TM modes may be treated using the dynamical equations. Coupling between the various modes is included through the effect of the mode ensemble on the electron trajectories. Hence, in order to complete the formulation, the electron orbit equations in the presence of the static and fluctuation fields must be specified. Since we deal with an amplifier model we choose to integrate in z and write the Lorentz force equations in the form

$$v_z \frac{d}{dz}\boldsymbol{p} = -e\delta\boldsymbol{E} - \frac{e}{c}\mathbf{v}\times(\boldsymbol{B}_0 + \boldsymbol{B}_w + \delta\boldsymbol{B}), \tag{5.124}$$

where the electric and magnetic fields

$$\delta\boldsymbol{E} = -\frac{1}{c}\frac{\partial}{\partial t}\delta\boldsymbol{A}, \quad \text{and} \quad \delta\boldsymbol{B} = \nabla\times\delta\boldsymbol{A} \tag{5.125}$$

are derivable from the vector potentials and consist of the sum of all the TE and/or TM modes included in any specific simulation. Substitution of the appropriate forms for the vector potentials shows that in the wiggler frame

$$\begin{aligned} v_3\frac{d}{dz}p_1 = &-\{\Omega_0[1 + k_w\varepsilon_0(z - z_0)] - k_w v_3 + 2\Omega_w I_1(\lambda)\sin\chi\}p_2 \\ &+ \Omega_w p_3 I_2(\lambda)\sin 2\chi - \frac{\varepsilon_0}{2}\Omega_0 p_3\lambda\sin\chi - \frac{m_e c}{2}\sum_{\text{TE modes}}\delta a_{ln} \\ &\times[(\omega - k_{ln}v_3)W_{ln}^{(-)} - 2\kappa_{ln}v_2 J_l(\kappa_{ln}r)\cos\alpha_{ln} + \Gamma_{ln}v_3 T_{ln}^{(+)}] \\ &- \frac{m_e c}{2}\sum_{\text{TM modes}}\delta a_{ln}\left[\left(\omega - \frac{k_{ln}^2 + \kappa_{ln}^2}{k_{ln}}v_3\right)W_{ln}^{(-)} + \Gamma_{ln}v_3 T_{ln}^{(+)}\right], \end{aligned} \tag{5.126}$$

$$\begin{aligned} v_3\frac{d}{dz}p_2 = &\{\Omega_0[1 + k_w\varepsilon_0(z - z_0)] - k_w v_3 + 2\Omega_w I_1(\lambda)\sin\chi\}p_1 \\ &- \Omega_w p_3[I_0(\lambda) + I_2(\lambda)\cos 2\chi] + \frac{\varepsilon_0}{2}\Omega_0 p_3\lambda\cos\chi + \frac{m_e c}{2}\sum_{\text{TE modes}}\delta a_{ln} \\ &\times[(\omega - k_{ln}v_3)T_{ln}^{(-)} - 2\kappa_{ln}v_1 J_l(\kappa_{ln}r)\cos\alpha_{ln} + \Gamma_{ln}v_3 W_{ln}^{(+)}] \\ &+ \frac{m_e c}{2}\sum_{\text{TM modes}}\delta a_{ln}\left[\left(\omega - \frac{k_{ln}^2 + \kappa_{ln}^2}{k_{ln}}v_3\right)T_{ln}^{(-)} + \Gamma_{ln}v_3 W_{ln}^{(+)}\right], \end{aligned} \tag{5.127}$$

and

$$\begin{aligned} v_3\frac{d}{dz}p_3 = &\Omega_w p_2[I_0(\lambda) + I_2(\lambda)\cos 2\chi] - \Omega_w p_1 I_1(\lambda)\sin 2\chi \\ &+ \frac{\varepsilon_0}{2}\Omega_0\lambda(p_1\sin\chi - p_2\cos\chi) - \frac{m_e c}{2}\sum_{\text{TE modes}}\delta a_{ln}[k_{ln}(v_1\, W_{ln}^{(-)} - v_2\, T_{ln}^{(-)}) \\ &+ \Gamma_{ln}(v_1\, T_{ln}^{(+)} + v_2\, W_{ln}^{(+)})] - \frac{m_e c}{2}\sum_{\text{TM modes}}\delta a_{ln} \end{aligned}$$

$$\times\left[\frac{k_{ln}^2+\kappa_{ln}^2}{k_{ln}}(v_1\,W_{ln}^{(-)}-v_2\,T_{ln}^{(-)})+\Gamma_{ln}(v_1\,T_{ln}^{(+)}+v_2\,W_{ln}^{(+)})\right.$$

$$\left.+\,2\omega\frac{\kappa_{ln}}{k_{ln}}J_l(\kappa_{ln}r)\cos\alpha_{ln}\right], \tag{5.128}$$

where $\Omega_0 \equiv eB_0/\gamma m_e c$, $\Omega_w \equiv eB_w(z)/\gamma m_e c$ is determined by the variation in the wiggler amplitude as given in equation (5.81), and

$$\Gamma_{ln} \equiv \frac{1}{\delta a_{ln}}\frac{d}{dz}\delta a_{ln} \tag{5.129}$$

defines the growth rate of the wave mode. Observe that this formulation of the Lorentz force equations includes the effects of tapered wiggler and axial solenoidal fields, and we neglect the terms of ε_0 for a uniform solenoidal field. In addition, we have that for either the TE or TM modes

$$v_3\frac{d}{dz}x = v_1\cos k_w z - v_2\sin k_w z, \tag{5.130}$$

$$v_3\frac{d}{dz}y = v_1\sin k_w z + v_2\cos k_w z, \tag{5.131}$$

and

$$\frac{d}{dz}\psi_{ln} = k_{ln} + lk_w - \frac{\omega}{v_3}. \tag{5.132}$$

Together, equations (5.110–13) for an ensemble of modes and equations (5.126–28) and (5.130–32) for an ensemble of particles completely specify the nonlinear problem which governs the self-consistent evolution of the wave modes and particle trajectories. As described for the case of the idealized one-dimensional formulation, the dynamical equations for the fields constitute a set of two second-order differential equations for the amplitude and phase of each mode, which can be converted into a set of four first-order differential equations. As a consequence, this formulation describes an initial value problem which consists of a set of $4N_m + 6N_p$ coupled first-order differential equations, where N_m and N_p denote the number of modes and electrons, respectively.

The first-order field equations

However, a simplification of the field equations can be made under the assumption that the second-order derivatives of the amplitude and phase (which is given by the derivative of the wavenumber) can be neglected. This is generally valid within the context of the implicit assumption of a slowly varying amplitude and wavenumber. The dynamical equations for the electromagnetic wave modes can be expressed in this manner as

$$k_{ln}^2 = \frac{\omega^2}{c^2} - \kappa_{ln}^2 - \frac{\omega_b^2}{c^2}\frac{H_{ln}}{\delta a_{ln}}\left\langle \frac{v_1\, T_{ln}^{(+)} + v_2\, W_{ln}^{(+)}}{|v_3|} \right\rangle, \tag{5.133}$$

$$2k_{ln}\frac{d}{dz}\delta a_{ln} = \frac{\omega_b^2}{c^2} H_{ln} \left\langle \frac{v_1\, W_{ln}^{(-)} - v_2\, T_{ln}^{(-)}}{|v_3|} \right\rangle, \tag{5.134}$$

for the TE modes, and

$$\left(1 + \frac{\kappa_{ln}^2}{k_{ln}^2}\right)\left(\frac{\omega^2}{c^2} - k_{ln}^2 - \kappa_{ln}^2\right) = \frac{\omega_b^2}{c^2}\frac{H_{ln}}{\delta a_{ln}}\left\langle \frac{v_1\, T_{ln}^{(+)} + v_2\, W_{ln}^{(+)}}{|v_3|} + 2\frac{\kappa_{ln}}{k_{ln}} J_l(\kappa_{ln} r)\sin\alpha_{ln} \right\rangle, \tag{5.135}$$

$$2\left(k_{ln} + \frac{\kappa_{ln}^2}{k_{ln}}\right)\frac{d}{dz}\delta a_{ln} = \frac{\omega_b^2}{c^2} H_{ln} \left\langle \frac{v_1\, W_{ln}^{(-)} - v_2\, T_{ln}^{(-)}}{|v_3|} + 2\frac{\kappa_{ln}}{k_{ln}} J_l(\kappa_{ln} r)\cos\alpha_{ln} \right\rangle, \tag{5.136}$$

for the TM modes. These two pairs of equations for the TE and TM modes represent an algebraic equation for the wavenumbers (equations (5.133) and (5.135)) and a first-order differential equation for the amplitudes (equations (5.134) and (5.136)) of each mode. In order to complete the formulation so as to specify the evolution of the phase, we define the relative phase of each mode in the form

$$\Delta\phi_{ln}(z) \equiv \int_0^z dz' \left[k_{ln}(z') - \sqrt{\left(\frac{\omega^2}{c^2} - \kappa_{ln}^2\right)} \right], \tag{5.137}$$

which describes the dielectric effect of the electron beam on the shift of the wavenumber from the vacuum value. The evolution of the relative phase is described by an additional first-order differential equation

$$\frac{d}{dz}\Delta\phi_{ln}(z) = k_{ln}(z) - \sqrt{\left(\frac{\omega^2}{c^2} - \kappa_{ln}^2\right)}. \tag{5.138}$$

This alternative formulation, therefore, consists of the simultaneous solution of $2N_m + 6N_p$ first-order differential equations, as well as the evaluation of an algebraic equation for the wavenumber of each mode. As a practical consequence, this results in a reduction in computational requirements with only a minimal loss in accuracy. Indeed, for all subsequent cases considered herein, the discrepancy between these two formulations of the problem is substantially less than 10%.

It should be remarked before we proceed to the discussion of numerical examples that the time-averaged Poynting flux P_{ln} for each mode is related to the field amplitudes and wavenumbers through the relations

$$P_{ln} = \frac{m_e^2 c^4}{8e^2}\omega k_{ln}\frac{R_g^2}{H_{ln}}\delta a_{ln}^2, \tag{5.139}$$

for the TE modes, and

$$P_{ln} = \frac{m_e^2 c^4}{8e^2}\omega\left(\frac{k_{ln}^2 + \kappa_{ln}^2}{k_{ln}}\right)\frac{R_g^2}{H_{ln}}\delta a_{ln}^2, \tag{5.140}$$

for the TM modes. Comparison of these expressions with equations (5.134) and (5.136) reveals that the dynamical equations describe the evolution of the time-averaged Poynting flux, which is a constant in the absence of the electron beam. Further, the source terms in the dynamical equations constitute a calculation of the average of $\delta \boldsymbol{J} \cdot \delta \boldsymbol{E}_{ln}$ for each mode.

5.3.2 The initial conditions

The set of coupled nonlinear differential equations can be solved numerically for an amplifier configuration in which a single wave of frequency ω is injected into the system at $z = 0$ in concert with the electron beam. The solution to this initial-value problem can be accomplished by a variety of different algorithms, including Adams–Moulton predictor/corrector and Runge–Kutta techniques. The advantage of the Adams–Moulton technique is that it is more stable than the Runge–Kutta algorithm; however, this occurs at the practical cost of a greatly increased memory requirement. In practice, it is found that the fouth-order Runge–Kutta–Gill technique leads to no serious numerical instabilities. Hence, either algorithm may be employed depending upon the available computational facilities. The averages in the dynamical equations for the wave modes are performed by means of an Nth-order Gaussian quadrature technique in each of the variables $(r_0, \theta_0, \psi_0, p_{z0}, \phi_0)$. In the absence of an energy spread, the number of degrees of freedom in the initial conditions reduces to (r_0, θ_0, ψ_0) and the choice of $N_r = N_\theta = N_\psi = 10$ (corresponding to 1000 particles) is typically made. The inclusion of an energy spread requires additional electrons in the numerical simulation. In this case we find that a reduction in the number of electrons corresponding to the radial positions to $N_r = 4$ does not result in a severe loss of accuracy, and we choose sixth- and fourth-order Gaussian quadratures to resolve the p_{z0} and ϕ_0 distributions. This gives a total number of 9600 electrons in the simulation.

The initial state of the electron beam is chosen to model the injection of a monoenergetic, uniform, axisymmetric electron beam with a flat-top density profile (i.e. $\sigma_\perp = \sigma_\parallel = 1$). The electron positions are chosen by means of the Gaussian algorithm within the ranges $-\pi \leqslant \psi_0 \leqslant \pi$, $0 \leqslant \theta_0 \leqslant 2\pi$, and $R_{\min} \leqslant r_0 \leqslant R_{\max}$. In the absence of an energy spread, the momenta are chosen so that $\boldsymbol{p}_{\perp 0} = 0$ and $p_{z0} = m_e c(\gamma_0^2 - 1)^{1/2}$, where γ_0 is the relativistic factor corresponding to the total beam energy. This model can treat a beam with either an annular or a solid cross-section. Within the context of this geometry, the plasma frequency is related to the total beam current, I_b, by means of the relation

$$\omega_b^2 = \frac{4e}{m_e v_{z0}} \frac{I_b}{R_{\max}^2 - R_{\min}^2}. \tag{5.141}$$

It is also required in order to self-consistently satisfy the condition necessary to the neglect of the beam space-charge waves in the high-gain Compton regime that

$$\frac{\omega_b^2}{\gamma_0^{1/2} c k_w} \ll \frac{v_\perp^2}{c^2} \gamma_z^3 \tag{5.142}$$

within the uniform wiggler region. The initial conditions on the radiation fields are chosen such that the initial amplitude of each TE or TM mode included in the simulation is chosen to correspond to an arbitrary input power via equations (5.139) or (5.140), and the initial value of the wavenumber is chosen to correspond to the vacuum value (i.e. $k_{ln}(z=0)=(\omega^2/c^2-\kappa_{ln}^2)^{1/2}$). Note also that the initial value of the growth rate of each mode is zero (i.e. $\Gamma_{ln}(z=0)=0$), which is consistent with the choice of a wiggler field amplitude which increases from zero adiabatically.

It is important to bear in mind that although the analysis is restricted to the high-gain Compton regime, collective effects are included through the dielectric response of the beam to the waveguide mode, and the analysis is not purely in the single-particle regime. Thus, while the wavenumbers are set initially to the vacuum dispersion relations, the system evolves into a fully self-consistent dielectrically loaded waveguide mode.

5.3.3 Numerical simulation for Group I orbit parameters

The first case under consideration is that of a wide-band 35 GHz amplifier operating in the TE_{11} mode. To this end, the wiggler field is assumed to be characterized by an amplitude and period of $B_w=2\,\text{kG}$ and $\lambda_w=1.175\,\text{cm}$, respectively, and an entry taper of $N_w=10$ wiggler periods is chosen for the smooth injection of the beam. The axial solenoidal magnetic field is chosen to be $B_0=1.3\,\text{kG}$. This corresponds to beam propagation on Group I trajectories comfortably far from the transition to orbital instability which occurs at $B_0\approx 3.9\,\text{kG}$. The electron-beam configuration is that of a solid (i.e. pencil) beam with an energy of 250 keV, a current of 35 A, and an initial radius $R_{max}=0.155\,\text{cm}$. These parameters ensure a relatively broad resonance band about 35 GHz in the TE_{11} mode for a waveguide radius of $R_g=0.36626\,\text{cm}$. Note that only one mode is included in this numerical example since all other modes are below cutoff in the vicinity of the TE_{11} resonance.

We first consider the case of an ideal beam with a vanishing axial energy spread (i.e. $\Delta p_\parallel=0$). The initial distribution in the axial phase space $(\psi_{ln}, d\psi_{ln}/dz)$ is shown in Fig. 5.15. Each dot in the figure represents a **phase sheet** composed of 100 electrons distributed throughout the cross-section of the beam. Each phase sheet, therefore, represents a cross-sectional slice of the beam, which is chosen initially as shown in Fig. 5.16. The circle shown in Fig. 5.16 represents the waveguide wall. Each phase sheet is chosen initially to be identical; however, the subsequent evolution of the electron trajectories in the presence of the electromagnetic fields is followed self-consistently. The positions in (r_0,θ_0,ψ_0) were chosen by means of a 10-point Gaussian weighting scheme, and the nonuniformity in the positions is compensated for by a nonuniform weighting of the electrons.

The evolution of the TE_{11} waveguide mode is illustrated in Fig. 5.17 in which we plot both the power (dashed line) and growth rate (solid line) of the mode as a function of axial position for an input signal of frequency $\omega/ck_w=1.3$ (corresponding to a frequency of 33.2 GHz) and a power of 10 W. As shown in

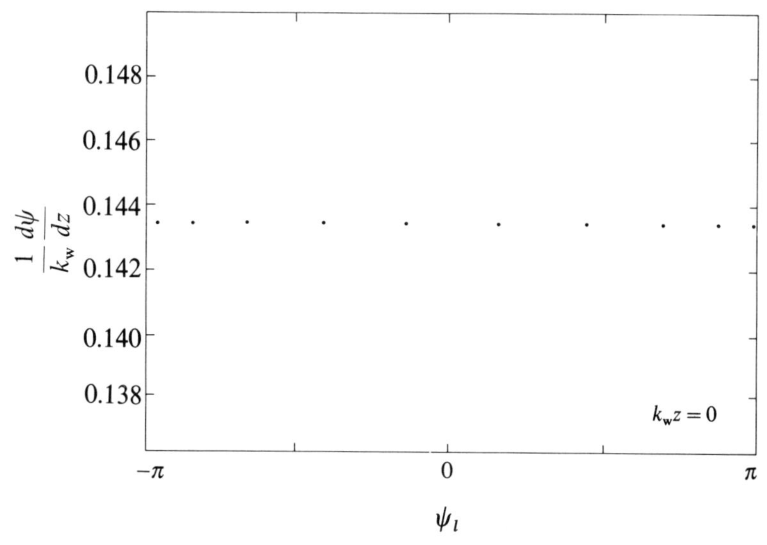

Fig. 5.15 Initialization of the axial phase space. Each point represents the superposition of 100 particles distributed throughout the cross-section of the beam.

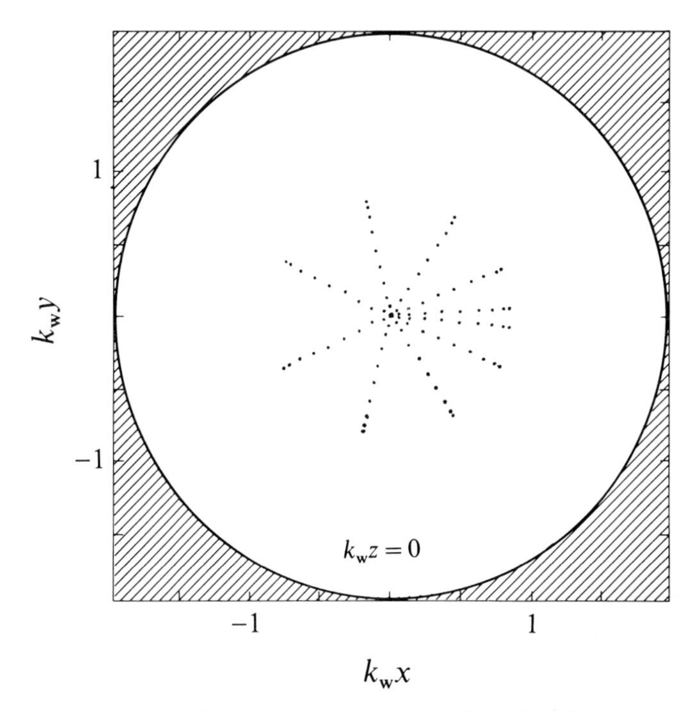

Fig. 5.16 Initialization of the cross-section of the beam.

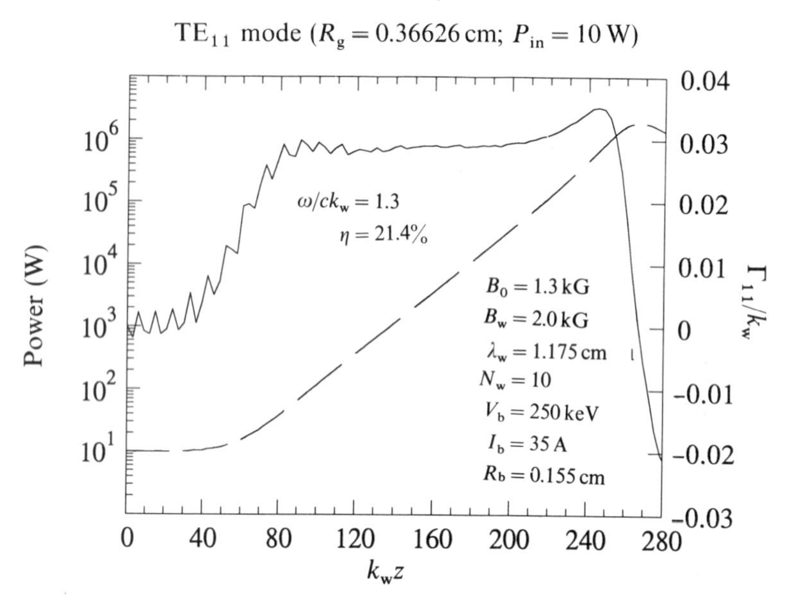

Fig. 5.17 Evolution of the radiation power (dashed line) and growth rate (solid line) of the TE_{11} mode as a function of axial position.

the figure, the growth of the wave mode was approximately exponential after an initial transient period of $k_w z \leqslant 80$ which includes the entry taper region. During the linear phase of the interaction the growth rate was $\Gamma_{11}/k_w \approx 0.029$, and a small increase is observed prior to saturation at $k_w z \approx 267$. The radiation power at saturation was approximately 1.87 MW for an overall efficiency of $\eta \approx 21.4\%$. A complete spectrum of the efficiency at saturation is shown in Fig. 5.18 as a function of frequency, in which the dots represent the results of the simulation. Observe that the peak efficiency occurs at $\omega_{max}/ck_w \approx 1.3$, with a bandwidth $\Delta\omega/\omega_{max} \approx 54\%$.

The consistency of the simulation procedure has been verified in several ways. The most fundamental is the requirement of energy conservation between the electrons and the wave, and agreement between the energy lost by the beam and that gained by the wave was found to be significantly better than 0.1% in all cases considered. In addition, the growth rate in the linear regime has been checked against the results of the three-dimensional linear theory described in Chapter 4. Although the linear theory represents an idealized model in which the entry taper is not included and all electrons execute the ideal helical trajectories, agreement between the nonlinear simulation and the linear theory is good. At the frequency corresponding to peak growth (i.e. $\omega/ck_w = 1.3$), the linear theory predicts a growth rate of $\Gamma_{lin}/k_w \approx 0.031$, which compares well with the simulation result of $\Gamma_{11}/k_w \approx 0.029$. Finally, an estimate of the efficiency based upon the *ad hoc* phase trapping model yields an estimate of $\eta \approx 19.1\%$, which is close to the 21.4% efficiency found in the simulation.

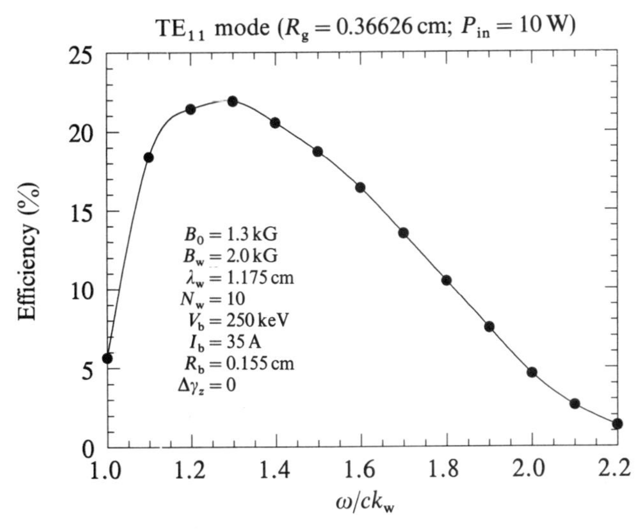

Fig. 5.18 Spectrum of the interaction efficiency of the TE_{11} mode as a function of frequency. The dots represent the results of the nonlinear simulation.

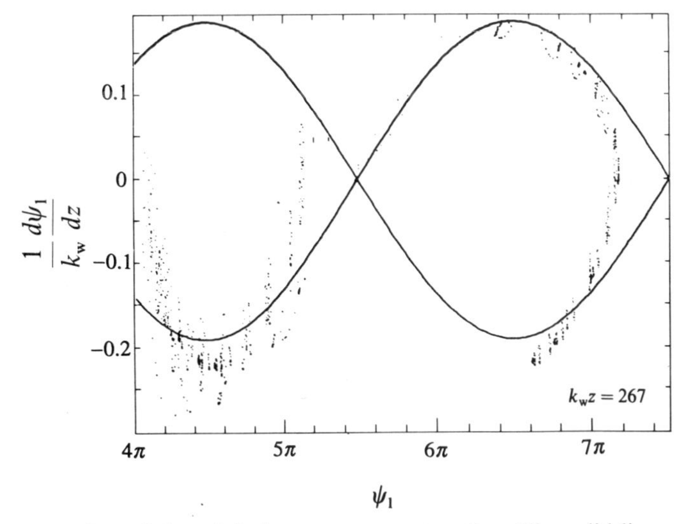

Fig. 5.19 Illustration of the axial phase space at saturation. The solid line represents an analytic approximation of the separatrix.

Saturation is by means of particle trapping in the ponderomotive wave. This is clearly shown in Fig. 5.19 in which we plot the axial phase space at saturation for the case of $\omega/ck_w = 1.3$. The solid lines in the figure represent *approximate* separatrices calculated under the assumption that all electrons are executing the ideal helical trajectories. Thus, the actual number of electrons trapped may differ slightly from that shown in the figure, but the essential conclusion remains valid. Note that the entire beam has not been trapped even at saturation.

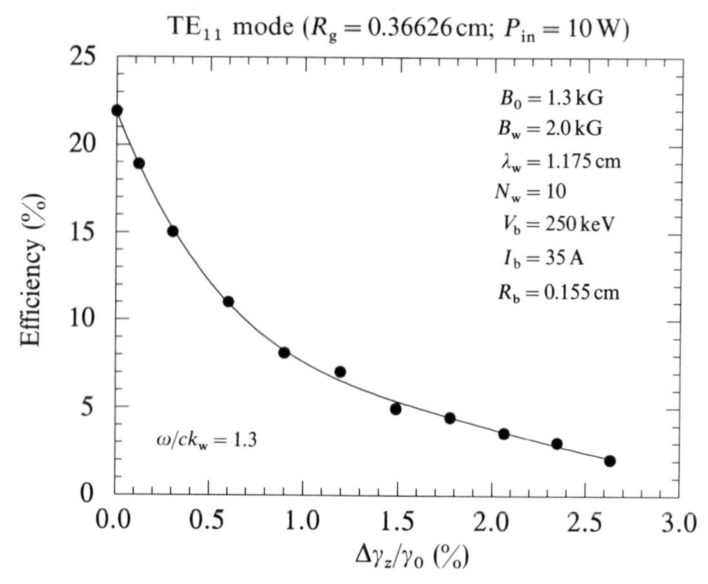

Fig. 5.20 Graph of the efficiency versus axial energy spread at the frequency corresponding to the maximum efficiency. The dots represent the results of the nonlinear simulation.

The effect of the *initial* axial energy spread upon the interaction at the frequency which corresponds to maximum efficiency is shown in Fig. 5.20, in which we plot the efficiency versus $\Delta\gamma_z/\gamma_0$. The decrease in the efficiency with the increase in the axial energy spread is quite rapid, and the efficiency is found to decrease by an order of magnitude as for an energy spread of 2.6%. In addition, the slope of this decrease is seen to change abruptly for $\Delta\gamma_z/\gamma_0 \approx 1.1$–$1.5\%$ corresponding to the point at which thermal effects become important. In order to understand this we observe that since $|\mathrm{Im}\,k|/k_w \approx 0.029$ and $\mathrm{Re}\,k/k_w \approx 0.88$ for this case, the critical axial velocity spread for which thermal effects become important is estimated to be $\Delta v/v \approx |\mathrm{Im}\,k|/[k_w + \mathrm{Re}\,k] \approx 1.5\%$. Using equation (5.122) to relate this axial velocity spread to the axial energy spread, we find that this corresponds to a $\Delta\gamma_z/\gamma_0 \approx 1.8\%$ which is in reasonable agreement with the results of the nonlinear simulation.

The detailed variation of the relative phase is shown in Fig. 5.21 as a function of axial position for a range of frequencies spanning the interaction spectrum, and the arrows indicate the positions at which the power saturates. Two features should be noted from the figure. The first is that it is possible to find a frequency which corresponds to a relatively small change in the relative phase over the course of the interaction. As shown in the figure, this corresponds to a frequency of $\omega/ck_w = 1.6$ at which the wavenumber is shifted relatively little by the dielectric effect of the beam in the wiggler. While this frequency is higher than the frequency of maximum saturation efficiency (i.e. $\omega/ck_w = 1.3$), it is slightly below the frequency of maximum growth rate, which occurs in the range of $\omega/ck_w \approx 1.7$–1.8

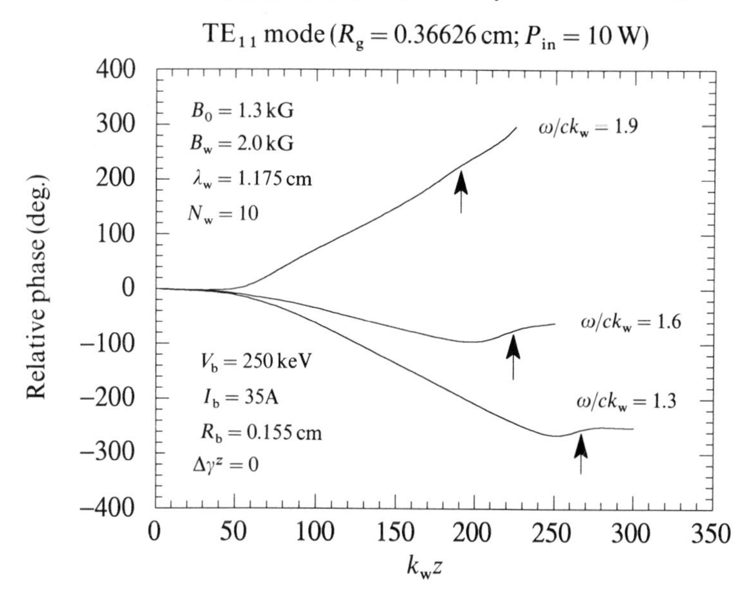

Fig. 5.21 Graph of the relative phase versus axial position at the three frequencies. The arrows indicate the saturation points for the interaction.

for the present choice of parameters. The second feature is that the frequency at which the phase remains relatively constant represents a transition point. Above this frequency, the relative phase increases monotonically with axial position reflecting an increase in the wavenumber with respect to the vacuum mode. In contrast, the relative phase decreases with axial position up to the point at which saturation occurs at lower frequencies. This decrease in the relative phase occurs because the wavenumber is downshifted with respect to the vacuum wavenumber in this regime.

On the basis of the rapid variation in the phase with axial position shown in Fig. 5.21 for the bulk of the interaction spectrum, it is to be expected that the phase observed at saturation can vary widely with beam voltage. Indeed, the typical variation in the phase at saturation with beam voltage across the unstable frequency spectrum ranges from a low of 44° per 1% variation in the beam voltage at $\omega/ck_w = 1.7$ to a high of 63° per 1% variation in the beam voltage at $\omega/ck_w = 1.3$.

The beam dynamics within the interaction region are illustrated by the evolution of the beam cross-section over the course of the interaction. The beam is spun up during the entry taper region ($k_w z \leqslant 62.8$), and the beam cross-section at a point just after the start of the uniform wiggler region is shown in Fig. 5.22. It is evident from this figure that, in comparison with the initial state shown in Fig. 5.16, the centre of the beam is displaced from the waveguide symmetry axis. The injection process yields orbits close to the ideal helical trajectories, and $k_w r_{centre} \approx 0.25$ for the beam which is in good agreement with the calculated value for the Group I orbit for these parameters. In addition, the beam has been

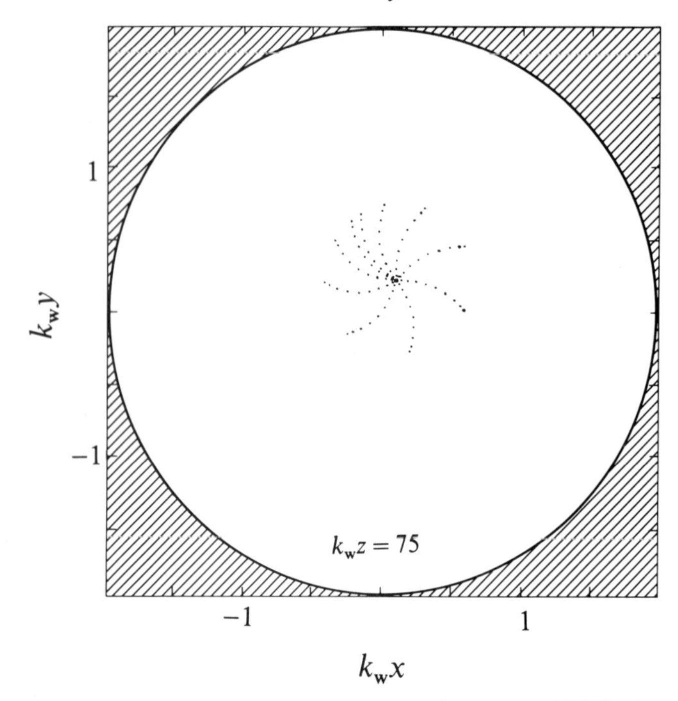

Fig. 5.22 Representation of the beam cross-section at $k_w z = 75$, which is shortly after the start of the uniform wiggler region.

compressed substantially due to the focusing effect of the helical wiggler. Finally, the beam executes rotational motion about the centroid which occurs due to the combined effects of a variation in the wiggler-induced transverse velocity across the beam and the betatron oscillation due to the transverse wiggler gradient.

The overall bulk motion of the beam is expected to display a periodicity at the wiggler period and to twist the beam into a helix about the axis of symmetry. Such a motion is observed as shown in Figs. 5.23–6, which display the beam cross-section as it evolves through one wiggler period from $k_w z = 150$ through 156. Note that this is within the linear (i.e. exponential growth) regime of the interaction, and substantial amplification of the signal has occurred.

The case of the TM_{11} mode can also be investigated. In order to facilitate comparison with the TE_{11} mode, all the parameters used in Fig. 5.17 have been retained. The only alteration is in the choice of the waveguide radius in order to shift the waveguide dispersion curve relative to the beam resonance line (i.e. $\omega \approx (k + k_w)v_\parallel$) so as to maintain the interaction in the same frequency range. We choose $R_g = 0.762\,23$ cm so that the waveguide cutoff is identical to that of the TE_{11} mode for $R_g = 0.366\,26$ cm. Numerical results indicate that the peak efficiency $\eta \approx 5.92\%$ occurs at a frequency $\omega/ck_w \approx 1.78$. This is significantly lower than the peak efficiency found for the TE_{11} mode, and is shifted relative to the TE_{11} mode peak (at $\omega/ck_w \approx 1.3$). The growth rate is also significantly lower,

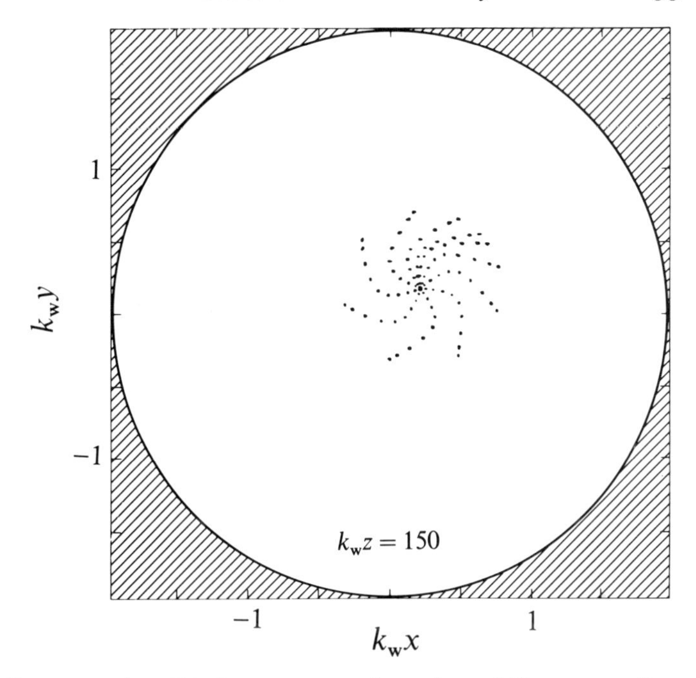

Fig. 5.23 Representation of the beam cross-section at $k_w z = 150$ corresponding to the linear phase of the interaction.

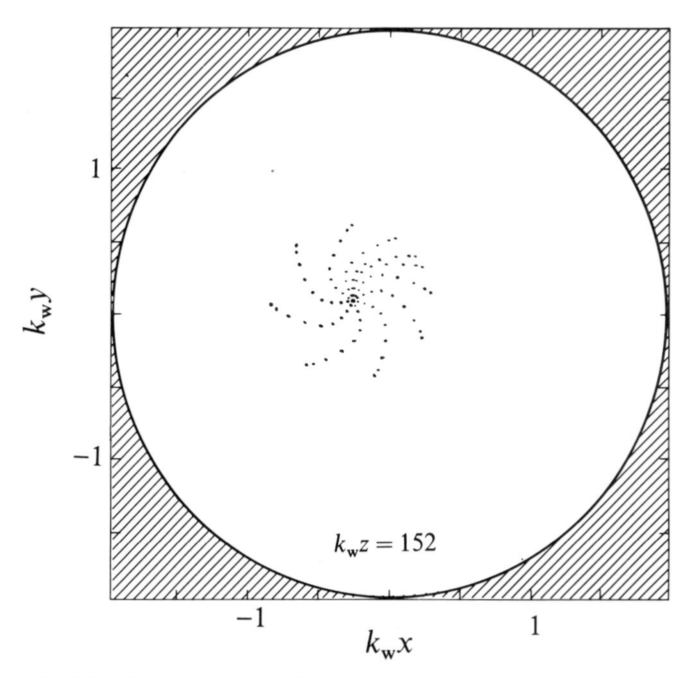

Fig. 5.24 Representation of the beam cross-section at $k_w z = 152$.

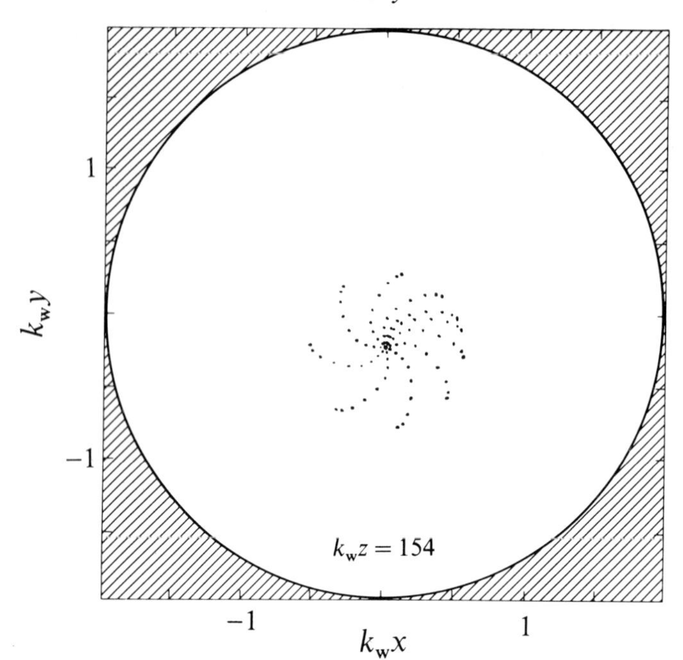

Fig. 5.25 Representation of the beam cross-section at $k_w z = 154$.

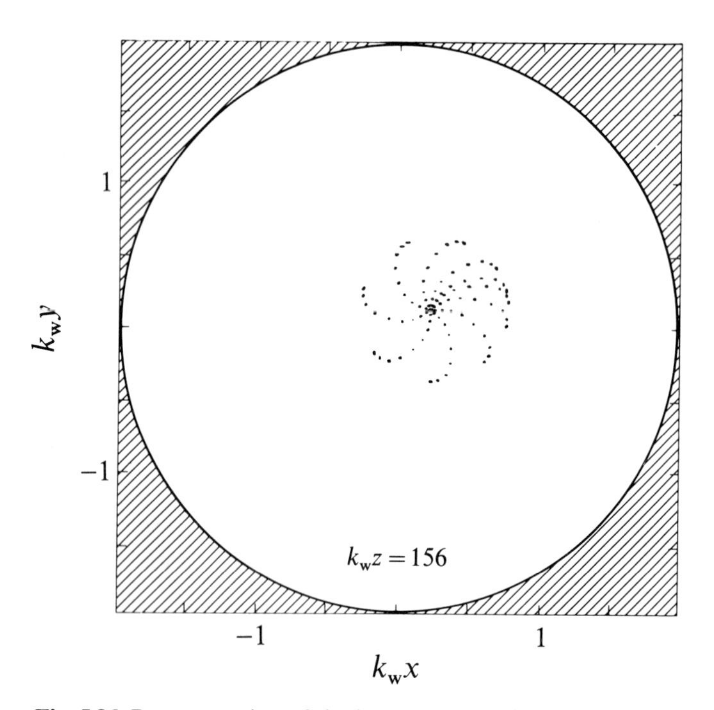

Fig. 5.26 Representation of the beam cross-section at $k_w z = 156$.

and we observe a growth rate of $\Gamma_{11}/k_w \approx 0.012$ at the frequency corresponding to peak efficiency.

5.3.4 Numerical simulation for Group II orbit parameters

The second case under consideration also corresponds to a 35 GHz amplifier, but for a set of parameters consistent with Group II orbits in the negative-mass regime [34]. In this case, we choose a wiggler field with an amplitude of $B_w = 1.0$ kG, a period of $\lambda_w = 3.0$ cm, and an entry taper of $N_w = 10$ wiggler periods. The beam is characterized by an energy of 1.0 MeV, a current of 50 A, and a beam radius of 0.2 cm. The sole interaction is with the TE_{11} mode of a waveguide with a radius $R_g = 0.5$ cm. For these parameters, the Group II negative-mass regime is found for axial solenoidal fields in the range $B_0 \approx 8.2$–12.0 kG. It is difficult to inject a beam for axial magnetic field levels that are extremely close to the magnetoresonance at $\Omega_0 \approx k_w v_\parallel$; hence, we choose to operate at $B_0 = 11.75$ kG which is close to the extreme high-field region of the negative-mass regime.

The results of a series of numerical simulations for these parameters are shown in Fig. 5.27 in which we plot the saturation efficiency as a function of frequency for an ideal beam in which $\Delta\gamma_z = 0$. It should be remarked that the transverse wiggler-induced velocity in this case $|v_w/v_\parallel| \approx 0.45$ is very large, and the beam is shifted very close to the waveguide wall. As a result, a small fraction of the beam is lost to the wall. The procedure followed when this occurs is to eject such particles from the simulation, and the efficiencies shown reflect this effect. The

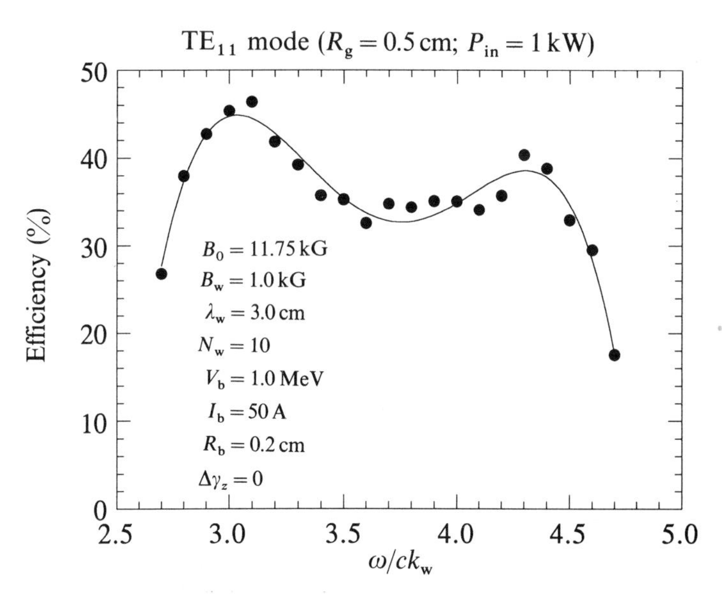

Fig. 5.27 Spectrum of the interaction efficiency as a function of the frequency for the TE_{11} mode corresponding to Group II negative-mass parameters.

peak efficiency obtained for this example was $\eta \approx 47\%$ at a frequency of $\omega/ck_w = 3.0$–3.1 (corresponding to approximately 31 GHz), with a bandwidth of approximately $\Delta\omega/\omega \approx 57\%$ extending over the range of $\omega/ck_w \approx 28$–45 GHz. It should be remarked for future reference that this efficiency has been obtained without recourse to a tapered wiggler field.

The small degree of variation in the efficiency about the line of best fit shown in the figure is difficult to account for precisely due to the complexity of the system of coupled equations. Some of the causes of the additional structure include: (1) variations in the number of particles which are lost to the wall and the axial positions at which they are lost, (2) the variation in the ponderomotive potential with the frequency at which the interaction occurs, and (3) variations in the evolution of the radial profile of the beam near saturation which have an effect upon both the effective filling factor and trapping fraction.

The effect of an axial energy spread upon the interaction is illustrated in Fig. 5.28 in which the efficiency is plotted as a function of $\Delta\gamma_z$ at the frequency of maximum efficiency (i.e. $\omega/ck_w = 3.1$). The efficiency decreases much more slowly with the axial energy spread than was found in Figs 5.19 and 5.20 corresponding to Group I orbit parameters. In this case, an axial energy spread of greater than 20% is required for the efficiency to decrease by an order of magnitude. The reason for this is that the transverse wiggler-induced velocity and, hence, the growth rate are extremely large this close to the magnetoresonance. Indeed, the growth rate and wavenumber corresponding to the case of $\omega/ck_w = 3.1$ were $|\mathrm{Im}\, k|/k_w \approx 0.045$ and $|\mathrm{Re}\, k|/k_w \approx 2.56$ which implies that thermal effects become important when

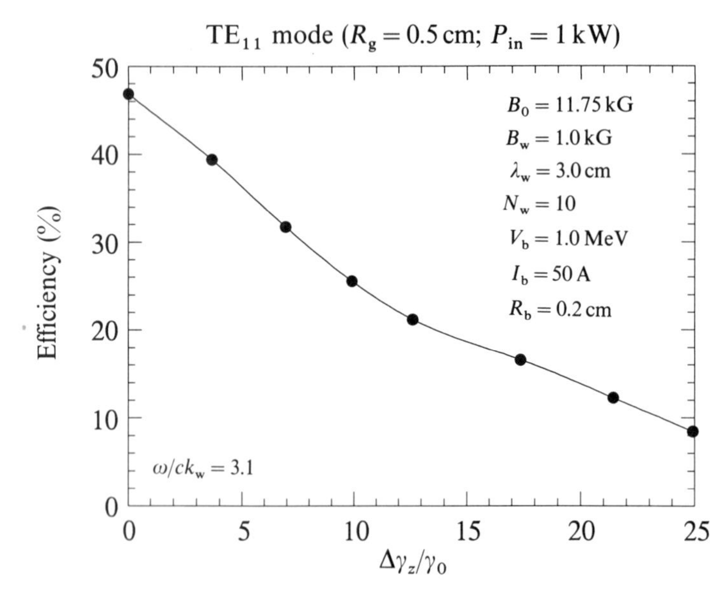

Fig. 5.28 Graph of the variation in the saturation efficiency with the axial energy spread for the Group II negative-mass regime.

$\Delta v_{\parallel}/v_{\parallel} \approx 1.26\%$. This is equivalent to an axial energy spread of $\Delta\gamma_z/\gamma_0 \approx 8.5\%$. As in the preceding case shown for the Group I orbit parameters, Fig. 5.28 shows that the decrease in the efficiency with the axial energy spread is monotonic but exhibits a discontinuity in the slope which corresponds to the onset of the thermally dominated regime.

The variation in the relative phase with axial position for a variety of frequencies ($\omega/ck_w = 3.1$, 3.5, 4.0 and 4.3) across the unstable spectrum is shown in Fig. 5.29 for the ideal beam limit. As in Fig. 5.21, the arows in the figure denote the points at which the power saturates. The sudden variations which are evident in the relative phase over the entry taper region ($k_w z \leqslant 62.8$) are associated with the rapid and substantial decrease in the axial velocity of the beam with the increase in the wiggler amplitude. The subsequent variation shows that the relative phase increases up to a point just prior to that at which the power saturates, after which the phase displays an oscillatory character.

It is important to note that the specific values for the relative phase at any given stage of the interaction are highly sensitive to the wave frequency. An alternative way of viewing this is to hold the frequency fixed but to vary the beam energy. In this way, the phase stability of the interaction against fluctuations in the beam voltage can be determined. For this example, the dependence of the relative phase on beam voltage (at the position corresponding to the saturation of the signal) varies with frequency. The best case (found for $\omega/ck_w = 3.1$) exhibits a 33° variation in the relative phase per 1% shift in the beam voltage, while the worst case shows a 60° variation in the relative phase. These estimates are comparable to those found in the preceding example for Group I orbit parameters.

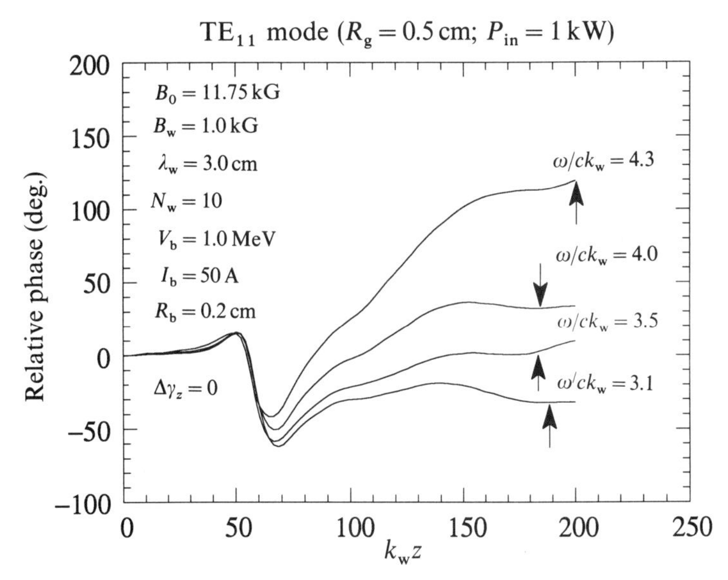

Fig. 5.29 Graph of the variation in the relative phase with the axial position for a variety of different frequencies in the Group II negative-mass regime.

The extremely high efficiencies and relative insensitivities to thermal effects which are evident in the Group II negative-mass regime raise the question of the essential limitations of this operating regime. On the one hand, the Group II negative-mass regime requires sufficiently strong axial magnetic fields that $\Omega_0 \geqslant ck_w$. As a consequence, the axial magnetic field increases in direct proportion to the beam voltage and in inverse proportion to the wiggler period. In practical terms, this prohibits the use of this operational regime for wavelengths in the submillimetre regime with the magnet technology presently available because the required beam voltages are too high. On the other hand, the negative-mass regime contracts to an extremely narrow range of axial magnetic fields (in the vicinity of the magnetoresonance) as the voltage decreases. This implies that it will be difficult to inject the beam on to stable helical trajectories if the beam voltage is too low. Since many applications also require a low beam voltage, the essential constraints in this limit are of great interest.

At low beam voltages, the narrow range of axial magnetic fields under which the negative-mass regime can be accessed implies that the interaction will be extremely sensitive to fluctuations in all the operating parameters. For operation in the vicinity of 35 GHz as described above, the low-energy limit on stable operation in this regime appears to be in the neighbourhood of 400 keV. However, for operation above this energy, this resonant interaction appears to be a realizable approach to high-efficiency free-electron laser amplifiers without the necessity of tapered wiggler fields. An example of the spectrum of efficiency versus frequency is shown in Fig. 5.30 for the case of a 400 keV electron beam. Observe that for

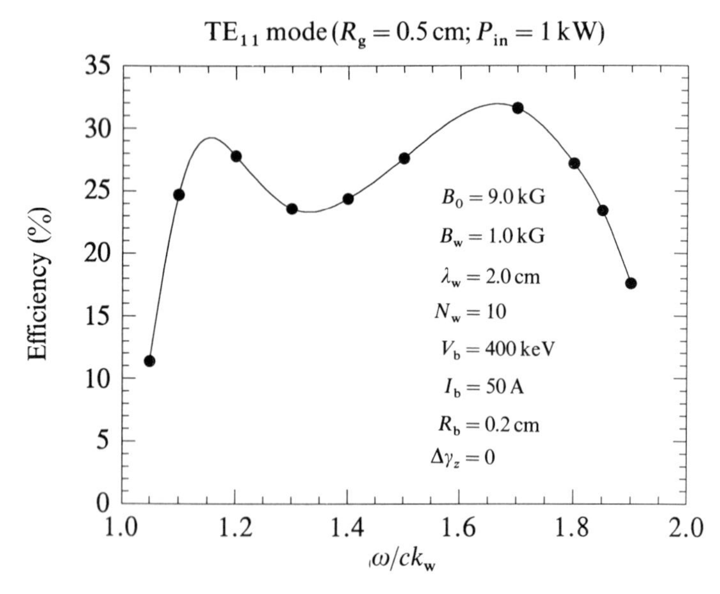

Fig. 5.30 Graph of the efficiency versus frequency for the Group II negative-mass regime corresponding to a beam energy of 400 keV.

the parameters indicated in the figure, the efficiency approaches 30% for frequencies in the range of $\omega/ck_w \approx 1.0$–3.0 (corresponding to 5–15 GHz).

5.3.5 Numerical simulation for the case of a tapered wiggler

The nonlinear formulation can also treat the case of efficiency enhancement by either a tapered wiggler or axial solenoidal field. As shown in equation (5.12) for an idealized one-dimensional model, the sense of the taper of the wiggler and/or axial magnetic fields depends upon whether the electrons' trajectories are of Group I or Group II. For small axial magnetic fields and Group I trajectories, both the wiggler field and axial magnetic fields must be tapered downward in order to achieve an efficiency enhancement. In contrast, for strong axial magnetic fields and Group II trajectories, the situation is more complicated and two distinct regimes are found. When $\Phi > 0$ (see equation (2.70)), the axial field must be tapered upward in order to achieve an efficiency enhancement, while the wiggler field must be tapered downward. In contrast, when $\Phi < 0$ in the negative-mass regime, efficiency enhancement is found when the axial field is tapered downward or the wiggler field is tapered upward. These conclusions are unaltered by three-dimensional effects.

The first case under consideration corresponds to Group I orbit parameters, and we choose a wiggler field with an amplitude $B_w = 1.0$ kG, a period $\lambda_w = 4.0$ cm, and an entry taper region $N_w = 6$ wiggler periods in length. The axial solenoidal field is assumed to be absent. An electron beam with an energy of 750 keV, a current of 200 A, and an initial beam radius of 0.5 cm is assumed to propagate through a waveguide of radius $R_g = 1.5$ cm. The resonant interaction is with the TE_{11} mode in the vicinity of $7.7 < \omega/ck_w < 9.3$. We consider the case in which $\omega/ck_w = 8.3$ (corresponding to 62.3 GHz). All other modes are either evanescent or nonresonant in this frequency range, and we may restrict the numerical analysis solely to the TE_{11} mode. The evolution of the power with axial position in the TE_{11} mode for a uniform and tapered wiggler (with $\varepsilon_w = -0.006$) is shown in Fig. 5.31 for the case of an ideal beam with $\Delta\gamma_z = 0$. In the case of a uniform wiggler with an input power of 10 W, the signal grows exponentially to saturation at $k_w z \approx 229$ with a power level of 10.7 MW. This corresponds to an efficiency of $\eta \approx 7.1\%$. In the tapered-wiggler cases, the signal is allowed to grow from an initial level of 10 W to a point just short of saturation before the taper is begun. It is found that the maximum efficiency enhancement is sensitive to the *start-point* of the taper. In practice, the taper must begin at the point at which the bulk of the electrons cross the separatrix on to trapped orbits. The ultimate efficiency enhancement is degraded if (1) the start point is chosen prior to the point at which the electrons become trapped, or (2) the uniform wiggler interaction has become saturated (i.e. if the electrons have undergone a rotation through half of the cycle in the ponderomotive wave). For the present case, the ideal start-point is located at $k_w z \approx 220$. In addition, the maximum efficiency is also dependent upon the slope of the taper, and too high a slope of the taper is as bad as too low. The evolution of the power for the tapered-wiggler case yielding the optimum

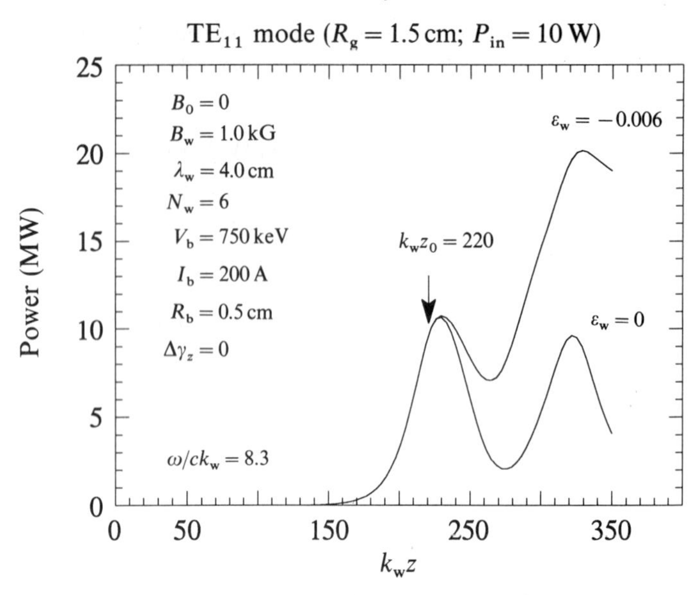

Fig. 5.31 Evolution of the radiation power in the TE_{11} mode versus axial position for Group I orbit parameters and uniform and tapered wigglers.

efficiency enhancement is shown in the figure. In this case the power can be almost doubled to approximately 20.2 MW for an efficiency of $\eta \approx 13.4\%$.

We now consider the case of Group II negative-mass orbit parameters for an axial magnetic field of $B_0 = 14.0$ kG, and a wiggler field of amplitude $B_w = 1.0$ kG, a period $\lambda_w = 3.0$ cm, and an entry taper region $N_w = 10$ wiggler periods. The electron-beam parameters for this case are an energy of 1.25 MeV, a current of 50 A, and an initial radius of 0.25 cm. As before, we consider the TE_{11} mode but with a waveguide radius of $R_g = 1.0$ cm, which results in amplification over the frequency range of $7.5 < \omega/ck_w < 8.5$. We choose $\omega/ck_w = 8.0$ (corresponding to 80 GHz) for purposes of illustration. The saturation efficiency for a uniform wiggler is found to be approximately $\eta \approx 5.47\%$ for a total power of 3.42 MW. For an input power of 10 W, the saturation point was found to be $k_w z = 253$. A near-optimum example of efficiency enhancements by means of a tapered wiggler is shown in Fig. 5.32 in which we plot the evolution of the power as a function of the axial position for both a uniform and a tapered wiggler ($\varepsilon_w = 0.001$ and $k_w z_0 = 247$). Observe that in the negative-mass regime the wiggler must, indeed, be tapered upward in order to enhance the efficiency. As shown in the figure, the efficiency may be enhanced to a level of approximately $\eta \approx 37.0\%$ for an output power of 23.1 MW, an enhancement of almost 700%.

The case of a tapered axial magnetic field is shown in Fig. 5.33 for the identical parameters illustrated in Fig. 5.32. In this case, however, we plot the evolution of the power with axial position for uniform fields and a tapered axial magnetic field with a normalized slope of $\varepsilon_0 = -0.0005$ and a start-point of $k_w z_0 = 247$.

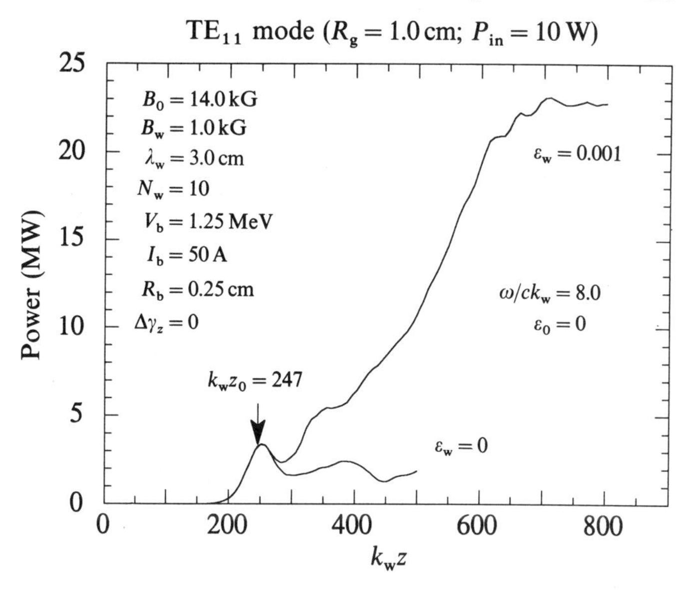

Fig. 5.32 Evolution of the radiation power in the TE$_{11}$ mode versus axial position for Group II negative-mass orbit parameters and uniform and tapered wigglers.

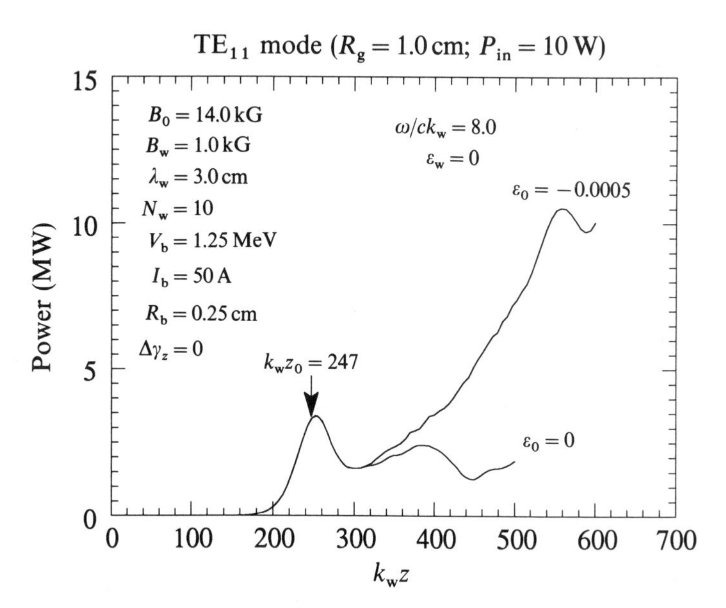

Fig. 5.33 Evolution of the radiation power in the TE$_{11}$ mode versus axial position for Group II negative-mass orbit parameters and uniform and tapered axial fields.

As is evident in the figure, the enhanced efficiency reaches a level of $\eta \approx 16.9\%$ for a maximum power of 10.6 MW. While this is substantially lower than that found for a tapered wiggler in Fig. 5.32, it should be observed that in this regime the wiggler must be tapered upward while the axial field must be tapered downward. As a result, there is a maximum efficiency enhancement possible for the tapered axial field associated with the point at which the axial magnetic field vanishes.

5.3.6 Comparison with experiment

The nonlinear formulation described herein has been compared with an intense-beam free-electron laser experiment conducted at the Massachusetts Institute of Technology with good results [51, 52]. This experiment operated in the superradiant regime in which the signal grows from noise, and employed a helical wiggler field with a period of 3.14 cm, an amplitude which could be varied up to almost 2 kG, and an entry taper region six wiggler periods in length. No axial magnetic field was imposed. The electron beam was produced by a Physics International Pulserad 110 A accelerator in concert with a five-stage multi-electrode field-emission electron gun to produce a beam with a pulse length of 30 ns, a maximum energy of up to 2.5 MeV, and a maximum current of up to 7 kA. The beam was then focused and apertured to produce a high-quality beam with a current of up to approximately 1 kA, which was then propagated through a drift tube with an inner diameter of 0.8 cm. The output of the free-electron laser was found to range in frequency from approximately 200 GHz to 500 GHz at power levels of the order of 18 MW in the neighbourhood of 470 GHz. Estimates of the beam quality were made on the basis of the electron-gun geometry, and produced estimates of the axial energy spread of approximately $\Delta\gamma_z/\gamma_0 \approx 0.22$–$0.40\%$.

Measurements of the growth rate were performed by means of a kicker magnet which dumped the electron beam to the wall at various points along the drift tube. Two examples are shown in Fig. 5.34 for wiggler field amplitudes of 1510 G and 1275 G, respectively, for a beam energy of 2.0 MeV, current 780 A, and equilibrium radius of approximately 0.4 cm. As shown in the figures, the gains for these two cases were in the neighbourhood of 70 ± 6 dB/m and 56 ± 3 dB/m, respectively. The polarization of the output radiation has been determined to be primarily in the fundamental TE_{11} mode of the waveguide.

The variation in the gain with axial energy spread for these two cases has been studied using the numerical simulation, and the results are shown in Fig. 5.35. These calculations were made using only the TE_{11} mode and with an initial beam radius of 0.4 cm. No phenomenological fit parameters are necessary. The maximum gain is found with a vanishing axial energy spread. For the case of $B_w = 1275$ G, the calculated gain is found to be approximately 62 dB/m for $\Delta\gamma_z/\gamma_0 = 0$, and decreases to 56 dB/m for $\Delta\gamma_z/\gamma_0 = 0.25\%$. Similarly, for $B_w = 1510$ G, the gains calculated for $\Delta\gamma_z/\gamma_0 = 0$ and $\Delta\gamma_z/\gamma_0 = 0.25\%$ are 70 dB/m and 64 dB/m, respectively. When compared with the measured gains, these results are consistent

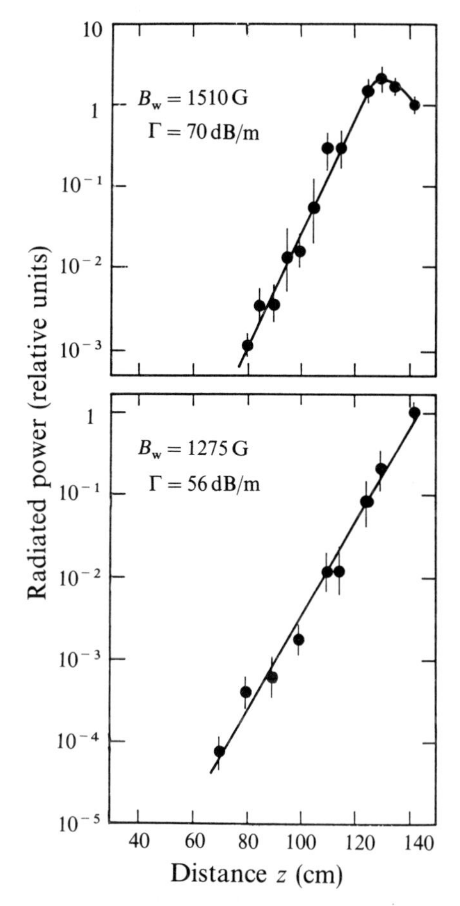

Fig. 5.34 Observations of the output power as a function of axial position for two wiggler fields (a) 1510 G, and (b) 1275 G with a beam energy of 2.0 MeV and current of 780 A [51, 52].

with the interpretation of an axial-beam energy spread of less than or of the order of 0.25%. This is also in excellent agreement with the estimates of beam quality to be obtained from the electron gun. Hence, the value of $\Delta\gamma_z/\gamma_0 = 0.25\%$ is used henceforth.

The measured output spectrum for the case of a 2.3 MeV, 930 A electron beam and a wiggler field amplitude of 1275 G is shown in Fig. 5.36, together with the results of the nonlinear simulation. Total measured output power is of the order of 18 MW for this example. In this case, the simulation included the TE_{11} and TM_{11} modes. Since the nonlinear formulation is restricted to single-frequency propagation, the numerical procedure used to generate the theoretical curve was to consider each frequency separately and to plot the combined output spectrum

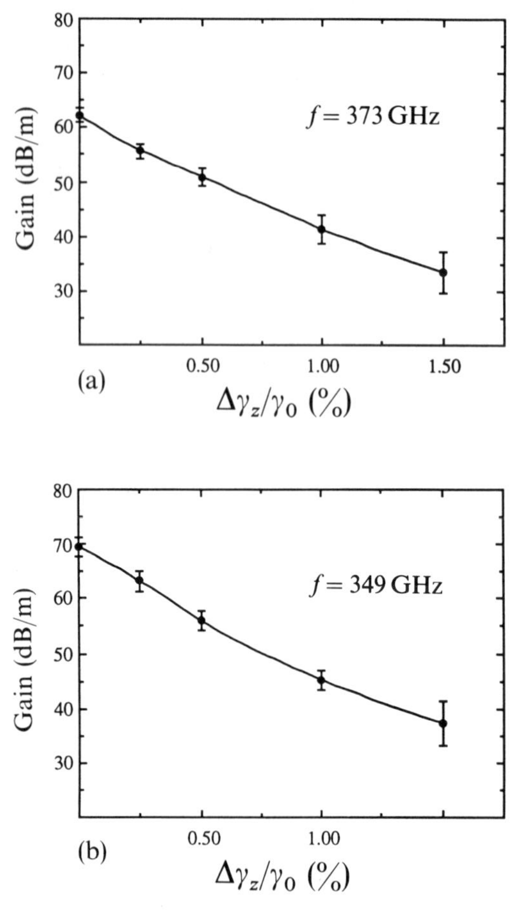

Fig. 5.35 Calculated gains versus axial energy spread for the two wiggler amplitudes shown in Fig. 5.33 [51, 52].

normalized to the peak power. As can be seen in the figure, the spectral agreement is excellent, and the measured and computed spectral peaks differ by less than 2%.

Finally, it should be observed that space-charge effects were not required to treat this intense-beam experiment, even though it should fall in the borderline region between the high-gain Compton and collective Raman regimes based upon the well-known criterion derived from the idealized one-dimensional formulation (equations (4.98) and (4.103)). We conclude, therefore, that three-dimensional effects imposed by the wiggler and bounded geometry of the waveguide act to reduce the effective plasma frequency and shift the experiment closer to the high-gain Compton regime.

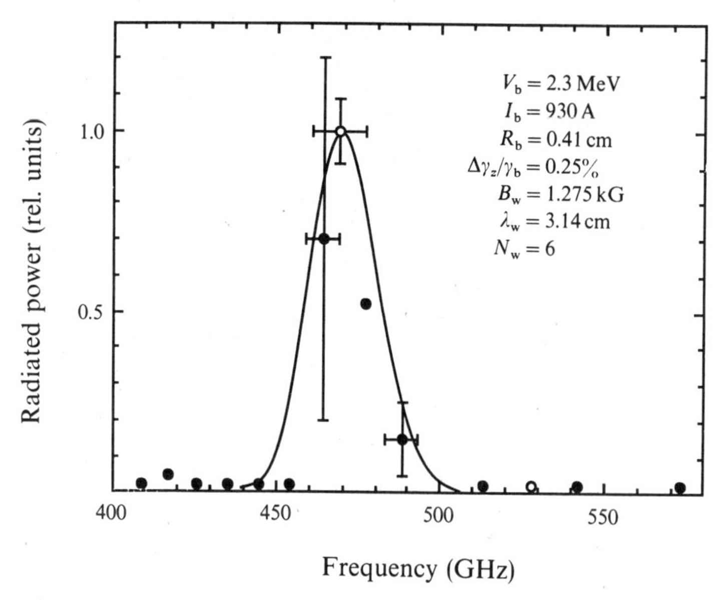

Fig. 5.36 The output spectrum for a 2.3 MeV, 930 A electron beam and an on-axis wiggler strength of 1275 G. The dots represent the measurements, and the curve is the result of the nonlinear simulation including the TE_{11} and TM_{11} modes [51, 52].

5.4 THREE-DIMENSIONAL ANALYSIS: PLANAR WIGGLERS

Planar wiggler configurations have been treated from a variety of standpoints. Due to their ease of construction and adjustment in comparison with helical wigglers (which can only be adjusted as to their bulk field magnitude), planar wigglers are employed in a wide variety of experimental configurations including both amplifiers and oscillators, and low- and high-gain systems. Low-gain oscillators will be treated in Chapter 9 which deals with oscillator configurations. Here we concentrate on high-gain amplifier configurations. Nonlinear amplifier analyses in both two and three dimensions based upon planar wiggler designs have been largely based upon an analysis in which the orbit equations are averaged over a wiggler period [29–31, 33, 39, 49, 53].

The formulation employed herein follows that described in references [42–6, and 54], and is similar to that described for helical wigglers in that the electron-orbit equations are not averaged over a wiggler period. Recall that the free-electron laser operates by means of the beating of the wiggler and radiation fields. In normal operation, it is the upper-frequency beat wave which results in the Doppler-upshifted interaction. In the case of helical wiggler geometries, the lower-frequency beat wave vanishes due to the symmetry of the configuration. This is not the case for planar geometries. In the planar geometry, the lower beat wave is present and results in an oscillation in both the power and phase with a period approximately equal to one half that of the wiggler period. Such effects are not included in descriptions which are based upon a wiggler-averaged orbit analysis. In most cases, the effects of the lower beat wave do not contribute substantially to the bulk evolution of the wave; however, the instantaneous values of the power and phase can be altered markedly by this effect. In the cases considered in this chapter, it will be shown that the overall magnitude of the oscillations in the power and phase of the signal can approach 10–20% of the overall bulk values for these quantities. In addition, the effect of the lower beat wave on the optical guiding of the signal will be discussed in Chapter 8.

5.4.1 The general configuration

The wiggler model we employ in the treatment of the three-dimensional nonlinear formulation is that given in equation (2.3)

$$\begin{aligned}\boldsymbol{B}_w(\boldsymbol{x}) = B_w(z)\Bigg\{\cos k_w z\Bigg[\hat{\boldsymbol{e}}_x \sinh\left(\frac{k_w x}{\sqrt{2}}\right)\sinh\left(\frac{k_w y}{\sqrt{2}}\right)\\ + \hat{\boldsymbol{e}}_y \cosh\left(\frac{k_w x}{\sqrt{2}}\right)\cosh\left(\frac{k_w y}{\sqrt{2}}\right)\Bigg]\\ - \sqrt{2}\hat{\boldsymbol{e}}_z \cosh\left(\frac{k_w x}{\sqrt{2}}\right)\sinh\left(\frac{k_w y}{\sqrt{2}}\right)\sin k_w z\Bigg\},\end{aligned} \tag{5.143}$$

which includes the effect of tapered pole faces for enhanced focusing. No axial-guide magnetic field is included in this formulation due to the deleterious drift associated with the combined planar wiggler and axial field. As in the case of the helical wiggler, we permit the amplitude to vary adiabatically in z as given in equation (5.81) in order to model both beam injection into the wiggler and tapered-wiggler efficiency enhancement.

The field equations

In treating the planar wiggler, we assume that the waveguide geometry corresponds to a loss-free rectangular waveguide, and the boundary conditions on the radiation fields at the waveguide wall are satisfied by expanding the field in terms of the vacuum waveguide modes. The vector potential in a rectangular waveguide bounded by $-a/2 \leqslant x \leqslant a/2$ and $-b/2 \leqslant y \leqslant b/2$ may be written in the form

$$\delta\boldsymbol{A}(\boldsymbol{x}, t) = \sum_{l,n=0}^{\infty}{}' \, \delta A_{ln}(z)\boldsymbol{e}_{ln}^{(1)}(x, y)\cos\alpha_{ln}(z, t), \tag{5.144}$$

for the TE modes, and

$$\begin{aligned}\delta\boldsymbol{A}(\boldsymbol{x}, t) = \sum_{l,n=1}^{\infty} \delta A_{ln}(z)\Bigg[\boldsymbol{e}_{ln}^{(2)}(x, y)\cos\alpha_{ln}(z, t)\\ + \frac{\kappa_{ln}}{k_{ln}}\sin\left(\frac{l\pi X}{a}\right)\cos\left(\frac{n\pi Y}{b}\right)\sin\alpha_{ln}(z, t)\Bigg],\end{aligned} \tag{5.145}$$

for the TM modes, where the phase for frequency ω and wavenumber $k_{ln}(z)$ is given by

$$\alpha_{ln}(z,t) = \int_0^z dz' k_{ln}(z') - \omega t, \tag{5.146}$$

the summation symbol $\sum'$ indicates that both l and n may not be zero,

$$\begin{aligned} \mathbf{e}_{ln}^{(1)}(x,y) = {} & \frac{n\pi}{\kappa_{ln} b}\hat{\mathbf{e}}_x \cos\left(\frac{l\pi X}{a}\right)\sin\left(\frac{n\pi Y}{b}\right) \\ & - \frac{l\pi}{\kappa_{ln} a}\hat{\mathbf{e}}_y \sin\left(\frac{l\pi X}{a}\right)\cos\left(\frac{n\pi Y}{b}\right), \end{aligned} \tag{5.147}$$

$$\begin{aligned} \mathbf{e}_{ln}^{(2)}(x,y) = {} & \frac{l\pi}{\kappa_{ln} a}\hat{\mathbf{e}}_x \cos\left(\frac{l\pi X}{a}\right)\sin\left(\frac{n\pi Y}{b}\right) \\ & + \frac{n\pi}{\kappa_{ln} b}\hat{\mathbf{e}}_y \sin\left(\frac{l\pi X}{a}\right)\cos\left(\frac{n\pi Y}{b}\right), \end{aligned} \tag{5.148}$$

where $X = x + a/2$ and $Y = y + b/2$, and the cutoff wavenumber of the mode is given by

$$\kappa_{ln} = \pi\sqrt{\left(\frac{l^2}{a^2} + \frac{n^2}{b^2}\right)}. \tag{5.149}$$

Observe that, unlike the case of a helical wiggler, the TE_{ln} and TM_{ln} modes have identical cutoff wavenumbers. As in the preceding cases, we assume implicitly that the amplitudes and wavenumbers of each mode vary slowly with respect to the wavelength.

The equations governing the evolution of the slowly varying amplitude and phase of each of the TE modes in a rectangular waveguide follow in a similar manner to that in the case of a cylindrical waveguide, and we assume the identical model for the source current and the electron-beam distribution function. The principal difference is that the orthogonalization is now over the x and y coordinates (rather than r and θ). The orthogonality integrals over the x-coordinate are

$$\int_0^a dx \cos\left(\frac{l\pi x}{a}\right)\cos\left(\frac{m\pi x}{a}\right) = \frac{a}{2}\delta_{l,m}[1 + \delta_{l,0}], \tag{5.150}$$

$$\int_0^a dx \sin\left(\frac{l\pi x}{a}\right)\sin\left(\frac{m\pi x}{a}\right) = \frac{a}{2}\delta_{l,m}[1 + \delta_{l,0}], \tag{5.151}$$

and similar relations hold for the y-orthogonalization. As a consequence, we find that the dynamical equations for each TE_{ln} mode are given by

$$\left[\frac{d^2}{dz^2} + \left(\frac{\omega^2}{c^2} - k_{ln}^2 - \kappa_{ln}^2\right)\right]\delta a_{ln} = 8\frac{\omega_b^2}{c^2}F_{ln}\left\langle \frac{\cos\alpha_{ln}}{|v_z|}\mathbf{e}_{ln}^{(1)}\cdot\mathbf{v}\right\rangle, \tag{5.152}$$

and

$$2k_{ln}^{1/2}\frac{d}{dz}\left(k_{ln}^{1/2}\,\delta a_{ln}\right)=-8\frac{\omega_b^2}{c^2}F_{ln}\left\langle\frac{\sin\alpha_{ln}}{|v_z|}\,\mathbf{e}_{ln}^{(1)}\cdot\mathbf{v}\right\rangle, \tag{5.153}$$

where the averaging operator is the same as that defined in equation (5.108) for a cross-sectional area $A_g = ab$, and $F_{ln} = 1/2$ whenever either $l = 0$ or $n = 0$, and unity otherwise. The dynamical equations for each TM_{ln} mode are governed by a similar set of equations

$$\left[\frac{d^2}{dz^2}+\left(1+\frac{\kappa_{ln}^2}{k_{ln}^2}\right)\left(\frac{\omega^2}{c^2}-k_{ln}^2-\kappa_{ln}^2\right)\right]\delta a_{ln}$$
$$=8\frac{\omega_b^2}{c^2}\left\langle\frac{\cos\alpha_{ln}}{|v_z|}\,\mathbf{e}_{ln}^{(2)}\cdot\mathbf{v}+\frac{v_z}{|v_z|}\frac{\kappa_{ln}}{k_{ln}}\sin\left(\frac{l\pi X}{a}\right)\sin\left(\frac{n\pi Y}{b}\right)\sin\alpha_{ln}\right\rangle, \tag{5.154}$$

and

$$2\left(k_{ln}+\frac{\kappa_{ln}^2}{k_{ln}}\right)^{1/2}\frac{d}{dz}\left[\left(k_{ln}+\frac{\kappa_{ln}^2}{k_{ln}}\right)^{1/2}\delta a_{ln}\right]$$
$$=-8\frac{\omega_b^2}{c^2}\left\langle\frac{\sin\alpha_{ln}}{|v_z|}\,\mathbf{e}_{ln}^{(2)}\cdot\mathbf{v}-\frac{v_z}{|v_z|}\frac{\kappa_{ln}}{k_{ln}}\sin\left(\frac{l\pi X}{a}\right)\sin\left(\frac{n\pi Y}{b}\right)\cos\alpha_{ln}\right\rangle. \tag{5.155}$$

These equations describe the evolution of the wavenumber and amplitude for each TE and TM mode, and must be integrated for each mode included in the simulation. A similar analysis has also been performed for the slow time-scale equations for the modes in a dielectric-lined rectangular waveguide [54].

The electron orbit equations

In order to complete the formulation, the trajectories of an ensemble of electrons must be determined using the Lorentz force equations in the aggregate fields. For the planar wiggler model and the TE and TM modes of the rectangular waveguide, we obtain

$$v_z\frac{d}{dz}p_x=\Omega_w\cosh\left(\frac{k_w x}{\sqrt{2}}\right)\left[\sqrt{2}\,p_y\sinh\left(\frac{k_w y}{\sqrt{2}}\right)\sin k_w z+p_z\cosh\left(\frac{k_w y}{\sqrt{2}}\right)\cos k_w z\right]$$
$$+m_e c\sum_{\text{TE modes}}\delta a_{ln}\left\{\frac{n\pi}{\kappa_{ln}b}(\omega-k_{ln}v_z)\cos\left(\frac{l\pi x}{a}\right)\sin\left(\frac{n\pi y}{b}\right)\sin\alpha_{ln}\right.$$
$$\left.+\sin\alpha_{ln}\cos\left(\frac{l\pi x}{a}\right)\left[\kappa_{ln}v_y\cos\left(\frac{n\pi y}{b}\right)+\Gamma_{ln}\frac{n\pi}{\kappa_{ln}b}v_z\sin\left(\frac{n\pi y}{b}\right)\right]\right\}$$
$$+m_e c\sum_{\text{TM modes}}\delta a_{ln}\frac{l\pi}{\kappa_{ln}a}\cos\left(\frac{l\pi x}{a}\right)\sin\left(\frac{n\pi y}{b}\right)$$
$$\times\left\{\sin\alpha_{ln}\left[\omega-\left(k_{ln}+\frac{\kappa_{ln}^2}{k_{ln}}\right)v_z\right]+\Gamma_{ln}v_z\cos\alpha_{ln}\right\}, \tag{5.156}$$

$$v_z \frac{d}{dz} p_y = -\Omega_w \sinh\left(\frac{k_w x}{\sqrt{2}}\right)\left[\sqrt{2}\, p_x \cosh\left(\frac{k_w y}{\sqrt{2}}\right) \sin k_w z + p_z \sinh\left(\frac{k_w y}{\sqrt{2}}\right) \cos k_w z\right]$$

$$- m_e c \sum_{\text{TE modes}} \delta a_{ln} \left\{ \frac{l\pi}{\kappa_{ln} a}(\omega - k_{ln} v_z) \sin\left(\frac{l\pi x}{a}\right) \cos\left(\frac{n\pi y}{b}\right) \sin \alpha_{ln}\right.$$

$$\left. + \cos \alpha_{ln} \cos\left(\frac{n\pi y}{b}\right)\left[\kappa_{ln} v_x \cos\left(\frac{l\pi x}{a}\right) + \Gamma_{ln} \frac{l\pi}{\kappa_{ln} a} v_z \sin\left(\frac{l\pi x}{a}\right)\right]\right\}$$

$$+ m_e c \sum_{\text{TM modes}} \delta a_{ln} \frac{n\pi}{\kappa_{ln} b} \sin\left(\frac{l\pi x}{a}\right) \cos\left(\frac{n\pi y}{b}\right)$$

$$\times \left\{ \sin \alpha_{ln}\left[\omega - \left(k_{ln} + \frac{\kappa_{ln}^2}{k_{ln}}\right) v_z\right] + \Gamma_{ln} v_z \cos \alpha_{ln}\right\}, \tag{5.157}$$

and

$$v_z \frac{d}{dz} p_z = -\Omega_w \cos k_w z\left[p_x \cosh\left(\frac{k_w x}{\sqrt{2}}\right) \cosh\left(\frac{k_w y}{\sqrt{2}}\right)\right.$$

$$\left. - p_y \sinh\left(\frac{k_w x}{\sqrt{2}}\right) \sinh\left(\frac{k_w y}{\sqrt{2}}\right)\right]$$

$$+ m_e c \sum_{\text{TE modes}} \delta a_{ln}(k_{ln} \sin \alpha_{ln} - \Gamma_{ln} \cos \alpha_{ln})$$

$$\times \left[\frac{n\pi}{\kappa_{ln} b} v_x \cos\left(\frac{l\pi x}{a}\right) \sin\left(\frac{n\pi y}{b}\right) - \frac{l\pi}{\kappa_{ln} a} v_y \sin\left(\frac{l\pi x}{a}\right) \cos\left(\frac{n\pi y}{b}\right)\right]$$

$$+ m_e c \sum_{\text{TM modes}} \delta a_{ln}\left[\left(k_{ln} + \frac{\kappa_{ln}^2}{k_{ln}}\right) \sin \alpha_{ln} - \Gamma_{ln} \cos \alpha_{ln}\right]$$

$$\times \left[\frac{l\pi}{\kappa_{ln} a} v_x \cos\left(\frac{l\pi x}{a}\right) \sin\left(\frac{n\pi y}{b}\right) + \frac{n\pi}{\kappa_{ln} b} v_y \sin\left(\frac{l\pi x}{a}\right) \cos\left(\frac{n\pi y}{b}\right)\right]$$

$$- \omega m_e c \sum_{\text{TM modes}} \delta a_{ln} \frac{\kappa_{ln}}{k_{ln}} \cos \alpha_{ln} \sin\left(\frac{l\pi x}{a}\right) \sin\left(\frac{n\pi y}{b}\right), \tag{5.158}$$

where $\Omega_w \equiv eB_w/\gamma m_e c$ and Γ_{ln} denotes the growth rate of the mode. In addition, we also integrate

$$\frac{d}{dz} x = \frac{v_x}{v_z}, \tag{5.159}$$

$$\frac{d}{dz} y = \frac{v_y}{v_z}, \tag{5.160}$$

and

$$\frac{d}{dz} \psi_{ln} = k_{ln} + k_w - \frac{\omega}{v_z}. \tag{5.161}$$

It should be remarked that these equations are implicitly slowly varying functions of axial position as long as the waves are near resonance (i.e. $\omega \approx (k + mk_w)v_z$, where m is an odd integer). Hence, no average over the wiggler period need be performed.

The first-order field equations

The second-order differential equations for the amplitude and phase of each mode can be simplified under the neglect of the second-order derivatives of the amplitude and phase. This yields algebraic equations for the wavenumber and amplitude

$$\left(\frac{\omega^2}{c^2} - k_{ln}^2 - \kappa_{ln}^2\right) = 8\frac{\omega_b^2}{c^2}\frac{F_{ln}}{\delta a_{ln}}\left\langle \frac{\cos \alpha_{ln}}{|v_z|} e_{ln}^{(1)}\cdot\mathbf{v} \right\rangle, \tag{5.162}$$

$$2k_{ln}\frac{d}{dz}\delta a_{ln} = -8\frac{\omega_b^2}{c^2}F_{ln}\left\langle \frac{\sin \alpha_{ln}}{|v_z|} e_{ln}^{(1)}\cdot\mathbf{v} \right\rangle, \tag{5.163}$$

of each TE mode. The analogous equations for the TM modes are

$$\left(1 + \frac{\kappa_{ln}^2}{k_{ln}^2}\right)\left(\frac{\omega^2}{c^2} - k_{ln}^2 - \kappa_{ln}^2\right)\delta a_{ln}$$
$$= 8\frac{\omega_b^2}{c^2}\left\langle \frac{\cos \alpha_{ln}}{|v_z|} e_{ln}^{(2)}\cdot\mathbf{v} + \frac{v_z}{|v_z|}\frac{\kappa_{ln}}{k_{ln}}\sin\left(\frac{l\pi X}{a}\right)\sin\left(\frac{n\pi Y}{b}\right)\sin \alpha_{ln} \right\rangle, \tag{5.164}$$

$$2\left(k_{ln} + \frac{\kappa_{ln}^2}{k_{ln}}\right)\frac{d}{dz}\delta a_{ln}$$
$$= -8\frac{\omega_b^2}{c^2}\left\langle \frac{\sin \alpha_{ln}}{|v_z|} e_{ln}^{(2)}\cdot\mathbf{v} - \frac{v_z}{|v_z|}\frac{\kappa_{ln}}{k_{ln}}\sin\left(\frac{l\pi X}{a}\right)\sin\left(\frac{n\pi Y}{b}\right)\cos \alpha_{ln} \right\rangle. \tag{5.165}$$

In addition, equation (5.138) is integrated for the relative phase of each mode. As in the case of the helical wiggler geometry, the discrepancy between this reduced set of dynamical equations and the complete set of second-order differential equations is under 10% for typical parameters of most free-electron lasers.

The initial conditions

The algorithms and initial conditions employed in the numerical simulation for the rectangular waveguide and planar wiggler configuration are identical to those described for the helical wiggler geometry.

5.4.2 Numerical simulation: single-mode limit

The particular example under consideration is that of a 35 GHz amplifier employing an electron beam with an energy of 3.5 MeV, a current of 800 A, and an initial radius of 1.0 cm which propagates through a waveguide characterized by $a = 9.8$ cm and $b = 2.9$ cm. In order to obtain peak growth rates in the vicinity of 35 GHz, we choose a wiggler field with an amplitude of $B_w = 3.72$ kG, a period of $\lambda_w = 9.8$ cm, and an entry taper region $N_w = 5$ wiggler periods in length. For purposes of illustration, we first consider the case of an ideal beam (i.e. $\Delta\gamma_z = 0$). For these parameters, there are three resonant modes: the TE_{01}, TE_{21} and TM_{21} modes. Before proceeding to the analysis of the complete multi-mode problem, we shall first consider the properties of the interaction solely in the presence of each of these modes. Bear in mind that these parameters are relevant to an experiment conducted at the Lawrence Livermore National Laboratory [5, 32, 39], and that a detailed discussion of the comparison of the results of the numerical simulation with the experiment will follow.

The initial electron beam distributions in axial phase space and in cross-section are similar to those used for the helical wiggler geometry. The axial phase space is shown in Fig. 5.37. Each dot in the figure denotes an entire phase sheet which represents a cross-sectional slice of the beam upon entry to the wiggler. The cross-sectional distribution of the beam is illustrated in Fig. 5.38 which represents the injection of a solid (i.e. pencil) beam with a flat-top density profile. As in the case of a helical wiggler, the nonuniform spacings between the electrons are

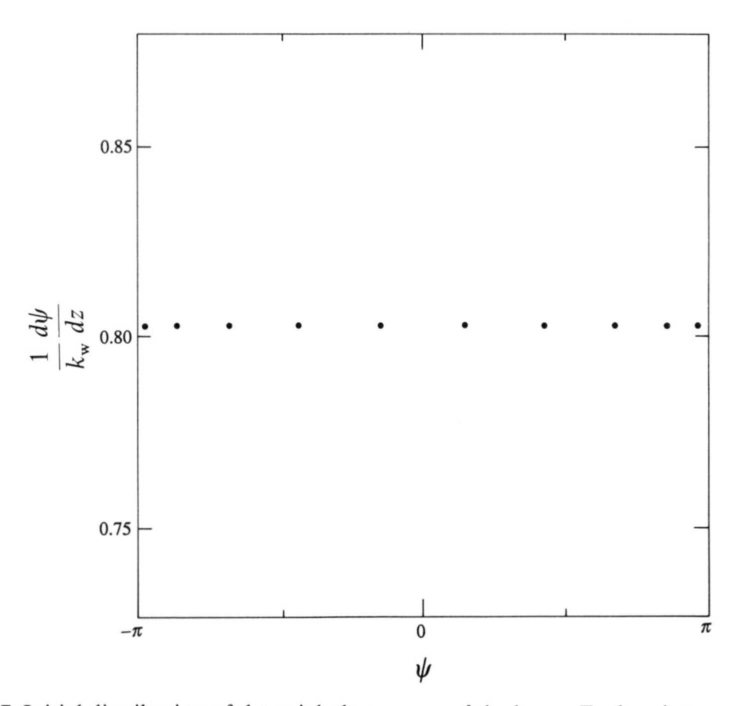

Fig. 5.37 Initial distribution of the axial phase space of the beam. Each point represents a phase sheet of electrons distributed throughout the cross-section of the beam.

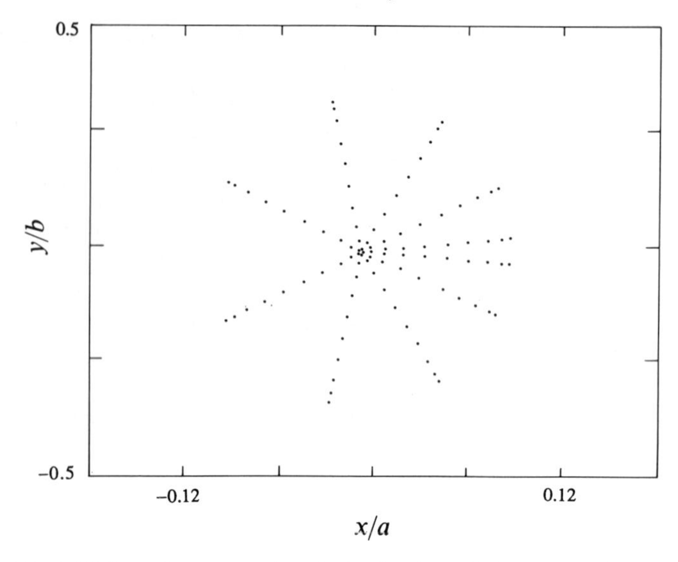

Fig. 5.38 Initial cross-sectional distribution of the electron beam.

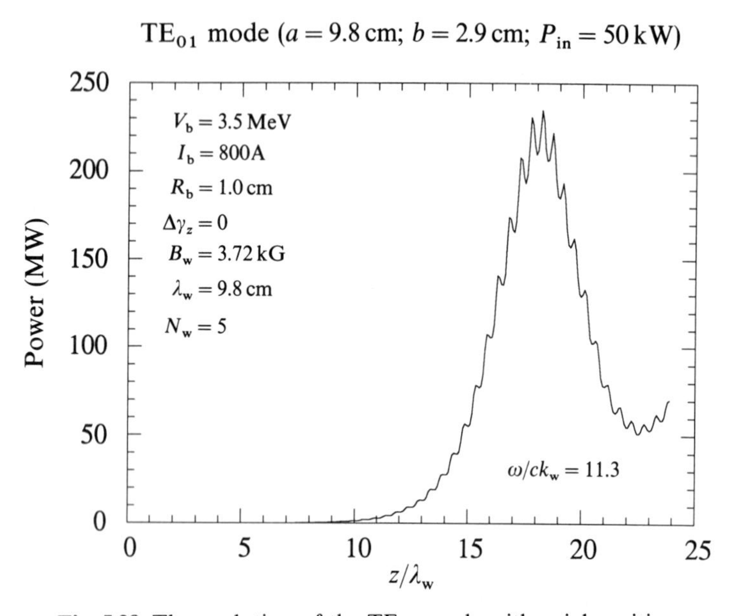

Fig. 5.39 The evolution of the TE_{01} mode with axial position.

artifacts of the Gaussian algorithm and are compensated for by nonuniform weights assigned to each electron.

The detailed evolution of the TE_{01} mode as a function of axial position is shown in Fig. 5.39 for the injection of a signal at a normalized frequency of $\omega/ck_w = 11.3$ (34.6 GHz) at a power level of 50 kW. As shown in the figure, saturation occurs at $k_w z \approx 115$ (1.79 m) at a power level of approximately 214 MW corresponding to an efficiency $\eta \approx 7.75\%$. Wave amplification occurs principally within the uniform wiggler region which is 1.30 m in length for this example. Hence, there is an average gain of 28 dB/m throughout the uniform wiggler region, which corresponds to an average normalized growth rate of $\Gamma_{01}/k_w \approx 0.05$.

One feature shown in the figure which merits discussion is the oscillation in the instantaneous power. This oscillation occurs at a period of $\lambda_w/2$, and is not generally found for helical wiggler configurations. In order to explain this oscillation, recall that the interaction in a free-electron laser arises due to the beating of the wiggler and radiation fields. The slowly varying ponderomotive wave which results in the bulk amplification of the signal corresponds to the upper beat wave, while the rapid oscillation shown in the figure is due to the lower beat wave. In order to illustrate this, observe that for the present planar configuration the bulk transverse velocity is aligned along the x-axis and varies approximately as

$$v_w \approx \frac{\Omega_w}{k_w} \hat{e}_x \sin k_w z. \tag{5.166}$$

The source terms contained in the dynamical equations are derived essentially from a calculation $\langle \boldsymbol{J} \cdot \delta \boldsymbol{E} \rangle$; hence, the principal wave–particle coupling is with the x-component of the radiation field. If we assume that $\delta E_x \approx \delta E_0 \sin(kz - \omega t)$, then it is evident that

$$\langle \boldsymbol{J} \cdot \delta \boldsymbol{E} \rangle \approx \frac{\Omega_w}{2k_w} \delta E_0 [\langle \cos \psi \rangle - \langle \cos(2k_w z - \psi) \rangle]. \tag{5.167}$$

The first average in the square brackets represents the upper beat wave while the second term represents the lower beat wave. Since amplification is found when the ponderomotive phase ψ is a slowly varying quantity, the lower beat wave describes an oscillation with a period half that of the wiggler period. Although the contribution from the lower beat wave provides no contribution to the bulk growth of the wave, the instantaneous magnitude of this contribution is comparable to that of the upper beat wave, and it does affect the instantaneous values of both the power and phase of the signal. This effect is not found in helical wiggler geometries because the wiggler-induced transverse velocity describes a helix with near-constant magnitude which drives an interaction with a circularly polarized wave. As a result, the symmetry of the helical interaction suppresses the lower beat wave.

A full spectrum of the TE_{01} mode is shown in Fig. 5.40 in which we plot the saturation efficiency and the distance to saturation as a function of frequency

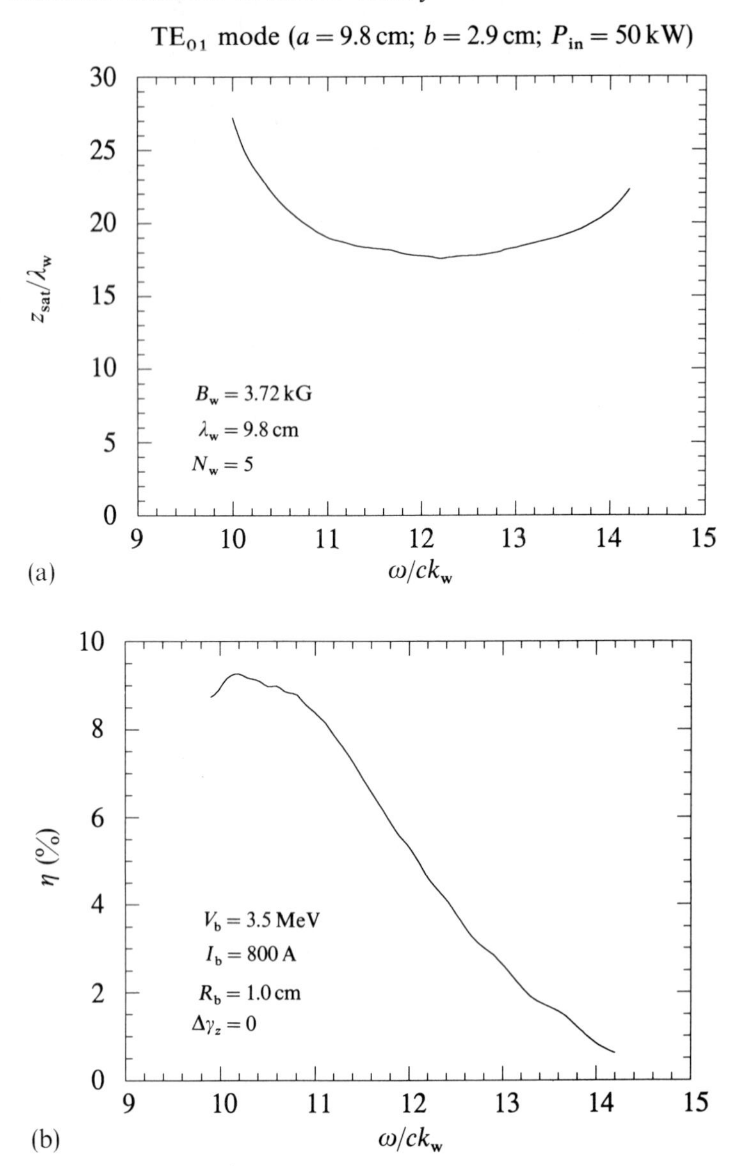

Fig. 5.40 Graphs showing the distance to saturation and the efficiency of the TE_{01} mode as a function of frequency.

within the unstable band. As shown in the figure, wave amplification is found for frequencies extending from $\omega/ck_w \approx 10$ (30.6 GHz) through $\omega/ck_w \approx 14.2$ (43.5 GHz) with a peak efficiency of the order of $\eta \approx 9.8\%$. The peak growth rate, by contrast (as measured by the shortest distance to saturation), occurs for $\omega/ck_w \approx 12.3$ (37.7 GHz).

The variation in the relative phase as a function of axial position is illustrated in Fig. 5.41 for $\omega/ck_w = 10.4$, 10.7, 11.0, 11.3 and 11.5. As is evident in the figure, the oscillation at one half the wiggler period due to the lower beat wave is also manifested in the evolution of the relative phase. The bulk variation (i.e. averaged over a wiggler period) shows the same qualitative behaviour as that found for a helical wiggler. Specifically, for frequencies at the low end of the gain band the relative phase decreases up to a point just prior to saturation (indicated in the figure by an arrow), after which the relative phase remains relatively constant. As the frequency increases, the variation in the relative phase decreases until a critical frequency is reached ($\omega/ck_w \approx 11$ for the example under consideration) at which the phase varies little over the course of the interaction. This frequency is typically found to be approximately 10% below the frequency of peak growth rate. For still higher frequencies, the bulk phase increases monotonically.

The axial phase space of the electron beam at saturation is illustrated in Fig. 5.42. The dashed lines represent an approximate separatrix calculated for electrons located at the centre of the beam; hence, many of the electrons which appear to be executing untrapped trajectories may instead be on trapped orbits at the edge of the beam. It should be remarked that this phase-space distribution is located in the lower half of the phase plane, which contrasts with the typical behaviour of the nonlinear pendulum equation in which the separatrix is centred at $d\psi/dz = 0$. The reason for this behaviour is that the single-particle oscillation in the axial velocity at half the wiggler period results in a large-amplitude oscillation in the axial phase of the electrons as well. Indeed the pendulum

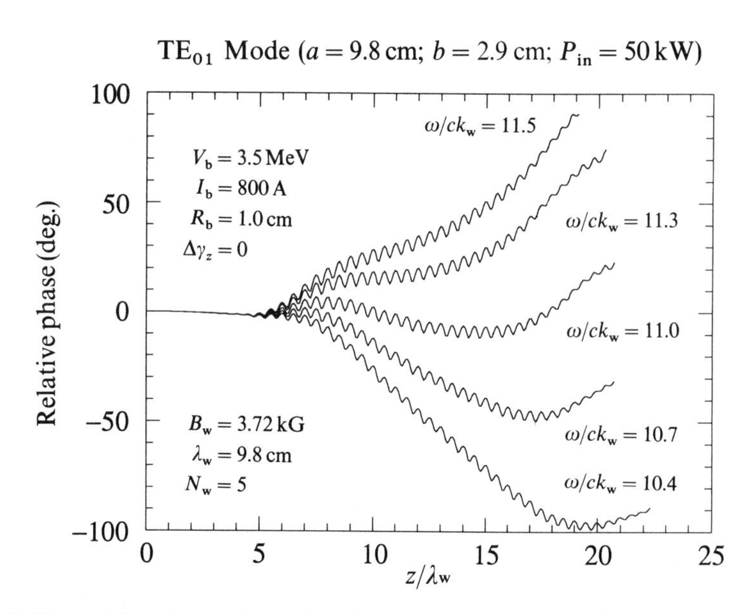

Fig. 5.41 Plots of the relative phase of the TE_{01} mode at a variety of frequencies within the gain band.

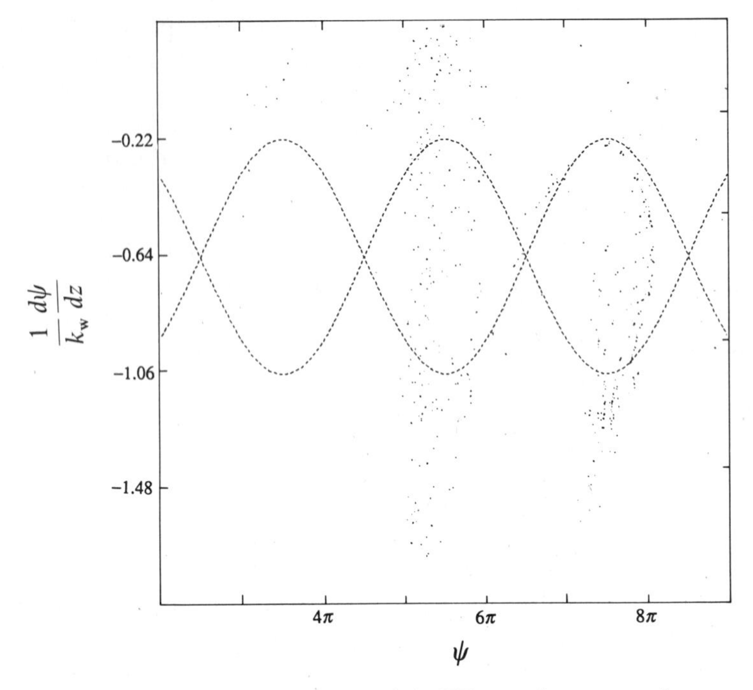

Fig. 5.42 The axial phase space of the TE_{01} mode at saturation.

equation for a planar wiggler takes the form

$$\frac{d^2}{dz^2}\psi_{ln} + K_{ln}^2 \cos\psi_{ln} = -\kappa^2 \sin 2k_w z, \tag{5.168}$$

where for the TE mode polarization

$$K_{ln}^2 \equiv \frac{\omega^2/c^2}{2\beta_\|^2\gamma_\|^2\gamma_0}\left(\frac{e\langle B_w\rangle}{m_e k_w c^2}\right)\left(\frac{e\langle \delta A_{ln}\rangle}{m_e c^2}\right), \tag{5.169}$$

and

$$\kappa^2 \equiv \frac{1}{2}k_w^2\left(\frac{\omega}{k_w v_\|}\right)\left(\frac{e\langle B_w\rangle}{m_e k_w c^2}\right)^2, \tag{5.170}$$

where $\langle B_w\rangle$ and $\langle \delta A_{ln}\rangle$ represent the average fields found at the centre of the beam, and $v_\|$ is the bulk axial velocity. The term in $\kappa^2 \sin 2k_w z$ arises from the oscillation in the axial velocity and generally dominates over the effect of the ponderomotive potential (i.e. $\kappa^2 \gg K_{ln}^2$). In principle, therefore, the effect of this term is to give rise to a general oscillation in the entire axial phase space of the electron beam.

The evolution of the cross-sectional distribution of the beam is illustrated in Figs 5.43–8. The cross-sectional distribution at the start of the uniform wiggler

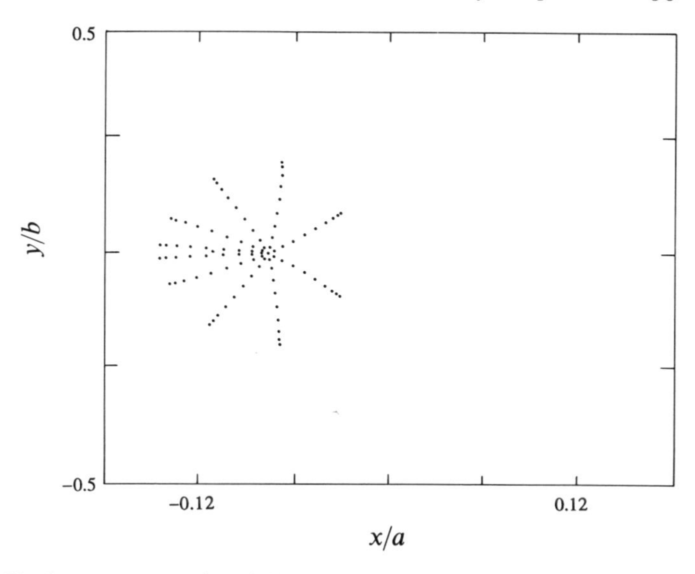

Fig. 5.43 The beam cross-sectional distribution at the start of the uniform wiggler region (i.e. $k_w z \approx 31$).

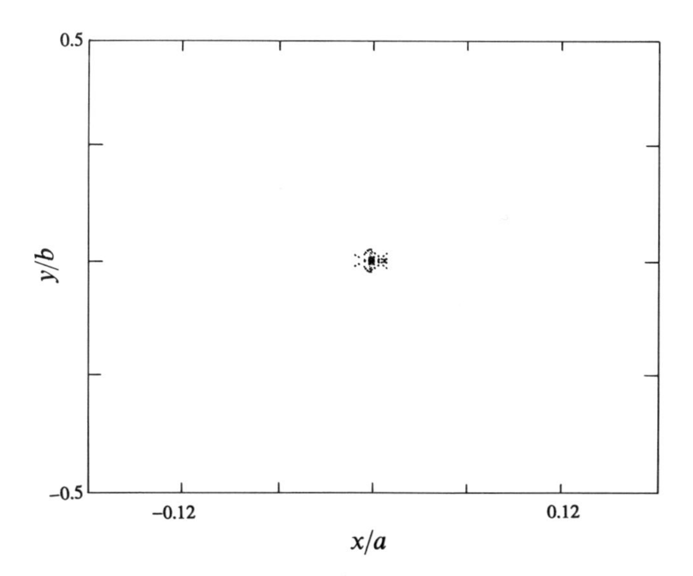

Fig. 5.44 The beam cross-sectional distribution at $k_w z = 36$.

region is shown in Fig. 5.43. The bulk motion exhibits four essential features. The first is the primary wiggler-induced oscillation which is aligned along the x-axis, and this is clearly shown in the figure. The second feature is that the transverse wiggler gradient has a focusing effect on the beam which results in a reduction in the maximum beam radius relative to the initial value. The third feature is

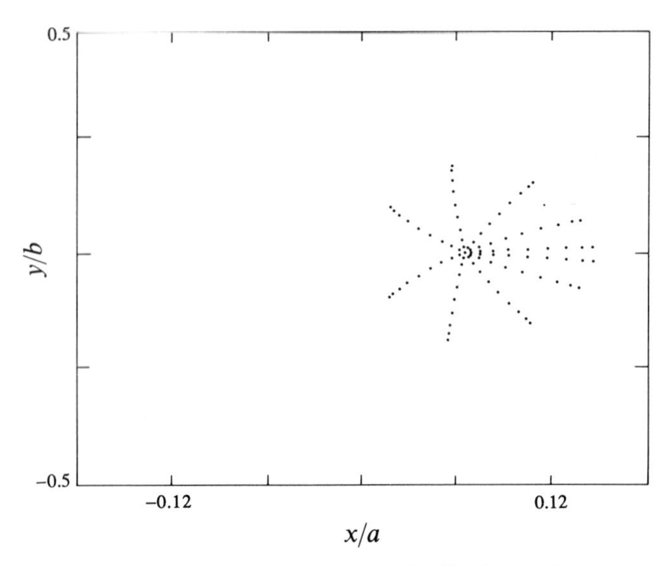

Fig. 5.45 The beam cross-sectional distribution at $k_w z = 41$.

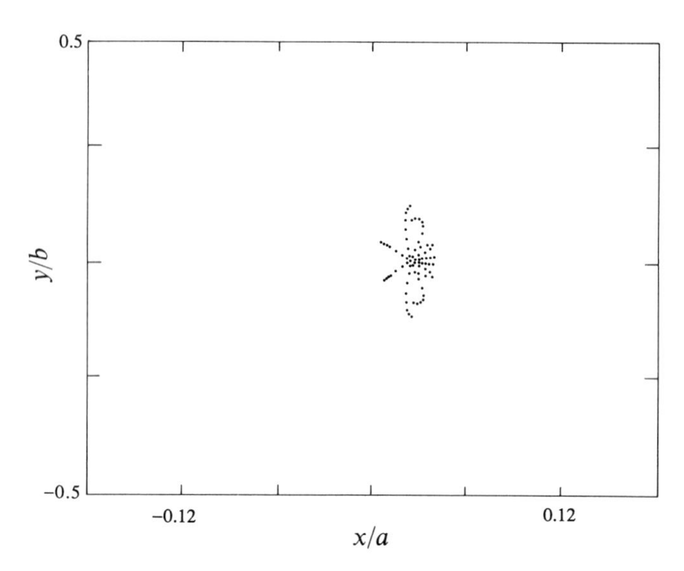

Fig. 5.46 The beam cross-sectional distribution at $k_w z = 46$.

that the transverse wiggler gradient introduces a betatron oscillation which causes a macroscopic scalloping of the beam envelope. The betatron oscillation for the present example occurs over a distance of $k_w \Delta z_b \approx 19.3$, and is illustrated in the sequence of Figs 4.43–7. In addition, on the microscopic level, the individual electrons come into a focus and out again on the opposite side of the beam. Lastly, the geometry of the wiggler tends to distort the overall beam into an elliptical

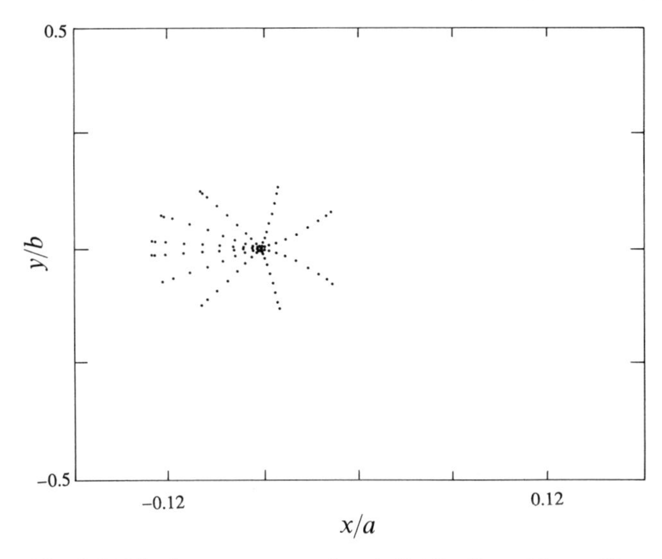

Fig. 5.47 The beam cross-sectional distribution at $k_w z = 51$.

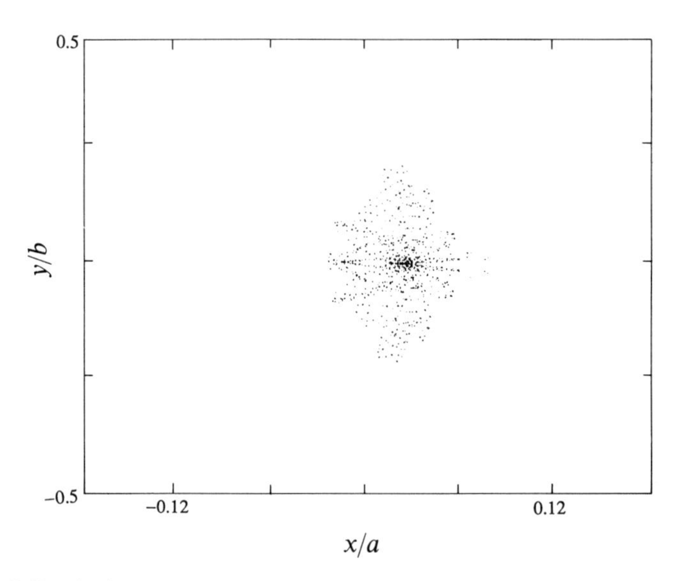

Fig. 5.48 The beam cross-sectional distribution at saturation for $k_w z \approx 115$.

cross-section. This is more clearly shown in Fig. 5.48, which depicts the cross-section of the beam at saturation.

The question of the effect of the process of injecting the electron beam into the wiggler can be addressed by varying the length of the entry taper region. The results of this analysis are shown in Fig. 5.49, in which the saturation efficiency

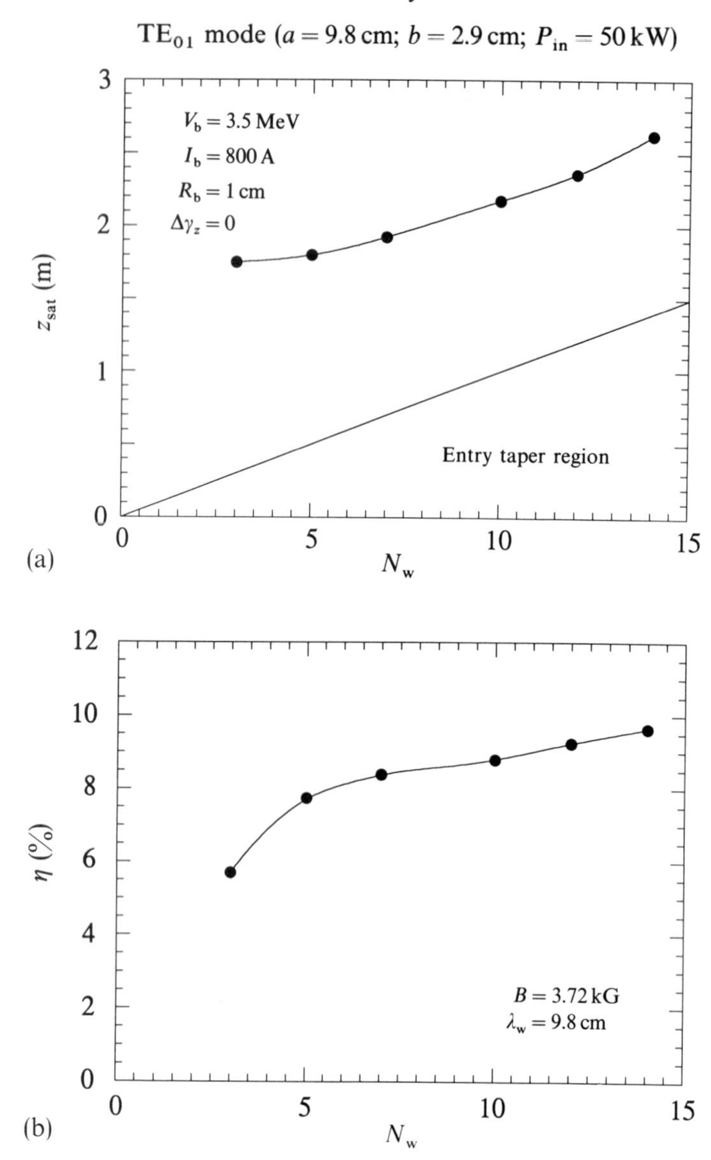

Fig. 5.49 Plot of the distance to saturation and the saturation efficiency of the TE$_{01}$ mode as functions of the length of the entry taper region.

and the distance to saturation are plotted as functions of the length of the entry taper region for $N_w \geqslant 3$. The minimum entry length of the entry taper region has been chosen to ensure that the fringing fields associated with the tapered wiggler amplitude may be neglected. The results indicate that the saturation efficiency increases markedly with the length of the entry taper region up to approximately

$N_w = 6$, after which the increase is more gradual. This increase in the efficiency can be attributed to a decrease in the effective axial energy spread induced by the injection process as the wiggler gradient becomes more adiabatic. It should also be noted that the increase in the distance to saturation is approximately linear for $N_w \geqslant 7$ and closely corresponds to the increase in the length of the entry taper. This implies that the length of the uniform wiggler region over which the bulk of the amplification occurs remains constant.

The evolution of the TE_{21} mode is shown in Fig. 5.50 in which the power is plotted versus axial position for the same parameters as shown for the TE_{01} mode. It is evident from the figure that the power saturates at $k_w z \approx 104$ at a power level of 194 MW for an efficiency of $\eta \approx 6.85\%$. In comparison with the TE_{01} mode, therefore, we conclude that the average growth rate of the TE_{21} mode is somewhat higher and the efficiency lower than for the lower-order mode.

A complete spectrum of the TE_{21} mode is shown in Fig. 5.51 in which the distance to saturation and the efficiency are plotted versus frequency. As shown in the figure, gain is found for frequencies ranging from $\omega/ck_w \approx 8.9$ through $\omega/ck_w \approx 14$, with a peak efficiency of approximately 12%. As a result, both the bandwidth and efficiency of the TE_{21} are greater than that found for the TE_{01} mode.

The evolution of the wave power with axial distance for the TM_{21} mode is shown in Fig. 5.52 for parameters identical to those used for the TE_{01} and TE_{21} modes. As shown in the figure, the power saturates at $k_w z \approx 237$ with a power level of approximately 68.5 MW and an efficiency of $\eta \approx 2.45\%$. This represents

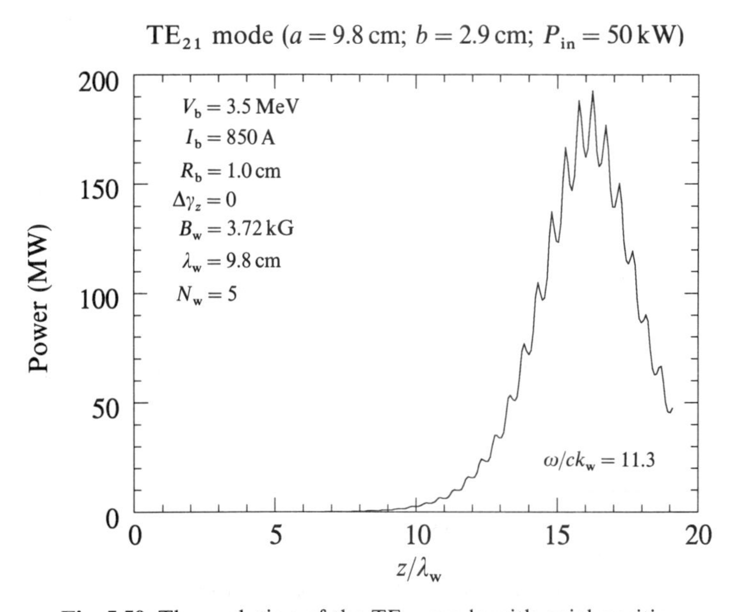

Fig. 5.50 The evolution of the TE_{21} mode with axial position.

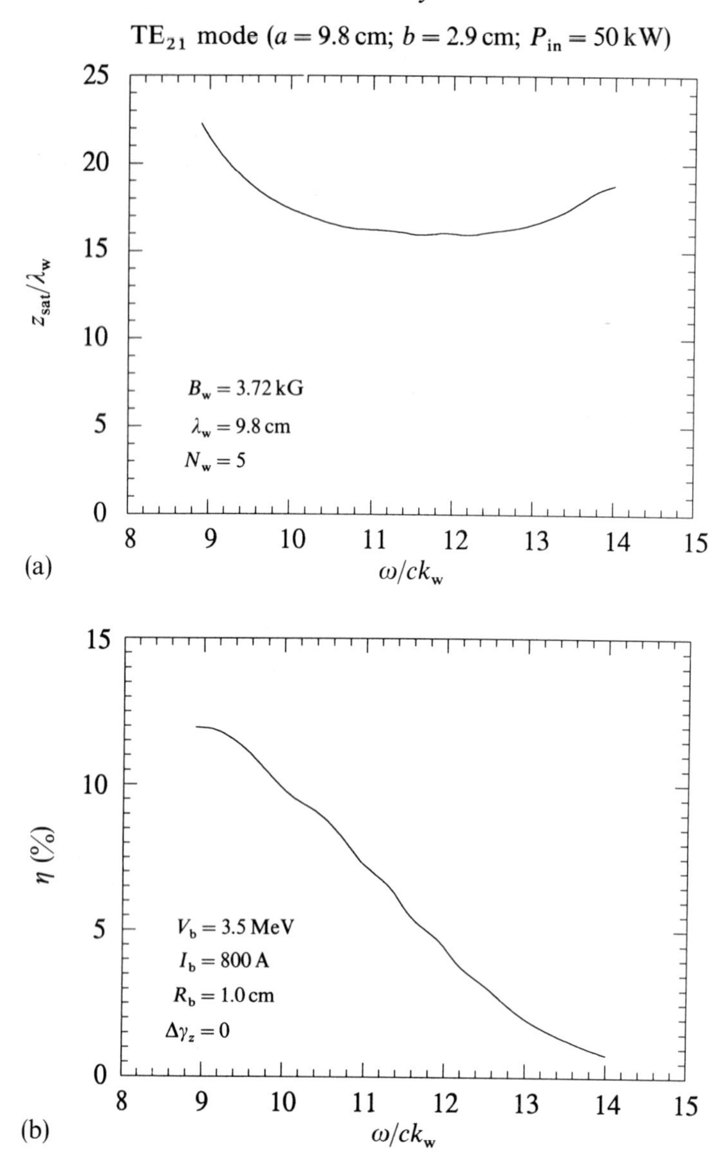

Fig. 5.51 Plot of the distance to saturation and the saturation efficiency of the TE$_{21}$ mode as functions of the length of the entry taper region.

a much lower growth rate and efficiency than found for either of the TE modes, despite the fact that the cutoff frequency and dispersion curves are degenerate for the TE$_{21}$ and TM$_{21}$ modes. The difference between the two modes lies in the transverse mode structure. As mentioned previously, the principal component of the wiggler-induced motion is aligned along the x-axis; hence, the wave–particle

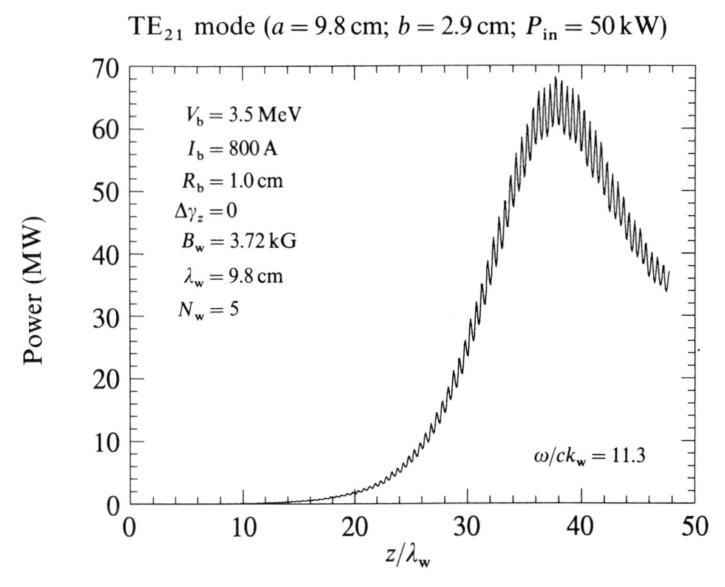

Fig. 5.52 The evolution of the TM_{21} mode with axial position.

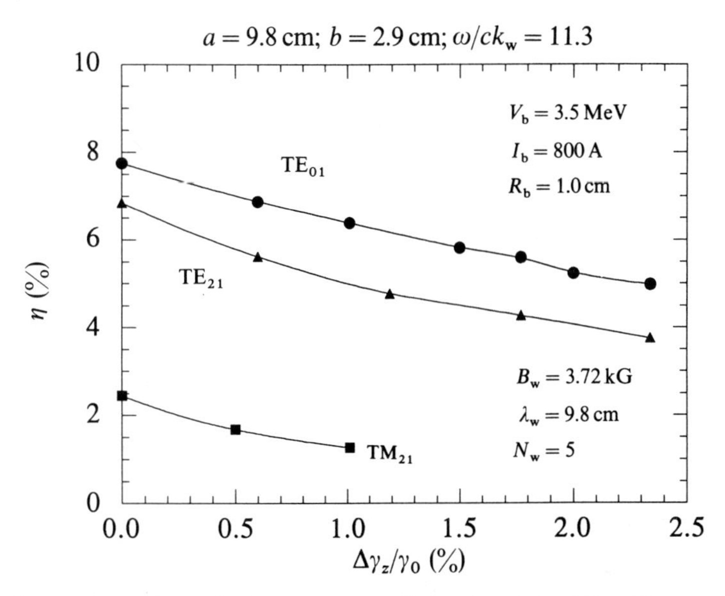

Fig. 5.53 Plot of the variation in the saturation efficiencies of the TE_{01}, TE_{21} and TM_{21} modes with the initial axial energy spread of the beam.

interaction is governed largely by the x-component of the electric field. Comparison of the mode structures for the TE and TM modes shows that for a given mode amplitude the ratio of the x-component of the electric field of the TM_{21} mode to that of the TE_{21} mode is approximately $2b/a \approx 0.59$. As a result, the wave–particle coupling is much weaker for the TM mode in the present configuration. A consequence of the weaker coupling of the TM mode is that the efficiency and gain are lower and the bandwidth narrower than for either TE mode.

The effect of an initial axial energy spread on each mode is shown in Fig. 5.53 in which the saturation efficiency is plotted versus the $\Delta\gamma_z/\gamma_0$ for the TE_{01}, TE_{21} and TM_{21} modes. Using the average growth rates found for each mode, the point at which thermal effects become important can be estimated to be $\Delta\gamma_z/\gamma_0 \approx 18\%$ for the TE_{01} and TE_{21} modes and $\Delta\gamma_z/\gamma_0 \approx 8\%$ for the TM_{21} mode. As a consequence, the cases shown in the figure are well short of the thermal-beam regime. The efficiency is found to decrease in an approximately linear fashion with increasing axial energy spread for each of these modes in correspondence with the results found for helical wiggler geometries. Note, however, that these results are obtained for the same model distribution, and the detailed variation of the efficiency with axial energy spread may vary with the choice of distribution. Be that as it may, it is found that the efficiency drops from $\eta \approx 7.75\%$ to $\eta \approx 4.98\%$ for the TE_{01} mode as the $\Delta\gamma_z/\gamma_0$ increases to approximately 2.3%. The efficiency of the TE_{21} mode drops from $\eta \approx 6.85\%$ to $\eta \approx 3.76\%$ for the same increase in the axial energy spread. The efficiency of the TM_{21} mode decreases from $\eta \approx 2.45\%$ to $\eta \approx 1.27\%$ as $\Delta\gamma_z/\gamma_0$ increases to only 1% in keeping with the weakened wave–particle coupling.

5.4.3 Numerical simulation: multiple modes

The multiple-mode analysis proceeds in the same fashion as that for single modes, except that all resonant modes are included simultaneously in the simulation. For the example under consideration this is restricted to the TE_{01}, TE_{21} and TM_{21} modes. The initializations of these modes are chosen to correspond to the injection of a signal at a frequency of $\omega/ck_w = 11.3$ which consists of 50 kW in the TE_{01} mode, 500 W in the TE_{21} mode, and 100 W in the TM_{21} mode. The detailed evolution of the total wave power with axial position for this example is shown in Fig. 5.54 for the case of an ideal beam. As seen in the figure, the total power saturates at a level of approximately 260 MW after a distance of $k_w z \approx 98$ (1.53 m). It is also evident that although the TE_{01} mode was overwhelmingly dominant upon the injection of the signal, it comprises only approximately 60% of the power at saturation. The remaining power is composed primarily of the TE_{21} mode (approximately 37%). This is due to the fact that the growth rate of the TE_{21} mode is higher than that of the TE_{01} mode at this frequency.

The phase variation of each of these modes is illustrated in Figs 5.55 and 5.56, where the arrow indicates the point at which the power saturates. It is evident from these figures that the bulk evolution (i.e. averaged over a wiggler period) of the relative phase of the TE_{01} mode increases monotonically with axial position

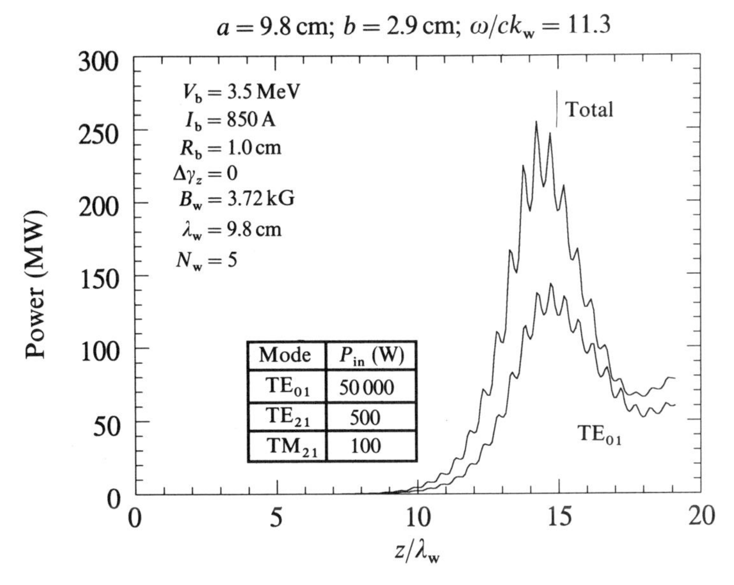

Mode	P_{in} (W)
TE_{01}	50 000
TE_{21}	500
TM_{21}	100

Fig. 5.54 Evolution of the wave power with axial position.

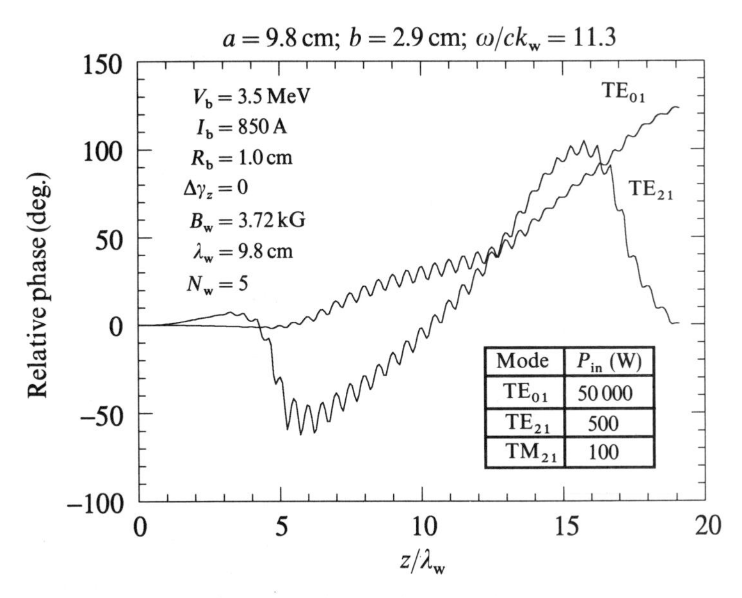

Mode	P_{in} (W)
TE_{01}	50 000
TE_{21}	500
TM_{21}	100

Fig. 5.55 The evolution of the relative phases of the TE_{01} and TE_{21} modes.

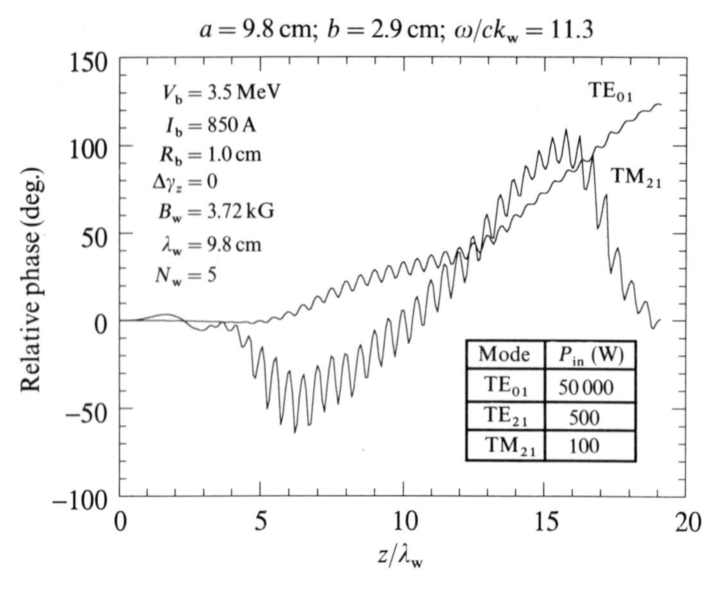

Fig. 5.56 The evolution of the relative phases of the TE_{01} and TM_{21} modes.

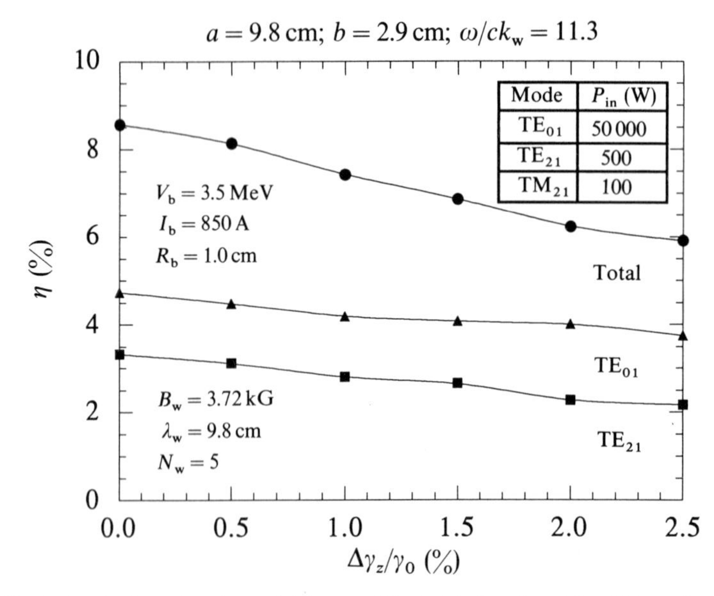

Fig. 5.57 Variation in the saturation efficiency of the total signal and the TE modes with the initial axial energy spread.

in much the same manner as in the single-mode example (see Fig. 5.41). In contrast, the relative phases of both the TE_{21} and TM_{21} modes are decidedly not monotonic and exhibit a decrease with axial position after a point somewhat beyond saturation. This is a multiple-mode effect since the relative phases of each of these modes also exhibit a monotonic increase with axial position in the single-mode analysis. Finally, we observe that the relative phase varies in an almost identical manner for the TE_{21} and TM_{21} modes. This is attributed to the fact that the vacuum dispersion curves for these modes are degenerate.

The effect of an initial axial energy spread on the saturation efficiency of the total signal as well as on the TE_{01} and TE_{21} modes is shown in Fig. 5.57. The TM_{21} mode is excluded from the figure because it composes a small fraction of the signal. As shown in the figure, the saturation efficiency is relatively insensitive to the axial energy spread over the range of $\Delta\gamma_z/\gamma_0 \leqslant 2.5\%$, and decreases from $\eta \approx 8.6\%$ at $\Delta\gamma_z = 0$ to $\eta \approx 5.9\%$ at $\Delta\gamma_z/\gamma_0 \approx 2.5\%$. The reason for this is, as demonstrated in the single-mode analysis, that the growth rate is large and the thermal regime occurs for $\Delta\gamma_z/\gamma_0 \approx 18\%$.

The saturation efficiency in the high-gain Compton regime scales as the cube root of the beam current at the frequency of maximum growth in the idealized one-dimensional model (see equation (1.15)). This type of scaling law is also found to be largely valid in three dimensions, as illustrated in Fig. 5.58. In this figure, the saturation efficiency is plotted as a function of beam current for an ideal beam ($\Delta\gamma_z = 0$) as well as for the case of $\Delta\gamma_z/\gamma_0 = 1.5\%$. Observe that in both cases the efficiency scales approximately as $I_b^{1/3}$.

Turning to the question of the enhancement of the efficiency by means of a tapered wiggler field, we plot the evolution of the power versus axial position in

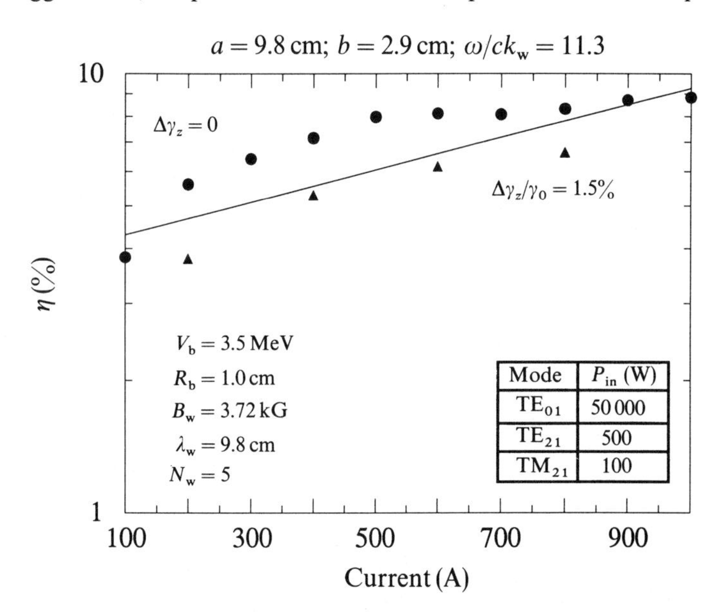

Fig. 5.58 The scaling of the saturation efficiencies with electron beam current.

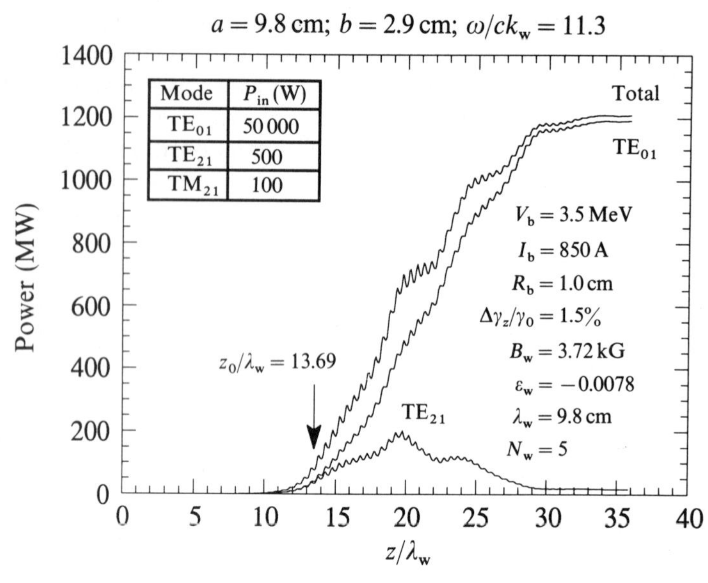

Fig. 5.59 The evolution of the total signal and the TE modes for a tapered wiggler.

Fig. 5.59 for parameters consistent with the uniform wiggler examples. In this case, however, we assume that the axial energy spread is $\Delta\gamma_z/\gamma_0 = 1.5\%$ and taper the wiggler downward with a slope $\varepsilon_w = -0.0078$. The saturation point for this case occurs at $k_w z \approx 96$ for a uniform wiggler, and the optimum start-taper point is found to be $k_w z \approx 86$. As is evident from the figure, the total power of the signal increases to a level of approximately 1.2GW for an efficiency of $\eta \approx 40.6\%$ if the wiggler field is tapered to zero. This represents an enhancement of more than 600% relative to the total efficiency for the uniform wiggler. In addition, it should be remarked that virtually the entire signal is in the TE_{01} mode; hence, we conclude that it is possible selectively to enhance a single mode through the tapered-wiggler interaction. Observe that the uniform wiggler interaction for these parameters yields a total efficiency of approximately 76.8% of which the TE_{01} mode comprises some 60% of the total. In contrast, the TE_{01} mode comprises some 99% of the signal at the end of the tapered wiggler interaction. Both the TE_{21} and TM_{21} (not shown in the figure) ultimately decay to extremely low intensities.

The phase variation of the TE_{01} mode in the tapered-wiggler multi-mode interaction is shown in Fig. 5.60. The relative phase follows the variation indicated in Figs 5.55 and 5.56 for the uniform wiggler region. Subsequent variation of the relative phase under the influence of the tapered wiggler shows a relative slowing of the increase up to an asymptotic level which is approximately 120° for the present example. This type of behaviour of the relative phase is consistently found for a tapered wiggler whether (1) the configuration is that of a helical or planar wiggler, or (2) the interaction is single-mode or multi-mode.

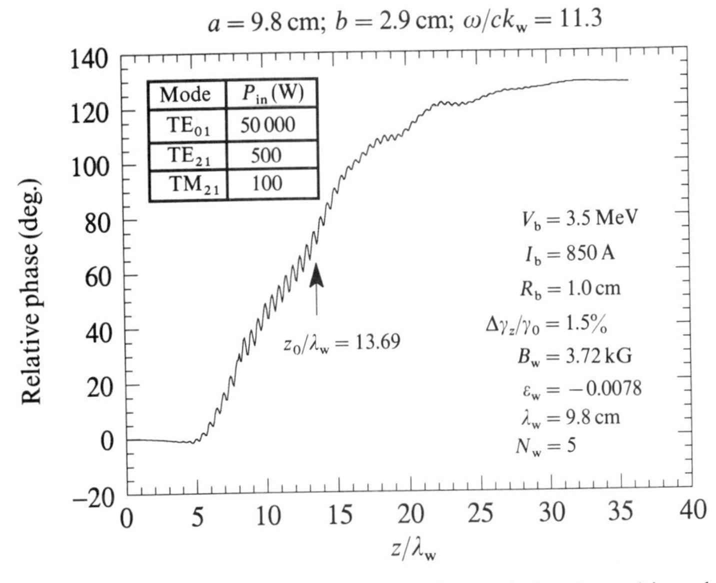

Fig. 5.60 The evolution of the axial phase of the TE_{01} mode for the multi-mode tapered-wiggler interaction.

5.4.4 Comparison with experiment

The aforementioned example nominally corresponds to an experiment conducted at the Lawrence Livermore National Laboratory [5, 32]. The principal differences between the analysis and the experiment are that in the experiment (1) the beam was injected into the wiggler through an entry taper region only one wiggler period in length, and (2) a quadrupole field was used to provide for enhanced focusing instead of parabolically tapered pole faces. Since the fringing fields associated with the wiggler field in the entry taper region are not included in the analytical model, it is invalid to apply the numerical simulation for $N_w = 1$. However, a choice of $N_w = 5$ is made as a compromise and gives good agreement with the experiment, subject to the additional assumption of an initial axial energy spread of 1.5%. This is within an upper bound of 2% on the axial energy spread established by means of an electron spectrometer measurement.

The experimental measurement for a uniform wiggler interaction resulted in a saturated power level of 180 MW over a length of 1.3 m. The simulation for this case gives a peak power of 204 MW and a saturated average power (i.e. averaged over the $\lambda_w/2$ oscillation due to the lower beat wave) of 185 MW. Given the experimental uncertainties in high-power measurements, the latter figure is more relevant for comparison and is in substantial agreement with the observations. In addition, the saturation length found in simulation (that is, the length of the uniform wiggler region plus one wiggler period for the entry taper) is 1.1 m, which is also in good agreement, and shows that the average growth rate from the nonlinear simulation is close to that found in the experiment. Note that the rapid oscillation due to the lower beat wave introduces a 10–20% uncertainty in the measurement of the power and phase. A comparison can also be made with the experiment for a tapered wiggler. In this case, the wiggler field was tapered downward by 55% (i.e. $\Delta B_w/B_w \approx 0.55$) over a length of 1.1 m in the experiment, and the efficiency was observed to increase to 34% for a total power of approximately 1 GW. This corresponds to the slope employed in the calculation shown in Figs 5.59 and 5.60. The calculation shown in these figures represents a tapering of the wiggler field to zero. However, the calculation is in substantial agreement with the experiment over a length of 1.1 m, for which a total power level of 1 GW is also recovered. The evolution of the relative phase is also in good agreement with the reported measurements of this quantity [39]. Thus, within the uncertainties imposed by the choices of the length of the entry taper region and the initial axial energy spread, the essential physics of the interaction are included within the formulation.

5.5 COLLECTIVE EFFECTS

The treatment of collective effects will be discussed for a helical wiggler/axial solenoidal magnetic field configuration in which the electron beam propagates through a cylindrical waveguide. Within the context of this formulation (discussed in [41]), the space-charge waves are treated in an approximate fashion by means of an expansion of the electrostatic potential in terms of the Gould–Trivelpiece modes of a fully filled cylindrical waveguide. The axial component of these electrostatic modes exhibits the same transverse-mode variation as the TM modes of the waveguide, and the eigenvalue (i.e. cutoff wavenumber) is also identical. Hence, the Gould–Trivelpiece modes [55] are orthogonal to the TE modes in the sense that there are no direct mode–mode coupling terms present in the dynamical equations for the fields, and the coupling occurs through the particle dynamics. In contrast, there are direct mode–mode coupling terms between the TM and space-charge modes. Because of this the admixture of TE and TM modes is neglected from the analysis, and the formulation is restricted to the collective interaction in the presence of TE modes only.

5.5.1 The inclusion of space-charge modes

When expressed in terms of the Gould–Trivelpiece modes of a fully filled cylindrical waveguide, the space-charge potential can be written in the form

$$\delta\Phi(\mathbf{x}, t) = \sum_{\substack{l=0\\n=1}}^{\infty} \delta\Phi_{ln}(z) J_l(\kappa_{ln} r) \cos\alpha_{ln}^{(sc)}, \tag{5.171}$$

where $\kappa_{ln} \equiv x_{ln}/R_g$ for $J_l(x_{ln}) = 0$ is the same cutoff wavenumber as obtained for a TM_{ln} mode, and the phase is given by

$$\alpha_{ln}^{(sc)} \equiv \int_0^z dz'\, k_{ln}^{(sc)}(z') + l\theta - \omega t \tag{5.172}$$

for a wave of frequency ω and wavenumber $k_{ln}^{(sc)}(z)$. Observe that the amplitude and wavenumber are implicitly assumed to vary slowly with respect to the wavelength. At the outset (i.e. at $z=0$), the wavenumbers of each mode satisfy a dispersion equation

$$(\omega - k_{ln}^{(sc)} v_{z0})^2 - \frac{\omega_b^2}{\gamma_0^3} = -\frac{\kappa_{ln}^2}{k_{ln}^{(sc)2}} \frac{(\omega - k_{ln}^{(sc)} v_{z0})^2}{(\omega - k_{ln}^{(sc)} v_{z0})^2 - \Omega_0^2} \times \left[(\omega - k_{ln}^{(sc)} v_{z0})^2 - \Omega_0^2 - \frac{\omega_b^2}{\gamma_0} \frac{(\omega - k_{ln}^{(sc)} v_{z0})}{\omega} \right], \qquad (5.173)$$

where γ_0 denotes the bulk initial relativistic factor for the beam, and v_{z0} is the initial axial velocity. Observe that this dispersion equation reduces to that of the space-charge modes in the idealized one-dimensional regime in the limit in which the wavenumber is much larger than the cutoff (i.e. $k_{ln}^{(sc)} \gg \kappa_{ln}$).

The Maxwell–Poisson equations for this configuration are of the form

$$\left(\nabla^2 - \frac{1}{c^2} \frac{\partial^2}{\partial t^2} \right) \delta \boldsymbol{A}(\boldsymbol{x}, t) - \frac{1}{c} \frac{\partial}{\partial t} \nabla \delta \Phi(\boldsymbol{x}, t) = -\frac{4\pi}{c} \delta \boldsymbol{J}(\boldsymbol{x}, t), \qquad (5.174)$$

and

$$\nabla^2 \delta \Phi(\boldsymbol{x}, t) = -4\pi \delta \rho(\boldsymbol{x}, t), \qquad (5.175)$$

where the vector potential is assumed to be given in terms of a superposition of TE modes of the vacuum waveguide (equation (5.84)), the source current is defined in equation (5.87), and the charge density is defined as

$$\delta\rho(\boldsymbol{x}, t) = -e n_b \iiint d^3 p_0 v_{z0} F_b(\boldsymbol{p}_0) \iint_{A_g} dx_0 dy_0 \sigma_\perp(x_0, y_0) \times \int_{-T/2}^{T/2} dt_0 \sigma_\parallel(t_0) \delta[x - x(z; x_0, y_0, t_0, \boldsymbol{p}_0)] \delta[y - y(z; x_0, y_0, t_0, \boldsymbol{p}_0)] \times \frac{\delta[t - \tau(z; x_0, y_0, t_0, \boldsymbol{p}_0)]}{|v_z(z; x_0, y_0, t_0, \boldsymbol{p}_0)|}. \qquad (5.176)$$

Due to the complexity of the numerical solution for Poisson's equation, it is more convenient to describe the evolution of the space-charge modes in terms of the z-component of Maxwell's equations (5.174). Since the z-component of the vector potential for the TE modes vanishes, this equation is

$$\sum_{l,n} J_l(\kappa_{ln} r) \left(\frac{d}{dz} \delta \Phi_{ln} \sin \alpha_{ln}^{(sc)} + k_{ln}^{(sc)} \delta \Phi_{ln} \cos \alpha_{ln}^{(sc)} \right) = \frac{4\pi}{\omega} \delta J_z. \qquad (5.177)$$

Orthogonalization of this equation in the azimuthal angle yields two equations

$$\sum_n J_l(\kappa_{ln} r) \left(\frac{d}{dz} \delta \Phi_{ln} \sin \alpha_{ln}^{(sc)} + k_{ln}^{(sc)} \delta \Phi_{ln} \cos \alpha_{ln}^{(sc)} \right) = \frac{4}{\omega} \int_0^{2\pi} d\theta \, \delta J_z \cos l\theta, \qquad (5.178)$$

and

$$\sum_n J_l(\kappa_{ln}r)\left(\frac{d}{dz}\delta\Phi_{ln}\cos\alpha_{ln}^{(sc)} - k_{ln}^{(sc)}\delta\Phi_{ln}\sin\alpha_{ln}^{(sc)}\right) = \frac{4}{\omega}\int_0^{2\pi} d\theta\,\delta J_z \sin l\theta, \quad (5.179)$$

where

$$\alpha_{ln}^{(sc)} \equiv \int_0^z dz'\,k_{ln}^{(sc)}(z') - \omega t. \quad (5.180)$$

Orthogonalization in r using equation (5.105) and the subsequent time average over a wave period yields the following dynamical equations governing the evolution of the amplitude and wavenumber of the space-charge modes

$$k_{ln}^{(sc)}\,\delta\varphi_{ln} = -2\frac{\omega_b^2}{\omega c}\frac{1}{J_{l+1}^2(x_{ln})}\langle J_l(\kappa_{ln}r)\cos\alpha_{ln}^{(sc)}\rangle, \quad (5.181)$$

and

$$\frac{d}{dz}\delta\varphi_{ln} = -2\frac{\omega_b^2}{\omega c}\frac{1}{J_{l+1}^2(x_{ln})}\langle J_l(\kappa_{ln}r)\sin\alpha_{ln}^{(sc)}\rangle, \quad (5.182)$$

where ω_b is the beam-plasma frequency corresponding to the ambient beam density, $\delta\varphi_{ln} \equiv e\delta\Phi_{ln}/m_e c^2$, and the average is defined as in equation (5.123) and makes use of the momentum space distribution in (5.120).

The transverse components of Maxwell's equations may be orthogonalized in the same manner as that employed previously for the TE modes in the high-gain Compton regime. In this case, the contributions due to the space-charge modes vanish, and equations (5.96) and (5.97) (or, alternatively, equations (5.133) and (5.134)) which govern the evolution of the TE modes are recovered. Hence, the dynamical equations which determine the evolution of the amplitude and wavenumber of each electromagnetic TE mode (equations (5.133) and (5.134)) and space-charge mode (equations (5.181) and (5.182)) are determined. The detailed coupling between each of these modes is governed by the orbit equations for the ensemble of electrons.

5.5.2 The electron orbit equations

The electron trajectories are governed by the complete Lorentz force equations in the combined magnetostatic, electromagnetic and electrostatic fields, and may be written as a generalization of equations (5.126–27)

$$\begin{aligned} v_3\frac{d}{dz}p_1 = &-\{\Omega_0[1 + k_w\varepsilon_0(z - z_0)] - k_w v_3 + 2\Omega_w I_1(\lambda)\sin\chi\}p_2 \\ &+ \Omega_w p_3 I_2(\lambda)\sin 2\chi - \frac{\varepsilon_0}{2}\Omega_0 p_3\lambda\sin\chi - \frac{m_e c}{2}\sum_{\text{TE modes}}\delta a_{ln} \\ &\times[(\omega - k_{ln}v_3)\,W_{ln}^{(-)} - 2\kappa_{ln}v_2 J_l(\kappa_{ln}r)\cos\alpha_{ln} - \Gamma_{ln}v_3\,T_{ln}^{(+)}] \\ &+ \frac{m_e c^2}{2}\sum_{sc\ \text{modes}}\kappa_{ln}\delta\varphi_{ln}(\bar{F}_{ln}^{(-)}\cos\bar{\psi}_{ln} - \bar{G}_{ln}^{(-)}\sin\bar{\psi}_{ln}), \end{aligned} \quad (5.183)$$

$$v_3 \frac{d}{dz} p_2 = \{\Omega_0[1 + k_w \varepsilon_0 (z - z_0)] - k_w v_3 + 2\Omega_w I_1(\lambda) \sin\chi\} p_1$$

$$- \Omega_w p_3 [I_0(\lambda) + I_2(\lambda)\cos 2\chi] + \frac{\varepsilon_0}{2}\Omega_0 p_3 \lambda \cos\chi + \frac{m_e c}{2} \sum_{\text{TE modes}} \delta a_{ln}$$

$$\times [(\omega - k_{ln} v_3)\, T_{ln}^{(-)} - 2\kappa_{ln} v_1 J_l(\kappa_{ln} r) \cos\alpha_{ln} + \Gamma_{ln} v_3\, W_{ln}^{(+)}]$$

$$- \frac{m_e c^2}{2} \sum_{sc\ \text{modes}} \kappa_{ln} \delta\varphi_{ln} (\bar{F}_{ln}^{(+)} \sin\bar{\psi}_{ln} - \bar{G}_{ln}^{(+)} \sin\bar{\psi}_{ln}), \tag{5.184}$$

and

$$v_3 \frac{d}{dz} p_3 = \Omega_w p_2 [I_0(\lambda) + I_2(\lambda)\cos 2\chi] - \Omega_w p_1 I_1(\lambda) \sin 2\chi$$

$$+ \frac{\varepsilon_0}{2}\Omega_0 \lambda (p_1 \sin\chi - p_2 \cos\chi) - \frac{m_e c}{2} \sum_{\text{TE modes}} \delta a_{ln}$$

$$\times [k_{ln}(v_1\, W_{ln}^{(-)} - v_2\, T_{ln}^{(-)}) + \Gamma_{ln}(v_1\, T_{ln}^{(+)} + v_2\, W_{ln}^{(+)})]$$

$$+ m_e c^2 \sum_{sc\ \text{modes}} J_l(\kappa_{ln} r) \left(\frac{d}{dz} \delta\varphi_{ln} \cos\alpha_{ln}^{(sc)} - k_{ln}^{(sc)} \delta\varphi_{ln} \sin\alpha_{ln}^{(sc)} \right). \tag{5.185}$$

Observe that these equations include the effects of both tapered axial and wiggler magnetic fields, and that

$$\bar{F}_{ln}^{(\pm)} \equiv J_{l-1}(\kappa_{ln} r) \cos(l-1)\chi \pm J_{l+1}(\kappa_{ln} r) \cos(l+1)\chi, \tag{5.186}$$

$$\bar{G}_{ln}^{(\pm)} \equiv J_{l-1}(\kappa_{ln} r) \sin(l-1)\chi \pm J_{l+1}(\kappa_{ln} r) \sin(l+1)\chi, \tag{5.187}$$

and

$$\bar{\psi}_{ln} \equiv \int_0^z dz' [k_{ln}^{(sc)}(z') + l k_w] - \omega t. \tag{5.188}$$

Observe that the appropriate cutoffs must be employed in the summations over the TE and space-charge modes. In addition, equations (5.130–32) for the coordinates must also be integrated, as well as an equation for the phase of each space-charge mode

$$\frac{d}{dz} \bar{\psi}_{ln} \equiv k_{ln}^{(sc)}(z') + l k_w - \frac{\omega}{v_z}. \tag{5.189}$$

As a result, the problem in its reduced form consists in the integration of equations for the amplitude and phase of each TE and space-charge mode included in the simulation and six equations per electron, as well as the evaluation of an algebraic equation for the wavenumber of each mode. It is important to bear in mind that the collective wave–particle resonance occurs for $k_{ln}^{(sc)} \approx k_{ln} + k_w$, so that the interaction is found when the space-charge mode has an azimuthal mode number which is one unit less than the TE mode. Specifically, the collective interaction of a TE_{11} mode and a space-charge mode will occur for an azimuthally symmetric space-charge mode (i.e. $l = 0$); however, an arbitrary number of radial eigenmodes may participate in the interaction.

5.5.3 Numerical examples

The initial conditions for the TE modes and the electrons considered in the collective regime are identical to those employed in the high-gain Compton limit, and are chosen to model the injection of an axi-symmetric beam into the waveguide in conjunction with an ensemble of TE modes of arbitrary initial powers. The space-charge modes describe the effect of shot noise on the beam, and must be initialized in a self-consistent fashion. For this purpose, we assume that the initial wavenumber is given approximately by

$$k_{ln}^{(sc)}(z=0) = \frac{\omega}{v_{z0}} + \frac{\omega_b}{\gamma_0^{3/2} v_{z0}} \sqrt{\left(1 - \frac{\kappa_{ln}^2 v_{z0}^2}{\omega^2}\right)}, \tag{5.190}$$

The initial amplitude of the space-charge modes is determined by means of the initial phase average

$$k_{ln}^{(sc)}(z=0)\delta\varphi_{ln}(z=0) = -2\frac{\omega_b^2}{\omega c}\frac{1}{J_{l+1}^2(x_{ln})}\langle J_l(\kappa_{ln} r)\cos\alpha_{ln}^{(sc)}\rangle, \tag{5.191}$$

which is a measure of the noise introduced by the discretization of the electron beam.

The first case we consider is that of the 35 GHz amplifier operating for Group I orbit parameters which was studied in the high-gain Compton regime and shown in Fig. 5.16. The specific parameters of this example involve the propagation of a 250 keV/35 A electron beam with an initial beam radius of 0.155 cm through a waveguide with a radius of 0.366 26 cm in the presence of a wiggler with an amplitude of 2.0 kG and a period of 1.175 cm and an axial magnetic field of 1.3 kG. The interaction for this case was with the TE_{11} mode. In view of the relatively low current, the high-gain Compton limit is expected to be an adequate approximation for this example. As shown in Fig. 5.16, a maximum efficiency of approximately 21.4% is found at a frequency of 33.2 GHz in the high-gain Compton regime. The collective effects upon this interaction are illustrated in Fig. 5.61, in which we plot the efficiency as a function of frequency for this case for both the high-gain Compton and collective Raman limits. The essential physics of the collective interaction are treated by means of a single space-charge mode with $l = 0$ and $n = 1$. It is evident from the figure that collective effects reduce both the bandwidth and efficiency of the interaction. However, as expected, the reductions are not large. In addition, a comparison of the linear growth rates found from the simulation and the linear theory described in Chapter 4 is shown in Fig. 5.62. The dashed line in the figure represents the linearized analysis, and good agreement is found near the centre of the gain band. The discrepancies which appear near the edges of the gain band result from the approximation imposed by the use of Gould–Trivelpiece modes for a fully filled waveguide which have the effect of underestimating the beam-plasma frequency by a small amount.

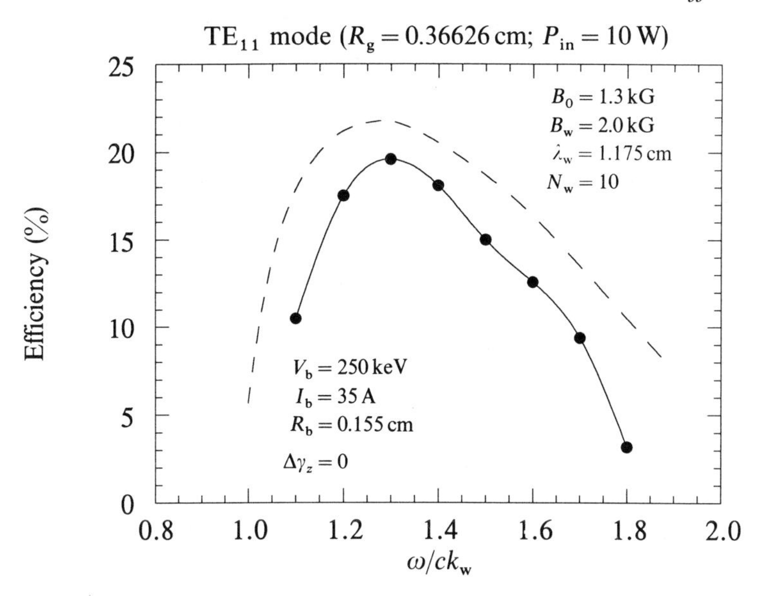

Fig. 5.61 Graph of the interaction efficiency versus frequency for both the high-gain Compton (dashed line) and collective Raman simulations of a 35 GHz amplifier operating on Group I orbit parameters.

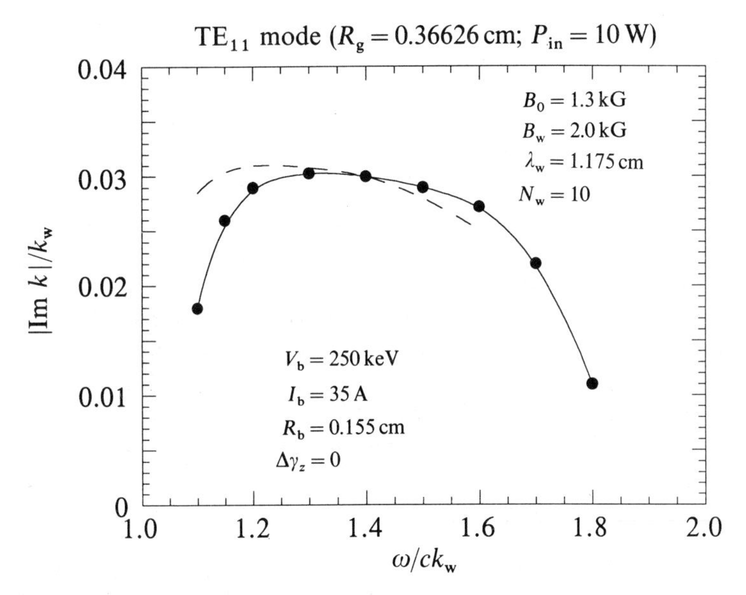

Fig. 5.62 Comparison of the growth rates found from simulation and the linearized theory (dashed line) for a 35 GHz amplifier operating on Group I orbit parameters.

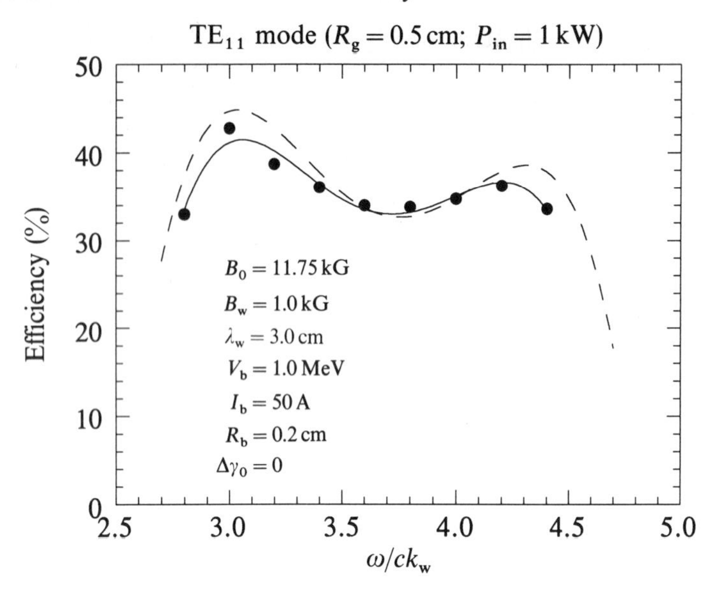

Fig. 5.63 Graph of the interaction efficiency versus frequency for both the high-gain Compton (dashed line) and collective Raman simulations of a 35 GHz amplifier operating on Group II negative-mass orbit parameters.

The collective effects upon the 35 GHz amplifier operating with Group II negative-mass orbit parameters have also been investigated. The magnetostatic fields for this example involve a wiggler field with an amplitude of 1.0 kG and a period of 3.0 cm, and an axial magnetic field of 11.75 kG. The electron beam is characterized by an energy of 1.0 MeV, a current of 50 A, and an initial beam radius of 0.2 cm. The interaction is with a single TE$_{11}$ mode in a waveguide of radius 0.5 cm. The results for the high-gain Compton regime simulation are shown in Fig. 5.63 in which the efficiency is plotted versus frequency, and a peak efficiency of the order of 47% is found at a frequency of 31 GHz. As in the previous case, the current for this example was chosen to be small enough so that the collective effect would not be important, and this is borne out by the simulation. The collective effects upon this case are adequately handled by inclusion of a single space-charge mode ($l = 0$ and $n = 1$), and the results of the simulation are shown in Fig. 5.63. As shown in the figure, the effect of the inclusion of the space-charge modes is to reduce both the efficiency and bandwidth of the interaction; however, these reductions are not large and the simulation in the high-gain Compton limit represents an adequate approximation for all practical purposes.

5.5.4 Comparison with experiment

An example of an experiment conducted at the Massachusetts Institute of Technology [56] in which collective effects were important is shown in Figs 5.64 and 5.65 in which we plot a comparison of the simulation with experimental

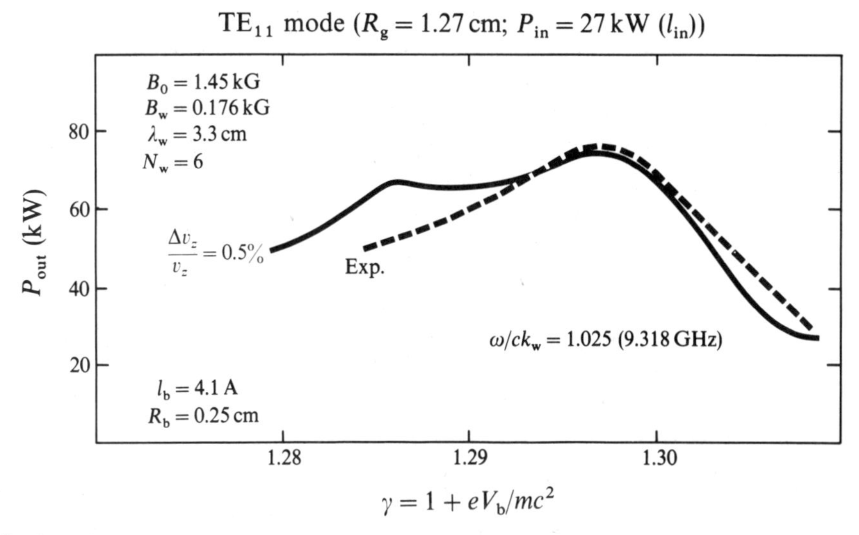

Fig. 5.64 Comparison of the output power versus beam energy from experiment and simulation.

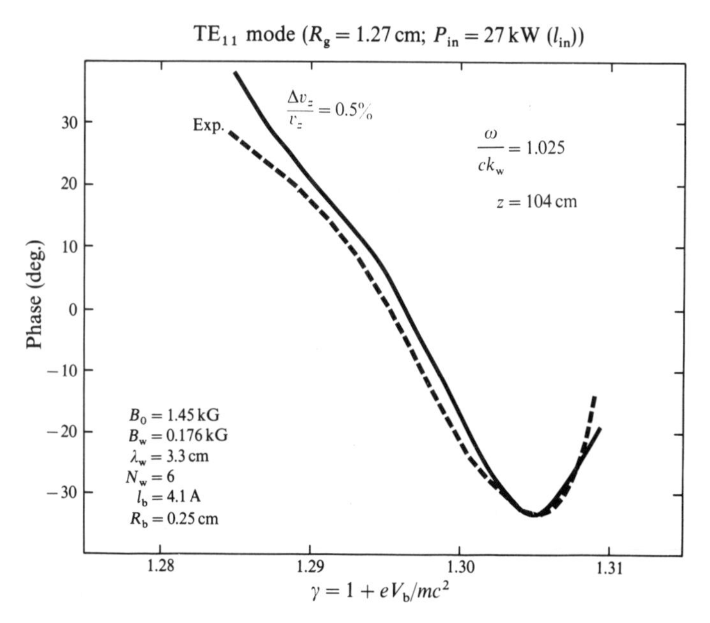

Fig. 5.65 Comparison of the phase of the output signal versus beam energy from experiment and simulation.

measurements of both the output power and phase of the signal as functions of beam energy. The electron beam in this experiment ranged in energy from about 143 to 159 keV with a current of approximately 4.1 A and an initial beam radius of 0.25 cm. The best estimate of the axial velocity spread of the beam in the experiment yielded a value of $\Delta v_z/v_0 \approx 0.5\%$, and this value was employed in the simulation. The electron beam propagated through a drift tube with a radius of 1.27 cm, and the interaction was with a single TE_{11} mode. This experiment employed a wiggler field with an amplitude of 0.176 kG, a period of 3.3 cm, and an entry taper region of six wiggler periods in length. The axial magnetic field for the specific case shown was 1.45 kG which corresponds to Group I orbit parameters. The experiment was operated as an amplifier with an input signal of 27 kW (linearly polarized) at a frequency of 9.318 GHz.

The output power is plotted as a function of beam energy in Fig. 5.64 in which the solid line represents the simulation and the dashed line represents the experimental measurements. The agreement between the simulation and the experiment is excellent for beam energies in the vicinity of the peak output power which corresponds to the centre of the gain band. However, it is evident that the simulation shows a secondary peak at lower beam energies. The reason for this peak stems from the approximate representation of the electrostatic potential which, as noted previously, underestimates the value of beam plasma frequency by a small amount. This has the effect of separating the two intersection points between the TE mode and space-charge mode dispersion curves which, in the present case, is enough to introduce two peaks corresponding to each intersection point. However, the phase of the output signal as calculated in the simulation, as plotted in Fig. 5.65 versus beam energy, is seen to be in good agreement with the measured value over the entire range of energies studied.

REFERENCES

1. Phillips, R. M. (1988) History of the ubitron, *Nucl. Instr. Meth.*, **A272**, 1.
2. Boehmer, H., Caponi, M. Z., Edighoffer, J., Fornaca, S., Munch, J., Neil, G. R., Saur, B. and Shih, C. (1982) Variable-wiggler free-electron laser experiment. *Phys. Rev. Lett.*, **48**, 141.
3. Warren, R. W., Newnam, B. E., Winston, J. G., Stein, W. E., Young, L. M. and Brau, C. A. (1983) Results of the Los Alamos free-electron laser experiment. *IEEE J. Quantum Electron.*, **QE-19**, 391.
4. Edighoffer, J., Neil, G. R., Hess, C. E., Smith, T. I., Fornaca, S. W. and Schwettman, H. A. (1984) Variable-wiggler free-electron laser oscillation. *Phys. Rev. Lett.*, **52**, 344.
5. Orzechowski, T. J., Anderson, B. R., Clark, J. C., Fawley, W. M., Paul, A. C., Prosnitz, D., Scharlemann, E. T., Yarema, S. M., Hopkins, D. B., Sessler, A. M. and Wurtele, J. S. (1986) High-efficiency extraction of microwave radiation from a tapered-wiggler free-electron laser. *Phys. Rev. Lett.*, **57**, 2172.
6. Sprangle, P., Tang, C. M. and Manheimer, W. M. (1979) Nonlinear formulation and efficiency enhancement of free-electron lasers. *Phys. Rev. Lett.*, **43**, 1932.
7. Kroll, N. M., Morton, P. L. and Rosenbluth, M. N. (1980) A variable parameter free-electron laser, in *The Physics of Quantum Electronics: Free-Electron Generators of Coherent Radiation*, Vol. 7 (ed. Jacobs, S. F., Pilloff, H. S., Sargent, M., Scully, M. O. and Spitzer, R.), Addison-Wesley, Reading, Massachusetts, p. 89.

8. Prosnitz, D., Szöke, A. and Neil, V. K. (1980) One-dimensional computer simulation of the variable wiggler free-electron laser, a variable parameter free-electron laser, in *The Physics of Quantum Electronics: Free-Electron Generators of Coherent Radiation*, Vol. 7 (ed. Jacobs, S. F., Pilloff, H. S., Sargent, M., Scully, M. O. and Spitzer, R.), Addison-Wesley, Reading, Massachusetts, p. 175.
9. Mani, S. A. (1980) Free-electron laser interaction in a variable pitch wiggler, a variable parameter free-electron laser, in *The Physics of Quantum Electronics: Free-Electron Generators of Coherent Radiation*, Vol. 7 (ed. Jacobs, S. F., Pilloff, H. S., Sargent, M., Scully, M. O. and Spitzer, R.), Addison-Wesley, Reading, Massachusetts, p. 589.
10. Louisell, W. H., Cantrell, C. D. and Wegener, W. A. (1980) Single-particle approach to free-electron lasers with tapered wigglers, a variable parameter free-electron laser, in *The Physics of Quantum Electronics: Free-Electron Generators of Coherent Radiation*, Vol. 7 (ed. Jacobs, S. F., Pilloff, H. S., Sargent, M., Scully, M. O. and Spitzer, R.), Addison-Wesley, Reading, Massachusetts, p. 623.
11. Brau, C. A. and Cooper, R. K. (1980) Variable wiggler optimization, a variable parameter free-electron laser, in *The Physics of Quantum Electronics: Free-Electron Generators of Coherent Radiation*, Vol. 7 (ed. Jacobs, S. F., Pilloff, H. S., Sargent, M., Scully, M. O. and Spitzer, R.), Addison-Wesley, Reading, Massachusetts, p. 647.
12. Slater, J. M., Adamski, J., Quimby, D. C., Grossman, W. M., Churchill, T. L. and Center, R. E. (1982) Tapered wiggler free-electron laser demonstration, in *Proceedings of the International Conference on Lasers '82* (ed. Powell, R. C.), STS Press, McLean, Virginia, p. 212.
13. Goldstein, J. C. (1984) Evolution of long pulses in a tapered wiggler free-electron laser, in *Free-Electron Generator of Coherent Radiation* (ed. Brau, C. A., Jacobs, S. F. and Scully, M. O.), Proc. SPIE 453, p. 2.
14. Freund, H. P. and Gold, S. H. (1984) Efficiency enhancement in free-electron lasers using a tapered axial guide field. *Phys. Rev. Lett.*, **52**, 926.
15. Kwan, T. J. T., Dawson, J. M. and Lin, A. T. (1977) Free-electron laser. *Phys. Fluids*, **20**, 581.
16. Taguchi, T., Mima, K. and Mochizuki, T. (1981) Saturation mechanism and improvement of conversion efficiency of the free-electron laser. *Phys. Rev. Lett.*, **46**, 824.
17. Freund, H. P. (1983) Nonlinear analysis of free-electron laser amplifiers with axial guide fields. *Phys. Rev. A*, **27**, 1977.
18. Colson, W. B. and Richardson, J. L. (1983) Multimode theory of free-electron lasers. *Phys. Rev. Lett.*, **50** 1050.
19. Antonsen, T. M. Jr and Levush, B. (1989) Mode competition and control in free-electron laser oscillators. *Phys. Rev. Lett.*, **62**, 1488.
20. Antonsen, T. M. Jr and Levush, B. (1989) Mode competition and suppression in free-electron laser oscillators. *Phys. Fluids B*, **1**, 1097.
21. Antonsen, T. M. Jr and Levush, B. (1990) Spectral characteristics of a free-electron laser with time-dependent beam energy. *Phys. Fluids B*, **2**, 2791.
22. Kishimoto, Y., Oda, H., Shiho, M., Odajima, K. and Maeda, H. (1990) Effect of electrostatic field on energy conversion efficiency in high current Raman regime free-electron laser. *J. Phys. Soc. Japan*, **59**, 118.
23. Kishimoto, Y., Oda, H. and Shiho M. (1990) Parasitic wave excitation by multimode coupling in a Raman-regime free-electron laser. *Phys. Rev. Lett.*, **65**, 851.
24. Bazylev, V. A. and Tulipov, A. V. (1990) Enhancement of the efficiency of a free-electron laser in a longitudinal magnetic field. *Sov. J. Quantum Electron.*, **20**, 115.
25. Iracane, D. and Ferrer, J. L. (1991) Stability of a free-electron laser spectrum in the continuous beam limit. *Phys. Rev. Lett.*, **66**, 33.
26. Kwan, T. J. T. and Snell, C. M. (1983) Efficiency of free-electron lasers with a scattered electron beam. *Phys. Fluids*, **26**, 835.
27. Lin, A. T., Lin, C. C., Taguchi, T. and Cheng, W. W. (1983) Nonlinear saturation of free-electron lasers around gyroresonance. *Phys. Fluids*, **26**, 3.

28. Parker, R. K., Jackson, R. H., Gold, S. H., Freund, H. P., Granatstein, V. L., Efthimion, P. C., Herndon, M. and Kinkead, A. K. (1982) Axial magnetic-field effects in a collective-interaction free-electron laser at millimeter wavelengths. *Phys. Rev. Lett.*, **48**, 238.
29. Prosnitz, D., Haas, R. A., Doss, S. and Galinas, R. J. (1981) A two-dimensional numerical model of the tapered wiggler free-electron laser, in *Physics of Quantum Electronics: Free-Electron Generators of Coherent Radiation*, Vol. 9 (ed. Jacobs, S. F., Pilloff, H. S., Sargent, M., Scully, M. O. and Spitzer, R.), Addison-Wesley, Reading, Massachusetts, p. 1047.
30. Sprangle, P. and Tang, C. M. (1981) Three-dimensional nonlinear theory of the free-electron laser. *Appl. Phys. Lett.*, **39**, 677.
31. Scharlemann, E. T. (1985) Wiggler plane focussing in linear wigglers. *J. Appl. Phys.*, **58**, 2154.
32. Orzechowski, T. J., Anderson, B. R., Fawley, W. M., Prosnitz, D., Scharlemann, E. T., Yarema, S. M., Hopkins, D. B., Paul, A. C., Sessler, A. M. and Wurtele, J. S. (1985) Microwave radiation from a high-gain free-electron laser amplifier. *Phys. Rev. Lett.*, **54**, 889.
33. Tang, C. M. and Sprangle, P. (1985) Three-dimensional numerical simulation of free-electron lasers by the transverse mode spectral method. *IEEE J. Quantum Electron.*, **QE-21**, 970.
34. Ganguly, A. K. and Freund, H. P. (1985) Nonlinear analysis of free-electron lasers in three dimensions. *Phys. Rev. A*, **32**, 2275.
35. Lin, A. T. and Lin, C. C. (1986) Mode competition in Raman free-electron lasers. *Nucl. Instr. Meth.*, **A250**, 1373.
36. Freund, H. P. and Ganguly, A. K. (1986) Nonlinear analysis of efficiency enhancement in free-electron laser amplifiers. *Phys. Rev. A*, **33**, 1060.
37. Freund, H. P. and Ganguly, A. K. (1986) Effect of beam quality on the free-electron laser. *Phys. Rev. A*, **34**, 1242.
38. Freund, H. P. and Ganguly, A. K. (1987) Phase variation in free-electron laser amplifiers. *IEEE J. Quantum Electron.*, **QE-23**, 1657.
39. Orzechowski, T. J., Scharlemann, E. T. and Hopkins, D. B. (1987) Measurement of the phase of the electromagnetic wave in a free-electron laser amplifier. *Phys. Rev. A*, **35**, 2184.
40. Ganguly, A. K. and Freund, H. P. (1988) High-efficiency operation of free-electron laser amplifiers. *IEEE Trans. Plasma Sci.*, **PS-16**, 167.
41. Ganguly, A. K. and Freund, H. P. (1988) Three-dimensional simulation of the Raman free-electron laser. *Phys. Fluids*, **31**, 387.
42. Freund, H. P., Bluem, H. and Chang, C. L. (1987) Three-dimensional analysis of free-electron laser amplifiers with planar wigglers. *Phys. Rev. A*, **36**, 2182.
43. Freund, H. P., Chang, C. L. and Bluem, H. (1987) Harmonic generation in free-electron lasers. *Phys. Rev. A*, **36**, 3218.
44. Freund, H. P., Bluem, H. and Chang, C. L. (1988) Three-dimensional simulation of free-electron lasers with planar wigglers. *Nucl. Instr. Meth.*, **A272**, 556.
45. Freund, H. P. (1988) Multimode nonlinear analysis of free-electron laser amplifiers in three dimensions. *Phys. Rev. A*, **37** 3371.
46. Bluem, H., Freund, H. P. and Chang, C. L. (1988) Harmonic content in a planar wiggler based free-electron laser amplifier. *Nucl. Instr. Meth.*, **A272**, 579.
47. Chang, S. F., Eldridge, O. C. and Sharer, J. E. (1988) Analysis and nonlinear simulation of a quadrupole wiggler free-electron laser at millimeter wavelengths. *IEEE J. Quantum Electron.*, **QE-24** 2309.
48. Freund, H. P., Bluem, H. and Jackson, R. H. (1989) Nonlinear theory and design of a harmonic ubitron/free-electron laser. *Nucl. Instr. Meth.*, **A285**, 169.
49. Chang, S. F., Joe, J. and Sharer, J. E. (1990) Nonlinear analysis of wiggler taper, mode competition, and space-charge effects for a 280 GHz free-electron laser. *IEEE Trans. Plasma Sci.*, **PS-18**, 451.

50. Wurtele, J. S., Chu, R. and Fajans, J. (1990) Nonlinear theory and experiment of collective free-electron lasers. *Phys. Fluids B*, **2**, 1626.
51. Kirkpatrick, D. A., Bekefi, G., DiRienzo, A. C., Freund, H. P. and Ganguly, A. K. (1989) A millimeter and submillimeter wavelength free-electron laser. *Phys. Fluids B*, **1**, 1511.
52. Kirkpatrick, D. A., Bekefi, G., DiRienzo, A. C., Freund, H. P and Ganguly, A. K. (1989) A high-power 600 μm wavelength free-electron laser. *Nucl. Instr. Meth.*, **A285**, 43.
53. Byers, J. and Cohen, R. H. (1988) A microwave free-electron laser code using waveguide modes. *Nucl. Instr. Meth.*, **A272**, 595.
54. Freund, H. P. (1991) Nonlinear theory of slow-wave ubitrons/free-electron lasers. *Nucl. Instr. Meth.*, **A304**, 555.
55. Krall, N. A. and Trivelpiece, A. W. (1986) *Principles of Plasma Physics*, San Francisco Press, San Francisco, p. 202.
56. Fajans, J., Wurtele, J. S., Bekefi, G., Knowles, D. S. and Xu, K. (1986) Nonlinear power saturation and phase in a collective free-electron laser amplifier. *Phys. Rev. Lett.*, **57**, 579.

6

Sideband instabilities

The growth of sidebands of the primary signal in free-electron lasers can occur after the bulk of the electron beam becomes trapped in the ponderomotive potential formed by the beating of the wiggler and radiation fields [1–31]. Trapped electrons execute an oscillatory bounce motion in the ponderomotive well, and the waves formed by the beating of this oscillation with the primary signal are referred to as the sidebands. The difficulties imposed by the growth of sidebands is that they can compete with and drain energy from the primary signal as well as considerably broaden the output spectrum. Sideband control, therefore, is an important consideration for free-electron laser configurations which operate in the trapped-particle regime. Examples of such systems include (1) tapered-wiggler configurations designed to trap the beam at an early stage of the interaction and then extract a great deal more energy from the beam over an extended interaction length, and (2) oscillators which run at sufficiently high power over an extended pulse time that the electron beam becomes trapped upon entry to the wiggler.

As a result, a great deal of effort has been expended on techniques of sideband suppression. One method of sideband suppression was employed in a free-electron laser oscillator at Columbia University [12, 14]. This experiment operated at a 2 mm wavelength in which the dispersion due to the waveguide significantly affected the resonance condition. As a consequence, it was found to be possible to shift the sideband frequencies out of resonance with the beam by the proper choice of the size of the waveguide. In order to understand this, observe that the phase of the sideband varies as [13, 14]

$$[(k + \Delta k) + k_w]z - (\omega + \Delta\omega)t \approx Kz, \tag{6.1}$$

where $(\Delta\omega, \Delta k)$ are the shifts in the angular frequency and the wavenumber due to the sideband, and K is the bounce period of the electrons in the ponderomotive potential which is given in equation (4.14) in the idealized one-dimensional limit. Noting that $z \approx v_\parallel t$ and $\omega \approx (k + k_w)v_\parallel$ for a bulk streaming velocity $v_\parallel$, it is clear that the shift in the angular frequency due to the sideband is given by

$$\Delta\omega \approx \frac{Kv_\parallel}{1 - v_\parallel/v_g}, \tag{6.2}$$

where $v_g = \Delta\omega/\Delta k$ is the group velocity of the wave. The sideband can be detuned and suppressed, therefore, when the group velocity of the wave matches the streaming velocity of the electron beam. The dispersion equation in a waveguide is given by $\omega^2 = c^2 k^2 + \omega_{co}^2$, where ω_{co} is the cutoff frequency. Therefore, the group velocity is given by

$$v_g = c\sqrt{\left(1 - \frac{\omega_{co}^2}{\omega^2}\right)}. \tag{6.3}$$

Equating this with the axial velocity of an electron in a helical wiggler, we find that the sideband can be suppressed if

$$\omega = \frac{\gamma\omega_{co}}{\sqrt{(1 + a_w^2)}}, \tag{6.4}$$

where $a_w \equiv eA_w/m_e c^2$.

In addition, work on an infrared free-electron laser oscillator at Los Alamos National Laboratory [16, 17, 23, 24] indicates that it is also possible to suppress sidebands by (1) using a Littrow grating to deflect the sidebands out of the optical cavity, or (2) changing the cavity length. Hence, it appears that sideband growth due to trapped electrons imposes no essential difficulties for the generation of coherent radiation in free-electron lasers at the present time.

In order to understand the basic nature of the sideband mechanism, we consider the process within the framework of an idealized one-dimensional model [4]. The purpose of the analysis is to examine the gain of the sideband modes which result from the trapped electron motion under a variety of conditions including the effect of a tapered wiggler field and the inclusion of a fluctuating space-charge field. More detailed analyses of the sideband instabilities in free-electron lasers include discussions of the linear kinetic theory of the instability in one [2, 7, 9–11, 30, 31] and two dimensions [15, 21, 22, 26, 27, 29] for a variety of trapped-particle distribution functions, as well as nonlinear analyses and simulations of the interaction [1, 3, 6, 19, 28].

6.1 THE GENERAL FORMULATION

The physical model under consideration consists of a relativistic electron beam propagating through an idealized one-dimensional helical wiggler represented in equation (2.5). The scattered electromagnetic waves are represented as transversely uniform circularly polarized waves in which

$$\delta \boldsymbol{A}(z,t) = \delta A(z)[\hat{\boldsymbol{e}}_x \cos\alpha(z,t) - \hat{\boldsymbol{e}}_y \sin\alpha(z,t)], \tag{6.5}$$

represents the large-amplitude primary signal, and

$$\delta \boldsymbol{A}_s(z,t) = \delta A_s(z)[\hat{\boldsymbol{e}}_x \cos\alpha_s(z,t) - \hat{\boldsymbol{e}}_y \sin\alpha_s(z,t)], \tag{6.6}$$

represents the sideband. The phase of each of these waves is represented in the form

$$\alpha(z,t) = \int_0^z dz' k(z') - \omega t, \tag{6.7}$$

and

$$\alpha_s(z,t) = \int_0^z dz' k_s(z') - \omega_s t + \theta_s, \tag{6.8}$$

where (ω, k) and (ω_s, k_s) are the angular frequency and wavenumber of the primary and sideband signals, respectively, and θ_s defines a phase shift of the sideband with respect to the primary. In addition, the fluctuating space-charge fields associated with each of these signals are

$$\delta\Phi(z,t) = \delta\Phi(z)\cos\phi(z,t), \tag{6.9}$$

for the primary signal, and

$$\delta\Phi_s(z,t) = \delta\Phi_s(z)\cos\phi_s(z,t), \tag{6.10}$$

for the sideband. The phases for the space-charge waves are defined as

$$\phi(z,t) = \int_0^z dz'\,\kappa(z') - \omega t + \delta, \tag{6.11}$$

and

$$\phi_s(z,t) = \int_0^z dz'\,\kappa_s(z') - \omega_s t + \delta_s, \tag{6.12}$$

where κ and κ_s denote the wavenumbers, and δ and δ_s represent the phase shifts of these waves with respect to the primary. Observe that the amplitudes and wavenumbers of each of these modes is implicitly assumed to vary slowly in comparison with the wavelengths.

Since $\delta \boldsymbol{A}$ and $\delta\Phi$ represent the primary signal which may have grown from either noise or a low-intensity drive signal or from a large-amplitude external coherent source, we assume that $|\delta \boldsymbol{A}_s| \ll |\delta \boldsymbol{A}| \ll |\boldsymbol{A}_w|$, and $|\delta\Phi_s| \ll |\delta\Phi|$. In addition, since the bounce frequency is, typically, much less than the resonant frequency of the interaction, we also assume that $|\omega_s - \omega| \ll \omega$.

The fields satisfy the Maxwell–Poisson equations (equations (5.19–21)). The analysis of the small-signal gain follows in essentially the same manner as that described in Chapter 5, except that the quasistatic assumption for the multi-wave system may be stated in the form that two electrons which enter the interaction region within an integral multiple of the beat period $T(\equiv 2\pi/|\omega - \omega_s|)$ execute identical trajectories. Hence, the dynamical equations which govern the evolution of the amplitudes and phases of the primary and sideband signals are

$$\left[\frac{d^2}{dz^2} + \left(\frac{\omega^2}{c^2} - k^2\right)\right]\delta a \cong -\frac{\omega_b^2}{c^2} v_{z0} a_w \left\langle \frac{\cos\psi}{\gamma v_z} \right\rangle, \tag{6.13}$$

$$\left[\frac{d^2}{dz^2} + \left(\frac{\omega_s^2}{c^2} - k_s^2\right)\right]\delta a_s \cong -\frac{\omega_b^2}{c^2} v_{z0} a_w \left\langle \frac{\cos\psi_s}{\gamma v_z} \right\rangle, \tag{6.14}$$

$$2k^{1/2}\frac{d}{dz}(k^{1/2}\delta a) \cong \frac{\omega_b^2}{c^2} v_{z0} a_w \left\langle \frac{\sin\psi}{\gamma v_z} \right\rangle, \tag{6.15}$$

$$2k_s^{1/2}\frac{d}{dz}(k_s^{1/2}\delta a_s) \cong \frac{\omega_b^2}{c^2} v_{z0} a_w \left\langle \frac{\sin\psi_s}{\gamma v_z} \right\rangle, \tag{6.16}$$

for the electromagnetic waves, and

$$(k+k_w)\begin{pmatrix}\delta\varphi_1\\ \delta\varphi_2\end{pmatrix}\cong -2\frac{\omega_b^2}{c^2}\frac{v_{z0}}{\omega}\begin{pmatrix}\langle\cos\psi\rangle\\ \langle\sin\psi\rangle\end{pmatrix}, \tag{6.17}$$

$$(k_s+k_w)\begin{pmatrix}\delta\varphi_{s1}\\ \delta\varphi_{s2}\end{pmatrix}\cong -2\frac{\omega_b^2}{c^2}\frac{v_{z0}}{\omega_s}\begin{pmatrix}\langle\cos\psi_s\rangle\\ \langle\sin\psi_s\rangle\end{pmatrix}, \tag{6.18}$$

for the space-charge modes, where $\delta a \equiv e\delta A/m_e c^2$, $\delta\varphi \equiv e\delta\Phi/m_e c^2$, $\delta\varphi_1 \equiv \delta\varphi\cos\delta$, $\delta\varphi_2 \equiv \delta\varphi\sin\delta$, $\delta\varphi_{s1} \equiv \delta\varphi_s\cos\delta_s$, and $\delta\varphi_{s2} \equiv \delta\varphi_s\sin\delta_s$. In addition, the ponderomotive phases are defined as

$$\psi = \psi_0 + \int_0^z dz'\left(k + k_w - \frac{\omega}{v_z}\right), \tag{6.19}$$

and

$$\psi_s = \frac{\omega_s}{\omega}\psi_0 + \int_0^z dz'\left(k_s + k_w - \frac{\omega_s}{v_z}\right) + \theta_s, \tag{6.20}$$

where $\psi_0 = -\omega t_0$. The source current and charge density used to derive the dynamical equations are given in equations (5.22) and (5.23), and ω_b denotes the plasma frequency associated with the ambient beam density. Finally, under the quasistatic assumption, Maxwell's equations have been averaged over the beat period to obtain the dynamical equations and the averages which appear are defined as

$$\langle(\cdots)\rangle \equiv \frac{1}{2\pi N}\int_{-N\pi}^{N\pi} d\psi_0\,\sigma(\psi_0)(\cdots), \tag{6.21}$$

where $N(\equiv \omega/|\omega-\omega_s|)$ describes the ratio of the beat period to the period of the primary signal and is assumed to be an integer which is much larger than unity, and $\sigma(\psi_0)$ denotes the distribution in the initial phases.

Observe that the time average has been performed over a much longer period than is necessary for a single-frequency amplifier model. The dynamical equations (6.13–18) can be integrated numerically to obtain the complete nonlinear evolution of the sideband instability in a manner analogous to that employed in Chapter 5 for the single-frequency amplifier model, and can be extended to include multiple sidebands as well, which are separated by the fixed angular frequency $\Delta\omega = |\omega - \omega_s|$. The distinction, however, arises from the time average which is performed, essentially, over a large number of periods of the primary signal. This means that a much larger number of electrons must be included in the description, as the formulation must include N beamlets of electrons (where a beamlet is defined as the group of electrons which enter the interaction region within one period of the primary signal).

The dynamical equations (6.13–16) simplify under the neglect of second-order derivatives of the amplitude and phase. For the purposes of the present analysis of the small-signal gain, therefore, we obtain the more tractable set of equations for the electromagnetic fields

$$\left(\frac{\omega^2}{c^2} - k^2\right)\delta a \cong -\frac{\omega_b^2}{c^2}\frac{v_w}{c}\langle\cos\psi\rangle, \tag{6.22}$$

$$\left(\frac{\omega_s^2}{c^2} - k_s^2\right)\delta a_s \cong -\frac{\omega_b^2}{c^2}\frac{v_w}{c}\langle\cos\psi_s\rangle, \tag{6.23}$$

$$2k\frac{d}{dz}\delta a \cong \frac{\omega_b^2}{c^2}\frac{v_w}{c}\langle\sin\psi\rangle, \tag{6.24}$$

and

$$2k_s\frac{d}{dz}\delta a_s \cong \frac{\omega_b^2}{c^2}\frac{v_w}{c}\langle\sin\psi_s\rangle. \tag{6.25}$$

The phases which appear in the dynamical equations are invariant under the transformation $\psi \to \psi + 2\pi$, $\theta_s \to \theta_s + 2\pi(\omega - \omega_s)/\omega$. As a consequence, $\psi(\psi_0 + 2\pi, \theta + 2\pi(\omega - \omega_s)/\omega, z) = \psi(\psi_0, \theta, z) + 2\pi$. This symmetry is also possessed by ψ_s. Hence, if it is also required that $\sigma(\psi_0 + 2\pi) = \sigma(\psi_0)$, then the averages in the dynamical equations can be decomposed into

$$\langle(\cdots)\rangle \cong \frac{1}{(2\pi)^2}\int_0^{2\pi} d\theta \int_{-\pi}^{\pi} d\psi_0\,\sigma(\psi_0)(\cdots), \tag{6.26}$$

for sufficiently large N.

6.2 TRAPPED ELECTRON TRAJECTORIES

Since the vector and scalar potentials are independent of the transverse coordinates, these coordinates are cyclic and the corresponding canonical momenta are constants of the motion. In view of this, the equation for the axial momentum takes the form

$$\frac{d}{dt}p_z = -m_e c\left[(\delta\varphi_1 \sin\psi - \delta\varphi_2\cos\psi)\frac{\partial}{\partial z}\psi + (\delta\varphi_{s1}\sin\psi_s - \delta\varphi_{s2}\cos\psi_s)\frac{\partial}{\partial z}\psi_s\right]$$
$$-\frac{m_e c^2}{2\gamma}\frac{d}{dz}\mu^2 - \frac{m_e c^2}{\gamma}a_w[(k + k_w)\delta a\sin\psi + (k_s + k_w)\delta a_s\sin\psi_s], \tag{6.27}$$

where $\mu \equiv 1 + a_w^2$, and terms of second order in the fluctuating fields have been neglected. The variation in the total electron energy is given by

$$\frac{d}{dt}\gamma = -v_z\left[(\delta\varphi_1\sin\psi - \delta\varphi_2\cos\psi)\frac{\partial}{\partial z}\psi + (\delta\varphi_{s1}\sin\psi_s - \delta\varphi_{s2}\cos\psi_s)\frac{\partial}{\partial z}\psi_s\right]$$
$$-\frac{1}{\gamma}a_w[\omega\,\delta a\sin\psi + \omega_s\delta a_s\sin\psi_s]. \tag{6.28}$$

Combination of these equations yields the equation for the axial velocity

$$\frac{d}{dt}v_z = -\frac{c^2\mu^2}{\gamma^3}[(k+k_w)(\delta\varphi_1\sin\psi - \delta\varphi_2\cos\psi) + (k_s+k_w)(\delta\varphi_{s1}\sin\psi_s - \delta\varphi_{s2}\cos\psi_s)]$$

$$-\frac{c^2}{2\gamma^2}\frac{d}{dz}\mu^2 - \frac{c^2}{\gamma^2}a_w\left[\left(k+k_w-\omega\frac{v_z}{c^2}\right)\delta a\sin\psi\right.$$

$$\left.+\left(k_s+k_w-\omega_s\frac{v_z}{c^2}\right)\delta a\sin\psi_s\right], \tag{6.29}$$

where terms of order $(\omega-\omega_s)/\omega$ have been ignored.

Substitution of the axial position for the time as the independent variable and the ponderomotive phase for the axial velocity, yields a nonlinear pendulum-like equation of the form

$$\frac{d^2}{dz^2}\psi \cong \frac{d}{dz}k_w - \frac{(k+k_w)c^2}{2\gamma_{ph}^2 v_{ph}^2}\frac{d}{dz}\mu^2 - \frac{\mu^2c^2(k+k_w)^2}{\gamma_{ph}^4 v_{ph}^2}a_w\delta a\left(\sin\psi + \varepsilon\frac{\omega}{\omega_s}\sin\psi_s\right)$$

$$+\frac{2\omega_b^2\mu^2 v_{z0}}{\gamma_{ph}^3 v_{ph}^3}\left[\left(\sin\psi\langle\cos\psi\rangle + \frac{\omega_s}{\omega}\sin\psi_s\langle\cos\psi_s\rangle\right)\right.$$

$$\left.-\left(\cos\psi\langle\sin\psi\rangle + \frac{\omega_s}{\omega}\cos\psi_s\langle\sin\psi_s\rangle\right)\right], \tag{6.30}$$

where $\varepsilon \equiv \delta a_s/\delta a$, $v_{ph} \equiv \omega/(k+k_w)$ is the phase velocity of the ponderomotive wave, and $\gamma_{ph} \equiv \mu(1 - v_{ph}^2/c^2)^{-1/2}$ is the relativistic factor corresponding to the ponderomotive frame. The inclusion of space-charge effects is made for the case of a diffuse electron beam in which $\omega_b \ll \omega$ and ω_s, so that we may write $\omega \approx ck$ and $\omega_s \approx ck_s$. Under the additional assumptions that $k_w \ll k$ and k_s, the pendulum equation reduces to

$$\frac{d^2}{dz^2}\psi \cong -\frac{K^2}{\cos\psi_{res}}\left(\sin\psi - \sin\psi_{res} + \varepsilon\frac{\omega_s}{\omega}\sin\psi_s\right)$$

$$+\delta K^2\left[\left(\sin\psi\langle\cos\psi\rangle + \frac{\omega_s}{\omega}\sin\psi_s\langle\cos\psi_s\rangle\right)\right.$$

$$\left.-\left(\cos\psi\langle\sin\psi\rangle + \frac{\omega_s}{\omega}\cos\psi_s\langle\sin\psi_s\rangle\right)\right], \tag{6.31}$$

where $K^2 \equiv 4k_w^2 a_w\delta a\cos\psi_{res}/\mu^2$, $\delta K^2 \equiv 2\mu^2\omega_b^2/\gamma^3c^2$, the resonant phase is determined by

$$\sin\psi_{res} \equiv -\frac{1}{2a_w\delta a}\left[\frac{a_w}{k_w}\frac{d}{dz}a_w - \frac{\mu^2}{2k_w^2}\frac{d}{dz}k_w\right], \tag{6.32}$$

and the phase of the sideband can be represented in the form

$$\psi_s \cong \frac{\omega_s}{\omega}\psi + \Delta kz + \theta_s, \tag{6.33}$$

for $\Delta k \equiv k_w(\omega-\omega_s)/\omega$.

The solution to the pendulum equation (6.31) is found by perturbation about the resonant phase $\psi \approx \psi_{res} + \delta\psi$, where $\delta\psi \ll \pi$. As a consequence,

$$\frac{d^2}{dz^2}\delta\psi \cong -K^2\delta\psi + \frac{K^2}{2}\tan\psi_{res}\delta\psi^2 + \frac{K^2}{6}\delta\psi^3 - \frac{\varepsilon K^2}{\cos\psi_{res}}\sin\left(\frac{\omega_s}{\omega}\psi_{res} + \Delta kz + \theta_s\right) + \delta K^2[\delta\psi - \langle\delta\psi\rangle]. \tag{6.34}$$

Observe that under the linearization of this equation $2\pi/K$ is the bounce period in the ponderomotive well, while $2\pi/\delta K$ is the oscillation (i.e. plasma) period in the space-charge wave. The perturbed phase is decomposed into $\delta\psi = \delta\psi^{(0)} + \delta\psi^{(1)}$ to denote the zeroth and first-order effects due to the sideband. As a result

$$\delta\psi^{(0)} \cong \alpha\cos K_a z + \frac{\alpha^2}{4}\tan\psi_{res}\left[1 - \frac{1}{3}\cos 2K_a z\right], \tag{6.35}$$

for small deviations from the resonant phase, where α is a constant fixed by the initial condition $\delta\psi^{(0)}(z=0) = \psi_0$ which implies that $\psi_0 - \psi_{res} = \alpha + \alpha^2\tan\psi_{res}/6$ for $|\alpha| \ll 1$, and

$$K_a \equiv K\left[1 - \frac{\alpha^2}{16}\left(1 + \frac{5}{3}\tan^2\psi_{res}\right)\right], \tag{6.36}$$

is the anharmonic bounce period in the ponderomotive potential. The first-order correction to this motion due to the sideband signal is given by

$$\begin{aligned}\delta\psi^{(1)} \cong & \frac{\varepsilon}{2\cos\psi_{res}}\frac{K}{\Delta K_\pm}\left[1 \pm \frac{\alpha^2 K}{16\Delta K_\pm}\left(1 + \frac{5}{3}\tan^2\psi_{res}\right)\right] \\ & \times\left\{\left[1 \pm \frac{\alpha^2 K}{8\Delta K_\pm}\left(1 + \frac{5}{3}\tan^2\psi_{res}\right)\right][\cos(\Delta K_\pm z + \phi_{res}) - \cos\phi_{res}]\sin Kz\right. \\ & \mp [\sin(\Delta K_\pm z + \phi_{res}) - \sin\phi_{res}]\cos Kz \\ & \left.\pm \frac{\alpha^2 K}{8\Delta K_\pm}\left(1 + \frac{5}{3}\tan^2\psi_{res}\right)Kz\sin\phi_{res}\sin Kz\right\} + \frac{\delta K^2}{2K^2}[\alpha - \langle\alpha\rangle]Kz\sin Kz,\end{aligned} \tag{6.37}$$

where $\phi_{res} \equiv \omega_s\psi_{res}/\omega + \theta_s$, $\Delta K_\pm \equiv \Delta k \pm K$, subject to the initial conditions $\delta\psi^{(1)}(z=0) = 0$ and $d\delta\psi^{(1)}(z=0)/dz = 0$. It is important to bear in mind that this solution requires that $\alpha^2 < \Delta K_\pm/K$, which implies that all electrons are deeply trapped with phases close to the resonant phase.

6.3 THE SMALL-SIGNAL GAIN

The gain per pass through a system of length L for the primary and sideband waves is found by integration of equations (6.24) and (6.25) to be

$$G(L) \cong \frac{\omega_b^2}{2\omega\gamma_0 c}\frac{a_w}{\delta a}\int_0^L dz\langle\sin\psi\rangle, \tag{6.38}$$

and

$$G_s(L) \cong \frac{\omega_b^2}{2\omega\gamma_0 c}\frac{a_w}{\delta a_s}\int_0^L dz\langle \sin\psi_s\rangle, \tag{6.39}$$

where γ_0 denotes the bulk energy of the beam, and we have assumed that $\omega \approx ck$ and $\omega_s \approx ck_s$.

In the case of the gain of the primary wave, we note that to the lowest nontrivial order the phase average is

$$\langle \sin\psi\rangle \cong \sin\psi_{res}\left[1-\frac{1}{2}\langle \delta\psi^{(0)2}\rangle - \langle \delta\psi^{(0)}\delta\psi^{(1)}\rangle\right]. \tag{6.40}$$

Because of the average over the relative phase between the primary and the sideband signals, only those components of $\delta\psi^{(1)}$ due to the space-charge potential can contribute to the gain to this order. In particular, the coupling between the primary and sideband waves cannot affect the gain of the primary to first order in ε. In addition, the space-charge effect is manifested by means of a beating between the bounce motion of the electrons in the ponderomotive potential and the space-charge wave itself. In the evaluation of these averages, the initial phase distribution is assumed to be of the form

$$\sigma(\psi_0) = \frac{\pi}{\Delta}H(\psi_0+\Delta-\psi_{res})H(\psi_{res}+\Delta-\psi_0), \tag{6.41}$$

in the range $-\pi \leqslant \psi_0 \leqslant \pi$, where $\Delta \ll 1$ and H denotes the Heaviside function. This distribution describes a tightly bunched, flat-topped beam about the resonant phase, and is depicted schematically in Fig. 6.1. As a consequence, the gain associated with this phase space distribution is

$$G(L) \cong \frac{\omega_b^2 L}{2\omega\gamma_0 c}\frac{a_w}{\delta a}\sin\psi_{res}\left\{1+\frac{\Delta^2}{12}\left[\frac{\sin KL}{KL}+\frac{2\sin 2KL}{3KL} + \frac{\delta K^2}{2K^2}\left(\cos 2KL - \frac{\sin 2KL}{2KL}\right)\right]\right\}. \tag{6.42}$$

Since the sideband is not expected to grow prior to the trapping of the beam, the phase-space distribution chosen reflects a beam for which the primary signal should be stable in the absence of a tapered wiggler, and this is reflected in the expression for the gain which vanishes when $\sin\psi_{res}=0$. In addition, the space-charge contribution vanishes in the limit in which $\Delta\to 0$. This conclusion is quite general and is not dependent upon the particular choice of the phase-space distribution. Indeed, it can be shown from equation (6.37) that the contribution from the space-charge potential vanishes when the spread in ψ_0 vanishes.

In the case of the sideband signal, the phase average is

$$\langle \sin\psi_s\rangle \cong \frac{\omega_s}{\omega}\langle \delta\psi^{(1)}\cos\delta\psi_s^{(0)}\rangle, \tag{6.43}$$

and the space-charge effect does not contribute to the interaction. Gain, therefore,

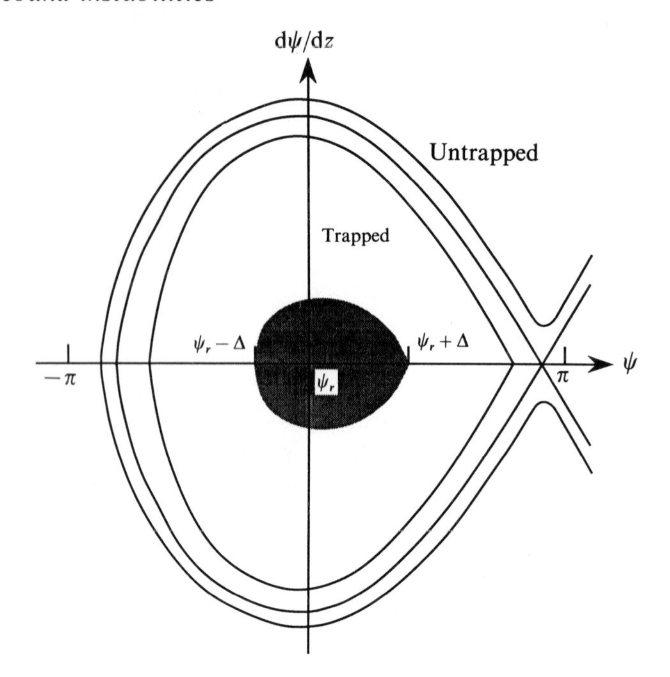

Fig. 6.1 Schematic representation of the phase-space distribution of the electron beam about the resonant phase in a tapered wiggler.

arises because the presence of the sideband induces an oscillation in the electron bounce motion, and is of the form

$$G_s(L) \cong \frac{\omega_b^2 KL^2}{16\omega_s \gamma_0 c \cos\psi_{res}} \frac{a_w}{\delta a}\left[1 \pm \frac{K\Delta^2}{48\Delta K_\pm}\right.$$
$$\left.\times\left(1+\frac{5}{3}\tan^2\psi_{res}\right)\left(1-\frac{\Delta K_\pm L}{2}\frac{d}{d\Theta_\pm}\right)\right]\left(\frac{\sin\Theta_\pm}{\Theta_\pm}\right)^2, \tag{6.44}$$

where $\Theta_\pm \equiv \Delta K_\pm L/2$. Observe that it is the lower sideband which is excited.

The preceding analysis applies principally to the case of amplifier configurations. However, it is the relative gain $R \equiv G_s(L)/G(L)$ which determines whether the sideband will grow from noise in an oscillator configuration.

$$R \cong \mp \frac{\omega KL}{4\omega_s \sin 2\psi_{res}}\left\{1 \pm \frac{\Delta^2}{24}\left[\frac{K}{2\Delta K_\pm}\left(1+\frac{5}{3}\tan^2\psi_{res}\right)\right.\right.$$
$$\left.\left.\mp\frac{\delta K^2}{K^2}\left(\cos 2KL - \frac{\sin 2KL}{2KL}\right)\right]\right\}\left(\frac{\sin\Theta_\pm}{\Theta_\pm}\right)^2. \tag{6.45}$$

If the relative gain is greater than unity for nearly equal loss rates, then the sideband will grow.

REFERENCES

1. Kwan, T. J. T. (1980) Trapped particle dynamics and efficiency optimization in free-electron lasers. *Phys. Fluids*, **23**, 1857.
2. Kroll, N. M. and Rosenbluth, M. N. (1980) Sideband instabilities in trapped particle free-electron lasers, in *Physics of Quantum Electronics: Free-Electron Generators of Coherent Radiation*, Vol. 7 (ed. Jacobs, S. F., Pilloff, H. S., Sargent, M., Scully, M. O. and Sitzer, R.), Addison-Wesley, Reading, Massachusetts, p. 147.
3. Lin, A. T. (1981) Saturation of sideband instabilities in a free-electron laser, in *Physics of Quantum Electronics: Free-Electron Generators of Coherent Radiation*, Vol. 9 (ed. Jacobs, S. F., Pilloff, H. S., Sargent, M., Scully, M. O. and Spitzer, R.), Addison-Wesley, Reading, Massachusetts, p. 867.
4. Freund, H. P., Sprangle, P. and Tang, C. M. (1982) Effect of fluctuating space-charge fields on sideband instabilities in free-electron lasers. *Phys. Rev. A.*, **25**, 3121.
5. Tang, C. M. and Sprangle, P. (1984) Semi-analytic formulation of the two-dimensional pulse propagation in the free-electron laser oscillator, in *Free-Electron Generators of Coherent Radiation* (ed. Brau, C. A., Jacobs, S. F. and Scully, M. O.), Proc. SPIE 453, pp. 11–24.
6. Goldstein, J. C. (1984) Evolution of long pulses in a tapered wiggler free-electron laser, in *Free-Electron Generators of Coherent Radiation* (ed. Brau, C. A., Jacobs, S. F. and Scully, M. O.), Proc. SPIE 453, p. 2.
7. Davidson, R. C. and McMullin, W. A. (1984) Detrapping stochastic particle instability for electron motion in combined longitudinal wiggler and radiation fields. *Phys. Rev. A*, **29**, 791.
8. Quimby, D. C., Slater, J. M. and Wilcoxon, J. P. (1985) Sideband suppression in free-electron lasers with multiple synchrotron periods. *IEEE J. Quantum Electron.*, **QE-21**, 979.
9. Davidson, R. C. (1986) Kinetic description of the sideband instability in a helical-wiggler free-electron laser. *Phys. Fluids*, **29**, 2689.
10. Colson, W. B. (1986) The trapped-particle instability in free-electron laser oscillators and amplifiers. *Nucl. Instr. Meth.*, **A250**, 168.
11. Davidson, R. C. and Wurtele, J. S. (1987) Single-particle analysis of the free-electron laser sideband instability for primary electromagnetic waves with constant phase and slowly varying phase. *Phys. Fluids*, **30**, 557.
12. Yee, F. G., Masud, J., Marshall, T. C. and Schlesinger, S. P. (1987) Power and sideband studies of a Raman free-electron laser. *Nucl. Instr. Meth.*, **A259**, 104.
13. Yu, S. S., Sharp, W. M., Fawley, W. M., Scharlemann, E. T., Sessler, A. M. and Sternbach, E. J. (1987) Waveguide suppression of the free-electron laser sideband instability. *Nucl. Instr. Meth.*, **A259**, 219.
14. Masud, J., Marshall, T. C., Schlesinger, S. P., Yee, F. G., Fawley, W. M., Scharlemann, E. T., Yu, S. S., Sessler, A. M. and Sternbach, E. J. (1987) Sideband control in a millimeter-wave free-electron laser. *Phys. Rev. Lett.*, **58**, 763.
15. Riyopoulos, S. and Tang, C. M. (1987) The free-electron laser sideband instability reconsidered. *Nucl. Instr. Meth.*, **A259**, 226.
16. Goldstein, J. C., Newnam, B. E. and Warren, R. W. (1988) Sideband suppression by an intracavity optical filter in the Los Alamos free-electron laser oscillator. *Nucl. Instr. Meth.*, **A272**, 150.
17. Warren, R. W. and Goldstein, J. C. (1988) The generation and suppression of synchrotron sidebands. *Nucl. Instr. Meth.*, **A272**, 155.
18. Sharp, W. M. and Yu, S. S. (1988) Validity of I-D free-electron laser sideband models. *Nucl. Instr. Meth.*, **A272**, 397.
19. Takeda, H. (1988) Analysis of sideband spectra from untapered undulator using analytical techniques. *Nucl. Instr. Meth.*, **A272**, 404.

20. Channell, P. J. (1988) Spectral evolution equation for the sideband instability in a free-electron laser oscillator. *Nucl. Instr. Meth.*, **A272**, 421.
21. Riyopoulos, S. and Tang, C. M. (1988) The structure of the sideband spectrum in the free-electron laser. *Phys. Fluids*, **31**, 1708.
22. Hafizi, B., Ting, A., Sprangle, P. and Tang, C. M. (1988) Development of sidebands in tapered and untapered free-electron lasers. *Phys. Rev. A*, **38**, 197.
23. Sollid, J. E., Feldman, D. W., Warren, R. W., Takeda, H., Gitomer, S. J., Johnson, W. J. and Stein, W. E. (1989) Sideband suppression in the Los Alamos free-electron laser using a Littrow grating. *Nucl. Instr. Meth.*, **A285**, 147.
24. Sollid, J. E., Feldman, D. W. and Warren, R. W. (1989) Sideband suppression for free-electron lasers. *Nucl. Instr. Meth.*, **A285**, 153.
25. Bhattacharjee, A., Cai, S. Y., Chang, S. P., Dodd, J. W. and Marshall, T. C. (1989) Sideband instabilities and optical guiding in a free-electron laser: experiment and theory. *Nucl. Instr. Meth.*, **A285**, 158.
26. Riyopoulos, S. (1990) Instability mechanisms in storage ring free-electron oscillators. *Nucl. Instr. Meth.*, **A296**, 485.
27. Sharp, W. M. and Yu, S. S. (1990) Two-dimensional Vlasov treatment of free-electron laser sidebands. *Phys. Fluids B*, **2**, 581.
28. Kishimoto, Y., Oda, H. and Shiho, M. (1990) Parasitic wave excitation by multimode coupling in a Raman-regime free-electron laser. *Phys. Rev. Lett.*, **65**, 851.
29. Sternbach, E. J. (1990) Coupled-wave theory of free-electron laser sidebands in a waveguide. *IEEE Trans. Plasma Sci.*, **PS-18**, 460.
30. Rosenbluth, M. N., Wong, H. V. and Moore, B. N. (1990) Sideband instabilities in free-electron lasers. *Phys. Fluids B*, **2**, 1635.
31. Elgin, J. N. (1991) Analysis of the sideband instability in the free-electron laser. *Phys. Rev. A*, **43**, 2514.

7

Coherent harmonic radiation

The electron-beam energy required for operation at a given frequency varies inversely as the square root of the harmonic number. For this reason, the mechanisms for the stimulated emission of harmonic radiation in free-electron lasers are of interest for many applications in which it is desired to reduce the beam energy requirement, and the concept has received widespread attention [1–30]. The principal difficulty with the interaction at the harmonics, however, is that the beam quality requirement associated with the tolerable axial energy spread of the beam increases with the harmonic number. In general, the free-electron laser resonance condition at the lth harmonic is $\omega \approx (k + lk_w)v_\parallel$; hence, thermal effects have a significant and deleterious impact on the interaction whenever $\Delta v_\parallel / v_\parallel \approx \mathrm{Im}\ k/(\mathrm{Re}\, k + lk_w)$. Because of this, great care in the design of the electron accelerator and beam optics is required in the design of harmonic free-electron lasers.

The mechanisms for coherent harmonic generation differ greatly between planar [1–3, 6–31] and helical [4, 5] wiggler configurations. In its most straightforward interpretation, the planar wiggler causes an oscillation in the axial velocity at a period one-half that of the wiggler period with an amplitude which depends upon the square of the wiggler-induced transverse velocity. This oscillation is the source of a resonant wave–particle interaction at the odd harmonics of the wiggler period. However, the strength of the harmonic interaction increases with the magnitude of the oscillation in the axial velocity which, in turn, depends upon the wiggler strength. As a result, the harmonic interaction in a planar wiggler is more sensitive to the wiggler strength than is the fundamental, and requires extremely larger-amplitude wigglers to be effective.

In contrast, the harmonic interaction in helical wiggler configurations does not depend upon an oscillatory axial velocity. In this case, the electron trajectories describe helices with uniform axial velocities, and the interaction is with a circularly polarized wave. Typically, the electron trajectory in a helical wiggler will vary in azimuth with $\theta \approx k_w z$. As a result, a sinusoidal wave which varies in phase as $\exp[ikz + il\theta - i\omega t]$, will resonate at the lth harmonic. For the case in which the drift tube acts as a waveguide in controlling the dispersion of the mode, this establishes a selection rule whereby the TE_{ln} and TM_{ln} modes resonate at the lth harmonic upshift. An important distinction between helical and planar wigglers in this regard is that the strength of the harmonic interaction is no more sensitive, in principle, to the wiggler amplitude than is the fundamental.

7.1 HELICAL WIGGLER CONFIGURATIONS

The linear gain in helical wiggler-based free-electron lasers can be treated using the stability analysis derived in Chapter 4 for realizable configurations in the presence of an axial magnetic field. Here, the discussion is confined to the analysis of the harmonic interaction in the low-gain regime [4]. It was shown in Chapter 4 that a selection rule exists whereby the azimuthal mode number in a cylindrical waveguide determines the harmonic number of the interaction. Specifically, the lth harmonic is resonant with the TE_{ln} or TM_{ln} modes.

7.1.1 Harmonic excitation in the low-gain regime

The small-signal gain for the various waveguide modes is given in equation (4.235), and can be used to determine the relative magnitudes of the various harmonics. The equation used to determine the ratio of the gain for the $TE_{l'n'}$ mode to that of the TE_{ln} mode for a cylindrical waveguide of radius R_g is given by

$$\frac{G_{l'n'}}{G_{ln}} \cong \frac{l'}{l}\frac{x_{l'n'}'^2}{x_{ln}'^2}\frac{x_{ln}'^2 - l^2}{x_{l'n'}'^2 - l'^2}\frac{J_l^2(x_{ln}')}{J_{l'}^2(x_{l'n'}')}\frac{J_{l'}'^2(\kappa_{l'n'}r_0)}{J_l'^2(\kappa_{ln}r_0)}, \tag{7.1}$$

where $J_l'(x_{ln}') = 0$, $\kappa_{ln}(\equiv x_{ln}'/R_g)$ is the cutoff wavenumber of the mode, and r_0 is the radius of the steady-state orbit. The ratio of the gain of the $TM_{l'n'}$ mode to that of the TM_{ln} mode is given similarly by

$$\frac{G_{l'n'}}{G_{ln}} \cong \frac{l'^3}{l^3}\frac{x_{l'n'}^2}{x_{ln}^2}\frac{J_l'^2(x_{ln})}{J_{l'}'^2(x_{l'n'})}\frac{J_{l'}^2(\kappa_{l'n'}r_0)}{J_l^2(\kappa_{ln}r_0)}, \tag{7.2}$$

where $J_l(x_{ln}) = 0$, and $\kappa_{ln}(\equiv x_{ln}/R_g)$ denotes the cutoff wavenumber of the TM mode. Analogous expressions can be derived for the ratios of the TE to TM modes or vice versa.

These expressions simplify considerably in the limit in which $|v_w/v_\parallel| \ll 1$ such that both $\kappa_{ln}r_0 \ll 1$ and $\kappa_{l'n'}r_0 \ll 1$. Since $k_w r_0 = |v_w/v_\parallel|$ for the steady-state trajectories, the ratio of the gain at the lth harmonic in the $TE_{l'n'}$ mode to the gain of the fundamental in the TE_{1n} mode is

$$\frac{G_{l'n'}}{G_{1n}} \cong \frac{l'}{(l'-1)!^2}\left(\frac{x_{l'n'}'}{2k_w R_g}\right)^{2l'-2}\left(\frac{v_w}{v_\parallel}\right)^{2l'-2}\frac{x_{l'n'}'^2}{x_{1n}'^2}\frac{x_{1n}'^2 - 1}{x_{1n'}'^2 - l'^2}\frac{J_1^2(x_{1n}')}{J_{l'}^2(x_{l'n'}')}. \tag{7.3}$$

The corresponding ratio for the TM modes is

$$\frac{G_{l'n'}}{G_{1n}} \cong \frac{l'}{(l'-1)!^2}\frac{x_{1n}^2}{x_{l'n'}^2}\left(\frac{x_{l'n'}}{2k_w R_g}\right)^{2l'}\left(\frac{v_w}{v_\parallel}\right)^{2l'}\frac{J_1'^2(x_{1n})}{J_{l'}'^2(x_{l'n'})}. \tag{7.4}$$

In this limit it is clear that the gain decreases rapidly with the harmonic number for a sufficiently weak wiggler. However, it is misleading to assume that this conclusion holds in general. The reason for this is that the argument of the Bessel functions, $\kappa_{ln}r_0$, is proportional to the ratio of the beam radius to the waveguide radius, and this can be appreciable even for relatively weak wigglers. Hence, it is possible to design free-electron lasers based upon helical wigglers in which the gain at the harmonics is comparable to that of the fundamental. A nonlinear simulation of such an example follows.

7.1.2 Numerical simulation of harmonic excitation

The nonlinear aspects of the harmonic interaction in a helical wiggler/axial magnetic field configuration can be investigated using the nonlinear formalism described in Chapter 5 in cylindrical waveguide geometry. In this case, the dynamical equations for the waveguide modes are given by equations (5.110) and (5.111), which are integrated for each waveguide mode in conjunction with the Lorentz force equations (5.126–28) and (5.130–32) for an ensemble of electrons. This formalism is capable of the treatment of an arbitrary harmonic interaction in both the linear and nonlinear regimes.

An example of a second harmonic interaction with the TE_{21} mode is illustrated in Fig. 7.1 in which we plot the efficiency as a function of normalized angular frequency [5]. The parameters of this case are similar to those examined for the fundamental TE_{11} interaction in Fig. 5.18 for Group I orbit parameters. Specifically, this example deals with the propagation of an electron beam with an energy of $V_b = 250\,\mathrm{keV}$, a current of $I_b = 35$ A, and an initial beam radius of $R_b = 0.155\,\mathrm{cm}$ through a waveguide with a radius of $R_g = 0.28\,\mathrm{cm}$ in the presence of a wiggler field with a magnitude of $B_w = 2.0\,\mathrm{kG}$ and a period of $\lambda_w = 1.175\,\mathrm{cm}$ and an axial solenoidal field of $B_0 = 1.3\,\mathrm{kG}$. The entry taper region is $N_w = 10$ wiggler periods in length. Note that the wiggler parameter $a_w (\equiv eA_w/m_e c^2) \approx 0.22$

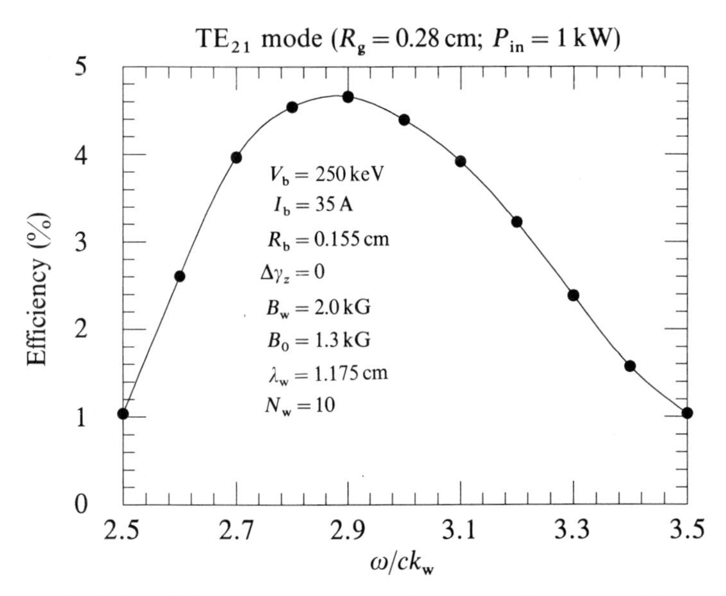

Fig. 7.1 Graph of the efficiency as a function of frequency for a second-harmonic TE_{21} mode interaction.

for this example. The only difference with the aforementioned TE_{11} mode analysis is that the waveguide radius has been increased in order to move the intersection points between the beam resonance line ($\omega \approx (k + 2k_w)v_\parallel$) and the TE_{21} mode dispersion curve close enough together to merge into a single broad gain band. As shown in the figure, a peak efficiency $\eta \approx 4.6\%$ is found at approximately $\omega/ck_w \approx 2.9$ (74 GHz). While the maximum efficiency of the second-harmonic TE_{21} mode interaction is substantially less than that found for the fundamental TE_{11} mode ($\eta \approx 21.4\%$), the growth rate of the TE_{21} mode at this frequency is approximately $\Gamma_{21}/k_w \approx 0.025$, which is comparable to that found for the fundamental TE_{11} mode interaction ($\Gamma_{11}/k_w \approx 0.029$ at $\omega/ck_w \approx 1.3$). In addition, observe that while the bandwidth of the second-harmonic interaction $\Delta\omega/\omega \approx 43\%$ is comparable to that of the fundamental, the frequency has been doubled.

7.2 PLANAR WIGGLER CONFIGURATIONS

The mathematical formalism necessary to treat harmonic excitation in planar wiggler configurations in both the linear and nonlinear regimes has been developed in Chapters 4 and 5. The linear theory of the interaction is described by the dispersion equation given in equation (4.180). This equation was derived subject to an idealized one-dimensional model, and is capable, principally, of treating the excitation of odd harmonics. However, beam thermal effects have been included in the analysis and the effect of an axial energy spread on the linear gain at the fundamental and harmonics can be studied. A nonlinear formulation of the planar wiggler interaction in three dimensions has also been presented in Chapter 5 (see equations (5.152–61)) for the case of a beam propagating through a rectangular waveguide in the presence of a planar wiggler generated by a magnet stack with parabolic pole faces for enhanced beam confinement. This formulation is capable of treating the fundamental and all harmonics, and will be applied in this chapter to the study of harmonic emission.

7.2.1 Harmonic excitation in the linear regime

The general dispersion equation which describes the linear growth rate for the fundamental and harmonics [31] in a planar wiggler has been derived in Chapter 4 (see equation (4.180)). This dispersion equation can be solved numerically for a case which illustrates the relative growth of the fundamental and the third harmonic. In general, strong harmonic amplification requires a relatively large oscillation in the axial velocity; hence, the growth rate at the harmonics increases rapidly with Ω_w/ck_w. Indeed, the growth rate at the harmonics as predicted by equation (4.180) can be larger than that at the fundamental when Ω_w/ck_w exceeds unity. However, this is an unjustifiable conclusion based upon the present type of formulation. It is important to bear in mind that the analysis cannot be applied for arbitrarily large values of this parameter because (1) the idealized one-dimensional model breaks down when the displacement of the electrons from the plane of symmetry becomes large, and (2) the Lagrangian time coordinate has been integrated in equation (4.143) under the assumption that the $v_w \ll v_\parallel$. Therefore, the analysis of cases in which Ω_w/ck_w is greater than unity requires a fully three-dimensional analysis. This will be discussed within the context of the nonlinear formulation of the interaction in planar wigglers derived in Chapter 5. However, in the analysis of the linear-dispersion equation, we restrict the numerical analysis to an example for which $\Omega_w/ck_w = 1$. This is a physically interesting case which is at the fringe of the range of validity of the formulation, and will serve to illustrate clearly the relationship of the harmonics to the fundamental. In addition, we shall assume that $\gamma_0 = 2.957$ and $\omega_b/ck_w = 0.1$. It should also be remarked that in order for the thermal effects to result in substantial growth at the even harmonics, $\Delta P/p_0$ must be of the order of unity. Since this is unreasonably high for any well-designed experiment, we shall ignore this effect henceforth, and concentrate on the emission at the odd harmonics.

The magnitude of the growth rate is plotted as a function of frequency in Fig. 7.2 for the fundamental and the third harmonic. The fundamental exhibits a peak growth rate of $|\mathrm{Im}\, k|/k_w \approx 0.065$ at a normalized frequency of $\omega/ck_w \approx 1.55$. In contrast, the magnitude of the growth rate at the third harmonic is $|\mathrm{Im}\, k|/k_w \approx 0.012$ at a frequency of $\omega/ck_w \approx 4.80$. Observe that both the magnitude and bandwidth of the harmonic are reduced relative to the fundamental.

The effect of the thermal spread on the fundamental and the third harmonic is shown in Fig. 7.3. Here we plot the normalized growth rate (defined as the ratio of the maximum growth rate for a specific value of $\Delta P/p_0$ to the maximum growth rate for $\Delta P/p_0 = 0$) as function of $\Delta P/p_0$ for the fundamental and third harmonic. Observe that thermal effects are expected to become important on

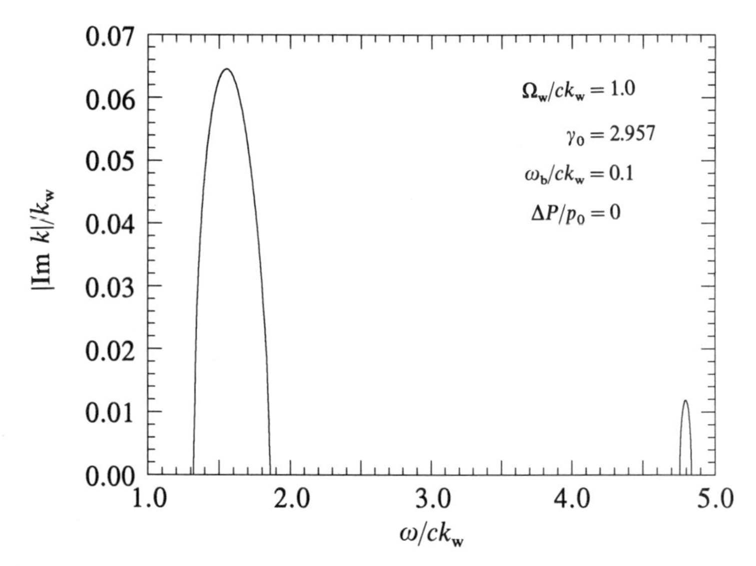

Fig. 7.2 Graph of the linear growth rates of the fundamental and third harmonics in a planar wiggler with $\Omega_w/ck_w = 1.0$, $\gamma_0 = 2.957$ and $\omega_b/\gamma_0^{1/2} ck_w = 0.1$ for an ideal beam.

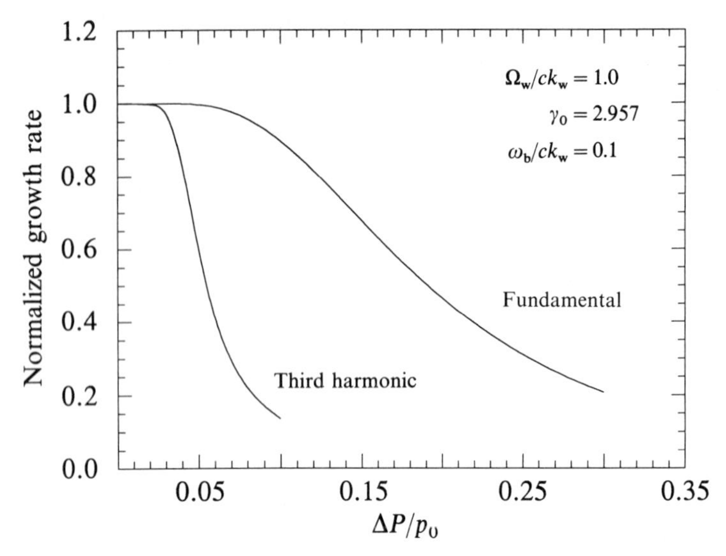

Fig. 7.3 Variation in the normalized growth rate with the axial momentum spread for the fundamental and third-harmonic for parameters consistent with Fig. 7.2.

the fundamental when $\Delta v_\parallel/v_\parallel \approx |\mathrm{Im}\,k|/(k_w + \mathrm{Re}\,k) \approx 0.025$, which corresponds to $\Delta P/p_0 \approx 22\%$ (recall that $\Delta v_\parallel/v_\parallel \approx \Delta P^2/2p_0^2$). This is in substantial agreement with the results shown in figure. For the case of the harmonics, thermal effects are expected to become important at a much reduced thermal spread; specifically, when $\Delta v_\parallel/v_\parallel \approx |\mathrm{Im}\,k|/[(1 + 2l)k_w + \mathrm{Re}\,k]$. For the third harmonic in the present example $\Delta v_\parallel/v_\parallel \approx |\mathrm{Im}\,k|/(3k_w + \mathrm{Re}\,k) \approx 0.0015$. This corresponds to $\Delta P/p_0 \approx 5.5\%$, which is also in good agreement with the calculation.

7.2.2 Numerical simulation of harmonic excitation

The nonlinear aspects of the harmonic interaction can be investigated using the formalism described in Chapter 5 for planar wigglers in a rectangular waveguide geometry [15, 16, 22]. In this case, the dynamical equations for the waveguide modes are given by equations (5.154) and (5.155) which are integrated for each waveguide mode in conjunction with the Lorentz force equations (5.156–61) for an ensemble of electrons. The wiggler model employed in this formulation is characterized by tapered pole faces (see equation (2.3)) to provide for enhanced beam confinement. Observe that no approximations are made in the orbit integrations; hence, unlike the aforementioned linear-stability analysis, the formalism is valid for arbitrarily large wiggler amplitudes in which $\Omega_w/ck_w > 1$.

The first example under consideration corresponds to parameters similar to those investigated for the fundamental interaction in Chapter 5. In particular, that of a 3.3 MeV, 100 A electron beam with an initial radius of 0.2 cm propagating through a rectangular waveguide with dimensions $a = 10.0$ cm and $b = 3.0$ cm in the presence of a wiggler with an amplitude $B_w = 4.2\,\text{kG}$ and a period $\lambda_w = 9.8\,\text{cm}$. The entry taper region into the wiggler is assumed to be $N_w = 10$ wiggler periods in length. This case represents an extremely strong wiggler parameter in which $a_w(\equiv eA_w/m_ec^2) \approx 3.84$. Amplification of a signal is found at frequencies in the vicinity of the odd harmonics. A spectrum of the output efficiency, η, of the TE_{01} mode is shown in Fig. 7.4 as a function of frequency for the fundamental through the seventh harmonic for an ideal beam with a vanishing axial energy spread. The doublet indicated in the neighbourhood of the fundamental ($\omega/ck_w \approx 7$) corresponds to the upper and lower intersection points between the TE_{01} mode dispersion curve and the beam resonance line ($\omega \approx (k + k_w)v_\parallel$). It is evident from the figure that there is a sharp decline in the efficiency between the fundamental and the third harmonic. However, the subsequent decrease in the efficiency with harmonic number is rather slow, and substantial power may be found at the higher harmonics. In addition, the decrease in the growth rates with harmonic number is rather slow as well.

It is difficult to estimate the precise growth rates at each harmonic because of (1) variations in the launching loss associated with the various harmonics, (2) the oscillations in the power and gain with a period half of $\lambda_w/2$ due to the lower beat wave, and (3) the betatron oscillations on the beam due to the wiggler gradients. However, the distance to saturation, z_{sat}, for a fixed input signal of 6 kW provides a measure of the average growth rate. For example, the peak growth rate is obtained at $\omega/ck_w \approx 7.25$ (22.2 GHz) for the fundamental (upper

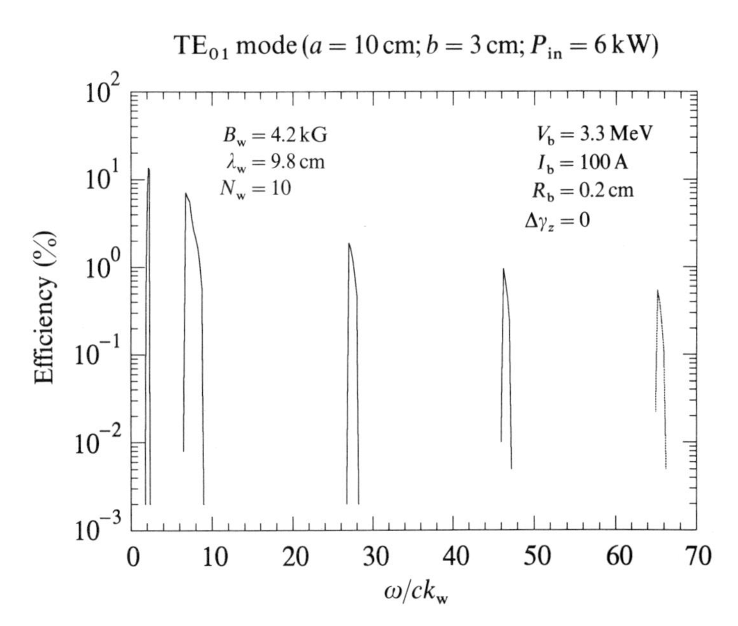

Fig. 7.4 Plot of the efficiency versus frequency at the fundamental and third, fifth and seventh harmonics.

intersection), and saturation occurs at $k_w z_{sat} \approx 184$ with an efficiency $\eta \approx 5.58\%$. Peak growth is found for $\omega/ck_w \approx 27.25$ (83.4 GHz) at the third harmonic, and saturation is found to occur at $k_w z_{sat} \approx 194$ with an efficiency $\eta \approx 1.61\%$. In view of the combination of the comparable saturation length and lower efficiency at the harmonic, it is clear that the gain per unit length at the harmonic is a substantial fraction of that at the fundamental. A similar situation is found at the fifth harmonic which displays a peak efficiency at $\omega/ck_w \approx 46.5$ (142 GHz), and for which saturation is found for $k_w z_{sat} \approx 205$ at an efficiency $\eta \approx 0.69\%$. The peak growth rate at the seventh harmonic is found for $k_w z_{sat} \approx 172$ at a frequency of $\omega/ck_w \approx 65.5$ (201 GHz) and an efficiency of $\eta \approx 0.39\%$.

The effect of an axial energy spread on the beam is illustrated in Fig. 7.5, in which we plot the normalized efficiency η/η_0 (where η_0 is defined as the efficiency in the limit in which $\Delta\gamma_z = 0$) as a function of the axial energy spread $\Delta\gamma_z/\gamma_0$. Recall that in the formulation under consideration, the electron-beam distribution is assumed to be monoenergetic but with a pitch-angle spread defined in equation (5.120). The principal conclusion to be drawn from the figure, as expected, is that the harmonic emission is far more sensitive to the thermal spread than the fundamental. While the efficiency decreases by an order of magnitude for $\Delta\gamma_z/\gamma_0 \approx 2\%$ at the fundamental, a corresponding decrease in the efficiency occurs for $\Delta\gamma_z/\gamma_0 \approx 0.7\%$ at the third harmonic. The decrease is even more rapid at still higher harmonics. The growth rates are also more sensitive to the thermal spread at the harmonics. The saturation length increases from $k_w z_{sat} \approx 184$ to $k_w z_{sat} \approx 194$

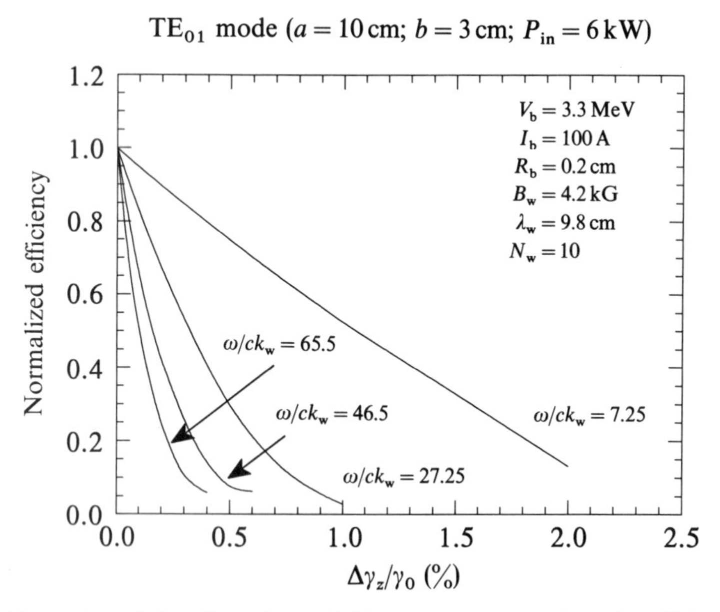

Fig. 7.5 Illustration of the effect of an axial beam energy spread on the efficiency at the fundamental and the harmonics.

at $\omega/ck_w = 7.25$ (fundamental) as the thermal spread increases to $\Delta\gamma_z/\gamma_0 = 1\%$. In contrast, the saturation length increases from $k_w z_{sat} \approx 194$ to $k_w z_{sat} \approx 232$ at $\omega/ck_w = 27.25$ (third harmonic) as the axial energy spread increases to $\Delta\gamma_z/\gamma_0 = 0.5\%$. At $\omega/ck_w = 65.5$ (seventh harmonic) the saturation length increases from $k_w z_{sat} \approx 172$ to $k_w z_{sat} \approx 224$ as the thermal spread increases to $\Delta\gamma_z/\gamma_0 = 0.27\%$. These conclusions dramatically illustrate the importance of beam quality on coherent harmonic emission.

The effect of betatron oscillations

The next example to be considered is similar to that shown in Figs 7.4 and 7.5. The principal difference is the thickness of the electron beam upon injection into the wiggler. This beam thickness has a significant impact on the betatron oscillations experienced by the beam which, in the case of the wiggler model, also has an impact on the harmonic spectrum. The specific parameters deal with the propagation of an electron beam with an energy of 3.5 MeV, a current of 800 A, and an initial beam radius of 1.0 cm through a waveguide with dimensions $a = 9.8$ cm and $b = 2.9$ cm in the presence of a wiggler with an amplitude $B_w = 3.72$ kG. The wiggler period and entry taper length are as previously with $\lambda_w = 9.8$ cm and $N_w = 5$. The mode of interest again is the TE_{01} mode, but with an input power level of 50 kW. This choice of parameters corresponds closely with the case examined in Chapter 5 for the fundamental interaction.

The spectrum of the saturation efficiency which results is shown in Fig. 7.6 for the fundamental through to the fifth harmonic. As expected, emission is found at the fundamental ($\omega/ck_w \approx 11.3$) and the odd harmonics ($\omega/ck_w \approx 40.3$ and 66.5). However, emission is also found in the vicinity of the even harmonics in the form of a frequency doublet. As shown in the figure, emission occurs for $\omega/ck_w \approx 22.0$ and 29.5, corresponding to the second harmonic, and at $\omega/ck_w \approx 52.3$ and 56.8, corresponding to the fourth harmonic. The efficiency at the second harmonic ($\eta \approx 0.27\%$) exceeds that found at the third harmonic ($\eta \approx 0.11\%$). As discussed previously, the efficiency decreases sharply between the fundamental and the nearest harmonic, but falls off slowly with harmonic number thereafter.

The source of the doublet emission in the vicinity of the even harmonics is the wiggler-plane focusing due to the parabolic pole face model for the wiggler. The spatial variation of the field in the wiggle direction induces a modified betatron oscillation in the x-component (i.e. the direction of the wiggler-induced transverse oscillation) of the velocity which is the source of oscillations in the axial velocity with periods $k \pm K_\beta$, where K_β denotes the modified betatron wavenumber. It is this oscillation which drives the even harmonic emission. It should be remarked that emission at the even harmonics can be expected to occur for any wiggler configuration which exhibits an inhomogeneity in the direction of the wiggler-induced transverse oscillation.

The modified betatron oscillations in a planar wiggler with parabolic pole faces has been discussed [16] to lowest order in the time domain in Chapter 2

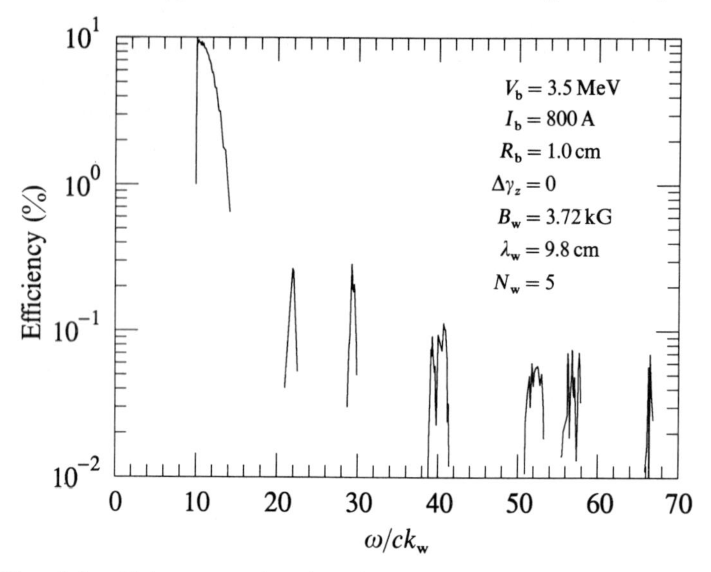

Fig. 7.6 Plot of the efficiency as a function of frequency at the fundamental through fifth harmonics. The doublets centred at $\omega/ck_w \approx 25$ and 55 correspond to emission at the second and fourth harmonics.

(see equation (2.143)). For the purposes of the present discussion, we treat the betatron oscillation in the spatial domain, and write the approximate orbit equations in the x–z-plane in the form

$$\frac{d}{dz}v_x = \Omega_w \cosh\left(\frac{k_w x}{\sqrt{2}}\right)\cos k_w z, \tag{7.5}$$

and

$$\frac{d}{dz}v_z = -\Omega_w \frac{v_x}{v_z}\cosh\left(\frac{k_w x}{\sqrt{2}}\right)\cos k_w z, \tag{7.6}$$

where $\Omega_w \equiv eB_w/\gamma m_e c$, and the y-variations are ignored. Solutions can be found which are of the form

$$x = x_\beta \cos K_\beta z - x_w \cos k_w z, \tag{7.7}$$

and

$$v_x = -x_\beta K_\beta v_z \sin K_\beta z + x_w k_w v_z \sin k_w z, \tag{7.8}$$

for which the axial velocity varies as

$$v_z \cong v_\parallel\left(1 + \frac{v_w^2}{4v_\parallel^2}\sin 2k_w z + \frac{v_\beta v_w}{4v_\parallel^2}[\cos(k_w + K_\beta)z - \cos(k_w - K_\beta)z]\right), \tag{7.9}$$

where $v_{\parallel}$ is the bulk axial velocity of an electron in a planar wiggler (2.105), x_w and v_w describe the amplitude of the fundamental wiggler-induced oscillation, and x_β and $v_\beta(\equiv -x_\beta K_\beta v_{\parallel})$ describe the amplitude of the modified betatron oscillations. Under the assumption that $|k_w x| \ll 1$, substitution of this solution into the orbit equations shows that the modified-betatron wavenumber is

$$K_\beta^2 \cong \frac{\Omega_w^2}{4v_{\parallel}^2}\left(1 + \frac{3}{16}\frac{v_w^2}{v_{\parallel}^2} + \frac{1}{8}k_w^2 x_\beta^2\right). \tag{7.10}$$

The analytic expression also predicts oscillations at $(2k_w \pm K_\beta)$ and $(k_w \pm 2K_\beta)$, although at lower amplitudes. The inhomogeneities in the y-direction, characteristic of the magnet itself, introduce an additional betatron oscillation which does not couple strongly to the axial velocity or the harmonic interaction.

In order to verify the form of this solution for the modified betatron oscillations, the single-particle equations of motion have been integrated numerically in z under the assumption of the same input taper as for the complete nonlinear analysis. The resulting velocities were then Fourier transformed over the uniform wiggler portion of their motion (i.e. $z > N_w \lambda_w$). In Figs 7.7 and 7.8 the results of the Fourier transforms of the x-components (a) and z-components (b) of the velocity are shown for initial conditions in which $y_0 = 0$ and $x_0 = 0$ and 0.3 cm, respectively. The different harmonic structure predicted by the analytic theory is contained in these spectra. Observe that the amplitude of the modified betatron doublet at $(k_w \pm K_\beta)$ in the axial velocity is relatively independent of the initial y-position of the electron. In addition, the amplitude of the doublet increases with the initial x-position. (N.B. The initial x-position is not identical with x_β.) This accounts for the lack of emission at the even harmonics in the previous example for which the initial beam radius was only 0.2 cm.

A comparison between the analytic approximation and the numerical results for the wavenumber of the modified oscillation is shown in Fig. 7.9 as a function of x_0 for $y_0 = 0$. It is evident that the analytic approximation holds for $x_0 \leqslant 0.35$ cm, after which it diverges rapidly from the numerical solution. As shown in Fig. 7.10, the magnitude of the velocity of the modified betatron oscillation increases in an approximately linear fashion with x_0.

Two difficulties associated with the design of a harmonic free-electron laser are (1) the suppression of the fundamental, and (2) the sensitivity of the harmonic interaction to the axial energy spread of the beam. One method of suppression of the fundamental is applicable when waveguide effects upon the wave dispersion are important is to design the system in such a manner that there is no intersection between the waveguide dispersion curve and the beam-resonance line $(\omega \approx (k + k_w)v_{\parallel})$. This allows the harmonic interactions to dominate. The beam-quality constraint can be addressed by the use of a tapered wiggler, in order to enhance the efficiency possible for a given axial energy spread.

In order to address these issues, we now consider the case of a third-harmonic interaction in which the fundamental resonance is below cutoff and, therefore, cannot interact with the beam. Further, this example deals with a single mode

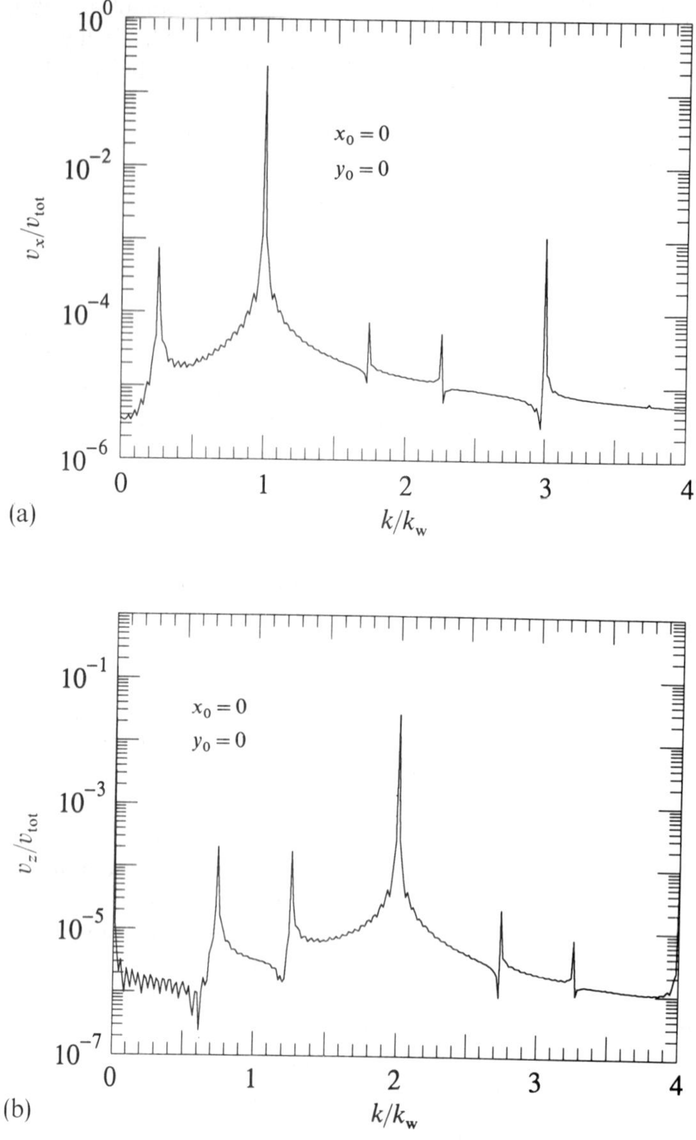

Fig. 7.7 Fourier decomposition of the x- and z-components of the electron velocities (normalized to the magnitude of the total velocity) for an electron which enters the wiggler at $x(z = 0) = 0$ and $y(z = 0) = 0$.

in which only the TE_{01} mode is resonant. The electron beam for this example is characterized by an energy of 55 keV, a current of 15 A, and an initial radius of 0.2 cm. The wiggler has an amplitude of 990 G, a period of 3.0 cm, and an entry taper region which is five wiggler periods in length. The waveguide dimensions

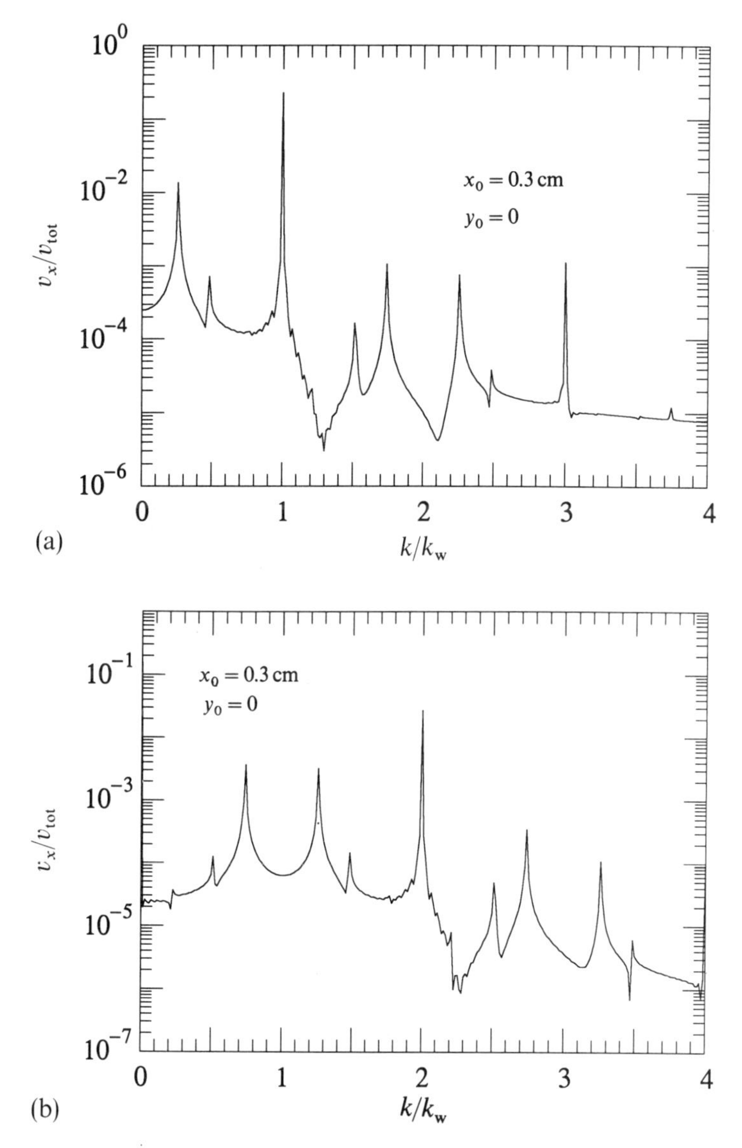

Fig. 7.8 Fourier decomposition of the x- and z-components of the electron velocities (normalized to the magnitude of the total velocity) for an electron which enters the wiggler at $x(z = 0) = 0.3$ cm and $y(z = 0) = 0$.

are $a = 3.2$ cm and $b = 1.58$ cm. We first consider the case of a uniform wiggler and investigate the sensitivity of the interaction to beam thermal effects. The spatial growth of a 3 kW injected signal at a frequency of 15.74 GHz is shown in Fig. 7.11 for the case of a beam with $\Delta\gamma_z = 0$. Saturation occurs with an efficiency of approximately 2.7%. This gives a net gain of 0.12 dB/cm over a total interaction

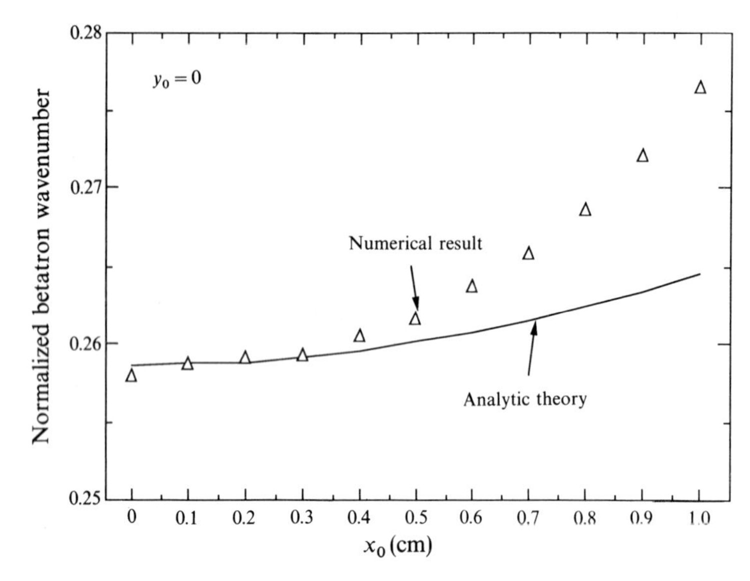

Fig. 7.9 Scaling of the modified betatron wavenumber with the initial x-coordinate of the electron.

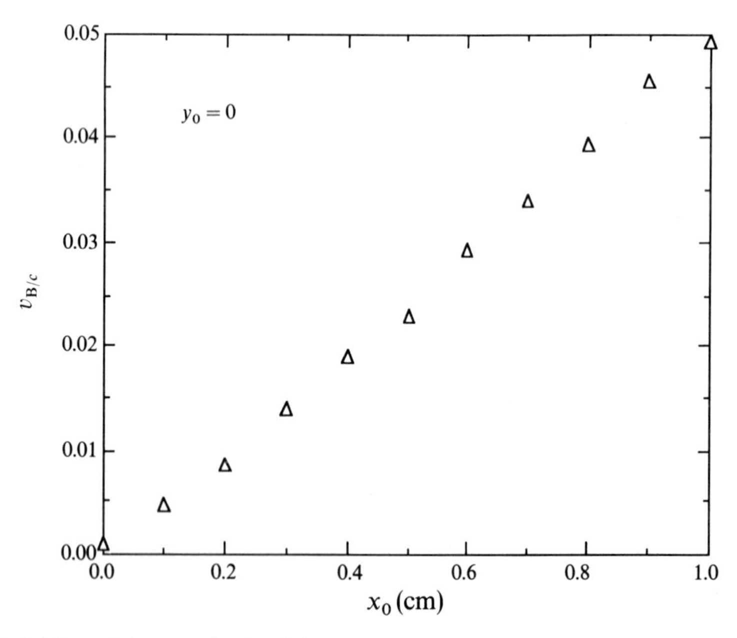

Fig. 7.10 Scaling of the magnitude of the velocity of the modified betatron oscillation with the initial x-coordinate of the electron.

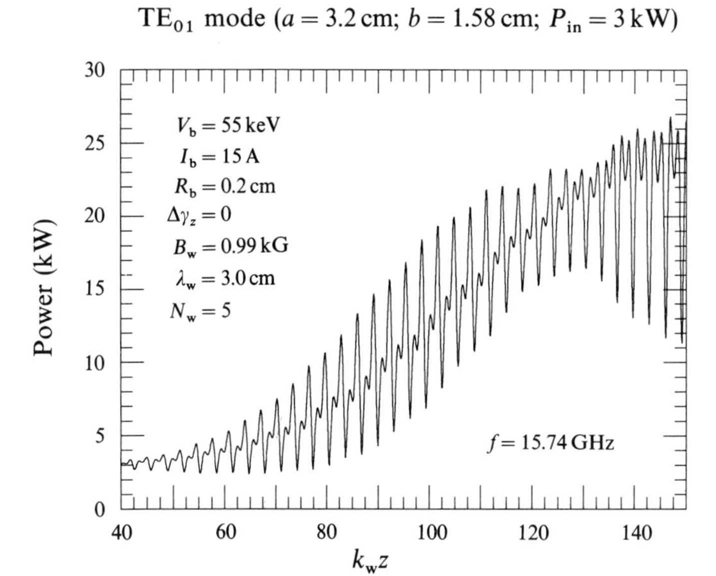

Fig. 7.11 The evolution of the power in the TE_{01} mode for a third-harmonic interaction.

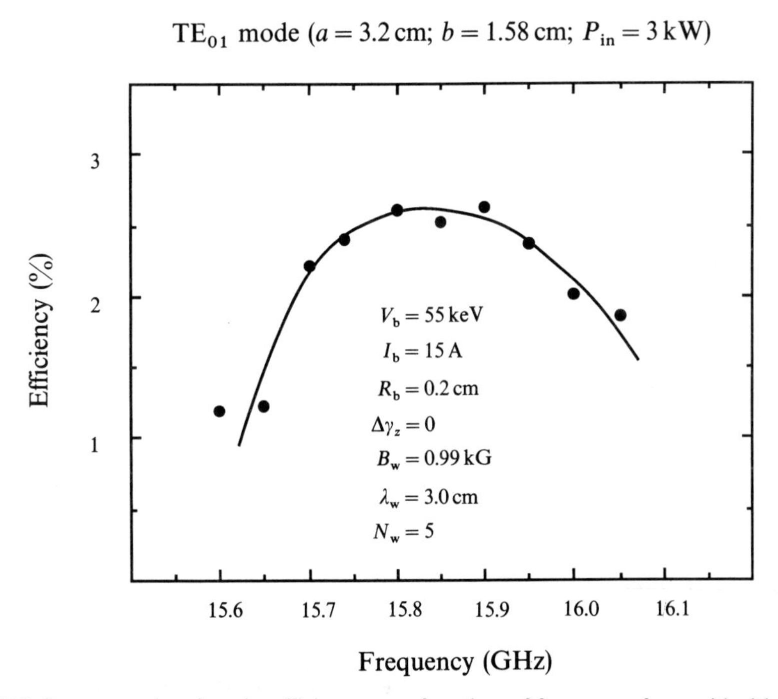

Fig. 7.12 Spectrum showing the efficiency as a function of frequency for an ideal beam.

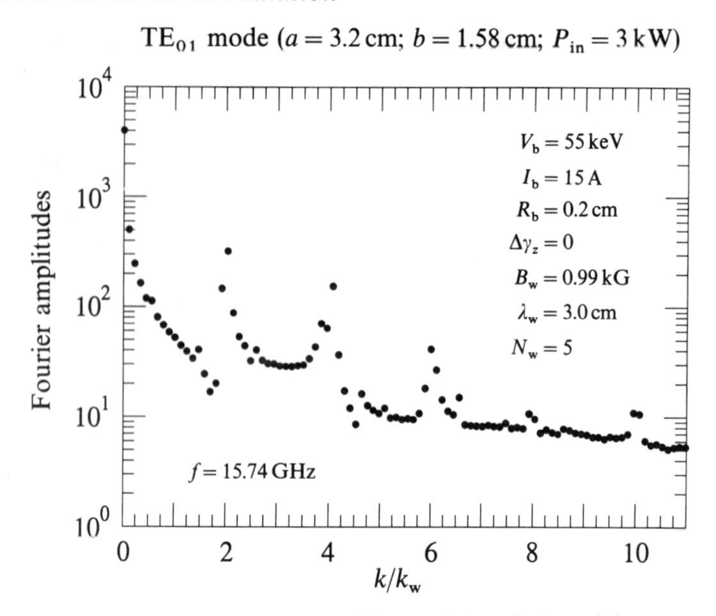

Fig. 7.13 The Fourier spectrum of the spatial evolution of the power.

length (i.e. including the entry taper region) of about 70 cm. This frequency corresponds to the central peak of the interaction spectrum. As shown in Fig. 7.12, resonance occurs from approximately 15.6 GHz through 16.1 GHz with a peak efficiency of the order of 2.8%.

It is evident from Fig. 7.11 that a substantial modulation is superimposed on the bulk growth of the signal. A Fourier transform of this spatial variation is shown in Fig. 7.13, and the modulation is evident with wavenumbers of $2k_w$, $4k_w$, $6k_w$, $8k_w$, $10k_w$, etc. This modulation is due to the effect of the lower beat wave in planar wiggler configurations which results in oscillations with a wavenumber of $2k_w$ at the fundamental. The interaction is similar at the harmonics, and the modulation at $4k_w$, $6k_w$, etc., derives from higher-order oscillations to the wiggler-induced velocity. Observe that such modulation of the signal at the even harmonics of the wiggler period occur, if at all, at a much lower level in the case of a helical wiggler because the symmetry acts to suppress the lower beat wave.

Returning to the question of the sensitivity of the interaction to thermal effects, we plot the decrease in the efficiency with increasing thermal (in this case through a pitch angle) spread in Fig. 7.14. As shown in the figure, the efficiency decreases by a factor of two as $\Delta p_{\|}/p_0$ increases to a value of about 0.045%, corresponding to a value of $\Delta\gamma_z/\gamma_0 = 0.01\%$. This is extremely difficult (if not impossible) to achieve given the current state of the art in electron-beam technology, and motivates our study of the effect of wiggler taper on the harmonic interaction.

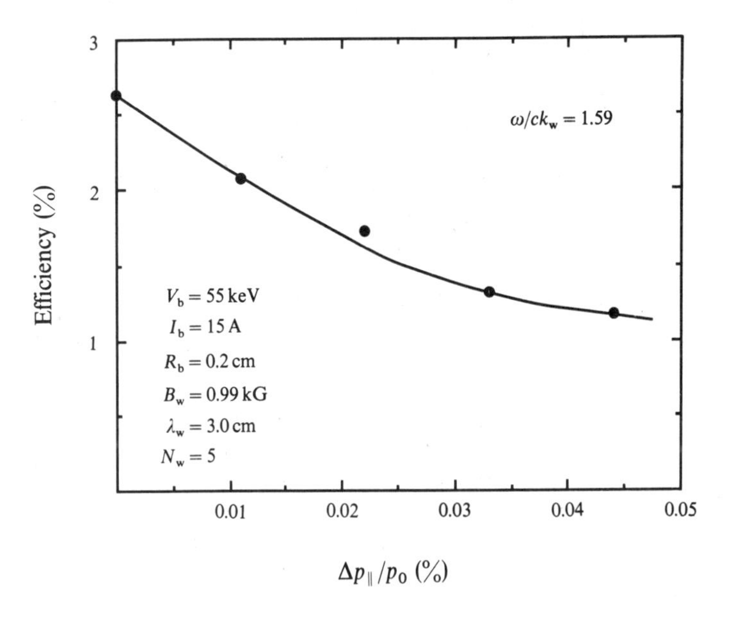

Fig. 7.14 Graph showing the decrease in efficiency with the increase in thermal spread.

Efficiency enhancement with a tapered wiggler

The effect of a tapered wiggler is shown in Fig. 7.15 in which we plot the maximum efficiency (i.e. if the field is tapered to zero) as a function of the scale length of the taper in the absence of thermal effects. As shown in the figure, the maximum obtainable efficiency is approximately 23% which is almost an order of magnitude greater than that found for the uniform wiggler. In addition, the optimal range for the taper itself is $0.0005 < |\varepsilon_w| < 0.0009$. The optimal range for the taper is governed by the requirements that the taper be rapid enough to maintain the axial beam velocity required for the wave–particle resonance but not so rapid that the beam becomes detrapped. By way of comparison, $\varepsilon_w = -0.0078$ in the experiment at LLNL described in Chapter 5, and simulations of the experiment indicate that the maximum extraction efficiency would occur for $\varepsilon_w \cong -0.002$. It appears, therefore, that harmonic experiments require slower tapers than at the fundamental. The reason for this is that the ponderomotive potential is directly proportional to the normalized wiggler amplitude $(\Omega_w/k_w c)$ for the fundamental, but varies as the cube of this factor at the third harmonic. As a result, the ponderomotive potential tends to be smaller at the harmonics and the effect of a given wiggler taper is larger. Indeed, the upper bound on the effective taper

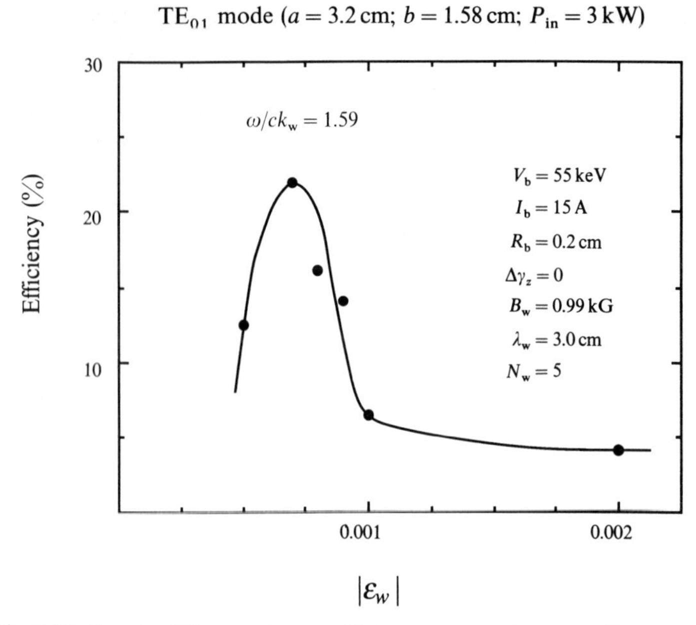

Fig. 7.15 Graph of the maximum efficiency versus the normalized taper.

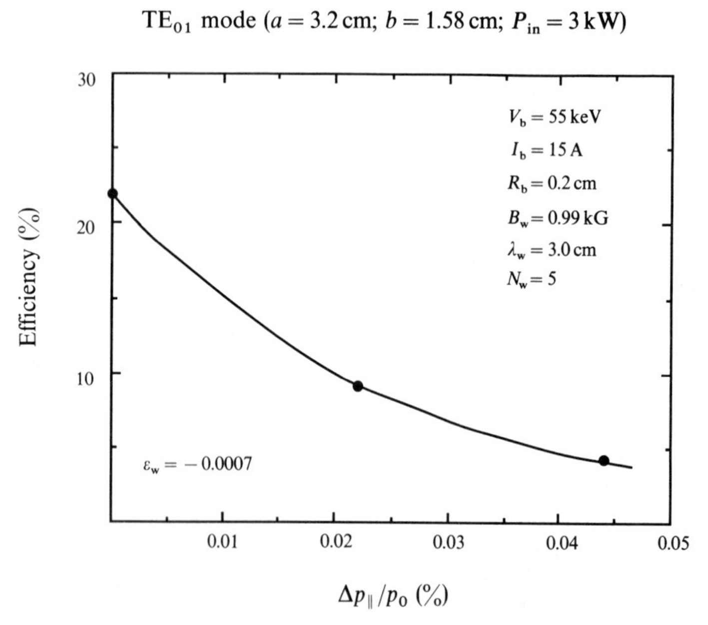

Fig. 7.16 Variation in the maximum efficiency of a third-harmonic tapered wiggler interaction with thermal effects.

shown in the figure for $\varepsilon_w \cong -0.001$ occurs because the effect of the taper has overwhelmed the ponderomotive force, and the particles have become detrapped.

The effect of an axial energy spread upon the third-harmonic tapered wiggler interaction is shown in Fig. 7.16, in which we plot the efficiency as a function of the axial momentum spread. As in the case of a uniform wiggler, the efficiency falls by approximately 60% as the axial momentum spread increases to 0.05%. However, the general levels of efficiency to be obtained for a comparable thermal spread are much greater than for the uniform wiggler.

7.2.3 The periodic position interaction

The conventional harmonic interaction in planar-wiggler free-electron lasers, as described above, occurs at the odd harmonics of the resonance frequency. These velocity harmonics are present even for ideal wigglers with perfect beam injection, and give rise to the periodic velocity instability of the free-electron laser. The even harmonic interaction has also been discussed, and can result from (1) betatron oscillations, and (2) beam-energy spread. There is an additional mechanism, however, which results in excitation of even harmonics without additional periodicity in the electron velocity. Rather, it depends on a synchronism in the electron position with respect to an antisymmetric radiation field. The interaction can occur with either the transverse or axial electric field. The transverse field is required to be odd in the direction of the wiggle motion with respect to the electron beam axis, and the axial field is required to be even for the respective interactions to occur. For a second-harmonic interaction, the radiation goes through two cycles as the electron beam traverses one wiggler period. For illustrative purposes, we consider this type of interaction at long wavelengths in which the electron beam interacts with modes supported by a rectangular waveguide. However, the fundamental process is also possible at optical wavelengths [11].

A brief discussion of both periodic position interaction types follows. For the transverse interaction, the on-axis electric field of the radiation mode is zero, and the field peaks off-axis. Considering only the central part of the beam, the essentials of the transverse interaction are shown in Fig. 7.17 in which the electron motion is greatly exaggerated and the transverse profile of the field is included (in this case, the TE_{11} rectangular waveguide mode). As seen in the figure, the electron will always be in either a decelerating or a zero electric field. Although a particle displaced from the horizontal centre of the beam will be in an accelerating field for a portion of the time, the bulk of the beam will be in a decelerating field most of the time, leading to a net amplification of the radiation. The axial interaction is also represented in Fig. 7.17, again for the central part of an on-axis beam. The transverse profile in this case represents the axial field of the TM_{11} mode. Here, even the central particle sees both an accelerating and a decelerating electric field. The particle is in a decelerating field on-axis where the field is at its maximum and the axial electron velocity is at a minimum and in an accelerating field off-axis where the field is reduced and the axial velocity is at a maximum. However, the

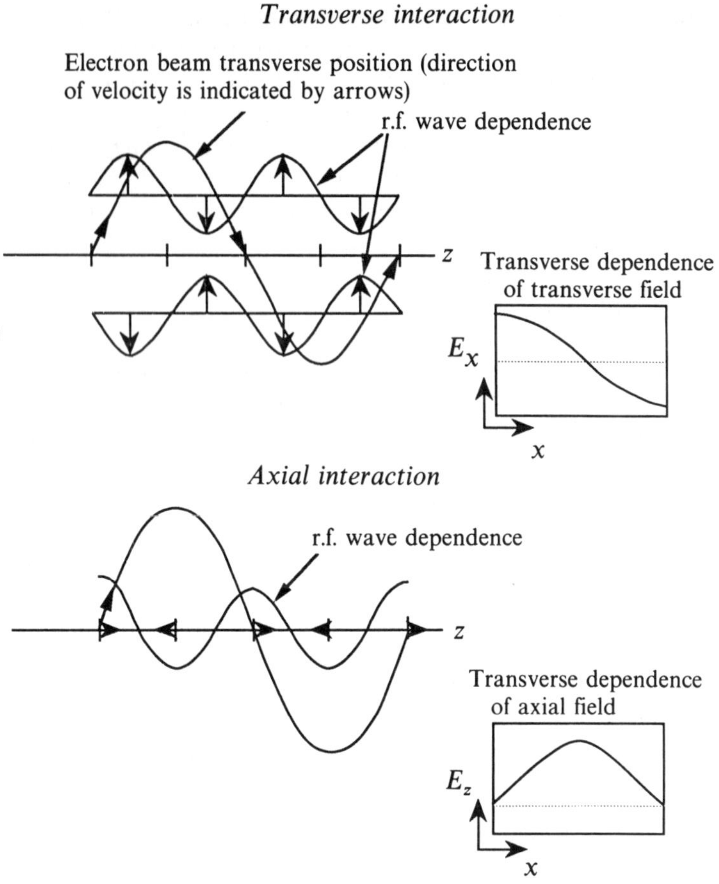

Transverse interaction

Electron beam transverse position (direction of velocity is indicated by arrows)

r.f. wave dependence

z

Transverse dependence of transverse field

E_x

x

Axial interaction

r.f. wave dependence

z

Transverse dependence of axial field

E_z

x

Fig. 7.17 Representation of the physical mechanism underlying the periodic position interaction.

transverse variation of the electric field is greater than the transverse variation of the axial velocity. This results in a stronger interaction on-axis, which, again, leads to net amplification.

Although the axial and transverse interactions have been considered separately in the preceding paragraph, it is difficult completely to separate the two interactions. The interaction can be treated using the nonlinear formulation described in Chapter 5 (equations (5.152–61)). In fact, numerical simulation of the interaction indicates that the overall performance at the second harmonic is improved when the two interactions are combined. Simulation also shows that the second-harmonic periodic position interaction can have a stronger growth rate than the fundamental interaction, and is a significantly stronger interaction than the third harmonic interaction for the current range of experimental parameters.

The periodic position interaction has been demonstrated recently in the laboratory [32], and we include a comparison of the experimental results with the nonlinear formulation for the interaction. The target frequency of the experiment was between 12 and 18 GHz. The experiment used a cylindrical electron beam tunable in energy from about 30 keV to 250 keV, with 100 keV as the nominal operating energy for the second-harmonic interaction. The wiggler consisted of a permanent-magnet assisted electromagnet with a period of 3 cm and an amplitude which was variable between 670 G and 1300 G, resulting in an a_w of 0.19–0.37. The pole pieces extended partially down the sides of the waveguide to provide wiggle-plane focusing. This geometry resulted in a very flat profile near the centre of the interaction region with the field rising sharply near the edges. The experiment operated as an amplifier and the interaction took place in an oversized waveguide with dimensions of 3.485 × 1.58 cm. The second-harmonic periodic position interaction occurred with the TE_{11} and TM_{11} modes since these are the lowest order waveguide modes with the necessary odd transverse symmetry.

Gain due to the second-harmonic periodic position interaction was measured at beam energies ranging from 78 keV to 106 keV and currents between 6 A and 10 A. This contrasts with energies in the range of 200–250 keV required for the fundamental interaction. Operation at frequencies between 12.5 GHz and 16.5 GHz was achieved through both energy and wiggler-field tuning, and the maximum observed gain was approximately 7 dB. It should be remarked that the experiment was not optimized for this interaction; hence, the interaction was not saturated at this value of gain with the available drive power.

The wiggler model used in the simulation of this experiment differed from the parabolic pole face model described previously (equation (2.3)), and was chosen to describe an inhomogeneity in the direction of the wiggle motion (denoted by the x-axis). The measured field for the experiment was quite uniform about the symmetry axis, and rose sharply toward the edges of the interaction region. A quartic profile of the field magnitude with x was found to provide a reasonable fit to this field. As such, the following wiggler model was employed

$$B_{w,x}(\mathbf{x}) = \left[\left(\sin k_w z - \frac{\cos k_w z}{k_w}\frac{d}{dz}\right)B_w(z)\right]\left[\sinh k_w y - \frac{Y(k_w y)}{2k_w^2}\frac{d^2}{dx^2}\right]\frac{1}{k_w}\frac{d}{dx}X(x), \tag{7.11}$$

$$B_{w,y}(\mathbf{x}) = \left[\left(\sin k_w z - \frac{\cos k_w z}{k_w}\frac{d}{dz}\right)B_w(z)\right]\left[\cosh k_w y - \frac{k_w y \sinh k_w y}{2k_w^2}\frac{d^2}{dx^2}\right]X(x), \tag{7.12}$$

$$B_{w,z}(\mathbf{x}) = B_w(z)\cos k_w z\left[\sinh k_w y - \frac{Y(k_w y)}{2k_w^2}\left(1 + \frac{1}{k_w^2}\frac{d^2}{dx^2}\right)\frac{d^2}{dx^2}\right]X(x), \tag{7.13}$$

where $B_w(z)$ describes the axial variation in the amplitude (see equation (5.14)), $X(x)$ denotes the variation in the wiggle plane, and $Y(k_w y) \equiv k_w y \cosh k_w y - \sinh k_w y$.

This field is not self-consistent in that it is divergence-free but not curl-free. However, the approximation is good as long as $B_w(z)$ and $X(x)$ vary slowly compared with λ_w.

The variation in the wiggler plane is described by a polynomial of the form

$$X(x) = 1 + \frac{1}{2}\left(\frac{x}{\alpha_x}\right)^{2m}, \tag{7.14}$$

where α_x denotes the scale length for variation of the field, and m is an integer. When $\alpha_x \to \infty$ this field reduces to that commonly used to describe a wiggler with flat pole faces. A comparison of the actual field with $X(x)$ as used in the code ($m = 2$ and $\alpha_x = 1.4938$) is shown in Fig. 7.18, and it is clear that the approximation gives a reasonable fit to the data.

The specific parameters used for comparison are an energy and current of 99.4 keV and 6.6 A with a beam radius of 0.4 cm. The wiggler was characterized by $B_w = 1.295$ kG, $\lambda_w = 3$ cm, an input taper of $N_w = 3$, and a total length of $34\lambda_w$. Both the TE_{11} and TM_{11} modes are included with an initial power split of 9 to 1 (TE to TM), and a total input power of 300 W. Figure 7.19 contains a comparison of the observations with numerical results over the unstable band for three different axial energy spreads: $\Delta\gamma_z/\gamma_0 = 0.0\%$, 0.025% and 0.05%. The experimental points over the entire frequency band fall, for the most part, between the curves representing an energy spread of 0.025% and 0.05%. This is in good agreement with the estimated energy spread in the experiment based upon electron-trajectory calculations of the gun geometry. It is important to bear in mind that the power has not saturated in any of these cases. The evolution of the power for a frequency of 14.4 GHz and an axial energy spread of 0.025% is seen from Fig. 7.20 to result in a gain of about 8 dB with a distance to saturation

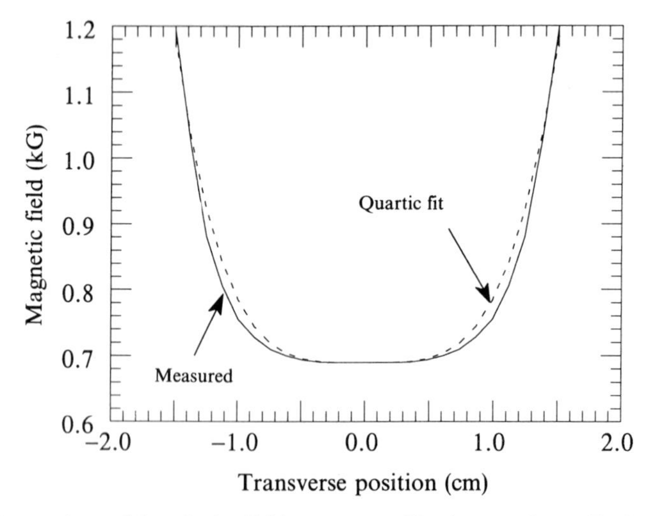

Fig. 7.18 Comparison of the wiggler field as measured in the experiment [32] and as used in simulation.

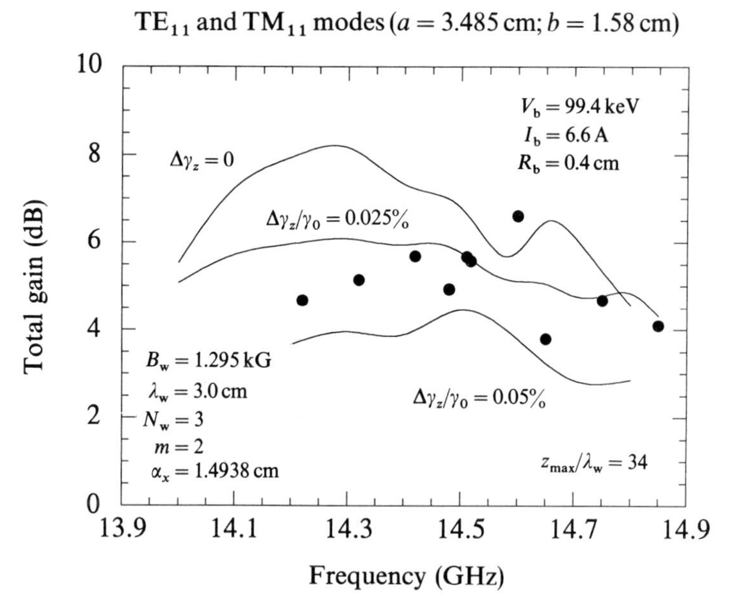

Fig. 7.19 Plot of the gain versus frequency as measured in the experiment (dots) and as determined from numerical simulation [32].

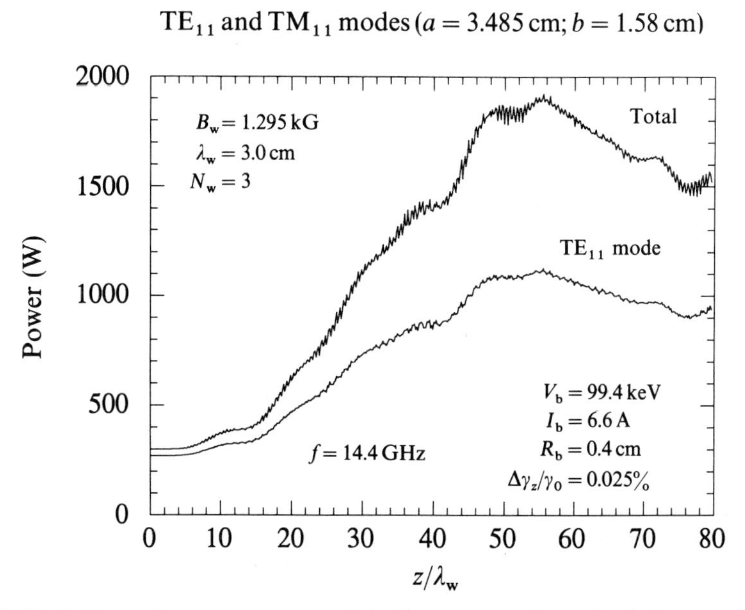

Fig. 7.20 Evolution of power versus axial distance as determined from numerical simulation [32].

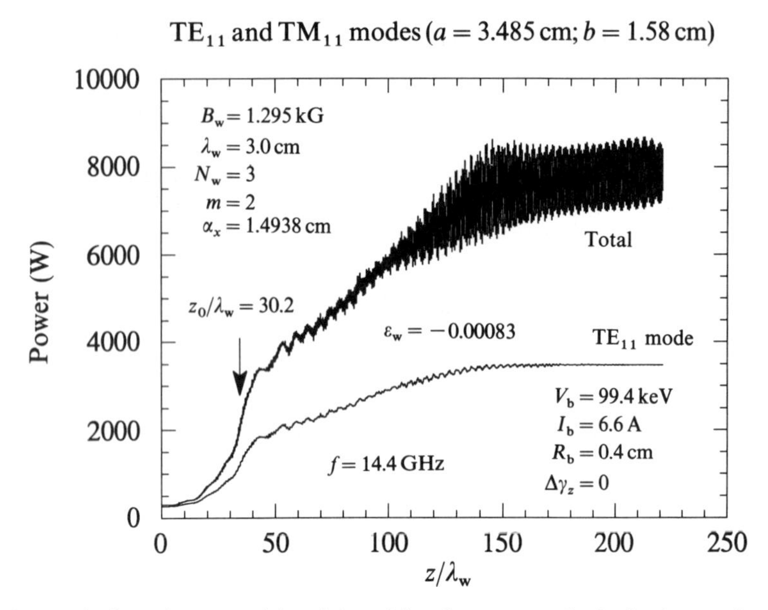

Fig. 7.21 Evolution of power with axial position for a tapered wiggler interaction [32].

of approximately $50\lambda_w$. In the absence of an energy spread, the gain for this case is approximately 10 dB with a saturation distance of $40\lambda_w$.

The effect of a wiggler field taper is shown in Fig. 7.21 for the case of an ideal beam ($\Delta\gamma_z = 0$) and a normalized taper of $\varepsilon_w = -0.000\,83$. The efficiency enhancement is extremely sensitive to the start-taper position, which must be chosen in the vicinity of the point at which the beam becomes trapped in the ponderomotive potential formed by the beating of the wiggler and radiation fields. For this example, the optimal start-taper point is $z_0 \approx 30.2$ cm. Only the total signal and the TE_{11} mode are shown in the figure, and the large oscillations in the total power are caused by the TM_{11} mode. It is evident that the saturated efficiency can be increased relative to that of the uniform-wiggler case for the present parameters by almost three-fold through the use of a tapered wiggler.

REFERENCES

1. Colson, W. B. (1981) The nonlinear wave equation for higher harmonics in free-electron lasers. *IEEE J. Quantum Electron.*, **QE-17**, 1417.
2. Coïsson, R. (1983) Generalized description of harmonic generation in a transverse optical klystron. *IEEE J. Quantum Electron.*, **QE-19**, 306.
3. Girard, B., Lapierre, Y., Ortega, J. M., Bazin, C., Billardon, M., Ellaume, P., Bergher, M., Velghe, M. and Petroff, Y. (1984) Optical frequency multiplication by an optical klystron. *Phys. Rev. Lett.*, **53**, 2405.
4. Freund, H. P., Johnston, S. and Sprangle, P. (1983) Three-dimensional theory of free-electron lasers with an axial guide field. *IEEE J. Quantum Electron.*, **QE-19**, 322.
5. Ganguly, A. K. and Freund, H. P. (1985) Nonlinear analysis of free-electron laser amplifiers in three dimensions. *Phys. Rev. A*, **32**, 2275.

6. Davidson, R. C. (1986) Kinetic description of harmonic instabilities in a planar wiggler free-electron laser. *Phys. Fluids*, **29**, 267.
7. Schmitt, M. J. and Elliot, C. J. (1986) Even harmonic generation in free-electron lasers. *Phys. Rev. A*, **34**, 4843.
8. Ortega, J. M. (1986) Harmonic generation in the VUV on a storage ring and prospects for Super-ACO at Orsay. *Nucl. Instr. Meth.*, **A250**, 203.
9. Kinkaid, B. M. (1986) Laser harmonic generation using the optical klystron and the effects of wiggler errors. *Nucl. Instr. Meth.*, **A250**, 212.
10. Ellaume, P. (1986) Theory of the optical klystron. *Nucl. Instr. Meth.*, **A250**, 220.
11. Schmitt, M. J. and Elliot, C. J. (1987) The effects of harmonic wiggler field components on free-electron laser operation. *IEEE J. Quantum Electron.*, **QE-23**, 1552.
12. Gover, A., Friedman, A. and Luccio, A. (1987) Three dimensional modelling and numerical analysis of super-radiant harmonic emission in a free-electron laser (optical klystron). *Nucl. Instr. Meth.*, **A259**, 163.
13. Elliot, C. J. and Schmitt, M. J. (1987) XUV harmonic enhancement by magnetic fields. *Nucl. Instr. Meth.*, **A259**, 177.
14. Prazeres, R., Lapierre, Y. and Ortega, J. M. (1987) Monte Carlo simulation of the harmonic generation in an optical klystron or undulator. *Nucl. Instr. Meth.*, **A259**, 184.
15. Freund, H. P., Chang, C. L. and Bluem, H. (1987) Harmonic generation in free-electron lasers. *Phys. Rev. A*, **36**, 3218.
16. Bluem, H., Freund, H. P. and Chang, C. L. (1988) Harmonic content in a planar wiggler based free-electron laser amplifier. *Nucl. Instr. Meth.*, **A272**, 579.
17. Prazeres, R., Ortega, J. M., Bazin, C., Bergher, M., Billardon, M., Couprie, M. E., Velghe, M. and Petroff, Y. (1988) Coherent harmonic generation in the vacuum ultraviolet spectral range on the storage ring ACO. *Nucl. Instr. Meth.*, **A272**, 68.
18. Carlson, K., Fann, W. and Madey, J. M. J. (1988) Spatial distribution of the visible coherent harmonics generated by the Mark III free-electron laser. *Nucl. Instr. Meth.*, **A272**, 92.
19. Hooper, B. A., Benson, S. V., Cutolo, A. and Madey, J. M. J. (1988) Experimental results of two stage harmonic generation with picosecond pulses on the Stanford Mark III free-electron laser. *Nucl. Instr. Meth.*, **A272**, 96.
20. Dutt, S. K., Friedman, A., Gover, A. and Pellegrini, C. (1988) Three dimensional simulation of a high harmonic transverse optical klystron. *Nucl. Instr. Meth.*, **A272**, 564.
21. Schmitt, M. J., Elliot, C. J. and Newnam, B. E. (1988) Harmonic power implications on free-electron laser mirror design. *Nucl. Instr. Meth.*, **A272**, 586.
22. Freund, H. P., Bluem, H. and Jackson, R. H. (1989) Nonlinear theory and design of a harmonic ubitron/free-electron laser. *Nucl. Instr. Meth.*, **A285**, 169.
23. Benson, S. V. and Madey, J. M. J. (1989) Demonstration of harmonic lasing in a free-electron laser, in *Proc. Int. Conf. Lasers '88* (ed. Sze, R. C. and Duarte, F. J.), STS Press, McLean, Virginia, p. 189.
24. Bamford, D. J. and Deacon, D. A. G. (1989) Measurement of the coherent harmonic emission from a free-electron laser oscillator. *Phys. Rev. Lett.*, **62**, 1106.
25. Benson, S. V. and Madey, J. M. J. (1989) Demonstration of harmonic lasing in a free-electron laser. *Phys. Rev. A*, **39**, 1579.
26. Warren, R. W., Haynes, L. C., Feldman, D. W., Stein, W. E. and Gitomer, S. J. (1990) Lasing on the third harmonic. *Nucl. Instr. Meth.*, **A296**, 84.
27. Bamford, D. J. and Deacon, D. A. G. (1990) Harmonic generation experiments on the Mark III free-electron laser. *Nucl. Instr. Meth.*, **A296**, 89.
28. Sharp, W. M., Scharlemann, E. T. and Fawley, W. M. (1990) Three-dimensional simulation of free-electron laser harmonics with FRED. *Nucl. Instr. Meth.*, **A296**, 335.
29. Bluem, H., Jackson, R. H., Pershing, D. E., Booske, J. H. and Granatstein, V. L. (1990) Final design and cold tests of a harmonic ubitron amplifier. *Nucl. Instr. Meth.*, **A296**, 37.
30. Latham, P. E., Levush, B., Antonsen, T. M. and Metzler, N. (1991) Harmonic operation of a free-electron laser. *Phys. Rev. Lett.*, **66**, 1442.

31. Freund, H. P., Davidson, R. C. and Kirkpatrick, D. A. (1991) Thermal effects on the linear gain in free-electron lasers. *IEEE J. Quantum Electron.*, **QE-27**, 2550.
32. Bluem, H., Jackson, R. H., Freund, H. P., Pershing, D. E. and Granatstein, V. L. (1991) Demonstration of a new free-electron laser harmonic interaction. *Phys. Rev. Lett.*, **67**, 824.

8
Optical guiding

Optical guiding during the course of the interaction in free-electron lasers refers to the self-focusing of the electromagnetic wave by the electron beam [1–30]. Optical guiding of the signal occurs by two related mechanisms referred to as gain and refractive guiding. Gain guiding describes the preferential amplification of radiation in the region occupied by the electron beam. Therefore, an optical ray will undergo amplification as long as it is coincident with the beam. If it propagates out of the beam, then the interaction will cease. Refractive guiding describes the focusing (or defocusing) of the radiation by means of the shift in the refractive index due to the dielectric response of the electron beam. In particular, if the wavenumber is shifted upward due to the interaction with respect to the vacuum state, then the phase velocity of the wave decreases and the beam acts as an optical guide. It should be remarked, however, that gain and refractive guiding are intimately linked and are not independent processes.

The process of refractive guiding is related to variation in the relative phase [30] described in Chapter 5, since this quantity measures the shift in the wavenumber due to the dielectric effect of the beam. As shown in the nonlinear simulations of both the helical and planar wiggler configurations, the relative phase decreases with axial position at the low-frequency portion of the gain spectrum. This decrease occurs because the dielectric shift induced by the beam reduces the wavenumber below that of the vacuum state, and corresponds to a defocusing of the signal. As the frequency increases, however, the downshift in the wavenumber decreases until a critical frequency is reached at which the relative phase remains approximately constant. This corresponds to a wavenumber which is comparable to the vaccum state, and for which there is no refraction of the signal. The frequency at which this is found is, typically, below the frequency of peak growth rate. For frequencies higher than the critical point, the relative phase increases with axial position, corresponding to the guiding of the signal.

The mechanism of optical guiding has been studied both by analytical means [2, 6, 9, 17, 18, 23, 27, 30] and numerical simulations [1, 3, 4, 8, 10, 11, 13–16, 20, 26, 28, 30]. These analyses indicate that the optical guiding of the beam can result in a focusing of the electromagnetic signal to smaller or larger areas depending upon the cross-section of the electron beam, as well as the steering of the radiation in which the centroid of the electromagnetic signal follows that of the electron beam. It is also important to note in this regard that the relative phase has been found to reach a saturation level in tapered wiggler configurations in both experiment and simulation [31, 32]. Beyond this saturation point, the relative phase varies little with axial position. As a consequence, the optical guiding mechanism is weakly operative in the presence of a tapered wiggler [28].

8.1 OPTICAL GUIDING AND THE RELATIVE PHASE

The physical basis of the optical-guiding mechanism can be best understood in terms of the behaviour of the relative phase, which is defined as the integrated difference between the wavenumber in the interaction region and the free-space wavenumber. This can be understood most clearly on the basis of the idealized one-dimensional analysis. We first consider the high-gain regime. Under the assumption that $|v_w/v_\parallel| \ll 1$, the dispersion equations for both the helical (4.83) and planar-wiggler (4.170) geometries can be expressed as

$$\left([\omega - (k + k_w)v_\parallel]^2 - \frac{\omega_b^2}{\gamma_0 \gamma_\parallel^2}\right)\left(\omega^2 - c^2 k^2 - \frac{\omega_b^2}{\gamma_0}\right) \cong \frac{v_w^2}{c^2}\frac{\omega_b^2}{\gamma_0}\omega c k_w, \tag{8.1}$$

where ω and k are the angular frequency and wavenumber of the electromagnetic wave, γ_0 denotes the relativistic factor corresponding to the bulk energy of the beam, $v_\parallel$ is the bulk axial velocity of the beam $\gamma_\parallel \equiv (1 - v_\parallel^2/c^2)^{-1/2}$, ω_b is the beam plasma frequency, and

$$\left|\frac{v_w}{c}\right| \equiv \begin{cases} \dfrac{\Omega_w}{ck_w}; & \text{Helical wiggler} \\[2ex] \dfrac{\Omega_w}{\sqrt{2}ck_w}; & \text{Planar wiggler,} \end{cases} \tag{8.2}$$

denotes the bulk wiggler-induced transverse velocity corresponding to $\Omega_w \equiv eB_w/\gamma_0 m_e c$ for a wiggler amplitude B_w and wavenumber k_w. In order to illustrate the refractive shift in the wavenumber, we transform the wavenumber in equation (8.1) to $\delta k = k - \omega/c$, which measures the shift from the vacuum wavenumber. Under the assumptions that $|\delta k| \ll k_0 \equiv (\omega^2 - \omega_b^2/\gamma_0)^{1/2}/c$, and $\omega_b/\gamma_0^{1/2} \ll \omega$, the dispersion equation can be written in the form of a cubic equation

$$(\delta k - k_0)\left[\delta k^2 v_\parallel^2 - 2\Delta\omega\,\delta k v_\parallel - \left(\Delta\omega^2 - \frac{\omega_b^2}{\gamma_0 \gamma_\parallel^2}\right)\right] \cong -\frac{v_w^2}{2c^2}\frac{\omega_b^2}{\gamma_0}k_w, \tag{8.3}$$

where $\Delta\omega \equiv (1 - v_\parallel/c)\omega - k_w v_\parallel$.

The solution to equation (8.3) for the real (solid line) and imaginary (dashed line) parts of δk as functions of frequency are shown in Fig. 8.1 for $\gamma_0 = 3.5$, $v_w/c = 0.05$, and $\omega_b/\gamma_0^{1/2} ck_w = 0.1$. The frequency corresponding to the peak growth rate for this choice of parameters is $\omega/ck_w \approx 21.6$. It is clear that the wavenumber is shifted downward from ω/c (i.e. $\delta k < 0$) for frequencies below $\omega/ck_w \leqslant 21.5$, which is below the frequency of peak growth. In contrast, for

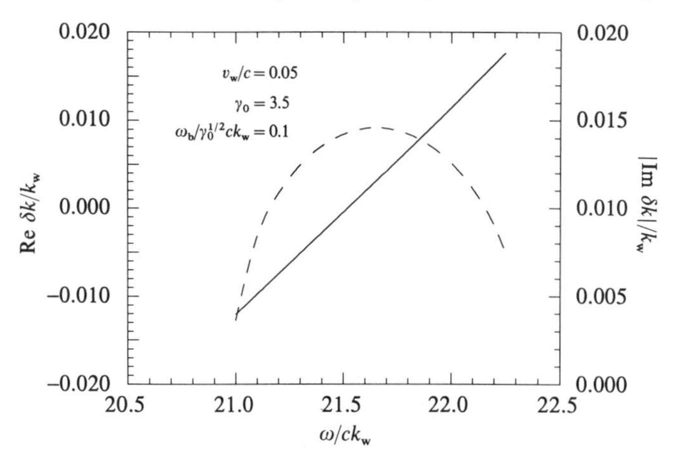

Fig. 8.1 Graph showing the growth rate (dashed line) and the shift in the wavenumber (solid line) as functions of frequency.

frequencies above this critical value, including at peak growth, the wavenumber is shifted upward from the vacuum wavenumber. Hence, the qualitative behaviour for the wavenumber in the idealized one-dimensional analysis is the same as that from the three-dimensional nonlinear formulation. The conclusion to be drawn from these results is that the refractive effect of the wave–particle interaction in the wiggler can either guide or defocus the electromagnetic wave depending upon the interaction frequency.

In order to understand the process of optical guiding in the low-gain regime, we consider an analysis of the evolution of the relative phase in an idealized one-dimensional model in which the vector potential of the optical signal is represented in the form

$$\delta \boldsymbol{A}(z,t) = \delta A(z)\hat{\boldsymbol{e}}_x \cos[\omega(z/c - t) + \Delta\phi(z)], \tag{8.4}$$

where $\Delta\phi$ denotes the relative phase, and both $\Delta\phi$ and the amplitude δA are assumed to be real. The wiggler field is assumed to be given by the idealized one-dimensional representation in equation (2.99). Substitution of this form for the vector potential into Maxwell's equations (5.19) yields an equation of the form

$$\delta k(z)\delta A(z)\cos[\omega(z/c - t)] + \frac{d}{dz}\delta A(z)\sin[\omega(z/c - t)] \cong \frac{2\pi}{\omega}\delta J_x, \tag{8.5}$$

where the source current is given in equation (5.22), $\delta k \equiv d(\delta\phi)/dz$ denotes the perturbed wavenumber, and it is assumed that $|\delta k| \ll \omega/c$. Note that second-order derivatives of the amplitude and phase have been neglected in equation (8.5). Multiplication by $\cos[\omega(z/c - t)]$ and subsequent averaging of the resulting equation over a wave period yield

$$\delta k \cong \frac{2}{\delta A}\int_0^{2\pi/\omega} dt\,\delta J_x \cos[\omega(z/c - t)], \tag{8.6}$$

where the source current is given by equation (5.22). Under the assumption that $v_x \approx v_w \cos k_w z$, therefore, the perturbed wavenumber becomes

$$\delta k \cong \frac{\omega_b^2}{2\omega c} \frac{v_w}{c} \frac{1}{\delta a} \langle \cos \psi \rangle, \tag{8.7}$$

where ψ denotes the ponderomotive phase, ω_b is the beam plasma frequency, $\delta a \equiv e\delta A/m_e c^2$ is the normalized amplitude, and the average is taken over the initial phase as defined in equation (5.41). Observe that the equation for the small-signal gain is obtained by multiplication by $\sin[\omega(z/c - t)]$.

The phase average may be determined by solution of the pendulum equation in the untrapped limit. The evolution of the ponderomotive phase in the combined electromagnetic and magnetostatic fields is governed by the nonlinear pendulum equation

$$\frac{d^2}{dz^2}\psi = K^2 \sin \psi, \tag{8.8}$$

where

$$K^2 \equiv \delta a \frac{v_w}{v_\parallel} \frac{\omega^2/c^2}{2\gamma_0 \gamma_\parallel^2 \beta_\parallel^3}. \tag{8.9}$$

In the linear regime, the solutions describe untrapped trajectories which may be determined by a perturbative solution to equation (8.8). In this case we expand $\psi = \psi_0 + \Delta k z + \delta\psi$, where ψ_0 is the initial phase,

$$\Delta k \equiv -\frac{\omega}{v_\parallel}\left(1 - \frac{v_\parallel}{c}\right) + k_w, \tag{8.10}$$

describes the mismatch parameter, and it is assumed that $|\delta\psi/\psi| \ll 1$. To lowest order in the perturbation, therefore, the pendulum equation can be expressed in the form

$$\frac{d^2}{dz^2}\delta\psi \cong K^2 \sin(\psi_0 + \Delta k z). \tag{8.11}$$

Subject to the initial conditions that $\delta\psi(z=0) = 0$ and $d\delta\psi(z=0)/dz = 0$, this equation has the solution

$$\delta\psi \cong -\frac{K^2}{\Delta k^2}[\sin(\psi_0 + \Delta k z) - \sin \psi_0 + \Delta k z \cos \psi_0]. \tag{8.12}$$

As a consequence,

$$\langle \cos \psi \rangle \cong \frac{K^2}{2\Delta k^2}[1 - \cos \Delta k z - \Delta k z \sin \Delta k z], \tag{8.13}$$

and

$$\langle \sin \psi \rangle \cong -\frac{K^2}{2\Delta k^2}[\sin \Delta k z - \Delta k z \cos \Delta k z]. \tag{8.14}$$

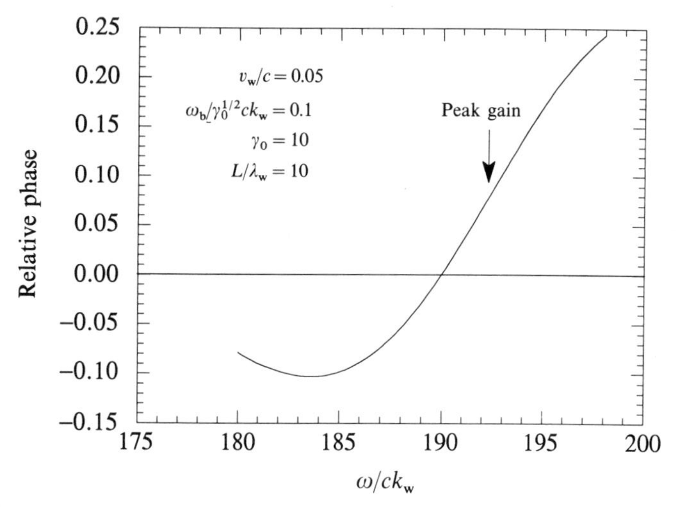

Fig. 8.2 The relative phase as a function of frequency within the gain band as calculated from the idealized one-dimensional model of the low-gain regime.

Hence, the perturbed wavenumber is given by

$$\delta k(z) \cong \frac{\omega_b^2}{8\gamma_0\gamma_\parallel^2 v_\parallel^2}\frac{v_w^2}{v_\parallel^2}\frac{\omega}{c\Delta k^2}[1 - \cos\Delta kz - \Delta kz\sin\Delta kz]. \tag{8.15}$$

The relative phase is found by integration of (8.15) over axial position; hence, the relative phase at the end of the wiggler (i.e. $z = L$) is given by

$$\delta\phi(L) \cong \frac{\omega_b^2 kL^3}{16\gamma_0\gamma_\parallel^2 v_\parallel^2}\frac{v_w^2}{2v_\parallel^2}\frac{4}{\Delta k^2 L^2}\left[1 + \cos\Delta kL - \frac{2}{\Delta kL}\sin\Delta kL\right]. \tag{8.16}$$

The relative phase has been evaluated for a choice of parameters in which $\gamma_0 = 10$, $v_w/c = 0.05$, $\omega_b/\gamma_0^{1/2}ck_w = 0.1$, and a wiggler which is $L = 10\lambda_w$ in length, and the results are illustrated in Fig. 8.2. The gain band for this choice of parameters ranges over $\omega/ck_w \approx 180$–200, and a peak gain of approximately 20% is found at $\omega/ck_w \approx 192$. As shown in the figure, the relative phase is positive at the high-frequency portion of the gain band which includes the frequency of peak gain.

8.2 THE SEPARABLE-BEAM LIMIT

The fundamentals of the optical guiding process can best be understood in terms of a separable-beam approximation [9, 17, 24, 27, 30] in which the wiggler-driven source current is decomposed into a product of functions depending upon radius and axial position. Hence, the electron beam is assumed to be cylindrically symmetric. The electron trajectories are treated in the context of an idealized one-dimensional approximation of the planar wiggler geometry (equation (2.99)). This model has the considerable advantage of allowing an analytic solution for the radiation spot size, and is generally valid as long as the beam radius is much less than either the spot size of the radiation or the wiggler period.

The idealized planar wiggler geometry implies that the interaction will be with a cylindrically symmetric plane-polarized wave in which the vector potential of the electromagnetic field is expressed in the form

$$\delta \mathbf{A}(r,z,t) = \frac{1}{2}\delta A(r,z)\hat{\mathbf{e}}_x \exp[i\omega(z/c - t)] + \text{c.c.}, \tag{8.17}$$

for an angular frequency ω and wavenumber ω/c. Substitution of this form of the vector potential into Maxwell's equations (5.92) yields a dynamical equation of the form

$$\left(\frac{1}{r}\frac{\partial}{\partial r}r\frac{\partial}{\partial r} + 2i\frac{\omega}{c}\frac{\partial}{\partial z}\right)\delta A(r,z) = -\frac{8\pi}{c}\delta J_x(r,z)\exp[-i\omega(z/c - t)], \tag{8.18}$$

where the source current is defined as

$$\delta J_x(r,z) = -en_b(r)v_{z0}\int_{-\infty}^{\infty} dt_0 \sigma(t_0)\frac{v_x}{|v_z|}\delta[t - \tau(z,t_0)], \tag{8.19}$$

where $n_b(r)$ notes the radial variation of the ambient beam density, $\tau(z,t_0)$ is the Lagrangian time coordinate defined in equation (5.24) of an electron which enters the interaction region at time t_0, and with initial axial velocity v_{z0}. As a consequence, averaging equation (8.18) over a wave period subject to the quasistatic approximation for the electron trajectories gives a dynamical equation of the form

$$\left(\frac{1}{r}\frac{\partial}{\partial r}r\frac{\partial}{\partial r} + 2i\frac{\omega}{c}\frac{\partial}{\partial z}\right)\delta a(r,z) = \frac{\omega_b^2(r)}{c^2}\beta_{z0}\frac{v_w}{v_\parallel}\langle\exp(-i\psi)\rangle, \tag{8.20}$$

where $\beta_{z0} \equiv v_{z0}/c$, $\omega_b^2(r) \equiv 4\pi e^2 n_b(r)/m_e$, $\psi \equiv (\omega/c + k_w)z - \omega t$ denotes the ponderomotive phase, $\delta a \equiv e\delta A/m_e c^2$, the orbits are given in the idealized approximation (equations (2.103) and (2.104) subject to the requirement that $|v_w/v_\parallel| \ll 1$, and the average is over the initial phase as defined in equation (5.41). The dynamical equation is solved for an electron beam with a Gaussian density profile

$$n_b(r) = n_{b0}\exp(-r^2/r_b^2), \tag{8.21}$$

where n_{b0} denotes the maximum density on-axis and r_b is the Gaussian beam radius.

The solution of the dynamical equation may be written in terms of a Green's function which satisfies the equation

$$\left(\frac{1}{r}\frac{\partial}{\partial r}r\frac{\partial}{\partial r} + 2i\frac{\omega}{c}\frac{\partial}{\partial z}\right)G(r,z-z') = \frac{\omega}{c}\exp(-r^2/r_b^2)\delta(z-z'). \tag{8.22}$$

Equations (8.20) and (8.22) can be combined to yield

$$\int_0^\infty dr\, r\int_0^{z'+0^+} dz\left[\delta a(r,z)\delta(z-z') - \frac{\omega_{b0}^2}{\omega c}\beta_{z0}\frac{v_w}{v_\parallel}G(r,z'-z)\langle\exp(-i\psi(z))\rangle\right]$$

$$= \frac{c}{\omega}\int_0^\infty dr \int_0^{z'+0^+} dz\left[\frac{\partial}{\partial r}\left(r\delta a(r,z)\frac{\partial}{\partial r}G(r,z'-z)\right)\right.$$

$$\left. - \frac{\partial}{\partial r}\left(rG(r,z'-z)\frac{\partial}{\partial r}\delta a(r,z)\right)\right]$$

$$- 2i\int_0^\infty drr \int_0^{z'+0^+} dz\left[\delta a(r,z)\frac{\partial}{\partial z}G(r,z'-z) + G(r,z'-z)\frac{\partial}{\partial z}\delta a(r,z)\right]. \quad (8.23)$$

The first integral on the right-hand side of equation (8.23) vanishes subject to the requirements that

$$\delta a(r=\infty,z) = G(r=\infty,z-z') = 0, \quad (8.24)$$

and

$$\frac{\partial}{\partial r}\delta a(r,z)\bigg|_{r=\infty} = \frac{\partial}{\partial r}G(r,z-z')\bigg|_{r=\infty} = 0. \quad (8.25)$$

The second integral on the right-hand side yields the homogeneous solution in the absence of the interaction. Hence, the solution in terms of this Green's function is

$$\delta a(r,z) = \delta a_h(r,z) + \frac{\omega_{b0}^2}{c^2}\beta_{z0}\frac{v_w}{v_\parallel}\frac{c}{\omega}\int_0^z dz'\,G(r,z'-z)\langle\exp[-i\psi(z')]\rangle, \quad (8.26)$$

where $\omega_{b0}^2 \equiv 4\pi e^2 n_{b0}/m_e$, and $\delta a_h(r,z)$ denotes the homogeneous solution which describes the evolution of the vector potential in the absence of the beam.

We now consider the low-gain regime. In order to describe the focusing or defocusing of the radiation due to the interaction, we define an *average* field in the form

$$\langle\delta a(r,z)\rangle_r \equiv \frac{2}{r_b^2}\int_0^\infty drr\exp(-r^2/r_b^2)\delta a(r,z). \quad (8.27)$$

This is a weighted average over the cross-section of the electron beam which is a measure of the effective field experienced by the electrons. In the low-gain regime, the ponderomotive phase is determined by the solution of the pendulum equation (8.8) in the untrapped regime. Utilization of this average vector potential in the definition of K^2 in equation (8.9), therefore, gives a phase average which varies as

$$\langle\exp[\pm i\psi(z)]\rangle \cong \frac{K^2}{2\Delta k^2}[1 - \exp(\pm i\Delta kz) \pm i\Delta kz\exp(\pm i\Delta kz)], \quad (8.28)$$

where

$$K^2 \cong \langle\delta a(r,z)\rangle_r \frac{v_w}{v_\parallel}\frac{\omega^2/c^2}{2\gamma_0\gamma_\parallel^2\beta_\parallel^3}. \quad (8.29)$$

It may be verified by substitution that the Green's function is

$$G(r,z-z') = \frac{1}{2i}\frac{\exp[r^2/r_b^2(1+2i(z-z')/z_b)]}{1+2i(z-z')/z_b}H(z-z'), \quad (8.30)$$

where $H(z)$ denotes the Heaviside function, and $z_b \equiv \omega r_b^2/c$. As a consequence, the average field can be expressed as

$$\langle \delta a(r,z)\rangle_r = \frac{\langle \delta a_h(r,z)\rangle_r}{S(z)}, \tag{8.31}$$

where

$$S(z) \equiv 1 + \left(\frac{v_w}{v_\parallel}\right)^2 \frac{\omega_{b0}^2 r_b^2}{16\gamma_0\gamma_\parallel^2 v_\parallel^2}\frac{\omega^2}{c^2\Delta k^2}\left\{1 + \ln\left(1 + i\frac{z}{z_b}\right) - \exp(-i\Delta kz)\right.$$
$$\left. - i[1 + \Delta k(z_b + iz)]\exp(-i\Delta kz)\int_0^{z/z_b} dx \frac{\exp(i\Delta k z_b x)}{1 + ix}\right\}. \tag{8.32}$$

If the homogeneous solution is given in terms of a Gaussian beam which is focused down to the minimum spot size at $z = 0$, then [33]

$$\delta a_h(r,z) = \delta a_0 \frac{\exp[r^2/r_0^2(1 + 2iz/z_0)]}{1 + 2iz/z_0}, \tag{8.33}$$

where δa_0 is real and denotes the initial amplitude of the field, r_0 denotes the Gaussian radius of the beam, and $z_0 \equiv \omega r_0^2/c$ denotes the Rayleigh length. As a consequence,

$$\langle \delta a_h(r,z)\rangle_r = \frac{\delta a_0}{1 + \dfrac{r_b^2}{r_0^2} + 2i\dfrac{z}{z_0}}. \tag{8.34}$$

An effective **focusing factor** may be defined in terms of this average field in the form

$$\mathscr{F}(z) \equiv \frac{|\langle \delta a(r,z)\rangle_r|^2}{\dfrac{2}{r_b^2}\displaystyle\int_0^\infty drr|\delta a(r,z)|^2}. \tag{8.35}$$

Since the homogeneous solution for the vector potential describes the propagation of the signal in the absence of the beam, the focusing factor measures the evolution of the radiation spot size relative to the diffraction of the Gaussian beam in free space.

The denominator can be determined by energy-conservation arguments. Returning to the dynamical equation for the field (8.20), we observe that

$$\delta a^*\left(\nabla_\perp^2 + 2i\frac{\omega}{c}\frac{\partial}{\partial z}\right)\delta a = \frac{\omega_{b0}^2}{c^2}\beta_{z0}\frac{v_w}{v_\parallel}\exp(-r^2/r_b^2)\delta a^*\langle \exp(-i\psi)\rangle, \tag{8.36}$$

and

$$\delta a\left(\nabla_\perp^2 - 2i\frac{\omega}{c}\frac{\partial}{\partial z}\right)\delta a^* = \frac{\omega_{b0}^2}{c^2}\beta_{z0}\frac{v_w}{v_\parallel}\exp(-r^2/r_b^2)\delta a\langle \exp\langle i\psi)\rangle. \tag{8.37}$$

Subtracting the second equation from the first, we find that

$$2i\frac{\omega}{c}\frac{\partial}{\partial z}|\delta a|^2 = \frac{\omega_{b0}^2}{c^2}\beta_{z0}\frac{v_w}{v_\parallel}\exp(-r^2/r_b^2)[\delta a^*\langle \exp(-i\psi)\rangle - \delta a\langle \exp(i\psi)\rangle]. \tag{8.38}$$

Observe that we have omitted the terms in $\delta a^* \nabla_\perp^2 \delta a - \delta a \nabla_\perp^2 \delta a^*$ since they will vanish upon integration over radius. Integration of (8.38) over both axial position and radius, therefore, yields

$$\frac{2}{r_b^2}\int_0^\infty drr|\delta a(r,z)|^2 \cong \frac{\delta a_0^2}{\left(1+\frac{r_b^2}{r_0^2}+4\frac{z^2}{z_0^2}\right)}\left[1+\frac{r_b^2}{r_0^2}+4\frac{z^2}{z_0^2}+G(z)\right], \tag{8.39}$$

where $G(z)$ is the expression for the power gain in the idealized one-dimensional formulation of the fundamental interaction in the low-gain regime (equation (4.159)). Combining equations (8.33–5) and (8.39), we find that the focusing factor over the total length of the system L may be approximated in the form

$$\mathscr{F}(L) \cong \frac{1}{\left[1+\frac{r_b^2}{r_0^2}+4\frac{L^2}{z_0^2}+G(L)\right]|S(L)|^2}. \tag{8.40}$$

Observe that in the limit in which the interaction vanishes (i.e. in the absence of either the beam or the wiggler field) the gain also vanishes and the focusing factor describes the free-space diffraction of the optical beam

$$\mathscr{F}_{fs}(L) \cong \frac{1}{1+\frac{r_b^2}{r_0^2}+4\frac{L^2}{z_0^2}}. \tag{8.41}$$

Hence, the effect of the interaction upon the diffraction of the optical beam can be effectively measured by a normalized focusing factor defined as

$$\mathscr{F}_n(L) \equiv \frac{\mathscr{F}(L)}{\mathscr{F}_{fs}(L)}. \tag{8.42}$$

Hence, when $\mathscr{F}_n(L) > 1$ the diffraction of the signal is slower than in the free-space limit, and the signal is effectively guided. Conversely, when $\mathscr{F}_n(L) < 1$, the diffraction is more rapid than in free space.

The specific example under consideration is that of a system for which $v_w/c = 0.05$, $\omega_{b0}/\gamma_0^{1/2}ck_w = 0.1$, and $\gamma_0 = 10$. Further, it is assumed that there are 10 wiggler periods within the interaction length (i.e. $L/\lambda_w = 10$), $k_w r_b = 0.1$, and that the optical beam is focused down to the beam radius upon the entry to the wiggler (i.e. $r_b = r_0$). Observe that this choice of parameters corresponds to that used in the calculation of the relative phase in the idealized one-dimensional regime shown in Fig. 8.2. The peak gain in the idealized one-dimensional limit for this choice of parameters over length L is of the order of 20% and occurs for $\omega/ck_w \approx 191.7$. The focusing factor at $z = L$ associated with this choice of parameters is shown in Fig. 8.3 for frequencies within the gain band. It is evident that the indicated diffraction of the radiation found on the basis of the calculation of the relative phase and the separable beam limits is qualitatively similar. That is that the diffraction is slower than the free-space value for the low-frequency

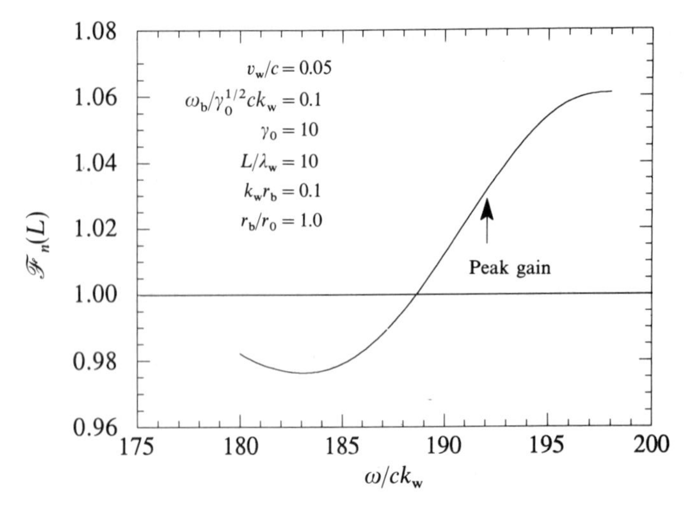

Fig. 8.3 Illustration of the normalized focusing factor for representative parameters in the low-gain regime.

portion of the gain band, but is faster than in free space at the edges of the gain band.

8.3 NUMERICAL SIMULATION

In order to study the optical-guiding process in more detail, the nonlinear formulation described in Chapter 5 to treat the geometry of an electron beam propagating through a rectangular waveguide in the presence of a planar wiggler is adapted to the treatment of an optical mode [29]. To this end, an optical mode, as represented by a Gaussian pulse, is expanded in terms of an ensemble of TE and TM modes of the empty waveguide. The dynamical equations for the waveguide modes are given by equations (5.154) and (5.155) which are integrated for each waveguide mode in conjunction with the Lorentz force equations (5.156–61) for an ensemble of electrons. The wiggler model employed in this formulation is characterized by tapered pole faces (see equation (2.3)) to provide for enhanced beam confinement. These equations describe both the gain and refractive guiding of the signal implicitly since they describe the evolution of the wavenumber and amplitude for each TE and TM mode.

Observe that no approximations are made in the orbit integrations in this formalism; hence, it is capable of treating the action of a lower beat wave upon the optical guiding of the signal. The lower beat wave introduces an oscillation at half the wiggler period which vanishes from the planar wiggler formulations when the orbits are averaged over a wiggler period, and makes only small contributions to the bulk properties of the system. In fact, this lower beat wave cancels out entirely for a helical wiggler. In contrast, the lower beat wave is present in planar configurations, and has been discussed in the nonlinear formulation presented in Chapter 5. In such cases, the instantaneous power and phase of the wave are seen to oscillate at half the wiggler period with amplitudes of up to 20% of the bulk values (i.e. the values obtained after averaging over a wiggler period). Since the relative phase measures the shift in the wavenumber from the vacuum state due to the interaction, it is related to the optical guiding of the signal. In this regard, the lower beat wave also results in an oscillation in the guiding of the wave as well as in the loss (or defocusing) of some fraction of the optical power during the course of the interaction.

The concept of optical guiding in a waveguide geometry requires some discussion. In this regard, it should be noted that the drift tube within which an electron beam propagates also constitutes a waveguide in that the electromagnetic field must satisfy the appropriate boundary conditions on the wall of the drift tube. As a result, unhindered diffractive spreading of an optical radiation pulse is not possible. However, the wave–particle interaction in the presence of the wiggler field constitutes a dielectrically loaded waveguide which alters the profile of the radiation with respect to that of the signal in the vacuum waveguide. These mechanisms are operative even in the presence of the waveguide, and are inextricably linked since the free-electron laser interaction results in a shift in the wavenumber which is comparable in magnitude to the growth rate. As a result, it may be expected that the profile of the electron beam will have an effect on that of the radiation field even in the presence of the waveguide. The free-electron laser interaction will cause each mode in the ensemble to either grow or decay at different rates; hence, the relative amplitudes of the modes will vary over the course of the interaction. This, in turn, will alter the radiation profile to describe the optical guiding of the signal. Note in this regard that refractive guiding is present within the context of the waveguide mode representation because variations of the wavenumber result in corresponding variations in the wave–particle resonance condition that alter the individual growth and/or damping rates of the various modes.

8.3.1 Gaussian-mode expansion

We treat the case of an amplifier configuration in which the output signal from a master oscillator is injected into the free-electron laser in synchronism with the electron beam. The injected radiation is assumed to be polarized linearly in the direction of the wiggler-induced oscillations in the electron beam (i.e. the x-axis in the current representation), and to be represented by a Gaussian which is focused to its minimum spot size upon entry to the wiggler. As a result, the cross-sectional variation of the electric field of the incident radiation can be represented in the form

$$\boldsymbol{E}_0 = \frac{\omega}{c} A_0 \hat{\boldsymbol{e}}_x \exp(-r^2/\rho_0^2) \sin(kz - \omega t), \tag{8.43}$$

at the entrance to the wiggler, where ρ_0 defines the area of the minimum spot size, and we assume that $\omega \approx ck$. It is necessary that the Gaussian spot size be much less than the waveguide dimensions in order to inject a large fraction of the laser signal into the waveguide. For simplicity, we shall assume that the entire fraction of the signal which is incident on the waveguide will be transmitted, so that the transmission coefficient

$$T = \mathrm{erf}\left(\frac{a}{\sqrt{2}\rho_0}\right)\mathrm{erf}\left(\frac{b}{\sqrt{2}\rho_0}\right), \tag{8.44}$$

is simply the ratio of the Poynting flux included within the waveguide cross-section to the total Poynting flux $P_0 = \omega^2 A_0^2/16c$. The advantage of using the waveguide modes to treat this problem (rather than, for example, the Gauss–Laguerre modes of an optical resonator) is that the drift tube within which the electron beam propagates constitutes a waveguide which imposes boundary conditions on the electromagnetic wave. These boundary conditions are satisfied implicitly by the waveguide modes.

The transmitted fraction of the incident signal is then decomposed into the various waveguide modes. These mode amplitudes are of the form [29]

$$\frac{\delta A_{ln}}{A_0} = 4F_{ln}C_{ln}\frac{\pi\rho_0^2}{ab}\frac{n\pi}{k_{ln}b} \tag{8.45}$$

for the TE_{ln} mode, and

$$\frac{\delta A_{ln}}{A_0} = 4C_{ln}\frac{\pi\rho_0^2}{ab}\frac{l\pi}{k_{ln}a} \tag{8.46}$$

for the TM_{ln} mode. Here

$$\begin{aligned} C_{ln} = {} & \cos\left(\frac{l\pi}{2}\right)\left\{\exp\left[-\left(\frac{l\pi\rho_0}{2a}\right)^2\right] - \frac{1}{\sqrt{\pi}}\exp\left(-\frac{a^2}{4\rho_0^2}\right)\cos\left(\frac{l\pi}{2}\right)\right. \\ & \left.\times \mathrm{Im}\left[Z\left(\frac{l\pi\rho_0}{2a} - i\frac{a}{2\rho_0}\right)\right]\right\} \times \sin\left(\frac{n\pi}{2}\right)\left\{\exp\left[-\left(\frac{n\pi\rho_0}{2b}\right)^2\right]\right. \\ & \left. + \frac{1}{\sqrt{\pi}}\exp\left(-\frac{b^2}{4\rho_0^2}\right)\sin\left(\frac{n\pi}{2}\right)\mathrm{Re}\left[Z\left(\frac{n\pi\rho_0}{2b} - i\frac{b}{2\rho_0}\right)\right]\right\}, \end{aligned} \tag{8.47}$$

and $Z(\zeta)$ is the plasma dispersion function [34]. Observe that this implies that the decomposition will include only those modes with even l and odd n. The total number of vacuum waveguide modes required to describe a Gaussian will vary with both the spot size of the Gaussian and the waveguide dimensions. For the parameters studied in the following section in which the radiation wavelength is 10.6 μm, $a/\rho_0 = 10$ and $b/\rho_0 = 3$, it will be shown that the decomposition can be accomplished adequately with as few as 21 modes.

For simplicity, the initial conditions are chosen to model the injection of a cold, pencil electron beam with a flat-topped density profile into the wiggler. The electrons are assumed to be propagating paraxially at the entry to the wiggler, but the subsequent evolution of the trajectories is followed in a self-consistent manner. Since this is a highly idealized beam with a vanishing emittance, no attempt has been made to *match* the beam into the wiggler (in the sense that the beam envelope remains constant). However, we do assume that the input laser signal is focused down to a spot size identical to that of the electron beam as it is on entry to the wiggler in order to minimize the effects of the bulk self-focusing and steering of the radiation signal. However, we are interested primarily in the question of the effect on the optical guiding of the radiation by the oscillations (at a period of $\lambda_w/2$) introduced by the lower beat wave. This is suitable for the purposes of the present analysis because this oscillation is a microscopic effect which is unaffected by the macroscopic structure of either the electron beam or the radiation field.

The specific parameters under consideration are as follows. The electron beam is characterized by an energy of 50 MeV, a current of 1 kA and an initial beam radius of 0.2 cm. The wiggler has an amplitude of 2.83 kG with a period of 7.2 cm and an entry taper region which is 10 wiggler periods long. This choice of beam and wiggler parameters provides for a resonance at a wavelength of 10.6 μm. The waveguide dimensions are $a = 2.0$ cm and $b = 0.6$ cm. The total signal power in the initial Gaussian is at a level of 20 MW, and is assumed to have been focused down to a spot size of 0.2 cm at the waist. Note that the overlap of the Gaussian beam and the waveguide imply that the transmission coefficient $T = 0.998$, which means that the injected signal is at a power level of 19.95 MW.

The decomposition of this signal into the TE and TM modes of the waveguide can be accomplished adequately with a relatively small number of modes. We find that for the present choice of parameters, if we include all modes for which $|\delta A_{ln}/A_0| > 0.005$ then 21 modes account for almost the entire input power at 99.8% of the injected signal. The magnitudes of the 21 highest mode amplitudes are shown in Table 8.1.

In general, there is an optimum range for the ratio between the Gaussian spot size and the waveguide dimensions which minimizes the number of modes required in the decomposition. If the spot size is comparable to the waveguide dimensions, then the fraction of power intercepted by the waveguide is small and the transmission coefficient is low. In this regime, a Gaussian approximation to the signal in the interaction regime is poor since the boundary conditions at the wall are not well satisfied. In addition, a large number of waveguide modes will be required to describe the transmitted power. In the opposite regime, a large number of waveguide modes is required when the Gaussian spot size is a negligibly small fraction of the waveguide dimensions. In this case, a Gaussian representation for the fields is preferable; however, it should be noted that the waveguide-mode representation is feasible even when the spot size is 10% of the larger and 30% of the smaller waveguide dimension.

A comparison of the shape of the superposition of the waveguide modes with the Gaussian is shown in Fig. 8.4 in which we plot the normalized E_x-component versus x (at $y = 0$) and y (at $x = 0$). The solid lines in the figure describe the Gaussian profile while the dots represent the waveguide mode superposition. This corresponds to the midplane of the waveguide, and it is clear

Table 8.1 Magnitudes of the highest-mode amplitudes for which $|\delta A_{ln}/A_0| \geqslant 0.005$.

		$\lvert\delta A_{ln}/A_0\rvert$	
l	n	TE Modes	TM Modes
0	1	0.161 01	
0	3	0.014 09	
2	1	0.250 18	
2	1		0.150 11
2	3	0.025 03	
2	3		0.005 01
2	5	0.006 78	
2	7	0.005 32	
4	1	0.138 91	
4	1		0.166 69
4	3	0.017 63	
4	3		0.007 05
6	1	0.064 33	
6	1		0.115 80
6	3	0.009 94	
6	3		0.005 96
8	1	0.025 53	
8	1		0.061 28
10	1	0.008 64	
10	1		0.025 91
12	1		0.008 89

that the mode selection is good. The principal discrepancy lies in the fact that the Gaussian does not satisfy the boundary conditions, and therefore is not negligible near the wall at $y/b = \pm 0.5$. The boundary condition is not satisfied at $x/a = \pm 0.5$ either; however, this is the long dimension of the waveguide and the Gaussian amplitude is negligible at this boundary. This illustrates the hazards of using a Gaussian mode in treating the dynamics of a free-electron laser. Even though the minimum waveguide dimension here is a factor of three longer than the Gaussian spot size, there is an appreciable discrepancy in the boundary condition at the wall. As a result, the use of a Gaussian mode to treat the radiation field in a free-electron laser must be approached with caution, and due consideration must be given to the relative sizes of the electron beam and the drift tube through which it propagates.

The normalized E_x-component away from the midplane is plotted in Fig. 8.5 versus x(at $|y/b| = 0.3$) and y (at $|x/a| = 0.3$). It is evident from the figure that the match between the waveguide- and Gaussian-mode representations versus x is good, but that the waveguide-mode representation diverges from the Gaussian in the description of the y-variation of the field. However, the field components are negligibly small in this regime, and the discrepancy is not significant. Indeed, it is the field match in this regime which diverges most rapidly as the mode number decreases. This match becomes increasingly poor as the number of waveguide

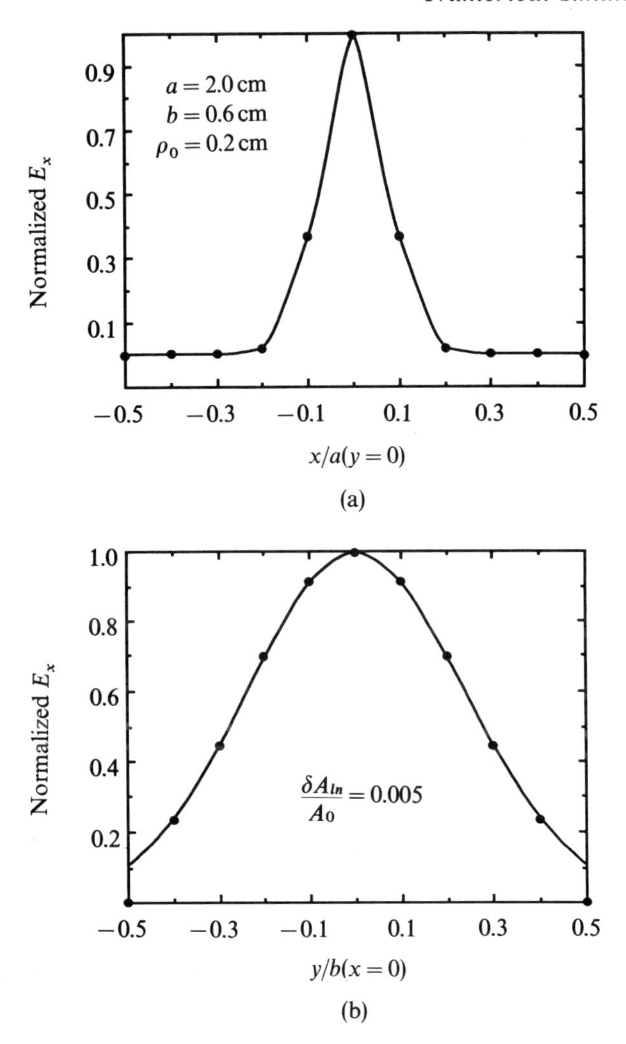

Fig. 8.4 Comparisons of the Gaussian optical profile (solid lines) with the waveguide superposition (dots) at the midplane of the waveguide.

modes included in the ensemble decreases further. This decomposition has been performed at the entry point to the wiggler where the Gaussian signal and electron beam overlap. However, while the electron beam is centred on the axis upon entry to the wiggler, the subsequent interaction with both the wiggler and radiation fields causes a complex motion which shifts both the centroid and envelope of the beam. The optical-mode pattern, in turn, will shift as it is guided by the microscopic and macroscopic effects of the interaction with the electron beam. Therefore, the discrepancies between the Gaussian- and waveguide-mode superposition away from the midplane have an important bearing on the question of

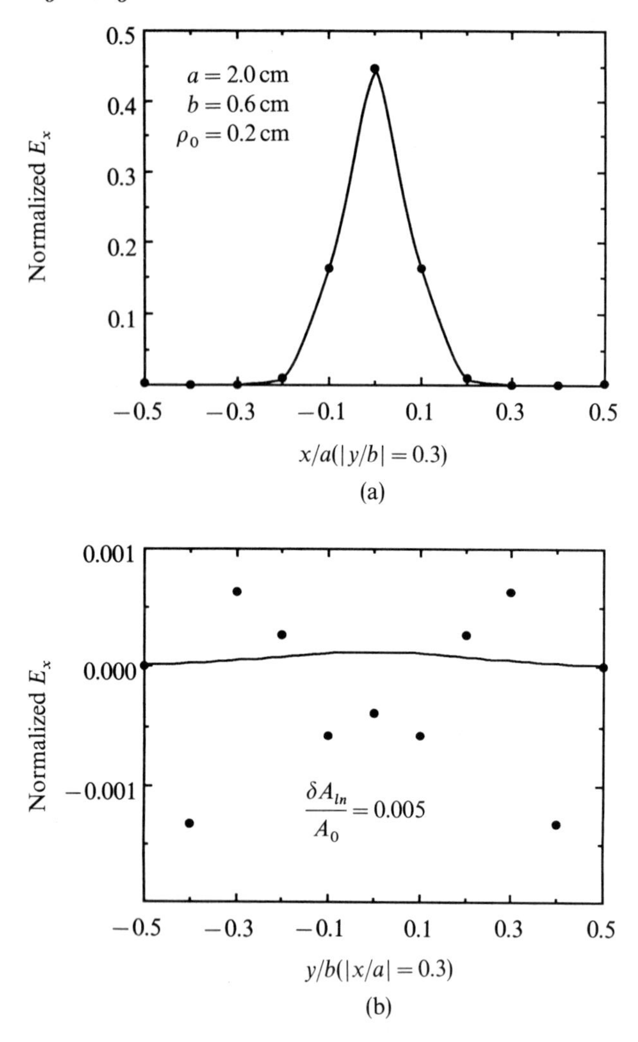

Fig. 8.5 Comparisons of the Gaussian optical profile (solid lines) with the waveguide superposition (dots) away from the midplane of the waveguide.

how many modes are needed to describe the interaction. It is clear that it is more complicated than just the analysis of the modal decomposition at the entry point.

8.3.2 Numerical simulation

The evolution of the total signal through the wiggler is shown in Fig. 8.6 in which we plot the total power in the 21 mode waveguide superposition versus axial position. Observe that the power saturates at $z/\lambda_w \approx 40$ with a power level of 808 MW for a total efficiency of 1.58%. The fluctuation introduced by the lower beat wave is clearly evident as an oscillation with an amplitude of

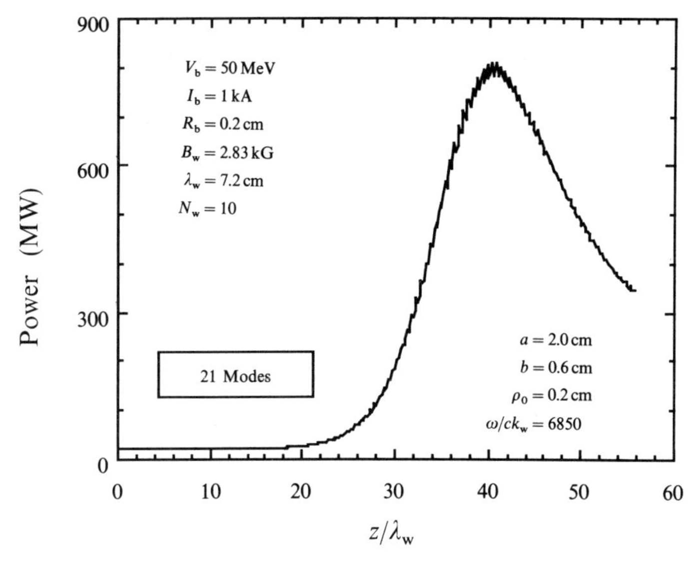

Fig. 8.6 Evolution of the total power in the superposition of waveguide modes as a function of axial position.

approximately 30 MW over a distance of half a wiggler period. A comparison of this result with similar simulation runs for both 8 and 13 mode decompositions (i.e. $|\delta A_{ln}/A_0| > 0.05$ and 0.01, respectively) is shown in Fig. 8.7 in which we plot the relative discrepancy in the power as a function of axial position between each

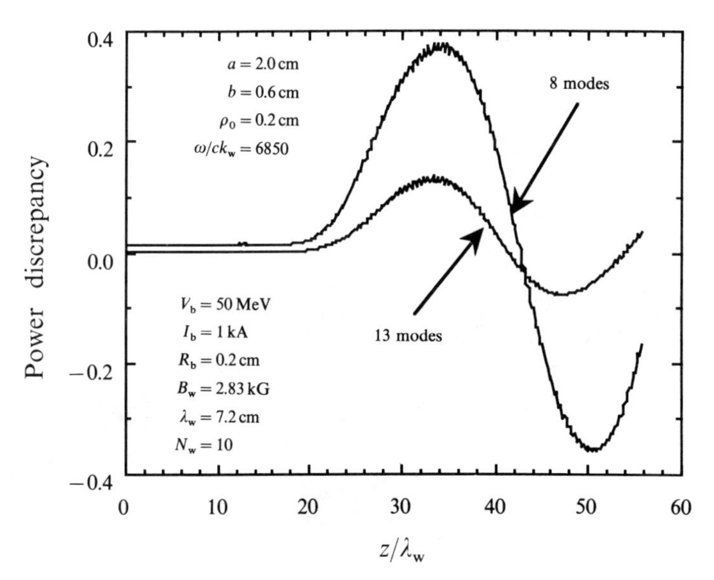

Fig. 8.7 Comparisons of the evolution of the total power in the superposition of both 8 and 13 waveguide modes with the 21-mode superposition as a function of axial position.

of these modal decompositions and the 21 mode case. Note that in both cases the discrepancy is relatively small at the point of saturation. However, the maximum discrepancy can approach 40% for the 8-mode decomposition compared with the 21-mode decomposition. As the mode number increases to 13 modes, the maximum discrepancy drops sharply to approximately 10%. The rapidity with which the relative discrepancy decreases with increases in the number of modes gives us confidence that the optical-mode representation is close to convergence with the 21-mode decomposition.

We now wish to address whether these oscillations in the power are associated with a corresponding oscillation in the cross-section of the radiation field. Perspective plots of the axial evolution of the radiation cross-section in the x- and y-directions are shown in Figs 8.8 and 8.9, respectively, over the entire length of the uniform interaction region (i.e. $10 < z/\lambda_w < 40$). It is clear from these figures that the radiation cross-section does indeed oscillate at one-half of a wiggler period in tune with the lower beat wave. The bulk self-focusing is visible in both

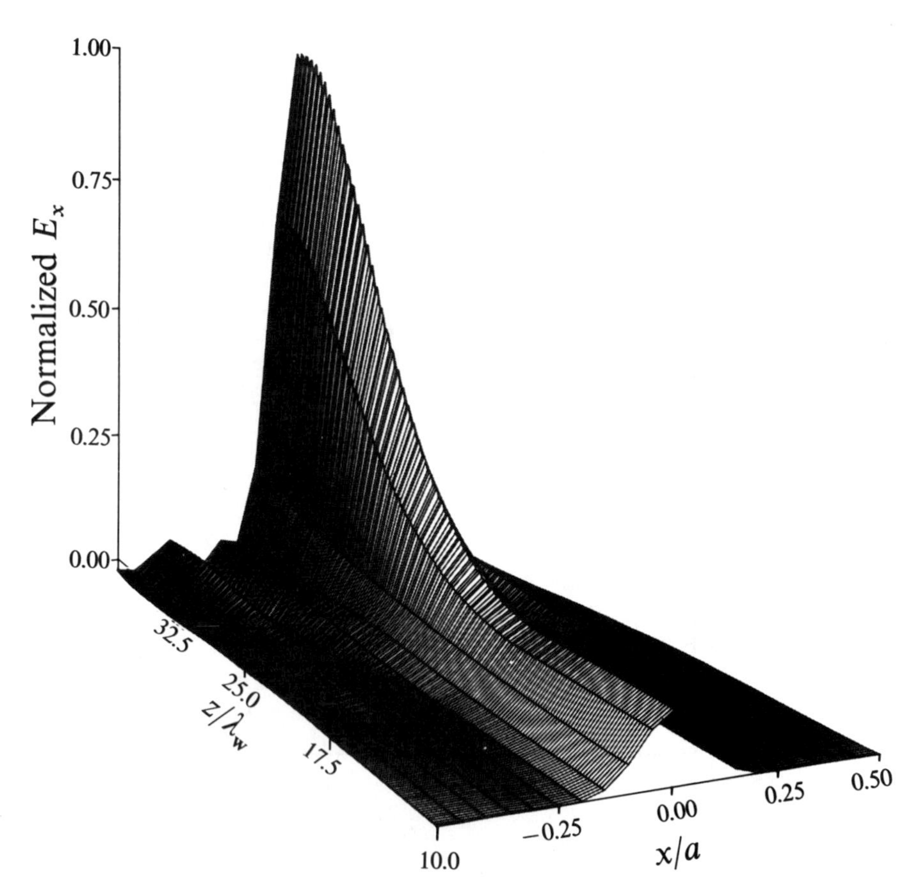

Fig. 8.8 Axial evolution of the cross-section in the x-direction in the $y = 0$ plane.

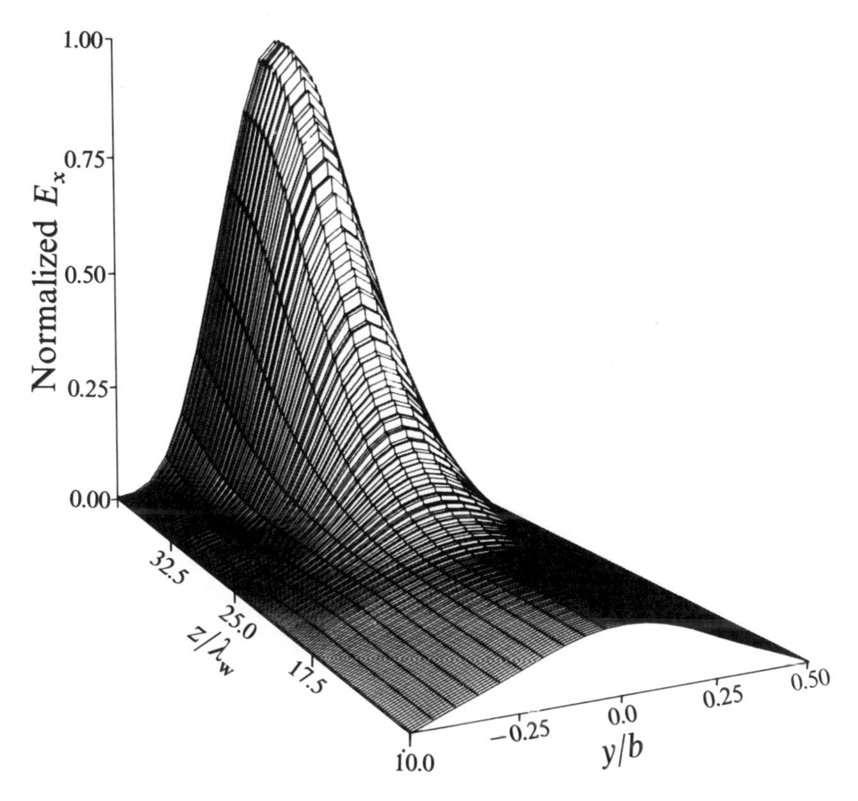

Fig. 8.9 Axial evolution of the cross-section in the y-direction in the $x = 0$ plane.

the x- and y-directions; however, it is much more evident in the x-direction. This is hardly surprising since this is the direction of the wiggler-induced motion. The bulk focusing occurs because the electron beam undergoes a focusing effect due to the transverse wiggler gradients. However, the additional focusing of the radiation field is negligible during the early phase of the interaction and does not seem to play a role until the power reaches several hundred megawatts. In addition, it does appear that **side lobes** grow in addition to the principal radiation beam. Thus, although the principal radiation beam is focused by the effect of the electron beam and wiggler field, these side lobes describe a defocusing effect which transfers energy away from the electron beam toward the drift-tube walls.

The side lobes are illustrated clearly by a contour plot of the normalized E_x-component of the field, which effectively describes the polarization vector of the electric field. The contours of the normalized E_x-component of the field at the entry to the wiggler are shown in Fig. 8.10 and describe a close match to the injected Gaussian signal. Note that the apparent noncircularity of the contour is accounted for by the fact that the x- and y-axes have been normalized to the waveguide dimensions and are shown at equal lengths. The contour at saturation is shown in Fig. 8.11, and the side lobes are clearly visible.

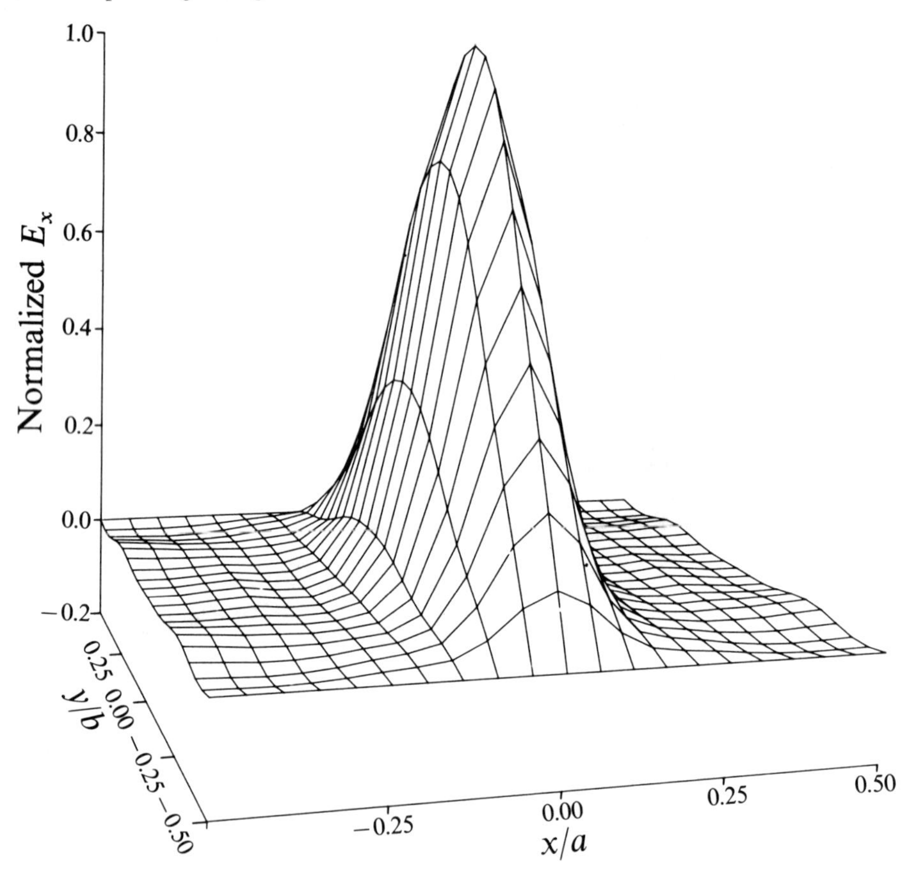

Fig. 8.10 Contour plot of the normalized E_x-component of the electric field at the entry to the wiggler.

8.3.3 Evolution of the electron beam

The evolution of the electron beam in this system describes (1) the effects of the wiggler-induced oscillation in the transverse and axial velocities, and (2) the effect of the wiggler gradients on the variation in the beam envelope due to betatron oscillations. The initial state of the electron beam is illustrated in Fig. 8.12 in which we plot the loading of the electron beam at $z = 0$. This reflects a pencil beam with a flat-top distribution and an initial radius of 0.2 cm. The particle positions shown in the figure reflect this uniform distribution. The positions in (r_0, θ_0) were chosen by means of a 10-point Gaussian algorithm in which the nonuniformity of the positions is compensated for by a nonuniformity in the weighting of the particles. As in the previous figures, the x- and y-coordinates in the plot are normalized to the corresponding dimensions of the waveguide; hence, the circular beam appears to be flattened in the x-direction.

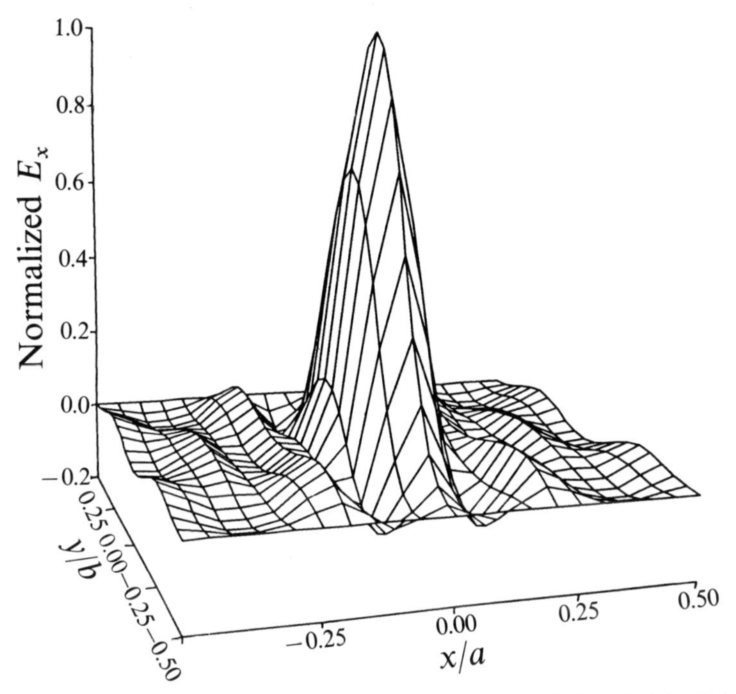

Fig. 8.11 Contour plot of the normalized E_x-component of the electric field at the saturation point of the interaction.

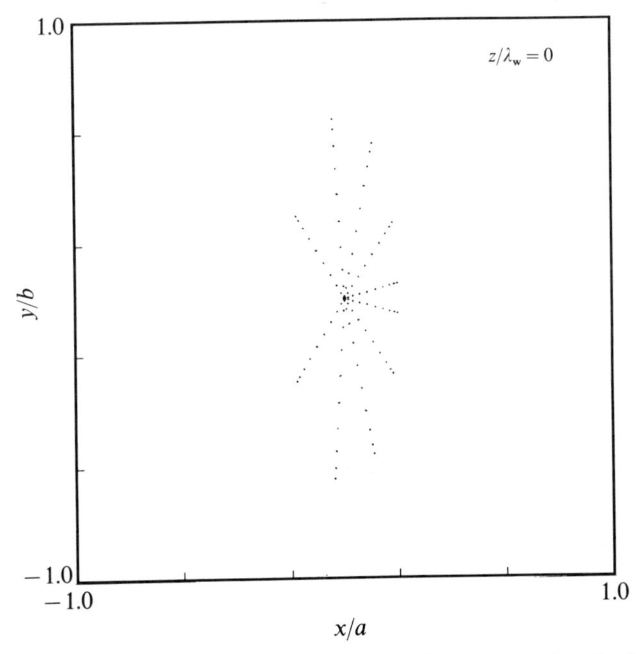

Fig. 8.12 Electron beam cross-section at the entry to the wiggler.

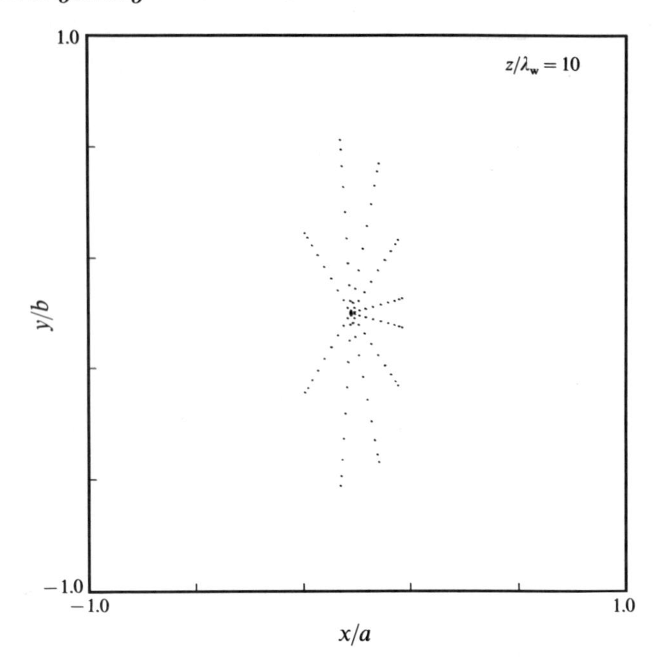

Fig. 8.13 Electron beam cross-section at $z/\lambda_w = 10$.

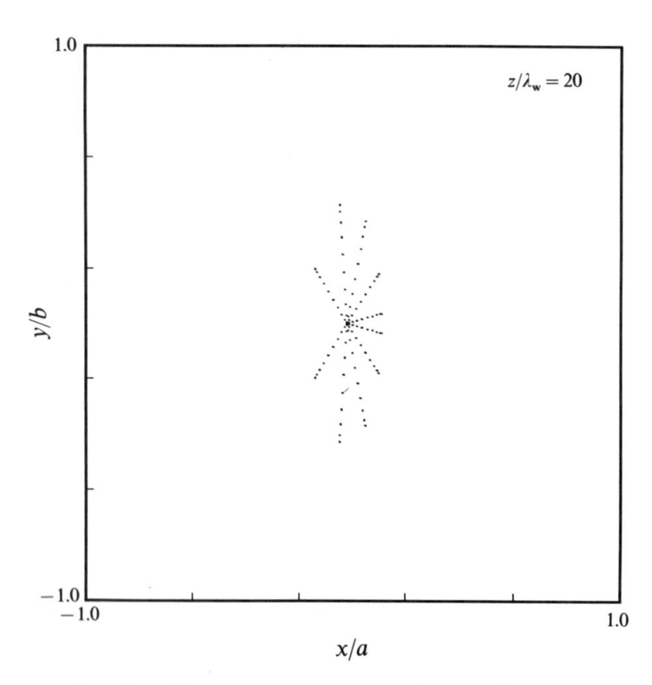

Fig. 8.14 Electron beam cross-section at $z/\lambda_w = 20$.

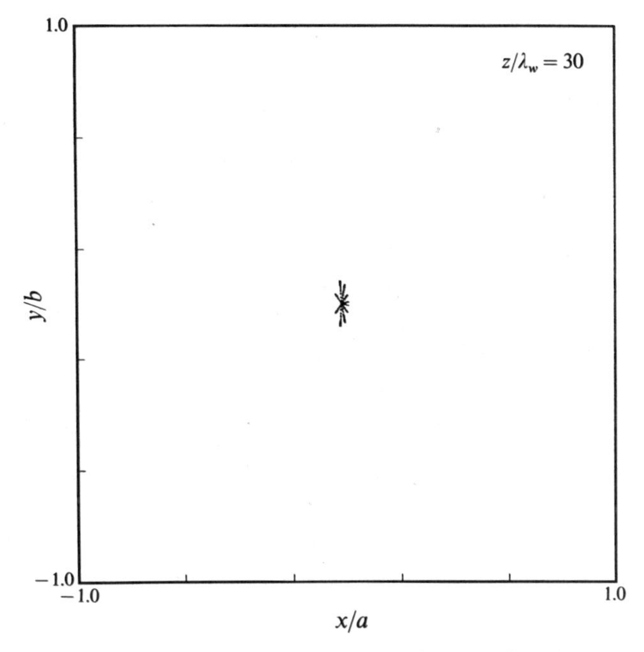

Fig. 8.15 Electron beam cross-section at $z/\lambda_w = 30$.

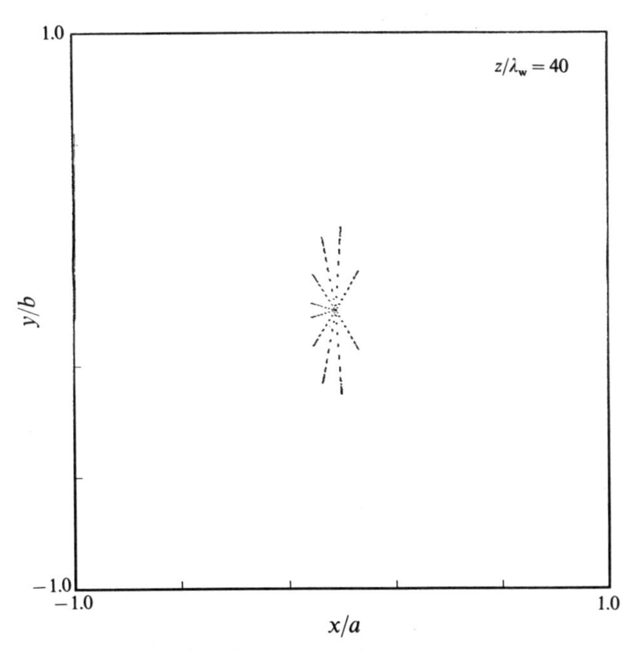

Fig. 8.16 Electron beam cross-section at $z/\lambda_w = 40$.

The inhomogeneity in the wiggler introduces a betatron oscillation which causes a scalloping in the beam envelope. The betatron frequency $\Omega_\beta = \Omega_w/\sqrt{2}\gamma$ yields a period, for the specific parameters studied, of approximately $73\lambda_w$. Since the saturation point occurs at $z \approx 40\lambda_w$, this permits somewhat more than one-half of a betatron oscillation during the course of the interaction at saturation. As a result of the betatron oscillation, the beam envelope goes through a focus at $z \approx 32\lambda_w$. The fact that the point of focus due to the betatron oscillation occurs prior to $z \approx 36.5\lambda_w$ is due to the fact that there is a 10-period entry taper over which the wiggler amplitude increases from zero to the constant level of 2.83 kG. The evolution of the beam envelope is illustrated in Figs 8.13–16 in which we plot the beam cross-section at $z/\lambda_w = 10$, 20, 30 and 40. Note that the magnitude of the oscillation at the wiggler period in this case is of the order of $|\Delta x| \approx 0.016$ cm, which is too small to resolve clearly on the scale of the figures.

It is significant that the radiation envelope does not reflect the betatron-induced scalloping of the electron-beam cross-section. The radiation spot size does not exhibit a focal point corresponding to that found for the electron beam. The reason for this is that, in this intense-beam/exponential-gain regime, the optical guiding due to both refractive and gain guiding is controlled predominantly by the microscopic interaction of the electrons and the radiation. Since the effect of the synchrotron–betatron coupling on both the growth rate and the dielectric response of the beam is weak in this example, the betatron oscillation has a minimal effect on the guiding of the signal.

REFERENCES

1. Scharlemann, E. T., Sessler, A. M. and Wurtele, J. S. (1985) Optical guiding in a free-electron laser. *Phys. Rev. Lett.*, **54**, 1925.
2. Moore, G. T. (1986) High-gain and large-diffraction regimes of the free-electron laser. *Nucl. Instr. Meth.*, **A250**, 381.
3. LaSala, J. E., Deacon, D. A. G. and Scharlemann, E. T. (1986) Optical guiding simulations for high-gain short-wavelength free-electron lasers. *Nucl. Instr. Meth.*, **A250**, 389.
4. Amir, A. and Greenzweig, Y. (1986) Three-dimensional free-electron laser gain and evolution of optical modes. *Nucl. Instr. Meth.*, **A250**, 404.
5. Luchini, P. and Solimeno, S. (1986) Optical guiding in a free-electron laser. *Nucl. Instr. Meth.*, **A250**, 413.
6. Xie, M. and Deacon, D. A. G. (1986) Theoretical study of free-electron laser active guiding in the small signal regime. *Nucl. Instr. Meth.*, **A250**, 426.
7. Gallardo, J. and Elias, L. R. (1986) Multimode dynamics in a free-electron laser with energy shift. *Nucl. Instr. Meth.*, **A250**, 438.
8. McVey, B. D. (1986) Three-dimensional simulations of free-electron laser physics. *Nucl. Instr. Meth.*, **A250**, 449.
9. Pantell, R. H. and Feinstein, J. (1987) Free-electron laser mode propagation at saturation. *IEEE J. Quantum Electron.*, **QE-23**, 1534.
10. Sprangle, P., Ting, A. and Tang, C. M. (1987) Analysis of radiation focusing and steering in the free-electron laser by use of a source dependent expansion technique. *Phys. Rev. A*, **36**, 2773.
11. Cai, S. Y., Bhattacharjee, A. and Marshall, T. C. (1987) Optical guiding in a Raman free-electron laser. *IEEE J. Quantum Electron.*, **QE-23**, 1651.

12. Luchini, P. (1987) More on optical guiding in a free-electron laser. *Nucl. Instr. Meth.*, **A259**, 150.
13. Warren, R. W. and McVey, B. D. (1987) Bending and focusing effects in a free-electron laser oscillator I: simple models. *Nucl. Instr. Meth.*, **A259**, 154.
14. McVey, B. D. and Warren, R. W. (1987) Bending and focusing effects in a free-electron laser oscillator II: numerical simulations. *Nucl. Instr. Meth.*, **A259**, 158.
15. Bhattacharjee, A., Cai, S. Y., Chang, S. P., Dodd, J. W. and Marshall, T. C. (1988) Observations of optical guiding in a Raman free-electron laser. *Phys. Rev. Lett.*, **60**, 1254.
16. Bourianoff, G., Moore, B., Rosenbluth, M., Waelbroeck, F., Waelbroeck, H. and Wong, H. V. (1988) Adaptive eigenmode expansion for 3-D free-electron laser simulations. *Nucl. Instr. Meth.*, **A272**, 340.
17. Antonsen, T. M. and Levush, B. (1988) Optical guiding in the separable beam limit. *Nucl. Instr. Meth.*, **A272**, 472.
18. Fruchtman, A. (1988) Optical guiding in a sheet-beam free-electron laser. *Phys. Rev. A*, **37**, 2989.
19. Whang, M. H. and Kuo, S. P. (1988) Self-focusing of laser pulses in magnetized relativistic electron beams. *Nucl. Instr. Meth.*, **A272**, 477.
20. Chen, Y. J., Solimeno, S. and Carlomusto, L. (1988) Optical guiding in a free-electron laser with full account of electron wiggling and 3-D propagation. *Nucl. Instr. Meth.*, **A272**, 490.
21. Gallardo, J. C., Dattoli, G., Renieri, A. and Hermsen, T. (1988) Integral equation for the laser field: multimode description of a free-electron laser oscillator. *Nucl. Instr. Meth.*, **A272**, 516.
22. Bhowmik, A., Bitterly, S., Cover, R. A., Kennedy, P. and Labbe, R. H. (1988) Transverse mode control in high gain free-electron lasers with grazing incidence, unstable ring resonators. *Nucl. Instr. Meth.*, **A272**, 524.
23. Xie, M., Deacon, D. A. G. and Madey, J. M. J. (1988) The guided mode expansion in free-electron lasers. *Nucl. Instr. Meth.*, **A272**, 528.
24. Antonsen, T. M. and Laval, G. (1989) Suppression of sidebands by diffraction in a free-electron laser. *Phys. Fluids B*, **1**, 1721.
25. Lontano, M., Sergeev, A. M. and Cardinali, A. (1989) Dynamical self-focusing of high-power free-electron laser radiation in a magnetized plasma. *Phys. Fluids B*, **1**, 901.
26. Bhattacharjee, A., Cai, S. Y., Chang, S. P., Dodd, J. W., Fruchtman, A. and Marshall, T. C. (1989) Theory and observation of optical guiding in a free-electron laser. *Phys. Rev. A*, **40**, 5081.
27. Metzler, N., Antonsen, T. M. and Levush, B. (1990) Nonlinear optical guiding in the separable beam limit. *Phys. Fluids B*, **2**, 1038.
28. Hafizi, B., Ting, A., Sprangle, P. and Tang, C. M. (1990) Effect of tapering on optical guiding and sideband growth in a finite-pulse free-electron laser. *Nucl. Instr. Meth.*, **A296**, 442.
29. Freund, H. P. and Chang, C. L. (1990) Effect of the lower beat wave on optical guiding in planar wiggler free-electron lasers. *Phys. Rev. A*, **42**, 6737.
30. Freund, H. P. and Antonsen, T. M. Jr (1991) The relationship between optical guiding and the relative phase in free-electron lasers. *IEEE J. Quantum Electron.*, **QE-27**, 2539.
31. Orzechowski, T. J., Scharlemann, E. T. and Hopkins, D. B. (1987) Measurement of the phase of the electromagnetic wave in a free-electron laser amplifier. *Phys. Rev. A*, **35**, 2184.
32. Freund, H. P. (1988) Multimode nonlinear analysis of free-electron laser amplifiers in three dimensions. *Phys. Rev. A*, **37**, 3371.
33. Yariv, A. (1967) *Quantum Electronics*, 2nd edn, Wiley, New York, p. 110.
34. Fried, B. D. and Conte, S. D. (1961) *The Plasma Dispersion Function*, Academic Press, New York.

9

Oscillator configurations

In the previous chapters it has been shown that an energetic electron beam propagating through an undulatory magnetic field is capable of amplifying electromagnetic radiation. That is, the combination of the beam and the undulatory field can be regarded as a medium with an inherent gain. The principal focus of the discussion in the previous chapters has been on the amplification of an injected signal with a specified frequency. However, the gain mechanism can also be used to form an oscillator by the feedback of a portion of the output signal. The radiation in an oscillator may be self-excited in the sense that radiation will grow spontaneously from noise in the oscillator if the gain the radiation experiences on traversing the interaction region exceeds the losses the radiation experiences on its return path to the input of the interaction region (including the portion of the radiation that is allowed to leave the oscillator as output). An oscillator may also be mode-locked by the injection of a large-amplitude signal with a frequency within the gain band.

Since the gain in the interaction region increases with beam current there is a threshold value of current (known as the start current) below which the gain is exceeded by the losses in the cavity and the radiation will not grow. If the beam current exceeds this threshold, spontaneously emitted noise in the electron beam will be amplified and will grow exponentially in time until the radiation in the oscillator reaches a saturation level. It will be shown in this and subsequent chapters that the characteristics of a free-electron laser oscillator are determined to a large degree by the ratio of the operating current to the start oscillation current.

There are a number of practical reasons for constructing an oscillator as opposed to an amplifier. First, it may be the case that no source is available in the frequency range of interest to drive an amplifier. Second, the amplification that the radiation may experience in one pass through the available interaction region in the frequency range of interest may be too small. An amplifier with a small gain is of little practical value. However, if most of the output signal can be fed back to the input a useful oscillator can be constructed. As a general rule, one finds that shorter wavelength devices which rely on relatively high-energy but low-current electron-beam sources fall into this category and tend to be oscillators because of their inherent low gain. In contrast, longer-wavelength devices which employ lower-energy but higher-current electron beams can drive amplifiers as well as oscillators.

As previously mentioned, the radiation in an oscillator is self-excited, whereas a signal with a specific frequency is injected in an amplifier. Thus, unless the oscillator is operated in a mode-locked manner, the frequency of the radiation which emerges from an oscillator is determined by the self-consistent action and reaction of the electron beam and the radiation. Consequently, it is more likely to be found that the radiation emerging from an oscillator is non-monochromatic (that is multi-frequency) than is that from an amplifier. This is particularly the case in the class of oscillators driven by pulsed electron beams, such as those produced by radio-frequency accelerators or storage rings, in which the emerging radiation has a temporal structure which also consists of pulses. In such systems, the radiation and beam pulses overlap within the interaction region. Due to the pulsed nature of the radiation, the emerging signal is necessarily non-monochromatic and the temporal length of the pulse determines the minimum possible width of the radiation spectrum. For long-pulse oscillators in which the electron beam can be regarded as continuous, in contrast, the spectrum may still be broad band due to other effects. In such cases the gain bandwidth (the range of frequencies over which a signal can be amplified) is typically larger than the spacing in frequency between the natural modes of the resonant cavity which is formed by the interaction region and the feedback loop. This spacing is the inverse of the transit time of radiation through the cavity. At least initially, therefore, one can expect a large number of different frequencies to be excited in the oscillator. Further, as we have seen in Chapter 6 where sidebands were discussed, the presence of a large-amplitude signal at one frequency can modify the electron trajectories and generate amplification at a second frequency, where previously (i.e. in the absence of the large-amplitude radiation signal) there was no amplification. The difference between the frequencies of these two competing signals can be much smaller than the bandwidth of the feedback path. Thus, it may be difficult to discriminate between these two signals in order to eliminate the undesired one. All told, the basic conceptual problem which is posed in the description of free-electron laser oscillators is that the radiation field is composed of a spectrum of frequencies which must be determined self-consistently along with the motion of the electrons, and which requires the development of a time-dependent formulation of the interaction.

9.1 GENERAL FORMULATION

As discussed in previous chapters, amplification of radiation occurs in a free-electron laser due to the bunching of and subsequent coherent radiation from electrons as they move in the combined wiggler and radiation fields. The electrons are bunched by a second-order axial force that results from the cross product of the first-order wiggle velocities and transverse magnetic fields associated with the radiation and the magnetic wiggler. This ponderomotive force appears as a beat wave, which has a phase velocity equal to the radiation frequency divided by the sum of the radiation and wiggler wave numbers ($\omega/(k_w + k)$). The most important physical processes then, are those which cause individual electrons to fall in and out of phase with the beat wave. These processes can be understood within the context of a simple one-dimensional model [1–3] in which the important dynamical variables are the energies and beat-wave phases of electrons, and the magnitude and phase of the radiation. Further, since the bunching force is a second-order effect it is safe to assume that the dynamical variables change slowly on the time scale associated with the traversal of one period of the wiggler. This leads to an additional simplification in that we can take an average over the short time scale associated with the traversal of one wiggler period. As shown in Chapter 5, this wiggler average removes the effect of the lower beat wave from the analysis of planar wiggler configurations. However, this is adequate for the purposes of determining the average properties of many free-electron lasers. The result is that the bulk energy and beat-wave phase of electrons are described by a pendulum equation in which the effective gravity is determined by the amplitude of the radiation. Further, the amplitude of the radiation evolves self-consistently in space and time in response to the current generated by the bunched electrons. This simple model has been studied for decades in connection with travelling wave tubes. The first section of this chapter is devoted to a simple derivation of these governing equations.

We begin our consideration of the free-electron laser oscillator equations with a discussion of the representation of the radiation field. For simplicity, we start with a one-dimensional model in which the vector potential that generates the radiation field can be represented in complex phasor notation as

$$\delta \boldsymbol{A}(z,t) = \delta A(z,t)\hat{\boldsymbol{e}}_p \exp[ikz - i\omega t] + \text{c.c.}, \tag{9.1}$$

where $\delta A(z,t)$ is a time- and space-dependent complex amplitude, ω and k are the central frequency and wavenumber of the radiation, $\hat{\boldsymbol{e}}_p$ is a polarization vector, and c.c. denotes the complex conjugate. It is assumed that the complex amplitude depends weakly on its arguments in the sense that the amplitude does not change significantly in a distance corresponding to a radiation wavelength or time corresponding to a period of the radiation. The polarization vector describes the relative orientations of the components of the radiation field that will be amplified. For example, a planar wiggler will excite plane-polarized radiation, while a helical wiggler excites circularly polarized radiation. The values of ω and k are chosen to be in the centre of the amplification band and are related by the dispersion relation for the propagation of radiation in the absence of an electron beam. For example, $\omega = ck$ for plane waves propagating in a vacuum. If the interaction occurs in a waveguide, the dispersion relation is modified to account for the cut-off frequency. It should be remarked that the exact values of the frequency and wavenumber are somewhat arbitrary. Different choices of ω and k will lead to differences in the solution for the amplitude but will presumably leave the actual vector potential unaffected. A precise definition of ω and k will be specified as appropriate examples are discussed.

Within the context of the one-dimensional model under consideration, the important coordinates of a particle are its phase relative to the ponderomotive beat wave formed by the beating of the wiggler and radiation fields

$$\psi = (k + k_w)z - \omega t, \tag{9.2}$$

and the electron's energy $\gamma m_e c^2$, where $\gamma \equiv (1 - v^2/c^2)^{-1/2}$ is the relativistic factor. As the electrons travel through the combined fields of the wiggler and the radiation their phases and energies evolve in time according to the pair of equations,

$$\frac{d}{dt}\psi = (k + k_w)v_z - \omega, \tag{9.3}$$

and

$$\frac{d}{dt}\gamma = -\frac{e}{m_e c}\mathbf{v}\cdot\delta\boldsymbol{E}, \tag{9.4}$$

where $\delta\boldsymbol{E}$ denotes the radiation electric field, and $\mathbf{v}$ is the electron velocity.

Consider the limit in which the interaction length L is many wiggler periods long (i.e. $\lambda_w \ll L$). One consequence of a long interaction region is that the spectrum of the radiation is narrow. This is evidenced by both the gain bandwidth (see equation (4.59) for the low-gain regime), and the bandwidth of spontaneous emission given by (3.24), both of which scale as λ_w/L. Thus, the fractional bandwidth scales as the reciprocal of the number of wiggler periods, and the assumption that the radiation amplitude varies slowly in time on the scale of the wave period is justified. A further consequence of this assumption is that the electric field in (9.4) may be evaluated by ignoring the time derivatives of the complex amplitude $\delta A(z, t)$

$$\delta\boldsymbol{E}(z, t) = \frac{i\omega}{c}\delta A(z, t)\hat{e}_p \exp(ikz - \omega t) + \text{c.c.} \tag{9.5}$$

The change in both phase and energy of an electron over one wiggler period is also expected to be small for a long interaction region. Recall that the efficiency will be a maximum and the saturation point will be found when an electron completes half of a synchrotron oscillation in the ponderomotive potential in the time it takes to traverse the interaction region. Alternatively, the ponderomotive phase will change by order unity on traversing the interaction region. Thus, the net change in phase in traversing one period of the wiggler will be small.

With the previous discussion in mind, equations (9.3) and (9.4) can be integrated by perturbation about the single-particle trajectories over a time equal to the travel time in traversing one period of the wiggler. In this case, it is assumed that the velocity (on the right-hand side of the equations) is given solely by its value in the presence of the wiggler field. Furthermore, the slow variation of the amplitude of the radiation vector potential can be ignored on the space and time scales associated with the traversal of one wiggler period. The average rate of change of phase and energy, therefore, can be expressed as

$$\frac{\Delta\psi}{\Delta T} = (k + k_w)v_\parallel - \omega, \tag{9.6}$$

and

$$\frac{\Delta\gamma}{\Delta T} = -\frac{i\omega}{c}\delta a \int_t^{t+\Delta T} \frac{dt'}{\Delta T} \mathbf{v}(z')\cdot\hat{\mathbf{e}}_p \exp(ikz' - i\omega t') + \text{c.c.}, \tag{9.7}$$

where $\delta a \equiv e\delta A/m_e c^2$ is the normalized vector potential,

$$\Delta T \equiv \int_z^{z+\lambda_w} \frac{dz'}{v_z(\gamma, z')} \tag{9.8}$$

is the travel time over one wiggler period, $v_\parallel \equiv \lambda_w/\Delta T$ is the wiggler-averaged axial velocity, and z' in equation (9.7) represents the position of a particle at time t'.

For the sake of simplicity, we will complete this derivation under the assumption of a helical wiggler. The discussion of the case of a planar wiggler will be given later. The helical wiggler case is easier to analyse due to the fact that the axial velocity of the beam in the wiggler is constant. In contrast, the axial velocity of an electron in a planar wiggler varies as it travels through the periodic magnetic field. While this complicates the analysis, the final equations for a wiggler-averaged orbit analysis of a planar wiggler configuration are formally identical to those for a helical wiggler. In the case of the idealized helical wiggler field given by equation (2.5), the single-particle electron velocity is given by

$$\mathbf{v}(z) = v_w[\hat{\mathbf{e}}_x \cos k_w z + \hat{\mathbf{e}}_y \sin k_w z] + v_\parallel \hat{\mathbf{e}}_z, \tag{9.9}$$

where $v_w = -\Omega_w/k_w$ is the wiggler-induced transverse velocity (see equations (2.7) and (2.8) for an idealized beam in which the canonical momenta vanish), and $v_\parallel$ is given by solution of equation (2.16). The appropriate polarization vector in this case describes a circularly polarized wave for which

$$\hat{\mathbf{e}}_p = \frac{1}{2}(\hat{\mathbf{e}}_x + i\hat{\mathbf{e}}_y), \tag{9.10}$$

and the time to traverse one wiggler period is $\Delta T = \lambda_w/v_\parallel$. Insertion of equations (9.9) and (9.10) into (9.7) allows the integral in equation (9.7) to be evaluated, yielding the average rate of change of energy

$$\frac{\Delta\gamma}{\Delta T} = -\frac{\omega}{2\gamma} a_w [i\delta a(z,t)\exp(i\psi) + \text{c.c.}], \tag{9.11}$$

where $a_w/\gamma \equiv v_w/c$.

Consider the energy dependence of the rate of change of phase given by equation (9.6). There exists a resonant energy $\gamma_r m_e c^2$ for which the phase is constant. At this energy, the average velocity of a particle is equal to the phase velocity of the beat wave, $v_\parallel = \omega/(k + k_w)$. Electrons with energies far from this resonance value will not interact strongly with the radiation and, hence, will not contribute to amplification. The width of the resonance can be estimated by consideration of the range of energies for which the total change in phase in traversing the interaction region will be of order unity; specifically,

$$|\gamma - \gamma_r| = \left[L(k + k_w) \frac{\gamma_r}{v} \frac{dv}{d\gamma_r} \right]^{-1}. \tag{9.12}$$

For a long interaction region, the relative deviation of an electron's energy from the resonant value will be small. As a consequence, a perturbation analysis of the dynamical equations which govern the phase and energy about the resonant energy may be performed in which $\gamma = \gamma_r + \delta\gamma$, where it is assumed that $|\delta\gamma| \ll \gamma_r$. The average rate of change of phase and energy are evaluated along the trajectory of a beam electron, and the time derivatives are expressed in Eulerian form as

$$\frac{\Delta\psi}{\Delta T} = \left(\frac{\partial}{\partial t} + v_{\parallel} \frac{\partial}{\partial z} \right) \psi = \omega_\gamma \delta\gamma, \tag{9.13}$$

and

$$\frac{\Delta\gamma}{\Delta T} = \left(\frac{\partial}{\partial t} + v_{\parallel} \frac{\partial}{\partial z} \right) \delta\gamma = -\frac{\omega}{2\gamma_r} a_w [i\delta a(z,t) \exp(i\psi) + \text{c.c.}], \tag{9.14}$$

where

$$\omega_\gamma \equiv (k + k_w) \frac{dv}{d\gamma_r} \cong \frac{\omega}{v_{\parallel}} \frac{dv_{\parallel}}{d\gamma_r} \tag{9.15}$$

is related to the derivative of the average velocity with respect to the relativistic factor. Observe that the Eulerian treatment describes the convective derivatives of the phase and energy along the electron trajectories. The importance of this will be seen in the later treatment of beam slippage.

Equations (9.13) and (9.14) determine the trajectory of a single electron in the combined wiggler and radiation fields. These two coupled first-order differential equations require two initial conditions for solution: specifically, the values of $\delta\gamma$ and ψ for an electron when it enters the interaction region. With multiple electrons, this information must be specified for them all. This is accomplished by definition of a distribution function $f_0(\delta\gamma, \psi, t)$ where $f_0(\delta\gamma, \psi, t) d\delta\gamma d\psi$ denotes the fraction of electrons which enter the interaction region at time t with energies and phases in the interval $\gamma\delta$ through $\delta\gamma + d\delta\gamma$ and ψ and $\psi + d\psi$ respectively. Thus, the normalization requirement on this distribution is chosen as

$$\int_{-\infty}^{\infty} d\delta\gamma \int_0^{2\pi} d\psi f_0(\delta\gamma, \psi, t) = 1. \tag{9.16}$$

Observe that if electrons enter the interaction region with entrance times distributed uniformly over a wave period, then it is appropriate to assume that the distribution function is independent of the phase.

Equations (9.13) and (9.14) are equivalent to the pendulum equation discussed in Chapter 4. These equations of motion can be derived from a Hamiltonian which has the form

$$H(\delta\gamma, \psi, z, t) = \frac{\omega_\gamma}{2}\delta\gamma^2 + \frac{\omega}{2\gamma_r}a_w[\delta a(z,t)\exp(i\psi) + \text{c.c.}]. \tag{9.17}$$

Hamilton's equations,

$$\frac{d}{dt}\psi = \frac{\partial H}{\partial \delta\gamma} \text{ and } \frac{d}{dt}\delta\gamma = -\frac{\partial H}{\partial \psi}, \tag{9.18}$$

then yield the dynamical equations (9.13) and (9.14), where the total time derivative follows the trajectory of an electron moving with the resonant velocity. If the radiation amplitude is independent of space and time then the Hamiltonian is conserved by the electron motion and one may draw the phase-space trajectories and separatrices as was done in Chapter 4. In the general case however, the field amplitude depends on both space and time, leading to extremely complicated electron motion.

The electron equations of motion can be replaced by the Vlasov equation which describes the evolution of the local distribution function $f(\delta\gamma, \psi, z, t)$

$$\left(\frac{\partial}{\partial t} + v_\parallel \frac{\partial}{\partial z}\right) f + \frac{\partial H}{\partial \delta\gamma}\frac{\partial f}{\partial \psi} - \frac{\partial H}{\partial \psi}\frac{\partial f}{\partial \delta\gamma} = 0. \tag{9.19}$$

The solution of the Vlasov equation must satisfy the boundary condition $f(\delta\gamma, \psi, z = 0, t) = f_0(\delta\gamma, \psi, t)$ representing the distribution of $\delta\gamma$ and ψ values for the injected beam. The particle equations of motion and the Vlasov equation describe exactly the same physics, and the particular choice of formulation is largely a matter of convenience.

It remains to determine the equations that describe the evolution of the radiation field. To do this, the specific form of the radiation vector potential is inserted into the one-dimensional wave equation (5.19). Observe that by writing the electric field in terms of the vector potential of an electromagnetic wave and not including a scalar potential, it is assumed implicitly that collective (i.e. Raman) effects may be neglected. For the specific choice of the vector potential in equation (9.1), the wave equation takes the form

$$\left(\frac{\partial}{\partial t} + v_g \frac{\partial}{\partial z}\right)\delta A(z,t) = \frac{2\pi i c}{\omega}\hat{e}_p^* \cdot \boldsymbol{J}(z,t)\exp(-ikz + i\omega t), \tag{9.20}$$

where $\boldsymbol{J}(z,t)$ is the current density, and $v_g = kc^2/\omega$ denotes the group velocity of the radiation. Note that only first-derivative terms of the complex amplitude have been retained due to the assumed slow space and time variation of the amplitude. For the plane-wave case developed here, $\omega = kc$ and the group velocity equals the speed of light. In a waveguide, ω and k satisfy a different dispersion relation which results in a group velocity which is below the speed of light. The distinction between the group velocity and the speed of light will be maintained here in the interest of generality even though they are identical for this particular example. The components of the current density which resonantly drive the radiation field can be computed by forming the product of the local beam charge density and the single-particle electron-velocity vector, $\boldsymbol{J}(z,t)=\rho(z,t)\mathbf{v}(z)$, where $\mathbf{v}(z)$ is given by equation (9.9) for the case of a helical wiggler under consideration.

The beam charge density $\rho(z,t)$ evolves into a nonuniform state in space and time owing to the bunching in phase of the beam particles as they interact with the radiation. The charge density can be visualized as consisting of clumps of charge moving at the resonant velocity $v_{\parallel}$, with a spatial wavenumber $k_b=\omega/v_{\parallel}$. This corresponds to a frequency of oscillation equal to that of the radiation. The current density, which is the product of the charge density and the electron velocity, will have a frequency equal to that of the radiation and a spatial variation with wavenumbers which are the sum and difference of the wiggler wavenumber and the wave number k_b of the charge density. According to the resonance condition ($\omega=(k+k_w)v_{\parallel}$), only the portion of the current density whose spatial wave number is the difference of the wiggler and charge-density wavenumbers will excite the radiation resonantly. This portion of the current density which interacts resonantly with the radiation can then be expressed in terms of an average of the factor $\exp(-i\psi)$ as follows

$$\boldsymbol{J}(z,t)\exp(-ikz+i\omega t)\rightarrow J(t_e)\frac{v_w}{v_{\parallel}}\hat{e}_p\langle\exp(-i\psi)\rangle, \tag{9.21}$$

where $J(t_e)$ is the current density entering the interaction region at time t_e. It is important to observe that the transverse current for the choice of a helical wiggler is precisely that required to excite the circularly polarized wave. The function $t_e(z,t)$ specifies the entrance time of the group of electrons which contribute to the high-frequency current at the point z and time t,

$$t_e(z,t)=t-z/v_{\parallel}. \tag{9.22}$$

This entrance time function will appear later in the discussion of slippage.

The average appearing in equation (9.21) can be represented as an average over the initial energies and entrance phases of beam particles injected at time t_e, as follows

$$\langle\exp(-i\psi)\rangle=\int_{-\infty}^{\infty}d\delta\gamma_i\int_0^{2\pi}d\psi_i f_0(\delta\gamma_i,\psi_i,t_e)\exp[-i\psi(\delta\gamma_i,\psi_i,t_e)], \tag{9.23}$$

where the subscript 'i' refers to the initial condition of the electron upon entry into the interaction region, $\psi(\delta\gamma_i,\psi_i,t_e)$ is the phase of the particle launched at time $t_e(z,t)$ with initial values of phase and energy deviation given by ψ_i and $\delta\gamma_i$, respectively. Equivalently, if the Vlasov equation has been solved, then the average can be taken over the distribution of energies and phases at the specific point (z,t)

$$\langle\exp(-i\psi)\rangle=\int_{-\infty}^{\infty}d\delta\gamma\int_0^{2\pi}d\psi f(\delta\gamma,\psi,z,t_e)\exp(-i\psi). \tag{9.24}$$

Once again, we observe that these two averages are equivalent and which is used depends largely on convenience. In terms of these averages, the final wave equation can now be written in the form

$$\left(\frac{\partial}{\partial t}+v_g\frac{\partial}{\partial z}\right)\delta A(z,t)=\frac{2\pi i}{\omega}\frac{c^2 a_w}{\gamma_r v_\parallel}J(t_e)\langle\exp(-i\psi)\rangle. \tag{9.25}$$

Solution of this equation requires specifying the complex radiation amplitude at the entrance of the interaction region, $\delta A(z=0,t)$.

In order to complete this formulation, an equation must be introduced which describes the feedback loop. The simplest model for this process is one in which the radiation propagates without dispersion through an optical system from the end of the interaction region at $z=L$ to the entrance at $z=0$, taking a time T_r and suffering an amplitude loss factor μ. The field amplitude at $z=0$, therefore, can be expressed as

$$\delta A(z=0,t)=\mu\exp[i(kL+\omega T_r)]\delta A(z=L,t-T_r), \tag{9.26}$$

where μ may be complex. The central frequency and wavenumber can be specified by the requirement that the product $R\equiv\mu\exp[i(kL+\omega T_r)]$ is real. In this case, the natural frequencies of the cavity formed by the feedback loop and the interaction region are given by

$$\omega_n=\omega+\frac{2\pi n+i\ln R}{T}, \tag{9.27}$$

where n is an integer (positive or negative) that labels the different longitudinal modes of the oscillator cavity, and

$$T\equiv T_r+\frac{L}{v_g}, \tag{9.28}$$

is the total round-trip time for radiation to circulate through the cavity. The imaginary part of the natural frequency ω_n describes the damping of the radiation in the cavity due to losses and output coupling. The power will decay exponentially with a characteristic time T_d given by

$$T_d\equiv\frac{T}{|\ln R^2|}. \tag{9.29}$$

The quality factor for the cavity is defined based upon this decay rate as

$$Q\equiv\omega T_d. \tag{9.30}$$

Due to the assumed narrowness of the spectrum and the fact that all frequencies are attenuated by the same factor R, the quality factor is essentially the same for all modes. Thus, in the present model there is no mechanism to control the frequency of oscillation other than the natural gain bandwidth of the free-electron laser amplification mechanism. Solutions of this governing system of equations (equations (9.13), (9.14) and (9.25) and the associated boundary conditions) describe a variety of different time-dependent phenomena which can occur in a free-electron laser oscillator. A discussion of these solutions for both continuous and pulsed electron-beam systems forms the bulk of the remaining discussion of free-electron laser oscillators.

9.2 PLANAR WIGGLER EQUATIONS

If the particular device under consideration has a planar wiggler, then the steps leading to the derivation of equations (9.13), (9.14) and (9.25) need to be modified. The important difference that arises in the consideration of planar wigglers is that the instantaneous axial velocity is no longer a constant as an electron propagates through one period of wiggler field. For example, in the planar wiggler field given by equation (2.99) the electrons will have a velocity $\mathbf{v}$ given by

$$\mathbf{v} = v_w \hat{\mathbf{e}}_x \cos k_w z + v_z \hat{\mathbf{e}}_z, \tag{9.31}$$

where the instantaneous z-dependent axial velocity is determined from energy conservation to be

$$v_z = c\left(1 - \frac{1}{\gamma^2} - \frac{v_w^2}{c^2}\cos^2 k_w z\right)^{1/2}, \tag{9.32}$$

and it has been assumed that the electrons are injected into the wiggler with no transverse canonical momentum. The wiggler velocity for this case is in the x-direction and, as a result, the radiation will be plane polarized with $\hat{\mathbf{e}}_p = \hat{\mathbf{e}}_x$.

The periodic modulations in the axial velocity cause the ponderomotive phase ψ given in (9.2) to have periodic modulations as well. Equation (9.6), which is obtained by averaging over the travel time through one period of the wiggler, represents the rate of change of the average value of the ponderomotive phase with the periodic modulations removed. Thus, it is convenient to write ψ as the sum of its slowly varying average part ψ_0, and the periodic modulations $\delta\psi$ with period equal to that of the wiggler. The periodic modulation is expressed in terms of the difference in the time taken to reach a particular point in the wiggler between an electron travelling at the instantaneous velocity and one travelling at the average velocity; hence

$$\delta\psi = -\omega \int_{z_0}^{z} dz' \left(\frac{1}{v_z} - \frac{1}{v_\parallel}\right), \tag{9.33}$$

where $v_\parallel$ denotes the average axial velocity defined in equation (2.105). The time integral appearing in equation (9.7) can now be evaluated as an integral over z',

$$\int_t^{t+\Delta T} \frac{dt'}{\Delta T} \mathbf{v}(z') \cdot \hat{\mathbf{e}}_p \exp(ikz' - i\omega t') = \frac{1}{2} v_w C_1 \exp(i\psi_0), \tag{9.34}$$

where C_1 is defined by the integral

$$C_1 \equiv \int_z^{z+\lambda_w} \frac{dz'}{v_z \Delta T} [1 - \exp(-2ik_w z')] \exp[i\delta\psi(z')]. \tag{9.35}$$

The modification to the equations of motion for the planar wiggler case, therefore, can be accounted for by replacing a_w by the product $a_w C_1$.

Similar considerations apply to the derivation of the wave equation (9.25). The periodic modulation of the electron phase must be accounted for in the excitation of the radiation. The result is that the factor a_w appearing in (9.25) is replaced by the product $a_w C_1^*$.

Finally, it is a straightforward procedure to incorporate the effects of higher harmonics in the radiation field in the present model. Let δa_l be the complex amplitude of the lth harmonic of the radiation having a frequency $l\omega$ and a wavenumber lk. These harmonics contribute to the rate of change of particle energy in equation (9.14) according to the substitution

$$\delta a \exp(i\psi) \rightarrow \sum_l \delta a_l C_l \exp(il\psi_0), \tag{9.36}$$

where a coefficient C_l is defined in analogy to equation (9.35),

$$C_l \equiv \int_z^{z+\lambda_w} \frac{dz'}{v_z \Delta T} [1 - \exp(-2ik_w z')] \exp[il\delta\psi(z')]. \tag{9.37}$$

The dependence of these coefficients on harmonic number has been discussed in Chapter 7. In particular, C_l is non-zero only for odd values of the harmonic number l and decreases with harmonic number. The wave equation must also be modified in the sense that a separate equation must be written for each harmonic amplitude δA_l, which takes the form,

$$\left(\frac{\partial}{\partial t} + v_g \frac{\partial}{\partial z}\right) \delta A_l = \frac{2\pi i\, c^2 a_w}{l\omega\ \ \gamma_r v_\parallel} C_l^* \exp(-il\psi_0). \tag{9.38}$$

The question of harmonic generation is important in free-electron laser oscillators because it is possible, by control of the optical feedback path, to cause the device to lase in a particular harmonic of the fundamental resonant frequency. In this way, shorter wavelengths can be generated without either increasing beam energy or decreasing the wiggler period.

9.3 CHARACTERISTICS-SLIPPAGE

In the preceding sections, a system of equations describing the interaction of the radiation and beam particles in a free-electron laser oscillator has been derived. The system consists of an ensemble of first-order differential equations where the time and space derivatives represent a convective time derivative following the trajectory of the beam electrons, equations (9.13), (9.14) and (9.19), or following the trajectory of the radiation, equation (9.25). It is usually the case that the radiation propagates faster than the beam. Thus, radiation which enters the interaction region at a particular time will overtake some group of electrons which have entered earlier. This is known as slippage, and its effects can be most easily understood by plotting the characteristics of the first-order partial differential equations in the z versus t plane [4].

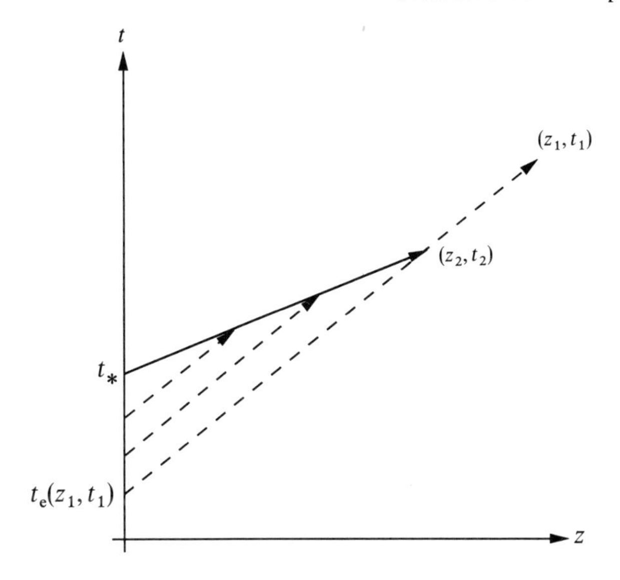

Fig. 9.1 Illustration of the characteristics in an oscillator configuration.

Figure 9.1 shows a plot of the relevant characteristics. The orbit equations can be integrated to find either the distribution function or the phases and energies of particles at a particular space-time point (z_1, t_1). The group of electrons which arrive at the point z_1 at time t_1 have moved along a trajectory $t = t_1 + (z - z_1)/v_\parallel$. This trajectory is known as a characteristic for the equation of motion since the convective derivative in equations (9.13), (9.14) and (9.19) can be written as a total derivative along this trajectory (shown as a dashed line in the figure). Electrons arriving at the space-time point (z_1, t_1) are only affected by the complex amplitude of the radiation $\delta a(z, t)$ evaluated for points on the characteristic trajectory. For example, one can write the solution of equation (9.14) as an integral along this characteristic

$$\delta\gamma = \delta\gamma_i + \frac{\omega a_w}{2\gamma_r}\int_0^{z_1}\frac{dz}{v_\parallel}[i\delta a(z,t)\exp(i\psi) + \text{c.c}], \tag{9.39}$$

where $\delta\gamma_i$ is the injected value of $\delta\gamma$, and z and t are related by the condition $t = t_1 + (z - z_1)/v_\parallel$.

To evaluate the integral in equation (9.39), the value of the radiation amplitude for points (z, t) on the characteristic must be determined. Let the space-time point (z_2, t_2) be a point on this characteristic where the complex amplitude of the radiation field is to be evaluated. To determine the field at this point, the wave equation (9.25) is integrated along its characteristics: that is, along a trajectory given by $t = t_2 + (z - z_2)/v_g$,

$$\delta A(z_2, t_2) = \delta A_e(t_2 - z_2/v_g) + \frac{2\pi i c^2 a_w}{\omega\gamma_r v_\parallel}\int_0^{z_2}\frac{dz}{v_g}J[t_e(z,t)]\langle\exp(-i\psi)\rangle, \tag{9.40}$$

where $\delta A_e(t)$ denotes the radiation amplitude upon entering the interaction region at time t. The characteristic for the wave equation is shown as a solid line in Fig. 9.1.

In order to perform the integral in (9.40), it will be necessary to know the high-frequency current density at all points along the wave-equation characteristic ending at (z_2, t_2). To determine this current density, it is necessary to integrate the equations of motion along a number of different characteristics of the type shown as dashed lines in Fig. 9.1. In this way, the value of the complex amplitude of the radiation field at (z_2, t_2) becomes dependent on the values of the radiation amplitude at the entrance of the interaction region for a range of times given by $t_* = t_2 - z_2/v_g > t > t_e(z_2, t_2) = t_2 - z_2/v_\parallel$. Different portions of the radiation pulse can influence each other in this way.

An alternative to writing the electron trajectories in the form of integrals along characteristics in the (z, t) plane is to express the equations of motion in Lagrangian, instead of Eulerian, coordinates. Basically, this amounts to expressing the dependence of the energies and phases of the particles as functions of position z and entrance time $t_e = t - z/v_\parallel$. With this change of variables, the convective derivative along the trajectory of a particle in the (z, t) plane becomes an ordinary derivative with respect to position z, with entrance time t_e held fixed,

$$v_\parallel \frac{d}{dz}\psi(z, t_e) = \omega_\gamma \delta\gamma(z, t_e), \tag{9.41}$$

and

$$v_\parallel \frac{d}{dz}\delta\gamma(z, t_e) = -\frac{\omega}{2\gamma_r} a_w [i\delta a(z, t_e + z/v_\parallel)\exp(i\psi) + \text{c.c}]. \tag{9.42}$$

The space and time dependence of the radiation-field amplitude in equations (9.41) and (9.42) have been left in terms of the Eulerian variables z and t. In order to continue to write the wave equation in Eulerian variables we must still express the high-frequency current density appearing in equations (9.25) and (9.40) in Eulerian variables

$$J(t_e)\langle\exp(-i\psi)\rangle = J[t_e(z,t)]\langle\exp\{-i\psi[z, t_e(z,t)]\}\rangle, \tag{9.43}$$

where $\psi(z, t_e)$ is the solution for the phase which satisfies the Lagrangian equations (9.41) and (9.42).

A particularly simple case is the one in which the group velocity of the radiation is equal to the beam velocity. In this case, the characteristics of the wave equation and the equations of motion coincide. Since the electrons and radiation enter the interaction region and propagate together, the complex amplitude of the radiation at any space-time point (z_2, t_2) is, therefore, dependent only on the radiation which entered the interaction region at the time $t_e = t_2 - z/v_\parallel$ and on the distribution function and current for the beam particles entering at that time. In this case different portions of the radiation pulse never communicate with one another. While it seems that the case of equal beam and radiation speeds is rather special, it turns out that for highly relativistic electron beams the beam velocity is close enough to the speed of light that in some cases the electron slippage can be neglected.

Figure 9.2 shows the system of characteristics for the determination of the amplitude of the radiation at the entrance of the interaction region $z = 0$ at the time t_3. The wavy line in the figure represents the return path of the radiation through the feedback loop. From the figure, it can be seen that the amplitude of the radiation entering the interaction region at time t_3 depends on the radiation that entered in the past between times $t_3 - T$ and $t_3 - T(1 + \varepsilon)$, where

$$\varepsilon = \frac{L}{T}\left(\frac{1}{v_\parallel} - \frac{1}{v_g}\right) = \frac{L}{L_c}\left(\frac{v_g}{v_\parallel} - 1\right) \tag{9.44}$$

is the slippage parameter which measures how much the electrons slip behind the radiation during a time equal to the round-trip travel time of radiation through the optical length L_c of the cavity, where $T = L_c/v_g$.

Also shown on the left-hand side of Fig. 9.2 are plots which characterize the injected electron current in a free-electron laser which is driven by a pulsed electron beam. In the plot, the electron pulses, known as micropulses, are shown arriving at times separated by the round-trip travel time of radiation in the cavity so that the returning radiation pulse overlaps with the arrival of the next beam pulse. It will be seen at a later stage of the discussion that due to an effect known as **laser lethargy** [5], the electron pulses actually should have a separation in arrival times which is slightly larger than the round-trip travel time. The width of the electron pulse is characterized by a time T_p. A measure of the importance of slippage is the parameter $\varepsilon T/T_p$ which is the ratio of the distance the electron pulse slips behind the radiation pulse in traversing the interaction length to the

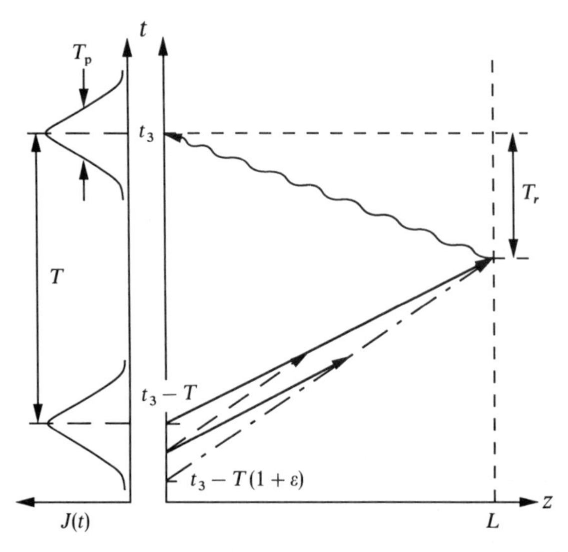

Fig. 9.2 Schematic illustration of the synchronism and slippage conditions between the electron beam pulses and the round-trip time of the radiation within the cavity.

spatial width of the electron pulse. This parameter is, typically, smaller than unity. In a free-electron laser oscillator which is driven by a continuous electron beam the relevant parameter is simply ε, which can be extremely small. Since it is through the process of slippage that different portions of the radiation pulse communicate with one another, the size of the slippage parameter determines the rate at which temporal coherence is established in the radiation pulse.

The integration in equations (9.39) and (9.40) can be performed for a relatively simple analytic formulation, known as the klystron model [4]. The coupling between the radiation field and the electrons is mediated by the wiggler field, as evidenced by the appearance of the constant a_w in both the equation of motion (9.14) and the wave equation (9.25). If the amplitude of the wiggler field were a function of position z, this would result in the coupling coefficient having a z-dependence as well. In principle, this would lead to an axial dependence of the resonant velocity $v_\parallel(\gamma_r)$ if the wiggler parameter were sufficiently large. This possibility will be neglected in the following discussion. In the klystron model, the axial dependence of the wiggler parameter is chosen to be of the form of two delta functions, one at the entrance of the interaction region and one at the exit

$$a_w(z) = a_{w0} L[\delta(z) + \delta(z - L)], \tag{9.45}$$

where a_{w0} is a constant amplitude giving the strength of the delta function.

The interaction in this model is identical to that in a conventional klystron. Electrons are impulsively kicked in energy by the radiation upon entry to the interaction region (i.e. at $z = 0$), and then ballistically bunch in phase while traversing the interaction region. The radiation is amplified by the coherently bunched particles as they leave the interaction region (i.e. $z = L$). While a delta function seems to be a rather radical dependence for the wiggler field, all that is required is that the distance over which the wiggler parameter is non-zero be short enough that the phase of individual particles remains relatively constant in the region of non-zero wiggler field. Further, the klystron model is only considered here for the insight it provides into the behaviour of the more complicated system.

The equations of motion and the wave equation for the given axial profile of the wiggler field strength can now be solved. An electron entering the interaction region with relativistic factor $\delta\gamma_i$ and phase ψ_i immediately receives a kick in energy. The relativistic factor for such an electron just after the first delta function is given by

$$\delta\gamma = \delta\gamma_i + \frac{\omega a_{w0} L}{\gamma_r v_\parallel} |\delta a_e(t)| \sin[\psi_i + \theta_e(t)], \tag{9.46}$$

where $\delta a_e(t) = |\delta a_e| \exp(i\theta_e)$ is the complex amplitude of the radiation entering the interaction region at time t. The beam electron then travels the length of the interaction region with a constant energy and a phase which increases linearly with time according to equation (9.13). When this electron arrives at the end of the interaction region, it has a phase given by

$$\psi(L, t) = \psi_i + p_i + |X(t_e)| \sin[\psi_i + \theta_e(t_e)], \tag{9.47}$$

where

$$p_i = \frac{L\omega_\gamma}{v_\parallel}\delta\gamma_i \tag{9.48}$$

is a normalized energy deviation,

$$X(t_e) = \frac{\omega\omega_\gamma a_{w0} L^2}{\gamma_r v_\parallel}\delta a_e(t_e) \tag{9.49}$$

is a normalized radiation amplitude, and $t_e(L, t)$ is the entrance time of the electron arriving at the exit at time t. Here, $X(t_e)$ plays the role of the bunching parameter in a klystron.

The complex amplitude of the radiation, which satisfies equation (9.25), will be constant as the radiation propagates through the interaction region except that it will exhibit discontinuous jumps at the points where the wiggler field is non-zero. If the electron beam has a uniform phase distribution upon entry to the interaction region, then there will be no jump at the entrance due to the fact that the phase average vanishes for a uniform distribution. In other words, the beam has yet to be bunched by the radiation. However, the radiation is amplified as it passes through the exit of the interaction region. Using equation (9.40), the complex amplitude of the radiation leaving the interaction region at time t can be expressed in terms of an average over the phases of particles at that point

$$\delta A(L,t) = \delta A_e(t - L/v_g) + \frac{2\pi i c^2 a_{w0} L}{\omega\gamma_r v_\parallel v_g} J(t_e)\langle \exp[-i\psi(L,t)]\rangle. \tag{9.50}$$

The average over injected phases and energies can now be written as

$$\langle \exp[-i\psi(L,t)]\rangle = iX(t_e)g(|X(t_e)|), \tag{9.51}$$

where

$$g(|X|) = -i\iint d\delta\gamma_i d\psi_i \frac{f_0(\delta\gamma_i, t_e)}{|X|}\exp[-i\psi_i - i\theta_e - ip_i - i|X|\sin(\psi_i + \theta_e)] \tag{9.52}$$

can be thought of as a complex, nonlinear gain function for the interaction region. The integral over ψ_i is readily carried out, yielding

$$g(|X|) = 2\pi\frac{J_1(|X|)}{|X|}\int d\delta\gamma_i f_0(\delta\gamma_i, t_e)\exp[-i(p_i - \pi/2)], \tag{9.53}$$

where J_1 is a regular Bessel function of the first kind. The remaining integral over energy can be carried out once the distribution of injected energies is specified. The important feature of the nonlinear gain function is that it decreases as the amplitude X of the radiation increases. This provides for the saturation of growth in the oscillator.

The derivation is now completed by applying the feedback boundary condition (9.26). It is reasonable to express the amplitude of the radiation in terms of the bunching parameter X since it appears in the argument of a Bessel function. When this is done one obtains a two-time-delay equation which describes the nonlinear, multi-frequency behaviour of radiation in a free-electron laser oscillator

$$X(t) = R\{X(t-T) + I_b(t')X(t')g[|X(t')|]_{t'=t-T(1+\varepsilon)}\}. \tag{9.54}$$

The normalized beam current $I_b(t')$ appearing in equation (9.54) is given by

$$I_b(t) = -\frac{2\pi e J(t) a_{w0}^2 L^3 \omega_\gamma}{m_e v_\parallel^3 v_g \gamma_r^2}. \tag{9.55}$$

The various terms in equation (9.54) can be understood with the aid of Fig 9.2. The first term on the right-hand side represents the contribution to the radiation arriving at the entrance of the interaction region which comes from the propagation of radiation along the characteristic indicated with a solid line in Fig. 9.2; that is, the direct path of radiation through the cavity. The second term represents the contribution from the path which is indicated by the dashed-dot line in the figure. This path represents the portion of the signal which is carried by the modulated beam in the interaction region. In the klystron model, it is only necessary to consider one such path since the electrons are affected by the radiation only as they just enter the interaction region. Observe that in the general case of a distributed interaction region, the contributions of the beam will depend on the radiation amplitude for all entrance times between $t - T$ and $t - T(1 + \varepsilon)$.

It is interesting to note that by introduction of normalized variables, an equation with very few parameters has been obtained. These parameters are the reflection coefficient R, the slippage parameter ε, the distribution of injected p_i values, and a normalized current I_b. The time can be normalized to T for a continuous beam or T_p for a pulsed beam. The same parameters can be used to describe the more general system from which the klystron model was derived [2, 6–9], and this will be shown in the next section where oscillator gain is discussed.

This concludes the general discussion of slippage in free-electron laser oscillators. However, extensive use of the basic ideas developed in this section (as well as the klystron model) will be used in subsequent sections and chapters.

9.4 OSCILLATOR GAIN

A free-electron laser oscillator may be characterized according to its single-pass gain. Low-gain oscillators are those in which the radiation amplitude grows by a small factor (for example 10%) over a single pass through the interaction region. High-gain oscillators are those in which the radiation grows by a factor of order unity or larger in traversing the interaction length. It is important to realize that the amplitude of the radiation in a low-gain oscillator can eventually grow large; and may be comparable to the power found in a high-gain oscillator. If almost all the radiation power is fed back to the input of the interaction region, the power can grow to large amplitude after many passes through the interaction region. What is important in both high- and low-gain oscillators is that the radiation losses over a single pass are lower than the gain.

Whether or not an oscillator with a particular set of parameters is a high-gain or low-gain device can be determined by calculating the maximum spatial growth rate using, for example, the theory of Chapter 5. For the sake of clarity and continuity of discussion, we shall briefly re-derive the gain in the Compton regime by an alternative method to that used in Chapter 5. If the spatial growth length in the infinite interaction region limit is much shorter than the actual length of the interaction region then the device must be of the high-gain type. To determine the gain for the Compton-regime model described by equations (9.13), (9.14) and (9.25), we consider the spatial dependence of the amplitude of the vector potential to be of the form

$$\delta A(z,t) = \delta \hat{A} \exp(i\delta k z), \tag{9.56}$$

where δk is a complex wavenumber whose imaginary part determines the spatial rate of growth of the radiation signal and whose real part determines the shift in wavenumber of the radiation from the vacuum case. It has been assumed here that the frequency of the radiation which is being amplified is ω and, thus, the complex amplitude $\delta A(z,t)$ has no explicit time dependence.

To determine the linearized spatial growth rate, it is convenient to use the Vlasov equation (9.19) rather than the individual particle equations of motion (9.13) and (9.14). Thus, the distribution function is expressed as a sum of the value at injection plus a small perturbation

$$f = f_0 + [\delta f \exp(i\delta k z) + \text{c.c.}], \tag{9.57}$$

where the amplitude of the small perturbation satisfies the linearized version of (9.25)

$$\left[i\delta k v_\parallel + \omega_\gamma \delta\gamma \frac{\partial}{\partial\psi} \right] \delta f - \frac{i\omega}{2\gamma_r} \delta\hat{a} \exp(i\psi) \frac{\partial}{\partial\delta\gamma} f_0 = 0, \tag{9.58}$$

where $\delta\hat{a} \equiv e\delta\hat{A}/m_e c^2$ is the normalized amplitude of the vector potential. Equation (9.58) can be readily solved for the perturbed distribution function in the form

$$\delta f = \frac{\omega a_w \delta\hat{a}}{2\gamma_r(\delta k v_\parallel + \omega_\gamma \delta\gamma)} \exp(i\psi) \frac{\partial}{\partial\delta\gamma} f_0. \tag{9.59}$$

The perturbed distribution function, which is proportional to the radiation amplitude is then inserted into the wave equation (9.25), yielding the following dispersion relation

$$v_g \delta k = -\frac{a_w^2 \omega_b^2}{4\gamma_r^2} \int \frac{d\delta\gamma}{\delta k v_\parallel + \omega_\gamma \delta\gamma} \frac{\partial f_0}{\partial\delta\gamma}, \tag{9.60}$$

where $\omega_b^2 \equiv -4\pi e J/(m_e v_\parallel)$ is the nonrelativistic beam plasma frequency. Solutions of equation (9.60) for δk give the local spatial growth rate. If this growth rate satisfies

$$\text{Im}(\delta k L) \ll 1, \tag{9.61}$$

then the device will operate in the low-gain regime.

To assess the gain for a particular device it is necessary to specify the distribution function for the injected beam. All injected-beam electrons have the same energy in the idealized limit, and the distribution is of the form

$$f_0 = \frac{1}{2\pi}\delta(\delta\gamma - \delta\gamma_0). \tag{9.62}$$

In this case, the integral over the energy in the dispersion relation can readily be performed to obtain the standard dispersion relation for travelling wave amplifiers [10, 11]

$$v_g\delta k = -\frac{a_w^2\omega_b^2\omega_\gamma}{4\gamma_r^2(\delta k v_\parallel + \omega_\gamma\delta\gamma)^2}. \tag{9.63}$$

The denominator on the right-hand side of equation (9.63) can be cast into a more familiar form using the definition of ω_γ in equation (9.15)

$$v_\parallel\delta k + \omega_\gamma\delta\gamma_0 = (k_w + k + \delta k)v_\parallel + (k + k_w)\delta v_\parallel - \omega, \tag{9.64}$$

where $(k + k_w)\delta v_\parallel = \omega_\gamma\delta\gamma$. The cubic dispersion relation that results is identical to equation (4.102). The spatial growth rate is maximum when the frequency is such that the average velocity of the beam matches the phase velocity of the ponderomotive wave. This occurs when $\delta\gamma_0 = 0$, in which case

$$(L\delta k)^3 = -\frac{a_w^2\omega_b^2\omega_\gamma L^3}{2\gamma_r^3 v_g v_\parallel^2} = -I_b, \tag{9.65}$$

where I_b is the normalized current density introduced in equation (9.54). As a consequence, the oscillator must be of the low-gain type if the normalized current density satisfies [2, 6–8]

$$I_b \ll 1. \tag{9.66}$$

The maximum spatial growth rate appearing in equation (9.65) was derived for an ideal, monoenergetic beam in an ideal wiggler. It is often the case that these assumptions are not satisfied and, as a result, the gain is reduced substantially from the ideal value. A discussion of the various causes for the reduction of gain by thermal effects can be found in Chapter 4 and will not be repeated here. However, it should be emphasized that these effects are important and often are the crucial factors limiting the gain in a particular device.

9.5 THE LOW-GAIN REGIME

It is possible to further reduce the governing system of equations (9.13), (9.14) and (9.25) for devices which operate in the low-gain regime [2, 4, 8]. In this case, the change in the amplitude of the radiation during each circuit of the cavity will be small. Hence, the time dependence of the radiation amplitude will be nearly periodic with a period equal to the round-trip time of radiation in the cavity. However, it is important to recognize that the radiation amplitude can still have large variations on the time scale associated with the round-trip travel time of radiation through the cavity. For example, in the case of an oscillator driven by a pulsed electron beam as illustrated in Fig. 9.2, it can be expected that the radiation entering the interaction region will develop into pulses which match the pulses of the beam current. Because the gain and losses are both small, the radiation pulses will vary little from pulse to pulse. That is, the radiation entering the interaction region will be nearly periodic in time with a period given by the separation in arrival times of successive beam pulses which must be nearly the same as the round-trip travel time T. The shape of the radiation pulses can change substantially but only after a large number of circuits of the cavity.

Consider one round trip of the radiation in the cavity by beginning with the wave equation integrated along its characteristic as is done in equation (9.40). In the low-gain limit, the second term on the right-hand side of (9.40), which represents the increase in amplitude of the radiation due to the beam, will be small compared to the first term. Thus, to a first approximation, the radiation amplitude in the interaction region can be expressed in terms of the amplitude of the radiation entering the interaction region

$$\delta A(z,t) \cong \delta A_e(t - z/v_g). \tag{9.67}$$

This space-time dependence can then be inserted into the equations of motion, which in Lagrangian coordinates become

$$v_\parallel \frac{d}{dz}\psi = \omega_\gamma \delta\gamma, \tag{9.68}$$

and

$$v_\parallel \frac{d}{dz}\delta\gamma = -\frac{\omega}{2\gamma_r} a_w\{i\delta a_e[t + z(v_\parallel^{-1} - v_g^{-1})]\exp(i\psi) + \text{c.c.}\}. \tag{9.69}$$

The solutions for the phases of particles may now be inserted in the wave equation (9.40) to determine the small gain that the radiation experienced on one transit of the interaction region

$$\delta A(L,t_2) = \delta A_e(t_2 - L/v_g) + \frac{2\pi i c^2 a_w}{\omega\gamma_r v_\parallel}\int_0^L \frac{dz}{v_g} J(t_e)\langle\exp[-i\psi(z,t_e)]\rangle, \tag{9.70}$$

where $t_e = t_e[z, t_2 + (z - L)/v_g] = t_2 - L/v_g - z(v_\parallel^{-1} - v_g^{-1})$. Note that the phase variable has been written in terms of its Lagrangian coordinates as discussed in (9.43).

The feedback boundary condition (9.26) can now be used to relate the radiation amplitude entering the cavity at time $t + T$ to the amplitude entering earlier

$$\delta A_e(t + T) = R\left[\delta A_e(t) - \frac{2\pi i c^2 a_w}{\omega\gamma_r v_\parallel}\int_0^L \frac{dz}{v_g} J(t_e)\langle\exp[-i\psi(z,t_e)]\rangle\right], \tag{9.71}$$

where it has been assumed that $t_2 = t + T - T_r$; hence, $t_e = t - z(v_\parallel^{-1} - v_g^{-1})$.

The similarity of equation (9.71) with the klystron model (9.53) is evident. In particular, the first term on the right-hand side of equation (9.71) represents the direct propagation of radiation through the cavity while the second term represents the contributions of the beam. This term depends on the complex amplitude of the radiation entering the cavity only for times $t > t' > t - \varepsilon T$.

As previously discussed, it is expected that the solutions of equation (9.71) will yield a radiation amplitude which is nearly periodic in time with a period equal to the arrival time of electron pulses which must be close to the round-trip time for the radiation. To describe this expected time dependence of the radiation pulse, it is convenient to introduce a new time variable t' defined by

$$t' = t - nT_a, \tag{9.72}$$

where T_a is the separation in arrival times of the beam pulses (known as micropulses) and n is the integer that puts t' in the range $-T_a/2 < t' < T_a/2$. Thus, the dependence of the radiation amplitude on the variable t' will determine the shape of the pulse and the dependence on the integer n will describe how the pulse evolves after many circuits of the resonator. In terms of a multiple time-scale perturbation theory, t' is described as a fast time variable and the index n as a slow time variable. The amplitude of the radiation entering the cavity as a function of time is written in the form

$$\delta A_e(t) = \delta A_e(n, t'), \tag{9.73}$$

and the amplitude of radiation entering a period T later is given by

$$\delta A_e(t + T) = \delta A_e(n + 1, t' + T_\delta), \tag{9.74}$$

where $T_\delta = T - T_a$ is the difference in round-trip time of the radiation and the separation in the arrival times of successive beam pulses. The time T_δ measures what is known as cavity detuning, and it will be found that it must be negative in order that arbitrarily small initial radiation amplitudes build up to saturation in a pulsed-beam oscillator [5].

The fact that the gain is small and the reflection coefficient R is close to unity allows considerable simplifications to be made. First, the low-gain assumption is that the radiation amplitude changes slowly with n. Second, T_δ may be assumed to be small in comparison with the temporal width of the electron pulse T_p. If T_δ is not small the radiation and beam pulses drift out of step before the interaction between the beam and the radiation can act to force synchronism. Therefore, a Taylor expansion in the dependence of the two arguments of δA_e yields a partial-differential-integral equation for the radiation amplitude

$$\left(\frac{\partial}{\partial n} + T_\delta \frac{\partial}{\partial t'}\right)\delta A_e = (R - 1)\delta A_e + \frac{2\pi i c^2 a_w}{\omega \gamma_r v_\parallel}\int_0^L \frac{dz}{v_g} J(t_e)\langle \exp[-i\psi(z, t_e)]\rangle \tag{9.75}$$

where we have assumed that $R \approx 1$ in the term containing the response of the beam.

At this point, it is convenient to introduce normalizations, as in the case of the klystron model. First, axial distance is normalized to the length of the interaction region, with the dimensionless distance given by $\xi = z/L$. Second, the energy deviation $\delta\gamma$ is normalized as in equation (9.48)

$$p = \frac{L\omega_\gamma}{v_\parallel}\delta\gamma, \tag{9.76}$$

where p is the normalized energy deviation. Note that the variable p can also be thought of as a normalized detuning parameter. In particular, p can also be expressed in terms of the average parallel velocity as

$$p = \frac{L}{v_\parallel}[(k + k_w)(v_\parallel + \delta v_\parallel) - \omega]. \tag{9.77}$$

Thus, p is the frequency detuning of a particle with energy $(\gamma + \delta\gamma)m_e c^2$ normalized to the time of flight through the interaction region. The choice of the letter p is motivated by the fact that this variable plays the role of a momentum in terms of the pendulum Hamiltonian.

The radiation amplitude is now normalized in the same way as the bunching parameter introduced in the klystron model was; specifically

$$X(n, t') = \frac{\omega\omega_\gamma a_w L^2}{\gamma_r v_\parallel}\delta a_e(n, t'), \tag{9.78}$$

where $\delta a_e \equiv e\delta A_e/m_e c^2$. With these normalizations, the Lagrangian equations of motion can be expressed as

$$\frac{\partial\psi}{\partial\xi} = p, \tag{9.79}$$

and

$$\frac{\partial p}{\partial\xi} = -\frac{1}{2}[iX(n, t_e + \varepsilon\xi T)\exp(i\psi) + \text{c.c.}], \tag{9.80}$$

where the slippage parameter ε has been introduced as defined by equation (9.44). The normalization for the radiation amplitude is understood by consideration of the discussion of the pendulum equation in Chapter 4. In particular, the normalized amplitude can be expressed as $X = (k_s L)^2$ where k_s is the synchrotron wavenumber for a particle at the bottom of the ponderomotive potential well of the beat wave.

Application of the preceding normalizations to equation (9.75) results in

$$\left(\frac{\partial}{\partial n} + T_\delta\frac{\partial}{\partial t'}\right)X(n, t') = (R - 1)X(n, t') - i\int_0^1 d\xi I_b(t_e)\langle\exp[-i\psi(\xi, t_e)]\rangle, \tag{9.81}$$

where $t_e = t' - \varepsilon\xi T$, and $I_b(t_e)$ is defined in equation (9.55).

The various terms on the left-hand side of equation (9.81) describe the slow evolution of the radiation pulse due to cavity detuning, and the terms on the right-hand side describe the effects of cavity losses and the gain due to interaction of the radiation with the electron beam. The real-time dependence of the amplitude of the radiation field entering the cavity can be obtained from the solution for $X(n, t')$ by replacing n by $t/T_a \approx t/T$ and t' by t. For example, in the absence of an electron beam the solution of (9.81) is

$$X(n, t') = X(0, t' - nT_\delta)\exp[-n(1-R)], \tag{9.82}$$

where $X(0, t')$ is periodic in t' with period T_a (the electron pulse arrival time) and represents the initial pulse shape. In the multiple time-scale variables n and t', the solution (9.82) represents a pulse which drifts to increasing values of t' (for $T_\delta > 0$) as the round-trip circuit index n increases. Furthermore, the amplitude of the pulse decreases with n due to cavity losses. The solution is illustrated schematically in Fig. 9.3. The decay of the radiation field is understood easily in that the power in the pulse decays exponentially in real time with characteristic time constant $T_d = T/[2(1-R)]$, which for R close to unity is identical to the decay rate predicted by equation (9.29). The drifting of the pulse in t' due to the cavity detuning is a consequence of having defined t' (equation (9.72)) in terms of the arrival time T_a of beam pulses, which is not necessarily the same as the round-trip time for radiation T. Equation (9.82) describes the solution in the absence of the electron beam. In this case, the radiation pulses arrive separated by the cavity round-trip time T. Thus, the radiation pulse appears to slip in the **beam pulse frame** by an amount $T_\delta = T - T_a$ on every round trip of the cavity. An alternative way to represent the solution (9.82) is to write the periodic function $X(0, t')$ in terms of a Fourier series with Fourier coefficients $X_m(0)$. Upon replacement of t' by t and n by t/T, equation (9.82) becomes

$$X(t/T, t) = \sum_{m=-\infty}^{\infty} X_m(0)\exp\{-i[2\pi m - i(1-R)]t/T\}, \tag{9.83}$$

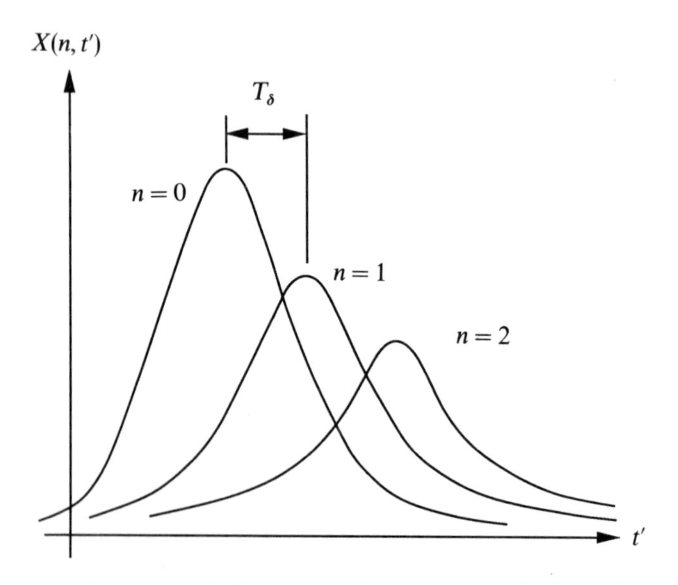

Fig. 9.3 Schematic illustration of a decaying radiation pulse.

which is seen to be a superposition of signals whose frequencies are those of the modes of the empty cavity given by equation (9.27).

Equation (9.81) does not represent the most compact form for the equation describing a low-gain oscillator. In principle, one could normalize the two time variables n and t'. The round-trip index could be normalized to the number of round trips of radiation occurring in the cavity decay time T_d. This would be accomplished by dividing (9.81) by $2(1-R)$. The normalized current would then enter only in combination with the quantity $2(1-R)$. Physically, this is a consequence of the fact that the important parameter measuring the current in a low-gain oscillator is the ratio of the current to the losses. Equivalently, the ratio of the operating current to the start current could be specified, which also depends only on the ratio of the current to the losses.

Finally, observe that the time variable t' that gives the shape of the radiation pulse has not been normalized. This is because the most appropriate normalization depends on whether a pulsed beam or continuous beam is driving the oscillator. In the case of a pulsed beam, the appropriate choice is to normalize t' to the width of the electron pulse T_p. Typically, this time is much shorter than the repetition time T_a; hence, the fact that the radiation amplitude is supposed to be periodic in t' with period T_a can be neglected and t' can be considered to run over all time. In the case of a continuous beam, the normalized current is constant and one can take $T_a = T$. The appropriate choice for the normalization of t' is, therefore, T. These two cases, that of a pulsed beam and that of a continuous beam, will be considered in more detail in subsequent sections.

9.6 LONG-PULSE OSCILLATORS

Oscillators driven by continuous electron beams are referred to as **long-pulse oscillators** and are capable in principle of oscillating at a single frequency. This is in contrast to devices driven by pulsed electron beams where the radiation field takes on a pulsed nature that, necessarily, consists of a spectrum of frequencies. What is meant by 'single frequency' is that solutions of equations (9.13), (9.14) and (9.25) can be found in which none of the quantities depend on the variable t. The fluctuating fields and current densities in a real device would have a more complicated time dependence which, nevertheless, is periodic with a period corresponding to the frequency ω. Hence, the real spectrum would consist of the frequency ω and all its harmonics. In the simple formulation described by equations (9.13), (9.14) and (9.25), the small components of the fields and current densities which are oscillating at harmonics of the basic frequency ω have been neglected. Therefore, these solutions are effectively single-frequency solutions.

The problem of finding the single-frequency states can be thought of as a nonlinear eigenvalue problem. Solutions of the governing equations, including the feedback boundary condition (9.26), exist only for specific values of the frequency ω and the amplitude of the radiation entering the interaction region. The solution of this nonlinear eigenvalue problem can be obtained by specification of ω and the initial amplitude of the radiation entering the interaction region. The equations in the interaction region can then be solved as if the device were an amplifier with a specified injected signal. Using the feedback boundary condition the magnitude and phase of the returning radiation can be found and compared with the assumed injected radiation. To find an eigensolution, the frequency and magnitude of the injected radiation must be adjusted to match the magnitudes and phases of the injected and returning radiation. The phase of the injected radiation is arbitrary in that changing it simply changes the phase of the radiation amplitude at any other point in the device by the same amount. In general, there are many different possible eigenfrequencies for a given set of beam, wiggler and feedback-path parameters. The separations between the frequencies in the present model are approximately uniform and equal to the inverse of the travel time of the radiation through the cavity. In the low-gain limit, the frequencies are nearly the same as those of the empty cavity given by equation (9.27). Thus, it is sometimes useful to think of these eigenfrequencies as corresponding to modes of the cavity.

While it may be clear from the above discussion that single-frequency states always exist in principle, it is not obvious whether it is possible to access these states in practice. For example, the single-frequency states may be unstable in the sense that radiation at other frequencies will spontaneously grow from noise even in the presence of a large-amplitude saturated signal. The sideband instability of Chapter 6 is an example of just such a situation. The problem of the stability of single-frequency states in an oscillator will be considered in detail in the next section.

A second reason preventing the attainment of true single-frequency states is the inherent noise present in any electron beam. There is always a minimum level of noise associated with discreteness of the charge of an electron. This is the same source of noise responsible for the spontaneous emission discussed in Chapter 3, and can cause spectral broadening in two ways. First, the radiation field may be essentially one of the single-frequency nonlinear eigenstates described in the preceding paragraphs, but, as was discussed, the precise phase of one of these states is an arbitrary constant. In the presence of noise, the phase becomes a time-dependent, random variable which gives rise to a finite spectral width for the radiation. This width is dependent on the amplitude of the noise. A second, and distinct, possibility is that the noise can excite other modes of the cavity which can achieve large amplitudes (depending on parameters). These two separate limits on the achievable spectral width will be determined later in this chapter.

A third impediment to the achievement of a true single-frequency state of the cavity is the finite time duration of any practical electron beam. One limit is the spectral width associated with the reciprocal of the temporal duration of the electron beam, although this width can be made very small. Further, even if the beam is of extremely long duration, its parameters may vary with time. This can lead to a situation in which a stable single-frequency state becomes unstable, and is replaced by a state (or states) with different frequencies. A more important limitation, however, is the slow rate at which coherence is established in devices with small slippage parameters. This effect was alluded to in Section 9.3 and will be considered in more detail in the following sections. The net result is that a device with a small slippage parameter will tend initially to oscillate over a relatively broad range of frequencies, and this range will narrow slowly but progressively over time. The rate of narrowing can be increased by various techniques that will be discussed.

Single-frequency states are difficult, if not impossible, to achieve in practice. However, their study for long-pulse oscillators is important since the interaction of these states (or their modification due to the temporal variation of a real beam pulse) provides a way of understanding the temporal behaviour of the radiation in a long-pulse oscillator.

9.6.1 Single-frequency states

The previous discussion regarding the problems of the existence and stability of single-frequency states and their sensitivity to noise applied to oscillators of both the high- and low-gain type. While the physical effects that were discussed are general and can be expected to occur independently of the oscillator gain they are most easily analysed, and understood, if one adopts the low-gain approximation. Therefore, in the next several sections we will concentrate on studying the solutions of the low-gain equations, in particular the normalized equations (9.79–81).

For the case of a continuous electron beam, the normalized current appearing in (9.81) can be assumed to be constant. Further, while the beam-pulse arrival time T_a has no specific meaning, the radiation amplitude will be nearly periodic, with a period equal to the round-trip time for radiation in the cavity T. Accordingly, we set $T_a = T$. The fast time dependence of the normalized radiation amplitude can then be expressed in terms of a Fourier series

$$X(n, t') = \sum_{m=-\infty}^{\infty} X_m(n) \exp(-i\Delta\omega_m t'), \tag{9.84}$$

where

$$\Delta\omega_m = \frac{2\pi m}{T}, \tag{9.85}$$

and the Fourier index m labels the mth mode of the cavity relative to the mode with frequency ω. That is, the Fourier amplitudes $X_m(n)$ are the amplitudes of the modes of the empty cavity whose real frequencies are approximately $\omega + \Delta\omega_m$. These mode amplitudes will change slowly with time (n denotes the round-trip index) due to the interaction of the modes with the electron beam and due to the output coupling and cavity losses. These effects are described by the two terms on the right-hand side of equation (9.81).

The modal representation for the normalized field amplitude can be inserted in the particle equations of motion (9.79) and (9.80). It is then necessary to integrate the equations of motion for an ensemble of initial conditions corresponding to a uniform distribution of phases, a distribution of momentum values p reflecting the energy dependence of the injected beam and, finally, a distribution of entrance times t_e. Since the radiation amplitude is periodic in the fast time variable t' with period T the solutions of (9.79) and (9.80) for the orbits will have the same periodicity. Therefore, it is only necessary to integrate the equations of motion for a distribution of entrance times t_e on an interval of length T.

As has been discussed in section 9.1, it is always possible to introduce a Vlasov equation to replace the individual particle equations,

$$\left\{\frac{\partial}{\partial\xi}+p\frac{\partial}{\partial\psi}-\frac{1}{2}[iX(n,t_e+\varepsilon\xi T)\exp(i\psi)+\text{c.c}]\frac{\partial}{\partial p}\right\}F(p,\psi,\xi,t_e)=0, \quad (9.86)$$

where $F(p,\psi,\xi,t_e)$ is the distribution function which satisfies the boundary condition

$$F(p,\psi,\xi=0,t_e)=F_0(p) \quad (9.87)$$

at the entrance of the cavity. Here $F_0(p)$ is the distribution function in normalized momentum p for the incoming beam, which satisfies a normalization condition

$$\int_0^{2\pi} d\psi \int_{-\infty}^{\infty} dp\, F_0(p)=1. \quad (9.88)$$

Comparison of the preceding with (9.16) along with the definition of normalized momentum p, (9.76) reveals that f_0 and F_0 differ only by a multiplicative constant, $L\omega_\gamma/v_\parallel$.

Once the orbits or the distribution function are calculated, the slow rate of change of the mode amplitudes with round-trip number is obtained from the mth Fourier component of equation (9.81)

$$\left[\frac{\partial}{\partial n}+(1-R)\right]X_m(n)=-iI_b\int_0^T\frac{dt'}{T}\int_0^1 d\xi\,\langle\exp[i\Delta\omega_m t'-i\psi(\xi,t_e)]\rangle. \quad (9.89)$$

Since the orbits are periodic in the entrance time variable t_e, the integral over t' in equation (9.89) can be transferred to an integral over t_e, resulting in the mode equation

$$\left[\frac{\partial}{\partial n}+(1-R)\right]X_m(n)=-iI_b\int_0^T\frac{dt_e}{T}\int_0^1 d\xi\,\langle\exp[i\Delta\omega_m(t_e+\varepsilon\xi T)-i\psi(\xi,t_e)]\rangle, \quad (9.90)$$

where the angular brackets imply an average over the distribution function of the incoming beam and can be evaluated as an average over particles as in (9.23), or as an average weighted by the local distribution function as (9.24).

The mode amplitude evolution equation (9.90) along with the equations of motion describe the nonlinear competition of the modes of a low-gain oscillator cavity. Before analysing the nonlinear saturated single-frequency solutions of this system, it is useful to examine the linear regime in which the radiation amplitude is small. It is more convenient in the linear regime to work with the Vlasov equation (9.86), which we then linearize by writing the distribution function as the sum of the injected distribution function plus a small perturbation

$$F = F_0(p) + [F_+(p, \xi, t_e)\exp(i\psi) + \text{c.c}]. \tag{9.91}$$

The perturbation satisfies the linearized Vlasov equation

$$\left[\frac{\partial}{\partial\xi} + ip\right]F_+ = \frac{i}{2}\sum_{m=-\infty}^{\infty} X_m \exp[-i\Delta\omega_m(t_e + \varepsilon\xi T)]\frac{\partial F_0}{\partial p}, \tag{9.92}$$

subject to the boundary condition that the perturbed distribution function vanishes at the entrance of the interaction region. Equation (9.92) can be integrated to obtain the perturbed distribution function

$$F_+ = \frac{i}{2}\int_0^{\xi} d\xi' \sum_{m=-\infty}^{\infty} X_m \exp[-ip(\xi - \xi') - i\Delta\omega_m(t_e + \varepsilon\xi' T)]\frac{\partial F_0}{\partial p}. \tag{9.93}$$

Substitution of the perturbed distribution function into the equation for the slow evolution of the mth mode amplitude (9.90) results in

$$\left[\frac{\partial}{\partial n} + (1 - R)\right]X_m = I_b G_m X_m, \tag{9.94}$$

where G_m is the linear, complex, normalized gain for the mth mode

$$G_m = \pi^2 \int_{-\infty}^{\infty} dp\, \Delta(p - \varepsilon\Delta\omega_m T)\frac{\partial F_0}{\partial p}, \tag{9.95}$$

where the resonance function $\Delta(p)$ is given by

$$\Delta(p) = -\frac{i}{\pi}\int_0^1 d\xi \frac{1 - \exp(-ip\xi)}{p} = \frac{1 - \cos p + i(\sin p - p)}{\pi p^2}. \tag{9.96}$$

A mode will grow if the product of the real part of the gain and the normalized current exceeds the cavity losses

$$\text{Re}\, G_m > \frac{1 - R}{I_b}. \tag{9.97}$$

For a cold beam with a monoenergetic distribution $F_0(p) = \delta(p - p_0)/2\pi$, the real part of the gain function reduces to that derived in Chapter 4 and displayed in equation (4.60),

$$G_m = -\frac{1}{2}\frac{\partial}{\partial p}\frac{1 - \cos p + i(\sin p - p)}{p^2}\bigg|_{p = p_0 - \varepsilon\Delta\omega_m T}, \tag{9.98}$$

this is, with $\Theta = -p/2$, $\text{Re}\, G_m = F(\Theta)/8$, where $F(\Theta)$ is defined in equation (4.60). The maximum value of G_m is approximately 0.068 and occurs for a value of its argument equal to approximately 2.6.

The solutions of equation (9.94) describe the initial growth of the cavity modes. The modes will grow with a temporal rate γ_m and have a frequency shift $\delta\omega_m$, where

$$\gamma_m = \frac{I_b}{T}\left[\text{Re}\, G_m - \frac{(1 - R)}{I_b}\right], \tag{9.99}$$

and

$$\delta\omega_m = -\frac{I_b}{T}\mathrm{Im}\,G_m. \tag{9.100}$$

The effective density of modes can be determined by the dependence of the gain function on the mode index m. In particular, for a monoenergetic beam, the argument of the gain function is the detuning

$$p_m = p_0 - 2\pi m\varepsilon \cong \frac{L}{v_\parallel}[(k_m + k_w)(v_\parallel + \delta v_\parallel) - (\omega + \Delta\omega_m)], \tag{9.101}$$

where $k_m = (\omega + \Delta\omega_m)/c$ is the spatial wavenumber of the mth mode, and $L(k_m + k_w)\delta v_\parallel / v_\parallel = p_0$ measures the deviation in velocity of the incoming beam from exact resonance with the $m = 0$ mode. As has been indicated in Chapter 4, the first positive peak of the gain function corresponds to values of the argument of G between approximately 0 and π. The number of modes which fall into this range is $N \approx (2\varepsilon)^{-1}$. Thus, the slippage parameter ε defined in equation (9.44) determines the mode density in a continuous-beam oscillator. With a small value of the slippage parameter, a large number of cavity modes will have positive growth rates and can be expected to compete for the free energy of the beam. Further, each mode is also characterized by a frequency shift which is induced by the beam. That is, the frequency of oscillation of the mode is not exactly equal to the oscillation frequency of a mode in the empty cavity, but rather is shifted relative to the corresponding empty-cavity frequency by a small amount (in the low-gain regime) which is proportional to the beam current and depends on detuning.

The gain function for arbitrary distribution functions is defined in equation (9.95). In general, as the spread in p values (parallel energy in (9.76) and (9.77)) increases, the gain is reduced and the range of frequencies (or detunings defined by (9.101) with p_0 being the mean value of p) for which the gain is positive increases. This transition occurs when the spread in p values is comparable to unity. In the case of a beam with an extremely large spread in p values, the integral in (9.95) can be performed by considering the derivative of the unperturbed distribution function to be constant over the range of p values for which the other factor Δ in the integrand is large; specifically, for $|p - \Delta\omega_m T| < \pi$. In this case, the factor $\Delta(p - \varepsilon\Delta\omega_m T)$ acts as a delta-function and the gain is proportional to the slope of the distribution function at the value of p which is resonant with the mode in question,

$$G_m \cong \frac{\pi}{2}\frac{\partial F_0}{\partial p}\bigg|_{p = \varepsilon\Delta\omega_m T} \tag{9.102}$$

Gain will be positive or negative depending upon the slope of the distribution function. A positive slope indicates that there are more electrons with energies greater than the resonant energy than with energies less than the resonant energy. As a result, energy is extracted from the beam and the radiation is amplified. A situation where the distribution function might be broad enough that (9.102) applies is the case of a storage-ring free-electron laser where the same group of electrons transits the interaction region many times, acquiring a large spread in parallel energies.

Equation (9.94) describes the small-signal growth of modes in a low-gain continuous-beam oscillator. If one of these modes grows and becomes dominant (i.e. it is the only mode with a substantial amplitude), then it can be regarded as a single-frequency state. Without loss of generality, we can label this mode the $m=0$ mode. The problem, therefore, is to analyse the characteristics of an oscillator in which the mode amplitude X_0 is sufficiently large that the gain function depends on the mode amplitude. Saturation will be achieved when the mode amplitude reaches a sufficient size that the power extracted from the beam balances the power lost from the cavity. The saturation process can be understood in the following way. In the low-gain single-mode regime, the Hamiltonian

$$H = \frac{p^2}{2} + \mathrm{Re}[X_0 \exp(i\psi)] \tag{9.103}$$

is conserved by electrons as they transit the interaction region. This is in contrast to the case of a high-gain amplifier or oscillator where the dependence of the radiation-field amplitude on axial distance destroys the constancy of the equivalent pendulum Hamiltonian. In spite of this difference, Fig. 4.1 also applies qualitatively to a low-gain oscillator. The efficiency at which energy is extracted from the beam is measured by the mean change in the momentum variable for the ensemble of electrons

$$\Delta p = \langle p_0 - p(\xi = 1) \rangle. \tag{9.104}$$

Recall that the momentum variable p is proportional to the energy deviation $\Delta\gamma$ according to equation (9.76). In terms of the dimensionless momentum change Δp, the average energy extracted per particle is

$$\Delta\gamma = \Delta p \frac{v_\parallel}{L\omega_\gamma} \cong \frac{\gamma \Delta p}{2\pi N_w} \frac{\beta_\parallel^2}{1+\beta_\parallel}, \tag{9.105}$$

where in the second equality N_w is the number of wiggler periods and we have evaluated the factor ω_γ defined in (9.15) for the case of a helical or a weak planar ($\gamma^2 \gg 1 + a_w^2$) wiggler. In the relativistic case this gives the standard estimate that the fractional change in energy scales as the reciprocal of the number of wiggler periods.

The dependence of Δp on the radiation amplitude for a specific value of $p_0 (= 2.6)$ is shown in Fig. 9.4. For small values of radiation amplitude, the dimensionless efficiency is a quadratic function of the radiation amplitude. As the radiation amplitude increases the dimensionless efficiency increases (but at a rate which is slower than predicted by the quadratic dependence in the linear regime) and reaches a maximum corresponding to the value of the radiation amplitude in Fig. 4.1(c). This maximum is seen to correspond to the case in which the bulk of the electrons has made half a synchrotron oscillation in the ponderomotive well.

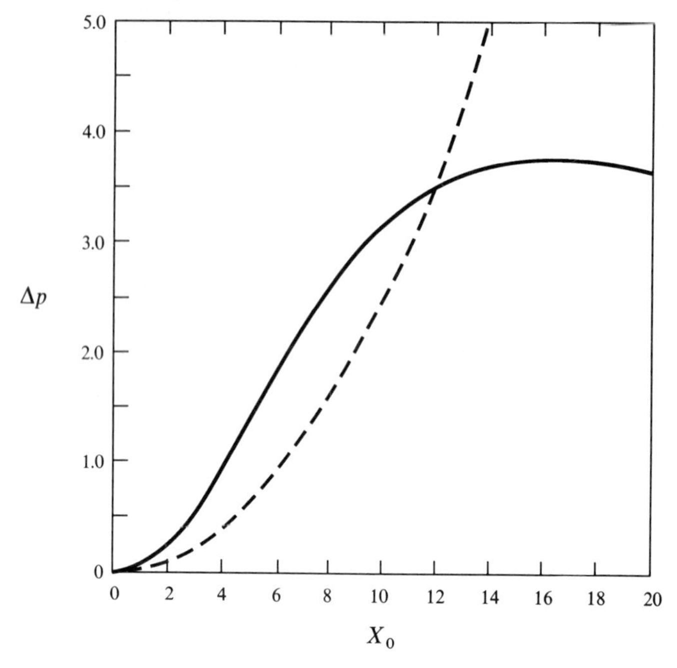

Fig. 9.4 Normalized energy extracted per electron as a function of radiation amplitude (solid line), and normalized energy radiated per electron (dashed line).

The process by which saturation is achieved is illustrated by the addition to the figure of the dashed line which represents the *per electron* losses. The rate at which energy is lost from the cavity is quadratic in the field amplitude. Thus the amount of energy per electron that must be extracted from the beam to maintain a given amplitude will appear as a parabola with a coefficient of proportionality which depends on the beam current and reflection coefficient. For currents above the start current, the energy extracted per electron exceeds that which is lost, and the mode grows. As the mode amplitude increases, the losses increase faster than the energy extracted per electron. Saturation of growth occurs when these two effects balance, as represented by the intersection of the solid and dashed curves in Fig. 9.4.

Curves similar to those of Fig. 9.4 could be plotted for different values of the detuning p_0. Instead, we show in Fig. 9.5 the level curves of Δp in the field amplitude versus detuning plane. For the range of parameters shown, the maximum dimensionless efficiency is $\Delta p \approx 5.5$ and occurs for a detuning of 5.2 and a field amplitude $X_0 = 18.1$. As the range of detunings and field amplitudes is increased other local maxima appear. However, as we shall soon see, these maxima occur for parameters for which one would not expect to find single-mode operation in an oscillator.

The efficiency of energy extraction is closely related to the single-mode nonlinear gain of the oscillator [12]. When a single mode is present ($m = 0$), equation (9.94)

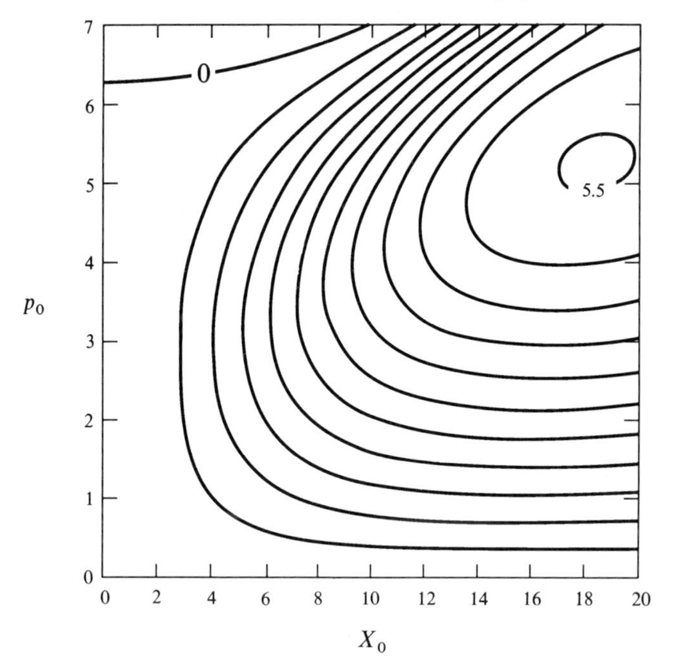

Fig. 9.5 Level curves of the normalized efficiency Δp in the normalized mode amplitude versus detuning plane. The separation between adjacent levels is $\Delta p = 0.5$.

can be generalized by the introduction of the nonlinear gain function G_{nl} defined by

$$G_{nl}(|X_0|, p_0) = -\frac{i}{X_0}\int_0^1 d\xi \langle \exp(-i\psi) \rangle. \tag{9.106}$$

That the gain depends only on the magnitude of the radiation amplitude can be seen by noting that a change in the phase of X_0 can be absorbed in the definition of the particle phase ψ both in (9.103) and (9.106). Since the average in equation (9.106) is taken over an ensemble of phases distributed over an interval of 2π, this change in phase leaves the nonlinear gain unaffected. The real part of the gain can now be related to the dimensionless efficiency by means of energy-conservation arguments, and we find using (9.80), (9.84) and (9.106) that

$$\Delta p = -\int_0^1 d\xi \left\langle \frac{dp}{d\xi} \right\rangle = |X_0|^2 \,\mathrm{Re}[G_{nl}(|X_0|, p_0)]. \tag{9.107}$$

The plot in Fig. 9.5 shows the efficiency as a function of mode amplitude and detuning. While the detuning is determined directly by the incoming beam, the mode amplitude is determined by power-balance considerations, as illustrated in Fig. 9.4. Specifically, the mode amplitude adjusts itself until the power extracted from the beam equals that dissipated in and radiated from the cavity. This power-balance constraint can be expressed by searching for steady-state solutions of equation (9.94) for the $m = 0$ mode using the nonlinear gain function defined in equation (9.106). In these steady-state solutions, the amplitude of the radiation is constant of the slow time scale (round-trip index n) and has a phase which decreases steadily with time representing the frequency shift $X_0 = |X|\exp(-i\delta\omega_0 Tn)$. The real and imaginary parts of equation (9.94) then give

$$(1 - R) = I_b \mathrm{Re}[G_{nl}(|X_0|, p_0)], \tag{9.108}$$

and

$$\delta\omega_0 T = -I_b \mathrm{Im}[G_{nl}(|X_0|, p_0)]. \tag{9.109}$$

The current required to maintain a particular mode amplitude can be obtained from the power balance relation (9.108). It is useful to compare this current with the minimum current necessary to initiate oscillations in a cavity with a given loss factor $1 - R$. This minimum current, known as the start current, is obtained by inserting the expression for the linear, small signal gain evaluated at the detuning $p_0 = 2.6$ in equation (9.98). Hence,

$$I_{st} = \frac{1 - R}{G_0(2.6)} \cong 15(1 - R). \tag{9.110}$$

The level curves of the current required to maintain a given mode amplitude given by equation (9.108) can be plotted versus mode amplitude and detuning. Such a plot appears in Fig. 9.6 where we have normalized the current to the start

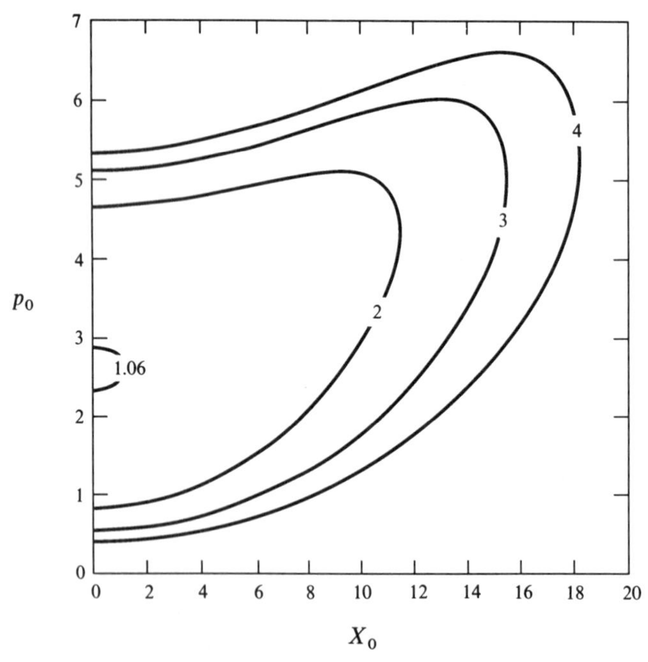

Fig. 9.6 Level curves of current ($\chi = I_b/I_{st}$) required to maintain power balance, plotted in the normalized mode amplitude v. detuning plane [15].

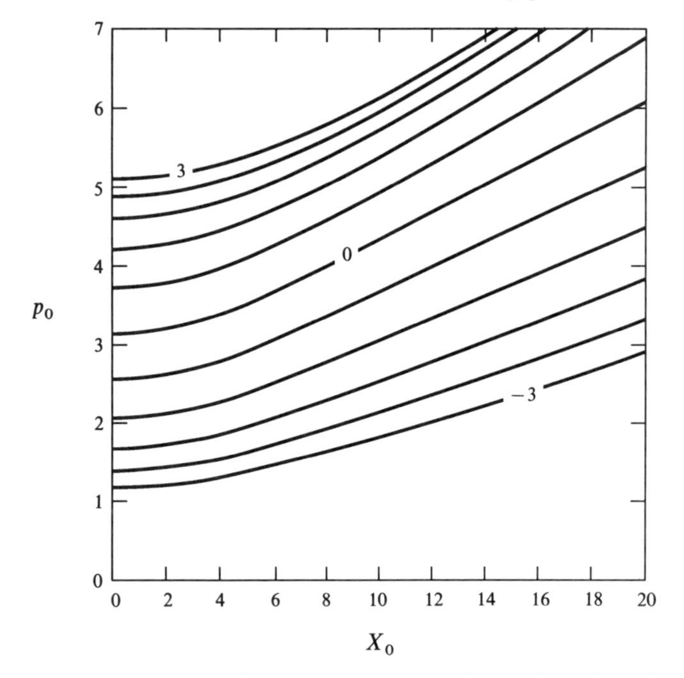

Fig. 9.7 Level curves of normalized frequency shift ($\delta\Omega = \delta\omega_0 T/2(1 - R)$) in the normalized mode amplitude v. detuning plane. The separation between adjacent levels is 0.6.

current. It is evident that a current about four times the start current is required to reach the maximum efficiency equilibrium. Further, for a given value of current in excess of the start current there is a range of detuning values for which single-frequency equilibrium states exist. If the slippage parameter is small, then according to equation (9.101) this range in detuning parameters translates into a large number of modes. Thus, with a given current in excess of the start current, there typically are a large number of possible single-frequency equilibria in a long-pulse oscillator. Which of these equilibria are stable and which can be accessed will be discussed in the subsequent sections.

Finally, we plot the level curves of the dimensionless frequency shift $\delta\Omega$ ($\equiv \delta\omega_0 T/2(1 - R)$) in the field amplitude versus detuning plane in Fig. 9.7. The frequency shift is normalized to the decay time of the radiation in the empty cavity T_d (i.e. the Q resonance width of the cavity). This frequency shift will be important when the effects of dispersion on mode competition are discussed in a later section.

9.6.2 Stability of single-frequency states

The single-frequency states of the previous section were found upon requiring that the radiation has a single spectral component. It was found that these states represented valid equilibrium solutions of the basic governing equations. In this section, the stability of these equilibrium states to perturbations will be examined. Clearly, due to the ubiquitous presence of noise, stability of a single-frequency state is necessary for the state ever to be reached in practice.

There are two principal considerations in the stability of single-frequency equilibria in the low-gain regime. The first is the stability of the equilibrium mode against perturbations in its amplitude and phase. The second consideration is the stability of the equilibrium mode against the introduction of other modes. The first consideration can be addressed relatively easily based upon information from the previous section regarding the characteristics of single-frequency equilibria. Since a free-running oscillator has no preferred phase of oscillation, the equilibrium state is neutrally stable to perturbations in its phase. Therefore, a small perturbation in the phase of the complex amplitude of the single-cavity mode will neither grow nor decay in time. This is manifested mathematically in the fact that the nonlinear gain function defined in (9.106) is independent of the phase of the radiation. Perturbations in the magnitude of the mode will decay if the rate at which power is radiated increases faster with field amplitude than does the rate at which power is extracted from the beam. For example (see Fig. 9.4), if the radiation amplitude is increased (due to the perturbation) to a point above the intersection of the power extracted per electron and losses curves, then the power lost from the cavity exceeds that extracted from the beam. As a result, the mode amplitude will decay back to the equilibrium point. We will see that this can, in general, be determined from the level curves of current plotted in Fig. 9.6.

The second stability consideration is whether the equilibrium state is stable with respect to the introduction of other modes. Specifically, can a mode other than the equilibrium state (which is excited by noise) grow exponentially in time? The answer to this question depends upon both the parameters of the equilibrium state, and the nearness in frequency to the equilibrium mode of the perturbing mode(s). In particular, there is a wide range of equilibrium states which are capable of suppressing competing modes [2, 13–15]. This mode suppression can be envisioned in the following manner. Assume that the current is several times the start current defined in equation (9.110), and the slippage parameter ε is small. In this case, a large number of modes have positive growth rates when all modes are small. However, if one mode is able to reach large amplitude, then the same nonlinear mechanism that reduces the gain of this mode at saturation also reduces the gain of the competing modes. The gain reduction of the competing modes is sufficient to cause the losses to exceed the gain for these modes, with the result that these other modes decay away with time.

It is important to bear in mind that the mode-suppression mechanism just described does not work for all the equilibrium states in the mode amplitude versus detuning plane of Figs 9.5, 9.6 and 9.7. Only those equilibria which fall within the triangular-shaped region of Fig. 9.8 are robustly stable. The triangular-shaped region has three fundamental boundaries, and equilibria outside these boundaries are unstable to the growth of other modes. There are basically two distinct types of instability that determine the three boundaries. The equilibria

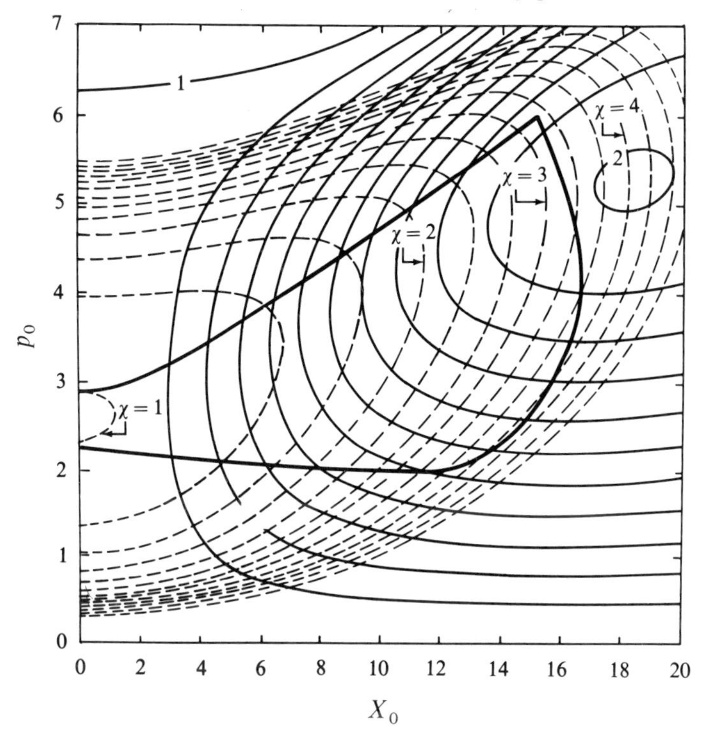

Fig. 9.8 Region of stable single-mode operation in the normalized mode amplitude v. detuning plane [15]. Only inside the triangular region is a single-frequency state stable to the introduction of other modes.

just outside the upper and lower boundaries are unstable to the growth of close-in-frequency modes. This instability is called a phase instability for reasons that will be given subsequently. The basic cause of the instability is the fact that if the equilibrium mode has a detuning p_0 which is too far from resonance (at which the linear growth is maximum), then a mode with a detuning which is closer to the resonance (and has a higher linear gain) will be able to grow. Equilibria which are just across the third boundary defining the right side of the triangle are unstable to the growth of modes which are predominantly of lower frequency than the equilibrium mode. This instability is called variously the sideband instability [3, 6, 16], the synchrotron instability [13], the trapped-particle instability [14], the overbunch instability [4, 15], or the spiking mode [17, 18] and can be explained as follows. Due to the phase-trapping nature of the saturation mechanism, the equilibrium mode is not able to extract any more beam energy once the bulk of the particles has completed about half of a synchrotron oscillation in the ponderomotive well. A lower frequency mode with a resonant detuning below that of the equilibrium mode can still extract energy from these particles. In fact it is now in a better position to trap the electrons in its ponderomotive well than it was in the absence of the equilibrium mode. This situation is illustrated schematically in Fig. 9.9, where the separatices corresponding to the equilibrium mode and the lower frequency satellite mode are superimposed. This figure is only schematic since, in principle, no separatrices exist if two modes are present. However, the separatrices of the individual modes give a qualitative picture of the phase-space trajectories of different electrons when more than one mode is present.

Based on Fig. 9.9, a mode with a resonant detuning lower than that of the equilibrium mode by about half the width of the trapped region of phase space for the mode at saturation would be most unstable. From the definition of detuning (9.101), this gives

$$\varepsilon\Delta\omega_m T \cong -(2|X_0|)^{1/2} \cong -5.7, \tag{9.111}$$

where we have substituted $|X_0| \approx 16$ as an estimate of the saturation amplitude. Interestingly, this estimate is identical to that of the frequency of the most unstable sideband mode obtained by the often-used theoretical argument that the sideband should be down-shifted by the synchrotron frequency. It is clear, however, that the threshold of the instability occurs for mode amplitudes such that particles execute only half a synchrotron oscillation on one transit of the interaction length, and, thus, the conventional argument is only qualitative. A third explanation of this instability based on the klystron model introduced in section 9.3 will be given in this section. This explanation is based on time-domain considerations and gives a different insight into the overbunch-sideband-spiking mode instability.

A final point which will be brought out in this section is that for equilibria within the stable triangle in parameter space, nearby modes decay slowly in time [15]. The slow temporal decay is a consequence of the fact that coherence in the

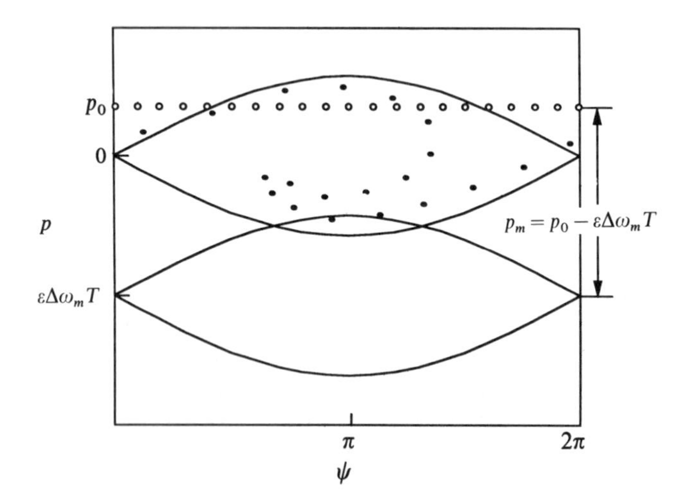

Fig. 9.9 Schematic phase-space plot illustrating the mechanism by which a lower-frequency sideband extracts energy from electrons trapped in the ponderomotive well of the carrier.

radiation pulse is established slowly for small slippage parameters. Hence, the smaller the slippage parameter, the closer in detuning neighbouring modes are to the equilibrium mode (see equation (9.101)) and the slower these modes decay. This important dependence of the damping rate on mode number is a consequence of the equal spacing in frequency of the cavity modes and, as a result, leads to the nonlinear interaction of modes being of the four-wave type [19]. Modifications to the frequencies of cavity modes which cause the separation in frequency of adjacent modes to become nonuniform (for example by including a dispersive element in the feed-back loop), therefore, can increase the rate at which coherence is established [20].

We begin the analysis by consideration of the stability of the equilibrium to perturbations of its own parameters. In particular, we write the slow time dependence of the mode amplitude $X_0(n)$

$$X_0(n) = [X_{00} + \delta X_0(n)] \exp(-i\delta\omega_0 nT), \tag{9.112}$$

where since only a single mode is present we may label it the $m = 0$ mode, X_{00} is the equilibrium amplitude which satisfies (9.108), $\delta\omega_0$ is the equilibrium frequency shift given by (9.109) with $|X_0|$ replaced by X_{00}, and δX_0 is a small complex perturbation. The real part of δX_0 gives the perturbation of the amplitude of the mode and the imaginary part gives the perturbation of the phase. This form of the radiation amplitude is then inserted into the equation for the slow time evolution (9.94) for the $m = 0$ mode with the gain function G_0 replaced by the nonlinear gain function defined in (9.106). The arguments of the gain function are the magnitude of the radiation amplitude, $|X_0|$ and the detuning p_0. Expansion of the gain function to first order in the perturbation yields

$$\left.\begin{aligned} G_{nl}(|X_0|, p_0) &\cong G_{nl}(X_{00}, p_0) + \delta X_{0,r} \frac{\partial G_{nl}(X_{00}, p_0)}{\partial X_{00}} \\ &\equiv G_{nl}(X_{00}, p_0) + \frac{\delta X_{0,r}}{X_{00}} G'_{nl}, \end{aligned}\right\} \tag{9.113}$$

where $\delta X_{0,r}$ is the real part of δX_0, and G'_{nl} is the normalized derivative of the nonlinear gain. The real and imaginary parts of (9.94) then give equations for the evolution of the real and imaginary parts of the perturbation,

$$\frac{\partial}{\partial n} \delta X_{0,r} \cong I_b \delta X_{0,r} \operatorname{Re} G'_{nl} \tag{9.114}$$

and

$$\frac{\partial}{\partial n} \delta X_{0,i} \cong I_b \delta X_{0,r} \operatorname{Im} G'_{nl}, \tag{9.115}$$

where $\delta X_{0,i}$ is the imaginary part of the perturbation to the mode amplitude.

There are two independent solutions to equations (9.114) and (9.115). The first solution has $\delta X_{0,r} = 0$ and $\delta X_{0,i} = \delta X_{0,i}(0)$, which is a constant. This corresponds to a neutrally stable perturbation of the radiation phase. The second solution is given by

$$\delta X_{0,r}(n) \cong \delta X_{0,r}(0)\exp(-n\gamma_a T), \tag{9.116}$$

and

$$\delta X_{0,i}(n) \cong \delta X_{0,i}(0) + I_b \operatorname{Im}(G'_{nl}) \int_0^n dn'\, \delta X_{0,r}(n'), \tag{9.117}$$

and represents stable or unstable perturbations in the amplitude of the radiation field with a damping rate $\gamma_a \equiv -(I_b/T)\operatorname{Re} G'_{nl}$. The solution will be stable as long as the gain decreases with radiation amplitude

$$\operatorname{Re}\left[\frac{\partial G_{nl}(X_{00}, p_0)}{\partial X_{00}}\right] \leqslant 0. \tag{9.118}$$

This is precisely the requirement that the rate at which power is lost from the cavity increases faster with radiation amplitude than the rate at which power is extracted from the beam.

Some of the equilibria represented by points in the amplitude versus detuning plane of Figs 9.5–8 are not stable according to the requirement (9.118). The unstable equilibria can be found by consideration of the level curves of current shown in Fig. 9.6. The functional dependence of the level curves of current in the mode amplitude versus detuning plane is given by equation (9.108). Differentiation of (9.108) with respect to mode amplitude yields

$$\frac{\partial I_b}{\partial X_{00}} G_{nl}(X_{00}, p_0) + I_b \operatorname{Re}\left[\frac{\partial G_{nl}(X_{00}, p_0)}{\partial X_{00}}\right] = 0. \tag{9.119}$$

Thus, if the current required to maintain an equilibrium increases with mode amplitude, then the gain at that mode amplitude will decrease with mode amplitude and the equilibrium will be a stable operating point. Examination of Fig. 9.6 shows that equilibria with relatively high detunings but low field amplitudes where the level curves of current have a positive slope correspond to currents which decrease with mode amplitude and, therefore, are unstable.

The stability of single-mode equilibria to perturbations in the amplitude of the equilibrium mode can be analysed in terms of derivatives of the gain function. In general, the stability of a nonlinear, single-mode equilibrium to perturbations by other modes must be analysed numerically. This is not the case, however, if one adopts the klystron model introduced in section 9.3. Consideration of the klystron model leads to the two-time-delay equation (9.54), and the same approximations leading to the low-gain equations of section 9.5 can be carried out on the klystron model. In particular, we introduce the two-time-scale variables t' and n defined in equation (9.72) and assume the current and losses are small. This enables us to cast equation (9.54) into the form of (9.81),

$$\left(\frac{\partial}{\partial n} + T_\delta \frac{\partial}{\partial t'}\right) X(n,t') = (R-1)X(n,t') + \{I_b(t'')X(n,t'')g[X(n,t'')]\}_{t''=t'-\varepsilon T}, \tag{9.120}$$

where g denotes the gain in the klystron model defined by equation (9.53).

Equations (9.81) and (9.120) are similar except for the terms which describe the effect of the beam. In the klystron model, the interaction with the beam is dependent on the radiation amplitude at the time a particular group of electrons enters the interaction region, $t' - \varepsilon T$. In the more general model described by equation (9.81), the interaction with the beam is dependent on the amplitude of the radiation entering the cavity for all times between t' and $t' - \varepsilon T$. A slight improvement to the klystron model is obtained by replacement of the slippage parameter ε by $\varepsilon/2$ to account for the fact that in the more general case some weighted average of the entering radiation between the times t' and $t' - \varepsilon T$ determines the response of the beam.

To study the stability of single-frequency states using the klystron model, we again take the normalized current I_b to be constant and set T_δ equal to zero. Further, the fast-time-scale dependence of the radiation field is expressed as the sum of the large amplitude equilibrium mode with $m = 0$ and an ensemble of small perturbing modes with $m \neq 0$,

$$X(n, t') = \exp(-i\delta\omega_0 nT)\left[X_{00} + \sum_{m \neq 0} \delta X_m(n)\exp(-i\Delta\omega_m t')\right], \quad (9.121)$$

where $\Delta\omega_m$ is defined in equation (9.85). The argument of the gain function in equation (9.120) is the magnitude of the radiation amplitude evaluated at the delayed time $t' - \varepsilon T$. For small amplitude of the perturbing modes this becomes,

$$|X(n', t' - \varepsilon T)| \cong X_{00} + \mathrm{Re}\left[\sum_{m \neq 0} \delta X_m(n)\exp(-i\Delta\omega_m(t' - \varepsilon T))\right]$$

$$\cong X_{00} + \frac{1}{2}\sum_{m \neq 0} [\delta X_m(n) + \delta X^*_{-m}(n)]\exp[-i\Delta\omega_m(t' - \varepsilon T)], \quad (9.122)$$

where in the second equality we have used the fact that $\Delta\omega_m = -\Delta\omega_{-m}$. It is important to note that the usual relation between the Fourier amplitudes of a real function, $\delta X_{-m} = \delta X^*_m$, is not satisfied in the present case since the time-dependent radiation amplitude $X(n, t')$ is complex. Physically, this is a result of the fact that the $\pm m$ modes are distinct, and there is no *a priori* reason why their amplitudes should be complex conjugates.

An evolution equation for the amplitude of the mth perturbing mode can be obtained by the following procedure. We substitute the expression for the radiation amplitude (9.122) into the nonlinear gain function g. This expression is expanded in powers of the field amplitude and subsequently inserted into the dynamical equation (9.120). In addition, the equilibrium relations (9.108) and (9.109) (with G_{nl} replaced by g) are used to eliminate the equilibrium terms. Finally, we multiply by $\exp(i\Delta\omega_m t')$ and integrate over t' to select the mth Fourier component

$$\left[\frac{\partial}{\partial n} - i\delta\omega_0 T + (1 - R)\right]\delta X_m(n)$$

$$= I_b\left\{g(X_{00})\delta X_m(n) + \frac{1}{2}g'(X_{00})[\delta X_m(n) + \delta X^*_{-m}(n)]\right\}\exp(-i\varepsilon\Delta\omega_m T). \quad (9.123)$$

An important consequence of equation (9.123) is that modes with equal and opposite mode numbers (i.e. modes which are equally spaced in frequency about the equilibrium mode) couple together in the presence of the equilibrium mode. This coupling vanishes as the amplitude of the equilibrium mode goes to zero, and [15, 19]

$$g'(X_{00}) \equiv X_{00}\frac{\partial g(X_{00})}{\partial X_{00}} \approx X_{00}^2. \tag{9.124}$$

The coupling can be thought of as the result of a nonlinear four wave mixing process [19]. In such a case, four modes with frequencies ω_1, ω_2, ω_3 and ω_4 will couple nonlinearly if the frequency matching criterion

$$\omega_1 + \omega_2 = \omega_3 + \omega_4 \tag{9.125}$$

is satisfied. In the present case, one can take the frequencies ω_1 and ω_2 to be those of the perturbing modes $(\omega + \Delta\omega_{\pm m})$ and the frequencies ω_3 and ω_4 to be that of the equilibrium mode ω. Because of this coupling, it is not proper to speak of the gain of a single perturbing mode in the presence of an equilibrium mode. Rather, one must always treat coupled pairs of modes. This coupling has important consequences regarding the damping rate of stable modes.

Since modes with equal and opposite mode numbers are coupled, it is convenient to write the evolution equations for both the mode amplitudes δX_m and δX^*_{-m}, and treat these mode amplitudes as two independent variables. The result is two coupled equations of the form of equation (9.123)

$$\frac{\partial}{\partial n}\delta X_m = I_b\left\{g\delta X_m[\exp(i\varepsilon\Delta\omega_m T) - 1] + \frac{g'}{2}(\delta X_m + \delta X^*_{-m})\exp(i\varepsilon T\Delta\omega_m)\right\}, \tag{9.126}$$

and

$$\frac{\partial}{\partial n}\delta X^*_{-m} = I_b\left\{g^*\delta X^*_{-m}[\exp(i\varepsilon\Delta\omega_m T) - 1] + \frac{g'^*}{2}(\delta X_m + \delta X^*_{-m})\exp(i\varepsilon T\Delta\omega_m)\right\}, \tag{9.127}$$

where we have used the equilibrium condition $-i\delta\omega_0 T + (1-R) = I_b g$ to simplify the coefficients in equations (9.126) and (9.127).

One limit in which the two coupled equations (9.126) and (9.127) are easily solved is the limit in which $\varepsilon\Delta\omega_m T \to 0$, which corresponds to the limit in which the perturbing modes are close in frequency to the equilibrium mode. In this case, 'close in frequency' means that the spacing between the perturbing modes and the equilibrium mode is a small fraction of the gain bandwidth. In this limit, the sum and difference of equations (9.126) and (9.127) yield

$$\frac{\partial}{\partial n}(\delta X_m + \delta X^*_{-m}) = I_b\,\mathrm{Re}(g')(\delta X_m + \delta X^*_{-m}), \tag{9.128}$$

and

$$\frac{\partial}{\partial n}(\delta X_m - \delta X^*_{-m}) = iI_b\,\mathrm{Im}(g')(\delta X_m + \delta X^*_{-m}). \tag{9.129}$$

Observe that these equations are formally identical to the pair of equations (9.114) and (9.115) which describe the stability of the equilibrium state against perturbations of its magnitude and phase. In particular, there are two solutions: one which is damped corresponding to perturbations of the magnitude of the radiation $\delta X_m + \delta X^*_{-m} \neq 0$, and one which is neutrally stable corresponding to a perturbation of the phase of the radiation $\delta X_m + \delta X^*_{-m} = 0$ and $\delta X_m - \delta X^*_{-m} \neq 0$. Thus, to lowest order, nearby modes which couple together in such a way that only the phase of the radiation field is perturbed are undamped and not suppressed. Observe that had the coupling between the $\pm m$ modes not been included in equations (9.126) and (9.127), then we would have predicted erroneously that all nearby modes were strongly damped in the presence of the equilibrium mode at a rate exactly half that at which the magnitude of the equilibrium mode decays when perturbed (as in (9.116)).

While the preceding result was derived for the klystron model, it is in fact more general. In particular, the lowest-order stability of nearby in-frequency modes was determined by taking the limit of a vanishingly small slippage parameter. If we let the slippage parameter vanish in equations (9.79–81), both the electron equations of motion and the source term representing the effect of the electron beam in (9.81) become identical to the single frequency equations leading to (9.103) and (9.106), except that the mode amplitude $X_0(n)$ is replaced by the fast time-dependent amplitude $X(n, t_e = t')$. This is precisely the situation discussed in section 9.3 when the case of equal beam and radiation speeds was considered. In the absence of slippage, electrons and photons move in synchronism through the interaction region. Therefore, the amplitude of the radiation entering the cavity at time $t + T$ depends only on the amplitude of the radiation entering at time t, and there is no communication between portions of the radiation pulse having different entrance times. The same situation effectively results with finite slippage if it is assumed that the radiation amplitude consists of a superposition of nearby in-frequency modes. The variation in the field amplitude on the time scale on which slippage is important (i.e. εT) is small for closely spaced modes, and slippage can be neglected to lowest order. Thus, in the limit of vanishing slippage, the radiation amplitude in the general case evolves according to (9.120) with T_δ set equal to zero and the klystron gain g replaced by the general nonlinear gain G_{nl},

$$\frac{\partial}{\partial n} X(n, t') = (R - 1) X(n, t') + I_b G_{nl}(|X(n, t')|, p) X(n, t'). \tag{9.130}$$

The nonlinear evolution of the field amplitude which satisfies an equation of the form of (9.120) with slippage parameter ε and detuning time T_δ set equal to zero can be described as follows. An arbitrary small initial radiation amplitude $X(0, t')$ will grow to saturation (with an equilibrium magnitude satisfying equations (9.108) and (9.109)). This magnitude at saturation is independent of the fast time variable t'. The phase of the complex amplitude at saturation, however, will be determined by the initial conditions, and will depend on the fast time variable t'. In the absence of slippage, there is no mechanism to bring the phases of different portions of the radiation pulse into coherence. Therefore, because the phase is time dependent, the radiation pulse will consist of a superposition of cavity modes expressed in the form of a Fourier series as in (9.84) [21].

The neutrally stable solution of (9.128) and (9.129) for the linear growth of near-by in-frequency satellite modes is modified when the effects of non-zero frequency separation are included. The growth or damping of the satellite modes in this case can be determined by looking for solutions of equations (9.126) and (9.127) in which the slow time dependence of the mode amplitudes is expressed in the form, $X_m(n) = X_{m0}\exp(\gamma_m nT)$ where γ_m is the complex growth rate. Substitution of this solution into equations (9.126) and (9.127) results in a quadratic dispersion relation for the growth rate

$$\left\{\gamma_m T - I_b\left[g(\exp(i\varepsilon t\Delta\omega_m) - 1) + \frac{g'}{2}\exp(i\varepsilon T\Delta\omega_m)\right]\right\}$$
$$\times\left\{\gamma_m T - I_b\left[g^*(\exp(i\varepsilon T\Delta\omega_m) - 1) + \frac{g'^*}{2}\exp(i\varepsilon T\Delta\omega_m)\right]\right\}$$
$$= \frac{I_b^2}{4}\exp(2i\varepsilon T\Delta\omega_m)|g'|^2. \tag{9.131}$$

Observe that in the limit in which $\varepsilon\Delta\omega_m T \to 0$, the previously discussed solutions for amplitude and phase perturbations are obtained.

Certain properties of the solutions of equation (9.131) can be determined from symmetry arguments. In particular, the equation is invariant under the combined operations of replacement of m by $-m$ and complex conjugation. Hence, $\gamma_{-m} = \gamma_m^*$, or if we write γ_m in terms of its real and imaginary parts, $\mathrm{Re}(\gamma_{-m}) = \mathrm{Re}(\gamma_m)$ and $\mathrm{Im}(\gamma_{-m}) = -\mathrm{Im}(\gamma_m)$. Thus, the growth or damping rate of the satellite pair is an even function of mode number and the frequency shift is an odd function of mode number. This implies that the complex growth rate of the solution (which in the limit of vanishing $\varepsilon\Delta\omega_m T$ corresponds to a perturbation of the phase of the radiation) must have the following Taylor expansion for small $\varepsilon\Delta\omega_m T$

$$\gamma_m \cong i\varepsilon\Delta\omega_m T\omega' - D(\varepsilon\Delta\omega_m T)^2, \tag{9.132}$$

to second order in $\varepsilon T\Delta\omega_m$, where the coefficients are obtained from the solution of (9.131) in ascending powers of $\varepsilon T\Delta\omega_m$. The stability of the satellite pair is determined by the sign of the coefficient D. If D is positive nearby satellite pairs are weakly stable, and if D is negative the pair are weakly unstable [15]. The situation in which the nearby pairs are weakly unstable is termed a phase instability. For the klystron model, the coefficients D and ω' can be determined analytically with some effort [15]. For the general case it can be shown that the frequency-shift coefficient ω' can be related to the equilibrium frequency shift (9.109) by

$$\omega' = \frac{d}{dp_0}\delta\omega_0(X_{00}, p_0), \tag{9.133}$$

where the derivative in (9.133) is performed by allowing X_{00} to vary in accordance with maintaining I_b constant in equation (9.108). The steps required to obtain this result will be presented when nonlinear gain narrowing is discussed. The important coefficient D which determines the stability of the satellite pair must be determined numerically in general. An order of magnitude estimate of D based on (9.131) indicates that it scales as I_b/T, which scales as the inverse of the cavity time T_d in equilibrium.

So far only the growth and damping of nearby in-frequency modes have been discussed. Recall that the term 'nearby in frequency' describes pairs of modes whose separation in frequency from the equilibrium mode is small compared with the gain bandwidth. Pairs of satellites which are displaced from the equilibrium mode by approximately the synchrotron frequency can also become unstable at large field amplitudes. To find this instability in the klystron model [4] we set $\varepsilon\Delta\omega_m T = \pi$, and impose the simplifying assumption that the nonlinear gain is real. From the definition of the klystron gain in equation (9.53), it is seen that the gain is real if the distribution of energies of the injected beam is centred about a detuning value $p_i = \pi/2$. This corresponds to the maximum gain in the klystron model. With these assumptions, the dispersion equation (9.131) yields two possible solutions

$$\gamma_m = -\frac{2gI_b}{T}, \tag{9.134}$$

and

$$\gamma_m = -\frac{I_b[2g + g']}{T}. \tag{9.135}$$

The solution given in (9.134) is always stable, while that in (9.135) may be either stable or unstable. The latter of these corresponds to the case in which $\delta X_m = \delta X^*_{-m}$; that is, when the phases of the two satellites are such that the magnitude of radiation is perturbed. The satellite pair will be unstable if

$$2g + X_{00}\frac{\partial g}{\partial X_{00}} < 0. \tag{9.136}$$

Hence, if the gain decreases too rapidly with mode amplitude, then sidebands with $\varepsilon\Delta\omega_m T = \pi$ become unstable. Furthermore, these sidebands produce perturbations in the magnitude of the radiation. Recalling the relationship between nonlinear gain and efficiency from equation (9.107), the condition (9.136) is seen to correspond to the field amplitude for maximum efficiency which can be expressed as

$$\frac{\partial}{\partial X_{00}}(gX_{00}^2) = \frac{\partial}{\partial X_{00}}\Delta p < 0. \tag{9.137}$$

Therefore, the sidebands become unstable when the equilibrium-mode amplitude exceeds that value at which the efficiency begins to decrease with field amplitude.

A physical picture of the instability [4] can be drawn using characteristic plots such as Fig. 9.10. The radiation amplitude shown on the left side of the figure has developed modulations with a period equal to twice the slippage time εT. In terms of modes, this is described by a superposition of the equilibrium mode and two

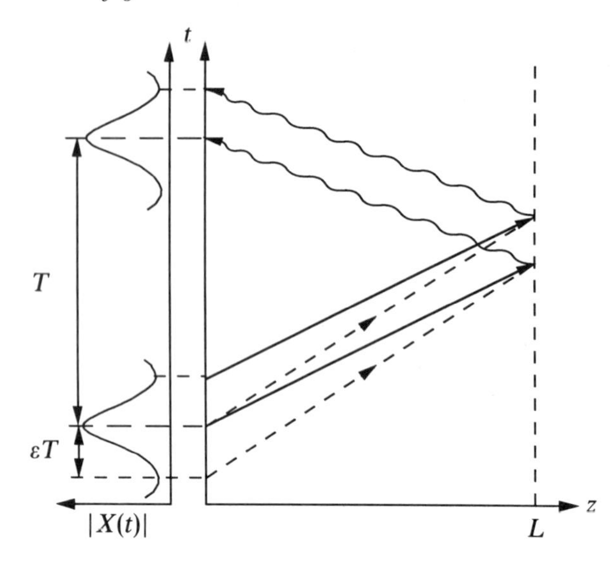

Fig. 9.10 Characteristic plots illustrating the mechanism by which nonlinear saturation of gain combined with slippage gives rise to the spiking mode.

satellites with frequencies displaced from the equilibrium mode by $\varepsilon\Delta\omega_m T = \pi$. The modulations grow because they are self-reinforcing if the gain decreases too rapidly with radiation amplitude. Electrons which enter the interaction region when the radiation amplitude is small produce a significantly larger gain than those entering when the radiation amplitude is large. In traversing the interaction region, the electrons slip behind the radiation by half a period of the modulation. Hence, the high-gain electrons then contribute to the high amplitude of the radiation on the next pass and the low-gain electrons contribute to the low-field amplitude thereby reinforcing the perturbation. This instability occurs when the gain decreases rapidly with field amplitude, and has been named the overbunch instability [22].

The klystron model affords us the opportunity to obtain a qualitative, if not quantitative, understanding of the stability of single-frequency states in a low-gain, continuous-beam oscillator. The determination of the stability of single-frequency states for the case of a distributed interaction region, as opposed to the klystron model, must proceed along numerical lines [15]. However, certain aspects of the klystron calculation can be employed. In particular, a radiation amplitude expressed in the form of equation (9.121) will turn equation (9.90) into an equation of the form of (9.123). Thus, a satellite mode with index m is driven by beam currents which are the responses to the presence not only of the mth satellite mode, but the $-m$th mode as well. The coefficients of proportionality must be determined numerically by integrating the equations of motion in the presence of each satellite. Once the coefficients are determined, the stability of the satellites is found by solving a quadratic equation of the form (9.131). This procedure has been carried out for a 50 × 50 grid of equilibria in the amplitude versus detuning plane [15]. For each equilibrium, the growth rates $\gamma_m(X_{00}, p_0)$ were calculated for 10 pairs of satellites and slippage parameter $\varepsilon = 0.1$. From this data, the boundary separating stable and unstable equilibria was interpolated as shown in Fig. 9.8. Examples of phase-unstable and overbunch-unstable equilibria are shown in Figs 9.11 and 9.12, where the growth rate of the most unstable of the two solutions of the quadratic equation corresponding to equation (9.131) is plotted versus detuning difference $\varepsilon\Delta\omega_m T$. As can be seen for the phase-unstable equilibrium, the most unstable modes are near to the equilibrium mode in frequency with growth rates that depend quadratically on detuning difference for small detuning differences. On the other hand, for the overbunch-unstable equilibrium, the unstable modes are displaced in detuning difference from the equilibrium mode.

In conclusion, the single-frequency states found in the previous section will be stable for equilibrium parameters that place them within the triangular region of Fig. 9.8. These stable equilibria are capable of suppressing parasitic modes, although this suppression becomes weak for pairs of parasitic modes equally spaced close in frequency to the equilibrium mode. The decay rate of these modes scales as the square of the detuning difference between the equilibrium mode and the satellites. The fact that these satellites decay slowly has two consequences. First, the slow decay implies that a single-frequency equilibrium will only be reached after a time sufficiently long so that all satellites decay [15, 21]. Second,

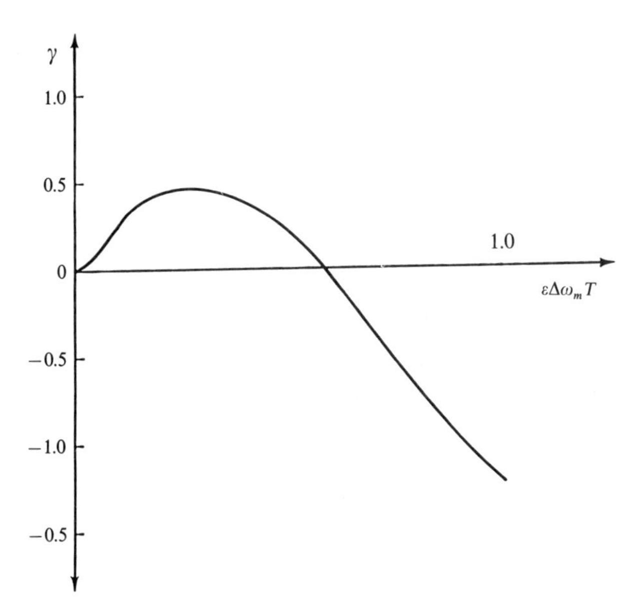

Fig. 9.11 Satellite growth rate versus detuning difference illustrating the phase instability. Equilibrium parameters are $X_0 = 6.0$ and $p_0 = 5.14$ [15].

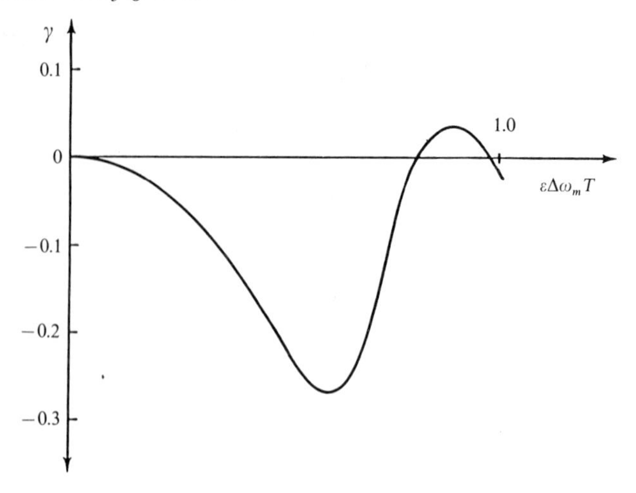

Fig. 9.12 Satellite growth rate versus detuning difference illustrating the overbunch instability. Equilibrium parameters are $X_0 = 14.8$ and $p_0 = 2.6$ [15].

noise fluctuations can excite the weakly damped satellites to an unexpectedly large amplitude which places a more stringent condition on the noise level required to reach a minimum bandwidth. Finally, other modes will not be suppressed for equilibrium parameters outside the stable region. In the case of a phase instability, the tendency is for nearby in-frequency modes to grow, and the unstable equilibrium mode is eventually replaced by one that is stable. In the case of the overbunch instability, the radiation amplitude develops modulations and single-frequency equilibria are not possible. This is known as the spiking mode, and it is expected to occur when the beam current exceeds the start current by a factor of between three and four (depending on the detuning of the equilibrium). For higher currents, the time dependence of the radiation field becomes progressively more complicated and eventually chaotic, as will be discussed in Chapter 11.

9.6.3 The effects of shot noise

So far our treatment of oscillators has neglected the effects of noise on the incoming electron beam. For example, in our discussion of long-pulse oscillators the beam is characterized by a current and a distribution function both of which are independent of time. When the parameters describing the beam become time dependent in a random way this can be considered as noise. Noise has both adverse and positive effects. On the positive side, it provides a seed signal which becomes amplified with time and allows the power in the oscillator to reach saturation without the need to inject power. On the negative side, it can degrade the performance of the oscillator from that which is expected based on time-independent beam parameters.

Classical noise results from the random variations of the injected beam over time. In the low-gain oscillator, the time dependence of the beam current enters the system of equations in three places with three distinct time scales. The shortest time scale is the reciprocal of the radiation frequency. Variations in beam current on this time scale result from the fact that the beam is composed of discrete electrons, which have random entrance times. This source of noise is responsible for the spontaneously emitted radiation that was discussed in Chapter 3 and will be discussed more extensively here. The noise may be included in the present theory by allowing the injected distribution function to be composed of a superposition of delta functions representing the discrete charges with random values of entrance phase ψ_i[23–6].

The next shortest time scale in a low-gain oscillator is the radiation round-trip time. Variation of the beam parameters on this time scale are particularly important for oscillators driven by pulsed electron beams [27]. If the beam pulses do not arrive at uniformly separated times, synchronism with the radiation pulses will be spoiled. If the pulse-to-pulse variations in the arriving beam are correlated over a time which is short compared with the cavity decay time, then they can be modelled by appropriate modification of the parameters of the incoming beam. For example, pulse-to-pulse variations in beam energy can be treated as an additional energy spread for a beam whose pulses are identical. Pulse-to-pulse variations in the arrival time of the beam effectively broaden the temporal width and lower the peak current of the beam micropulse.

Finally, the longest time scale is the cavity decay time. Variations of the beam parameters over this time scale can lead to mode-hopping [28]. In this case, a stable single-frequency state may have established itself for a given set of beam parameters. However, if the beam parameters change, then this mode may become unstable and be replaced by another mode or modes. This will be the case if the beam energy varies, since the free-electron laser resonance is particularly sensitive to energy. Fractional changes in beam energy of order $1/N_w$ (where N_w is the length of the interaction region measured in wiggler periods) are sufficient to change the detuning of a mode and move it out of the gain bandwidth.

The shot noise associated with the discreteness of the electron charge is the most straightforward to calculate and probably represents a minimum to the noise that would be encountered in a realistic beam. For this reason, we will focus our attention on it here. It will be seen that shot noise can be modelled by adding a random source term to the right-hand side of equation (9.90). The random source term can be taken to be a Gaussian white-noise process with respect to the round-trip index n. It is characterized by a matrix of correlation functions between the amplitudes of the source terms driving modes with different indices m [23]. If the mode amplitudes are all small (as they would be during the start-up phase of the oscillator), this matrix will be diagonal; that is, the source terms driving different modes are uncorrelated. The diagonal terms will be found to be proportional to the rate of spontaneous emission of radiation into each mode of the cavity. A consequence of the independence of the source terms is that each mode will grow independently during the initial start-up phase of the oscillator and, at any instant of time before saturation, the mode amplitudes will have a Gaussian probability distribution characterized by a mode number-dependent temperature. Thus, there will be large statistical fluctuations in the amplitudes of individual modes even though the expected amplitude of a mode (the mode temperature) depends smoothly on mode numbers [29].

In the nonlinear regime, shot noise will cause the phase of the signal to diffuse in round-trip time n if conditions of the previous section are satisfied for the existence of stable single-frequency equilibria. This phase diffusion with time leads to a broadened spectrum in the form of a Lorentzian distribution with a width proportional to the spontaneous emission rate [25].

It is convenient to begin the analysis with equation (9.90) which describes the evolution of the amplitude of the mth cavity mode in a low-gain oscillator. The time integral and average on the right-hand side of this equation represent the contribution to the growth of the amplitude of the mth mode of an ensemble of electrons with a uniform distribution of entrance times t_e and phases ψ_i. Such a uniform distribution approximates the situation when a large number of electrons enter the device at random times. The average in (9.90) is more properly thought of as a sum over the N_T electrons that entered the device during a time period T. Given that the electrons enter at random times, the right-hand side will have an average component and a random component which fluctuate with iteration number n. Let us define the exponent inside the average of (9.90) to be some phase,

$$\theta_m(t_e, \psi_i) = \Delta\omega_m(t_e + \varepsilon\xi T) - \psi(\xi, t_e, \psi_i), \tag{9.138}$$

where we have expressed explicitly the dependence of θ_m on the initial entrance phase and entrance time of the particle and suppressed writing the dependence of θ_m on the axial distance, initial energy and the amplitudes of the modes $X_m(n)$.

The average over entrance phases and entrance times in (9.90) is now replaced by a sum over entrance times and entrance phases of N_T individual particles

$$Z_m = \frac{1}{N_T} \sum_{j=1}^{N_T} \int_0^1 d\xi \exp[i\theta_m(t_{ej}, \psi_{ij})]. \tag{9.139}$$

The quantity Z_m is a complex random variable which is the sum of a large number N_T of individual random variables. If we assume that the entrance times and entrance phases for the individual electrons are random variables which are independent and identically distributed, then by the central limit theorem Z_m will be a Gaussian random variable characterized only by a mean and a variance which describes the random nature of the electron beam. The mean value of Z_m, denoted by $\bar{Z}_m$, is found by averaging over the entrance times and entrance phases of all N_T particles under the assumption that the entrance times and entrance phases of different particles are independent and identically distributed and that the distribution is uniform with respect to entrance time and entrance phase. Hence

$$\bar{Z}_m = \frac{1}{(2\pi T)^{N_T}} \int_0^{2\pi} d\psi_{i1} \ldots d\psi_{iN_T} \int_0^T dt_{e1} \ldots dt_{eN_T} Z_m(\psi_{i1}, \ldots, \psi_{iN_T}, t_{e1}, \ldots, t_{eN_T}). \tag{9.140}$$

Evaluation of these integrals yields a mean value for Z_m of

$$\bar{Z}_m = \frac{1}{2\pi T}\int_0^{2\pi} d\psi_i \int_0^T dt_e \int_0^1 d\xi \exp[i\theta_m(t_e, \psi_i)]. \tag{9.141}$$

As expected, this gives precisely the right-hand side of (9.90). The variance of $|Z_m|^2$ is obtained by forming the square of the magnitude of Z_m given by (9.139),

$$|Z_m|^2 = \frac{1}{N_T^2}\sum_{j,j'=1}^{N_T}\int_0^1 d\xi \exp[i\theta_m(t_{ej}, \psi_{ij})]\int_0^1 d\xi \exp[-i\theta_m(t_{ej'}, \psi_{ij'})], \tag{9.142}$$

and averaging over the entrance times and entrance phases of all the electrons. The double sum in equation (9.142) results in two types of terms when averaged. There are N_T diagonal terms for which $j = j'$, and $N_T(N_T - 1)$ terms for which particles $j \neq j'$. Under the assumption that N_T is large, this results in

$$\overline{|\delta Z_m|^2} = \overline{|Z_m|^2} - |\bar{Z}_m|^2 = \frac{1}{N_T}\int_0^{2\pi}\frac{d\psi_i}{2\pi}\int_0^T \frac{dt_e}{T}\left|\int_0^1 d\xi \exp[i\theta_m(t_e, \psi_i)]\right|^2. \tag{9.143}$$

Thus, the root-mean-square value of the random fluctuation of Z_m from its mean value scales as $N_T^{-1/2}$ which is small when the number of electrons is large.

The random component of the source term in (9.90) will fluctuate with round-trip index n. Each time the round-trip index increases by one, a new group of N_T electrons will have passed through the interaction region. This group will be uncorrelated with the previous groups of electrons. Thus, the effective correlation time in round-trip index n is unity. Since field amplitudes in the low-gain limit can change only after many round-trip times, a correlation time of one round trip is essentially the same as a correlation time of zero. In other words, the noise term can be taken to represent white noise

$$\overline{\delta Z_m(n)\delta Z_m^*(n')} = \delta(n - n')\overline{|\delta Z_m|^2}. \tag{9.144}$$

In principle there can be correlations between the source terms (i.e. on the right-hand side of equation (9.90)) driving different modes. That is, the expectation value of the cross term $\delta Z_n^* \delta Z_m$ may be non-zero [23]. The cross-correlation function between the sources for two modes labelled m and n can be defined in analogy to (9.144),

$$\overline{\delta Z_n^* \delta Z_m} = \frac{1}{N_T}\int_0^{2\pi}\frac{d\psi_i}{2\pi}\int_0^T\frac{dt_e}{T}\int_0^1 d\xi \exp[-i\theta_n(t_e, \psi_i)]\int_0^1 d\xi \exp[i\theta_m(t_e, \psi_i)]. \tag{9.145}$$

Examining equation (9.138) for the phase θ_n, we see that the cross-correlation between the noise signal driving the two modes will be non-zero if the phase ψ depends periodically on the entrance time t_e with a frequency equal to the difference of the frequencies of the two modes. This requires that a large-amplitude mode with this difference frequency be present in the cavity to modify the electron trajectories. If the amplitudes of all modes are small, as they are during the linear start-up phase of the oscillator, then the cross-correlations vanish and each mode is excited by its own independent white-noise source. On the other hand, if the mode amplitudes are large then there will be correlations between the noise sources driving each mode. Another instance in which the noise sources are cross correlated is the case of a pulsed-beam oscillator [23]. In this case the current in (9.90) is time dependent and should be taken inside the integral over entrance time. This same factor would then appear, but squared, inside the integral in (9.145). The time dependence of the beam current would then induce a correlation between the noise signals driving two different modes even when the mode amplitudes are small.

The simplest case to analyse is the case in which all modes are assumed to be small. This was done in Chapter 3 in which spontaneous emission was studied in detail. The amplitude of the fluctuations expressed in (9.143) can be calculated by substitution of the expression for the free-streaming phase of an electron in the interaction region in terms of its initial phase and entrance time

$$\psi(\xi, \psi_i, t_e) = \psi_i + p_0 \xi. \tag{9.146}$$

Evaluation of (9.143) then gives the spontaneous emission noise amplitude in terms of the dimensionless detuning $p_m = p_0 - \varepsilon \Delta\omega_m T = p_0 - 2\pi\varepsilon m$, for which

$$\overline{|\delta Z_m|^2} = \frac{1}{N_T}\left|\frac{\exp(ip_m) - 1}{ip_m}\right|^2 = \frac{1}{N_T}\frac{\sin^2\left(\frac{p_m}{2}\right)}{p_m^2}. \tag{9.147}$$

Examination of the effect of noise on the evolution of individual mode amplitudes requires solution of the set of equations represented by (9.90) and either the particle equations (9.79) and (9.80) or the Vlasov equation (9.86) supplemented by the addition of the random source representing the fluctuations in the variables Z_m. These equations can be solved analytically in two limits: (1) the limit of small mode amplitudes (i.e. the start-up phase), and (2) the limit of a single mode. In both these cases, correlations between different modes can be neglected. In the general case, computer simulation of the system is required with specific realizations for the random variables. A technique for performing such a simulation which models the noise correctly without requiring the solution for the trajectories of an unduly large number of electrons will be given at the end of this section.

We now consider the case of the evolution of a single mode in the presence of a white-noise source. Because of the presence of a random-noise source, the complex amplitude of the mode X_m itself becomes a random variable. At a particular time, labelled by the round-trip index n, there is a probability distribution function which gives the probability of observing the mode to have a given amplitude. Since the amplitude is a complex variable, we must introduce the joint probability distribution function for either the real and imaginary parts of the amplitude, or its magnitude and phase. It is convenient at this point to introduce the two-dimensional vector $\boldsymbol{X}$ whose components are the real and imaginary parts of the mode amplitude X_m. This permits the use of vector notation to write the probability of observing the vector mode amplitude $\boldsymbol{X}$ in the small two-dimensional area

dX^2 centred at $\boldsymbol{X}$

$$dP = P(\boldsymbol{X}, n)d^2\boldsymbol{X}. \tag{9.148}$$

Because the mode amplitudes change slowly with time, and because the noise is white noise, the evolution of the probability distribution function is governed by the Fokker–Planck equation

$$\frac{\partial}{\partial n}P(\boldsymbol{X}, n) = \frac{\partial}{\partial \boldsymbol{X}}\cdot\left[-\frac{\overline{\Delta \boldsymbol{X}}}{\Delta n}P(\boldsymbol{X}, n) + \frac{\partial}{\partial \boldsymbol{X}}\cdot\left(\frac{1}{2}\frac{\overline{\Delta \boldsymbol{X}\Delta \boldsymbol{X}}}{\Delta n}P(\boldsymbol{X}, n)\right)\right], \tag{9.149}$$

where the coefficients in (9.149) give the expected rates of drift and diffusion of the mode amplitude vector $\boldsymbol{X}$. Derivations of the Fokker–Planck equation can be found in standard textbooks on kinetic equations [30]. The coefficients in (9.149) are most easily written when the vector $\boldsymbol{X}$ is represented in polar coordinates (X, φ),

$$\frac{\partial}{\partial n}P(\boldsymbol{X}, n) = \frac{1}{X}\left[\frac{\partial}{\partial X}(X\Gamma_X) + \frac{\partial}{\partial \varphi}\Gamma_\varphi\right], \tag{9.150}$$

where the fluxes are given by

$$\left.\begin{aligned}
\Gamma_X &= X[(1-R) - I_b \operatorname{Re} G_{nl}]P + \frac{1}{X}\left[\frac{\partial}{\partial X}(XPD_{XX}) + \frac{\partial}{\partial \varphi}(PD_{\varphi X}) - PD_{\varphi\varphi}\right] \\
\Gamma_\varphi &= -X\,\delta\omega_0 P + \frac{1}{X}\left[\frac{\partial}{\partial X}(XPD_{X\varphi}) + \frac{\partial}{\partial \varphi}(PD_{\varphi\varphi}) + PD_{\varphi X}\right],
\end{aligned}\right\} \tag{9.151}$$

and the components of the diffusion tensor in polar coordinates are

$$\left.\begin{aligned}
D_{XX} &= \frac{I_b^2}{2N_T}\int_0^{2\pi}\frac{d\psi_i}{2\pi}\int_0^T\frac{dt_e}{T}\left[\int_0^1 d\xi \sin\theta_m(\psi_i, t_e)\right]^2 \\
D_{\varphi\varphi} &= \frac{I_b^2}{2N_T}\int_0^{2\pi}\frac{d\psi_i}{2\pi}\int_0^T\frac{dt_e}{T}\left[\int_0^1 d\xi \cos\theta_m(\psi_i, t_e)\right]^2 \\
D_{X\varphi} = D_{\varphi X} &= \frac{I_b^2}{2N_T}\int_0^{2\pi}\frac{d\psi_i}{2\pi}\int_0^T\frac{dt_e}{T}\int_0^1 d\xi \sin\theta_m(\psi_i, t_e) \\
&\quad\times\int_0^1 d\xi \cos\theta_m(\psi_i, t_e).
\end{aligned}\right\} \tag{9.152}$$

In evaluating the average drift term (the first terms in equations (9.151)), we have used the result that the expected value of the variable Z_m given by (9.141) can be expressed in terms of the nonlinear gain function defined in (9.106). This introduces the real part of the gain and the frequency shift through equation (9.109). The diffusive terms (in particular the elements of the diffusion tensor) are evaluated by solution for the trajectories of electrons in the presence of a real mode amplitude $X_m = X$. Equation (9.150) is expected to apply if many modes are present (but all are small), or if only the mth mode is present. Observe that when the mode amplitude is small in the linear regime, the diffusion tensor is diagonal and can be expressed in terms of the normalized spontaneous emission rate (9.147)

$$D_d = \frac{I_b^2}{4N_T} \frac{\sin^2\left(\frac{p_m}{2}\right)}{p_m^2} \equiv \frac{1}{2} S_m, \tag{9.153}$$

where D_d denotes the diagonal elements of the diffusion tensor, and S_m is used to denote the source term which is the normalized rate of spontaneous emission.

In the linear regime, we can look for solutions of equation (9.150) of the form of a Gaussian probability distribution function with a time-varying mode temperature

$$P(X, n) = \frac{1}{2\pi T_m(n)} \exp\left(-\frac{X^2}{2T_m(n)}\right). \tag{9.154}$$

Insertion of the assumed form of the distribution function into equation (9.150) results in a consistency relation determining the growth with time of the mode temperature

$$\left(\frac{d}{dn} + \gamma_m T\right) T_m = S_m, \tag{9.155}$$

where γ_m is the growth rate in the linear regime given by (9.99). If the macroscopic beam parameters such as current and voltage are constant in time, equation (9.155) is readily integrated to obtain

$$T_m(n) = \frac{S_m}{2\gamma_m T} [\exp(2\gamma_m n T) - 1]. \tag{9.156}$$

For times shorter than the growth time $\gamma_n n T \ll 1$, the mode temperature increases linearly with time reflecting the initial buildup of energy in the mode due to the constant rate of spontaneous emission (the exponential growth is unimportant on this time scale). Initially, the temperature increases by an amount S_m in the time it takes the radiation to circuit the cavity once. For times greater than the growth time $\gamma_n n T \gg 1$, the temperature grows exponentially due to gain with an effective initial amplitude proportional to the spontaneous emission rate. This effective initial amplitude corresponds to a temperature $T_m(0) = S_m/(2\gamma_m T)$ which is roughly the temperature achieved at the end of the initial phase after a number $(2\gamma_m T)^{-1}$ of round trips through the cavity by the radiation. If the macroscopic beam parameters vary with time, equation (9.155) can be solved numerically to determine the mode temperature in the linear regime. An example of this will be given in the section on nonlinear gain narrowing.

In addition to provide a seed signal to start the oscillator, shot noise also causes fluctuations in the amplitude and phase of a mode at saturation, which may be determined quantitatively from equation (9.150). We assume that a single mode is present and calculate the time-asymptotic probability distribution function for the amplitude of this mode. When expressed in polar coordinates, equation (9.150) describes the drift and diffusion of the amplitude and phase of the cavity mode. Due to the diffusion of the phase, we can assume that the probability distribution function becomes independent of the phase of the mode asymptotically in time. The dependence of the distribution function on the magnitude of the mode is obtained by averaging equation (9.150) over phase (the polar angle of $\mathbf{X}$) and integrating once with respect to mode amplitude under the assumption that a steady state has been reached

$$\Gamma_X = X[(1-R) - I_b \operatorname{Re} G_{nl}]P + \frac{1}{X}\left[\frac{\partial}{\partial X}(XPD_{XX}) - PD_{\varphi\varphi}\right] = 0, \quad (9.157)$$

where we have set the constant of integration to zero by virtue of the requirement that the distribution function must vanish as $X \to \infty$. Equation (9.157) can be integrated once more with respect to the magnitude of the mode amplitude to find the time-asymptotic probability distribution function

$$P = \frac{C}{D_{XX}} \exp\left[-\int_0^X dX' \frac{X'[(1-R) - I_b \operatorname{Re} G_{nl}] + (D_{XX} - D_{\varphi\varphi})/X'}{D_{XX}}\right], \quad (9.158)$$

where C is a constant chosen to normalize the probability distribution function to unity. The probability distribution function described by (9.158) will peak at values of mode amplitude X where the integrand in the exponent of P vanishes. If the noise is weak (i.e. $X^2(D_{\varphi\varphi} - D_{XX}) \ll 1 - R$ and $I_b \operatorname{Re} G_{nl}$), then the peak will occur at the value of the mode amplitude which gives saturation in the absence of noise (9.108) (i.e. $1 - R = I_b \operatorname{Re} G_{nl}(X_0)$). The nonlinear coefficients appearing in (9.158) can be evaluated by Taylor expansion around this value. Specifically, the probability distribution becomes a Gaussian peaked at the saturated mode amplitude X_0

$$P(X, n \to \infty) = \frac{1}{2\pi^{3/2} X_0 \Delta X} \exp\left\{-\frac{[X - X_0]^2}{(\Delta X)^2}\right\}, \quad (9.159)$$

where the width ΔX is proportional to the spontaneous emission rate

$$(\Delta X)^2 = \frac{2D_{XX}}{\gamma_a T} = \frac{\Phi_A(X_0, P_0)}{N_*}, \quad (9.160)$$

and the quantity γ_a appearing in (9.160) is defined in (9.116) and represents the damping rate for perturbations of the mode amplitude away from equilibrium. It must be positive for the saturated equilibrium to be stable. The noise-induced fluctuations in the mode amplitude are inversely proportional to this rate. Further, equation (9.160), along with the expression for the diffusion tensor (9.152), shows that the relative fluctuations in the amplitude of a saturated mode are inversely proportional to the square root of the number of electrons which pass through the interaction region during this relaxation time. This dependence is more concisely expressed in the second equation (9.160) where equation (9.152) has been used to eliminate D_{XX}, and the equilibrium relation (9.108) (i.e. $1 - R = I_b \operatorname{Re} G_{nl}(X_0, p)$) has been used to eliminate the current. This allows the expression for the noise-induced amplitude fluctuations to be expressed in terms of $N_* = N_T/[2(1-R)]$, which is the number of electrons passing through the device in a cavity decay time, $T_d = T/[2(1-R)]$, and a form factor

$$\Phi_A(X_0, P_0) = \frac{\displaystyle\int_0^{2\pi} \frac{d\psi_i}{2\pi} \int_0^T \frac{dt_e}{T} \left[\int_0^1 d\xi \sin\theta_0(\psi_i, t_e) \right]^2}{2\mathrm{Re}[G(X_0, p_0)]\mathrm{Re}[X_0 \partial G(X_0, p_0)/\partial X_0]}, \tag{9.161}$$

which depends only on the normalized mode amplitude and the detuning for the nonlinear saturated mode. Level curves of the form factor Φ_A are plotted in Fig. 9.13. Only those values inside the stable triangle should be considered since only for these parameters is single-mode operation possible. The dominant dependence of the amplitude fluctuations is therefore determined by N_*, the number of electrons passing through the device in a cavity decay time. Since this is typically a large number, the relative fluctuation in the amplitude of a large saturated mode due to shot noise is expected to be quite small and will have a negligible effect on the spectrum of the radiation.

The dominant effect of shot noise on the spectral width of a saturated mode is due to the diffusion in time of the phase of the mode. To calculate this effect we introduce the power spectrum, which is the time Fourier transform of the two-time correlation function of the complex mode amplitude

$$S(v) = \int_{-\infty}^{\infty} dnT \exp(ivnT)\overline{X_m(n)X_m^*(0)}. \tag{9.162}$$

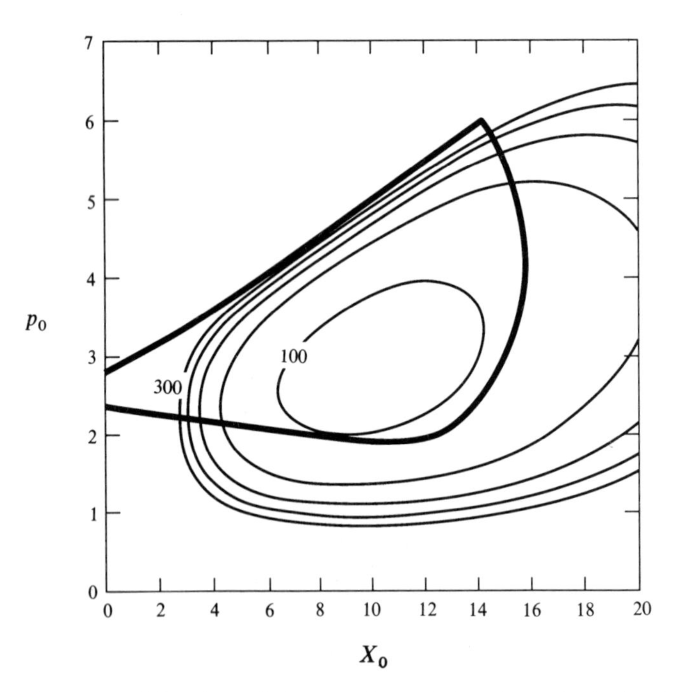

Fig. 9.13 Level curves of the form factor Φ_A determining the noise-induced amplitude fluctuations for a nonlinearly saturated mode. The separation between adjacent levels is 50.

We must find the expected value of the product of the complex mode amplitude and its complex conjugate (i.e. the two-time correlation function) evaluated at two different times, 0 and nT, in order to evaluate the power spectrum. Under the assumption that the fluctuations in the magnitude of the complex amplitude are small, the time dependence of the correlation function is dominated by the diffusion of the phase of the mode with time. The correlation function may be determined by assuming that the magnitude of the amplitudes of the mode at the two times are the same and equal to the expected value of the amplitude at saturation. The expected dependence on the phase difference between the complex amplitudes of the mode at the two different times can be determined by solving the Fokker–Planck equation (9.149) subject to the initial condition that the phase of the mode at time $n=0$ is known. Since the probability distribution will be narrowly peaked around the saturation amplitude X_0, equation (9.150) can be integrated over the magnitude of $\boldsymbol{X}$ to obtain an equation for the phase dependence of the amplitude integrated probability distribution function

$$\frac{\partial}{\partial n}\bar{P}(\varphi, n) = \frac{\partial}{\partial \varphi}\left[-\delta\omega_0 T(X_0, p_m)\bar{P} + \Delta\nu T\frac{\partial}{\partial \varphi}\bar{P}\right], \tag{9.163}$$

where

$$\bar{P}(\varphi, n) = \int_0^\infty dX\, X\, P(X, \varphi, n), \tag{9.164}$$

and

$$\Delta\nu = \frac{D_{\varphi\varphi}}{TX_0^2} = \frac{\Phi_B}{T_d N*} \tag{9.165}$$

is the spontaneous emission-induced diffusion coefficient for the phase. The diffusion tensor is evaluated at the saturated mode amplitude X_0.

The diffusion rate $\Delta\nu$ will be shown to determine the width of the spectrum for the saturated mode. The second equality in (9.165) shows that the width scales inversely with the cavity decay time and the number of electrons transiting the device in the cavity decay time. Again, there is a form factor

$$\Phi_B(X_0, P_0) = \frac{\displaystyle\int_0^{2\pi}\frac{d\psi_i}{2\pi}\int_0^T\frac{dt_e}{T}\left[\left[\int_0^1 d\xi \cos\theta_0(\psi_i, t_e)\right]\right]^2}{8X_0^2\{\mathrm{Re}[G(X_0, p_0)]\}^2}, \tag{9.166}$$

which depends only on the normalized parameters of the saturated equilibrium. The level curves of Φ_B are shown in Fig. 9.14.

The analysis continues by solution of the diffusion equation (9.163) subject to the condition that the initial phase is specified as φ_0; hence

$$\bar{P} = \frac{1}{\sqrt{(2\pi|n|\Delta\nu T)}}\exp\left[-\frac{(\varphi - \varphi_0 + \delta\omega_0 nT)^2}{4|n|\Delta\nu T}\right], \tag{9.167}$$

which shows the phase to be both drifting in time with an average rate equal to

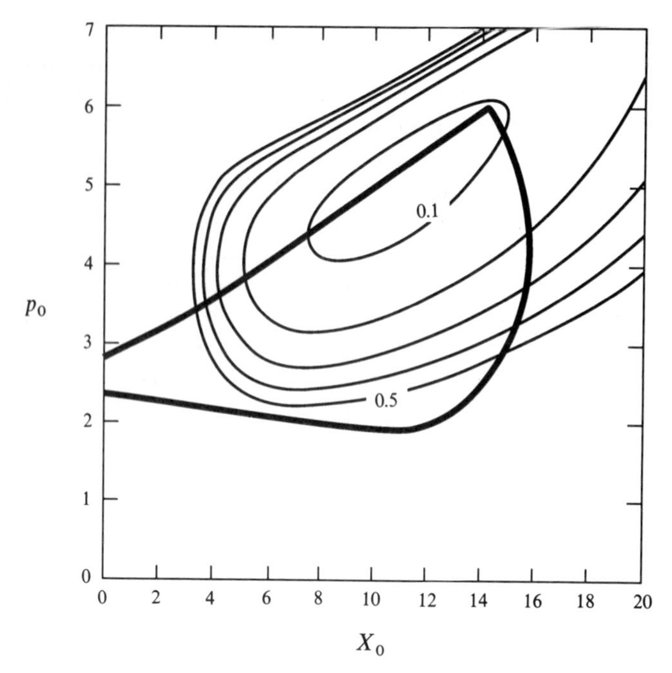

Fig. 9.14 Level curves of the form factor Φ_B determining the noise-induced spectral width for a single nonlinearly saturated mode. The separation between adjacent levels is 0.1.

the frequency shift defined for the saturated mode amplitude, and to be diffusing due to the shot noise. The correlation function, that is the integrand in (9.162), is then obtained by integration of the amplitude-averaged probability distribution function over all angles weighted by the initial and final mode amplitudes. Fourier transformation in the time domain then yields the power spectrum

$$S(\nu) = \frac{2\Delta\nu X_0^2}{(\nu - \delta\omega_0)^2 + \Delta\nu^2}. \tag{9.168}$$

Thus, the power spectrum is peaked at a frequency $\delta\omega_0$ corresponding to the shifted frequency of a saturated mode given by (9.109) and has a Lorentzian shape with a width given by $\Delta\nu$. This power spectrum was defined for the time dependence of the mode amplitudes on the slow time scale. The actual spectrum will, of course, be shifted by the empty cavity frequency of the mode in question $\omega_m = \omega + \Delta\omega_m$.

The spectral width defined in equation (9.165) assumes that only one mode is present in the time-asymptotic state. However, it was shown in the last section that nearby modes are weakly damped with a damping rate that scales as the square of the frequency difference between the mode in question and the large-amplitude equilibrium mode. Equation (9.132) expresses this damping rate in terms of a coefficient D (not to be confused with the elements of the Fokker–Planck diffusion tensor defined in (9.152)) which is roughly of the order of the inverse of the cavity decay time T_d. Using the previous calculations as a guide (in particular, equation (9.160)) which states that the r.m.s. level of fluctuations in a mode scale as the normalized spontaneous emission rate divided by the damping rate, we arrive at the estimate

$$|X_m|^2 = \frac{S_m}{DT(\varepsilon\Delta\omega_m T)^2} \approx \frac{\Phi_C(X_0, p_0)}{N*(2\pi\varepsilon m)^2}, \tag{9.169}$$

where Φ_C is a form factor analogous to Φ_A and Φ_B and we have used $\Delta\omega_m T = 2\pi m$. Recall that equation (9.132) corresponds to the damping rate of a coupled pair of modes; however, this will not affect the order-of-magnitude estimate given here. If the index m corresponds to the nearest-neighbour pair of modes ($\Delta\omega_m T = 2\pi$), then this pair of modes can be excited to large amplitude even if the number of electrons transiting in a cavity decay time is large. A noise level consistent with

$$\varepsilon^2 N_* < 1 \tag{9.170}$$

is required in order that the single-mode regime be achieved. Thus, if the slippage parameter is too small, then the single-mode spectral width calculated in (9.165) will not apply. Instead, the final asymptotic state of the oscillator will be one in which the shot noise excites a number of modes to large amplitude. This number is estimated by using equation (9.169) to determine the mode number for which the energy in the noise-excited satellites is comparable to that of the equilibrium mode. That is, if a large number of modes are excited but the spectrum is still narrower than the gain bandwidth, we expect $\Sigma_m |X_m|^2 = |X_0|^2$, where X_0 is the saturated amplitude for a single mode with the corresponding central detuning. Thus, the number of excited modes can be estimated from (9.169) by finding the value of m such that $m|X_m|^2 \approx |X_0|^2$. The resulting spectral width is given by

$$\Delta\omega_m T \cong \frac{1}{\varepsilon^2 N*}. \tag{9.171}$$

To summarize, shot noise can be responsible for the initial excitation of modes in the cavity which then grow due to their positive gain. Modes are excited with a Gaussian probability distribution function prior to saturation which is characterized by a time-dependent mode temperature. Due to the statistical nature of the noise-excitation process, a given mode will have 100% fluctuations in its amplitude from shot to shot. Equivalently, nearby modes in the same shot will have large fluctuations in their amplitudes. If the beam parameters are such that a stable single-frequency equilibrium is possible, the spectrum will tend to approach the single-frequency state asymptotically in time. However, it will ultimately be limited by noise. This limit may be either of the form of a broadening of a single mode (9.165) [25], or the excitation of a number of modes (9.171). The process of saturation and spectral narrowing which leads to the time-asymptotic single-mode state will be discussed in the next section.

Finally, we conclude this section with the description of a numerical technique for incorporating noise realistically into particle simulations of free-electron laser oscillators [29]. As previously noted, electrons do not enter the interaction region with phases and entrance times uniformly distributed over the intervals $(0, 2\pi)$ and $(0, T)$. Rather, a large number of electrons enter with random entrance times, and phases which are independent and identically distributed. This is the source of the noise. To account for the noise, it would be impractical to distribute the phases and entrance times randomly and independently since this would require simulating the trajectories of too many electrons in order to achieve a low enough level of noise to be comparable with that in an experiment. Instead, one can ascribe a weight $W_j = 1 + \delta W_j$ to each particle in the sum in (9.139), and simulate the trajectories of N_s electrons where $N_s \ll N_T$. The electrons included in the simulation can then be distributed uniformly on a grid of entrance phases and entrance times. If the δW_j were zero ($W_j = 1$), then there would be no noise. To adjust the level of noise in the simulation to be equal to that in an experiment, one can allow the δW_j to be random variables that are independent and distributed identically with a zero mean and a variance given by

$$\overline{\delta W_j^2} = \Delta n \frac{N_s}{N_T}, \tag{9.172}$$

where Δn is the time step used in the integration of the mode amplitude equation. The noise in the simulation will be the result of the addition of a large number N_s of random variables. Hence, the noise will be Gaussian white noise with identical statistics to that expected from experimental shot noise. The mode amplitudes calculated in this way will represent a single shot, and a number of simulations will be required to obtain statistical averages. Examples of the start-up phase of a long-pulse oscillator using this technique will be presented in the next section.

9.6.4 Linear and nonlinear spectral narrowing

In the previous sections, it has been shown that both the spectrum of spontaneous emission and the intrinsic gain have frequency bandwidths that correspond to modes whose dimensionless detunings fall in the range $0 \leqslant p_m \leqslant \pi$. This is evidenced by the dependence of both the rate of spontaneous emission (9.147) and the gain (9.98) on the dimensionless detuning parameter. This range of dimensionless detunings translates, using the definitions appearing in equations (9.101) and (9.44), into a fractional frequency bandwidth given by

$$\frac{\Delta\omega_m}{\omega} \cong \frac{\pi}{\varepsilon\omega T} = \frac{1}{2N_w}, \tag{9.173}$$

where we use N_w here to denote the number of wiggler periods in the interaction region. Thus, in an oscillator which is initiated by shot noise, modes in this frequency range will be excited. If the level of noise is small, then the radiation in the cavity will have to be greatly amplified over many passes before saturation is achieved. During this time, the modes with the largest growth rates (that is, the highest gains) will outgrow modes with smaller gain, so that the spectrum of modes by the time saturation is achieved will be narrower than that which was excited initially by noise. This process, which is common to lasers, is known as gain narrowing.

The amount of gain narrowing that is expected to occur can be quantified using equation (9.156) for the mode temperature (the expected squared amplitude of the mode X_m) in the linear regime. For large times nT, the dominant dependence of the mode temperature on the mode number is through the dependence of the growth rate appearing in the exponent. If the growth rate is expanded about the detuning value p_{max}, which gives the maximum growth rate (i.e. $p_{max} \approx 2.6$ for a monoenergetic beam), then the dependence of the mode temperature on detuning is of the form of a Gaussian

$$T_m \cong T_{max} \exp\left[-\frac{(p_m - p_{max})^2}{2\delta p^2} \right], \tag{9.174}$$

where T_{max} is the amplitude of the largest mode and δp is the width in the dimensionless detuning of the spectrum

$$\delta p^2 = \frac{1}{nT} \left| \frac{d^2 \gamma_m}{dp_m^2} \right|^{-1}. \tag{9.175}$$

Equation (9.175) shows that the spectral width starts approximately at the gain bandwidth and decreases steadily with time. This form of linear gain narrowing is expected to proceed at least until saturation occurs, at which time the factor nT can be estimated from (9.156) as

$$nT \cong \frac{1}{2\gamma_m} \ln \left| \frac{\gamma_m T T_{sat}}{S_m} \right|, \tag{9.176}$$

where T_{sat} is the mode temperature corresponding to a saturated amplitude, and S_m is the spontaneous emission rate. Therefore, the larger the saturated amplitude compared with the emission rate, the more growth is required to reach saturation, and the narrower the spectrum as saturation is approached.

Once the saturation process begins, the mode amplitudes interact nonlinearly and the formulae describing linear gain narrowing are no longer valid. It will be shown, however, that formula (9.175) is approximately valid with the spectral width decreasing with time but with the second derivative of the linear growth rate being replaced by the coefficient D which appears in the formula (9.132) for the damping rate of nearby in-frequency modes in the presence of a large-amplitude equilibrium mode. Note that the coefficient D is the second derivative with respect to detuning of the damping rate of nearby satellite modes. Since the coefficient D is a frequency which scales as the inverse of the cavity decay time T_d (see the discussion following (9.133)), we can write the expression for the time-dependent spectral width as

$$\delta p^2 \cong \frac{T_d}{nT}, \tag{9.177}$$

which, when expressed as a fractional bandwidth (as in (9.173)), becomes [15, 21]

$$\frac{\Delta\omega}{\omega} \cong \frac{1}{2N_w}\left(\frac{T_d}{nT}\right)^{1/2}. \tag{9.178}$$

Therefore, many cavity decay times are required for the spectrum to become significantly narrower than the gain bandwidth.

Recall that the slow decay of the coupled pairs of modes described by (9.132) is a consequence of the fact that coherence in the radiation pulse is established only by the slippage of the beam electrons with respect to the radiation. Thus, if the slippage is small a long time is required before all temporal portions of the radiation pulse are oscillating in phase. The rate at which coherence is established can be increased if some other mechanism can be found to provide communication between different portions of the radiation pulse. An example is the addition of a dispersive element to the radiation feedback path [20]. The effect of a dispersive element on the rate of spectral narrowing will be discussed at the end of this section.

A second method of achieving a narrow bandwidth in a relatively short time is to **seed** the oscillator with a signal at the desired frequency. This process is often referred to as **mode-locking.** That is, if radiation is injected into the cavity at a level significantly larger than the noise level, and if the gain at the frequency of the injected radiation is near the peak of the gain spectrum, then the radiation spectrum at saturation will be essentially single frequency [31].

An example of a free-electron laser in which gain narrowing and mode competition have been important is the experiment at the University of California at Santa Barbara [28, 31–3]. The slippage parameter in this experiment was of the order of $\varepsilon \approx 1.7 \times 10^{-3}$, the cavity decay time was 0.2 μs, and the duration of the electron pulse was about 2.75 μs. Thus, a large number of modes were excited initially. These modes grew and saturated, but there was insufficient time for coherence to be established. Indeed, measurements have shown that in a large number of cases the spectrum was multi-moded at the end of the shot [33]. In a smaller number of cases (roughly 29% of the discharges), a single mode appeared to dominate. The criterion for single-mode domination was that of the 20 or so modes whose frequencies fell in the range of the detector, the power in one mode exceeded by a factor of two the power in any other mode. This is most likely due to the statistical fluctuations in the spontaneous emission process discussed in the previous section.

A complicating factor in the Santa Barbara experiment was that the beam energy decreased in time. Thus, the modes that dominated at the end of the pulse were not necessarily the ones that grew initially [28]. It will be shown in this section that the time dependence of the beam energy has a dramatic effect on the efficiency of energy extraction. Specifically, the peak efficiency can vary by over a factor of five, depending on whether the beam voltage increases or decreases over time. An increasing beam voltage is preferred because it allows the high-efficiency single-mode equilibrium occurring at a high detuning parameter ($p_0 \approx 5.14$) to be accessed.

We now describe the results of a numerical simulation of the Santa Barbara experiment [29]. The equations which were solved numerically were the electron equations of motion (9.79) and (9.80)

$$\frac{d\psi}{d\xi} = p, \tag{9.79}$$

and

$$\frac{dp}{d\xi} = -\frac{1}{2}[iX(n, t_e + \varepsilon\xi T)\exp(i\psi) + \text{c.c.}], \tag{9.80}$$

for an ensemble of electrons, with the dimensionless field amplitude represented as a superposition of cavity modes as in (9.84),

$$X(n, t') = \sum_{m=-\infty}^{\infty} X_m(n)\exp(-i\Delta\omega_m t'). \tag{9.84}$$

The equations of motion were solved at each time step in round-trip index n for an ensemble of particles launched on a uniform grid of initial phases $(0, 2\pi)$ and entrance times $(0, T)$ with an initial momentum (detuning parameter for the central mode) p_0 appropriate for a monoenergetic beam. The expression for the detuning parameter is given in (9.101)

$$p_m = p_0 - 2\pi m\varepsilon \cong \frac{L}{v_\parallel}[(k_m + k_w)(v_\parallel + \delta v_\parallel) - (\omega + \Delta\omega_m)]. \tag{9.101}$$

The evolution of the mode amplitudes with slow time (round-trip index n) is governed by equation (9.90),

$$\begin{aligned}\left[\frac{\partial}{\partial n} + (1 - R)\right]X_m(n) &= -iI_b Z_m \\ &= -\frac{iI_b}{T}\int_0^T dt_e \int_0^1 d\xi \langle \exp[i\Delta\omega_m(t_e + \varepsilon\xi T) - i\psi(\xi, t_e)]\rangle.\end{aligned} \tag{9.90}$$

To model spontaneous noise accurately in the simulation, the entrance phase and entrance time averages in equation (9.90) are evaluated as sums over randomly weighted particles are given by (9.139) and (9.172)

$$Z_m = \frac{1}{N_s}\sum_{j=1}^{N_s} W_j \int_0^1 d\xi \exp[i\theta_m(t_{ej}, \psi_{ij})], \tag{9.139}$$

where N_s is the number of simulation particles traversing the interaction region per slow time step, and W_j denotes the particle weights.

Simulation parameters were chosen to match the conditions of the experiment as closely as possible. Recall that the slippage parameter in the experiment was very small with $\varepsilon \approx 1.7 \times 10^{-3}$. Using this value in a simulation would require keeping approximately 750 modes (to span the gain bandwidth). Instead, a value of $\varepsilon = 5.0 \times 10^{-3}$ was chosen, and 251 modes were retained in equations (9.90), (9.79) and (9.80). Thus, each mode in the simulation can be thought of as representing three modes in the experiment. With this number of modes it was necessary to solve the equations of motion for 300 different entrance times and 15 different entrance phases, yielding $N_s = 15 \times 300 = 4500$ electrons per step.

The beam voltage in the Santa Barbara experiment decreased linearly with time at a rate of approximately 2 kV/μs. This is modelled by allowing the detuning parameter p_0 to be a function of the slow time variable n through the dependence of the injected parallel velocity on beam energy (c.f. equation (9.101)). Evaluation of the formula for the normalized current for the parameters of the experiment requires making assumptions on the structure of the modes in the cavity. In the low-gain limit, these can be taken to be the same as the vacuum modes in the cavity. The result is that the actual current is found to be about three times the start current. This is consistent, as will be seen, with the observed rate of increase of power in the experiment. The final important parameter is the noise level, which is modelled using the technique described in the last section. The relevant experimental number which must be supplied for the simulation is the number of electrons traversing the interaction region in a given time which is determined by the beam current (≈ 1 A).

The normalized efficiencies Δp (defined in equation (9.104)) versus slow time $\tau_s = nT/T_d$ (where T_d is the decay time of the empty cavity, from (9.29)) are shown for three different runs in Fig. 9.15. Using equation (9.105), it is seen that a value of $\Delta p = 2\pi$ corresponds (for a relativistic beam) to the standard estimate $\Delta\gamma/\gamma \approx 1/2N_w$. The three curves of Fig. 9.15 correspond to different dependences of beam energy on time. The curve labelled (a) corresponds to a falling beam

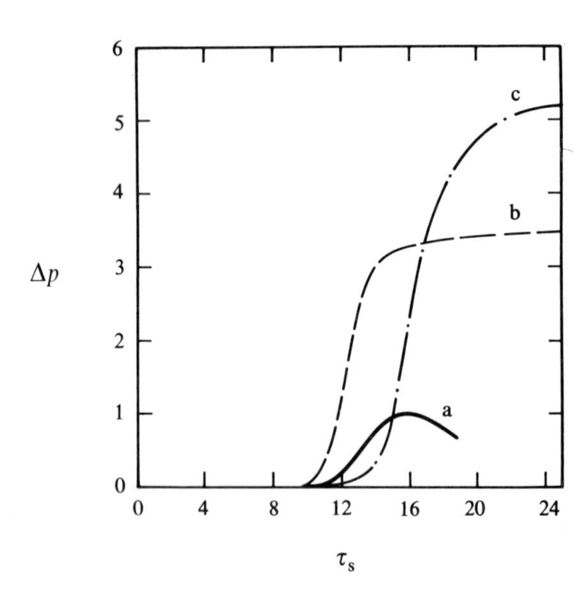

Fig. 9.15 Normalized efficiency versus time for three simulations appropriate to the parameters of the UCSB continuous beam experiment. The three cases correspond to: (a) a falling beam voltage, (b) a constant beam voltage, and (c) a rising beam voltage [29].

voltage as in the experiment, the curve labelled (b) corresponds to a constant beam voltage, and the curve labelled (c) corresponds to a beam voltage that rises with time at a rate opposite to that of case (a). It is evident that the time history of the beam voltage has a large effect on the efficiency of energy extraction. Note that for the conditions of the experiment shown in curve (a), the efficiency is about 16% of the much-quoted standard value.

Figure 9.16 shows the mode amplitudes at various times during the simulations illustrated in Fig. 9.15. In particular, Figs. 9.16(a) and 9.16(b) show the spectrum at $\tau_s = 15$ of case (a) corresponding to a falling beam voltage. The realizations differ in the random numbers that generated the incoherent noise, and illustrate the statistical fluctuations predicted to occur when modes are excited by incoherent noise. In particular, there are 100% fluctuations in the magnitude of a given mode from realization to realization as well as 100% fluctuations in the magnitude of neighbouring modes. Plots of the efficiency versus time for these two realizations are virtually identical. The statistical fluctuations in mode amplitudes are further illustrated by a comparison of the expected amplitudes of the modes just prior to saturation given by the solution of equation (9.155), including the appropriate time dependence of the beam energy. The simulation results for one realization of the incoherent noise are shown in Fig. 9.17. The spectrum is approximately Gaussian, as predicted by equation (9.174).

For the three cases of increasing, uniform and decreasing voltage, the spectra at saturation are peaked at different mode numbers. The central detunings are given by $p_m = 5.67$, 2.60 and 1.16 for these cases, respectively. If these spectra were single-moded with the same detunings, Figs 9.5 and 9.6 could be used to predict the efficiency. With the given value of the normalized current (about three times the start current) and the three values of the detuning from the simulation, the predicted efficiencies at saturation are $\Delta p = 5.0$, 3.4 and 1.0, which is in agreement with the simulations. Thus, as far as efficiency is concerned the

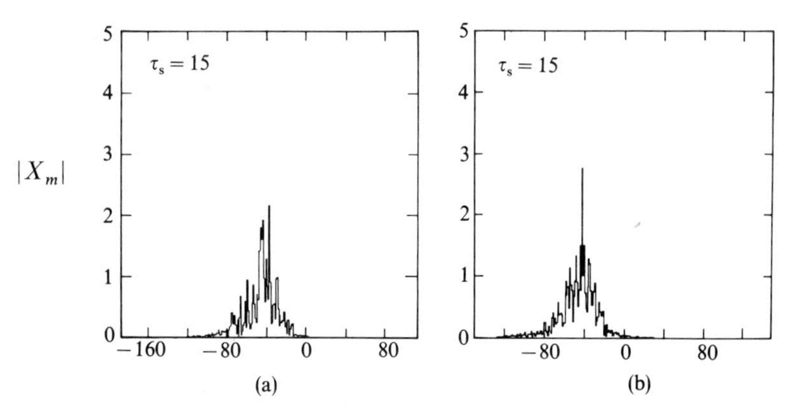

Fig. 9.16 Spectra at saturation for two different realizations of the UCSB experiment. The difference is in the random number **seed** used to generate the spontaneous noise [29].

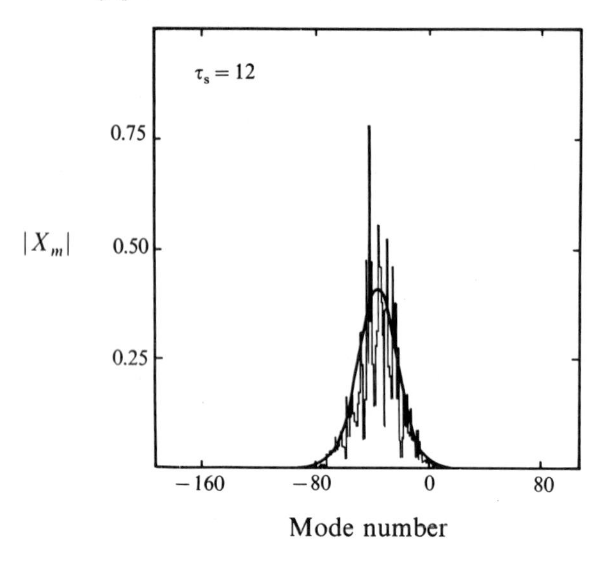

Fig. 9.17 Spectrum of mode amplitudes just prior to saturation for one realization of the UCSB experiment. Superimposed is the expected value of the mode amplitudes as predicted by (9.174) [29].

simulations are behaving as if the spectra were single-moded. This is to be expected since a significant amount of gain narrowing has occurred during the exponential-growth phase.

The reason that the spectrum is peaked at different detunings depending on the time history of the beam voltage can be understood by examination of equation (9.101) for the dimensionless detuning. Due to the time dependence of the beam voltage, the detuning of the central mode p_0 and, hence, the detunings of all modes become a function of time. The mode which will experience the greatest amount of linear gain will be that mode whose detuning spends the longest time in the vicinity of the maximum instantaneous growth rate $p_{max} = 2.6$. If the voltage and detunings are rising, then the detuning of the mode with the maximum total gain will be higher than p_{max} when it reaches saturation, while if the voltage is falling the detuning will be lower than p_{max}. Once saturation is achieved the detuning of the dominant mode will still continue to change with beam voltage. As a result, the detuning of the dominant mode will be pulled out of the stable triangle of Fig. 9.8 and other modes whose frequencies are more in resonance with the beam will grow. This process of mode hopping [28, 29] will continue as long as the beam voltage is changing and a monochromatic spectrum will never be achieved.

A continuation of the simulation with constant-beam parameters results in a spectrum which progressively narrows with time [34]. Figure 9.18 shows spectra from a simulation in which the beam energy becomes constant in time after an initial transient. Spectra are shown for two different times after the power has

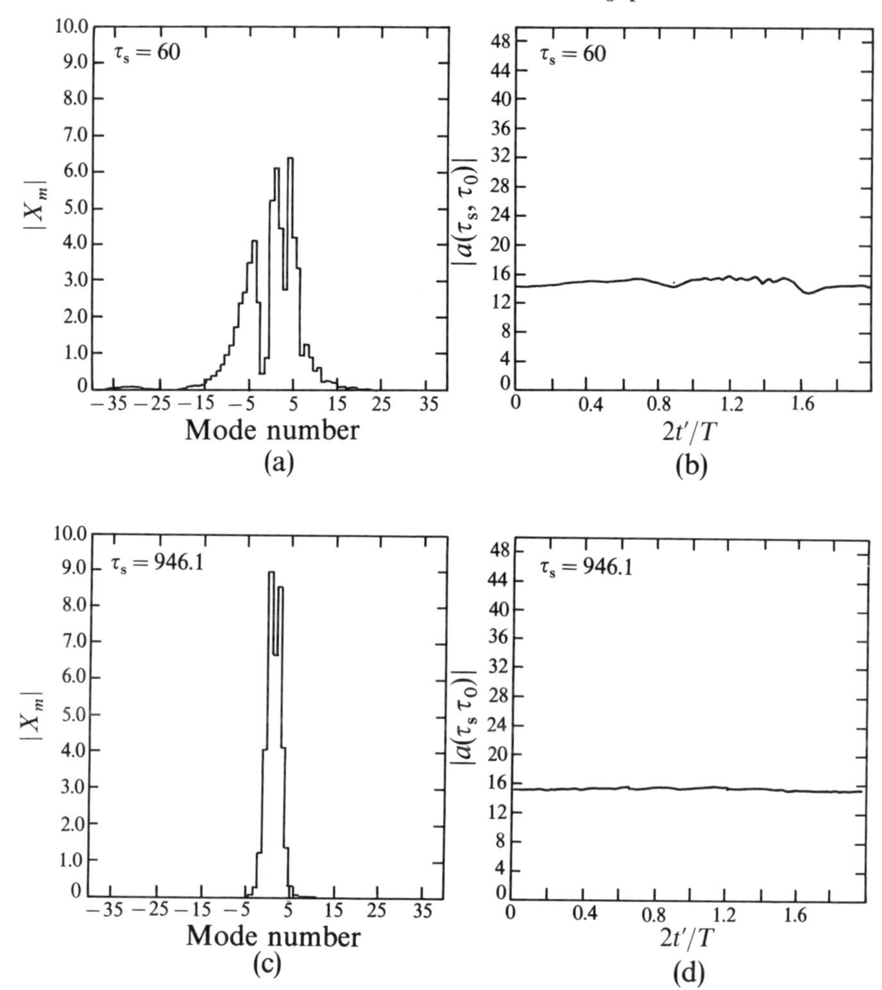

Fig. 9.18 Spectrum of modes at constant voltage showing nonlinear spectral narrowing [34].

saturated. It is evident that the spectrum narrows very slowly with time. In particular, it is still multi-moded after about 1000 cavity decay times. It has been verified [34] that the width of the spectrum was governed by (9.179). Also shown are reconstructions of the fast-time dependence of the magnitude of the radiation field corresponding to the two spectra. Note the absence of modulations in the magnitudes, in accordance with the zero slippage prediction. A quantitative analysis of nonlinear spectral narrowing proceeds by assuming that as the radiation grows and approaches saturation the spectrum has already narrowed linearly to some degree so that the width of the spectrum is considerably less than the gain bandwidth when nonlinear effects come into play. Thus, the radiation spectrum will peak around a mode with a dimensionless detuning parameter which we label p_0. In the simplest case of a beam whose parameters are constant in time, this will correspond to the detuning which gives maximum linear gain (i.e. $p_0 \approx 2.6$ for a monoenergetic beam). As has been seen, if the beam parameters vary with time the detuning will correspond to the value which maximizes the growth of the modes over the time history of the beam. The nonlinear evolution of the radiation pulse which consists of a narrow spectrum of modes peaked about a mode detuning p_0 is governed by equation (9.130) when the effects of slippage are ignored. As discussed, the evolution of the complex amplitude of the radiation field in this case is to approach a constant magnitude asymptotically, as described by equation (9.108) but with phase depending on the fast-time variable t', leading to a non-monochromatic spectrum. Ultimately, the variations in phase will diminish as coherence in different portions of the radiation pulse is established, but this involves the action of slippage which is not included in (9.130).

To examine the way in which slippage eventually establishes coherence in the radiation pulse, we return to consideration of the basic low-gain system of equations (9.79–81) and treat these equations by expansion in powers of the slippage parameter ε. Equation (9.130) is the result of the lowest-order expansion if we set the slippage parameter to zero. This equation leads to an asymptotic solution in which the magnitude of the radiation is independent of the fast-time variable t', reflecting local power balance, and the phase advances with round-trip index at a constant rate

$$\frac{1}{T}\frac{\partial \phi(n,t')}{\partial n} = -\delta\omega_0, \tag{9.179}$$

where $X(n,t') = X_{00}\exp[i\phi(n,t')]$ is the complex amplitude of the radiation. The value of the amplitude X_{00} and the frequency shift $\delta\omega_0$ are determined by the nonlinear equilibrium equations (9.108) and (9.109) for given values of the beam current and central detuning. The lack of coherence in the absence of slippage is reflected in the fact that equation (9.179) does not require that the phase be independent of the fast-time variable t', but only that it advances at the same rate everywhere in the radiation pulse. Keeping higher-order terms in the slippage parameter will lead to an evolution equation for the phase which couples together the values of the phase at different times t' in the radiation pulse

$$\frac{1}{T}\frac{\partial \phi}{\partial n} = -\delta\omega_0 - C_1 \varepsilon T\frac{\partial \phi}{\partial t'} - \frac{C_2}{2}(\varepsilon T)^2\left(\frac{\partial \phi}{\partial t'}\right)^2 + C_3(\varepsilon T)^2\frac{\partial^2 \phi}{\partial t'^2}, \tag{9.180}$$

where the coefficients C_1, C_2 and C_3 depend on the magnitude of the radiation X_{00} and the central detuning p_0. That the evolution equation must be of this form can be deduced from the way in which the slippage parameter enters equations (9.79–81). In particular, it always appears multiplied by the product of the round-trip time T and the dimensionless axial distance ξ, where $0 \leqslant \xi \leqslant 1$. Further, this product always appears additively in the fast-time argument of the radiation amplitude and beat-wave phase. Thus, in an expansion in ε such as

(9.180) each power of ε must be accompanied by a derivative with respect to the fast-time variable. Physically, the slippage induces a coupling between neighbouring in-time phases of the radiation amplitude. Equation (9.180) represents all possible terms in an expansion up to and including the ε^2 term. Direct calculation of the coefficients appearing in equation (9.180) would be tedious and, in the end, requires numerical evaluation. It is possible, however, to deduce the coefficients on the basis of some general considerations.

The coefficients in equation (9.180) can be determined based on our knowledge of the equilibrium and stability of single-frequency states. For example, suppose the mode phase depends linearly on the fast-time variable

$$\phi = \phi_0(n) - \Delta\omega_m t'. \tag{9.181}$$

This variation of the phase corresponds to a single-frequency state with a detuning parameter $p_m = p_0 - \varepsilon\Delta\omega_m T$. Thus, the rate at which the phase advances with respect to the slow time index n must be of the form

$$\frac{1}{T}\frac{\partial\phi}{\partial n} = -\,\delta\omega_0(I_b, p_0 - \varepsilon\Delta\omega_m T), \tag{9.182}$$

where we have written the frequency shift explicitly in terms of the beam current as opposed to the mode amplitude since current is held fixed and the mode amplitude adjusts itself according to equation (9.108) to maintain power balance. Insertion of equation (9.181) into (9.180) allows the coefficients C_1 and C_2 to be identified in terms of the derivatives of the frequency shift with respect to detuning at constant current by expansion of the frequency shift to second order in the slippage parameter

$$\begin{aligned} C_1 &= -\frac{d}{dp_0}\delta\omega_0 = -\,\delta\omega_0' \\ C_2 &= -\frac{d^2}{dp_0^2}\delta\omega_0 = -\,\delta\omega_0''. \end{aligned} \tag{9.183}$$

To determine the final coefficient, consider the case in which the phase is constant in the fast-time variable except for a small sinusoidal perturbation

$$\phi = \phi_0(n) + \phi_m \cos(\Delta\omega_m t' + \alpha_m). \tag{9.184}$$

Since only the phase of the radiation is perturbed, this corresponds to a large equilibrium mode perturbed by two small satellites having $X_m + X^*_{-m} = 0$. Based on the results of the previous section concerning the stability of single-frequency equilibria, the perturbation ϕ_m must decay with round-trip time n at a rate given by (9.132). Therefore, the coefficient C_3 must be the same as the coefficient D in (9.132). Further, using (9.183) we can deduce relation (9.133). As before, the coefficient C_3 must be determined numerically for each equilibrium and will only be positive within the triangular region of Fig. 9.8 corresponding to stable single-frequency equilibria.

Having deduced the coefficients in the nonlinear evolution equation (9.180) we now turn to its solution. This equation is a form of Burger's equation and, consequently, we look for a solution which is expressed in the form

$$\phi(n,t') = -\delta\omega_0 nT - \frac{2C_3}{C_2}\ln[1+u(n,t')], \tag{9.185}$$

where insertion of (9.185) in (9.180) yields a linear differential equation for the function $u(n,t')$

$$\frac{1}{T}\frac{\partial u}{\partial n} = -C_1 \varepsilon T \frac{\partial u}{\partial t'} - C_3(\varepsilon T)^2 \frac{\partial^2 u}{\partial t'^2}. \tag{9.186}$$

Thus, we have converted a nonlinear equation for the phase into a linear equation for the function u. Representing the function $u(n,t')$ in the form of a Fourier series in the fast-time variable t', the solution of equation (9.186) can be written as

$$u(n,t') = \sum_{m=-\infty}^{\infty} u_m(0)\exp(-i\Delta\omega_m t' + \gamma_m nT), \tag{9.187}$$

where $u_m(0)$ are the initial Fourier amplitudes of the function $u(0,t')$, and the complex growth rate γ_m is given by (9.132). Interestingly, the time dependence of the Fourier components of $u(n,t')$ is precisely the same as that predicted by linear theory for the small perturbing amplitudes in the presence of a large equilibrium mode (equation (9.132)). Complete solution of the problem thus requires specifying the initial fast-time dependence of function $u(0,t')$, which is determined by the spontaneous emission. However, it is clear that the function u will eventually decay to a constant independent of t' at a rate determined by the damping coefficient C_3. Consequently, the phase given by equation (9.185) becomes independent of the fast-time variable, indicating that all temporal portions of the radiation pulse ultimately oscillate coherently.

The slow rate of spectral narrowing predicted by the previous analysis is a consequence of the fact that radiation propagates in this model without dispersion, and only the slippage of the beam provides communication between different temporal portions of the radiation pulse. The communication between different portions can be increased, along with the rate of spectral narrowing, by adding a dispersive element to the path of the radiation [20]. The effect of weak dispersion is to cause the frequencies of the longitudinal modes of the cavity to become nonuniformly spaced. That is, the natural frequencies will have a dependence on mode number of the form

$$\omega_m = \omega + \frac{m}{T}(2\pi + \alpha m), \tag{9.188}$$

where the coefficient α measures the amount of dispersion and may be either positive or negative. If we assume the dispersion is small, then its effect will only be apparent after a time corresponding to many circuits of the radiation through the cavity. Therefore, the dispersion can be included by modifying the slow-time evolution of the radiation amplitude as given by (9.81) [4],

$$\left[\frac{\partial}{\partial n} - i\alpha\left(\frac{T}{2\pi}\right)^2 \frac{\partial^2}{{}^2t'^2} + (1-R)\right] X(n,t') = -iI_b \int_0^1 d\xi \langle \exp(-i\psi) \rangle. \quad (9.189)$$

It is clear that in the absence of a beam the solutions of (9.189) give precisely the dispersion modified frequencies indicated in (9.188).

The effect of dispersion on the rate of spectral narrowing can now be assessed by calculating the rate of decay of small perturbing satellites in the presence of a large equilibrium mode. In order to accomplish this, the effect of slippage is neglected so that the beam response can be modelled as in equation (9.130). Specifically, the beam provides a nonlinear gain dependent on both the amplitude of the radiation field and the detuning of the equilibrium mode. The radiation field is then written as a superposition of a large-amplitude equilibrium mode and small perturbing satellites as in (9.121). The evolution of the small-amplitude satellites is governed by an equation of the form of (9.123) except that slippage is set to zero and dispersion is added. This gives the equation

$$\left[\frac{\partial}{\partial n} - i(\delta\omega_0 T + \alpha m^2) + (1-R)\right]\delta X_m = I_b\left[G\delta X_m + \frac{1}{2}G'(\delta X_m + \delta X^*_{-m})\right]. \quad (9.190)$$

Again, it is seen that the beam couples modes which are equally spaced in mode number about the equilibrium mode. Writing the corresponding evolution equation for the mode amplitude δX^*_{-m} as in (9.127) and assuming an exponential slow-time dependence we arrive at a quadratic dispersion relation replacing (9.131)

$$\left[\gamma_m T + i\alpha m^2 - \frac{I_b}{2}G'\right]\left[\gamma_m T - i\alpha m^2 - \frac{I_b}{2}G'^*\right] = \frac{I_b^2}{4}|G'|^2. \quad (9.191)$$

The solutions for the complex growth rate of the coupled pairs of satellites is found to be

$$\gamma_m T = \frac{I_b}{2}\operatorname{Re} G' \pm \frac{1}{2}\left[(I_b \operatorname{Re} G')^2 - 4\alpha m^2\left(1 - \frac{I_b \operatorname{Im} G'}{\alpha m^2}\right)\right]^{1/2}, \quad (9.192)$$

and both solutions are damped provided

$$1 > \frac{I_b}{\alpha m^2}\operatorname{Im} G'. \quad (9.193)$$

Notice that if the dispersion vanishes, then one of the two solutions is strongly damped (corresponding to a perturbation of the magnitude of the radiation field), and the other solution is marginally stable (corresponding to perturbations of the phase). Dispersion causes a coupling of amplitude and phase perturbations which results in both solutions of the quadratic equation being strongly damped. Thus, a dispersive element can be used to hasten the spectral narrowing. In the limit in which (9.193) is strongly satisfied

$$\gamma_m T = \frac{I_b}{2}\operatorname{Re} G' \pm i\alpha m^2, \quad (9.194)$$

which is the damping rate one would obtain had one neglected the coupling between modes with opposite signs of the mode index m [35] in equation (9.190). In this case, dispersion has modified the frequencies of the cavity to the point that the modes are no longer effectively coupled. Specifically, they no longer satisfy the four-wave matching condition given by equation (9.125). Dispersive elements have been used to hasten the coherence in the radiation pulse in storage ring free-electron lasers [36], and have the advantage over filters in that the tunability of the free-electron laser with beam energy is not compromised by the dispersive element.

9.7 REPETITIVELY PULSED OSCILLATORS

The basic operating wavelength of a free-electron laser scales inversely with the energy of the electron beam. In order to operate at shorter wavelengths, one approach is to increase the energy of the beam. The electron beams in most high-energy accelerators, however, are not continuous in time but rather consist of a series of micropulses of a short duration with a fixed time between pulses. The series of micropulses is known as a macropulse. As a result it is of fundamental importance to understand the effect of the pulsed nature of the electron beam on the operation of the oscillator in which it is being used [5, 17, 18, 37–41]. When pulsed beams are used to drive an oscillator, the radiation develops a temporal structure that more or less matches that of the electron beam. It is necessary, therefore, to synchronize the time of flight of radiation in the cavity with the arrival of successive beam micropulses, thus providing for temporal overlap of the radiation with the beam pulses. The degree of synchronism that is required, as well as the temporal structure of the resulting radiation pulse, will depend on the amount of electron slippage, the duration of the micropulse, and the gain and losses.

Since the radiation emerging from an oscillator driven by a pulsed electron beam also consists of pulses, the spectrum will be broader than that which could be achieved with a continuous electron beam. If the duration of the radiation pulses is the same as the beam micropulse T_p, then the minimum spectral width will be determined by the Fourier transform limit, $2\pi/T_p$. By comparison, the gain bandwidth is approximately $2\pi/\varepsilon T$. Thus, pulses which are of a duration T_p which is longer than the slippage time εT can have spectra which are narrower than the gain bandwidth. This can be visualized in the time domain using Fig. 9.2 and imagining that the micropulse duration T_p is much greater than the slippage time εT. In this case, the behaviour of the radiation in the centre of the pulse is to some degree independent of the shape of the pulse, and only dependent on the instantaneous beam current and energy. As a result, in a limited sense, the centre of the pulse can be treated as if the electron beam were continuous. Thus, concepts that have been developed in the treatment of continuous-beam oscillators can be applied to the understanding of the more complex pulsed-beam case if the duration of the micropulse is greater than the slippage time. In particular, the conditions for starting oscillations from noise, the mechanism for saturation of the radiation by phase trapping of electrons, and the criteria for the onset of sidebands (overbunch instability) in pulsed-beam oscillators may be inferred from the continuous-beam result.

9.7.1 Cavity detuning

There are important differences, however, in the behaviour of pulsed- and continuous-beam oscillators even when the micropulse duration is long compared with the slippage time. These differences relate to the interaction of slippage and cavity detuning in determining the dynamics of the radiation pulse [5]. Cavity detuning relates to the difference between the round-trip time of the radiation in the cavity T and the arrival time of electron-beam pulses T_a, and is characterized by the difference of these two times $T_\delta = T - T_a$. It will be shown in this section that for normal slippage $\varepsilon > 0$, it is necessary to have a negative value of the cavity detuning time T_δ in order for coherent oscillations to grow from arbitrarily small noise levels [5, 38–42]. This requirement is attributed to an effect known as **laser lethargy** [5], and is analogous to a similar requirement on the conditions for the existence of absolute instability in a medium that supports two waves which when coupled together are unstable [43, 44]. The requirement is that if the medium is of finite size, then the group velocities of the two waves must be directed oppositely. If the group velocities are not directed oppositely then a small initial disturbance will grow in time due to the instability, but at the same time it will escape convectively from the medium, so that ultimately the perturbations decay to zero. If noise is present continually then perturbations will grow to a size which is determined by the noise level and the gain a perturbation experiences before being carried out of the unstable region. On the other hand, if the group velocities are directed oppositely, an inherent feedback loop exists which allows an arbitrarily small initial perturbation to grow to saturation. This is known as an absolute instability.

To see how these ideas apply to the interplay of slippage and cavity detuning we redraw Fig. 9.2 in the variables n and t', introduced in equation (9.72), corresponding to the round-trip index and the time within the radiation pulse. Figure 9.19, therefore, attempts to illustrate the way in which the field amplitude at time $t' = t*$ and round-trip index $n + 1$ is determined by the field amplitude at the earlier round-trip index n and fast-time variable t'. The two cases shown in the figure correspond to positive cavity detuning and negative cavity detuning, respectively. The various lines in the figure correspond to terms in equation (9.81) which determine the evolution of the complex radiation amplitude $X(n, t')$. The dashed lines represent the contribution to the rate of change of the radiation amplitude with round-trip index due to the interaction with the beam. Due to electron slippage, the effect of the beam on the radiation amplitude at fast time $t' = t*$ is determined by the radiation amplitude the previous round trip for fast time in the range $t* > t' > t* - \varepsilon T$. Recall that t' is defined such that events occurring at the same value of t' and successive values of the round-trip index n are actually separated in time by T_a, the time between arrival of successive beam pulses. The solid lines show the contribution to the radiation amplitude from the

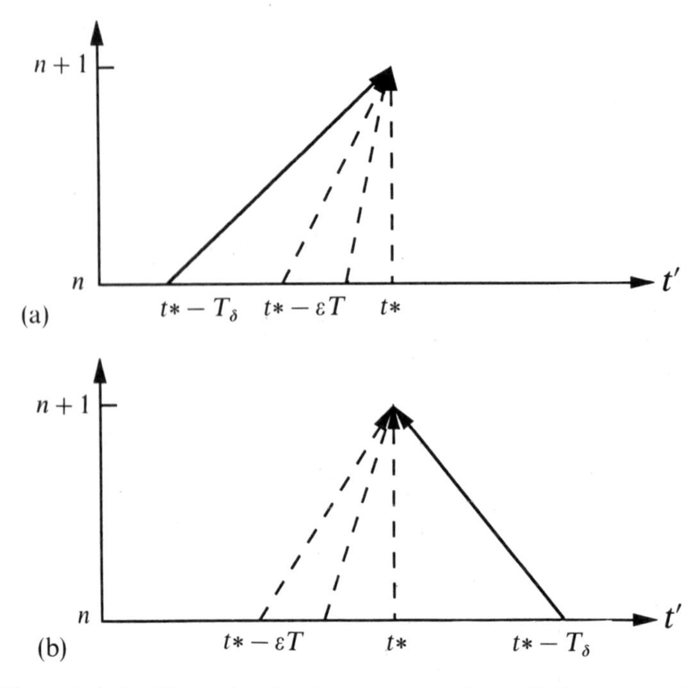

Fig. 9.19 Characteristics illustrating the dependence of the radiation amplitude entering the cavity at a time $t*$ on the radiation amplitude entering the cavity with the pervious micropulse. For case (a) cavity detuning is positive and no absolute instability is possible, and for case (b) cavity detuning is negative and absolute instability is possible.

direct propagation of radiation through the cavity. Due to cavity detuning, the amplitude of the radiation at time $t*$ is determined by the radiation entering one round trip earlier at a time $t* - T_\delta$. This is also illustrated in Fig. 9.3. As is evident, the effect of slippage is such that information from the interaction with the beam always propagates to larger values of t'. If the cavity detuning is positive, information resulting from the direct path of the radiation also propagates to larger values of t'. In this case, both signal paths have positive group velocities, if one thinks of t' as a space-like variable and n as a time-like variable. If the electron pulse is of a finite duration in t', disturbances will grow but also convect out of the electron pulse. In this case, no self-sustaining oscillations can develop. On the other hand, if the cavity detuning is negative, information from the direct path of the radiation is propagated toward negative values of t' and the group velocities of the two signal paths are directed oppositely. As a result, self-sustaining oscillations can grow to saturation from arbitrarily small noise levels.

The effect of cavity detuning on the start current of a pulsed-beam oscillator is illustrated in Fig. 9.20 which plots the start current for a monoenergetic beam normalized to the minimum start current for the corresponding continuous beam (cf. equation (9.110)) versus the normalized cavity detuning parameter. The calculation applies to the case in which the slippage time is much shorter than

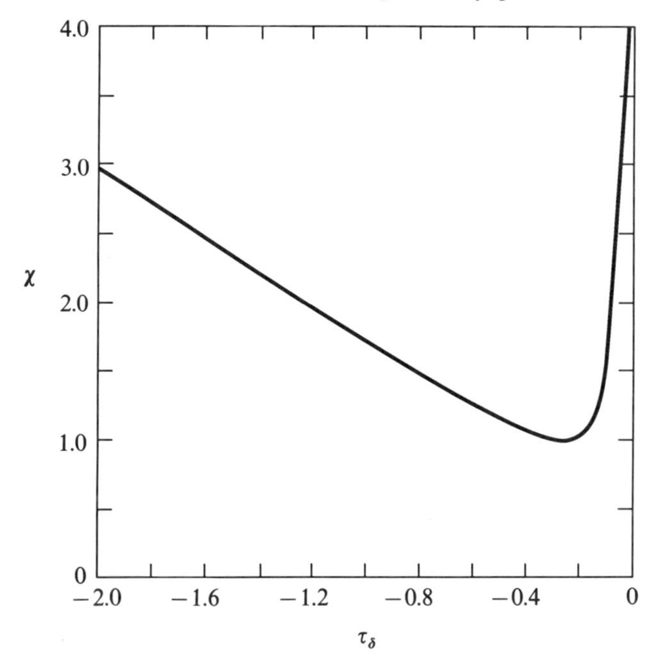

Fig. 9.20 The peak current required to start oscillations in a monoenergetic pulsed beam normalized to the corresponding start current for a continuous beam as a function of normalized cavity detuning time $\tau_\delta = T_\delta/[\varepsilon T(1 - R)]$. The calculation has assumed that the duration of the micropulse T_p is much longer than the slippage time εT.

the micropulse duration and, as a result, the shape of the beam micropulse is not important. The steps required to obtain this plot will be detailed subsequently. The plot shows that there is an optimum cavity detuning time T_δ (which is negative) for which the start current is the same as in a continuous beam. Actually, the start current for the pulsed beam will be slightly higher than the corresponding continuous beam if corrections due to the finite value of the pulse duration time are retained. Further, the start current is infinite for positive values of cavity detuning and becomes large for large negative values of cavity detuning.

9.7.2 Supermodes

A theoretical treatment of the small-signal growth of radiation in pulsed oscillators as described in the previous paragraph follows. To determine the conditions for the existence of absolute instability one linearizes the equations of motion and evaluates the last term on the right-hand side of equation (9.81) describing the interaction of the radiation with the beam. The result can be represented in the form of a convolution of the radiation amplitude with a kernel and is written

[37, 39, 40]

$$\left(\frac{\partial}{\partial n} + T_\delta \frac{\partial}{\partial t'}\right) X(n,t') = (R-1)X(n,t') + \int_{t'-\varepsilon T}^{t'} dt'' K(t'', t'-t'')X(n,t''). \tag{9.195}$$

The kernel $K(t'', t'-t'')$ depends not only on the difference time $t'-t''$ accounting for slippage but also on the time argument t'' due to the temporal dependence of the injected beam. If the beam parameters are independent of time, the kernel will depend only on the difference time, and solutions of the form $X(n,t') \sim \exp(-i\Delta\omega t')$ can be found (as discussed in the sections on continuous-beam oscillators). The time scale for the dependence of the kernel on the difference time is the slippage time εT. The time scale for the explicit dependence of the kernel on t'' is the characteristic time of duration of the electron micropulse T_p. Solutions of this linear integral equation corresponding to exponential dependence on the round-trip index n

$$X(n,t') = \hat{X}(t')\exp(\gamma_{sm} n T) \tag{9.196}$$

are known as supermodes [37, 39, 40], and in general must be determined numerically.

Analytic progress can be made when the duration of the micropulse is much longer than the slippage time. In this case the explicit dependence of the kernel on the time variable t'' is relatively weak and one can expect the radiation field locally to be of the form of a complex exponential

$$\hat{X}(t') \cong \hat{X} \exp\left[-i\int^{t'} dt'' \Delta\omega(t'')\right]. \tag{9.197}$$

That is, one can use a **WKB** representation for the radiation amplitude. Equation (9.197) is then inserted in the integral equation with the parameters of the beam evaluated at time t'. The contribution of the beam is then formally identical to that which would be found for a monochromatic radiation signal interacting with a continuous electron beam with the instantaneous parameters of the micropulse. We can thus express the response of the beam in terms of a linear gain function of the form derived in (9.95). The resulting dispersion equation determining the instantaneous frequency of the radiation including the effect of cavity detuning follows from (9.195)

$$\gamma_{sm} T - iT_\delta \Delta\omega(t') = R - 1 + I_b(t')G[t', -\varepsilon T\Delta\omega(t')], \tag{9.198}$$

where $I_b(t')$ denotes the time-dependent electron-beam current, and $G[t', -\varepsilon T\Delta\omega(t')]$ is the instantaneous gain defined by analogy with (9.95),

$$G[t', -\varepsilon T\Delta\omega(t')] = \pi^2 \int dp\,\Delta(p - \varepsilon T\Delta\omega)\frac{\partial}{\partial p}F_0(p,t'), \tag{9.199}$$

where $\Delta(p)$ is defined in equation (9.96), and G is time dependent due to the fact that the beam-distribution function may have a time dependence within the micropulse. For example, the gain would be given by equation (9.98) with the parameter p_0 a function of the fast-time variable t' for a beam which at any instant is monoenergetic but whose energy depends on time.

Equation (9.198) is insufficient by itself to determine the complex growth rate of the radiation on the round-trip time scale. In fact, for given values of t' and the fast-time frequency shift $\Delta\omega$ there will always be solutions for the slow-time growth rate γ_{sm}. What is required is some way of enforcing the boundary condition that the field amplitude vanishes far outside of the beam micropulse as $|t'| \to \infty$. Within the context of WKB theory, the problem of enforcing the boundary conditions at infinity reduces to the problem of finding two turning points: specifically, two values of t' (t_1 and t_2) where two distinct solutions for the fast-time frequency ($\Delta\omega_1$ and $\Delta\omega_2$) coalesce. The correct slow-time growth rate is then determined by the quantization condition that the integral of the difference of the two fast-time frequencies between the two turning points is given by

$$\left(m + \frac{1}{2}\right)\pi = \int_{t_1}^{t_2} dt' [\Delta\omega_1(t') - \Delta\omega_2(t')], \tag{9.200}$$

where m is a positive integer labelling the various normal modes (supermodes). The problem is further complicated by the fact that, in general, the two turning points may occur for complex values of t'. Thus, even though the solution can be written in a compact form, implementation of this solution still requires computation.

The lowest-order solution (i.e. the ground state) with $m = 0$ can be found by noting that the quantization condition requires that the two turning points be close to each other on the time scale of the duration of the electron micropulse. Thus, not only must there be a coalescence of the two solutions for the fast-time frequency $\Delta\omega$, but also the two turning points must coalesce. Thus, to lowest order, the complex growth rate can be determined by solving equation (9.198) along with the subsidiary conditions that the time t' and frequency shift $\Delta\omega$ are such that derivatives of (9.198) with respect to time and frequency shift vanish

$$\frac{dI_b(t')}{dt'} G(t', -\varepsilon T\Delta\omega) + I_b(t') \frac{\partial}{\partial t'} G(t', -\varepsilon T\Delta\omega) = 0, \tag{9.201}$$

and

$$\varepsilon T I_b(t') \frac{\partial G}{\partial \tilde{p}}\bigg|_{\tilde{p} = -\varepsilon T\Delta\omega} = iT_\delta. \tag{9.202}$$

These latter two conditions ensure that the dispersion equation is solved for parameters which give simultaneous coalescence of both the turning points and the fast-time frequencies. It is now a good idea to count the number of equations and unknowns. There are three complex equations (9.198), (9.201) and (9.202) along with three complex unknowns: the location of the coalesced turning points t', the complex frequency at the turning points $\Delta\omega$, and the complex growth rate on the round-trip time scale γ_{sm}. Thus, we have the right number of knowns and unknowns to calculate the round-trip time growth rate.

To simplify matters further, assume that the distribution function for the injected beam is independent of time and only the beam current varies within the micropulse. Thus, the gain function is independent of time and equation (9.201) requires that the coalesced turning points occur for real values of t' at the maximum of the beam current. Since the coalesced turning points occur for real t', the normalized beam current appearing in (9.198) and (9.202) is real and equal to the maximum beam current. As a consequence equation (9.202) now determines implicitly the value of the complex frequency shift $\Delta\omega$. This value may then be inserted into (9.198) to determine the complex supermode growth rate γ_{sm}. This procedure was followed for the case of a monoenergetic beam and the value of the beam current required to initiate absolute instability ($\mathrm{Re}\,\gamma_{sm} = 0$) is plotted versus cavity detuning in Fig. 9.20.

As pointed out previously, the minimum starting current predicted by this method is the same as the minimum starting current in a continuous-beam oscillator. However, this minimum occurs only for a specific value of cavity detuning. This optimum detuning can be found from (9.198) and (9.202) by the following argument. The real part of equation (9.202) requires that the derivative of the real part of the gain with respect to frequency shift vanish. This certainly occurs for the real frequency shift that maximizes the gain. At this frequency shift the start current ($\mathrm{Re}\ \gamma_{sm} = 0$) can be obtained from the real part of (9.198),

$$I_b(t') = \frac{1 - R}{\mathrm{Re}[G(t', -\varepsilon T\Delta\omega)]}, \tag{9.203}$$

where t' and $\Delta\omega$ give the maximum current and gain. This expression is identical to the corresponding condition for starting oscillations in a continuous-beam oscillator (9.97). The optimum value of the cavity detuning which produces the minimum start current is then obtained from the imaginary part of equation (9.202).

The structure of the supermode on the fast time-scale variable t' predicted by this theory will tend to be Gaussian, as will be seen. At first one might expect that the radiation pulse predicted by this theory would be peaked on the maximum current, since this is the location of the coalesced turning points in t'. It must be remembered, however, that the coalesced frequency shift $\Delta\omega_c$ is complex. Thus, depending on the sign of the imaginary part of $\Delta\omega_c$ at the coalescence points, the radiation peak will be either advanced or retarded with respect to the peak of the beam current. Generally, the more negative the detuning, the more retarded the peak is. The special case of optimum detuning predicts a pulse whose peak coincides with that of the beam current.

At this point it must be recalled that the present method of solution has relied on the approximation that the micropulse duration is much longer than the slippage time. When this is not the case, the full integral equation (9.195) must be solved and one can expect the start current to be higher when higher-order finite pulse duration effects are included. This can be seen when one calculates the first-order corrections to the starting current or growth rate implied by the quantization condition (9.200). Rather than perform the integral in (9.200), we will convert the local dispersion relation (9.198) to the corresponding second-order differential equation by expanding about the fast time $t' = t_{\max}$, and frequency shift $\Delta\omega = \Delta\omega_c$ implied by simultaneous solution of (9.203) and (9.204). We again assume the beam-distribution function is time independent (only the beam current varies) and we define $\tau = t' - t_{\max}$, where $I_b(t_{\max})$ is the maximum beam current. To obtain a differential equation we expand the frequency shift about the coalescence frequency shift by substitution of

$$\Delta\omega = \Delta\omega_c - i\frac{\partial}{\partial\tau}, \tag{9.204}$$

and expand equation (9.198) to second order in τ and its derivative. The result is a harmonic oscillator equation of the form [36, 42]

$$\left[T_1^2\frac{d^2}{d\tau^2} - \frac{\tau^2}{T_2^2} - 2\Lambda\right]\hat{X}(\tau) = 0 \tag{9.205}$$

where the coefficients appearing in (9.205) are defined by

$$T_1^2 = -(\varepsilon T)^2 \frac{1}{G}\frac{\partial^2 G}{\partial p^2}\bigg|_{p = -\varepsilon T\Delta\omega_c} \approx (\varepsilon T)^2, \tag{9.206}$$

$$T_2^2 = -\frac{1}{I_b}\frac{\partial^2 I_b}{\partial t'^2}\bigg|_{t' = t_{\max}} \approx T_p^{-2}, \tag{9.207}$$

and

$$\Lambda = \frac{\gamma_{sm} T - iT_\delta \Delta\omega_c - (1 - R)}{I_b(t_{\max}) G(t_{\max}, -\varepsilon T\Delta\omega_c)} - 1. \tag{9.208}$$

Note that we have indicated the order of magnitude of the coefficients T_1 and T_2. In particular, T_1 scales as the slippage time εT and T_2 scales as the pulse duration T_p.

The solution of the harmonic-oscillator equation is given in terms of Hermite polynomials

$$\hat{X}_m(\tau) = H_m\left(\frac{\tau}{\sqrt{(T_1 T_2)}}\right)\exp\left(-\frac{\tau^2}{2T_1 T_2}\right), \tag{9.209}$$

and

$$\Lambda_m = -\frac{T_1}{T_2}\left(m + \frac{1}{2}\right), \tag{9.210}$$

where the index m labels various supermodes by the number of their temporal nodes. The lowest-order solution has a start current which is elevated slightly above that predicted on the basis of (9.198), (9.201) and (9.202) from $\Lambda_0 = 0$. As can be seen, this difference is small when pulse duration exceeds the slippage time. Also of interest is the temporal width of the solution $\sqrt{(T_1 T_2)}$, which is the geometric mean of the slippage time and the pulse duration. For the case of optimum detuning this is the width of the supermode. For cavity detunings which are not optimum, the radiation pulse is peaked away from the maximum of the beam current. The width of the radiation peak is determined by the WKB representation of the solution (9.197) expanded about the fast time t', where the imaginary part of the local frequency shift vanishes. This width also scales as the geometric mean of the slippage and pulse-duration times. As the current is raised above the start value, an increasing number of supermodes are driven unstable. The fastest-growing mode is the lowest-order one which has the narrowest temporal width. Thus, as radiation approaches saturation it will be somewhat narrower than the pulse duration and will correspond to a spectral width that is the geometric mean of the gain bandwidth and the pulse-duration Fourier-transform limit. The shape of the saturated radiation pulse depends on the nature of the saturation mechanism. In storage-ring free-electron lasers, where the electron beam is recirculated continually through the interaction region, saturation is achieved at low field amplitudes by a process in which the quality of the beam is degraded continually until the linear gain exactly balances losses [45, 46]. In this case, the supermode structure is likely to remain intact [36, 42, 45, 46]. In free-electron lasers driven by linear accelerators, saturation is achieved by phase trapping at large field amplitude, and the radiation pulse shape is altered significantly from that in the small-signal phase [14, 17, 18, 27, 41, 42].

It is important to emphasize at this point that the previous discussion of supermodes focuses on the question of whether the conditions of cavity detuning, slippage and gain are such that an absolute instability is present. That is, will radiation grow to saturation if the noise level is arbitrarily small? In practice, the noise level is not arbitrarily small and radiation can grow to a substantial level even if it is only convectively unstable. Noise that is excited at some point in the micropulse will grow as it convects out of the pulse. If the gain during the time when the disturbance in the pulse is large, then a small but non-zero noise signal can amplify to a significant level. The gain can be estimated for various parameters from (9.198)

$$N_e \cong T_p \operatorname{Im} \Delta\omega, \tag{9.211}$$

where N_e is the number of exponentiations and T_p is the pulse duration. Under the assumption that the detuning time T_δ is large compared with the slippage time εT, the imaginary part of the fast-time frequency is determined by gain and losses

$$\operatorname{Im} \Delta\omega = \frac{1}{T_\delta}\{I_b \operatorname{Re}[G(t_{max}, -\varepsilon T\Delta\omega)] - (1 - R)\}, \tag{9.212}$$

where the gain is evaluated with a real frequency shift. Thus, the number of exponentiations N_e scales as

$$N_e \cong \frac{T_p}{T_\delta}\{I_b \operatorname{Re}[G(t_{max}, -\varepsilon T\Delta\omega)] - (1 - R)\}, \tag{9.213}$$

which for long pulses can be a large number even if the conditions for absolute instability are not satisfied. In this limit one can think of cavity detuning as providing an additional loss mechanism where the fraction of radiation which is lost on each round trip is the ratio of the cavity detuning time to the micropulse duration time. We caution the reader that the simple estimate given above applies only if the cavity detuning time is much longer than the slippage time and much shorter than the pulse duration. In cases where both these inequalities cannot be satisfied it is necessary to resort to numerical solution to determine more precise estimates of the amount of radiation growth.

9.7.3 Spiking mode and cavity detuning

If the maximum beam current in the micropulse exceeds either of the thresholds discussed in the previous section, then radiation can grow to large amplitude and saturate. The resulting radiation pulse will have a time dependence determined by the shape of the beam micropulse as well as by other parameters, such as cavity detuning. Due to the time dependence, the radiation must always have a non-zero spectral width.

The characteristics of the spectrum depend sensitively on whether the conditions for the excitation of the overbunch instability are satisfied. The overbunch instability was discussed in Section 9.6 in the context of continuous-beam oscillators. It occurs when the radiation amplitude becomes sufficiently large that electrons are trapped in the ponderomotive well and execute about half a synchrotron oscillation on transit through the interaction region. The instability manifests itself by the appearance of sidebands in the spectrum which correspond to temporal oscillations of the magnitude of the radiation amplitude whose period is roughly the slippage time. Thus, the sidebands are displaced from the original signal by, roughly, the gain bandwidth. If the overbunch instability is not present, then the spectrum can be much narrower than the gain bandwidth and its width can approach the Fourier-transform limit determined by the duration of the beam micropulse.

In the case of a continuous-beam oscillator, the conditions for the excitation of the overbunch instability can be determined with great precision and are displayed in Fig. 9.8. A similarly detailed calculation of the conditions for the onset of this instability in the case of a pulsed-beam oscillator is not available due to the added complexity introduced by the time dependence of the beam within a micropulse and the important effect of cavity detuning. There has been, however, extensive numerical simulation of pulsed-beam oscillators which tends to confirm that the threshold for the onset of the overbunch stability is correlated with the peak radiation amplitude exceeding a value for which electrons can complete half a synchrotron oscillation [14, 17, 18, 27, 41, 47]. This rule is not expected to be hard and fast due to complicated interaction of slippage and cavity detuning. Recall that the instantaneous linear gain given by equation (9.198) can be positive, yet if the cavity detuning is not optimized small signals will not grow significantly before being convected out of the beam pulse. Similarly, the conditions for the onset of the overbunch instability may be satisfied locally at some point in the radiation pulse but, due to a large value of cavity detuning, the instability can be suppressed. Indeed, it has been found that cavity detuning is effective in suppressing the overbunch instability [17, 18].

The features of the onset of the overbunch instability can be illustrated with the use of the klystron model introduced previously. In particular, Figs 9.21 and 9.22 show examples of the radiation profile obtained by numerical solution of equation (9.120) which may be rewritten as

$$\left[\frac{\partial}{\partial n} + T_\delta \frac{\partial}{\partial t'} + (1-R)\right] X(n,t') = I_b(t'')X(n,t'')g[|X(n,t'')|]|_{t''=t'-\varepsilon T}. \quad (9.120)$$

For the particular simulations shown the time dependence of the beam current was chosen to be a Gaussian with a width T_p

$$I_b(t'') = I_{bo}\exp\left(-\frac{t''^2}{T_p^2}\right), \quad (9.214)$$

and the gain function g defined in equation (9.53) was taken to be, simply,

$$g = 1 - |X|^2, \quad (9.215)$$

which has the required feature to produce the overbunch instability; specifically, that the gain decreases with radiation amplitude. This requirement is discussed in Section 9.6 and illustrated in Fig. 9.10. The boundary in current versus cavity detuning for the excitation of the overbunch instability obtained by solution of equation (9.120) for a large number of parameters is shown in Fig. 9.23, along with the curve giving the start-oscillation current. It can be seen that as the magnitude of the cavity detuning is increased both the start current and the

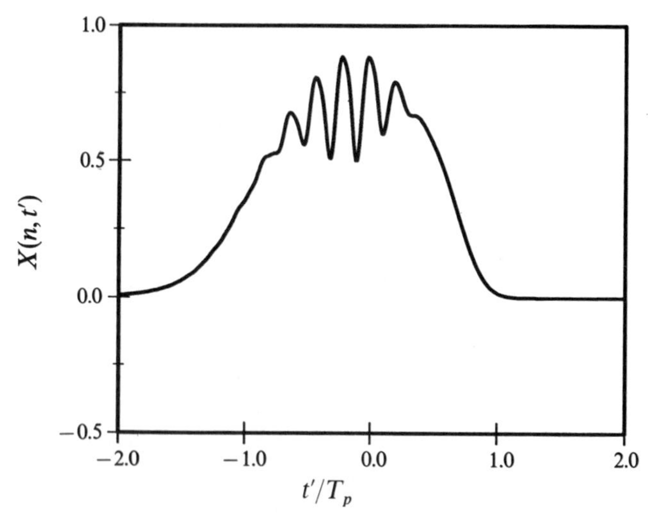

Fig. 9.21 The time-dependence of the radiation pulse showing the spiking mode. The magnitude of the radiation pulse is plotted versus t'_0 for a specific value of the cavity detuning τ_δ. ($\tau_\delta = -2$; $I_{bo} = 2.4$.)

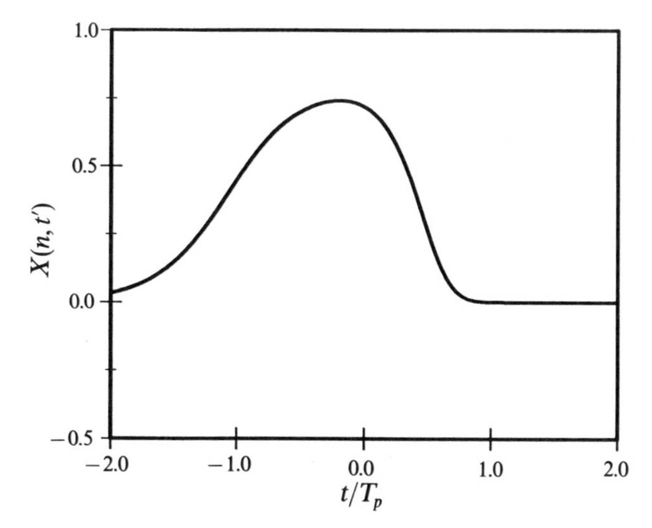

Fig. 9.22 The time-dependence of the radiation pulse showing the absence of the spiking mode. The magnitude of the radiation pulse is plotted versus t'_0 for the same parameters as Fig. 9.21 except that the magnitude of the cavity detuning τ_δ has been increased. ($\tau_\delta = -3$; $I_{b0} = 2.4$.)

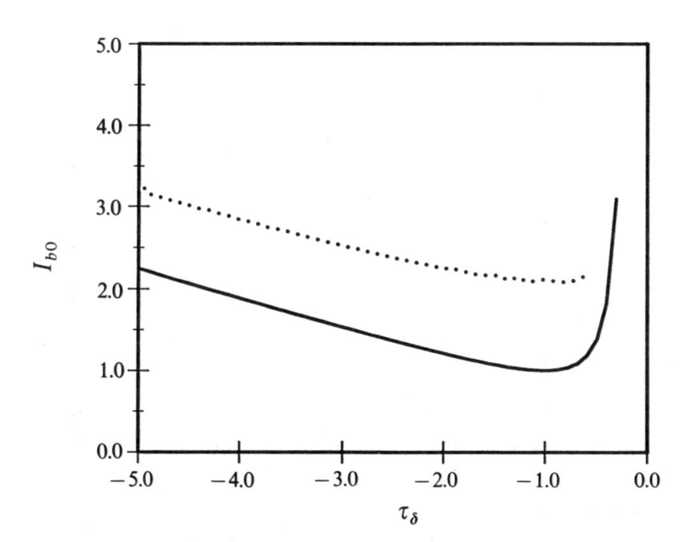

Fig. 9.23 The peak current required to initiate the spiking mode in the klystron (dotted line) as a function of the normalized cavity detuning time $\tau_\delta = T_\delta/[\varepsilon T(1 - R)]$. Also shown is the start current versus normalized detuning (solid line).

overbunch threshold current increase. Thus, stable operation at high current can be achieved by increasing the magnitude of the cavity detuning.

9.8 MULTI-DIMENSIONAL EFFECTS

Our treatment of oscillators has thus far been strictly one-dimensional in that all quantities have been assumed to depend only on the axial coordinate and time. Of course, in a real device the radiation field, the wiggler field and the electron beam will have significant variations in the directions transverse to the axis of the beam. The resulting multi-dimensional effects have been discussed in extensive detail for the case of amplifiers in the early chapters of this book. In particular, the linear and nonlinear interaction of a beam with multiple transverse modes in a waveguide has been considered in Chapters 4 and 5 while the related problem of diffraction and optical guiding for the case of unconfined radiation was discussed in Chapter 8.

The basic physical effects which govern the transverse dependence of radiation profiles in amplifiers also apply to oscillators. The problem is more complicated in an oscillator, however, due to the fact that the radiation circulates continually through the interaction region. In an amplifier, the transverse distribution of the radiation field that enters the interaction region is determined by the launching structure more or less independently of the interaction of the radiation with the electron beam. The radiation is then amplified and the transverse distribution modified as it propagates down the interaction region. In an oscillator, the amplified radiation that leaves the interaction region is fed back through an optical system and returns to the entrance of the interaction region. Thus, the transverse dependence of the radiation entering the oscillator must be determined self-consistently with the interaction.

If the interaction is strong in the sense that the device is of the high-gain type, the modification of the transverse dependence of the radiation profile by the beam can be significant and must be accounted for in the design of experiments. Sophisticated computer codes which self-consistently determine the transverse and axial dependences of the radiation field have been written for this task [48–52]. If the device is of the low-gain type, then the transverse dependence of the radiation field can be assumed to be the same as that of the eigenmodes of the empty cavity, and the one-dimensional theory presented here can be adopted with certain modifications. The remainder of this section will address the issue of incorporation of multi-dimensional effects into the low-gain model of an oscillator.

The first modification to the one-dimensional theory introduced in Section 9.1 is that the full three-dimensional variation of the vector potential must included. Thus, equation (9.1) for the vector potential is rewritten

$$\delta \boldsymbol{A}(\boldsymbol{x}, t) = \delta A(\boldsymbol{x}, t)\hat{\boldsymbol{e}}_p \exp(ikz - \omega t) + \text{c.c.} \tag{9.216}$$

The equations for the rate of change of the beat-wave phase and the electron energy (equations (9.3) and (9.4)) are formally the same, except that the full three-dimensional dependence of the radiation electric field must be included. As a consequence, it is necessary also to introduce equations describing the transverse location of the beam electrons since electrons at different transverse locations experience different radiation fields. Within the approximations leading to the wiggler-period averaged equations (9.6) and (9.7), it is now assumed that the average transverse location of an electron does not change much during the time it takes the electron to travel one wiggler period. Thus, in equation (9.14), the normalized radiation vector potential is evaluated at the average transverse location of an electron

$$\left(\frac{\partial}{\partial t}+v_{\|}\frac{\partial}{\partial z}\right)\delta\gamma=-\frac{\omega}{2\gamma_r}a_w[i\delta a(\boldsymbol{x}_{0\perp},z,t)\exp(i\psi)+\text{c.c.}]. \tag{9.217}$$

This position, which we label $x_{0\perp}$, is sometimes termed the **oscillation centre** since the motion of the electron in the wiggler consists of periodic oscillations about the average trajectory $\boldsymbol{x}_{0\perp}(t)$. By replacement of the exact transverse position of an electron with the average transverse position, we assume implicitly that variations of quantities such as the wiggler field strength and the radiation field strength are small over a distance corresponding to the displacement of the electrons in the wiggler field. Note that this assumption eliminates the possibility of including periodic position interactions of the type described in Chapter 7. The equations for energy and phase must now be supplemented by an equation for the transverse location of the oscillation centre. The simplest case is the one in which the transverse motion is determined solely by the strong focusing of the wiggler field and corresponds to betatron oscillations with betatron wave number k_β

$$\left(\frac{\partial}{\partial t}+v_{\|}\frac{\partial}{\partial z}\right)^2\boldsymbol{x}_{0\perp}=-k_\beta^2v_{\|}^2\boldsymbol{x}_{0\perp}. \tag{9.218}$$

A further simplification is the case in which the amplitude of the betatron oscillations is so small that the transverse dimension of the beam is much smaller that the scale length over which the radiation vector potential varies. In this case, all electrons sample only the radiation field on-axis and it is not necessary to follow the transverse motion of the electron at all.

Inclusion of transverse dependence in the radiation field results in a modification to the wave equation (9.25) to account for diffraction; specifically, the transverse Laplacian must be added to the left-hand side of equation (9.25) and the source term on the right-hand side now acquires transverse dependence due to the fact that the beam is of finite size,

$$\left(\frac{\partial}{\partial t}+c\frac{\partial}{\partial z}-\frac{ic}{2k}\nabla_\perp^2\right)\delta A(\boldsymbol{x}_\perp,z,t)=\frac{2\pi i\,c^2a_w}{\omega\;\gamma_rv_{\|}}\langle J(t_e)\exp(-i\psi)\rangle. \tag{9.219}$$

Here we have replaced v_g by c because we are treating the transverse dependence explicitly and equation (9.219) determines self-consistently the group velocity. The new system of equations is now capable of describing all the effects of optical guiding discussed in Chapter 8. Finally, to treat an oscillator, the equation describing the optical feedback path must be modified. In the onc-dimensional case, this equation was a simple algebraic relation between the radiation leaving the interaction region at time $t - T_r$ and the radiation entering the interaction region at time t (equation (9.26)). When multi-dimensional effects are included, this algebraic relation is replaced by a complicated integral relation describing the transformation of the transverse profile of the radiation by the system of mirrors (and/or waveguides) composing the feedback loop. We will not discuss this problem in detail here, however, as there have been several publications describing numerical methods for treating these effects [48–52]. It will be seen that in the low-gain limit the effects of the optical feedback path enter into the determination of the empty cavity modes, and that once these are found the effect of the interaction can be treated by perturbation.

We proceed with a discussion of the low-gain limit by consideration of solutions of (9.219) under the assumption that the right-hand side is small. The development parallels that given in Section 9.5 for the one-dimensional case. In the absence of a beam, the solution of the wave equation should yield a radiation field whose structure can be described by a superposition of empty cavity modes. Inclusion of the beam in the low-gain regime will cause the amplitudes of the various cavity modes to evolve in time as in the one-dimensional limit described by equation (9.90). We assume for simplicity that only the lowest-order transverse eigenmodes of the cavity are excited, and that these modes have a transverse dependence which is Gaussian with a minimum spot size σ_0 occurring at an axial location z_0. The radiation field in the interaction region to which the electrons respond can then be written by analogy with (9.67)

$$\delta A(\boldsymbol{x}_\perp, z, t) = \delta A_e(t - z/c)u(\boldsymbol{x}_\perp, \sigma), \tag{9.220}$$

where the functions $u(\boldsymbol{x}_\perp, \sigma)$ and $\sigma(z)$ describe the transverse dependence of the radiation field

$$u(\boldsymbol{x}_\perp, \sigma) = \frac{\sigma_0}{\sigma(z)}\exp[-x_\perp^2/2\sigma(z)], \tag{9.221}$$

and

$$\sigma(z) = \sigma_0 + \frac{i(z - z_0)}{k} \tag{9.222}$$

determines the local spot size. The field given by equation (9.221) is then inserted into the electron equations of motion to calculate the interaction with the electron beam. This leads to equations of the form of (9.68) and (9.69), except that the factor $u(\boldsymbol{x}_{0\perp}, \sigma)$ multiplies the radiation amplitude. These equations are normalized subsequently to yield the multi-dimensional counterparts of (9.79) and (9.80)

$$\frac{\partial \psi}{\partial \xi} = p, \tag{9.223}$$

and

$$\frac{\partial p}{\partial \xi} = -\frac{1}{2}[iu(x_{0\perp}, \sigma)X(n, t_e + \varepsilon\xi T)\exp(i\psi) + \text{c.c.}], \tag{9.224}$$

where $\xi = z/L$ is substituted into equation (9.223). These equations are to be solved for an ensemble of electrons with initial conditions appropriate to the incoming beam. In addition, the equations for the trajectories of the oscillation centres (equation (9.219)) must also be solved with a distribution of initial conditions describing the transverse positions and angular divergences of the electrons in the beam.

The solutions for the phases and transverse positions of the electrons must now be inserted in the wave equation (9.119) to determine the gain that the radiation experiences on one transit of the interaction region. The interaction with the beam produces a small increase in the amplitude of the radiation whose transverse dependence is described by the function $u(\boldsymbol{x}_\perp, \sigma)$. To find this small increase, we multiply the wave equation by u^*, integrate over transverse coordinates and along the radiation characteristics in the (z, t) plane and obtain an expression analogous to (9.70) of the one-dimensional case

$$\delta A(L,t) = \delta A_e(t - L/c) + \frac{2\pi i\, ca_w}{\omega\ \gamma_r v_\parallel} \frac{\int_0^L dz \int d^2 x_\perp u^* \langle J \exp(i\psi)\rangle}{\int d^2 x_\perp |u|^2}. \tag{9.225}$$

The transverse average appearing on the right-hand side of (9.225) can be expressed in terms of an average over the initial coordinates of the beam electrons entering the interaction region and an effective current density equal to the total beam current divided by the minimum spot area

$$\frac{\int d^2 x_\perp u^* \langle J \exp(-i\psi)\rangle}{\int d^2 x_\perp |u|^2} = J_{\text{eff}} \langle u^* \exp(-i\psi)\rangle, \tag{9.226}$$

where $J_{\text{eff}} = I_{total}/(\pi\sigma_0)$, and the average is over the initial distribution of the injected electrons.

Thus, it is seen that multi-dimensional effects can be included in the one-dimensional formulation in the low-gain regime by introducing transverse profile functions $u(\boldsymbol{x}_\perp, \sigma)$ into the equations of motion and the wave equation, and by defining an effective one-dimensional current density. This leads to the following general conclusions in regard to the choice of experimental parameters based on the requirement that the profile functions remain of order unity for all electrons in the interaction region. First, the transverse size of the beam must be kept smaller than the minimum spot size in order that all electrons in the beam interact strongly with the radiation field. Second, the interaction length L must not be much greater than the Rayleigh length so that radiation does not diffract significantly over a distance corresponding to the interaction length. However, if the gain is high this latter condition can be relaxed due to the optical guiding effect discussed in Chapter 8.

9.9 STORAGE-RING FREE-ELECTRON LASERS

The high-energy electrons circulating in storage rings are suitable for driving free-electron laser oscillators. In these devices, the same groups of electrons circulate continually around the ring with their energy determined by a balancing of gain due to an accelerating r.f. field and loss due to synchrotron radiation. Thus, when a wiggler is added to the storage ring, the properties of the beam entering the interaction region vary on each pass around the ring in response to the interaction of the beam with the radiation on the previous pass. This situation for the beam is now analogous to that of the radiation in an oscillator and introduces another degree of complexity in the design and analysis of these devices.

The interaction with the radiation acts primarily to alter the energy of the electrons in the beam, and this leads to a modification of the interaction of the electrons with the radiation on its subsequent passes through the interaction region. The modifications of the interaction can be divided into two basic classes: (1) modifications due to the sensitivity of the performance of the storage ring to changes in beam energy, and (2) modifications due to the sensitivity of the basic free-electron laser interaction to the energy of the entering beam. Examples of the first class relate to the dispersion and energy acceptance of the ring. As the electron energy changes, the path followed through the ring also changes, which is known as dispersion. If this change is comparable to the spot size of the radiation, then the interaction of the electron with the radiation will be weakened if the radiation profile remains fixed. Accompanying the change in path for the electron is a change in travel time around the ring. If this change in travel time becomes comparable to the period of the accelerating field, then the electron will fall out of phase with the accelerating field and will no longer be accelerated. This determines the energy acceptance. Both these effects serve to remove an electron from interaction with the radiation once its energy has changed by a specific amount. The importance of these effects depends on details of the storage ring and will not be pursued here.

The second, and perhaps more fundamental, class of modifications relates to the effect of changes in the energy of electrons on the free-electron laser interaction itself. The interaction with the radiation lowers, on average, the energy of electrons and thus produces an amplification of the radiation. However, the energies of different electrons are changed by different amounts. Thus, the interaction also leads to an increase in the spread of energies at the same time the average is lowered. The spread in energy of the beam degrades the interaction on its subsequent passes around the ring. Energy spread degrades the interaction in two ways. First, it lowers the gain directly since with a spread all electrons cannot have the optimum detuning. Second, a spread in energy leads to an increase in the length of the micropulse and a lowering of the peak beam current [46, 53].

The rate at which energy is extracted from the beam and the rate at which the energy spread of the beam is increased are related directly in the small radiation amplitude limit by Madey's second theorem [54],

$$\langle \gamma_f - \gamma_i \rangle = \frac{1}{2}\frac{\partial}{\partial \gamma_i}\langle (\gamma_f - \gamma_i)^2 \rangle, \tag{9.227}$$

where γ_f and γ_i are the final and initial relativistic factors for electrons transiting the interaction region and the average is over an ensemble of electrons entering the interaction region with a relativistic factor γ_i and a uniform distribution of entrance phases. Relation (9.227) is a consequence of the fact that the average rate of energy extraction is second order in the radiation field strength, whereas the magnitude in the change in energy for an electron is first order. That is, in a group of electrons entering the interaction region with a uniform distribution of entrance phases, half the electrons are accelerated initially by the radiation field and half are decelerated, leading to a spread in energies. To first order in the radiation field the net change in energy averages to zero. Due to the first-order acceleration and deceleration, electrons become bunched in phase as they transit the interaction region. The slightly higher-energy electrons catch up to the lower-energy ones. This leads to a net change in energy which is second order in the field strength and which satisfies the Madey theorem. The Madey theorem will be derived in this section by solution of the Vlasov equation (9.86) in powers of the radiation field. The development leads to what is known as a quasilinear theory [55,56] of the evolution of the energy-distribution function for the beam electrons [46, 57, 58].

The increase in energy spread of the beam due to its interaction with the radiation leads to a limit, known as the Renieri limit [45, 46], on the average amount of power that can be extracted from the beam. In the absence of the free-electron laser interaction, a spread in energy of the beam is damped by the combined effects of synchrotron radiation and r.f. acceleration. Since the rate of synchrotron radiation increases strongly with energy, while the rate of acceleration is approximately independent of energy (in the typical range of interest), the competition of the two causes all electrons to tend to the energy at which the rate of increase of energy due to acceleration and the rate of decrease due to radiation balance. Synchrotron radiation scales as the second power of energy and magnetic field strength; therefore, the rate at which deviations in energy decay is given by [59]

$$\nu_s = \frac{P_s(2 + D_b)}{\gamma m_e c^2}, \tag{9.228}$$

where P_s is the power radiated by an electron in the ring, the factor of 2 comes from the energy dependence of the synchrotron radiation, and the factor D_b accounts for the variation of magnetic field strength with the variation of the path (dispersion) implied by the variation in energy. The Renieri limit states that the maximum average power extracted from the beam by the free-electron laser interaction is given by [45, 46]

$$\bar{P} = \frac{P_s}{2N_w}, \tag{9.229}$$

where N_w denotes the number of wiggler periods in the interaction region. This power, while a fraction of the synchrotron radiation power, is concentrated in a narrow range of frequencies and in a well-collimated beam and is thus still of practical interest.

Finally, with the available current in storage rings it is frequently difficult to obtain a significant gain with an interaction region whose length is not too long as to be incompatible with operation of the storage ring. As a result, there has been developed a wiggler configuration known as the optical klystron [60, 61] which is capable of achieving enhancements in gain above that predicted for the conventional uniform wiggler. The basic operation of the optical klystron can be explained in terms of the klystron model introduced in Section 9.3. This will be discussed briefly at the end of this chapter.

We begin our analysis of the dynamics of the radiation and the electron beam in a storage-ring oscillator by making a number of simplifying assumptions. First, we assume that a one-dimensional, low-gain theory, as developed in Section 9.5, applies. This imposes the requirement that the size of the electron beam must be less than the minimum spot size of the radiation, and that the interaction length must be shorter than the Rayleigh length. Further, we assume that the duration of the micropulses in the storage ring is much longer than the slippage time, and neglect details of the pulsed nature of the electron beam. Recall from (9.203) that the start oscillation current for a pulsed-beam device is the same as that in a continuous-beam device in this limit at the optimum cavity detuning. Thus, the theory will parallel that of a continuous-beam oscillator as developed in Section 9.6, except that we now must account for the recirculation of the beam electrons through the interaction region.

The basic equations governing the dynamics of the beam particles are the Vlasov equation (9.86) along with boundary condition (9.87), which must now be modified to describe the recirculation of the electron beam. The distribution function for injected electrons will vary in time as it is affected by the radiation. This variation will be ascribed to the slow-time variable n, which measures the number of round trips of the radiation through the cavity. We shall assume (1) that the radiation field in the interaction region is weak in the sense that electrons will suffer only small changes in energy on each pass through the device, (2) that the beam electrons return to the entrance of the interaction region after a time T_{sr} which depends weakly on the energy of the electrons, and (3) that during this time the energy of each electron is constant. Actually, electrons are reaccelerated and lose energy to synchrotron radiation during their trip around the ring, but this effect can easily be added later in the analysis. The normalized energies and phases of electrons entering the interaction region at round-trip index n can now be expressed in terms of the same quantities for electrons leaving the interaction region at a round-trip index N_{sr} earlier, where $N_{sr} = T_{sr}/T$. In particular, the energies are unchanged, and the phases have advanced by an amount $\Delta\psi$, where

$$\Delta\psi = \omega T_{sr}(\gamma) \cong \frac{2\pi L_{sr}(\gamma)}{\lambda} \tag{9.230}$$

is the change in phase accumulated in the time T_{sr} it takes for electrons to return to the interaction region. The second equality in (9.230) expresses the phase change in terms of the path length L_{sr} for the returning electrons, assuming they travel at the speed of light, and λ is the wavelength of the radiation.

Equation (9.230) allows for the possibility that returning electrons maintain phase coherence with the radiation. If the variations in the phase change from electron to electron are small compared with unity, then the entering beam is effectively **prebunched** by its previous encounters with the radiation. This could lead, in principle, to large enhancements in gain [62]. The requirements for phase coherence are quite stringent in that variations of the return time with energy are large enough that $\Delta\psi$ varies considerably over the range of energies of interest [63]. In this case it is appropriate to assume that phase coherence is lost by the returning electrons and the distribution function for electrons entering the interaction region can be taken to be independent of entrance phase, with an energy dependence given by the phase average of the distribution function for electrons leaving the interaction region on the previous circuit of the ring. Further, consistent with our neglect of the pulsed nature of the electron beam, we assume that the entering distribution function is independent of entrance time during the time interval T_p (i.e. the pulse duration) and equal to the time average of the distribution function for electrons leaving the interaction region.

An evolution equation for the phase- and entrance-time-averaged distribution function entering the interaction region at round-trip index n is then obtained by integrating the Vlasov equation (9.86) over all phases and entrance times within a micropulse, and over axial distance through the interaction region,

$$\bar{F}(p, n + N_{sr}) = \bar{F}(p, n) + \frac{1}{2}\frac{\partial}{\partial p}\int_0^1 d\xi \int_0^{2\pi}\frac{d\psi}{2\pi}\int_0^{T_p}\frac{dt_e}{T_p}$$

$$\times\,[iX(n, t_e + \varepsilon\xi T)\exp(i\psi) + \text{c.c.}]F(p, \psi, \xi, t_e). \tag{9.231}$$

To solve (9.231) we must still supply on the right-hand side the detailed, phase-dependent distribution function for electrons in the interaction region. To do this, we make what is known as the quasilinear approximation [55, 56] which assumes that the distribution function is perturbed only slightly by the radiation on each pass of the electrons through the interaction region. This is valid if the amplitude of the radiation is small enough that the usual phase-trapping effects which cause nonlinear saturation are not operative. Instead, the energies of electrons, as measured by the dimensionless momentum p, change only by small amounts on each pass through the interaction region. After many passes, a large change in the distribution function can occur, and the nature of this change is diffusive. On each pass through the interaction region, an electron receives a small kick in normalized momentum. Kicks on subsequent passes are uncorrelated due to the loss of phase coherency by the electrons during their trip around the ring. Thus, the normalized momentum of each electron executes a random walk leading to a diffusion equation. The quasilinear diffusion equation is obtained by calculating the first-order correction to the distribution function, as one would do in linear theory, and inserting it in (9.231). Specifically, we represent the radiation field during the time interval T_p in terms of a Fourier series (as in equation (9.84)), except that the frequency separation defined in (9.85) will now be based on the pulse-duration time T_p instead of the round-trip travel time of the radiation T. The distribution function is then written as the sum of the injected distribution and a small perturbation proportional to the radiation field, as in (9.91). The perturbation satisfies the linearized Vlasov equation (9.92) and can be written as an integral as in (9.93). Substitution of the perturbed distribution function into (9.231) and performing the averages over phase and entrance time then yields the quasilinear diffusion equation

$$\bar{F}(p, n + N_{sr}) = \bar{F}(p, n) + \frac{\partial}{\partial p} D_{QL}(p) \frac{\partial}{\partial p} \bar{F}(p, n), \tag{9.232}$$

where the diffusion coefficient D_{QL} obtained after a number of integrations is second order in the radiation amplitude

$$D_{QL}(p) = \frac{\pi}{2} \sum_{m=-\infty}^{\infty} |X_m(n)|^2 \,\mathrm{Re}[\Delta(p - \varepsilon\Delta\omega_m T)]. \tag{9.233}$$

The resonance function $\Delta(p)$ appearing in equation (9.233) is defined in (9.96), and it appears in the expressions for the gain of the radiation (9.95). It is referred to as a resonance function since it is positive, peaked at zero argument, and has unit area. Equation (9.233) states that the contribution of each component of the radiation spectrum to the diffusion coefficient is peaked at a normalized energy corresponding to the resonance with the beat wave for that frequency. The width of the resonance, as measured in normalized energy, is unity, corresponding to the range of energies for which the Doppler-shifted beat-wave frequency is of the order of the inverse of the transit time through the interaction region. If the dependence of the spectral amplitudes on the mode index m is weak or, equivalently, if the spectrum is broader than the gain bandwidth, then the sum in (9.233) can be replaced by an integral and the resonance function treated as a delta function.

Madey's second theorem [54] can now be derived by comparing the quasilinear equation (9.232) with the Fokker–Planck equation [30] for the distribution function

$$\bar{F}(p, n + N_{sr}) = \bar{F}(p, n) + \frac{\partial}{\partial p}\left[\frac{\partial}{\partial p}\left(\frac{\Delta\dot{p}^2}{2}\bar{F}(p, n)\right) - \Delta\dot{p}\bar{F}(p, n)\right], \tag{9.234}$$

where $\Delta\dot{p}$ and $\Delta\dot{p}^2$ are the average rate of drift and diffusion of the normalized energy on each pass through the interaction region. In order that the Fokker–Planck equation reduces to the quasilinear equation it is required that

$$\Delta\dot{p} = \frac{1}{2}\frac{\partial}{\partial p}\Delta\dot{p}^2. \tag{9.235}$$

Expressing the normalized energy in terms of the initial and final relativistic factors then gives (9.227). That the above relationship is a consequence of the underlying Hamiltonian dynamics has been discussed by Manheimer and Dupree [64] and by Krinsky *et al.* [65].

Madey's theorem, as derived from the quasilinear equation, does not depend on the detailed form of the diffusion coefficient, but only on the order of the diffusion coefficient and the derivatives with respect to normalized energy appearing in the quasilinear equation. In particular, the diffusion coefficient is placed between the two p derivatives, not outside. That this must be the case follows from conservation of particles and phase-space area. Specifically, in order that the total number of electrons be conserved, the change in the distribution function must be expressed as the derivative of a flux. This fixes the first derivative on the left of the diffusion coefficient. In addition, the flow in phase space implied by the Vlasov equation is incompressible. Thus, if the injected distribution function were independent of normalized energy in a particular range of energies, then there could be no resulting change in the distribution function in this range of energies. This fixes the second derivative with respect to p in that it must act directly on the distribution function. Finally, there can be only two derivatives with respect to p since the quasilinear equation is second order in the radiation field strength and the derivative with respect to normalized energy in the original Vlasov equation is multiplied by the radiation field strength.

The evolution of the distribution function due to the interaction of the electrons and the radiation is described as quasilinear flattening. Specifically, the diffusion in energy of the electrons causes the distribution function to become independent of energy in the range of normalized energies which are resonant with the radiation. This leads to a progressive reduction in the gain given by (9.95) since the gain is proportional to the average of the slope of the distribution function weighted by the resonance function. Suppose that only a single spectral component is present. The distribution function satisfying the quasilinear equation will then tend monotonically toward a constant for values of normalized momentum p between the first two zeros of the resonance function $p - \varepsilon\Delta\omega_m T = \pm 2\pi$. Thus, the beam acquires a spread in energies corresponding to $\delta\gamma \approx \gamma/2N_w$ (where N_w denotes the number of wiggler periods in the interaction region). Further, in the absence of other effects, specifically reacceleration and synchrotron cooling, the gain falls below the level required to overcome losses and the radiation decays away.

The basic time scale over which quasilinear flattening occurs is governed by the rate of buildup of the radiation field; hence, the flattening will occur over a time which scales as the inverse of the initial exponentiation rate of the radiation. Typically, this rate is much greater than the rate at which synchrotron radiation and acceleration combine to restore the beam to its initial state (9.228). Thus, on the initial flattening time scale, the effects of acceleration and synchrotron radiation can be neglected. Ultimately, the distribution function is determined by the balancing of the diffusive effects of the radiation and the damping effect of the synchrotron radiation and acceleration. In addition, the balance of these effects determines the average rate at which power can be extracted from the beam by the free-electron laser interaction.

An order-of-magnitude estimate of the average radiation power extracted from the beam going to the free-electron laser interaction can be obtained as follows. In equilibrium, the average rate of increase of the normalized energy spread due to the free-electron laser interaction is balanced by synchrotron cooling which occurs at a rate given by (9.228). Thus, for a typical electron we have

$$\nu_s T_{sr} p^2 \cong |\Delta \dot{p}^2| \cong |\Delta \dot{p}|, \tag{9.236}$$

where the second approximation follows from Madey's theorem along with the knowledge that the gain is saturated when the distribution function acquires a spread in p values of order unity. Finally, since the spread in p values is of order unity we have $|\Delta \dot{p}| = \nu_s T_{sr}$, where $|\Delta \dot{p}|$ is the typical energy lost to the free-electron laser interaction per trip around the ring. Restoring units to the normalized energy, we arrive at the Renieri limit (9.229).

The above calculation applies in steady state. However, a weakly damped relaxation oscillation cycle is often excited due to the large disparity in time scales between growth of radiation and the damping of energy deviations by synchrotron radiation [53, 66, 67]. This cycle is characterized by bursts of radiation which increase the spread in energies of the electron beam beyond that which is necessary to suppress growth of the radiation. The radiation then decays to a low level while the beam cools due to synchrotron radiation. Beam cooling continues for a time even after the gain of the radiation becomes positive, since modes do not grow to finite amplitude immediately. As a result, radiation then grows again to a large amplitude, and the bursting process is repeated. The power level during one of the bursts can be many times the average, thus temporarily exceeding the Renieri limit. Further, it has been demonstrated that the timing of the bursts has been controlled by periodically changing the overlap of the radiation and beam pulses [53].

9.10 OPTICAL KLYSTRONS

In the low-gain regime the one-dimensional theory predicts a single-pass gain for the radiation that scales as the third power of the length of the interaction region. In particular, in the small signal-limit the growth of the radiation with each pass through the interaction is given by (9.94). The gain is thus determined by the product I_b and G_m, where I_b is a normalized current given by equation (9.55), and G_m is a normalized gain factor which accounts for the energy distribution of the incoming beam. The dependence on interaction length enters through the definition of the normalized current. Physically, the third power of the length enters due to the fact that there are three spatial derivatives in the governing equations: one in the wave equation and two in the equations of motion. Thus, if one doubles the length, keeping the radiation amplitude fixed, the first-order change in an electron energy will also double since the radiation electron field acts to accelerate electrons over a longer distance. This causes the degree of phase bunching to quadruple since both the variation of particle energy and the length of the interaction region have doubled. Finally, the amplification of the radiation will increase by a factor of eight due to the four-fold increase in the bunched current, and the factor of two increase in the length over which amplification occurs.

In a short-wavelength device, the third power of the length is reduced to only the second power by multi-dimensional effects. Recall from equation (9.226) that when the finite transverse size of the beam and radiation envelopes are included in the analysis, the effective current density is the total current divided by the minimum spot size of the radiation. The minimum spot size is constrained by the requirement that the Rayleigh length be of the order of or less than the interaction length $L/k\sigma_0$. This effectively reduces by one power the length dependence of the gain. It is still advantageous to design a wiggler which is long to increase gain if the energy spread of the available beam permits. However, there may be physical limitations in the case of a storage ring simply due to the available space between bending magnets [53].

A method of increasing the gain without increasing the overall length of the interaction region is to use the optical klystron configuration [60, 61, 68–71]. The optical klystron works by effectively increasing the bunching length for the electrons. To see this, we note that the normalized current is proportional to the factor

$$\frac{\omega_\gamma L}{v_\parallel} = -\frac{d}{d\gamma}\left(\frac{\omega L}{v_\parallel}\right), \tag{9.237}$$

where ω_γ is defined in equation (9.15). Hence, the degree of bunching (and, hence, the gain) is proportional to the derivative with respect to the energy of the phase change of an electron in traversing the interaction region. For electrons travelling along the same path at nearly the speed of light this derivative is small. However, it can be enhanced greatly if high-energy electrons travel on a shorter path than low-energy electrons. This is accomplished by constructing an interaction region consisting of two sections of undulator separated by a drift region with a dispersive magnetic field. This is the optical klystron. The dispersive magnetic field imparts an impulse to the electron in the direction transverse to the axis of the beam, and causes the path of the electron to veer off axis and then to return. Due to the relativistic energy dependence of the electron mass, the path followed by a high-energy electron will be shorter than that followed by a low-energy electron. For small deviations from the axis, the energy dependence of the time of flight through the drift region can be expressed as [53, 70]

$$T_d(\gamma) = \frac{L_d}{v_\parallel} + \frac{1}{2}\int_0^{L_d} \frac{dz}{v_\parallel}\left[\int_0^z \frac{dz'}{v_\parallel}\frac{eB(z')}{\gamma m_e c}\right]^2, \tag{9.238}$$

where L_d is the length of the drift region and $B(z')$ is the magnitude of the dispersive magnetic field. Differentiating this expression with respect to energy and assuming highly relativistic motion one finds that the effective bunching length of the drift region is now given by

$$L_{\text{eff}} = L_d + \int_0^{L_d} dz\left[\int_0^z dz' \frac{eB(z')}{\gamma m_e c}\right]^2. \tag{9.239}$$

The operating characteristics of the optical klystron can be understood at a basic level by studying the klystron model of Section 9.3 provided a number of parameters are replaced by values which are appropriate to the use of a dispersive drift section. In particular, the normalized momentum p_i (9.48) is now defined in terms of the energy dependence of the phase change in transitting the dispersive section

$$p_i = \frac{L_{\text{eff}}\,\omega_\gamma}{v_\parallel}\,\delta\gamma_i. \tag{9.240}$$

The normalized radiation amplitude $X(t_e)$ (see equation (9.49)) measures the strength of the kick in energy that an electron receives in the bunching section. Thus it is normalized by the strength a_{w1} and length L_1 of the wiggler in the buncher, the spot size in the buncher (σ_0/σ_1) and the energy dependence of the phase change through the drift region

$$X(t_e) = \frac{\sigma_0 a_{w1}\, L_1\, L_{\text{eff}}\,\omega\omega_\gamma}{\sigma_1 \gamma_r v_\parallel}\,\delta a_e. \tag{9.241}$$

The change in amplitude of the radiation as expressed in (9.50) is proportional to the strength a_{w2} and length L_2 of the wiggler as well as the spot size of the radiation in the output section (σ_0/σ_2). This leads again to the klystron equation (9.54) except that the normalized current is now given by

$$I(t) = \frac{2e\sigma_0 a_{w1} a_{w2}\, L_1\, L_2\, L_{\text{eff}}\,\omega_\gamma}{\sigma_1 \sigma_2^* \gamma_r^2 m_e c v_\parallel^3}\, I_T(t). \tag{9.242}$$

Finally, the slippage parameter is defined such that εT is the difference in time for the propagation of radiation and electrons through the interaction region,

$$\varepsilon = \frac{L}{L_c}\left[\frac{T_d(\gamma_r)c}{L} - 1\right], \tag{9.243}$$

where $T_d(\gamma_r)$ is the mean travel time for electrons through the drift section.

The small-signal gain for an arbitrary distribution of injected electrons is obtained from (9.54) and (9.53). For a monoenergetic beam interacting with a mode with frequency shifted from exact resonance by an amount $\Delta\omega$, the gain has a purely sinusoidal dependence on frequency shift

$$\left|\frac{X(t-T)}{X(t)}\right| = \left|1 + \frac{I_b}{2}\exp\left(i\frac{\pi}{2} + i\varepsilon\Delta\omega T\right)\right|. \tag{9.244}$$

The klystron model has vanishingly small buncher and output sections and this leads artificially to the appearance of gain over an infinite range of frequencies. If the non-zero length of the buncher and output sections are accounted for, then the frequency dependence of the gain is modulated by an envelope function with width scaling as the gain band width of the individual wiggler sections [70].

The effect of energy spread on gain in the optical klystron is particularly easy to evaluate since (9.53) reveals that the gain is proportional to the Fourier transform of the distribution function. Thus, for a Gaussian spread in energies with a width Δ_γ,

$$f(\delta\gamma_i) = \frac{1}{2\pi^{3/2}\Delta\gamma}\exp\left[-\left(\frac{\delta\gamma_i}{\Delta\gamma}\right)^2\right], \tag{9.245}$$

the gain function g is reduced by a factor h from its value for a monoenergetic beam, where

$$h = \exp\left[-\left(\frac{L_{\text{eff}}\omega_y\Delta\gamma}{v_\parallel}\right)^2\right]. \tag{9.246}$$

The requirement that this factor not be too small sets an upper limit on the effective length of the dispersive section for a given beam quality. In addition, saturation of the gain occurs when the energy spread of the electron beam acquires a width $\Delta\gamma$ such that the reduced gain implied by (9.246) balances losses. In conclusion, use of the optical klystron is an effective way to increase gain in cases in which physical space limitations as opposed to beam quality limit the usable length of a conventional undulator.

REFERENCES

1. Colson, W. B. and Ride, S. K. (1980) The free-electron laser, Maxwell's equations driven by single particle currents, in *Physics of Quantum Electronics: Free-Electron Generators of Coherent Radiation*, Vol. 7 (ed. Jacobs, S. F., Pilloff, H. S., Sargent, M., Scully, M. O. and Spitzer, R.), Addison-Wesley, Reading, Massachusetts, p. 377.
2. Bogomolov, Ya. L., Bratman, V. L., Ginzburg, N. S., Petelin, M. I. and Yunakovsky, A. D. (1981) Nonstationary generation in free-electron lasers. *Opt. Commun.*, **36**, 209.
3. Kroll, N. M., Morton, P. L. and Rosenbluth, M. N. (1981) Free-electron lasers with variable parameter wigglers. *IEEE J. Quantum Electron.*, **QE-17**, 1436.
4. Ginzburg, N. S. and Petelin, M. I. (1985) Multifrequency generation in free-electron lasers with quasi-optical resonators. *Int. J. Electronics*, **59**, 291.
5. Al-Abawi, H., Hopf, F. A., Moore, G. T. and Scully, M. O. (1979) Coherent transients in the free-electron laser: laser lethargy and coherence brightening. *Opt. Commun.*, **30**, 235.
6. Yu, S. S., Sharp, W. M., Fawley, W. M., Scharlemann, E. T., Sessler, A. M. and Sternbach, E. J. (1987) Waveguide suppression of the free-electron laser sideband instability. *Nucl. Instr. Meth.*, **A259**, 219.
7. Colson, W. B. and Blau, J. (1988) Parameterizing physical effects in free-electron lasers. *Nucl. Instr. Meth.*, **A272**, 386.
8. Levush, B. and Antonsen, T. M. Jr (1988) Regions of stability of free-electron laser oscillators. *Nucl. Instr. Meth.*, **A272**, 375.
9. Colson, W. B. (1990) Classical free-electron laser theory, in *The Laser Handbook: Free-Electron Lasers*, Vol. 6 (ed. Colson, W. B., Pellegrini, C. and Renieri, A.), North Holland, Amsterdam, p.115.
10. Pierce, J. R. (1950) *Traveling Wave Tubes*, Van Nostrand, New York.
11. Collin, R. E. (1966) *Foundations for Microwave Engineering*, McGraw-Hill, New York.
12. Hopf, F. A., Meystre, P., Moore, G. T. and Scully, M. O. (1978) Nonlinear theory of free-electron devices, in *Physics of Quantum Electronics: Novel Sources of Coherent Radiation*, Vol. 5 (ed. Jacobs, S. F., Sargent, M. and Scully, M. O.), Addison-Wesley, Reading, Massachusetts, p. 41.

13. Colson, W. B. and Freedman, R. A. (1983) Synchrotron instability for long pulses in free-electron lasers. *Opt. Commun.*, **46**, 37.
14. Colson, W. B. (1986) The trapped particle instability in free-electron laser oscillators and amplifiers. *Nucl. Instr. Meth.*, **A250**, 168.
15. Antonsen, T. M. Jr and Levush, B. (1989) Mode competition and suppression in free-electron laser oscillators. *Phys. Fluids B*, **1**, 1097.
16. Masud, J., Marshall, T. C., Schlesinger, S. P., Yee, F. G., Fawley, W. M., Scharlemann, E. T., Yu, S. S., Sessler, A. M. and Sternbach, E. J. (1987) Sideband control in a millimeter-wave free-electron laser. *Phys. Rev. Lett.*, **58**, 763.
17. Warren, R. W., Goldstein, J. C. and Newnam, B. E. (1986) Spiking mode operation for a uniform-period wiggler. *Nucl. Instr. Meth.*, **A250**, 19.
18. Warren, R. W., Sollid, J. E., Feldman, D. W., Stein, W. E., Johnson, W. J., Lumpkin, A. H. and Goldstein, J. C. (1989) Near-ideal lasing with a uniform wiggler. *Nucl. Instr. Meth.*, **A285**, 1.
19. Nusinovich, G. S. (1980) The mode interaction in free-electron lasers. *Sov. Phys. Tech. Phys.*, **6**, 848.
20. Stanford, E. R. and Antonsen, T. M. Jr (1991) The effect of dispersion on mode competition in free-electron laser oscillators. *Nucl. Instr. Meth.*, **A304**, 659.
21. Antonsen, T. M. Jr and Levush, B. L. (1989) Mode competition and control in free-electron laser oscillators. *Phys. Rev. Lett.*, **62**, 1488.
22. Ginzburg, N. S., Kuznetsov, S. P. and Fedoseeva, T. M. (1978) Theory of transients in backward wave tubes. *Radiofiz.*, **21**, 1037.
23. Sprangle, P., Tang, C. M. and Bernstein, I. B. (1983) Initiation of a pulsed beam free-electron laser. *Phys. Rev. Lett.*, **50**, 1775.
24. Kim, K. J. (1986) An analysis of self amplified spontaneous emission. *Nucl. Instr. Meth.*, **A250**, 396.
25. Becker, W., Gea-Banacloche, J. and Scully, M. O. (1986) Intrinsic linewidth of a free-electron laser. *Phys. Rev. A*, **33**, 2174.
26. Friedman, A., Gover, A., Kurizki, G., Ruschin, S. and Yariv, A. (1988) Spontaneous and stimulated emission from quasi-free electrons. *Rev. Mod. Phys.*, **60**, 471.
27. Warren, R. W. and Goldstein, J. C. (1988) The generation and suppression of synchrotron sidebands. *Nucl. Instr. Meth.*, **A272**, 155.
28. Elias, L. R., Hu, R. J. and Ramian, G. J. (1985) The UCSB electrostatic accelerator free-electron laser: first operation. *Nucl. Instr. Meth.*, **A237**, 203.
29. Antonsen, T. M. Jr and Levush, B. (1990) Spectral characteristics of a free-electron laser with time-dependent beam energy. *Phys Fluids B*, **2**, 2791.
30. Liboff, R. L. (1969) *Introduction to the Theory of Kinetic Equations*, Wiley, New York.
31. Amir, A., Hu, R. J., Kielmann, F., Mertz, J. and Elias, L. R. (1988) Injection locking experiment at the UCSB free-electron laser. *Nucl. Instr. Meth.*, **A272**, 174.
32. Elias, L. R., Ramian, G. J., Hu, J. and Amir, A. (1986) Observation of single mode operation in a free-electron laser. *Phys. Rev. Lett.*, **57**, 424.
33. Danly, B. G., Evangelides, S. G., Chu, R., Temkin, R. J., Ramian, G. J. and Hu, J. (1990) Direct spectral measurements of a quasi-cw free-electron laser. *Phys. Rev. Lett.*, **65**, 2251.
34. Levush, B. and Antonsen, T. M. Jr (1989) Nonlinear mode competition and coherence in low gain free-electron laser oscillators. *Nucl. Instr. Meth.*, **A285**, 136.
35. Kimmel, I. and Elias, L. R. (1988) Long-pulse free-electron lasers as sources of monochromatic radiation. *Nucl. Instr. Meth.*, **A272**, 368.
36. Litvenenko, V. N. and Vinokurov, N. A. (1991) Lasing spectrum and temporal structure in storage ring free-electron lasers: theory and experiment. *Nucl. Instr. Meth.*, **A304**, 66.
37. Dattoli, G. and Renieri, A. (1979) Classical multimode theory of the free-electron laser. *Lett. Nuovo Cimento*, **59B**, 1.
38. Colson, W. B. (1982) Optical pulse evolution in the Stanford free-electron laser and in a

tapered undulator, in *Physics of Quantum Electronics: Free-Electron Generators of Coherent Radiation*, Vol. 8 (ed. Jacobs, S. F., Moore, G. T., Pilloff, H. S., Sargent, M., Scully, M. O. and Spitzer, R.), Addison-Wesley, Reading, Massachusetts, p. 457.

39. Dattoli, G., Hermsen, T., Renieri, A., Torre, A. and Gallardo, J. C. (1988) Lethargy of laser oscillations and supermodes in free-electron lasers: I. *Phys. Rev. A*, **37**, 4326.
40. Dattoli, G., Hermsen, T., Mezi, L., Renieri, A. and Torre, A. (1988) Lethargy of laser oscillations and supermodes in free-electron lasers: II–quantitative analysis. *Phys. Rev. A*, **37**, 4334.
41. Goldstein, J. C., Newnam, B. E., Warren, R. W. and Sheffield, R. L. (1986) Comparison of the results of theoretical calculations with experimental measurements from the Los Alamos free-electron laser oscillator experiment. *Nucl. Instr. Meth.*, **A250**, 4.
42. Elleaume, P. (1985) Storage ring free-electron laser theory. *Nucl. Instr. Meth.*, **A237**, 28.
43. Kroll, N. M. (1965) Excitation of hypersonic vibrations by means of photoelastic coupling of high intensity light waves to elastic waves. *J. Appl. Phys.* **36**, 34.
44. Pesme, D., Laval, G. and Pellat, R. (1973) Parametric instabilities in bounded plasmas. *Phys. Rev. Lett.*, **31**, 203.
45. Renieri, A. (1979) Storage ring operation of a free-electron laser: the amplifier. *Nuovo Cimento*, **53B**, 160.
46. Dattoli, G. and Renieri, A. (1980) Storage ring operation of a free-electron laser: the oscillator. *Nuovo Cimento*, **59B**, 1.
47. Goldstein, J. C. (1984) Evolution of long pulses in a tapered wiggler free-electron laser, in *Free-Electron Generators of Coherent Radiation* (ed. Brau, C. A., Jacobs, S. F. and Scully, M. O.), Proc. SPIE 453, Bellingham, Washington, p. 2.
48. Colson, W. B. and Richardson, J. L. (1983) Multimode theory of free-electron laser oscillators. *Phys. Rev. Lett.*, **50**, 1050.
49. Goldstein, J. C., McVey, B. D., Carlsten, B. E. and Thode, L. E. (1989) Integrated numerical modeling of free-electron laser oscillators. *Nucl. Instr. Meth.*, **A285**, 192.
50. Goldstein, J. C., McVey, B. D., Tokar, R. L., Elliot, C. J., Schmidt, M. J., Carlsten, B. E. and Thode, L. E. (1989) Simulation codes for modeling free-electron laser oscillators, in *Modeling and Simulation of Laser Systems* (ed. Bullock, D. L.), Proc. SPIE 142, Bellingham, Washington, p. 28.
51. Riyopoulos, S., Sprangle, P., Tang, C. M. and Ting, A. (1988) Reflecting matrix for optical resonators in free-electron laser oscillators. *Nucl. Instr. Meth.*, **A272**, 543.
52. Iracane, D. and Ferrer, J. L. (1990) An optical basis equation for solving the time-dependent Schrödinger equation: simulation of guiding and multifrequency mechanisms. *Nucl. Instr. Meth.*, **A296**, 417.
53. Deacon, D. A. G. and Ortega, J. M. (1990) The storage ring free-electron laser, in *The Laser Handbook: Free-Electron Lasers*, Vol. 6 (ed. Colson, W. B., Pellegrini, C. and Renieri, A.), North Holland, Amsterdam, p. 345.
54. Madey, J. M. J. (1979) Relationship between mean radiated energy, mean squared radiated energy, and spontaneous power spectrum in a power series expansion of the equation of motion in a free-electron laser. *Nuovo Cimento*, **50B**, 64.
55. Vedenov, A. A., Velikov, E. P. and Sagdeev, R. Z. (1961) Nonlinear oscillations of rarified plasma. *Nucl. Fusion*, **1**, 82.
56. Drummond, W. E. and Pines, D. (1962) Nonlinear stability of plasma oscillations. *Nucl. Fusion Suppl.*, **3**, 1049.
57. Taguchi, T., Mima, K. and Mochizuki, T. (1981) Saturation mechanism and improvement of conversion efficiency of the free-electron laser. *Phys. Rev. Lett.*, **46**, 824.
58. Ginzburg, N. S. and Shapiro, M. A. (1982) Quasilinear theory of multimode free-electron lasers with an inhomogeneous frequency broadening. *Opt. Commun.*, **40**, 215.
59. Edwards, D. A. and Syphers, M. J. (1989) An introduction to the physics of particle accelerators, in *Physics of Particle Accelerators*, Vol. 1 (ed. Month, M. and Dienes, M.), American Institute of Physics Conference Proceedings #184, New York, p. 2.

60. Vinokurov, N. A. and Skrinsky, A. N. (1979) *Optical Range Klystron Oscillator using Ultrarelativistic Electrons*, Preprint 77-59 of the Institute of Nuclear Physics, Novosibirsk.
61. Vinokurov, N. A. and Skrinsky, A. N. (1977) *On Ultimate Power of the Optical Klystron Installed on Electron Storage Ring*, Preprint 77-67 of the Institute of Nuclear Physics, Novosibirsk.
62. Deacon, D. A. G. and Madey, J. M. J. (1980) Isochronous storage ring laser: a possible solution to the electron heating problem in recirculating free-electron lasers. *Phys. Rev. Lett.*, **44**, 449.
63. Van Steenbergen, A. (1990) Accelerators and storage rings for free-electron lasers, in *The Laser Handbook: Free-Electron Lasers*, Vol. 6 (ed. Colson, W. B., Pellegrini, C. and Renieri, A.), North Holland, Amsterdam, p. 417.
64. Manheimer, W. M. and Dupree, T. H. (1968) Weak turbulence theory of velocity space diffusion and nonlinear Landau damping of waves. *Phys. Fluids.*, **11**, 2709.
65. Krinsky, S., Wang, J. M. and Luchini, P. (1982) Madey's gain spread theorem for the free-electron laser and the theory of stochastic processes. *J. Appl. Phys.*, **53**, 5453.
66. Billardon, M., Elleaume, P., Ortega, J. M., Bazin, C., Bergher, M., Velghe, M., Petroff, Y., Deacon, D. A. G., Robinson, K. E. and Madey, J. M. J. (1983) First operation of a storage ring free-electron laser. *Phys. Rev. Lett.*, **51**, 1652.
67. Elleaume, P. (1984) Macrotemporal structure of free-electron lasers. *J. Phys.*, **45**, 997.
68. Elleaume, P. (1982) Optical klystron spontaneous emission and gain, in *Physics of Quantum Electronics: Free-Electron Generators of Coherent Radiation*, Vol. 8 (ed. Jacobs, S. F., Moore, G. T. , Pilloff, H. S., Sargent, M., Scully, M. O. and Spitzer, R.), Addison-Wesley, Reading, Massachusetts, p. 119.
69. Shih, C. C. and Caponi, M. Z. (1983) An optimized multicomponent wiggler design for a free-electron laser. *IEEE J. Quantum Electron.*, **QE-19**, 369.
70. Elleaume, P. (1990) Free-electron laser undulators, electron trajectories and spontaneous emission, in *The Laser Handbook: Free-Electron Lasers*, Vol. 6 (ed. Colson, W. B., Pellegrini, C. and Renieri, A.), North Holland, Amsterdam, p. 91.
71. Csonka, P. L. (1980) High-modulation electron beams in storage rings. *Particle Accelerators*, **11**, 45.

10

Electromagnetic-wave wigglers

The physical mechanism in the free-electron laser depends upon the propagation of an electron beam through a periodic magnetic field. Both incoherent and coherent radiation result from the undulatory motion of the electron beam in the external fields which permits a wave–particle coupling to the output radiation. Coherent radiation depends upon the stimulated emission due to the ponderomotive wave formed by the beating of the radiation and wiggler fields. The wiggler field itself may be either magnetostatic or electromagnetic in nature. Although the bulk of experiments as of this time have relied upon magnetostatic wigglers with either helical or planar polarizations, the fundamental principle has also been demonstrated in the laboratory using a large-amplitude electromagnetic wave to induce the requisite undulatory motion in the electron beam [1, 2].

The basic difference between magnetostatic and electromagnetic-wave wigglers lies in the frequency of the output radiation, which depends upon both the wiggler period and the beam energy in both cases. In the case of a magnetostatic wiggler, the wavelength of the output radiation scales as $\lambda = \lambda_w/2\gamma_b^2$ where λ_w denotes the wiggler period and γ_b is the bulk relativistic factor of the beam. In contrast, the wavelength of the output radiation for an electromagnetic-wave wiggler scales as $\lambda \approx \lambda_w/4\gamma_b^2$. As a result, for fixed wiggler periods and beam energies, the electromagnetic-wave wiggler will produce shorter output wavelengths. As a consequence, electromagnetic-wave wigglers become attractive alternatives to magnetostatic wigglers for the production of short wavelengths when the electron-beam energy is constrained.

Several different configurations have been proposed, and analysed, to make use of electromagnetic-wave wigglers [3–19]. The earliest of these is the simplest, and involves nothing more than the use of a large-amplitude radiation pulse from some convenient source which is launched in synchronism with the electron beam. An interesting variant on this concept makes use of an external radiation source of moderate to high intensities to **pump up** a resonant cavity to extremely high intensities prior to the injection of an electron beam. In this concept the electromagnetic wave constitutes a standing wave of the cavity, and contrasts with the earlier design which employs a travelling electromagnetic wave. One proposal for such a design would make use of a long-pulse or CW gyrotron oscillator [15, 17] as a source of radiation at wavelengths below approximately 1 cm to pump up the resonant cavity. This would permit the generation of infrared radiation with wavelengths in the range of 300 μm with relatively modest beam energies of the order of 1 MeV. The most ambitious concept is the two-stage free-electron laser [10–12], which is a variant on an oscillator configuration. In this concept, an electron beam propagates through a magnetostatic wiggler located within a resonant cavity of some kind. The radiation generated by this means will itself act as an electromagnetic-wave wiggler to generate still shorter-wavelength radiation on the electron beam. Indeed, this mechanism can operate within any long-pulse, high-intensity oscillator design. The principal difficulty with this concept as a source of short-wavelength, high-intensity radiation is that if the interaction in either stage reaches sufficiently high efficiencies, then the electron beam quality can be degraded and can **quench** both stages of the interaction.

Whatever the source of the electromagnetic wave, however, the principal difficulty is the same as that found with short-period magnetostatic wigglers, specifically, that in order for the oscillatory motion of the electron beam to reach the large amplitudes necessary to achieve high gains and efficiencies a large-amplitude signal must be generated. At the present time, it is still an open question as to which is the most advantageous configuration for this type of alternative wiggler design. In this chapter, therefore, we shall analyse a relatively simple configuration which consists of a uniform circularly polarized electromagnetic wave propagating antiparallel to the electron beam. Three basic issues will be addressed. First, the single-particle trajectories will be treated for a system which includes an axial magnetic field [8]. This is to provide for enhanced confinement of the electron beam. Note in this regard, that the typical helical magnetostatic wiggler will act to confine the electron beam against defocusing due to self-field effects through the transverse gradients of the field. In addition, planar wigglers can also be designed with parabolic pole faces for enhanced confinement. However, it appears to be difficult to tailor a large-amplitude electromagnetic wave for such a purpose, and some additional means to ensure beam confinement will be necessary. Second, the small-signal gain for such a combined electromagnetic-wave wiggler/axial magnetic field configuration will be addressed [13]. Finally, while tapered magnetostatic wigglers are relatively easy to construct, tapered electromagnetic-wave wigglers present technical difficulties in design due to problems in mode control. For example, the coupling coefficient between modes in a tapered waveguide depends upon the slope of the taper. Difficulties may ensue, therefore, if the slope of the taper is comparable to the coupling coefficient mediating the free-electron laser interaction. Hence, an alternative efficiency-enhancement scheme for electromagnetic-wave wiggler configurations is described which employs a tapered axial magnetic field [14].

10.1 SINGLE-PARTICLE TRAJECTORIES

The specific configuration to be investigated consists in the propagation of an electron beam counter to that of a circularly polarized electromagnetic wave. As such, the electromagnetic wave acts as a wiggler which induces an undulatory motion on the beam. In order to study the electron trajectories under the combined influence of an axial solenoidal magnetic field and a large-amplitude circularly polarized electromagnetic wave, we assume that the electromagnetic wave is approximately uniform in the transverse direction and write the electric and magnetic fields of the wave as

$$\boldsymbol{B}_w = B_w[\hat{\boldsymbol{e}}_x \cos(k_w z + \omega_w t) + \hat{\boldsymbol{e}}_y \sin(k_w z + \omega_w t)], \tag{10.1}$$

and

$$\boldsymbol{E}_w = -\frac{\omega_w}{ck_w} B_w[\hat{\boldsymbol{e}}_x \sin(k_w z + \omega_w t) - \hat{\boldsymbol{e}}_y \cos(k_w z + \omega_w t)], \tag{10.2}$$

where the subscript 'w' is used throughout to denote quantities associated with the electromagnetic-wave wiggler, B_w denotes the amplitude of the magnetic field, and ω_w and k_w denote the angular frequency and wavenumber of the wiggler. Observe that the Poynting vector for these fields is

$$\boldsymbol{S}_w = -\frac{1}{4\pi}\frac{\omega_w}{ck_w} B_w^2 \hat{\boldsymbol{e}}_z, \tag{10.3}$$

which demonstrates that the electromagnetic-wave wiggler describes a backwards-propagating wave for $\omega_w > 0$ and $k_w > 0$. It should be remarked that a magnetostatic wiggler is recovered in the limit in which ω_w vanishes.

The orbit equations for an electron in the combined fields are given by

$$\frac{d}{dt}\mathbf{v} = -\frac{e}{\gamma m_e}\left[\left(\boldsymbol{I} - \frac{1}{c^2}\mathbf{vv}\right)\cdot \boldsymbol{E}_w + \frac{1}{c}\mathbf{v} \times (B_0 \hat{\boldsymbol{e}}_z + \boldsymbol{B}_w)\right], \tag{10.4}$$

and

$$\frac{d}{dt}\gamma = -\frac{e}{m_e c^2}\mathbf{v}\cdot \boldsymbol{E}_w, \tag{10.5}$$

where $\boldsymbol{I}$ is the unit dyadic. For convenience, we transform to the rotating wiggler-frame

$$\hat{\boldsymbol{e}}_1 \equiv \hat{\boldsymbol{e}}_x \cos(k_w z + \omega_w t) + \hat{\boldsymbol{e}}_y \sin(k_w z + \omega_w t), \tag{10.6}$$

$$\hat{\boldsymbol{e}}_2 \equiv -\hat{\boldsymbol{e}}_x \sin(k_w z + \omega_w t) + \hat{\boldsymbol{e}}_y \cos(k_w z + \omega_w t), \tag{10.7}$$

$$\hat{\boldsymbol{e}}_3 \equiv \hat{\boldsymbol{e}}_z. \tag{10.8}$$

In this frame, the orbit equations take the form

$$\frac{d}{dt}v_1 = -\left[\Omega_0 - k_w(v_3 + v_p) - \Omega_w \frac{v_1 v_p}{c^2}\right]v_2, \tag{10.9}$$

$$\frac{d}{dt}v_2 = [\Omega_0 - k_w(v_3 + v_p)]v_1 - \Omega_w(v_3 + v_p) + \Omega_w \frac{v_2^2 v_p}{c^2}, \tag{10.10}$$

$$\frac{d}{dt}v_3 = \Omega_w\left(1 + \frac{v_3 v_p}{c^2}\right)v_2, \tag{10.11}$$

and

$$\frac{d}{dt}\gamma = -\gamma\frac{v_2 v_p}{c^2}\Omega_w, \tag{10.12}$$

where $\Omega_{0,w} \equiv eB_{0,w}/\gamma m_e c$ and $v_p \equiv \omega_w/k_w$ denotes the phase velocity of the electromagnetic-wave wiggler. Observe that these equations are analogous to equations (2.10–12) for the magnetostatic helical wiggler.

The steady-state orbits for this configuration are found under the requirement that the total energy of the electron remains constant (i.e. $d\gamma/dt = 0$). This implies that $v_2 = 0$, which in turn means that v_1 and v_3 are constant as well. Denoting the constant energy and axial velocity by γ_0 and $v_\parallel$, we find that the constant transverse velocity is given by [8]

$$v_1 = v_w \equiv \frac{\Omega_w(v_\parallel + v_p)}{\Omega_0 - (\omega_w + k_w v_\parallel)}, \tag{10.13}$$

where now $\Omega_{0,w} \equiv eB_{0,w}/\gamma_0 m_e c$. The assumption of a electromagnetic-wave wiggler which is uniform in the transverse direction requires that $v_w \ll v_\parallel$. Since the energy is a constant for the steady-state trajectories v_w and $v_\parallel$ are related through

$$v_\parallel^2 + v_w^2 = (1 - \gamma_0^{-2})c^2. \tag{10.14}$$

The dispersion relation between ω_w and k_w is determined in a self-consistent fashion by the dielectric properties of the medium. In this case, we assume that the beam is cold and uniform in the transverse direction. Hence, the dispersion equation which relates the frequency and wavenumber is

$$\omega_w^2 - c^2 k_w^2 - \frac{\omega_b^2(\omega_w + k_w v_\parallel)}{\gamma_0(\omega_w - \Omega_0 + k_w v_\parallel)} = 0. \tag{10.15}$$

Equations (10.13–15) are sufficient to determine v_w, $v_\parallel$ and k_w for fixed values of B_0, B_w, γ_0 and ω_w.

The orbital stability of these steady-state trajectories is determined by a straightforward perturbation analysis. We write $v_1 = v_w + \delta v_1, v_2 = \delta v_2, v_3 = v_\parallel + \delta v_3$, and $\gamma = \gamma_0 + \delta\gamma$ and find that the orbit equations are

$$\frac{d}{dt}\delta v_1 = -\left[\Omega_0 - k_w(v_\parallel + v_p) - \Omega_w\frac{v_w v_p}{c^2}\right]\delta v_2, \tag{10.16}$$

$$\frac{d}{dt}\delta v_2 = [\Omega_0 - k_w(v_\parallel + v_p)]\,\delta v_1 - (\Omega_w + k_w v_w)\delta v_3 - \frac{v_w}{\gamma_0}k_w(v_\parallel + v_p)\delta\gamma, \tag{10.17}$$

$$\frac{d}{dt}\delta v_3 = \Omega_w\left(1 + \frac{v_\parallel v_p}{c^2}\right)\delta v_2, \tag{10.18}$$

and

$$\frac{d}{dt}\delta\gamma = -\gamma_0\frac{v_p}{c^2}\Omega_w\delta v_2, \tag{10.19}$$

correct to first order in the perturbations. If we take the derivative of equation (10.17) and substitute the values for the derivatives of δv_1, δv_3 and $\delta\gamma$ from equations (10.16), (10.18) and (10.19), then we obtain

$$\left(\frac{d^2}{dt^2} + \Omega_r^2\right)\delta v_2 = 0, \tag{10.20}$$

and

$$\frac{d}{dt}\left(\frac{d^2}{dt^2} + \Omega_r^2\right)\begin{pmatrix}\delta v_1\\ \delta v_3\\ \delta\gamma\end{pmatrix} = 0, \tag{10.21}$$

where

$$\Omega_r^2 \equiv [\Omega_0 - (\omega_w + k_w v_\parallel)]\left\{\Omega_0\left[1 + \frac{v_w^2}{c^2}\frac{c^2 k_w^2 - \omega_w^2}{(\omega_w + k_w v_\parallel)^2}\right] - (\omega_w + k_w v_\parallel)\right\}. \tag{10.22}$$

Orbital instability occurs whenever $\Omega_r^2 < 0$.

Solution of equations (10.13–15) must take cognizance of the properties of the dispersion equation (10.15) which exhibits three distinct branches. A schematic illustration of these branches is shown in Fig. 10.1. Two branches of interest exist corresponding to backwards-propagating waves in the first quadrant. These are (1) an electromagnetic electron cyclotron wave which is found over a restricted range of wavenumbers given approximately $0 < k_w < \Omega_0/v_\parallel$, and (2) the electromagnetic escape mode which is found for frequencies above the cutoff frequency

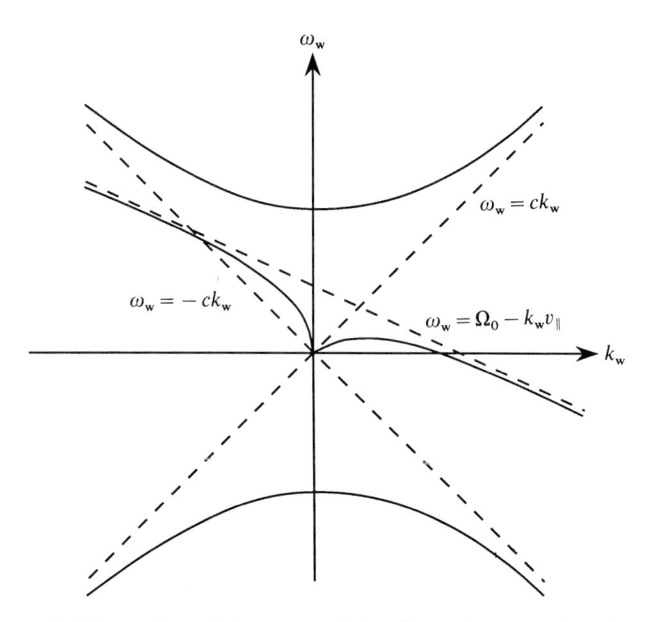

Fig. 10.1 Schematic illustration of the roots of the dispersion equations for the electromagnetic wave.

$\omega_w \geqslant \omega_{co}$, where

$$\omega_{co} \equiv \frac{\Omega_0}{2}\left[1 + \sqrt{\left(1 + \frac{4\omega_b^2}{\gamma_0\Omega_0^2}\right)}\right]. \tag{10.23}$$

The character of the steady-state orbits is dependent upon the choice of wave mode. As in the case of the magnetostatic helical wiggler, two classes of orbit are found. The generalization of Group I trajectories occurs when $\Omega_0 < \omega_w + k_w v_\parallel$ and corresponds to waves on the electromagnetic escape branch. In contrast to the Group I trajectories in a magnetostatic wiggler, however, these orbits are stable. In order to understand this, observe that the orbital stability criterion for these trajectories $[\Omega_r^2 > 0]$ requires that

$$\Omega_0\left(1 + \frac{v_w^2}{c^2}\frac{c^2k_w^2 - \omega_w^2}{(\omega_w + k_w v_\parallel)^2}\right) < (\omega_w + k_w v_\parallel). \tag{10.24}$$

This condition is trivially satisfied since waves on the escape branch are supraluminous (i.e. $\omega_w > ck_w$). The generalization of the Group II trajectories of the magnetostatic wiggler occurs when $\Omega_0 > \omega_w + k_w v_\parallel$, and corresponds to waves on the electromagnetic electron cyclotron branch. In this case, the stability criterion is

$$\Omega_0\left(1 + \frac{v_w^2}{c^2}\frac{c^2k_w^2 - \omega_w^2}{(\omega_w + k_w v_\parallel)^2}\right) > (\omega_w + k_w v_\parallel). \tag{10.25}$$

These orbits are stable as well since the waves on this branch are subluminous. Hence, in contrast to the case of a helical magnetostatic wiggler, all orbits in a circularly polarized electromagnetic-wave wiggler are stable. An example of the

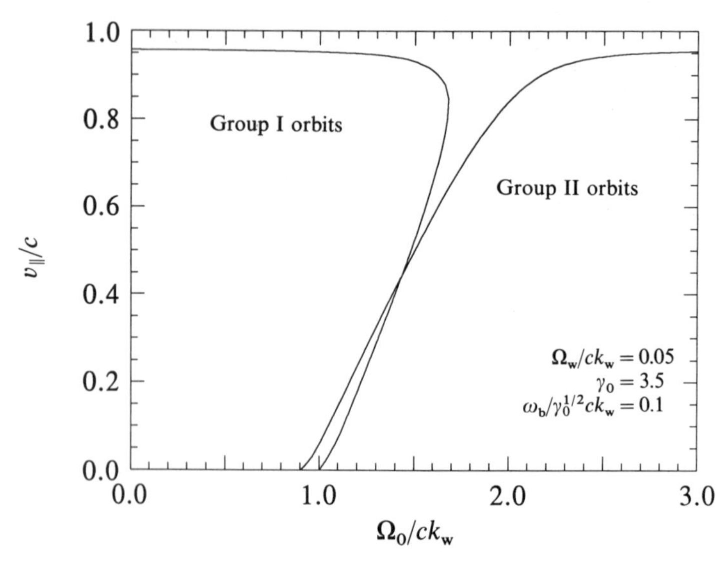

Fig. 10.2 The Group I and Group II trajectories found in the electromagnetic-wave wiggler.

variation of the axial velocity of each orbit group with the axial magnetic field is shown in Fig. 10.2 for $\gamma_0 = 3.5$, $\Omega_w/ck_w = 0.05$, and $\omega_b/\gamma_0^{1/2}ck_w = 0.1$. The Group I trajectories for these parameters represent a multi-valued function of the axial magnetic field and occur for $\Omega_0/ck_w \leqslant 1.68$. In contrast, the Group II trajectories are single-valued and occur for $\Omega_0/ck_w \geqslant 0.94$.

10.2 THE SMALL-SIGNAL GAIN

The small-signal gain is due to the beating of the wiggler and radiation fields and is governed by a nonlinear pendulum equation. In the present case, the interaction gives rise to the stimulated scattering of a right-hand circularly polarized wave propagating parallel to the electron beam. Within the context of the idealized one-dimensional model, the vector potential of the scattered wave of angular frequency ω and wavenumber k takes the form

$$\delta \boldsymbol{A}(z,t) = \delta A[\hat{\boldsymbol{e}}_x \cos(kz - \omega t) - \hat{\boldsymbol{e}}_y \sin(kz - \omega t)], \tag{10.26}$$

where the amplitude and wavenumber are assumed to be slowly varying functions of axial position. In the rotating wiggler frame, the electric and magnetic fields corresponding to this vector potential are

$$\delta \boldsymbol{E}(z,t) = -\frac{\omega}{c}\delta A[\hat{\boldsymbol{e}}_1 \sin\psi(z,t) + \hat{\boldsymbol{e}}_2 \cos\psi(z,t)], \tag{10.27}$$

and

$$\delta \boldsymbol{B}(z,t) = k\delta A[\hat{\boldsymbol{e}}_1 \cos\psi(z,t) - \hat{\boldsymbol{e}}_2 \sin\psi(z,t)], \tag{10.28}$$

where

$$\psi(z,t) \equiv (k + k_w)z - (\omega - \omega_w)t \tag{10.29}$$

denotes the generalized ponderomotive phase.

The nonlinear pendulum equation is obtained by a perturbation analysis of the orbit equations to first order in the radiation fields about the steady-state trajectories. Hence, writing $v_1 = v_w + \delta v_1, v_2 = \delta v_2, v_3 = v_\parallel + \delta v_3$, and $\gamma = \gamma_0 + \delta\gamma$ we obtain

$$\frac{d}{dt}\delta v_1 = -\left[\Omega_0 - k_w(v_\parallel + v_p) - \Omega_w \frac{v_w v_p}{c^2}\right]\delta v_2 + \frac{e\delta A}{\gamma_0 m_e c}\left[\omega\left(1 - \frac{v_w^2}{c^2}\right) - kv_\parallel\right]\sin\psi, \tag{10.30}$$

$$\frac{d}{dt}\delta v_2 = [\Omega_0 - k_w(v_\parallel + v_p)]\delta v_1 - (\Omega_w + k_w v_w)\delta v_3 - \frac{v_w}{\gamma_0}(\omega_w + k_w v_\parallel)\delta\gamma + \frac{e\delta A}{\gamma_0 m_e c}(\omega - kv_\parallel)\cos\psi, \tag{10.31}$$

$$\frac{d}{dt}\delta v_3 = \Omega_w\left(1 + \frac{v_\parallel v_p}{c^2}\right)\delta v_2 + \frac{e\delta A}{\gamma_0 m_e c}\frac{v_w^2}{c^2}\omega \sin\psi, \tag{10.32}$$

and

$$\frac{d}{dt}\delta\gamma = -\gamma_0\frac{v_p}{c^2}\Omega_w\delta v_2 + \frac{e\delta A}{m_e c}\frac{v_w^2}{c^2}\omega\sin\psi. \tag{10.33}$$

Equations (10.30–3) are a straightforward generalization of equations (10.16–19). Differentiating equation (10.31) with respect to time, we find that

$$\left(\frac{d^2}{dt^2} + \Omega_r^2\right)\delta v_2 \cong -\frac{e\delta A}{\gamma_0 m_e c}\sin\psi\left[(\omega - kv_\parallel)(\omega_w - \Omega_0 + k_w v) + \frac{v_w^2}{c^2}\frac{\Omega_0(\omega\omega_w + c^2 kk_w)}{\omega_w + k_w v_\parallel}\right], \tag{10.34}$$

where terms in $d\psi/dt$ have been neglected. Hence,

$$\frac{d}{dt}\left(\frac{d^2}{dt^2} + \Omega_r^2\right)\delta v_3 \cong -\frac{e\delta A}{m_e c}\frac{v_w}{v_\parallel}\frac{\omega - \omega_w}{\gamma_0\gamma_\parallel^2}\sin\psi \times\left[\Omega_r^2 + \gamma_\parallel^2\Omega_0\left(\frac{v_w}{v_p + v_\parallel}\right)^2\left(1 + \frac{v_\parallel v_p}{c^2}\right)^2(\omega_w - \Omega_0 + k_w v_\parallel)\right], \tag{10.35}$$

where $\gamma_\parallel^2 \equiv (1 - v_\parallel^2/c^2)^{-1}$. Under the assumption that the phase and the perturbed velocity vary more slowly than Ω_r, we obtain the pendulum equation

$$\frac{d^2}{dz^2}\psi \cong \delta a\frac{v_w}{v_\parallel}\frac{(\omega - \omega_w)^2 c}{\gamma_0\gamma_\parallel^2 v_\parallel^3}\Phi_{em}\sin\psi, \tag{10.36}$$

where $\delta a \equiv e\delta A/m_e c^2$,

$$\Phi_{em} \equiv 1 - \left(1 + \frac{v_\parallel v_p}{c^2}\right)^2\frac{\gamma_\parallel^2\left(\dfrac{v_w}{v_p + v_\parallel}\right)^2\Omega_0}{\left[1 + \left(\dfrac{v_w}{v_p + v_\parallel}\right)^2(1 - v_p^2/c^2)\right]\Omega_0 - (\omega_w + k_w v_\parallel)}, \tag{10.37}$$

and we have made use of the relation

$$\frac{d}{dt}\delta v_3 \cong \frac{v_\parallel^3}{\omega - \omega_w}\frac{d^2}{dz^2}\psi. \tag{10.38}$$

Observe that Φ_{em} is the counterpart of Φ (equation (2.24)) for the magnetostatic helical wiggler, and reduces to this function in the limit in which ω_w vanishes. Equation (10.36) governs the axial bunching of the electron beam in the presence of the wiggler and radiation fields.

The behaviour of Φ_{em} as a function of the axial magnetic field is shown in Fig. 10.3 for parameters corresponding to the Group I and II orbits in Fig. 10.2. The essential character of this function is similar to that of Φ for the magnetostatic wiggler (see Fig. 2.3). The principal differences arise due to the facts that (1) the Group II orbits do not extend to the $\Omega_0 = 0$ limit, and (2) there is no orbital

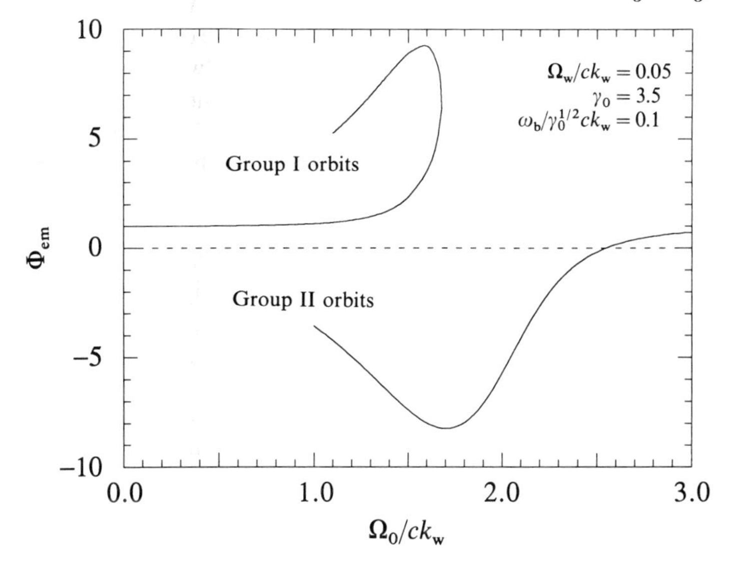

Fig. 10.3 Plot of Φ_{em} as a function of the axial magnetic field.

instability for the electromagnetic-wave wiggler. The latter implies that, although Φ_{em} can become large, it displays no singularity.

The small-signal gain is determined from Maxwell's equations in the idealized one-dimensional limit. The source current is assumed to represent an ideal monoenergetic electron beam, and can be expressed as

$$\delta \boldsymbol{J}(z,t) = -en_b v_{z0} \int_{-\infty}^{\infty} dt_0 \sigma(t_0) \frac{\boldsymbol{p}(t,t_0)}{p_z(t,t_0)} \delta[t - \tau(z,t_0)], \tag{10.39}$$

where v_{z0} is the initial axial electron velocity, n_b is the ambient beam density, $\sigma(t_0)$ is the distribution in entry times (t_0) at which the electrons cross the $z=0$ plane, $\boldsymbol{p}(t,t_0)$ is the electron momentum at time t for a particle which crossed the $z=0$ plane at time t_0, and $\tau(z,t_0)$ is the Lagrangian time coordinate defined in equation (4.27). Substitution of the representations of the source current and vector potential into equation (4.28) yields two coupled equations for the evolution of the amplitude and wavenumber

$$\left[\frac{d^2}{dz^2} + \left(\frac{\omega^2}{c^2} - k^2\right)\right]\delta a \cong \frac{\omega_b^2}{c^2}\frac{v_w}{c}\int_{-\infty}^{\infty} dt_0 \sigma(t_0) \cos\psi\, \delta[t - \tau(z,t_0)], \tag{10.40}$$

and

$$2k^{1/2}\frac{d}{dz}(k^{1/2}\delta a) \cong -\frac{\omega_b^2}{c^2}\frac{v_w}{c}\int_{-\infty}^{\infty} dt_0 \sigma(t_0) \sin\psi\, \delta[t - \tau(z,t_0)], \tag{10.41}$$

where ω_b denotes the beam plasma frequency.

Under the quasi-static assumption for a large-amplitude electromagnetic-wave wiggler, electrons which enter the interaction region within an integral number of beat-wave periods $T(\equiv 2\pi/(\omega - \omega_w))$ execute identical trajectories. Hence, both $\tau(z, t_0) = \tau(z, t_0 + 2\pi N/T)$ and $\psi[z, \tau(z, t_0)] = \psi[z, \tau(z, t_0 + 2\pi N/T)]$ for integer N. Making use of this symmetry, we find that upon averaging equations (10.40) and (10.41) over a beat-wave period

$$\left[\frac{d^2}{dz^2} + \left(\frac{\omega^2}{c^2} - k^2\right)\right]\delta a \cong \frac{\omega_b^2}{c^2}\frac{v_w}{c}\langle\cos\psi\rangle, \tag{10.42}$$

and

$$2k^{1/2}\frac{d}{dz}(k^{1/2}\delta a) \cong -\frac{\omega_b^2}{c^2}\frac{v_w}{c}\langle\sin\psi\rangle, \tag{10.43}$$

where the average

$$\langle(\ldots)\rangle \equiv \frac{1}{2\pi}\int_0^{2\pi} d\psi_0\sigma(\psi_0)(\ldots) \tag{10.44}$$

is over the initial phase $\psi_0 \equiv -\omega t_0$. The dynamical equations (10.42) and (10.43) simplify under the assumptions (1) that the wave frequency is much greater than the beam plasma frequency, and (2) that second-order derivatives of the amplitude and phase may be neglected. In this case, we obtain $\omega \approx ck$ and

$$2k\frac{d}{dz}\delta a \cong -\frac{\omega_b^2}{c^2}\frac{v_w}{c}\langle\sin\psi\rangle. \tag{10.45}$$

The gain in power over a length L is expressed as

$$G(L) \equiv 2\frac{\delta A(z=L) - \delta A(z=0)}{\delta A(z=0)}, \tag{10.46}$$

and can be obtained by integration of equation (10.45). Under the assumption of low gain (i.e. $G(L) \ll 1$) we obtain

$$G(L) \cong -\frac{\omega_b^2}{kc^2}\frac{v_w}{c}\frac{1}{\delta a(z=0)}\int_0^L dz\langle\sin\psi\rangle. \tag{10.47}$$

The small-signal gain is obtained by integration of equation (10.47) under the assumption that the electrons are on untrapped trajectories. This has been discussed within the context of the low-gain limit for a realizable helical magnetostatic wiggler in Chapter 4. The untrapped trajectories in the electromagnetic-wave wiggler are of the form $\psi = \psi_0 + \Delta kz + \delta\psi$, where

$$\Delta k \equiv k + k_w - \frac{\omega - \omega_w}{v_\parallel} \tag{10.48}$$

and the perturbed phase satisfies

$$\frac{d^2}{dz^2}\delta\psi \cong K^2\sin(\psi_0 + \Delta kz), \tag{10.49}$$

where

$$K^2 \equiv \frac{c(\omega - \omega_w)^2}{\gamma_0\gamma_\parallel^2 v_\parallel^3}\frac{v_w}{v_\parallel}\delta a\Phi_{em}. \tag{10.50}$$

Therefore, the solution for the phase subject to the initial conditions $\delta\psi(z=0)=0$ and $d\delta\psi(z=0)/dz=0$ is

$$\psi \cong \psi_0 + \Delta kz + \frac{K^2}{\Delta k^2}[\sin\psi_0 - \sin(\psi_0 + \Delta kz) + \Delta kz\cos\psi_0], \tag{10.51}$$

which, in turn, implies that

$$\langle \sin\psi \rangle \cong -\frac{K^2}{2\Delta k^2}(\sin\Delta kz - \Delta kz\cos\Delta kz). \tag{10.52}$$

As a result, the small-signal gain takes on a form [13] which is analogous to that found for the magnetostatic wiggler (4.59)

$$G(L) \cong -\frac{\omega_b^2 L^3 k v_w^2}{8\gamma_0\gamma_\parallel^2 v_\parallel^3 v_\parallel^2}\left(\frac{\omega-\omega_s}{\omega}\right)^2 \Phi_{em}F(\Theta_{em}), \tag{10.53}$$

where $\Theta_{em} \equiv \Delta kL/2$ and F is the spectral function defined in equation (4.60). It is a straightforward matter to demonstrate that this expression for the small-signal gain reduces to the result for a helical magnetostatic wiggler in the limit in which the ω_w vanishes.

The extrema for the gain occur for $\Theta_{em} \approx \pm 1.3$ at which $F(\Theta_{em}) \approx \mp 0.54$, which corresponds to the frequencies

$$\omega \cong \left(1 - \frac{v_\parallel}{c}\right)^{-1}\left[\omega_w + k_w v_\parallel\left(1 \mp \frac{2.6}{k_w L}\right)\right]. \tag{10.54}$$

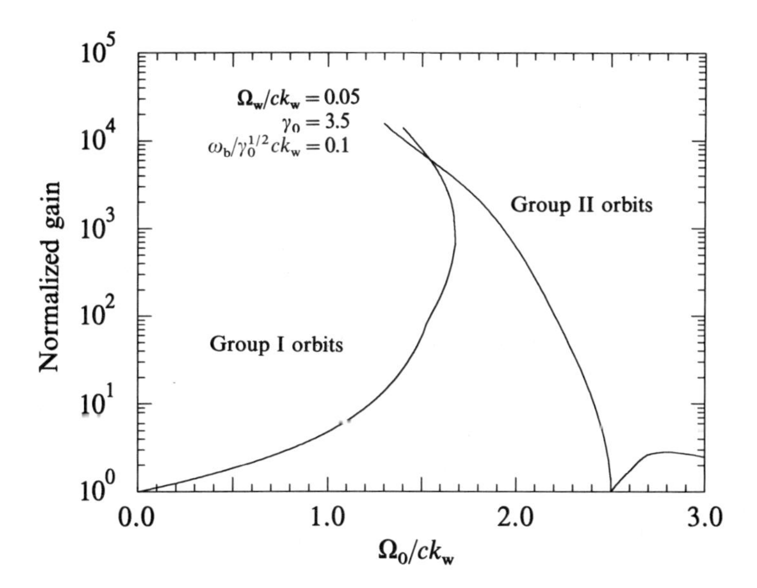

Fig. 10.4 Graph of the normalized value of the peak gain for Group I and Group II orbits corresponding to the orbits illustrated in Fig. 10.2.

Note that for $\Phi_{em} > 0(<0)$ the maximum gain is found to correspond to $\Theta_{em} \approx 1.3(-1.3)$; consequently, the effect of the axial magnetic field is to cause a relative phase shift between the Group I and II classes of trajectory. In either case, however, the maximum gain is given approximately by

$$G_{max}(L) \cong 0.034\left(1+\frac{v_{\parallel}}{c}\right)\frac{\omega_b^2}{\gamma_0 c^2 k_w^2 v_{\parallel}^2}(k_w L)^3 \frac{(\omega_w + ck_w)^2}{ck_w(\omega_w + k_w v_{\parallel})}|\Phi_{em}|. \quad (10.55)$$

The resonant enhancement in both v_w and Φ_{em} due to the axial magnetic field can result in a substantial enhancement in the gain. As an example, observe that $v_w/c \approx 0.052$ and $\Phi_{em} = 1.0$ in the absence of an axial magnetic field for the parameters shown in Figs 10.2 and 10.3. This results in a maximum gain of $G_{max}(L) \approx 3.69 \times 10^{-6}(k_w L)^3$. In contrast, $v_w/c \approx 0.53$ and $\Phi_{em} \approx 1.68$ when $\Omega_0/ck_w = 1.68$ near resonance. As a consequence, the maximum gain is enhanced by several orders of magnitude and we find that $G_{max}(L) \approx 2.48 \times 10^{-3}(k_w L)^3$. A more detailed variation of the maximum gain as a function of the axial magnetic field is shown in Fig. 10.4 in which the maximum gain (normalized to the value of $G_{max}(L)$ for $B_0 = 0$) is plotted versus Ω_0/ck_w.

10.3 EFFICIENCY ENHANCEMENT

It has been noted that there are practical difficulties in the control of the tapering of an electromagnetic-wave wiggler for the purpose of efficiency enhancement. As a result, it may be advantageous to adopt an alternate scheme of efficiency enhancement through the tapering of the axial magnetic field. This technique has been discussed for magnetostatic wigglers in Chapter 5, in which it was shown that a tapered axial field is equivalent to a tapered wiggler for the purpose of efficiency enhancement. In order to formulate the problem for an electromagnetic-wave wiggler, it is assumed that the axial guide field is uniform for $z \leqslant z_0$ and displays a linear taper thereafter. Hence, we write [14]

$$B_0(z) = \begin{cases} B_0; & z \leqslant z_0 \\ B_0[1 + \kappa_0(z - z_0)]; & z > z_0, \end{cases} \quad (10.56)$$

where B_0 is the amplitude of the axial field in the uniform-field region and $\kappa_0(\equiv B_0^{-1} dB_0(z)/dz)$ represents the normalized scale length for variation of the axial field.

The response of the electron beam to the tapered axial field can be determined by a first-order perturbation analysis about the steady-state trajectories in κ_0. The equations for the perturbations in δv_1, δv_3 and $\delta\gamma$ that result are unchanged and are given by equations (10.30), (10.32) and (10.33). The effect of the tapered axial field is upon the equation for δv_2, which becomes

$$\frac{d}{dt}\delta v_2 = [\Omega_0 - (\omega_w + k_w v_{\parallel})]\delta v_1 - (\Omega_w + k_w v_w)\delta v_3 - \frac{v_w}{\gamma_0}(\omega_w + k_w v_{\parallel})\delta\gamma$$
$$+ v_w \Omega_0 \kappa_0 (z - z_0) + \frac{e\delta A}{\gamma_0 m_e c}(\omega - k v_{\parallel})\cos\psi. \quad (10.57)$$

As a result of this, the nonlinear pendulum equation in the presence of the tapered axial magnetic field is

$$\frac{d^2}{dz^2}\psi \cong \delta a \frac{v_w}{v_\parallel} \frac{(\omega - \omega_w)^2 c}{\gamma_0 \gamma_\parallel^2 v_\parallel^3} \Phi_{em}(\sin\psi - \sin\psi_{res}), \tag{10.58}$$

where

$$\sin\psi_{res} \equiv \frac{\kappa_0}{k + k_w} \frac{\gamma_0 \beta_\parallel}{\delta a} \frac{v_\parallel + v_p}{v_w} \frac{\Phi_{em} - 1}{\Phi_{em}} \left(1 + \frac{v_p v_\parallel}{c^2}\right)^{-1}, \tag{10.59}$$

and $\beta_\parallel \equiv v_\parallel / c$. Equation (10.56) describes the trapping of electrons in the ponderomotive potential formed by the electromagnetic-wave wiggler and the radiation field, and the term in $\sin\psi_{res}$ describes the bulk acceleration or deceleration due to the tapered axial magnetic field.

In order to determine the effect of the tapered field upon the efficiency of the interaction, the gain must be calculated using equation (10.47) under the assumption that the taper begins at the point at which the bulk of the electron beam has become trapped. Hence, the phase average is performed for $\psi \approx \psi_{res}$, and the enhancement in the gain over an additional length L becomes

$$\Delta G(L) = 2\frac{\delta A(z = z_0 + L) - \delta A(z = z_0)}{\delta A(z = z_0)}$$

$$\cong \kappa_0 L \frac{\omega_b^2}{c^2 k_w^2} \frac{\gamma_0 \beta_\parallel^2}{\delta a^2(z_0)} \frac{k_w(\omega_w + k_w v_\parallel)}{k(\omega - \omega_w)} \frac{1 - \Phi_{em}}{\Phi_{em}} \left(1 + \frac{v_p v_\parallel}{c^2}\right)^{-1}. \tag{10.60}$$

The efficiency enhancement associated with this enhancement in the gain is calculated from the ratio of the increase in the Poynting flux to the electron beam power flux. Since the Poynting flux increases by an amount $\Delta P = (\omega k / 2\pi) \Delta G(L) \delta A^2(z_0)$, the efficiency enhancement is [14]

$$\Delta\eta(L) \cong \kappa_0 L \beta_\parallel \frac{k(\omega_w + k_w v_\parallel)}{k_w(\omega - \omega_w)} \frac{1 - \Phi_{em}}{\Phi_{em}} \left(1 + \frac{v_p v_\parallel}{c^2}\right)^{-1}. \tag{10.61}$$

In order to estimate the efficiency enhancements which are possible by this mechanism, we consider parameters consistent with those discussed previously for the uniform field. Specifically, $\gamma_0 = 3.5$, $\Omega_w / ck_w = 0.05$, $\omega_b / \gamma_0^{1/2} ck_w = 0.1$, and we assume that $\lambda_w = 1$ cm. We first consider Group I orbits and assume that $\Omega_0 / ck_w = 0.5$, which corresponds to an axial magnetic field of 18.7 kG. For this choice of parameters, the orbit parameters are: $v_\parallel / c \approx 0.956$, $v_w / c \approx 0.067$ and $\Phi_{em} \approx 1.018$. In addition, the frequency of the electromagnetic-wave wiggler is 30.2 GHz and the maximum gain for the interaction occurs at a resonant frequency $\omega / ck_w \approx 44.6$ and a wavelength of 220 μm. As a consequence, the efficiency enhancement per unit length $\Delta\eta(L)/L \approx -0.035\,\kappa_0$. Thus, if the axial magnetic field is tapered downward to zero, then it is possible to obtain an efficiency enhancement of 3.5% of the initial electron beam power. As the magnetic resonance is approached more closely, the efficiency enhancement can be increased. To see this, note that if we double the axial magnetic field to 37.5 kG, then $\omega / ck_w \approx 41.7$

(for $\lambda \approx 240\,\mu$m), $v_{\parallel}/c \approx 0.953$ and $\Phi_{em} \approx 1.117$. Hence, the efficiency enhancement per unit length increases to $\Delta\eta(L)/L \approx -0.10\,\kappa_0$ and an efficiency enhancement of the order of 10% is possible if the field is tapered to zero. This efficiency enhancement can be increased still further as the axial field approaches the magnetic resonance more closely; however, the resonant wavelength of the interaction will also increase. In short, quite appreciable enhancements in the interaction efficiency for the electromagnetic-wave wiggler are possible by means of a tapered axial magnetic field.

REFERENCES

1. Granatstein, V. L., Sprangle, P., Parker, R. K., Pasour, J. A., Herndon, M. and Schlesinger, S. P. (1976) Realization of a relativistic mirror: electromagnetic backscattering from the front of a magnetized relativistic electron beam. *Phys. Rev. A*, **14**, 1194.
2. Granatstein, V. L., Schlesinger, S. P., Herndon, M., Parker, R. K. and Pasour, J. A. (1977) Production of megawatt submillimeter pulses by stimulated magneto-Raman scattering. *Appl. Phys. Lett.*, **30**, 384.
3. Sprangle, P., Granatstein, V. L. and Baker, L. (1975) Stimulated scattering from a magnetized relativistic electron beam. *Phys. Rev. A*, **12**, 1697.
4. Sprangle, P. and Drobot, A. T. (1979) Stimulated backscattering from relativistic unmagnetized electron beams. *J. Appl. Phys.*, **50**, 2652.
5. Lin, A. T. and Dawson, J. M. (1980) Nonlinear saturation and thermal effects on the free-electron laser using an electromagnetic pump. *Phys. Fluids*, **23**, 1224.
6. Hiddleston, H. R. and Segall, S. B. (1981) Equations of motion of a free-electron laser with an electromagnetic pump field and an axial electrostatic field. *IEEE J. Quantum Electron.*, **QE-17**, 1488.
7. Hiddleston, H. R., Segall, S. B. and Catella, G. C. (1982) Gain-enhanced free-electron laser with an electromagnetic pump field, in *Physics of Quantum Electronics: Free-Electron Generators of Coherent Radiation*, Vol. 9 (ed. Jacobs, S. F., Moore, G. T., Pilloff, H. S., Sargent, M., Scully, M. O. and Spitzer, R.), Addison-Wesley, Reading, Massachusetts, p. 849.
8. Freund, H. P., Kehs, R. A. and Granatstein, V. L. (1985) Electron orbits in combined electromagnetic wiggler and axial guide magnetic fields. *IEEE J. Quantum Electron.*, **QE-21**, 1080.
9. Goldring, A. and Friedland, L. (1985) Electromagnetically pumped free-electron laser with a guide magnetic field. *Phys. Rev. A*, **32**, 2879.
10. Pasour, J. A., Sprangle, P., Tang, C. M. and Kapetanakos, C. A. (1985) High-power two-stage free-electron laser oscillator operating in the trapped particle mode. *Nucl. Instr. Meth.*, **A237**, 154.
11. Segall, S. B., Curtin, M. S. and Von Laven, S. A. (1986) Key issues in the design of a two-stage free-electron laser. *Nucl. Instr. Meth.*, **A250**, 316.
12. Kimel, I., Elias, L. R. and Ramian, G. (1986) The University of California at Santa Barbara two-stage free-electron laser. *Nucl. Instr, Meth.*, **A250**, 320.
13. Freund, H. P., Kehs, R. A. and Granatstein, V. L. (1986) Linear gain of a free-electron laser with an electromagnetic-wave wiggler and an axial guide magnetic field. *Phys. Rev. A.*, **34**, 2007.
14. Freund, H. P. (1987) Efficiency enhancement in free-electron lasers driven by electromagnetic-wave wigglers. *IEEE J. Quantum Electron.*, **QE-23**, 1590.
15. Danly, B. G., Bekefi, G., Davidson, R. C., Temkin, R. J., Tran, T. M. and Wurtele, J. S. (1987) Principles of gyrotron powered electromagnetic wigglers for free-electron lasers. *IEEE J. Quantum Electron.*, **QE-23**, 103.

16. Gea-Banacloche, J., Moore, G. T., Schlicher, R. R., Scully, M. O. and Walther, H. (1987) Soft X-ray free-electron laser with a laser undulator. *IEEE J. Quantum Electron.*, **QE-23**, 1558.
17. Tran, T. M., Danly, B. G. and Wurtele, J. S. (1987) Free-electron lasers with electromagnetic standing-wave wigglers. *IEEE J. Quantum Electron.*, **QE-23**, 1578.
18. Sharma, A. and Tripathi, V. (1988) A whistler pumped free-electron laser. *Phys. Fluids*, **31**, 3375.
19. Hofland, R. and Pridmore-Brown, D. C. (1989) Optically-pumped free-electron laser with electrostatic reaccelerations. *Nucl. Instr. Meth.*, **A285**, 276.

11

Chaos in free-electron lasers

The past two decades have seen the emergence of a new field of science known as chaos or nonlinear dynamics. Researchers in this field are concerned with the basic properties of the solutions of systems of nonlinear equations. This interest stems from the fact that almost every physical system can be described at some level of approximation by a system of nonlinear equations. While scientists working in different fields have been aware for a long time of the ubiquity of nonlinear effects, it has been only recently that this study has become a scientific field of its own. The development of this field has led to several general conclusions about nonlinear systems. On the one hand, even the simplest deterministic nonlinear systems can exhibit behaviour which is complicated and appears to be random. This behaviour has been termed chaos. On the other hand, the chaotic behaviour of much more complicated systems often seems to follow the same rules as that of the simple systems. Thus, there is order in the chaos.

Before going into a description of chaos and nonlinear dynamics as it applies to free-electron lasers, it is useful to introduce some basic concepts from that field. The discussion given here will be elementary and qualitative. A reader who is interested in the detailed points of nonlinear dynamics should read one of the several books that have been written on the subject [1–3]. First, what is chaos? A system which is chaotic exhibits extreme sensitivity to initial conditions. Small differences in the initial conditions for the system ultimately lead to large differences in the system's variables. For a chaotic system, the rate at which small differences become large differences is exponential with time. This implies that precise prediction of the future values of the system variables after a certain time is almost impossible. A small error in the initial conditions will ultimately lead to a large error in the predicted value of a system variable, and increasing the accuracy of the initial conditions by an order of magnitude results in only a modest increase in the length of time over which the solution is accurate. In chaotic systems, the sensitivity to initial conditions is measured by the Lyapunov exponent, which is the time-average rate of exponential divergence of two solutions of the system equations whose initial conditions differ by an infinitesimal amount.

There is an important distinction between the phase-space trajectories of a deterministic system which is chaotic and trajectories of a system which is driven externally by random noise. For deterministic systems, the trajectories may still be restricted to certain regions of phase space. For example, the motion of a particle in a two-dimensional time-independent potential is described as a trajectory in a four-dimensional phase space. Since the Hamiltonian is conserved, for a given initial energy the particle trajectory is constrained to a three-dimensional volume in the four-dimensional phase space. The nature of the motion in this three-dimensional volume depends on whether or not the trajectory is chaotic. A nonchaotic trajectory is characterized by the existence of another constant of motion which further restricts the trajectory to lie on a two-dimensional surface. For example, if the potential is circularly symmetric, then angular momentum is conserved. Such a trajectory is said to be integrable, as is the region of phase space filled with such trajectories. A chaotic trajectory in this system is not constrained to a surface but will eventually fill up a three-dimensional volume. This volume might be bounded by surfaces on which the trajectories are non-chaotic. Thus, the chaotic trajectory is constrained to stay within a certain sub-region of the available phase space. For the two-degrees-of-freedom Hamiltonian problem under discussion, the two-dimensional surfaces on which nonchaotic trajectories are constrained to lie are known as KAM (Kolmogorov, Arnold and Moser) surfaces [1, 2]. The KAM theorem states that a small perturbation to the system will destroy only a small number of these surfaces. For example, if we start with a circularly symmetric potential, then angular momentum is conserved, and all trajectories lie on two-dimensional surfaces. Suppose we now add a small nonsymmetric perturbation to the potential. Angular momentum is no longer conserved, but this does not mean that every trajectory becomes chaotic. Most trajectories will still be nonchaotic with a constant of motion which differs slightly from the angular momentum. A relatively small fraction of the trajectories will have no such constant, and will chaotically visit a small three-dimensional region of phase space. As the size of the nonsymmetric perturbation is increased, the regions of phase space which are filled by chaotic trajectories increase and, correspondingly, the integrable regions will decrease in size.

A different type of constraint on trajectories in phase space occurs for non-Hamiltonian or dissipative systems. For such systems, the phase-space volume is not conserved. Specifically, if one started with a group of initial conditions lying on a closed surface in phase space, then the volume of phase space inside that surface would not be constant in time as it is for Hamiltonian systems. For a purely dissipative system, the volume inside the surface will progressively shrink with time. As a consequence, there is a tendency for the solutions for different initial conditions to become attracted to the same trajectory. Such a trajectory is known as an **attractor**. A simple example is a damped pendulum. In the absence of any external forcing, all initial conditions ultimately lead to the state where the pendulum is motionless in the bottom of its potential well. The attractor is a point in the two-dimensional phase space of the pendulum. If the pendulum is driven by a weak time-varying sinusoidal force, then the attractor will be a limit cycle corresponding to steady oscillation of the pendulum. In this case, the attractor is one-dimensional. If the amplitude of the driving force is increased, the motion becomes chaotic and nearby trajectories diverge exponentially in time at the same time that phase-space volume contracts. This leads to what are known as **strange attractors** [3], which have fractional dimension. The attracting trajectory of the chaotic system does not fill up all of the available phase space, in fact it consists of a geometrically complex set of points which has measure zero.

In general, physical systems are characterized by a number of parameters. For example, a free-electron laser oscillator is characterized by the electron-beam current, the wiggler field strength, the mirror reflectivity, etc. The behaviour of the system can vary as the parameters of the system are changed. A system may exhibit chaotic behaviour for some values of the parameters but not for others. What is frequently of interest is the way in which the system behaviour changes as a parameter is varied. In particular, how does the transition from nonchaotic to chaotic behaviour occur? This is referred to as the route to chaos, and it is believed that the route to chaos in most systems falls into one of a few different classes.

In the present chapter, we shall discuss these two types of chaos. Specifically, Hamiltonian chaos which arises due to self-electric and self-magnetic fields in the single-particle trajectories of electrons propagating through a helical magnetostatic wiggler, and dissipative chaos in multimode free-electron laser oscillators. The study of chaos in free-electron lasers is relatively new, and these two aspects should not be thought of as the only sources of chaos in free-electron lasers. However, the onset of chaos in both cases acts to degrade the efficiency or spectral purity of the free-electron laser, and it appears likely that the study of chaos in coherent radiation sources is useful, primarily, as a guide to which operating regimes to avoid.

11.1 CHAOS IN SINGLE-PARTICLE ORBITS

One example of chaos which has recently been discussed in the literature appears due to the effect of the self-electric and self-magnetic fields of the electron beam on the dynamics of the single-particle trajectories in free-electron lasers [4–8]. These analyses have treated helical wiggler configurations in the presence of an axial solenoidal magnetic field (see section 2.1), and found that the effect of the self-fields can induce chaos in the single-particle trajectories in the vicinity of the gyroresonance. As we demonstrated in Chapters 4 and 5, the growth rates and saturation efficiencies of the free-electron laser interaction are enhanced in the vicinity of the gyroresonance (where the Larmor period associated with the axial magnetic field is comparable to the wiggler period). Hence, the ultimate impact of the orbital chaos induced by the self-fields can be to degrade the interaction.

However, caution must be used in the interpretation of this conclusion due to practical limitations imposed by the injection of the electron beam into the wiggler. As discussed in section 5.1, it becomes progressively more difficult to inject the electron beam onto well-behaved steady-state trajectories as the gyroresonance is approached (see Figs 5.2 and 5.3). Hence, as shown in Figs 5.7 and 5.8, it is already evident that the gain and efficiency decrease in the vicinity of the gyroresonance. Therefore, the relative impact on the electron orbits of (1) the onset to chaos due to self-fields, and (2) the injection process must be evaluated before a definitive conclusion can be reached on this issue.

The analysis we present here, unlike most treatments of classical chaos, is developed from the particle orbit equations rather than a Hamiltonian formalism. In addition, in the interests of simplicity and clarity, we limit the discussion to the idealized one-dimensional limit. The inclusion of wiggler inhomogeneities in the context of a fully three-dimensional analysis [4–8] requires immediate recourse to numerical methods which obscures the essential physics of the interaction.

11.1.1 The equilibrium configuration

The external magnetostatic fields are given by the idealized wiggler field in the idealized one-dimensional limit (equation (2.5)) in conjunction with an axial solenoidal field; hence

$$\boldsymbol{B}_{ext} = B_0\hat{\boldsymbol{e}}_z + B_w(\hat{\boldsymbol{e}}_x \cos k_w z + \hat{\boldsymbol{e}}_y \sin k_w z), \tag{11.1}$$

which can be represented in terms of a vector potential of the form

$$\boldsymbol{A}_{ext} = B_0 x\hat{\boldsymbol{e}}_y - \frac{B_w}{k_w}(\hat{\boldsymbol{e}}_x \cos k_w z + \hat{\boldsymbol{e}}_y \sin k_w z). \tag{11.2}$$

The steady-state orbits in this idealized limit are of the form

$$\mathbf{v}_0 = v_w(\hat{\boldsymbol{e}}_x \cos k_w z + \hat{\boldsymbol{e}}_y \sin k_w z) + v_\parallel \hat{\boldsymbol{e}}_z, \tag{11.3}$$

where the magnitudes of the transverse and axial velocities are determined by solution of equations (2.15) and (2.16). Recall that the idealized steady-state solutions are valid as long as $|v_w/v_\parallel| \ll 1$. More general solutions can be found in terms of the elliptic functions (see equations (2.31), (2.37) and (2.38)). The steady-state trajectories exhibit a resonant enhancement in the transverse velocity when $\Omega_0 \approx k_w v_\parallel$, and orbital instability is found in the vicinity of the gyroresonance for Group I trajectories in which $\Omega_0 < k_w v_\parallel$.

We are interested in the effect of the self-electric and self-magnetic fields of the electron beam upon these trajectories. We shall assume for this purpose that the electron beam has a flat-top density profile of the form

$$n_b(r) = \begin{cases} n_{b0};\, 0 \leqslant r \leqslant r_b \\ 0; \quad r > r_b, \end{cases} \tag{11.4}$$

as well as a flat-top current profile $\boldsymbol{J}_b(r) = -en_b(r)v_b\hat{\boldsymbol{e}}_z$, where n_{b0} is the uniform ambient beam density, and v_b denotes the bulk axial velocity of the beam. Observe that the axial beam velocity is assumed to be constant, and no attempt is made herein to obtain a self-consistent solution for the axial electron velocity. In addition, the effect of the transverse wiggle-motion of the beam on the self-fields is neglected (as a higher-order effect) in the idealized treatment.

Within the context of this beam configuration, the self-electric field within the beam (i.e. $r \leqslant r_b$) can be shown to be [9]

$$\boldsymbol{E}_s = -\frac{m_e\omega_b^2}{2e}(x\hat{\boldsymbol{e}}_x + y\hat{\boldsymbol{e}}_y), \tag{11.5}$$

where $\omega_b^2 \equiv 4\pi e^2 n_{b0}/m_e$. This is associated with the scalar potential

$$\Phi_s = \frac{m_e \omega_b^2}{4e} r^2. \tag{11.6}$$

The self-magnetic field and associated vector potentials within the beam are

$$\boldsymbol{B}_s = \frac{m_e \omega_b^2}{2e} \beta_b (y\hat{\boldsymbol{e}}_x - x\hat{\boldsymbol{e}}_y), \tag{11.7}$$

and

$$\boldsymbol{A}_s = \frac{m_e \omega_b^2}{4e} \beta_b r^2 \hat{\boldsymbol{e}}_z, \tag{11.8}$$

where $\beta_b \equiv v_b/c$. The question of chaos in the single-particle trajectories is analysed by solution of the orbit equations subject to the external magnetostatic fields and self-fields.

11.1.2 The orbit equations

The orbit equations for an electron in combined idealized helical wiggler and axial guide magnetic fields are given in equations (2.10–12) in the absence of self-electric and self-magnetic fields. The inclusion of self-fields results in equations of the form

$$\frac{d}{dt} p_x = -\Omega_0 p_y + \Omega_w p_z \sin k_w z + m_e \frac{\omega_b^2 x}{2}(1 - \beta_b \beta_z), \tag{11.9}$$

$$\frac{d}{dt} p_y = \Omega_0 p_x - \Omega_w p_z \cos k_w z + m_e \frac{\omega_b^2 y}{2}(1 - \beta_b \beta_z), \tag{11.10}$$

and

$$\frac{d}{dt} p_z = -\Omega_w (p_x \sin k_w z - p_y \cos k_w z) + \frac{\omega_b^2}{2\gamma c} \beta_b (x p_x + y p_y), \tag{11.11}$$

in rectangular coordinates, where $\Omega_{0,w} \equiv eB_{0,w}/\gamma m_e c$ are the relativistic cyclotron frequencies associated with the axial and wiggler fields, $\gamma \equiv (1 + p^2/m_e^2 c^2)^{1/2}$, and $\beta_z \equiv v_z/c$. It is follows from these equations that

$$\Gamma = \gamma - \frac{\omega_b^2 r^2}{4c^2} \tag{11.12}$$

is a constant of the motion which is related to the total energy. In addition, the canonical momenta in the presence of the self-fields are

$$P_x = p_x + m_e \frac{\hat{\Omega}_w}{k_w} \cos k_w z, \tag{11.13}$$

$$P_y = p_y - m_e \hat{\Omega}_0 x + m_e \frac{\hat{\Omega}_w}{k_w} \sin k_w z \tag{11.14}$$

and

$$P_z = p_z - m_e \frac{\omega_b^2}{4c} \beta_b r^2, \tag{11.15}$$

where $\hat{\Omega}_{0,w} \equiv eB_{0,w}/m_e c$ are the nonrelativistic cyclotron frequencies associated with the axial and wiggler fields. Observe that the self-fields appear explicitly only in the expression for P_z.

11.1.3 The canonical transformation

It will now prove to be convenient to make the canonical transformation [4] $(x, y, z, P_x, P_y, P_z) \to (\varphi, \psi, z', P_\varphi, P_\psi, P_{z'})$,

$$x = \sqrt{\left(\frac{2P_\varphi}{m_e \hat{\Omega}_0}\right)} \sin(\varphi + k_w z') - \sqrt{\left(\frac{2P_\psi}{m_e \hat{\Omega}}\right)} \cos(\psi - k_w z'), \tag{11.16}$$

$$y = -\sqrt{\left(\frac{2P_\varphi}{m_e \hat{\Omega}_0}\right)} \cos(\varphi + k_w z') + \sqrt{\left(\frac{2P_\psi}{m_e \hat{\Omega}_0}\right)} \sin(\psi - k_w z'), \tag{11.17}$$

$$z = z', \tag{11.18}$$

$$P_x = \sqrt{(2m_e \hat{\Omega}_0 P_\varphi)} \cos(\varphi + k_w z'), \tag{11.19}$$

$$P_y = \sqrt{(2m_e \hat{\Omega}_0 P_\psi)} \cos(\psi - k_w z'), \tag{11.20}$$

and

$$P_z = P_{z'} - k_w P_\varphi + k_w P_\psi. \tag{11.21}$$

Note that ψ as used in the canonical transformation should not be confused with the ponderomotive phase. In terms of these canonical coordinates, the orbit equations can be expressed as

$$\frac{d}{dt}\varphi = (\Omega_0 - k_w v_z) - \frac{\Omega_w}{k_w}\sqrt{\left(\frac{m_e \hat{\Omega}_0}{2P_\varphi}\right)} \cos\varphi - \frac{\omega_b^2}{2\hat{\Omega}_0}(1 - \beta_b \beta_z)\left[1 - \sqrt{\left(\frac{P_\psi}{P_\varphi}\right)} \sin(\varphi + \psi)\right], \tag{11.22}$$

$$\frac{d}{dt}\psi = k_w v_z - \frac{\omega_b^2}{2\hat{\Omega}_0}(1 - \beta_b \beta_z)\left[1 - \sqrt{\left(\frac{P_\varphi}{P_\psi}\right)} \sin(\varphi + \psi)\right], \tag{11.23}$$

$$\frac{d}{dt}P_\varphi = -\frac{\Omega_w}{k_w}\sqrt{(2m_e \hat{\Omega}_0 P_\varphi)} \sin\varphi - \frac{\omega_b^2}{\hat{\Omega}_0}\sqrt{(P_\varphi P_\psi)}(1 - \beta_b \beta_z)\cos(\varphi + \psi), \tag{11.24}$$

$$\frac{d}{dt}P_\psi = -\frac{\omega_b^2}{\hat{\Omega}_0}\sqrt{(P_\varphi P_\psi)}(1 - \beta_b \beta_z)\cos(\varphi + \psi), \tag{11.25}$$

and

$$\frac{d}{dt}P_{z'} = 0. \tag{11.26}$$

Hence, $P_{z'}$ is also a constant of the motion.

11.1.4 Integrable trajectories

These equations of motion yield the steady-state trajectories discussed in Chapter 2 in the limit in which the self-fields can be neglected. In order to understand this, observe that when $\omega_b \to 0$, P_ψ is a constant of the motion (denoted by $P_{\psi 0}$), and that $d\psi/dt = k_w v_z$. If we require as well that $dP_\varphi/dt = 0$ for steady-state trajectories, then it follows that $P_\varphi = P_{\varphi 0}$ (constant), $v_z = v_\parallel$ (constant), and $\varphi = \varphi_0$ (constant) such that $\cos\varphi_0 = \pm 1$. Since φ is also constant for the steady-state trajectories, this requires from equation (11.22) that

$$\sqrt{\left(\frac{2P_{\varphi 0}}{m_e\hat{\Omega}_0}\right)} = \pm\frac{v_w}{k_w v_\parallel}, \tag{11.27}$$

where v_w denotes the wiggler-induced transverse velocity given in equation (2.15), and the '$\pm$' sign refers to the Group II and I trajectories respectively in order to insure that $P_{\varphi 0}$ remains real. Note also that the energy-conservation requirement implies that v_w and $v_\parallel$ are related via equation (2.16). The values of $P_{\psi 0}$ and ψ are determined by the initial conditions on the steady-state orbit. In particular,

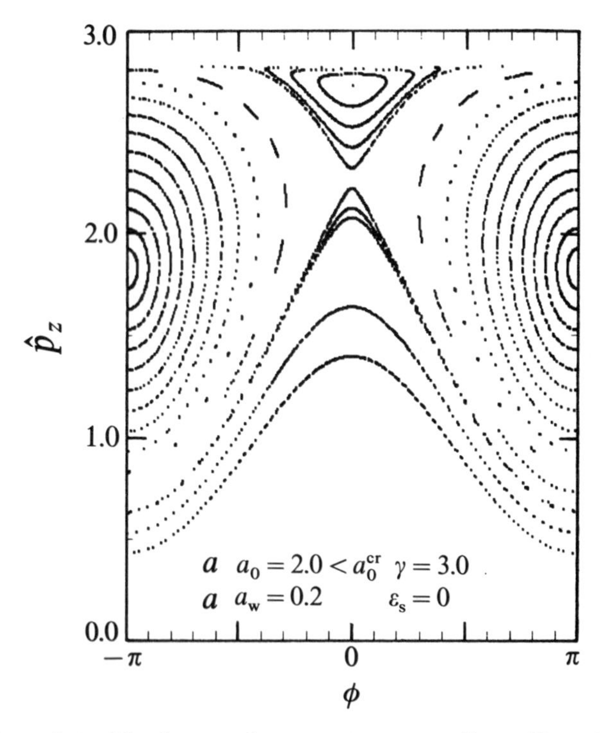

Fig. 11.1 Contour plots of the electron phase space corresponding to Group I trajectories in the integrable limit in which the self-fields are neglected [4].

$\psi = \psi_0 + k_w z$, where (we assume that $z(t=0)=0$)

$$x(t=0) = -\sqrt{\left(\frac{2P_{\psi 0}}{m_e \hat{\Omega}_0}\right)} \cos\psi_0. \tag{11.28}$$

In addition, we must also have that

$$y(t=0) = -\sqrt{\left(\frac{2P_{\varphi 0}}{m_e \hat{\Omega}_0}\right)} + \sqrt{\left(\frac{2P_{\psi 0}}{m_e \hat{\Omega}_0}\right)} \sin\psi_0. \tag{11.29}$$

Together, equations (11.27–9) specify the values of $P_{\varphi 0}$, $P_{\psi 0}$ and ψ_0 corresponding to the steady-state trajectories in the absence of self-fields.

The characteristic electron phase space corresponding to the equations of motion in the limit in which the self-fields vanish is shown in Figs 11.1 and 11.2 in both the Group I and Group II regimes [4] for $\gamma = 3.0$ and $a_w = 0.2$ (recall that $a_w \equiv eB/m_e c^2 k_w$). Since both P_φ and P_ψ are constants, the phase space can be described by a normalized axial momentum

$$\hat{p}_z \equiv \frac{1}{m_e c}(P_{z'} - P_\varphi + P_\psi), \tag{11.30}$$

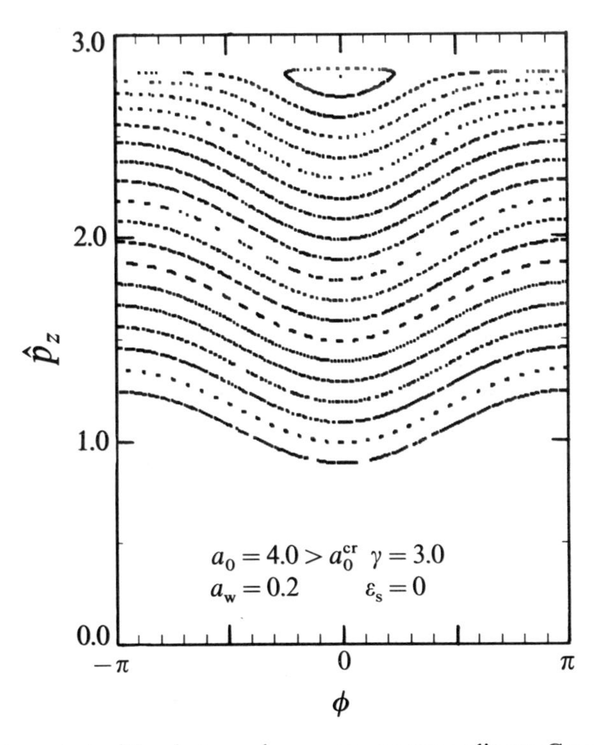

Fig. 11.2 Contour plots of the electron phase space corresponding to Group II trajectories in the integrable limit in which the self-fields are neglected [4].

and φ. The case corresponding to Group I parameters for a weak axial magnetic field shown in Fig. 11.1 corresponds to $a_0(\equiv eB_0/m_e c^2 k_w) = 2.0$. The family of curves here describes the general trajectories, and the elliptic and hyperbolic fixed points evident in the figure correspond to the stable and unstable steady-state trajectories. The strong axial magnetic field (Group II) case is shown in Fig. 11.2, for which $a_0 = 4.0$. Observe that only an elliptic fixed point is evident for this case since, as discussed in Chapter 2, there are no unstable Group II trajectories in the idealized one-dimensional limit. In the absence of the self-fields, no chaos is evident in either the Group I or Group II trajectories.

11.1.5 Chaotic trajectories

Numerical integration of the complete equations of motion (11.22–6) is required to demonstrate the onset of chaos. The first case under consideration (Fig. 11.3) corresponds to the Group I trajectories shown in Fig. 11.1 [4]. Here it is assumed that $\Gamma = 3.0$, $a_w = 0.2$, $a_0 = 2.0$, and that $\omega_b^2/c^2 k_w^2 = 0.16$. In addition, the initial value of P_ψ is $k_w P_\psi(t=0)/m_e c = 0.0625$, while the initial axial momentum varies. The existence of chaotic orbits associated with the unstable Group I trajectories is evident in Figure 11.3.

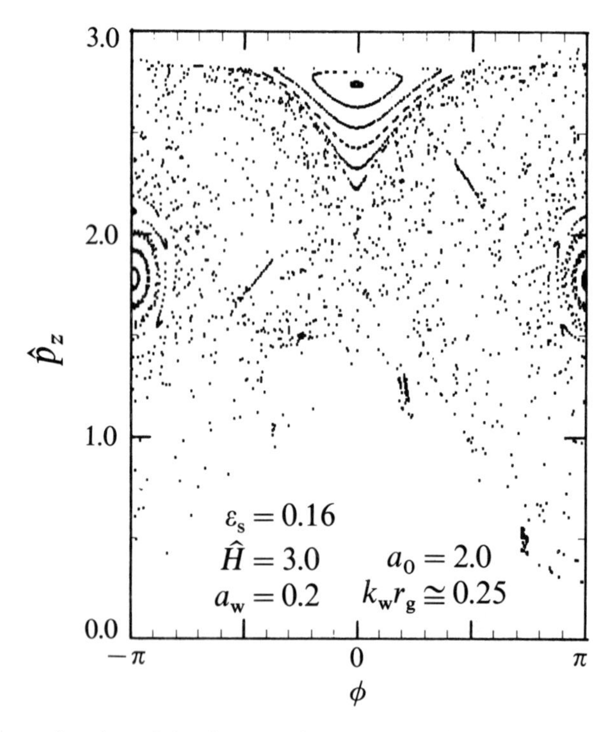

Fig. 11.3 Surface of section of the electron phase space at $\psi = 0$ corresponding to Group I trajectories in the chaotic regime [4].

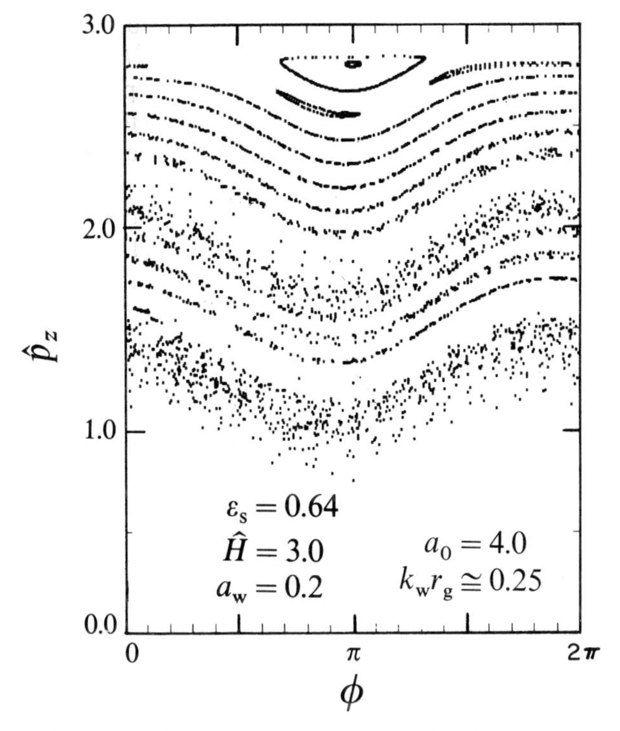

Fig. 11.4 Surface of section of the electron phase space at $\psi = 0$ corresponding to Group II trajectories in the chaotic regime [4].

The case corresponding to the Group II parameters is shown in Fig. 11.4 (this corresponds to the integrable limit shown in Fig. 11.2), where $\Gamma = 3.0$, $a_w = 0.2$, $a_0 = 4.0$, $\omega_b^2/c^2k_w^2 = 0.64$ and $k_w P_\psi(t=0)/m_e c = 0.125$. The existence of chaotic orbits in this case is also evident, although higher values of self-fields are required since these orbits are stable in the absence of the self-fields. In this regard it should be observed that unstable trajectories are found for Group II orbits in the absence of self-fields when three-dimensional effects are included (see Chapter 2), and that the conclusion that stronger levels of self-fields are required for the onset of chaos in Group II orbits does not hold in the three-dimensional case.

The above analysis is restricted to the special case of an idealized helical wiggler in which the self-fields are described under the assumption that the axial velocity of the beam is constant and the transverse velocity can be neglected. The natural generalization of this model is to employ the self-consistent axial velocity in the expressions for the self-fields. The results of this generalization [5] yield results which are qualitatively similar to those described for constant axial velocities as long as the densities are not too high.

While the self-fields are necessary in order to obtain chaotic trajectories for a helical wiggler, this is not the case for a planar wiggler. In the planar wiggler geometry, chaotic orbits have been obtained due merely to the presence of an axial solenoidal field [5].

11.2 CHAOS IN FREE-ELECTRON LASER OSCILLATORS

The physics underlying the operation of a free-electron laser is classical, and one can expect that nonlinear dynamics and chaos theory might play an important role in many aspects of the study of these devices [10]. In this section we discuss the nonlinear dynamics of electrons in the combined fields of the radiation and wiggler. This leads to the study of Hamiltonian motion with one or more degrees of freedom depending on the complexity of the model under consideration. If one regards the one-degree-of-freedom description as the ideal, then the question which is raised is: does the inclusion of higher dimensional effects destroy the ideal picture? For example, are electrons expelled from the beam due to the spatial inhomogeneities of a realistic wiggler field [4, 5] or detrapped from the ponderomotive well by multifrequency fields [11]? While questions of beam confinement are important, the focus of nonlinear dynamics theory is often time-asymptotic features of the electron motion, such as KAM surfaces, Lyapunov exponents and Arnold diffusion. In practice, an individual electron spends only a limited time in the interaction region (except for storage-ring free-electron lasers) and, thus, the practical issue of beam confinement is not necessarily tied directly to the important issues of nonlinear dynamics.

A second type of nonlinear problem [12] concerns the self-consistent action and reaction of the electrons and the radiation in a free-electron laser oscillator. Here, the radiation bounces back and forth between mirrors, gaining power while interacting with the electron beam and losing power due to output coupling and losses. New electrons are injected continually (except in storage rings where they are reaccelerated) and lose energy (on average) to the field before exiting the interaction region. In any case, due to its open nature the free-electron laser oscillator is a nonlinear dissipative system as opposed to a Hamiltonian one. The oscillator can, in principle, be run continuously and thus presents itself as a suitable system for the study of nonlinear dynamics and chaos. Indeed conventional laser oscillators have already been studied extensively in this light.

11.2.1 Return maps

The simplest example of a nonlinear dissipative system is the one-dimensional return map, and it is not hard to construct a model of a free-electron laser oscillator which reduces to such a map. To do this, suppose that the electron beam is so energetic that we can assume that it moves at essentially the same speed as the radiation. Further, let us represent the time dependence of the radiation field entering the interaction region as

$$\delta\hat{a} = \delta a_e(t)\exp(-i\omega t) + \text{c.c.}, \tag{11.31}$$

where ω is the central frequency of the radiation and $\delta a_e(t)$ is a complex time-dependent amplitude. When the radiation enters the interaction region it is met by a group of electrons which has yet to be affected by the radiation. Under the assumption of equal beam and radiation speeds, the radiation and electrons move together without slippage and interact, resulting in an amplification of the radiation. The radiation that emerges from the interaction region will then depend on the radiation that entered and the beam current. We write this dependence in terms of a nonlinear gain function

$$\delta a_0(t) = [1 + I_b g(|\delta a_e(t - T_L)|)]\delta a_e(t - T_L), \tag{11.32}$$

where $\delta a_0(t)$ is the complex amplitude of the radiation emerging from the interaction region, I_b is a parameter representing the beam current, T_L is the time of flight through the interaction region, and the nonlinear (and possibly complex) gain function g depends only on the amplitude of the radiation due to the fact that the entering electrons are uncorrelated with the radiation phase.

The radiation that emerges from the interaction region is then reflected by a system of mirrors back to the entrance of the interaction region, during which time a fraction of power is removed as output and a fraction is lost due to dissipation. Let the reflection coefficient R denote these loss processes, and write the field at the entrance of the interaction region at time t in terms of the field at the exit at time $t - T_R$ where T_R is the time required for the radiation to return from the exit to the entrance of the interaction region, $\delta a_e(t) = R\delta a_0(t - T_R)$. Combining these equations we thus arrive at a return map [12]

$$\delta a_e(t + T) = R[\delta a_e(t) + I_b g(|\delta a_e(t)|)]\delta a_e(t), \tag{11.33}$$

where $T = T_R + T_L$ is the total round-trip time for the radiation in the cavity. Such a map can of course be derived more rigorously. For example, (11.33) is precisely of the same form as the delay equation derived in Chapter 9 for the klystron model (9.54) in the absence of slippage. Further, as was discussed in Section 9.3, for the special case of negligibly small slippage, the radiation in an oscillator will evolve according to an equation of the form of (11.33).

The return map in equation (11.33) is slightly different from that which is traditionally studied, since it involves a continuous and complex amplitude $\delta a_e(t)$. The latter issue is easily resolved simply by taking the magnitude of both sides of equation (11.33). The result is a map for the magnitude $|\delta a_e(t)|$ in terms of the magnitude $|\delta a_e(t - T)|$. This map can be advanced independently of the phase of $\delta a_e(t)$. By examining the phase of each side of equation (11.33), we obtain

$$\arg[\delta a_e(t)] = \arg[\delta a_e(t - T)] - \Omega|\delta a_e(t - T)|, \tag{11.34}$$

where $\Omega = -\arg[R(1 + I_b g)]$, from which the phase of $\delta a_e(t)$ is advanced. This phase does not affect the evolution of the amplitude in the present model.

The behaviour of simple return maps of the form described is well known [13]. As the parameter I_b (i.e. the current) is changed, one observes the following behaviour. For small I_b, the gain is insufficient to overcome losses and the solution $\delta a_e(t) = 0$, corresponding to no oscillation, is reached asymptotically in time. Above a threshold I_{st}, known as the start current, the $\delta a_e(t) = 0$ solution becomes unstable. In the simple case where g is real, the start current is given by $I_{st} = (1-R)/[Rg(0)]$. If the current exceeds the start current by a moderate amount, the solution asymptotes in time to one in which the amplitude of $\delta a_e(t)$ is constant, $|\delta a_e| = \delta a_0$, and the phase advances by an amount $-\Omega(a_0)$ on each iteration. In this equilibrium, the saturated nonlinear gain just balances losses at all times. In the special case of real g, the equilibrium $\delta a_e = \delta a_0$ is determined by the transcendental equation

$$I_b = \frac{1-R}{Rg(\delta a_0)}. \tag{11.35}$$

At a current $I_b = I_2$, this equilibrium becomes unstable and the sequence of a values oscillates between two values. For real g the critical current I_2 is given by

$$RI_2 \delta a_0 \frac{\partial g}{\partial \delta a_0} = -2. \tag{11.36}$$

Higher values of current I_b above I_2 lead to the familiar sequence of period doublings until a critical value I_∞ of current is reached at which the sequence becomes chaotic. One can then construct bifurcation diagrams and calculate the **winding number** [10]

$$\omega = \lim_{N\to 1} \frac{1}{N} \sum_{n+1}^{N} \Omega[\delta a_e(nT)]. \tag{11.37}$$

The bifurcation diagrams appear to be the same as one would obtain for the logistics map.

11.2.2 Electron slippage

While the study of the return map is appealing because of its simplicity, it cannot be taken too seriously as a model of a free-electron laser. This is because the solutions of the return map eventually violate the assumptions under which the map was derived [12]. The map is a prescription for advancing a continuous time function $\delta a_e(t)$. One must initially specify $\delta a_e(t)$ on the interval $0 < t < T$, and from the map determine $\delta a_e(t)$ at a subsequent time. However, adjacent time slices do not affect each other. Thus, if our initial condition corresponded to a random complex function $\delta a_e(t)$ (as one would expect due to spontaneous noise), the time-asymptotic solution for currents in the range $I_{st} < I_b < I_2$ is

$$\delta a_e(t) = \delta a_0 \exp[-i\Omega t + i\phi(t)] \tag{11.38}$$

where $\phi(t)$ is some arbitrary, random, periodic function of time which depends on the initial conditions. That is, the radiation comes to an equilibrium where its amplitude is a constant, δa_0, but its phase is arbitrary. This effect is responsible for the initial, but erroneous, claim that the free-electron laser at the University of California at Santa Barbara was operating in a single mode [14]. There the constancy of the amplitude of the signal was argued to imply that only a single frequency was present. Subsequent theory [15, 16] demonstrated that the field could be of the form of (11.38) and, owing to the time dependence of $\phi(t)$, many frequencies could still be present. A later measurement of the spectrum [17] confirmed this prediction.

A more serious problem occurs for currents $I_b > I_2$, in which case the sequence of $|\delta a_e(t - nT)|$ values might be, say, a period two sequence. If this is the case, it is impossible for $\delta a_e(t)$ to remain a continuous function of time. It will necessarily develop discontinuities at some sequence of times separated by T.

The assumption that becomes strained under these circumstances is that the beam and radiation move at the same speed. While electrons are in the interaction region they continually slip behind the radiation, and by this process information is communicated between different time slices of the radiation field. Observe that even if one slows down the radiation to match the speed of the beam, then dispersion will be introduced and again the return map will not be valid. The electron slippage is characterized by the parameter

$$\varepsilon = \frac{L}{L_c}\left(\frac{c}{v_z} - 1\right), \tag{11.39}$$

where L and L_c are the length of the interaction region and twice the separation between the mirrors, c and v_z are the speeds of the radiation and the beam, and ε is typically small. In terms of our simple map, the result of including electron slippage is that the term representing the amplification of the radiation in equation (11.33) is not simply dependent on $\delta a_e(t)$, but depends in a complicated manner on all values of $\delta a_e(t')$ in the interval $t - \varepsilon T < t' < t$. This was illustrated by Fig. 9.2 in the discussion of slippage in Chapter 9.

A simple way of incorporating slippage into the model has been proposed [12]. It corresponds to consideration of an interaction region which consists of two parts: a prebunching region at the entrance, and a power-extraction region at the exit. This is the klystron model which is discussed in Chapter 9. The simple result is that the return map becomes a two-time delay equation,

$$\delta a_e(t + T) = R[\delta a_e(t) + I_b g(|\delta a_e(t - \varepsilon T)|) a(t - \varepsilon T)]. \tag{11.40}$$

Introduction of slippage modifies the results of the return map in two important ways. First, for cases where the current I_b is only moderately above the start current, the final asymptotic state is one given by equation (11.38) but with $\phi(t)$ constant. That is, a single-frequency equilibrium is established. The time T_c to reach this coherent state depends on ε as [15]

$$T_c \approx \frac{T_d}{\varepsilon^2}, \tag{11.41}$$

where $T_d = -T/\ln R^2$ is the decay time of radiation in the empty cavity. As seen in Chapter 9, this formula applies to more realistic models of a free-electron laser as well.

Of greater concern from the point of view of this discussion is that the presence of slippage leads to a new route to chaos. For a current I_b between three and four times the start current (depending on parameters), the single-frequency equilibrium solution becomes unstable to the overbunch/sideband/spiking mode [18, 19] as shown in Chapter 9. The unstable solution is characterized by modulations of the amplitude $\delta a_e(t)$ with a period $2\varepsilon T$. The values of current for which the instability appears are those values for which the equilibrium amplitude is such that $d(\ln g)/d(\ln \delta a_0) \leqslant -2$. In terms of our simple model, one can understand the onset of this instability in the following way [12]. If the dependence of the gain on the radiation amplitude becomes too strong, then the constant-amplitude states become unstable to a perturbation which changes sign every εT units of time. This perturbation reinforces itself constructively. Namely, a large value of the radiation field at time t produces a smaller amplification for the field at time $t + \varepsilon T$. This produces a smaller value of $\delta a_e(t + \varepsilon T)$, which then results in a large amplification and a larger value for the field $\delta a_e(t + 2\varepsilon T)$, and so on.

Another feature of this instability is that it can occur even in low-gain oscillators. If we consider situations where the beam current is weak $I_b \ll 1$, then to develop sustained oscillations one must have a reflection coefficient R close to unity. In such a case, the field is approximately periodic in time with period T. That is, if viewed over a relatively small number of periods $\delta a_e(t)$ would be periodic. Changes in $\delta a_e(t)$ occur only over a long time of the order $T_d = -T/\ln R^2$. In this case, the two-time delay map (11.40) can be transformed to a partial differential delay equation. Let us ascribe the nearly periodic behaviour of $\delta a_e(t)$ to two time variables, $\delta a_e(t) = \delta a_e(t_0, t_s)$, where $\delta a_e(t_0, t_s)$ is periodic in the variable t_0 with period T and changes slowly with t_s on the time scale T. This multiple time-scale expansion is identical to that introduced in Chapter 9 where the round-trip index n was used as the slow-time variable and the amplitude of the radiation was assumed to be periodic in t' with period T. Insertion of this form into the two-time delay equation gives

$$\delta a_e(t_0, t_s + T) = R[\delta a_e(t_0, t_s) + I_b g(|\delta a_e|)\delta a_e(t_0 - \varepsilon T, t_s - \varepsilon T)]. \quad (11.42)$$

We now assume that R is close to unity and I_b is small so that δa_e depends weakly on t_s. Thus, in the last term on the right-hand side we neglect the slippage delay in the t_s dependence of δa_e. Further, we Taylor expand the t_s dependence of δa_e on the left-hand side to obtain

$$T\frac{\partial}{\partial t_s}\delta a_e(t_0, t_s) = (R-1)\delta a_e(t_0, t_s) + I_b g(|\delta a_e|)\delta a_e(t_0 - \varepsilon T, t_s). \quad (11.43)$$

Introducing the scaled time τ, where $t_s = T\tau/[2(1-R)]$, we obtain an equation with a single parameter $I' = I_b/[2(1-R)]$

$$\frac{\partial}{\partial \tau}\delta a_e(t_0, \tau) = -\frac{1}{2}\delta a_e(t_0, \tau) + I' g(|\delta a_e|)\delta a_e(t_0 - \varepsilon T, \tau). \quad (11.44)$$

The parameter I' reflects the fact that only the ratio of the current to the losses is of importance in the low-gain limit.

The sequence of period doublings associated with the simple return map requires a current $I_b > I_2$, which is many times the start current in the limit $R \to 1$. This sequence is not permitted in the low-gain model. On the other hand, the

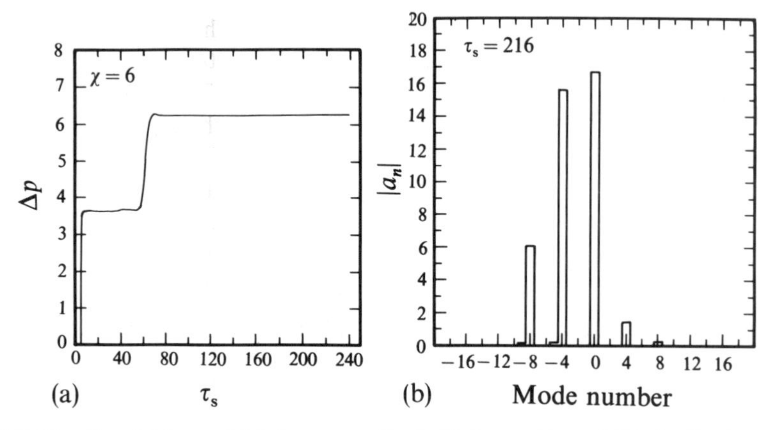

Fig. 11.5 The time history of (a) the normalized efficiency Δp, and (b) the final spectrum from a simulation with 21 modes $\varepsilon = 0.25$, and a current six times the start current [20].

overbunch instability that arises due to finite slippage typically occurs for currents only several times the start current and is thus more likely to be observed and can be studied with the low-gain model.

Just above the threshold current for the overbunch instability, steady periodic modulations of the amplitude and phase of the radiation signal appear. The spectrum in this case consists of a discrete set of narrow lines spaced by an amount $\Delta f = 1/2\varepsilon T$ corresponding to the period of the overbunch instability. As current is increased, the temporal modulations develop into spikes. The spectrum in this case is still composed of discrete lines; however, the amplitudes of the components displaced from the fundamental increase. An example of such a spectrum appears in Fig. 11.5 where the magnitudes of the complex amplitudes of a set of modes are shown for a numerical simulation with a current about six times the start current. The system of equations which were solved by computer are the same as those which were described in section 9.6.4 except that a slippage parameter $\varepsilon = 0.25$ was used. Note that the spectrum consists of a set of modes which have a separation of four modes between each large mode. In the final state, the magnitudes of these modes are constant and the phases are locked in such a way that the slow-time frequency shift for each mode is proportional to mode number. As the current is raised, different types of behaviour have been observed. In some numerical simulations, the amplitudes of the large modes became time dependent [12] in a periodic fashion, implying that the amplitude of the radiation was quasi-periodic with two frequencies. At higher currents the quasi-periodic behaviour makes a transition to chaos. This is illustrated in Fig. 11.6 where the time dependence of the efficiency and the final (nonstationary) spectrum are obtained from a computer run with a current 10 times the start current [20]. However, the exact route to chaos is not yet known.

An alternative route [10] to chaos is predicted by the map (11.40). As the current is raised, the radiation goes through a sequence of transitions from (1) single frequency (constant amplitude) to (2) spiking mode (periodic amplitude), and (3)

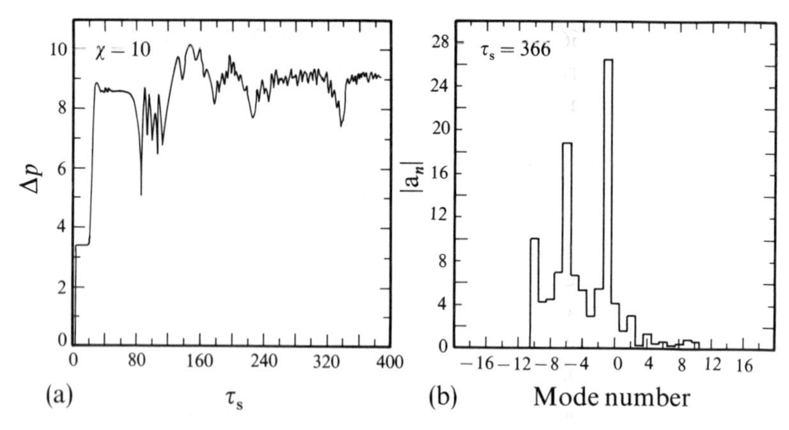

Fig. 11.6 The time history of (a) the normalized efficiency Δp and (b) the final spectrum from a simulation with 21 modes $\varepsilon = 0.25$, and a current ten times the start current [20].

through a sequence of period doublings in which the spikes acquire structure on successively longer time scales given roughly by $T_m = 2^m \varepsilon T$. For higher currents, the periodic behaviour gives way to quasi-periodic behaviour and eventually chaos.

Figure 11.7 shows a bifurcation diagram for the map (11.40) with $\varepsilon = 1/8$, $g = 1 - |\delta a_e|^2$ and $R = 0.9$. The time asymptotic values of sequences of field values $\delta a(m\varepsilon T)$ are plotted in Fig. 11.7(a) versus beam current I_b, with m an integer. The Lyapunov exponents corresponding to Fig. 11.7(a) are shown in Fig. 11.7(b). The start current is $I_{st} = 0.11$. It is seen that at $I_b = 0.22$, the steady solution bifurcates to a period two solution. Two additional bifurcations occur at higher current, producing a period eight orbit. At a current $I_b = 0.2559$, the period eight becomes unstable, but is unable to become period 16 due to the size of ε. Instead, it becomes quasiperiodic. This is evident on an expanded version of Fig. 11.5(b) where the Lyapunov exponent is zero for a range of currents $0.2559 < I_b < 0.2566$.

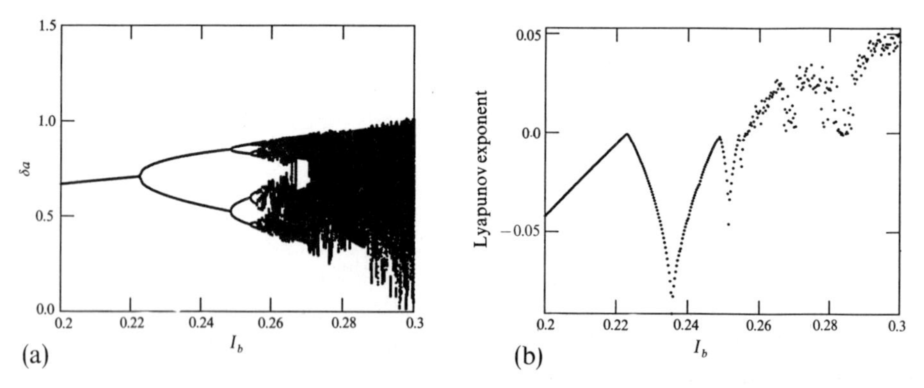

Fig. 11.7 Superimposed sequences of $\delta a(T)$ versus beam current from the map in (11.39) (a), and the corresponding values of the most positive Lyapunov exponents (b).

For currents above 0.2566, the Lyapunov exponent becomes positive – indicating chaos. At still higher currents, there are bands of currents where the solution is again quasiperiodic with zero Lyapunov exponent.

11.2.3 Pulsed injection

Up to now the beam current I_b has been assumed to be a time-independent parameter. However, free-electron lasers are often driven by accelerators such as r.f. linacs in which the beam consists of a stream of micropulses of duration T_p and separation T_a. The separation between micropulses T_p and the cavity round-trip time T are made nearly (but not exactly) to coincide. The difference in these times $T_\delta = T - T_a$ (often measured in micrometres assuming speed of light propagation) is called the cavity detuning.

There has been a great deal of study of free-electron lasers of this type (see Chapter 9) theoretically, numerically and experimentally. It has been found that at sufficiently large amplitude δa_0, the radiation spontaneously develops spikes with a width determined by the slippage. However, the cavity detuning constitutes an extra parameter, which is effective in suppressing this modulational instability [21, 22]. Due to the extra parameter in this case, as well as the complication of the current pulse waveform, the specific route to chaos in such devices has yet to be determined.

The new effects of pulsed beams and cavity detuning can be incorporated into the derivation of the low-gain model by assuming that the radiation field is periodic in the fast-time variable t_0 with period T_a as opposed to the cavity round-trip time T. The result is

$$\left(\frac{\partial}{\partial \tau} + \frac{T_\delta}{2(1-R)}\frac{\partial}{\partial t_0} + \frac{1}{2}\right)\delta a_e(t_0, \tau) = I'(t_0 - \varepsilon T) g[|\delta a_e(t_0 - \varepsilon T, \tau)|]\delta a_e(t_0 - \varepsilon T, \tau). \tag{11.45}$$

The field $\delta a_e(t_0, \tau_1)$ is now periodic with period T_a; it will be non-zero for times which are coincident with the electron micropulse. Now assume that the current is peaked at $t_0 = 0$ with a width T_p characterizing the micropulse. If T_p is much smaller than T_a then it is reasonable to rescale the fast time variable to T_p. That is, define $t_{0'} = t_0/T_p$, $\varepsilon' = \varepsilon T/T_p$ and $\tau_\delta = T_\delta/[2T_p(1-R)]$, and equation (11.45) becomes

$$\left(\frac{\partial}{\partial \tau} + \tau_\delta \frac{\partial}{\partial t_0'} + \frac{1}{2}\right)\delta a_e(t_0', \tau) = I_b \delta a_e g(|\delta a_e|)|_{t_0' - \varepsilon'}, \tag{11.46}$$

where $\delta a_e(t_0', \tau)$ is periodic in t_0' with period T/T_p which can be taken to be infinity. Preliminary investigations of this model show that it is relatively successful in predicting the onset of spiking in pulsed free-electron lasers. However, the accuracy of the differential time-delay equation in the high-current regime has not been checked.

11.2.4 Chaos in storage rings

It has recently been shown that time dependence of the radiation power in a storage ring can exhibit chaotic behaviour when a parameter is varied periodically in time [23, 24]. Recall from the discussion of storage-ring free-electron lasers in Chapter 9 that the mutual interaction of the radiation and the beam gives rise to a weakly damped relaxation oscillation in which periodic bursts of radiation are excited. A large burst of radiation degrades the beam to the extent that the cavity losses exceed the gain. This effectively turns off the radiation. With the radiation off, the beam cools down until the net gain is again positive and a new burst appears. If the ring is stable the relaxation oscillations eventually decay and, in the final state, the laser power is continuous in time. This relaxation oscillation has been modelled successfully with the following simple set of equations [25]

$$\left.\begin{aligned} \frac{dI}{dt} &= \frac{I(g-p)}{T}, \\ \frac{d\sigma^2}{dt} &= -2\nu_s(\sigma^2-\sigma_0^2)+\alpha I, \\ g &= g_0\exp[-k(\sigma^2-\sigma_0^2)], \end{aligned}\right\} \tag{11.47}$$

where I is the intensity of the radiation in a micropulse, g and p are the gain and losses per time T, σ^2 is the normalized energy spread in the beam, σ_0^2 is the equilibrium energy spread in the absence of radiation, ν_s is the synchrotron radiation damping rate, and αI represents the rate of heating of the beam due to the interaction with the radiation. The final equation expresses the dependence of the gain on energy spread.

Equations (11.47) represent a two-dimensional, autonomous (time does not appear explicitly on the right-hand side), dissipative system. Such a system cannot be chaotic, and the solutions of this system eventually settle down to an equilibrium in which the energy spread and the radiation intensity are constants: a zero-dimensional attractor. In this respect, the system is rather like a damped nonlinear pendulum. In order to make the system chaotic, one of the parameters is varied periodically in time. This is analogous to a damped driven pendulum [26]. In the experiment, the gain is modulated periodically in time by periodically modulating the frequency of the r.f. accelerating field. This changes the arrival times of the beam micropulses and, hence, the effective cavity detuning. The result is that the gain is modulated at twice the frequency at which the r.f. field is modulated (this assumes that in the absence of modulation the cavity detuning has been optimized for maximum gain). The modulation is accounted for in the model by multiplying the gain by a time-dependent factor

$$g = g_0\exp[-k(\sigma^2-\sigma_0^2)](1+a\sin^2\Omega t). \tag{11.48}$$

The above system of equations was found successfully to model the results of a series of experiments performed on the Super-ACO storage ring [23, 24]. In particular, both the experiment and the model predicted period doubling in the bursting mode of the radiation. In addition, chaos in the temporal behaviour of the radiation was observed for approximately the same parameters as in the experiment.

To summarize, as the current in a free-electron laser oscillator is raised, the radiation field progresses from being single frequency to broadband and chaotic. Simple model maps have been proposed which predict how the transition to chaos occurs. Certain features of these maps agree with numerical simulation codes; however, detailed correspondence has yet to be verified. Experiments with pulsed beams are most likely to exhibit the behaviour described here, but these experiments contain an additional parameter further complicating the picture. Finally, the practical question of whether chaos in these devices needs to be understood rather than simply avoided, needs to be addressed.

REFERENCES

1. Lichtenburg, A. and Liberman, M. (1983) *Regular and Stochastic Motion*, Springer, New York.
2. Sagdeev, R. Z., Usikov, D. A. and Zaslavky, G. M. (1988) *Nonlinear Physics: From the Pendulum to Turbulence and Chaos*, Harwood Academic Publishers, Chur, Switzerland.
3. Ott, E. (1981) Strange attractors and chaotic motions of dynamical systems. *Rev. Mod. Phys.*, **53**, 655.
4. Chen, C. and Davidson, R. C. (1991) Chaotic particle dynamics in free-electron lasers. *Phys. Rev. A.*, **43**, 5541.
5. Michel, L., Bourdier, A. and Buzzi, J. M. (1991) Chaos electron trajectories in a free-electron laser. *Nucl. Instr. Meth. A*, **304**, 465.
6. Spindler, G. and Renz, G. (1991) Chaotic behaviour of electron orbits in a free-electron laser near magnetoresonance. *Nucl. Instr. Meth.*, **A304**, 492.
7. Michel-Lours, L., Bourdier, A. and Buzzi, J. M. (1993) Chaotic electron trajectories in a free-electron laser with a linearly polarized wiggler. *Phys. Fluids B*, **5**, 965.
8. Bourdier, A. and Michel-Lours, L. (1994) Identifying chaotic electron trajectories in a helical wiggler free-electron laser. *Phys. Rev. E*, **49**, 3553.
9. Davidson, R. C. (1990) *Physics of Nonneutral Plasmas*, Addison-Wesley, Reading, MA.
10. Antonsen, T. M. Jr (1991) Nonlinear dynamics of radiation in a free-electron laser, in *Nonlinear Dynamics and Particle Acceleration* (ed. Ichikwa, Y. H. and Tajima, T.), AIP Conference Proceedings No. 230, New York, p. 106.
11. Riyopoulos, S. and Tang, C. M. (1988) Chaotic electron motion caused by sidebands in free-electron lasers. *Phys. Fluids*, **31**, 3387.
12. Ginzburg, N. S. and Petelin, M. I. (1985) Multi-frequency generation in free-electron lasers with quasi-optical resonators. *Int. J. Electronics*, **59**, 291.
13. Fiegenbaum, M. J. (1978) Quantitative universality for a class of nonlinear transformations. *J. Stat. Phys.*, **19**, 25.
14. Elias, L. R., Ramian, G. J., Hu, J. and Amir, A. (1986) Observation of single-mode operation in a free-electron laser. *Phys. Rev. Lett.*, **75**, 424.
15. Antonsen, T. M. and Levush, B. (1989) Mode competition and suppression in free-electron laser oscillators. *Phys. Fluids B*, **1**, 1097.
16. Antonsen, T. M. and Levush, B. (1989) Mode competition and control in free-electron laser oscillators. *Phys. Rev. Lett.*, **62**, 1488.
17. Danly, B. G., Evangelides, S. G., Chu, T. S., Temkin, R. J., Ramian, G. and Hu, J.

(1990) Direct spectral measurements of a quasi-cw free electron laser. *Phys. Rev. Lett.*, **65**, 2251.

18. Bogomolov, Ya. L., Bratman, V. L., Ginzburg, N. S., Petelin, M. I. and Yunakovsky, A. D. (1981) Nonstationary generation in free-electron lasers. *Opt. Comm.*, **36**, 209.
19. Colson, W. B. and Freedman, R. A. (1983) Synchrotron instability for long pulses in free-electron lasers. *Opt. Commun.*, **46**, 37.
20. Levush, B. and Antonsen, T. M. Jr (1989) Effect of nonlinear mode competition on the efficiency of low gain free-electron laser oscillator. *IEEE J. Quantum Electron.*, **1061**, 2.
21. Warren, R. W., Sollid, J. E., Feldman, D. W., Stein, W. E., Johnson, W. J., Lumpkin, A. H. and Goldstein, J. C. (1989) Near-ideal lasing with a uniform wiggler. *Nucl. Instr. Meth.*, **A285**, 1.
22. Warren, R. W. and Goldstein, J. C. (1988) The generation and suppression of synchrotron sidebands. *Nucl. Instr. Meth.*, **A272**, 155.
23. Billardon, M. (1990) Storage-ring free-electron laser and chaos. *Phys. Rev. Lett.*, **65**, 713.
24. Billardon, M. (1991) Chaotic behavior of the storage-ring free-electron laser. *Nucl. Instr. Meth.*, **A304**, 37.
25. Ellaume, P. (1984) Macrotemporal structure of free-electron lasers. *J. Phys.*, **45**, 997.
26. Grebogi, C., Ott, E. and Yorke, J. A. (1987) Chaos, strange attractors, and fractal basin boundaries in nonlinear dynamics. *Science*, **238**, 585.

12
Wiggler imperfections

The free-electron laser operates by the coherent axial bunching of electrons in the ponderomotive wave formed by the beating of the wiggler and radiation fields. The interaction is extremely sensitive to the axial energy spread of the electron beam, and an energy spread of one percent or less is sufficient to cause substantial reductions in the efficiency due to the detuning of the wave-particle resonance. A related effect is caused by random imperfections in the wiggler field. Planar wigglers can easily exhibit a random rms fluctuation of 0.5% from pole to pole [1]. This yields a velocity fluctuation which causes a phase jitter that also detunes the wave-particle resonance. In this chapter, we explore the effects of wiggler imperfections on free-electron laser performance, and compare the effects of wiggler imperfections with those of an axial energy spread.

The effects of random wiggler imperfections have been studied using a random walk model for the electron trajectories and their effects upon both spontaneous emission [2] and the linear gain [3,4]. Nonlinear modeling of wiggler field imperfections has been based [4–7] upon the inclusion of an analytic model of the random walk in a wiggler-period averaged formalism of the electron trajectories. In contrast to these approaches, we adapt the nonlinear formalism described in Chapter 5 to treat the effect of wiggler imperfections [8,9]. No average over a wiggler period is performed in this approach, and no explicit assumption of the random walk is included. Instead, this formalism relies upon a model of the imperfections in the wiggler field, and the evolution of the electron trajectories, as well as the growth of the radiation field, is then determined self-consistently by integration of the coupled nonlinear differential equations for the electrons and the fields.

Consideration of the effects of wiggler imperfections shows that any perturbation induced in one of the pole pieces of a planar wiggler will induce a series of correlated changes in the field over several adjacent wiggler periods. This effect has been measured in the laboratory on a prototype planar wiggler design [10]. Here, an error was introduced by reducing the gap spacing between one set of pole pieces. An axial scan of the on-axis field showed that the error propagated through ± 1 wiggler period (± 2 pole pieces for this design) with an increase in amplitude at the adjacent poles of approximately 55% and at the next poles of approximately 10%. The amplitude and extent of these correlations are dependent upon the detailed design of any given wiggler, and can be substantial. Thus, the question of the nature of 'random' imperfections in wiggler magnets requires further study. As a first step, a continuous mapping of random field variations from pole to pole has been included.

Two specific examples are discussed. The first is at a relatively long wavelength of approximately 8 millimeters and corresponds to the 35 GHz example discussed in section 5.4. As such, the discussion makes use of the nonlinear formalism developed in Chapter 5 directly. The second example corresponds to a short wavelength free-electron laser. For this purpose, the nonlinear formalism developed in Chapter 5 is extended to treat an ensemble of optical modes.

12.1 THE WIGGLER MODEL

The wiggler configuration we employ for both the long and short wavelength examples is that of the parabolic-pole-face planar wiggler introduced in equations (2.3) and (5.142). This wiggler model provides enhanced focusing of the electron beam with respect to a planar wiggler with flat pole faces, and was shown to provide good agreement with experiment [see section 5.4.4]. The variation in the wiggler amplitude in the axial direction will be assumed to contain both systematic and random components and we shall write $B_w(z) = B_{w_0}(z) + \Delta B_w(z)$, where $B_{w_0}(z)$ denotes the systematic variation and $\Delta B_w(z)$ denotes the random variation.

The systematic variation in the wiggler amplitude describes both the adiabatic entry taper as well as any amplitude tapering for efficiency enhancement. In this chapter, however, we shall ignore any tapering for efficiency enhancement and assume that the systematic variation in the wiggler amplitude solely includes the adiabatic entry taper as

$$B_{w_0}(z) = \begin{cases} B_w \sin^2\left(\dfrac{k_w z}{4N_w}\right); 0 \leqslant z \leqslant N_w\lambda_w \\ \\ B_w \qquad\qquad\qquad ; N_w\lambda_w < z \end{cases} \tag{12.1}$$

where B_w is the systematic amplitude of the uniform wiggler, λ_w is the wiggler period, k_w [$\equiv 2\pi/\lambda_w$] is the wiggler wavenumber, and the adiabatic entry taper is over N_w wiggler periods.

The random component of the amplitude is chosen at regular intervals using a random number generator, and a continuous map is used between these points. Since a particular wiggler may have several sets of pole faces per wiggler period, the interval is chosen to be $\Delta z = \lambda_w/N_p$, where N_p is the number of pole faces per wiggler period. Hence, a random sequence of amplitudes $\{\Delta B_n\}$ is generated, where $\Delta B_n \equiv \Delta B_w(n\Delta z)$. The only restriction is that $\Delta B_w = 0$ over the entry taper region (i.e., $\Delta B_n = 0$ for $0 \leqslant n \leqslant 1 + N_p N_w$) to ensure a positive amplitude. The variation in $\Delta B_w(z)$ between these points is given by

$$\Delta B_w(n\Delta z + \delta z) = \Delta B_n + [\Delta B_{n+1} - \Delta B_n]\sin^2\left(\frac{\pi}{2}\frac{\delta z}{\Delta z}\right), \tag{12.2}$$

where $0 \leqslant \delta z \leqslant \Delta z$. It is important to note that it is possible to model the effects of pole-to-pole variations in specific wiggler magnets with this formulation.

Before turning to a detailed numerical analysis of the effect of wiggler imperfections, we consider the effect of the wiggler fluctuations on the variations in the axial energy along the electron trajectory. It is clear that the effect of axial variations in the wiggler field will be to cause an oscillation in the axial velocity of the electrons. It is possible to show that the fluctuation in the axial electron velocity Δv_z caused by some given fluctuation in the wiggler field ΔB_w is given by

$$\frac{\Delta v_z}{v_z} = -\left(\frac{v_w}{v_z}\right)^2 \frac{\Delta B_w}{B_w}, \tag{12.3}$$

where v_z is the bulk axial velocity, v_w is the wiggler-induced transverse velocity, and

$$\left(\frac{v_w}{v_z}\right)^2 = \frac{a_w^2/2}{\gamma_0^2 - 1 - a_w^2/2}, \tag{12.4}$$

for a planar wiggler. This can be related to a fluctuation in the axial energy via

$$\frac{\Delta \gamma_z}{\gamma_0} = 1 - \left[1 + 2(\gamma_0^2 - 1)\left|\frac{\Delta v_z}{v_z}\right|\right]^{-1/2}. \tag{12.5}$$

Together equations (12.3) and (12.5) describe the relation between the fluctuation in the wiggler field amplitude and the corresponding axial energy fluctuation of the electrons.

12.2 THE LONG WAVELENGTH REGIME

Recall that the nonlinear formulation developed in Chapter 5 includes the simultaneous integration of a slow-time-scale formulation of Maxwell's equations for an ensemble of TE and TM modes as well as the complete Lorentz force equations for an ensemble of electrons. The wiggler model includes an adiabatic entry taper which describes the injection of the beam into the wiggler. As a result, the initial conditions on the electron beam are specified at the entrance to the wiggler, and the subsequent evolution of the electromagnetic field and the electron beam are integrated in a self-consistent manner. Thermal effects are included under the assumption that the electron beam is initially monoenergetic but with a pitch angle spread to describe the axial energy spread.

The waveguide configuration employed is that of the rectangular geometry described in section 5.4 in which the electron beam propagates through a loss-free rectangular waveguide with the dimensions $-a/2 \leqslant x \leqslant a/2$ and $-b/2 \leqslant y \leqslant b/2$. The equations which govern the evolution of the individual TE and TM modes of this structure are given in equations (5.151–154).

The specific example under consideration is one in which a 3.5 MeV/850 A electron beam with an initial radius of 1.0 cm propagates through a rectangular waveguide ($a = 9.8$ cm, $b = 2.9$ cm) in the presence of a wiggler with $B_w = 3.72$ kG, $\lambda_w = 9.8$ cm and $N_w = 5$. Hence, the wiggler parameter $a_w = 3.404$. Resonant interaction occurs with the TE_{01}, TE_{21} and TM_{21} modes at frequencies of 30–40 GHz, and the efficiency decreases with increasing frequency across this band. For an ideal beam (with a vanishing thermal energy spread for which $\Delta\gamma_z = 0$) and wiggler ($\Delta B_w = 0$) the efficiency falls off from a maximum $\eta = 12.4\%$ at 30 GHz to a minimum of $\eta = 3.6\%$ at 40 GHz. A frequency of 34.6 GHz is selected for the comparison since this was the operational frequency used in the experiment.

The variation in the wiggler imperfection-induced axial energy fluctuation versus the wiggler fluctuation given in equations (12.3–5) is shown in Fig. 12.1 for this case. It is evident that the axial energy fluctuation increases rapidly with the magnitude of the wiggler fluctuation. In particular, a wiggler fluctuation of 5% results in a corresponding fluctuation in the axial energy of the beam of approximately 22%. One important question which we shall address in the numerical analysis of wiggler imperfections is whether such wiggler-induced fluctuations in the axial energy of the beam will have a comparable effect on the free-electron laser interaction as a thermal axial energy spread.

In the case of an ideal wiggler (i.e., $\Delta B_w = 0$) discussed in section 5.3.3 the initial drive powers in the modes are chosen to be 50 kW in the TE_{01} mode, 500 W in the TE_{21} mode and 100 W in the TM_{21} mode. The evolution of the power as a function of axial distance is shown in Fig. 5.53 for the choice of an initial axial energy spread of $\Delta\gamma_z/\gamma_0 = 0$. Observe that this refers to the thermal energy spread and not the induced energy fluctuation due to wiggler imperfections. We find that saturation occurs after a distance of 153 cm at a total power level in all modes of 260 MW for an efficiency of 8.6%. The effect of the axial energy spread is illustrated in Fig. 5.57 in which the extraction efficiency is plotted as a function of

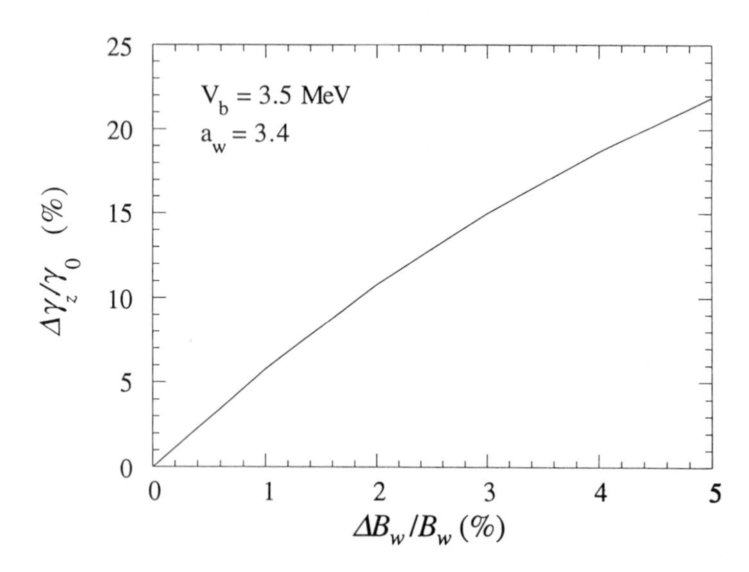

Fig. 12.1 Plot of the variation in the axial electron field as a function of the amplitude of the wiggler function.

$\Delta\gamma_z/\gamma_0$. As shown, the efficiency decreases gradually with the axial energy spread (due to the relatively high a_w) for $\Delta\gamma_z/\gamma_0 \leqslant 2.5\%$, at which point the efficiency has fallen to approximately 5.9%.

Random wiggler fluctuations can take many different forms for a fixed rms value. It is most natural to consider a random fluctuation which is relatively uniform over the interaction region (i.e., $\langle \Delta B_w \rangle = 0$); however, other configurations are possible. For example, fluctuations where the wiggler field is always greater (or less) than the systematic value for B_w are possible, as is one in which ΔB_w is very large over a small range and zero elsewhere. These are only limited examples, and a thorough analysis necessitates a large number of simulations with different random wiggler fluctuations to obtain adequate statistics. Typically, we find that a choice of approximately 35 different wiggler fluctuation distributions is required in order for the mean efficiency to converge to within 1%.

The effect of random wiggler errors is shown in Fig. 12.2 where the efficiency is plotted versus the rms wiggler variation (for $\Delta\gamma_z = 0$) for $N_p = 1$. This describes random wiggler variations at intervals of the wiggler period. The dots represent the average efficiency over the *ensemble* of random fluctuations, and the error bars denote the standard deviation. As shown, the average efficiency is relatively insensitive to wiggler errors for $(\Delta B_w/B_w)_{rms} \leqslant 5\%$, although the standard deviation increases with the rms error. For this example, the effect of a given $(\Delta B_w/B_w)_{rms}$ is much more benign than for a comparable $\Delta\gamma_z/\gamma_0$. We note from Fig. 12.1 that a wiggler fluctuation of 5% would correspond to a fluctuation in the axial energy of approximately 22%.

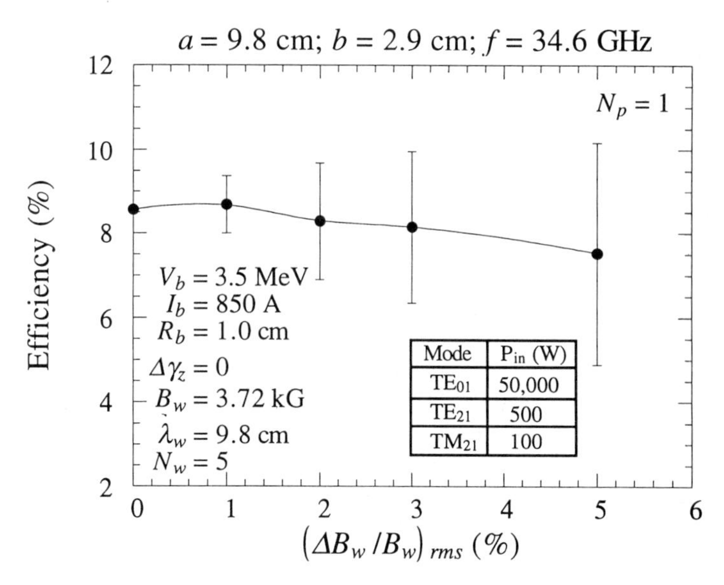

Fig. 12.2 Plot of the variation in the ensemble-averaged efficiency as a function of the rms wiggler fluctuation for $N = 1$. The error bars denote the standard deviation.

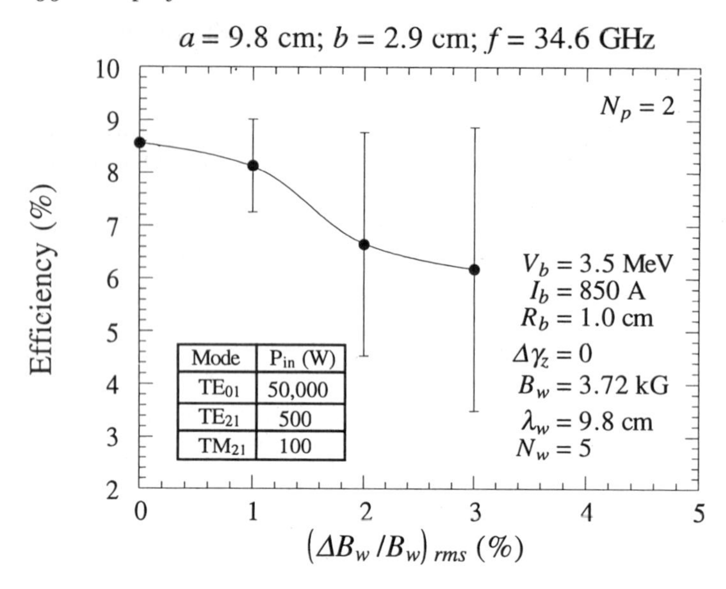

Fig. 12.3 Plot of the variation in the ensemble-averaged efficiency as a function of the rms wiggler fluctuation for $N = 2$. The error bars denote the standard deviation.

The interval over which the random wiggler imperfections occur is dependent upon a particular wiggler design; specifically, upon the number of pole pieces per wiggler period. This can be important because the effect of wiggler imperfections varies somewhat with the period over which the imperfections occur. This is illustrated in Fig. 12.3 in which the average efficiency is plotted versus the rms wiggler variation for $N_p = 2$, which describes random wiggler fluctuations at half-wiggler-period intervals. In comparison with Fig. 12.2, the efficiency falls somewhat more rapidly in this case and drops from 8.6% to 6.2% as the rms wiggler fluctuation increases to 3%. However, the effect of decreasing the interval between the random fluctuations does not result in a monotonic variation in the average efficiency. This is shown in Fig. 12.4 in which the variation in the average efficiency is plotted versus N_p for $(\Delta B_w/B_w)_{rms} = 3\%$.

Observe that the efficiency increases relative to the ideal wiggler case for particular wiggler fluctuations. In order to understand this, recall that the efficiency varies across the frequency band. This tuning can also be accomplished by variations in the wiggler magnitude, and an increase or decrease in the mean B_w can result in an increase or decrease in the efficiency as long as the chosen frequency remains in the resonant bandwidth of the interaction. Another way in which the wiggler fluctuation can affect the efficiency is if the field exhibits a bulk taper either upwards or downwards over the interaction region. A downward or upward taper increases or decreases the efficiency. In fact, it is this average up- or down-taper which accounts for the extrema in simulation.

In order to illustrate this, consider the case for which $(\Delta B_w/B_w)_{rms} = 3\%$ and $N_p = 1$. Wiggler error distributions which give rise to $\eta = 10.3\%$ and 5.9%

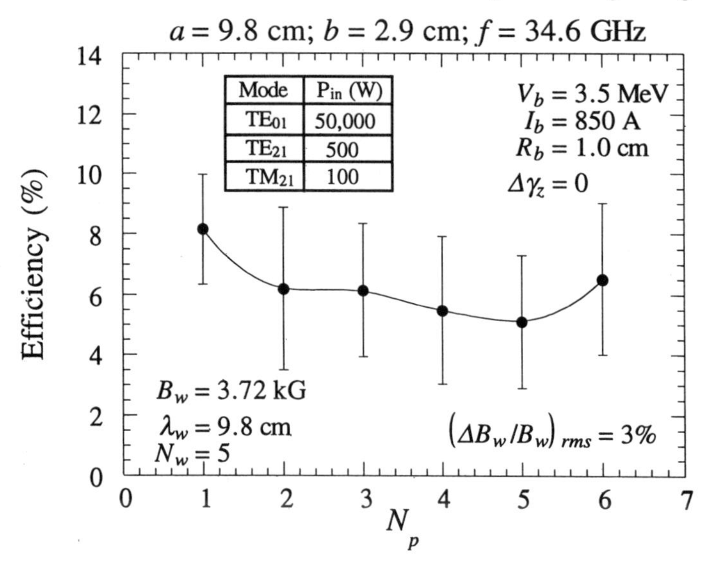

Fig. 12.4 Variation in the average efficiency versus the interval between random wiggler fluctuations.

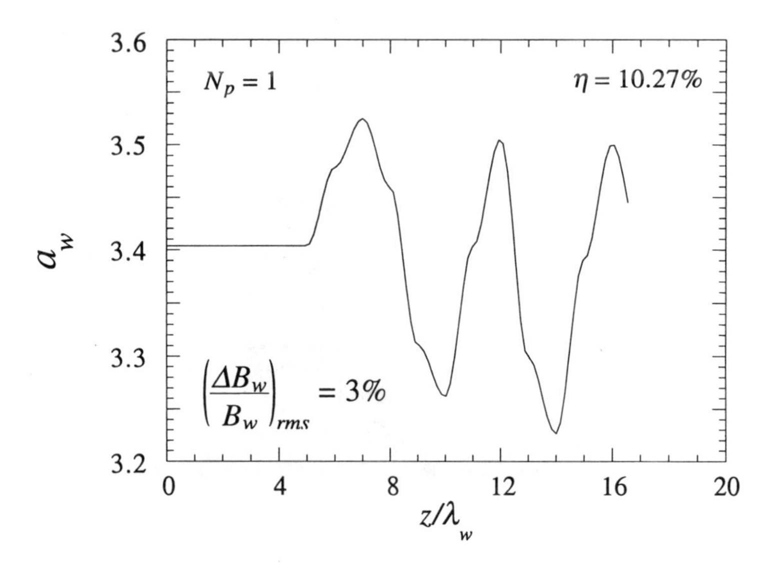

Fig. 12.5 Wiggler fluctuation distribution exhibiting a down-taper resulting in the highest efficiency in the ensemble.

(compared to $\eta = 8.6\%$ for an ideal wiggler) respectively are shown in Figs 12.5 and 12.6. The average a_w for each of these cases is close to the systematic value of 3.404; however, the field exhibits a downward taper in Fig. 12.5 and an upward taper in Fig. 12.6.

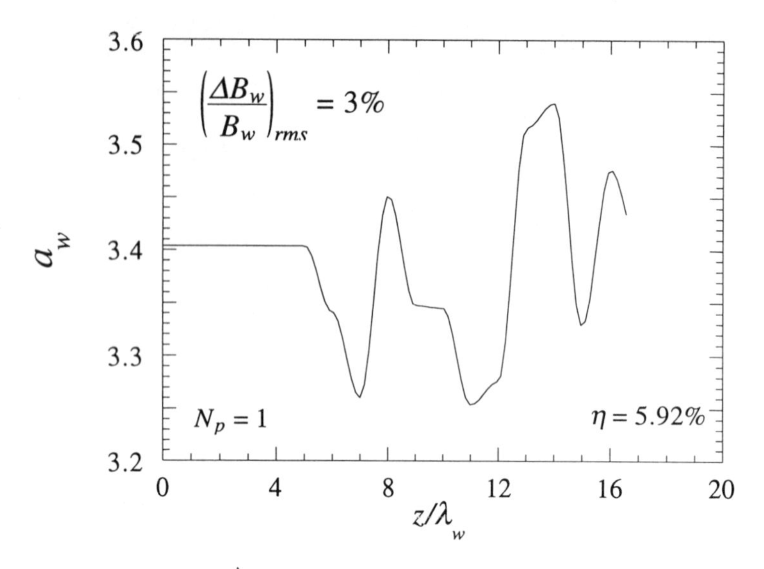

Fig. 12.6 Wiggler fluctuation distribution exhibiting an average up-taper which resulted in the lowest efficiency in the ensemble.

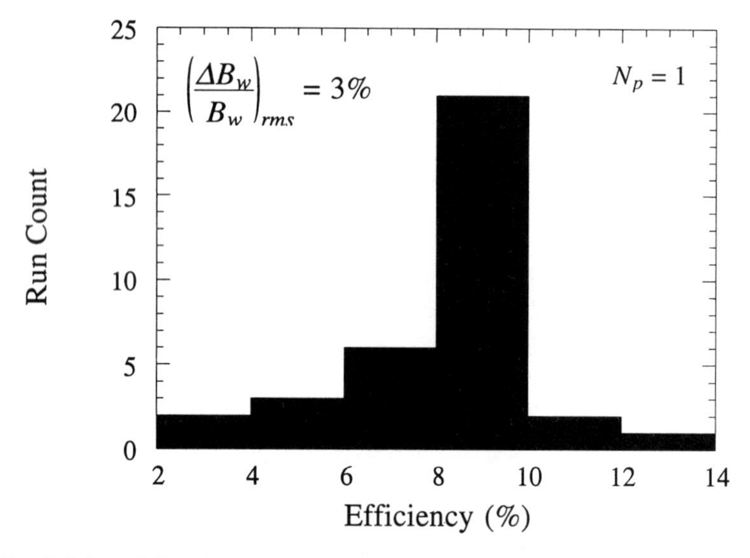

Fig. 12.7 Probability histogram including 35 different wiggler fluctuation distributions.

In general, the statistical distribution of the efficiency differs from the normal distribution, and the standard deviation must be used with some caution. For example, 35 runs were generally required to obtain adequate statistics, and the probability histogram is shown in Fig. 12.7 for $(\Delta B_w/B_w)_{rms} = 3\%$ and $N_p = 1$.

Here, the skewness ≈ -0.41 and the kurtosis ≈ 0.92, indicating a distribution skewed below the mean and more peaked than the normal distribution.

12.3 THE SHORT WAVELENGTH REGIME

In treating the effect of wiggler imperfections in short wavelength free-electron lasers, we turn to a nonlinear formulation based upon Gaussian optical modes. In this section, we continue to examine the wiggler imperfections in a planar wiggler configuration and use the same parabolic-pole-face model wiggler with systematic and random variations in the amplitude used in section 12.1; hence, it is most convenient to employ the Gauss–Hermite modes [11]. In contrast, the Gauss–LaGuerre modes would be most convenient in the treatment of a helical wiggler configuration. As in the preceding nonlinear analyses, the electron dynamics are treated using the complete three-dimensional Lorentz force equations for the magnetostatic and electromagnetic fields; however, since the wavelength of interest here is less than or of the order of several microns, the collective Raman effects due to the beam space-charge waves are neglected.

The Gauss–Hermite modes constitute a complete basis set which is consistent with the planar symmetry imposed by the wiggler geometry. It should be noted, however, that Gaussian optical modes must be used with some caution because the drift tube in which the electron beam propagates also constitutes a waveguide, and the Gaussian modes do not rigorously satisfy the boundary conditions on the drift tube wall. As a result, the analysis must be restricted to cases where the radiation spot size is much smaller than the radius of the drift tube. Since the radiation will be guided by the interaction with the electron beam, this condition is equivalent to the requirement that the electron beam radius be much less than the drift tube radius.

12.3.1 The nonlinear formulation

The vector potential of the Gauss–Hermite modes can be expressed as

$$\delta\mathbf{A}(\mathbf{x},t)=\sum_{l,n=0}^{\infty}\delta A_{l,n}(z)\frac{w_0}{w}\exp(-r^2/w^2)H_n\left(\frac{\sqrt{2}y}{w}\right)\left\{H_l\left(\frac{\sqrt{2}x}{w}\right)\hat{\mathbf{e}}_x\sin\varphi_{l,n}\right.$$
$$-\frac{\sqrt{2}}{k_{l,n}w}\hat{\mathbf{e}}_z\left[\frac{\sqrt{2}x}{w}\frac{z}{z_0}H_l\left(\frac{\sqrt{2}x}{w}\right)\sin\varphi_{l,n}+\left(\frac{\sqrt{2}x}{w}H_l\left(\frac{\sqrt{2}x}{w}\right)\right.\right.$$
$$\left.\left.\left.-H_l'\left(\frac{\sqrt{2}x}{w}\right)\right)\cos\varphi_{ln}\right]\right\} \tag{12.6}$$

where H_n denotes the Hermite polynomials, w_0 denotes the spot size at the radiation waist, and for frequency and wavenumber $(\omega, k_{l,n})$ the phase is given by

$$\varphi_{l,n}=\int_0^z dz'k_{l,n}(z')+\frac{k_0r}{2R}-(l+n+1)\tan^{-1}\left(\frac{z}{z_0}\right)-\omega t \tag{12.7}$$

In addition, $k_0 \equiv \omega/c$ is the free-space wavelength, $w^2 \equiv w_0^2(1 + z^2/z_0^2)$, $R(z) \equiv z(1 + z_0^2/z^2)$, and $z_0 \equiv k_0 w_0^2/2$ is the Rayleigh length. Observe that the amplitude and wavenumber of each mode is allowed to vary slowly in z to describe the growth of the wave as well as the dielectric effect of the beam on the dispersion. The Poynting flux for each mode can be written as

$$P_{l,n} = \frac{2^{l+n} l! n!}{16} \omega k_{l,n} w_0^2 \delta A_{l,n}^2. \tag{12.8}$$

It should be remarked that this representation is correct to first order in $(k_{l,n} w)^{-1} \approx \lambda/w$, where λ denotes the wavelength; hence, this representation is valid only as long as the spot size is much greater than the wavelength. Observe as well that these modes approximate TEM modes only as long as $\lambda \ll w$. For all cases of interest in this paper, this inequality is satisfied, and it will prove convenient to use the TEM approximation for the field.

The dynamical equations which govern the evolution of the amplitude and wavenumber of each mode are found by substitution of the mode representation into Maxwell's equations after averaging the equations over a wave period and orthogonalization in the transverse mode structure. The procedure is formally equivalent to that described for long wavelength free-electron lasers described in Chapter 5, and results in equations of the form [12]

$$\left[\frac{d^2}{dz^2} + \left(\frac{\omega^2}{c^2} - k_{l,n}^2\right)\right] \delta a_{l,n} = \frac{4\omega_b^2}{c^2} \frac{1}{2^{l+n} l! n!} \frac{w_0}{w} \left\langle \frac{v_x}{|v_z|} \exp(-r^2/w^2) H_l\left(\frac{\sqrt{2}x}{w}\right) H_n\left(\frac{\sqrt{2}y}{w}\right) \sin \varphi_{l,n} \right\rangle \tag{12.9}$$

and

$$2k_{l,n}^{1/2} \frac{d}{dz}(k_{l,n}^{1/2} \delta a_{l,n}) = \frac{4\omega_b^2}{c^2} \frac{1}{2^{l+n} l! n!} \frac{w_0}{w} \left\langle \frac{v_z}{|v_z|} \exp(-r^2/w^2) H_l\left(\frac{\sqrt{2}x}{w}\right) H_n\left(\frac{\sqrt{2}y}{w}\right) \cos \varphi_{l,n} \right\rangle \tag{12.10}$$

where $\delta a_{l,n} \equiv e\delta A_{l,n}/m_e c^2$, ω_b is the beam plasma frequency, $\mathbf{v}$ is the instantaneous electron velocity, e and m_e are the electronic charge and rest mass, and c is the speed of light *in vacuo*. Here, the averaging operator is defined for the same electron distribution described in equations (5.120–5.122), and can be written as

$$\langle (\cdots) \rangle \equiv \frac{A}{4\pi^2 w_0^2} \int_0^{2\pi} d\phi_0 \int_0^{p_0} dp_{z_0} \beta_{z_0} \exp[-(p_{z_0} - p_0)^2/\Delta p_z^2] \times \int_{A_b}\int dx_0 dy_0 \sigma_\perp(x_0, y_0) \int_{-\pi}^{\pi} d\psi_0 \sigma_\parallel(\psi_0)(\cdots) \tag{12.11}$$

where A_b denotes the initial cross sectional area of the beam, $\phi_0 \equiv \tan^{-1}(p_{y_0}/p_{x_0})$, $\beta_{z_0} \equiv v_{z_0}/c$, $\psi_0(\equiv -\omega t_0$, where t_0 is the injection time) is the initial ponderomotive phase, $\sigma_\parallel(\psi_0)$ and $\sigma_\perp(x_0, y_0)$ describe the initial beam distributions in phase and cross section, and A is the normalization constant defined in equation (5.121).

These equations for the amplitude and phase of each of the Gauss–Hermite modes included in the ensemble of modes are integrated simultaneously with the three-dimensional Lorentz force equations for an ensemble of electrons. As such, the procedure is capable of treating the self-consistent injection of the beam into the wiggler, emittance growth due to the inhomogeneities in the wiggler and radiation fields, betatron oscillations, wiggler imperfections and optical guiding of the radiation.

12.3.2 The numerical analysis

The short wavelength free-electron laser parameters under study here corresponds to the proposed Linac Coherent Light Source (LCLS) at the Stanford Linear Accelerator Center in the X-ray spectrum [13]. Due to the lack of sources to drive a MOPA in this spectral band, the device would operate in the SASE mode using an energy of 15 GeV. The beam pulses would be compressed in the axial direction to achieve a peak current of 5 kA, and would have a radial extent of only 16 μm. The proposed wiggler would achieve a 16 kG amplitude at a period of 2.7 cm, and it is assumed for this study that the entry taper region is 10 wiggler periods in length. This implies a resonant wavelength in the neighborhood of 1.4 Å. It shall also be assumed that the initial power is 10 kW in the TEM_{00} mode, and that the initial spot size matches the beam radius at 16 μm.

The issue of quantum mechanical effects should be discussed for this configuration if only for the purpose of dismissing them. Quantum mechanical effects can be neglected if the spreading of the electron wave packet over the length of the wiggler is less than the radiation wavelength. As shown in equation (1.28) this can be formulated as

$$\Delta z = \frac{\lambda_c L}{\gamma_0 \lambda_w} \ll \lambda \tag{12.12}$$

where Δz denotes the spreading of the electron wave packet. For the parameters of interest to the LCLS, the spreading of the wave packet over a wiggler length of 30 m is $\Delta z \approx 9.2 \times 10^{-4}$ Å which is approximately three orders of magnitude less than the 1.4 Å wavelength. At this length of wiggler, therefore, quantum mechanical effects can be neglected from the treatment. However, this is the closest that any operational or proposed free-electron laser has approached to the regime where quantum mechanical effects are important, and if a wiggler of 100 m or more in length were required, then quantum mechanical effects might become important.

For this case, it is found that adequate convergence of the mode superposition is achieved using 38 modes. A plot of the power as a function of axial position is shown in Fig. 12.8 for an ideal beam at a wavelength of 1.4323 Å. The interaction saturates at a power of approximately 98 GW for an efficiency of 0.13% over an interaction length of 30 m. This contrasts with a saturated power level of about

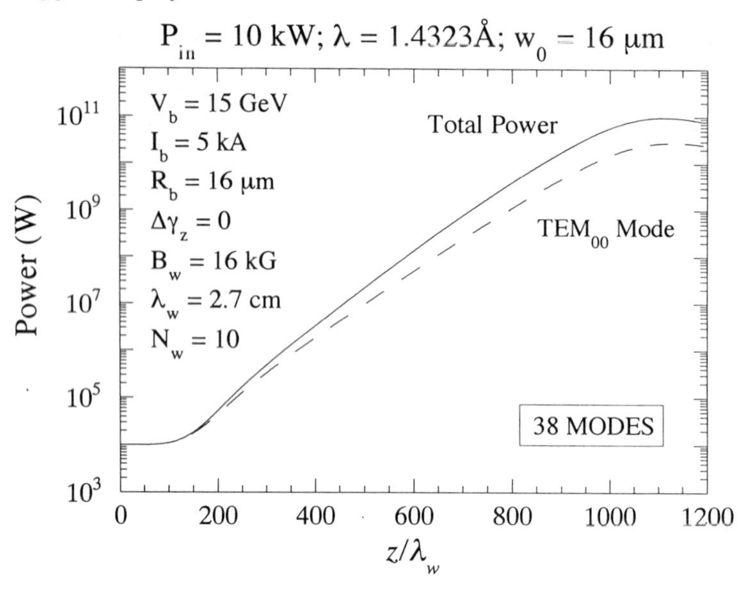

Fig. 12.8 Evolution of the power versus axial position.

73 GW over a saturation length of 52 m which is found using the TEM_{00} mode alone. As a consequence, as in the infrared example, while the TEM_{00} mode is dominant, it constitutes only about 27% of the total power at saturation. The relative mode amplitudes at saturation are shown in Fig. 12.9 and are normalized to the power in the TEM_{00} mode. It is clear that, as in the infrared example, the dominant modes are the TEM_{00}, TEM_{02}, TEM_{20}, TEM_{04}, and TEM_{40}, but substantial amounts of power are contained in the higher order modes.

As might be expected, the interaction for the X-ray free-electron laser is extremely sensitive to the axial energy spread. The average growth rate associated with the interaction in Fig. 12.8 is $|\mathrm{Im}\, k|/k_w \approx 1.15 \times 10^{-3}$, which implies that the transition to the thermal regime is found for $\Delta\gamma_z/\gamma_0 \approx 0.05\%$. A plot of the variation in the extraction efficiency and the saturation length at a wavelength of 1.4323 Å spanning the thermal transition regime is shown in Fig. 12.10. As shown in the figure, the extraction efficiency drops by more than half as the axial energy spread increases to 0.1%, and by 39% at the thermal transition. The growth rate also decreases, and the saturation length increases from about 30 m for an ideal beam to 39 m at the high end of this range. As such, it is necessary to keep the beam emittance as small as possible to remain far below the thermal transition for optimal performance, and the achievement of near-peak efficiencies requires that $\Delta\gamma_z/\gamma_0 \leqslant 0.015\%$.

The effect of wiggler imperfections is studied for an ideal beam as in the previous section for the long wavelength free-electron laser. Here we choose to

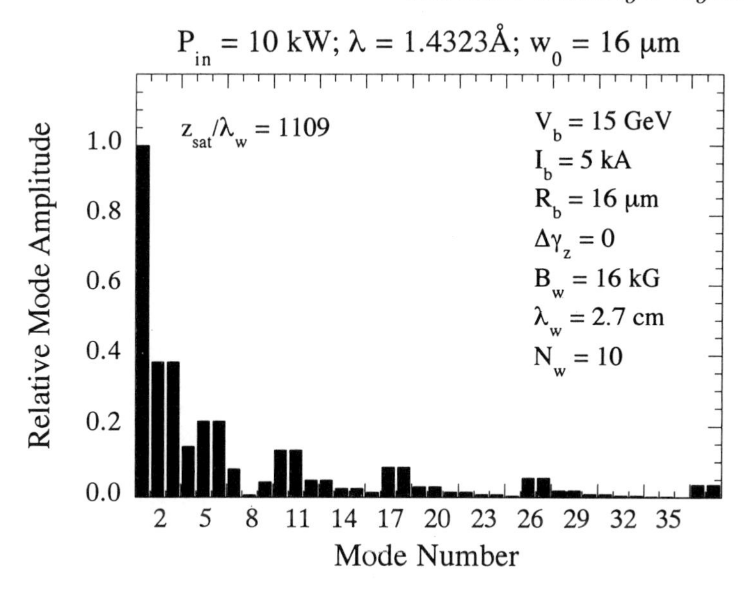

Fig. 12.9 Mode decomposition at saturation.

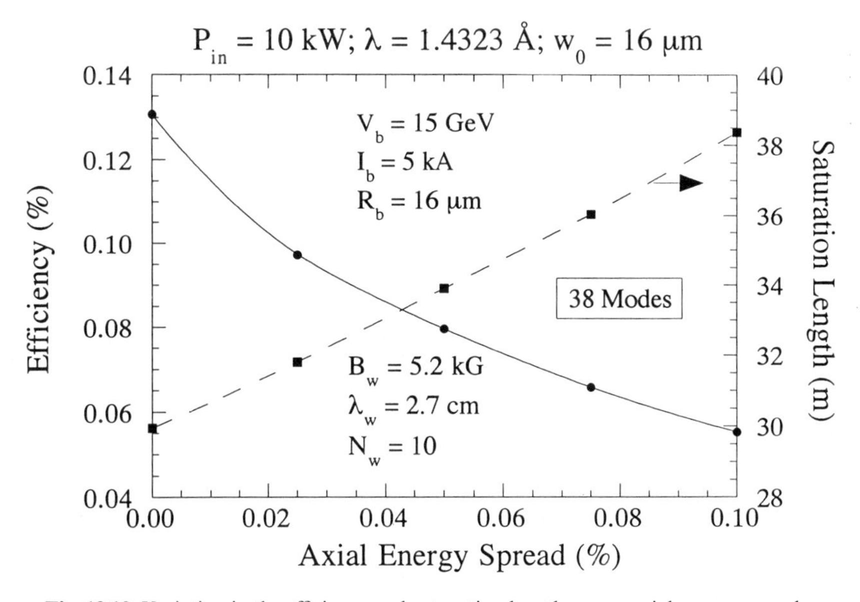

Fig. 12.10 Variation in the efficiency and saturation length versus axial energy spread.

study the case in which the random imperfections occur at intervals of half of the wiggler period; hence, $N_p = 2$. The variation in the average extraction efficiency versus $(\Delta B_w/B_w)_{rms}$ is shown in Fig. 12.11 in which the average is over 35 randomly chosen wiggler fluctuation distributions and the error bars denote the

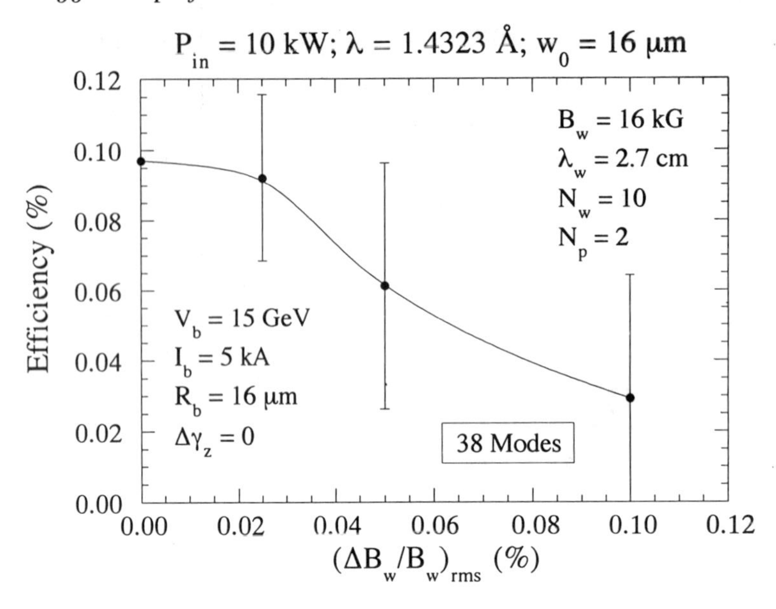

Fig. 12.11 Variation in the efficiency versus the rms wiggler fluctuation.

standard deviation. Evidently, the efficiency falls by a factor of approximately 70% as $(\Delta B_w/B_w)_{rms}$ increases to 0.1%.

The choice of $(\Delta B_w/B_w)_{rms} = 0.1\%$ corresponds to an induced axial energy spread of $\Delta\gamma_z/\gamma_0 \approx 0.8\%$. This shows far less of an effect on the interaction due to axial energy variations than was found for the thermal energy spread. As shown in Fig. 12.10, the thermal axial energy spread results in a decline of approximately 43% in the efficiency as the thermal energy spread increases to 0.1%. A comparable value of the axial energy spread induced by wiggler imperfections is found for $(\Delta B_w/B_w)_{rms} \approx 0.012\%$. As seen in Fig. 12.10, however, this value of the rms wiggler imperfections results in a negligible decline in the efficiency. Hence, as found in the long wavelength example, the axial energy spread induced by wiggler imperfections is more benign than that arising from beam thermal spread.

12.4 SUMMARY

In summary, a self-consistent analysis of the effect of random wiggler errors on the saturation efficiency of the free-electron laser has been presented in which no *a priori* assumption of a random walk of the electron orbits has been imposed. For the specific parameters under study, the results indicate that the effects of random wiggler errors are relatively more benign than the effects of beam energy spread, and some error configurations chosen at random were found to result in efficiency enhancements due to effective increases.

One final issue that merits discussion is the relevance of a statistical analysis of wiggler errors to any specific free-electron laser experiment. Consider the construction of a planar wiggler from an assembly of permanent magnets. Such a collection of permanent magnets can be expected to exhibit a fluctuation in magnetization of as much as $\pm 5\%$. If the wiggler is constructed using a random selection process, then the present analysis may be expected to characterize the statistical properties of *a large number of wigglers constructed in this way* and the performance of any specific wiggler can be expected to fall within the range determined by the statistical analysis. Wigglers are not constructed by a random selection process, however, but by sorting the permanent magnets to either minimize or optimize the field fluctuations. In either case, however, the performance of a free-electron laser is a deterministic process which is governed by the electron trajectories in the specific wiggler employed in the device. The formulation described herein is capable of modeling specific free-electron lasers by the relatively simple expedient of using measured values for the pole-to-pole fluctuations $\{\Delta B_w\}$ of any given wiggler.

REFERENCES

1. Hoyer, E., Chan, T., Chin, J. Y. G., Halbach, K., Kim, K. J., Winick, H. and Yang, J. (1983) The beam line VI Rec-Steel Hybrid Wiggler for SSRL. *IEEE Trans. Nucl. Sci.* **NS-30**, 3118 (1985).
2. Kincaid B. M. (1985) Random errors in undulators and their effects on the radiation spectrum. *J. Opt. Soc. Am. B*, **2**, 1294.
3. Marable, W. P., Esarey, E. and Tang C. M. (1990) Vlasov theory of wiggler field errors and radiation growth in a free-electron laser. *Phys. Rev. A*, **42**, 3006.
4. Yu, L. H., Krinsky, S., Gluckstern, R. L. and van Zeijts, J. B. J. (1992) Effect of wiggler errors on free-electron laser gain. *Phys. Rev. A*, **45**, 1163.
5. Elliot, C. J. and McVey B. (1987) Analysis of undulator field errors for XUV free-electron lasers, in *Undulator Magnets for Synchrotron Radiation and Free Electron Lasers* (eds. Bonifacio, R., Fonda, L. and Pellegrini, C.), World Scientific, Singapore, p. 142.
6. Shay, H. D. and Scharlemann, E. T. (1988) Reducing sensitivity to errors in free-electron laser amplifiers. *Nucl. Instr. Meth.*, **A272**, 601.
7. Marable, W. P., Tang, C. M. and Esarey, E. (1991) Simulation of free-electron lasers in the presence of correlated magnetic field errors. *IEEE J. Quantum Electron.*, **QE-27**, 2693.
8. Freund, H. P. and Jackson, R. H. (1992) Self-Consistent Analysis of Wiggler Field Errors in Free-Electron Lasers. *Phys. Rev. A*, **45**, 7488.
9. Freund, H. P. and Jackson, R. H. (1993) Self-Consistent Analysis of the Effect of Wiggler Field Errors in Free-Electron Lasers. *Nucl. Instr. Meth.*, **A331**, 461.
10. Bluem, H. (1990) *Generation of Harmonic Radiation from a Ubitron/FEL Configuration*, Ph.D. Thesis, University of Maryland.
11. Marcuse D. (1982) *Light Transmission Optics*, Van Nostrand, New York, Chapter 6.
12. Freund, H. P. (1995) Nonlinear theory of short wavelength free-electron lasers. *Phys. Rev. E.*, **52**, 5401.
13. Winick, H. *et al.* (1993) *Proceedings of the 1993 IEEE Particle Accelerator Conference* (IEEE Cat. No. 93CH3279-7), p. 1445.

13
The reversed-field configuration

Thus far, the primary consideration in the treatment of free-electron lasers with the combination of helical wigglers and axial guide fields has been devoted to configurations in which the guide field is directed parallel to the wiggler, since the bulk of free-electron laser experiments conducted to date have employed this geometry. However, the physical mechanism of the free-electron laser will also operate when the axial guide field is directed anti-parallel to the wiggler field. We shall refer to this as a **reversed-field** configuration.

One difference which arises from the reversed-field geometry is that the wiggler-induced transverse velocity is reduced relative to that found in the absence of a guide field. Hence, the linearized gain for the interaction is reduced relative to typical values expected for the Group I and II regimes. However, this does not necessarily imply that the saturation efficiency in the reversed-field case is also reduced. Indeed, one experiment which studied both orientations of the guide field found higher efficiencies for the reversed-field orientation [1, 2]. In this chapter, we shall discuss the physics of the reversed-field interaction including both consideration of the single-particle trajectories and a detailed analysis of the aforementioned reversed-field experiment. The nonlinear formulation used in the analysis is discussed in section 5.5.

13.1 SINGLE-PARTICLE TRAJECTORIES

The basic orbital dynamics of electrons in reversed-field configurations follow routinely from the discussion in Chapter 2. The steady-state orbits for such a configuration are governed by equations (2.53) and (2.54) as in the case in which the axial magnetic field is directed parallel to the wiggler field. The principal distinction which obtains in the reversed-field configuration is that there is no resonant enhancement in the wiggler-induced transverse velocity v_w when the Larmor period associated with the axial field is comparable to the wiggler period. As a result, there is only one class of trajectory for this configuration and it is a stable one. An example of the variation in the axial velocity as a function of the magnitude of the reversed axial field is shown in Fig. 13.1 in which we plot $v_\parallel/c$ versus the magnitude of the axial field for a beam energy of 750 keV, and a wiggler field with a magnitude of 1.47 kG and a period of 3.18 cm. As shown in the figure, the axial velocity is relatively constant over the entire range of guide fields

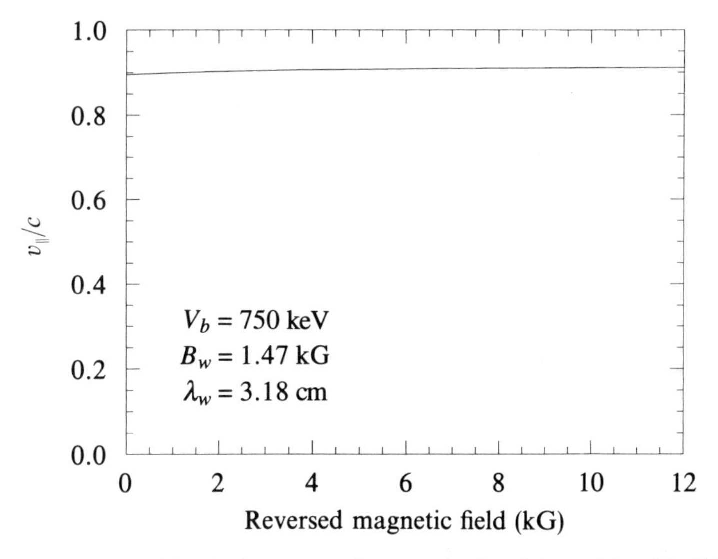

Fig. 13.1 Plot of the axial velocity versus the magnitude of the axial guide field in the reversed-field orientation.

shown. Note that the parameters used in the figure correspond to a reversed-field free-electron laser amplifier experiment [1,2] which will be analysed in detail.

However, the electron dynamics in the reversed-field configuration are far more complex than is indicated by the study of the steady-state trajectories. The reason for this is that the inhomogeneity in the wiggler field introduces a sinusoidal driving term to the electron orbit equations [3–5]. This term arises from the fact that an electron on a helical orbit centred off the axis of symmetry experiences a sinusoidally varying wiggler field which acts to drive the electrons at a period close to the wiggler period. Hence, this effect becomes important for electron beams which are large enough that a substantial fraction of the electrons are located relatively far from the symmetry axis. In addition, there is a resonant enhancement in this effect when the Larmor period associated with the axial field is comparable to the wiggler period. We shall discuss this **anti-resonance** in more detail in the comparison of the simulation with the aforementioned experiment, since the orbital irregularities introduced by this effect have a significant impact on the output of the free-electron laser.

The fundamental physics of this effect of the wiggler inhomogeneity and the anti-resonant enhancement process can be understood by means of a relatively simple treatment of the single-particle orbit dynamics [4]. In order to accomplish this, we turn to a guiding-centre formalism developed in Chapter 2. In this treatment, the electron position and velocity are written as $\boldsymbol{x} = \boldsymbol{x}_c + \boldsymbol{x}_{osc}$, where the subscript c denotes the guiding-centre position and osc denotes the various oscillatory motions. Under the assumption that the guiding-centre position is fixed (i.e. $\mathbf{v} = \mathbf{v}_{osc}$), expansion of the orbit equations about the guiding-centre position results in the following equations for the electron velocity (where we drop the subscript *osc* for convenience)

$$\left.\begin{aligned} \frac{d}{dt}v_1 &= -(\Omega_0 - k_w v_3)v_2 + \Omega_w v_3 I_2(\lambda_c)\sin 2\chi_c, \\ \frac{d}{dt}v_2 &= (\Omega_0 - k_w v_3)v_2 - \Omega_w v_3[I_0(\lambda_c) + I_2(\lambda_c)\cos 2\chi_c], \\ \frac{d}{dt}v_3 &= \Omega_w v_2[I_0(\lambda_c) + I_2(\lambda_c)\cos 2\chi_c] - \Omega_w v_1 I_2(\lambda_c)\sin 2\chi_c, \end{aligned}\right\} \tag{13.1}$$

where (r_c, θ_c) denote the guiding-centre position in cylindrical coordinates, and $\lambda_c \equiv k_w r_c$ and $\chi_c \equiv \theta_c - k_w z$. If we now expand about the steady-state orbits via $v_1 = v_w + \delta v_1, v_2 = \delta v_2$ and $v_3 = v_\| + \delta v_3$, where

$$v_w \equiv \frac{\Omega_w v_\|}{\Omega_0 - k_w v_\|} I_0(\lambda_c), \tag{13.2}$$

then the equations for the perturbations are

$$\left.\begin{aligned} \frac{d}{dt}\delta v_1 &= -(\Omega_0 - k_w v_\|)\delta v_2 + \Omega_w v_\| I_2(\lambda_c)\sin 2\chi_c, \\ \frac{d}{dt}\delta v_2 &= (\Omega_0 - k_w v_\|)\delta v_1 - \frac{v_w}{v_\|}\Omega_0 \delta v_3 - \Omega_w v_\| I_2(\lambda_c)\cos 2\chi_c, \\ \frac{d}{dt}\delta v_3 &= \Omega_w \delta v_2 I_0(\lambda_c) - \Omega_w v_w I_2(\lambda_c)\sin 2\chi_c. \end{aligned}\right\} \tag{13.3}$$

Note that we have also neglected terms which vary as $\delta v I_2(\lambda_c)$ under the assumption that $\lambda_c < 1$ as well. In this representation, the electrons execute a helical trajectory centred on the guiding-centre. In addition, this representation is quasi-idealized in the sense that the transverse velocity includes three-dimensional effects only in the inclusion of the $I_0(\lambda_c)$ function which describes the effect of the off-axis increase in the magnitude of the field at the guiding-centre. These equations may be reduced to a set of second-order differential equations

$$\left(\frac{d^2}{dt^2} + \Omega_r^2\right)\begin{bmatrix} \delta v_1 \\ \delta v_2 \\ \delta v_3 \end{bmatrix} = v_\| \Omega_w I_2(\lambda_c) \begin{bmatrix} (\Omega_0 - 3k_w v_\|)\cos 2\chi_c \\ (\Omega_0 - 3k_w v_\| + \beta_w^2 \Omega_0)\sin 2\chi_c \\ -\beta_w(\Omega_0 - 3k_w v_\|)\cos 2\chi_c \end{bmatrix}, \tag{13.4}$$

where $\beta_w \equiv v_w/v_\|$, and Ω_r is given by equation (2.21). The particular solutions of these equations are

$$\begin{bmatrix} \delta v_1 \\ \delta v_2 \\ \delta v_3 \end{bmatrix} = \frac{\Omega_w v_\parallel I_2(\lambda_c)}{(\Omega_0 + k_w v_\parallel)(\Omega_0 - 3k_w v_\parallel) + \beta_w^2 \Omega_0 (\Omega_0 - k_w v_\parallel)} \times \begin{bmatrix} (\Omega_0 - 3k_w v_\parallel)\cos 2\chi_c \\ [(1+\beta_w^2)\Omega_0 - 3k_w v_\parallel]\sin 2\chi_c \\ -\beta_w(\Omega_0 - 3k_w v_\parallel)\cos 2\chi_c \end{bmatrix}. \tag{13.5}$$

In order to demonstrate the anti-resonant nature of this effect, consider the limit in which $\beta_w^2 \ll 1$ for which these particular solutions reduce to the simpler form

$$\begin{bmatrix} \delta v_1 \\ \delta v_2 \\ \delta v_3 \end{bmatrix} = \frac{\Omega_w v_\parallel I_2(\lambda_c)}{\Omega_0 + k_w v_\parallel} \begin{bmatrix} \cos 2\chi_c \\ \sin 2\chi_c \\ -\beta_w \cos 2\chi_c \end{bmatrix}, \tag{13.6}$$

which clearly shows the anti-resonant enhancement in the perturbation when $\Omega_0 \approx -k_w v_\parallel$. However, since the particular solution also depends upon $I_2(\lambda_c)$, this effect will not become appreciable unless the electron guiding-centre is located relatively far from the symmetry axis.

Note, however, that this effect of the inhomogeneity in the wiggler field, even near the anti-resonance, is not as serious a problem for beam transport as the Group I and II orbital instabilities which occur when the axial guide field is oriented parallel to the wiggler field.

13.2 THE EXPERIMENTAL DESCRIPTION

The basic configuration used in the experiment was that of an amplifier in which a weakly relativistic electron beam was injected into a cylindrical waveguide in the presence of both a helical wiggler field and an axial guide solenoidal field. The wave–particle interaction was with the fundamental TE_{11} mode of the waveguide at a frequency of 33.39 GHz, which corresponds to the frequency of the magnetron used to drive the amplifier.

The electron beam was generated by a Physics International Pulserad 110A by means of field emission from a graphite cathode, and the beam energy used in the experiment was 750 keV ($\pm$ 50 keV). The quality (i.e. the emittance and energy spread) of the beam delivered to the interaction region was controlled by scraping the beam with a shaped graphite anode. This technique was originally pioneered at the Naval Research Laboratory for use in a free-electron laser experiment driven by the VEBA accelerator [6]. In the experiment of present interest, the shaped anode–cathode geometry results in a beam with a radius of 0.25 cm (corresponding to the radius of the anode aperture) and an axial energy spread estimated to be $\Delta\gamma_z/\gamma_0 \approx 1.5\%$. This energy spread corresponds to a normalized root-mean-square beam emittance of $\varepsilon_n \lesssim 4.4 \times 10^{-2}$ cm rad.

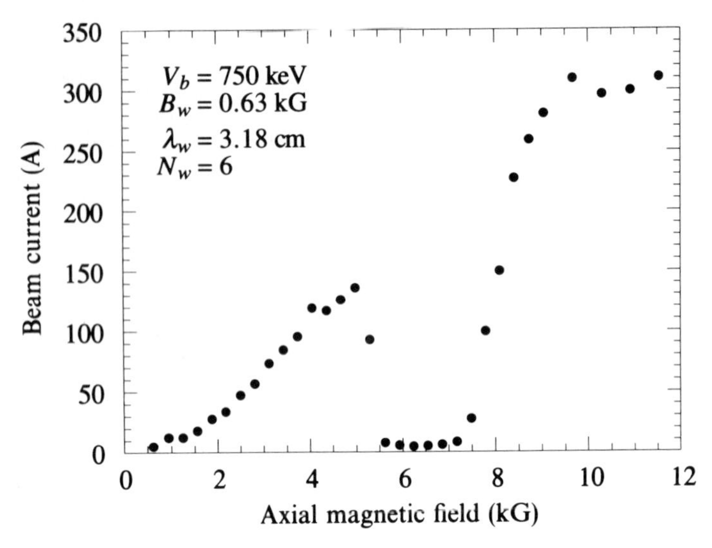

Fig. 13.2 Current propagation as a function of the axial guide field for the parallel orientation of the field.

The current available using this configuration was of the order of 300 A ($\pm$ 30 A) as delivered to the interaction region at the entrance to the wiggler. However, the amount of current which could be propagated through the wiggler/guide field configuration varied based upon the stability of the electron trajectories. Current propagation data indicated quite different results depending upon the orientation of the axial guide field. The results found for the current propagating through the wiggler as a function of the magnitude of the axial guide field for orientations in which the guide field is directed parallel and anti-parallel (referred to as the reversed-field configuration) to the wiggler are shown in Figs 13.2 and 13.3 respectively (data courtesy of Conde and Bekefi). In each case, the amount of current which could be propagated generally increased with increases in the magnitude of the guide field. An exception to this, however, was found when the guide field was oriented parallel to the wiggler in the vicinity of the magnetic resonance at which the Larmor period associated with the guide field is comparable to the wiggler period. In this case, the orbital instability in the electron trajectories described in Chapter 2 prevents propagation of the beam. In addition, the increase in the propagated current levelled off in the field-reversed case when the Larmor and wiggler periods were comparable. For convenience, we shall refer to this as the **anti-resonance**.

The explanation for this anti-resonant effect is that the inhomogeneity in the wiggler field introduces a sinusoidal driving term to the electron orbit equations discussed in the previous section which arises from the fact that an electron on a helical orbit centred off the axis of symmetry experiences a sinusoidally varying wiggler field which acts to drive the electrons at a period close to the wiggler

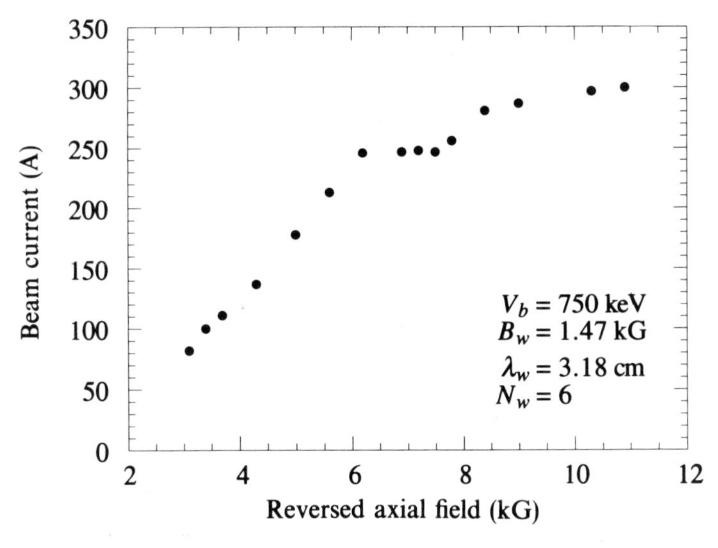

Fig. 13.3 Current propagation as a function of the axial guide field for the reversed-field configuration.

period. Hence, this effect becomes important for electron beams which are big enough that a substantial fraction of the electrons are located relatively far from the symmetry axis. In addition, it is resonant for axial fields close to the anti-resonance. We shall discuss this in more detail during the comparison of the simulation with the power measurements, since the orbital irregularities introduced by this effect have a significant impact on the growth of the signal.

The wiggler field is produced by a bifilar helix with a period of $\lambda_w = 3.18$ cm, a length of $50\lambda_w$ and an adiabatic entry taper which is six wiggler periods in length. The wiggler amplitude was continuously adjustable up to an amplitude of approximately 1.8 kG. The axial guide field could be adjusted up to a maximum amplitude of almost 12 kG.

The beam propagated through a cylindrical waveguide of 0.51 cm in radius, which provided for a wave–particle resonance with the fundamental TE_{11} mode in the vicinity of 35 GHz. The free-electron laser was operated as an amplifier, and a magnetron which produced approximately 17 kW ($\pm 10\%$) at a frequency of 33.39 GHz was used as a driver. Since the output from the magnetron was linearly polarized, this corresponded to approximately 8.5 kW in the right-hand-circularly-polarized state which was capable of interacting with the helical wiggler geometry.

The output from the amplifier showed the greatest efficiency for the field-reversed configuration. In this case, a peak power of 61 MW for a conversion efficiency of 27% was found for a wiggler field magnitude of approximately 1.47 kG and an axial magnetic field of 10.92 kG. The current which could be propagated in these fields was near the maximum of 300 A. The output power for the field-reversed configuration also showed a severe decrease in the vicinity of the anti-resonance, dropping by more than three orders of magnitude. The power observed when the axial magnetic field was oriented parallel to the wiggler was much less than for the field-reversed configuration, and showed a maximum measured power of approximately 4 MW. Details of the output power spectra will be presented in the comparison with the theoretical results.

13.3 THE EXPERIMENTAL COMPARISON

The experiment has been operated with the axial magnetic field oriented both parallel and anti-parallel to the wiggler field, and we shall discuss the comparison with each of these regimes separately. Features common to both regimes, however, involve the choice of various system parameters as well as the initialization of the modes included in the simulation. The basic analysis for the nonlinear Raman regime is described in section 5.4, and a more detailed discussion can be found in references [4] and [7].

Features common to all cases studied derive from the geometry of the system. Specifically, we take the waveguide radius to be $R_g = 0.51$ cm, the wiggler period to be $\lambda_w = 3.18$ cm and the wiggler entry taper as $N_w = 6$ wiggler periods in length. In addition, while the beam current varies with the magnitude of the axial guide field, the beam energy is 750 keV and the radius is fixed at the aperture of the anode to $R_b = 0.25$ cm.

Since the frequency of the amplifier experiment is fixed by the 33.39 GHz magnetron, the beam energy of 750 keV and the waveguide radius of 0.51 cm ensure that a resonant interaction is possible only with the fundamental TE_{11} mode of the guide. Since the magnetron produces approximately 17 kW with a linear polarization, we assume that only half of this power is available with the correct circular polarization to interact with the beam. Hence, the initial power of the TE_{11} mode is chosen to be 8.5 kW. The collective Raman interaction in a free-electron laser couples the TE_{11} mode, in principle, with each of the Gould–Trivelpiece modes having an azimuthal mode number of $l = 0$. In practice, however, we find that inclusion of only the lowest order radial mode is required to give reasonable agreement with the experiment. Note that the axial electric field of this mode has the same transverse variation as the TM_{01} waveguide mode. Hence, the following simulations have been performed using only one waveguide mode and one Gould–Trivelpiece mode.

13.3.1 The reversed-field configuration

The first case we consider is that of a reversed-field configuration in which the nominal experimental magnetic field parameters were an axial field magnitude of 10.92 kG and a wiggler field of 1.47 kG. The transmitted current for these field parameters was 300 A ($\pm$ 10%), and the axial energy spread of the beam is assumed to be 1.5% as indicated in the experiment. These parameters represent the case of the peak power observed in the experiment of 61 MW.

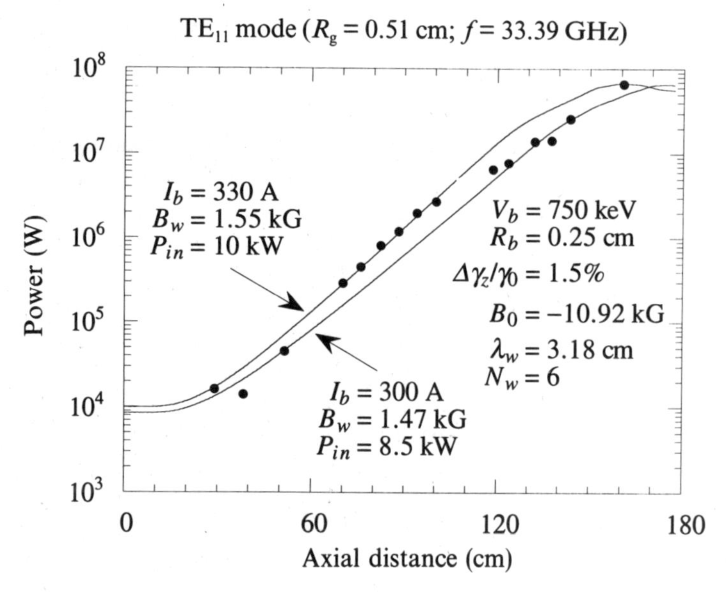

Fig. 13.4 The evolution of the power with axial position as determined from the theory and from the experiment (dots) for a reversed-field configuration.

The comparison of the theoretical formulation and simulation and the experiment is shown in Fig. 13.4 in which we plot the power as a function of axial position, and in which the dots represent the power as measured in the experiment. As shown in the figure, the theoretical analysis was performed for two sets of parameters. The first corresponds to the nominal experimental values given above, and the second corresponds to the upper limits on (1) the current, (2) the wiggler field and (3) the input power (due to the experimental uncertainties) of 330 A, 1.55 kG and 10 kW respectively. As is evident in the figure, the agreement between the experimental measurements and the theory is good, and virtually all the data points fall between these two curves. The saturated power for these two choices of the current, wiggler field and input power differ only marginally and are close to the 61 MW measured in the experiment. The principal difference is in the saturation length which is due to a small discrepancy in the growth rates for these two cases.

The interaction efficiency in this case is approximately 27%, and is comparable to that which is expected due to the phase trapping of the electron beam in the ponderomotive wave formed by the beating of the wiggler and radiation fields. The efficiency estimated by this technique represents the energy lost by the electrons as the axial velocity decreases by an amount $\Delta v_z = 2(v_z - v_{ph})$, where $v_{ph} = \omega/(k + k_w)$ is the phase velocity of the ponderomotive wave. Here (ω, k) are the angular frequency and wavenumber of the electromagnetic wave, and k_w is the wiggler wavenumber. The phase trapping estimate for the efficiency can be expressed in the form

$$\eta \approx 2\gamma_{\parallel}^{2} \frac{v_{\parallel}}{c} \left(\frac{v_{\parallel}}{c} - \frac{\omega/c}{k + k_w} \right). \tag{13.7}$$

The inclusion of Raman effects in this estimate is accomplished by the choice of the appropriate frequency and wavenumber in the phase velocity of the ponderomotive wave. The theoretical formulation includes collective Raman effects, and for the example shown results in a normalized wavenumber of $k/k_w \approx 2.98$ for the TE_{11} mode at a frequency of 33.39 GHz (i.e. $\omega/ck_w \approx 3.5$). As a consequence, the phase velocity of the ponderomotive wave is $v_{ph}/c \approx 0.879$. Note that this wavenumber differs from the normalized wavenumber for a vacuum TE_{11} mode for which $k/k_w \approx 2.985$, and that the dielectric loading of the waveguide due to the interaction in either the Raman or Compton regime is included in the theory. In addition, the axial electron velocity for the steady-state orbit in this field configuration is $v_z/c \approx 0.911$ which gives $\gamma_z^2 \approx 5.89$. As a result, the estimate of the phase trapping efficiency is approximately 34%. It is important to recognize that estimates such as this must be employed with caution, and should be taken in the present case to indicate the possibility of high efficiency operation. However, it should also be noted that while the estimate is higher than that found in either the simulation or the experiment, it does not include the effects of an axial energy spread.

The variation in the output power over an interaction length of 150 cm as a function of the magnitude of the reversed magnetic field is shown in Fig. 13.5.

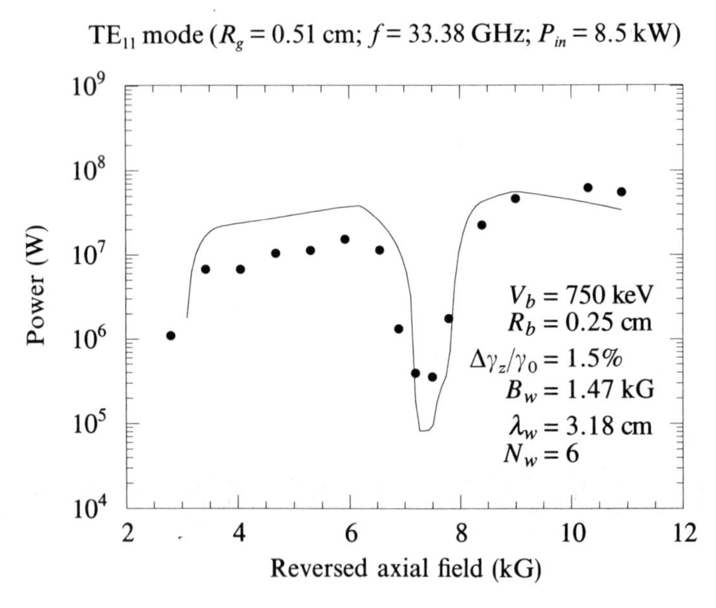

Fig. 13.5 The variation in the output power versus the magnitude of the reversed axial field as measured in the experiment (dots) over an interaction length of 150 cm and in simulation.

Again, the dots represent the experimentally measured power and the curve is the result from the theory. The current used in the simulation for each value of the axial field corresponds to the transmitted current shown in Fig. 13.3. Agreement between the experiment and theory is good across the entire range studied. Of particular note, however, is the sharp decrease in the output power in the vicinity of the anti-resonance at axial field magnitudes between approximately 7 and 8.5 kG.

The source of this anti-resonant decrease in the interaction efficiency is the irregularities introduced into the electron trajectories by the transverse inhomogeneity in the wiggler. For this particular example, the radius of the wiggler-induced motion (i.e. the radius of the helical steady-state trajectory) is approximately 0.04 cm. However, the beam radius is 0.25 cm in this experiment. As a consequence, the electrons at the outer regions of the beam are quite sensitive to the wiggler inhomogeneity, and experience a sinusoidally varying wiggler field during the course of their trajectories. These effects are implicitly included in the theoretical formulation, and we can illustrate their effect on the electron beam by examining the orbits of selected electrons in the simulation.

The first case we shall consider is that of an electron which is located near the centre of the beam upon entry to the wiggler for a reversed axial field of 7.2 kG, which is in the centre of the anti-resonance region. The evolution of the trajectory in the transverse plane is shown in Fig. 13.6, in which the jaggedness is a artifact introduced into the figure by plotting only every tenth point in the integration.The orbit shown in the figure exhibits the expected **spin-up** of the electron trajectory due to the adiabatic injection into the wiggler field, and the electron

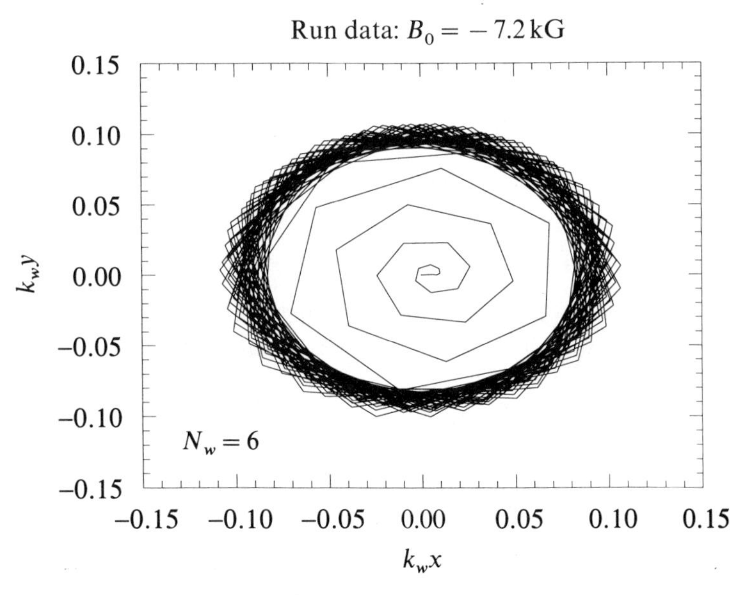

Fig. 13.6 The cross-sectional evolution of the trajectory of an electron injected near the centre of the beam for an axial field in the vicinity of the anti-resonance.

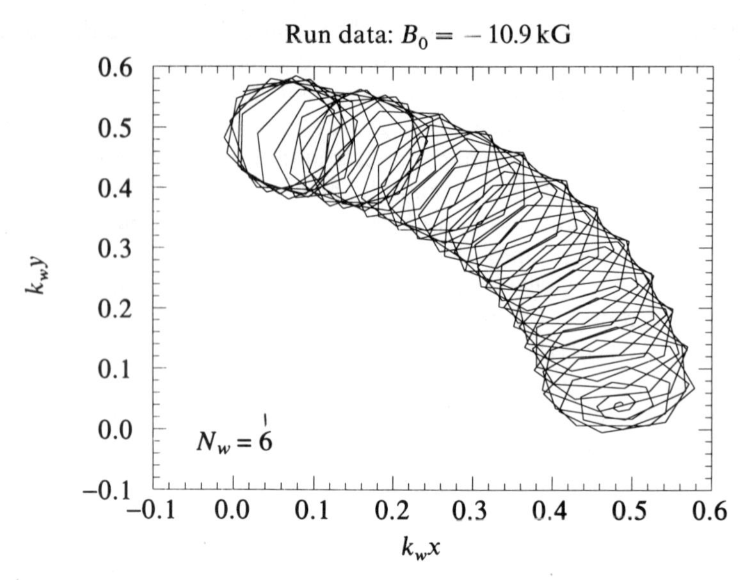

Fig. 13.7 The cross-sectional evolution of the trajectory of an electron injected at the edge of the beam for an axial field far from the anti-resonance.

executes a near-helical steady-state orbit upon transition to the uniform-wiggler region (i.e. after the six wiggler periods of the entry taper region). The principal characteristics of such an orbit are the regular wiggler-induced transverse velocity which mediates the interaction, and a near-uniform axial velocity which permits the resonant wave–particle interaction to occur over an extended interaction length. This behaviour is also found for electrons at the centre of the beam for axial fields away from the anti-resonance. The differences occur principally for the edge electrons.

In order to show the nature of these differences, we focus on a characteristic electron which is initially located at the edge of the beam ($x \approx 0.25$ cm and $y \approx 0$) upon entry to the wiggler. The cross-sectional evolution of the trajectory of such an electron for a reversed axial field of 10.92 kG is shown in Fig. 13.7. This case is not in the vicinity of the anti-resonance, and the orbit illustrates several features. The predominant feature is the aforementioned spin-up of the electron due to the bulk wiggler motion. However, the electron also executes slower motion which corresponds to betatron and Larmor motion due to the wiggler inhomogeneity which manifests as a guiding-centre drift in the counterclockwise direction. This orbit is fairly regular, and does not result in any significant degradation in the interaction efficiency.

However, the situation is quite different for an edge electron in the vicinity of the anti-resonance. The cross-sectional evolution of such a trajectory is shown in Fig. 13.8 for an axial field magnitude of 7.2 kG. The orbit in this case exhibits the initial spin-up due to the bulk wiggler action, but subsequently undergoes what

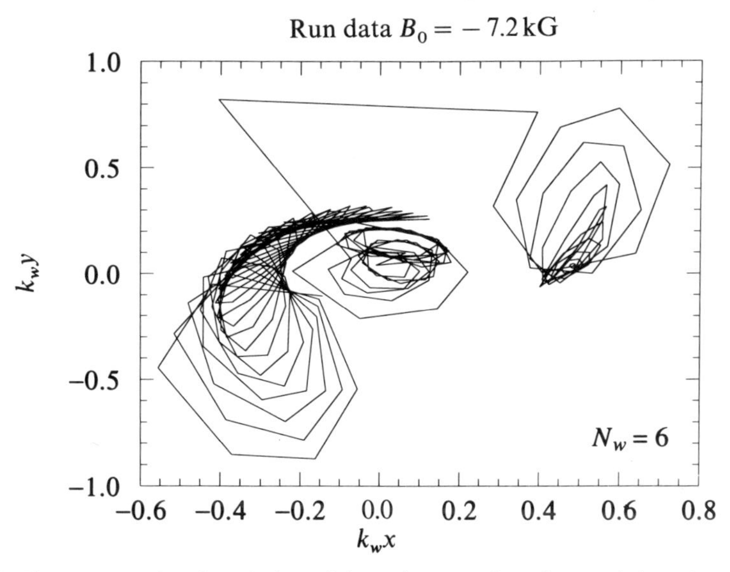

Fig. 13.8 The cross-sectional evolution of the trajectory of an electron injected at the edge of the beam for an axial field in the vicinity of the anti-resonance.

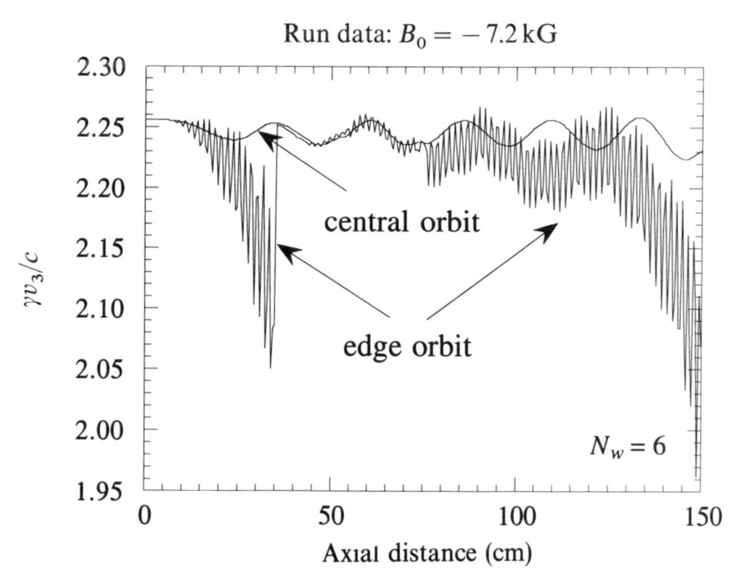

Fig. 13.9 The evolution of the axial momentum versus axial position for electrons injected at the edge of the beam and the centre of the beam in the vicinity of the anti-resonance.

appears to quite irregular motion. The effect of this motion on the axial momentum is shown in Fig. 13.9 in which we plot the axial momentum versus axial position for the central and the edge electrons. It is clear from the figure that the axial momentum exhibits regular oscillations about the bulk value for the central electron, but not for the edge electron. In the latter case, the motion exhibits far more structure reflecting the betatron and Larmor motions as well as sudden transitions. Similar behaviour is found for the transverse components of the momentum as well. These rapid and large variations in the axial momentum are the major cause of the degradation in the interaction efficiency, since they act to disrupt the resonant wave–particle interaction.

Finally, with regard to the issue of beam transport in the reversed-field configuration, we remark that the entire beam was able to propagate through the wiggler for $B_0 = -10.92\,\text{kG}$, which is far from the anti-resonance. However, substantial beam loss was found in the vicinity of the anti-resonance. For example, approximately 60% of the beam was found in the simulation to be lost to the waveguide wall for $B_0 = -7.2\,\text{kG}$. Such massive beam loss is a contributing factor to the low saturation efficiencies found near the anti-resonance.

13.3.2 The Group I and II regimes

The agreement between the simulation and the experimental measurements in the Group I and II regimes is not as good as that found for the reversed-field configuration. In the cases in which the axial guide field is oriented parallel to the wiggler field, we find that a much larger energy spread than that estimated for the experiment is required in order to account for the measured power levels. Indeed, we find that the assumption of an energy spread of 1.5% results in efficiencies comparable to that found for the reversed-field configuration, and we note that the efficiency predicted in simulation does not vary appreciably for either orientation of the magnetostatic fields for comparable beam energy spreads. In order to account for the Group I data we must assume an energy spread of approximately 6.25% to account for the measured power. The Group II data are more difficult to explain. The assumption of a comparable energy spread of 6.4% results in reasonable agreement for the power at the end of the interaction region, but not for the detailed evolution of the signal (i.e. the launching loss and the instantaneous growth rate) during the course of the interaction.

We have no definitive explanation for this discrepancy, but merely suggest that there might be some misalignment or other beam transport problem from the gun to the wiggler which is exacerbated by the orientation of the axial guide field. A possible source for such a discrepancy could be the existence of irregularities in the wiggler field upstream from the entrance due to the sudden termination of the coils. Such irregularities might give rise to orbital instabilities for a parallel alignment of the wiggler and axial guide fields which result in enhanced emittance growth. A detailed evaluation of these suggestions, however, can only be accomplished by means of a thorough analysis of the experimental configuration.

With all of this in mind, we plot the evolution of the beam transmission and power versus axial position as measured in the experiment and as determined with the simulation in Fig. 13.10 for wiggler and axial guide fields of 0.63 kG and 4.06 kG respectively, and for an axial energy spread of 6.25%. These fields correspond to Group I operation, and the initial current at the entrance to the

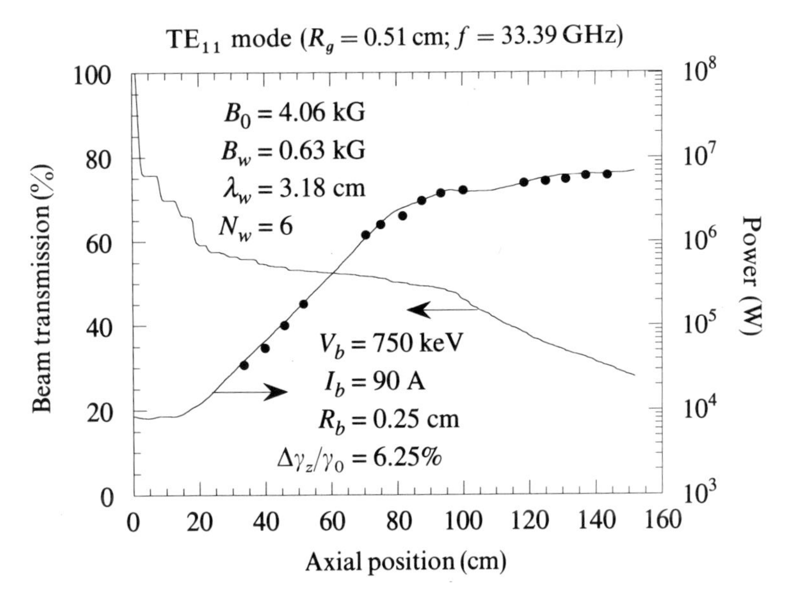

Fig. 13.10 Plot of the evolution of the transmitted beam and power as seen both in simulation and in the experiment (as indicated by the dots) versus axial position for Group I parameters.

wiggler is assumed to be 90 A. As shown in the figure, the simulation is in substantial agreement with the experiment for the presumed energy spread of 6.25% as regards both the linear growth rate and the saturation efficiency.

It is evident from the figure that a substantial fraction of the beam is lost in the entry taper region, but that little beam is lost thereafter until the electro-magnetic power reaches a level of approximately 4–5 MW, after which beam loss occurs at a faster rate. We conclude from this that the saturation mechanism in this case is not phase trapping in the ponderomotive wave, as in the reversed-field example at $B_0 = -10.92$ kG. Rather, saturation occurs in this case through loss of the beam to the waveguide wall.

The Group II data are more difficult to explain than the Group I case. The assumption of a comparable energy spread of 6.4% results in reasonable agreement for the power at the end of the interaction region, but not for the detailed evolution of the signal (i.e. the launching loss and the instantaneous growth rate) during the course of the interaction.

Comparison of the simulation with experiment in the case of wiggler and guide fields of 0.63 kG and 10.92 kG respectively and a transmitted current of 300 A is shown in Fig. 13.11 in which we plot the evolution of the transmitted current and the power both from the simulation and from the experiment (shown by the dots). This case shows rough agreement in the case of a 6.4% energy spread for the power (≈ 4 MW) and saturation efficiency but not the growth rate. In addition, the launching loss observed in the experiment is much higher than that seen in

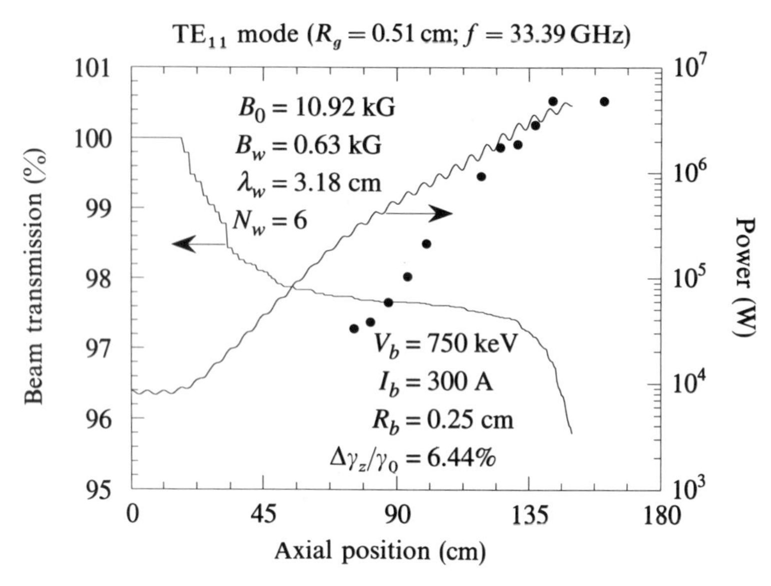

Fig. 13.11 Plot of the evolution of the transmitted beam and power as seen both in simulation and in the experiment (as indicated by the dots) versus axial position for Group II parameters.

simulation, as evidenced by the fact that there is negligible growth in the observed power until after an axial position of 70 cm after the wiggler entrance. Beam loss is not found in simulation to be a major factor for this case until a power level of aproximately 4 MW is reached, after which beam loss occurs at a very rapid rate. Hence, beam loss appears to be the saturation mechanism for this case as well as for the Group I cases.

In general, therefore, we find that particle loss provides a significant, if not dominant, contribution to the saturation mechanism for the Group I and Group II regimes for this experiment.

13.4 CONCLUSIONS

The fundamental conclusion based upon this analysis is that high efficiencies are possible based upon the reversed-field geometry. Indeed, the efficiency found in the experiment described in this chapter compares favourably with the efficiency found in the tapered-wiggler experiment discussed in section 5.3 which was also conducted at 35 GHz, but employed a 3.5 MeV/850 A electron beam with a planar wiggler with an amplitude of 3.72 kG and a period of 98 cm. However, the efficiency found in the reversed-field regime is comparable to those also predicted for the Group I and II regimes in sections 5.3.3 and 5.3.4, and is consistent with the expected efficiency calculated from phase trapping arguments. Hence, the unexplained feature of the experiment is, rather, why the Group I and II geometries exhibited lower than expected efficiencies. At the present time, there is no unequivocal explanation for this. A definitive answer to this question must rest on the detailed beam transport system used in the experiment which, it is speculated, might give rise to unexpectedly high beam energy spreads.

REFERENCES

1. Conde, M. E. and Bekefi, G. (1991) Experimental study of a 33.3 GHz free-electron laser amplifier with a reversed axial guide magnetic field. *Phys. Rev. Lett.*, **67**, 3082.
2. Conde, M. E. and Bekefi, G. (1992) Amplification and superradiant emission from a 33.3 GHz free-electron laser with a reversed axial guide magnetic field. *IEEE Trans. Plasma Sci.*, **20**, 240.
3. Chu, K. R. and Lin, A. T. (1991) Harmonic gyroresonance of electrons in combined helical wiggler and axial guide magnetic fields. *Phys. Rev. Lett.*, **67**, 3235.
4. Freund, H. P. and Ganguly, A. K. (1992) Nonlinear simulation of a high-power collective free-electron laser. *IEEE Trans. Plasma Sci.*, **20**, 245.
5. Bazylev, V. A., Bourdier, A., Gouard, Ph. and Buzzi, J. M. (1993) Electron trajectories in a free-electron laser with a reversed axial guide field. *Phys. Rev. E*, **48**, 3959.
6. Parker, R. K., Jackson, R. H., Gold, S. H., Freund, H. P., Granatstein, V. L., Efthimion, P. C., Herndon, M. and Kinkead, A. K. (1982) Axial magnetic field effects in a collective-interaction free-electron laser at millimeter wavelengths. *Phys. Rev. Lett.*, **48**, 238.
7. Freund, H. P. (1993) Beam transmission in a high-power collective free-electron laser. *Phys. Fluids B*, **5**, 1869.

14

Collective effects in free-electron lasers

Collective effects in free-electron lasers can be classified into two categories. The first category is the collective Raman regime that was previously discussed within the context of the small-signal linearized gain theory in Chapter 4, and in the nonlinear analysis in Chapter 5. The fundamental mechanism for this process is the induced scattering of a negative-energy beam space-charge wave and the wiggler to produce a daughter wave. The second category includes the self-electric and self-magnetic fields induced by the steady-state charge and current distributions of the non-neutral electron beam.

As formulated in the linearized gain theory in Chapter 2, the Raman regime is found when the beam density is sufficiently high that the space-charge potential exceeds the ponderomotive potential formed by the beating of the wiggler and radiation fields. The criterion for this condition is given in inequality (4.98), which requires relatively high beam currents to be satisfied. Comparison of the linearized theory with experiments in the Raman regime was made in section 4.2.3. The nonlinear formulation of the Raman regime necessitated the explicit inclusion of space-charge waves in the formulation, which appeared in sections 5.1 and 5.2 in the idealized one-dimensional analysis, and in section 5.5 in three dimensions. Comparisons of the nonlinear analysis with Raman free-electron laser experiments appeared in section 5.5 as well as in Chapter 13. However, the criteria for the importance of Raman effects can be more subtle than is implied in the criterion in (4.98). In addition to this criterion, it is also necessary to consider (1) the relative magnitudes of the Raman frequency shift and the interaction bandwidth, and (2) the damping of the space-charge waves [1]. The former condition is related to the previously stated Raman criterion (4.98).

The self-electric and self-magnetic fields are manifested in different ways from the Raman interaction. The self-electric field due to the non-neutral charge distribution of the beam results in a variation in the kinetic energy of the electrons across the beam. This so-called **space-charge depression**, therefore, acts to induce a thermal energy spread across the beam which can degrade the interaction. The self-magnetic field can act to induce either paramagnetic or diamagnetic modifications to the magnetostatic fields (i.e. both the wiggler and axial guide fields) which can also modify the gain and saturation efficiencies of the interaction.

This chapter is devoted to a discussion and analysis of both categories of collective effects in free-electron lasers by means of comparison of several intense beam free-electron laser experiments with the theory. Four experiments are used to analyse the collective effects for both helical and planar wiggler geometries. These experiments have been discussed in Chapters 5 and 13. It is interesting to note that of these four experiments, Raman effects are found to be unimportant in the two with the highest currents, but to be essential to the understanding of the interaction in the lower current experiments.

14.1 EXAMINATION OF THE RAMAN REGIME

As we discussed in Chapters 4 and 5, the free-electron laser operates subject to two mechanisms. In the Compton regime, the electron beam interacts with the ponderomotive potential formed by the beating of the wiggler and radiation fields. For high currents, the electrostatic potential due to the beam space-charge waves is dominant over the ponderomotive potential, and the interaction proceeds by stimulated Raman scattering of the negative-energy space-charge wave off the wiggler. Of course, there is also an intermediate regime in which both of these mechanisms are operative. The transition between the Compton and Raman regimes has been discussed in terms of a linear theory of the gain in free-electron lasers in Chapter 4, and the Raman–Compton criteria are stated in (4.95) and (4.98) in the idealized one-dimensional limit. Here, we explore the nature of the Raman interaction in more depth by studying the importance of space-charge effects in a selection of free-electron lasers experiments [2–5].

14.1.1 Raman criteria

For convenience, we restate the condition required for the dominance of the Raman regime as

$$\frac{\omega_b}{\gamma_0^{1/2} c k_w} \gg \frac{\gamma_z^3}{16} \frac{v_w^2}{c^2}, \tag{14.1}$$

where $\omega_b^2 \equiv 4\pi e^2 n_b/m_e$, n_b is the ambient beam density, γ_0 is the relativistic factor corresponding to the bulk beam energy and $\gamma_z^2 \equiv (1 - v_z^2/c^2)^{-1}$ for a bulk axial velocity v_z. In addition, $v_w \equiv -\Omega_w/k_w$ is the transverse wiggler velocity, where $\Omega_w \equiv eB_w/\gamma_0 m_e c$ for a wiggler amplitude B_w, and k_w is the wiggler wavenumber for a period λ_w. For a planar wiggler, the root-mean-square wiggler amplitude must be used in v_w.

This criterion is derived by examining the relative importance of the ponderomotive and space-charge forces on the linear growth rate in the idealized one-dimensional regime, and is a measure of the dominance of the space-charge forces. However, this criterion must be used with some caution in characterizing free-electron laser experiments, and its application to real systems is clouded by several factors. Firstly, the boundary conditions imposed by the drift tube walls reduce the effective plasma frequency. Secondly, the bulk characteristics of the electron orbits are modified by wiggler inhomogeneities, beam thermal effects and the use of an axial guide magnetic field. Planar wiggler configurations introduce further difficulties since, in contrast to a helical wiggler, the axial and transverse electron velocities are oscillatory. Due to these difficulties, a full three-dimensional nonlinear analysis is often required to characterize space-charge effects in any given experiment.

In addition to the above-mentioned criterion for the importance of space-charge effects, a second criterion required for space-charge effects to be important is that the Raman frequency shift be comparable to or greater than the linewidth of the free-electron laser. The physical interpretation of this criterion is that the wiggler must be long enough for several plasma oscillations during the course of the interaction. Of course, realistic three-dimensional effects can be expected to modify this condition as well.

Finally, a third criterion required for space-charge effects to play an important role is that Landau damping of the space-charge waves due to the thermal spread of the beam must be small. In general, Landau damping of space-charge waves is important for wavelengths less than the Debye length.

14.1.2 Experimental summary

In order to elaborate on the importance of space-charge effects in free-electron lasers, we shall consider four experiments. The operational frequencies of these experiments extend from 9 to 500 GHz, and the beam parameters are currents ranging from 4 to 900 A and energies ranging from 150 keV to 3.5 MeV. Three of the experiments [2, 4, 5] employed a helical wiggler and two also used an axial guide field [2, 5]. Of these two, one used a guide field oriented parallel with the wiggler [2], while the other used a reversed-guide field orientation [5]. The remaining experiment used a planar wiggler configuration [3]. Thus, these experiments cover a wide range of parameter space. Note that each of these experiments has already been discussed in Chapters 4, 5 and 13. It is interesting to observe that only two of these experiments were unequivocally in the Raman regime, and that these were the two with the lowest currents. The other two experiments were in the borderline regime. A summary of each of these experiments is given in Table 14.1. The experiments are analysed using the three-dimensional non-linear formulations described in Chapter 5 for both helical and planar wiggler geometries.

14.1.3 Comparison with theory

The first experiment [2] was conducted as an amplifier driven by a travelling wave tube. As shown in Table 14.1, criterion (1) places this experiment in the Raman regime. Further, the wiggler length of approximately 150 cm permitted five to six plasma oscillations over the course of the interaction; hence, the Raman frequency shift is expected to be relatively large as well. Finally, the low energy spread achieved in the experiment ($\Delta\gamma_z/\gamma_0 \approx 0.3\%$) minimized the effect of Landau

Table 14.1 A summary of the operational parameters for the experiments under consideration

	V_b (MeV)	I_b (A)	R_b (cm)	B_w (kG)	λ_w (cm)	B_0 (kG)	P (MW)	η (%)	f (GHz)	f_b (GHz)	$\frac{\omega_b}{\gamma_0^{1/2} c k_w}$	$\frac{\gamma_z^3}{16} \frac{v_w^2}{c^2}$	Raman?
Fajans et al. [2]	0.155	4.1	0.25	0.176[a]	3.3	1.45	0.08	12	9.318	0.72	0.069	0.0012	Yes
Orzechowski et al. [3]	3.6	850	1.0	3.6[b]	9.8	N/A	185	6	34.6	2.2	0.25	0.170	Borderline
Kirkpatrick et al. [4]	2.3	930	0.41	1.275[a]	3.14	N/A	18	0.8	470	5.5	0.25	0.040	Borderline
Conde and Bekefi [5]	0.75	300	0.25	1.47[a]	3.18	−10.92	61	27	33.4	5.2	0.35	0.005	Yes

[a] Helical.
[b] Planar.

damping. Note that the current of 4.1 A in this experiment was the lowest in the group. A comparison of the variation in the power and phase of the output signal with beam energy from the experiment and as determined with the nonlinear simulation is shown in Figs 5.64 and 5.65. The phase measurement is equivalent to a tuning curve which is the most sensitive test of the space-charge effect (as opposed to absolute power measurements). It is evident in the figures that substantial agreement exists between the experiment and the nonlinear simulation; hence, we conclude that the collective interaction is treated correctly in this formulation.

That this experiment is in the Raman regime has also been demonstrated by observation [6]. The variation in the gain as a function of current for this experiment is shown in Fig. 14.1. At currents below approximately 0.5 A, the gain is sufficiently small that thermal effects are important (i.e. $\mathrm{Im}\,k/(\mathrm{Re}\,k + k_w) \leqslant \Delta v_z/v_z \approx 0.5\%$) and the gain falls rapidly with current. However, it is evident that, for currents above this value, the gain scales as the fourth root of the current. As discussed in section 4.1 (Fig. 4.6) this is only possible in the Raman regime. Note, that as in the case of Fig. 4.16, the data are in close agreement with theory, and the line corresponding to the gain for $\Delta\gamma_\parallel/\gamma_\parallel = 0$ was computed using the theory presented in section 4.2.2.

The second experiment is the ELF experiment at Lawrence Livermore National Laboratory [3] which operated as a 35 GHz amplifier driven by a 50 kW magnetron with a planar wiggler and a rectangular waveguide. This experiment has been extensively analysed in section 5.4. As shown in Table 14.1, the experiment is transitional between the Compton and Raman regimes on the basis of the idealized criterion, and it might be expected that space-charge effects play some role. However, the linewidth determined from the nonlinear simulation as shown in Fig. 5.40 is approximately 15 GHz, which is much greater than the plasma frequency (≈ 2.2 GHz). In addition, the axial energy spread ($\Delta\gamma_z/\gamma_0 \leqslant 2\%$) yields a Debye length of approximately 0.8 cm. This is comparable to the space-charge

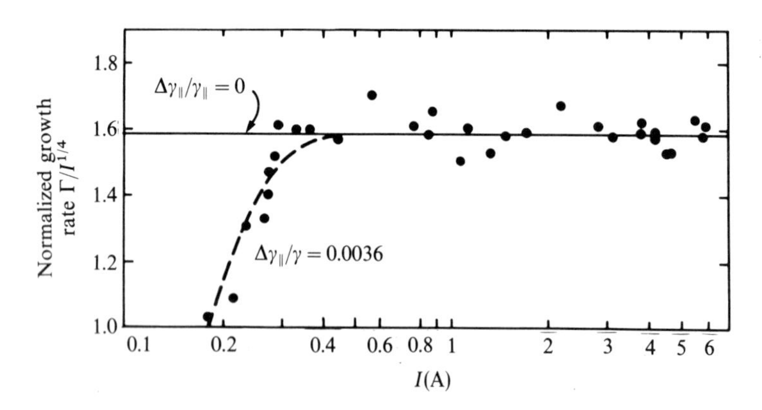

Fig. 14.1 Scaling of the gain versus beam current in the first experiment [2]. (From reference [6].)

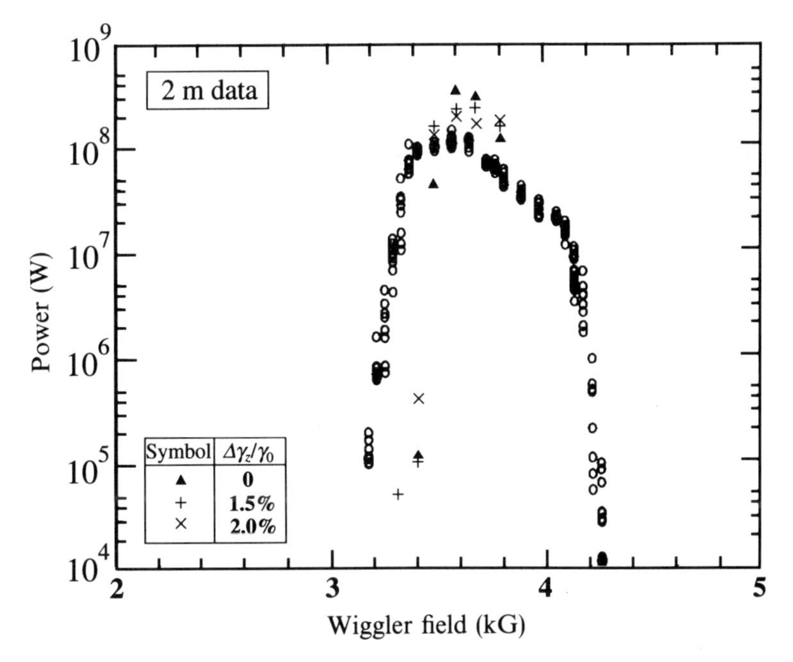

Fig. 14.2 Comparison of the tuning curves for the ELF experiment [3] shown as circles and the simulation over a 2 m interaction length.

wavelength of about 0.8 cm; hence, the space-charge waves are strongly damped [7]. The issue, therefore, is not whether this experiment is in the Raman regime (it is not), but the extent to which space-charge effects were important. This question can be addressed by comparison of the nonlinear theory with a detailed experimental spectrum.

Such a comparison is shown in Fig. 14.2 where we plot output power after 2 m as calculated by the nonlinear simulation (for three choices of the axial energy spread) with an experimentally measured tuning curve. The data here are taken from reference [8]. Note that (1) since the experiment was for a 34.6 GHz amplifier driven by a magnetron, the tuning is accomplished by varying the magnetic field, and (2) saturation was found to occur over a length of 1.4 m. In view of the latter point, a detailed comparison of the spectral width is not valid because sidebands are expected to result in spectral broadening after saturation, and the nonlinear analysis as described in Chapter 5 does not include sidebands in the formulation. Be that as it may, the agreement between the spectral peak predicted by the simulation and that found in the experiment is excellent and does not vary greatly with the choices of axial energy spread. The agreement between the peaks from the simulation and the experiment is to within approximately 30 G, which is well within the experimental uncertainty. As a result, we conclude that space-charge effects did not play a role in this experiment.

The third experiment was for a superradiant amplifier (i.e. the signal grew from noise) employing a helical wiggler without an axial field [4]. This experiment has been analysed in some depth in section 5.3.6. As shown in Table 14.1, the experiment is expected to be in the borderline Raman–Compton regime. However, the linewidth of 50 GHz was much greater than the plasma frequency. The beam energy spread of $\Delta\gamma_z/\gamma_0 \approx 0.25\%$ was consistent with an analysis of the gun geometry and with gains predicted with the simulation. Hence, the Debye length here is approximately 0.09 cm, while the space-charge wavelength is approximately 0.06 cm. As a result, we expect that space-charge waves will be damped, and that Raman effects will be unimportant. This conclusion is supported by the simulation. The output power versus frequency from the simulation and the experiment is shown in Fig. 5.36. Evidently, the simulation accurately reproduces the observed spectrum. It should be noted that the space-charge waves can be disabled in the simulation, and that the predicted spectrum is unaffected by the inclusion of space-charge waves.

The fourth experiment [5] has been extensively discussed in Chapter 13 and operated as an amplifier. As shown in Table 14.1, this experiment is expected to be in the Raman regime. In addition, the wiggler length of approximately 150 cm permitted 28–29 plasma oscillations during the course of the interaction. Finally, the Debye length for this experiment was approximately 0.14 cm while the wavelength of the space-charge waves was approximately 0.80 cm. Hence, Landau damping of the space-charge waves is not important. Hence, we expect this experiment to be in the Raman regime, and this conclusion is supported by the simulations which show good agreement with the experiment. Maximum power of approximately 61 MW was found for a reversed-guide-field orientation. A comparison of the evolution of the power versus axial position as determined in the experiment and computed with the nonlinear simulation is shown in Fig. 13.4. Two curves from the simulation are shown corresponding to the nominal experimental parameters ($I_b = 300$ A, $B_w = 1.47$ kG and $P_{in} = 8.5$ kW) as well as the upper limits due to experimental uncertainties. It is evident from the figure that the simulation is in agreement with the experiment to within the experimental uncertainty. Note that this agreement cannot be obtained if the space-charge modes are disabled.

In conclusion, the idealized Raman criterion (4.98) must be used with caution since three-dimensional effects alter the relationship between the ponderomotive and space-charge potentials. In addition, two other criteria must be considered. Specifically, (1) the Raman shift in the resonance condition must be greater than the free-electron laser linewidth, and (2) Landau damping of the space-charge waves must be small in order for space-charge effects to be important.

14.2 SELF-FIELD EFFECTS IN FREE-ELECTRON LASERS

Self-field effects have been treated in one-dimensional analyses of free-electron lasers both in linear theory [9, 10] and by means of a nonlinear particle-in-cell simulation [10]. Here, we describe a nonlinear model of self-fields in three dimensions [11]. Of course, the most general treatment of the self-fields arises in the context of a fully three-dimensional particle-in-cell simulation of the free-electron laser. In many cases, however, this poses an unacceptably high computational burden. Instead, we have constructed a model of the self-fields which is incorporated into the slow time-scale formulation for both helical and planar wiggler geometries described in Chapter 5. The results from this nonlinear formulation including the self-fields is compared with three intense-beam experiments using a planar wiggler at the Lawrence Livermore National Laboratory [3], and helical wigglers at the Massachusetts Institute of Technology [5] and the Naval Research Laboratory [12, 13].

14.2.1 The self-fields

An electron beam in a physically realizable free-electron laser is born, accelerated and possibly transported for some distance before it enters the wiggler. Various focusing schemes are often employed to transport the beam to the wiggler which can rely upon external magnetic (typically either solenoids or magnetic quadrupoles) or electric fields. In addition, many experiments employ some form of beam scraping (sometimes referred to as emittance selection) to ensure a beam with a small axial velocity spread [3, 14]. Hence, the electron beam, and therefore the self-fields, can exhibit a complex structure at the entrance to the wiggler, and a complete treatment of the initial conditions and self-fields in the electron beam in a free-electron laser would require a full-scale particle-in-cell simulation for each specific configuration including the accelerator and beam transport system as well as the free-electron laser. Instead, we adopt a simpler approach here and develop a model of the self-fields based upon the treatment of an electron beam derived from an idealized model of a beam with uniform, azimuthally symmetric profiles in both the density and velocity. This describes the case of the injection of a uniform parallel-propagating beam.

In such a case, the beam density is given by $n_b(r) = n_b$ for $r \leqslant R_b$ and is zero otherwise (where R_b denotes the beam radius) and the self-electric field, $\boldsymbol{E}^{(s)}$, is determined by Poisson's equation

$$\frac{1}{r}\frac{\partial}{\partial r}(rE_r^{(s)}) = -4\pi e n_b(r), \tag{14.2}$$

which has the solution

$$\boldsymbol{E}^{(s)} = -\frac{m_e}{2e}\omega_b^2 r\hat{\boldsymbol{e}}_r, \tag{14.3}$$

where e and m_e are the electronic charge and mass, and $\omega_b^2 \equiv 4\pi e^2 n_b/m_e$ is the square of the beam-plasma frequency. Energy conservation for this configuration is given by the sum of the kinetic and potential (due to the self-electric field) energies. Within the beam, the Lorentz force equations yield

$$\frac{d}{dt}\left(\gamma - \frac{\omega_b^2}{4c^2}r^2\right) = 0, \tag{14.4}$$

where γ is the relativistic factor. This results in a space-charge depression in the kinetic energy across the beam which may be expressed as

$$\gamma(r) = \gamma_0 + \frac{\omega_b^2}{4c^2}(r^2 - R_b^2). \tag{14.5}$$

In equation (14.5) γ_0 denotes the kinetic energy at the edge of the beam or, alternatively, the total energy.

The lowest-order representation for the self-magnetic field is obtained under the assumption that the beam propagates paraxially with $\mathbf{v} = v_z\hat{\mathbf{e}}_z$ for $r \leqslant R_b$ and is zero otherwise. This assumption requires that the space-charge depression across the beam be small. Observe that the space-charge depression in the kinetic energy at the centre of the beam depends upon $\omega_b^2 R_b^2/4c^2$ and is proportional to the total beam current (I_b) through

$$\frac{\Delta\gamma_{self}}{\gamma_0} \approx 5.88 \times 10^{-5} \frac{I_b(\text{A})}{\sqrt{(\gamma_0^2 - 1)}}. \tag{14.6}$$

For the specific cases under consideration here $\Delta\gamma_{self}/\gamma_0 < 1\%$, and the assumption of a uniform axial velocity provides a good approximation for the self-magnetic field. In this case, the self-magnetic field is determined by Ampère's law,

$$\frac{1}{r}\frac{\partial}{\partial r}(rB_\theta^{(s)}) = -\frac{4\pi e}{c}n_b(r)v_z, \tag{14.7}$$

hence

$$\boldsymbol{B}^{(s)} = -\frac{m_e}{2e}\omega_b^2\beta_z r\hat{\mathbf{e}}_\theta, \tag{14.8}$$

where c is the speed of light *in vacuo* and $\beta_z \equiv v_z/c$.

It is important to observe that several implicit assumptions underlie this approximation for the self-fields in the wiggler. The first is that even while the beam density is assumed to be a uniform flat-top profile with a circular cross-section upon entry to the wiggler, it does not necessarily remain either uniform or circular during the course of the interaction. The second is that the self-magnetic field has been derived under the additional assumption of uniform paraxial motion of the beam. However, the effect of the wiggler is to induce a bulk transverse wiggle motion and a velocity shear due to the wiggler field gradients and these distortions to the beam current can affect the self-magnetic field in the wiggler. In order to estimate the magnitude of this effect, consider the inclusion of the lowest-order wiggler-induced motion in an idealized one-dimensional helical wiggler in the source current for the self-magnetic field.

Using the idealized one-dimensional representation of a helical wiggler given in equation (2.5) and the helical steady-state trajectories described in section 2.1.1, Ampère's law can be writen in the form

$$\left.\begin{aligned} \frac{1}{r}\frac{\partial}{\partial\theta}B_z^{(s)} - \frac{\partial}{\partial z}B_\theta^{(s)} &= -\frac{4\pi e}{c}n_b(r)v_w\cos(k_w z-\theta)\,, \\ \frac{\partial}{\partial z}B_r^{(s)} - \frac{\partial}{\partial r}B_z^{(s)} &= -\frac{4\pi e}{c}n_b(r)v_w\sin(k_w z-\theta)\,, \\ \frac{1}{r}\frac{\partial}{\partial r}(rB_\theta^{(s)}) - \frac{1}{r}\frac{\partial}{\partial\theta}B_r^{(s)} &= -\frac{4\pi e}{c}n_b(r)v_\parallel\,. \end{aligned}\right\} \tag{14.9}$$

Equations (14.9) have solutions of the form

$$\left.\begin{aligned} B_r^{(s)} &= \hat{B}_r(r)\cos(k_w z-\theta)\,, \\ B_\theta^{(s)} &= \hat{B}_\theta(r)\sin(k_w z-\theta)+\bar{B}_\theta(r)\,, \\ B_z^{(s)} &= \hat{B}_z(r)\sin(k_w z-\theta)\,. \end{aligned}\right\} \tag{14.10}$$

Substitution of (14.10) into (14.9) yields

$$k_w r\hat{B}_\theta + \hat{B}_r = \frac{4\pi e n_b}{c} r v_w\,, \tag{14.11}$$

$$k_w\hat{B}_r - \frac{d}{dr}\hat{B}_\theta = \frac{4\pi e n_b}{c} v_w\,, \tag{14.12}$$

$$\frac{d}{dr}(r\bar{B}_\theta) + \left[\frac{d}{dr}(r\hat{B}_\theta) - \hat{B}_r\right]\sin(k_w z-\theta) = -\frac{4\pi e n_b}{c} r v_w\,. \tag{14.13}$$

The last of these equations implies that

$$\frac{d}{dr}(r\bar{B}_\theta) = -\frac{4\pi e n_b}{c} r v_w\,, \tag{14.14}$$

and

$$\frac{d}{dr}(r\hat{B}_\theta) = \hat{B}_r\,, \tag{14.15}$$

which can also be obtained by elimination of $\hat{B}_z$ from equations (14.11) and (14.12). Equation (14.14) can be integrated immediately to give

$$\bar{B}_\theta = \frac{m_e}{2e}\omega_b^2 r\frac{v_w}{c}. \tag{14.16}$$

The requirement that the divergence of this field vanishes yields

$$\frac{1}{r}\frac{d}{dr}(r\hat{B}_r) - \frac{1}{r}\hat{B}_\theta + k_w\hat{B}_z = 0. \tag{14.17}$$

Elimination of $\hat{B}_z$ and $\hat{B}_r$ using equations (14.11) and (14.15) gives

$$\frac{d^2}{dr^2}\hat{B}_\theta + \frac{3}{r}\frac{d}{dr}\hat{B}_\theta - k_w^2\hat{B}_\theta = -\frac{m_e}{e}\frac{\omega_b^2}{c}k_w v_w. \tag{14.18}$$

The simplest solution to equation (14.18) is

$$\hat{B}_\theta = \frac{\omega_b^2}{k_w^2 c^2}\frac{k_w v_\parallel}{\Omega_0 - k_w v_\parallel}B_w, \tag{14.19}$$

which implies that

$$\hat{B}_r = \frac{\omega_b^2}{k_w^2 c^2}\frac{k_w v_\parallel}{\Omega_0 - k_w v_\parallel}B_w, \tag{14.20}$$

and

$$\hat{B}_z = 0. \tag{14.21}$$

Finally, this results in a correction to the self-magnetic field of the form

$$\boldsymbol{B}^{(s)} = -\frac{m_e}{2e}\omega_b^2\beta_\parallel r\hat{\boldsymbol{e}}_\theta + \frac{\omega_b^2}{\gamma_0 k_w^2 c^2}\frac{k_w v_\parallel}{\Omega_0 - k_w v_\parallel}\boldsymbol{B}_w, \tag{14.22}$$

where $\Omega_0 \equiv eB_0/\gamma_0 m_e c$. The second term can describe either diamagnetic ($\Omega_0 < k_w v_\parallel$) or paramagnetic ($\Omega_0 > k_w v_\parallel$) corrections to the wiggler field [9] but is negligible as long as $k_w R_b > 2|v_w/v_\parallel|$, where v_w is the wiggler-induced transverse velocity. Hence, equations (14.3) and (14.8) represent a reasonable approximation for the self-fields in the wiggler as long as the transverse electron displacement due to the wiggler is less than the beam radius.

14.2.2 The nonlinear formulation

The analysis employed is that presented in Chapter 5 which describes intense beams propagating through both rectangular and cylindrical waveguides in the presence of both planar and helical wigglers. Note that the Gould–Trivelpiece modes are included only in the helical wiggler–cylindrical waveguide configuration since, as discussed earlier in the chapter, the collective Raman effects are unimportant for the planar wiggler experiment under consideration [3].

Within the context of this formulation, each mode interacts resonantly with the electrons and is coupled via the Lorentz force equations in the combined static and fluctuating fields which include the self-electric and -magnetic fields of the beam. The description of the self-fields given in equations (14.3) and (14.8) is modified to allow for the motion of the beam centroid in the wiggler and uses an average axial velocity in the self-magnetic field. As a result, the self-electric and magnetic fields are represented as

$$\boldsymbol{E}^{(s)} = -\frac{m_e}{2e}\omega_b^2[(x - \langle x\rangle)\hat{\boldsymbol{e}}_x + (y - \langle y\rangle)\hat{\boldsymbol{e}}_y], \tag{14.23}$$

and

$$\boldsymbol{B}^{(s)} = -\frac{m_e}{2e}\omega_b^2\langle\beta_z\rangle[(y-\langle y\rangle)\hat{\boldsymbol{e}}_x - (x-\langle x\rangle)\hat{\boldsymbol{e}}_y]. \tag{14.24}$$

Given these self-fields, as well as the external fields for the axial guide and wiggler fields, the Lorentz force equations take the form

$$v_z\frac{d}{dz}\boldsymbol{p} = -e(\boldsymbol{E}^{(s)} + \delta\boldsymbol{E}) - \frac{e}{c}\mathbf{v}\times(B_0\hat{\boldsymbol{e}}_z + \boldsymbol{B}_w + \boldsymbol{B}^{(s)} + \delta\boldsymbol{B}), \tag{14.25}$$

where $\delta\boldsymbol{E}$ and $\delta\boldsymbol{B}$ represent the aggregate electric and magnetic fields from each TE, TM and Gould–Trivelpiece mode. The magnetostatic fields employed are given by equation (2.1) for the helical wiggler and equations (7.11–13) for the planar wiggler. Observe that this form for the planar wiggler was used in Chapter 13 to model the effects of wiggler imperfections in the ELF experiment [3].

14.2.3 The numerical analysis

The initial state of the electron beam is chosen to model the injection of a monoenergetic, uniform, axi-symmetric electron beam with a flat-top density profile for $r_0 \leqslant R_b$. The effect of the self-electric field on the initial kinetic energy mirrors the space-charge depression, where γ_0 describes the total energy which is the initial kinetic energy at the edge of the beam. As a result, we scale the initial momentum

$$\boldsymbol{p}_0(r_0) \cong \boldsymbol{p}_0\sqrt{\left(1 + \frac{\Delta\gamma(r_0)[2\gamma_0 + \Delta\gamma(r_0)]}{\gamma_0^2 - 1}\right)}, \tag{14.26}$$

where

$$\Delta\gamma(r_0) \equiv \frac{\omega_b^2}{4c^2}(r_0^2 - R_b^2) \tag{14.27}$$

and $|\boldsymbol{p}_0|^2/m_e^2c^2 = \gamma_0^2 - 1$. Thus, the initial kinetic energy of the particles increases with radius from the beam centre.

It is important to observe here that no attempt is made to match the beam into the wiggler in order to achieve a beam envelope with a relatively constant radius throughout the wiggler. We treat a simpler model in which a paraxially propagating beam is injected into the wiggler, and the subsequent motion is calculated for the assumed electrostatic, magnetostatic (including the self-magnetic) and electromagnetic fields.

The effect of the self-electric and -magnetic fields is studied for parameters consistent with two 35 GHz amplifier experiments corresponding to planar and helical wiggler configurations and with a 16 GHz amplifier experiment using a helical wiggler. The planar wiggler experiment is the ELF experiment conducted at Lawrence Livermore National Laboratory [3]. The first helical wiggler experiment is for the reversed-field free-electron laser located at the Massachusetts

Institute of Technology [5]. These experiments have been analysed earlier in this chapter and need no further description.

The second helical wiggler experiment is located at the Naval Research Laboratory [12, 13] and employs a 250 keV/100 A electron beam with an initial beam radius of 0.4 cm. The wiggler field has a period of 2.54 cm, an entry taper region of five wigger periods in length, an exit taper of three wiggler periods in length and an amplitude variable up to 500 G. The axial guide field can be varied up to a field of 3.2 kG. This is an amplifier experiment at frequencies in the range of 12–18 GHz which employs a waveguide with a radius of 0.815 cm; hence, the primary interaction is with the TE_{11} mode. It should be noted that although the beam parameters were very different in these experiments, the magnitudes of the space-charge depression across the beam in each case are similar and we find that $\Delta\gamma_{self}/\gamma_0 \approx 0.64\%$ in the Lawrence Livermore National Laboratory experiment, 0.78% in the Massachusetts Institute of Technology experiment (at a current of 300 A) and 0.53% in the Naval Research Laboratory experiment.

The planar wiggler configuration

The amplifier experiment at Lawrence Livermore National Laboratory was overmoded and the power was predominantly injected into the TE_{01}, TE_{21} and TM_{21} modes, although the TE_{01} mode was dominant. Recall that the experimental results indicated that saturation occurred at a power level of approximately 180 MW over a length of 1.3 m (including the entry taper). The bulk of the output signal was found to be in the TE_{01} mode, but there was also substantial power in the TE_{21} and TM_{21} modes as well. Recall also that the experimental measurement of the axial energy spread on the beam yields an upper bound on the axial energy spread of approximately 2%.

In the absence of the self-fields, the formulation described in Chapter 5 provided close agreement with the experimental observations for the choice of $\Delta\gamma_z/\gamma_0 = 1.5\%$ which is within the bound set by the experimental measurements. The plot of the growth of the signal versus axial distance for these parameters (Fig. 12.2) shows both the total power and the power in the TE_{01} mode. Agreement between the simulation and the experimental measurement of the saturated power is good. The peak saturated power found in the simulation is approximately 190 MW, which falls to approximately 180 MW when averaged over the lower beat wave. The saturation length is found to be approximately 1.45 m, as compared to 1.3 m in the experiment.

The interaction efficiency is relatively insensitive to the initial axial energy spread for $\Delta\gamma_z/\gamma_0 \leqslant 2\%$ in the absence of the self-fields. However, this is not the case when the self-fields are included in the simulation, since the self-fields and the initial axial energy spread both act to increase the spreading of the beam. Recall that the axial energy spread is due to a pitch-angle spread; hence, increases in the axial energy spread imply increases in the transverse momenta of the beam. As a result, the combined effects of the self-fields and an increasing axial energy

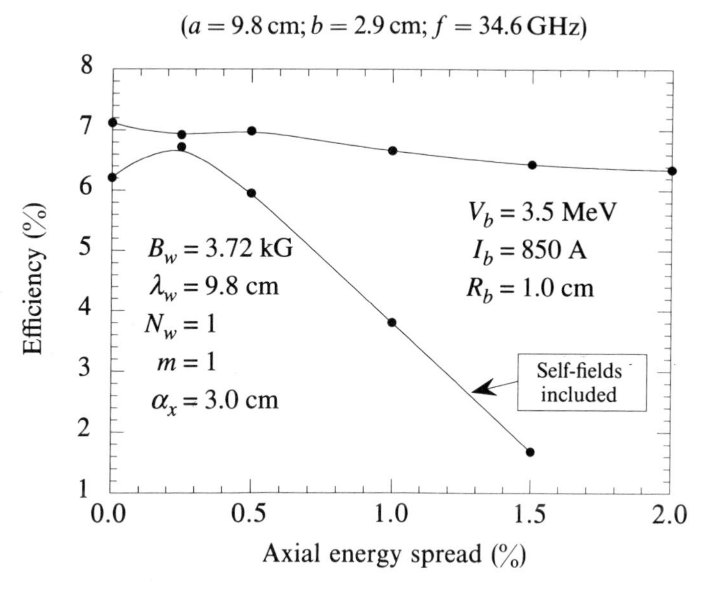

Fig. 14.3 Plot of the efficiency versus axial energy spread with and without the inclusion of self-fields.

spread act to enhance beam loss to the waveguide walls as well as to decrease the coupling between the beam and the waveguide modes.

This is illustrated clearly in Fig. 14.3 in which we plot the variation in the efficiency as a function of the initial axial energy spread both with and without the inclusion of the self-fields. Note that the wiggler field model used in this illustration differs from that used in Chapter 5. Here, we employ the planar wiggler model described in equations (7.11–13). As shown in the figure, the efficiency decreases from about 7.12% to 6.35% as the axial energy spread varies up to 2% without the inclusion of the self-fields. In constrast, when self-fields are included, the efficiency falls off rapidly for $\Delta\gamma_z/\gamma_0 > 0.5\%$. Note that the initial increase in the efficiency with the axial energy spread for $\Delta\gamma_z/\gamma_0 \leqslant 0.25\%$ is due to the shift in the tuning of the interaction with changes in the energy spread. This is due to the fact that the increase in the axial energy spread effectively reduces the average streaming velocity of the beam. This shifts the gain band, and can increase the efficiency at fixed frequencies, although the maximum efficiency across the gain band decreases.

It is clear from Fig. 14.3 that the power measured in the experiment can be recovered from the simulation with the self-fields for $\Delta\gamma_z/\gamma_0 \leqslant 0.5\%$. This is within the experimental uncertainty. The general conclusion to be drawn from this is that the effect of the self-fields on the interaction can be significant, but that in this case they are smaller than the effect of the uncertainties in the initial axial energy spread despite the use of a high-current beam.

The evolution of the power as a function of axial distance subject to the inclusion of the self-fields is shown in Fig. 14.4 for the choice of an initial axial

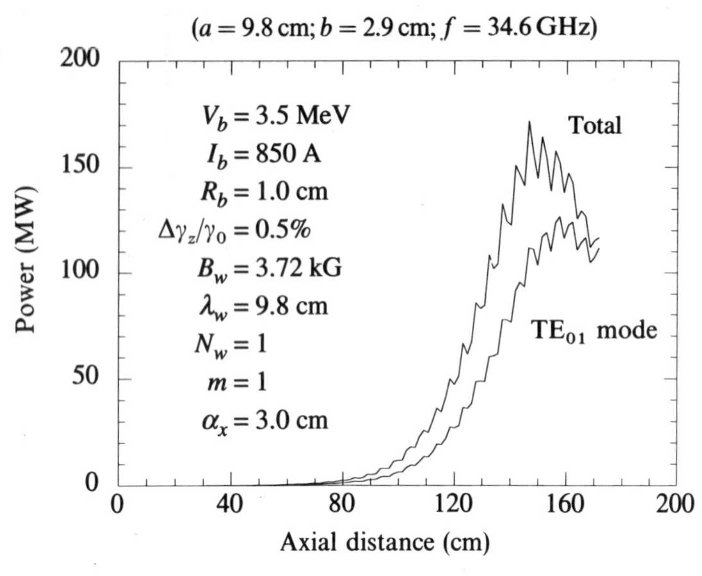

Fig. 14.4 Plot of the evolution of the power with distance subject to the inclusion of the self-fields.

energy spread of 0.5%. This is chosen for illustrative purposes since the saturated power in the TE_{01} mode is relatively unchanged in comparison with the results shown in Fig. 12.2. In this case, a peak saturated power level of approximately 175 MW (falling to $\approx$ 165 MW when averaged over the lower beat wave) was found over a saturation length of about 1.45 m. Hence, the result of the inclusion of the self-fields is a reduction of approximately 8% in the total saturated power, and somewhat less of a reduction in the growth rate. Note that a magnetic quadrupole field was included in the experiment for additional beam focusing but is not included in this wiggler model; hence, the effects of the self-fields seen in simulation may be greater than in the experiment. However, given the experimental uncertainties in the power measurements and the fact that only an upper bound of 2% is known regarding the initial axial energy spread of the beam, the results from simulation either with or without the self-fields are consistent with the experimental measurements.

It is important to observe that the saturated power in the TE_{01} mode both with and without the self-fields is of the order of 125 MW; hence, the reduction in the total saturated power is due largely to a decline in the power in the TE_{21} (and to a lesser extent the TM_{21}) mode. A detailed analysis of the electron dynamics is required in order to explain why the TE_{21} and TM_{21} modes are more sensitive to the self-fields for these parameters. In either case, the initial cross-section of the electron beam is chosen to model the injection of a cylindrical pencil beam.

In the absence of the self-fields, the beam undergoes complex motion which includes the bulk wiggler-induced transverse oscillation, betatron oscillations due to wiggler gradients, and responses to the electromagnetic fields. Figures 14.5–14.8

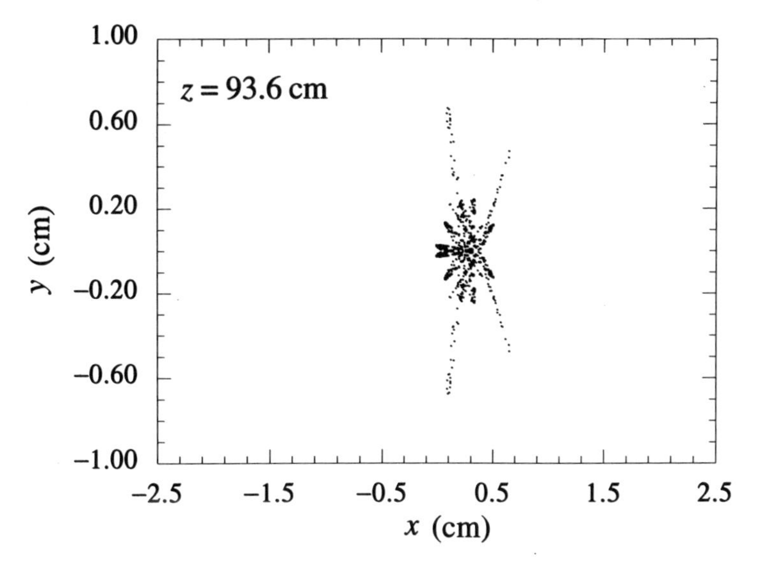

Fig. 14.5 Beam cross-section at $z = 93.6$ cm without self-fields.

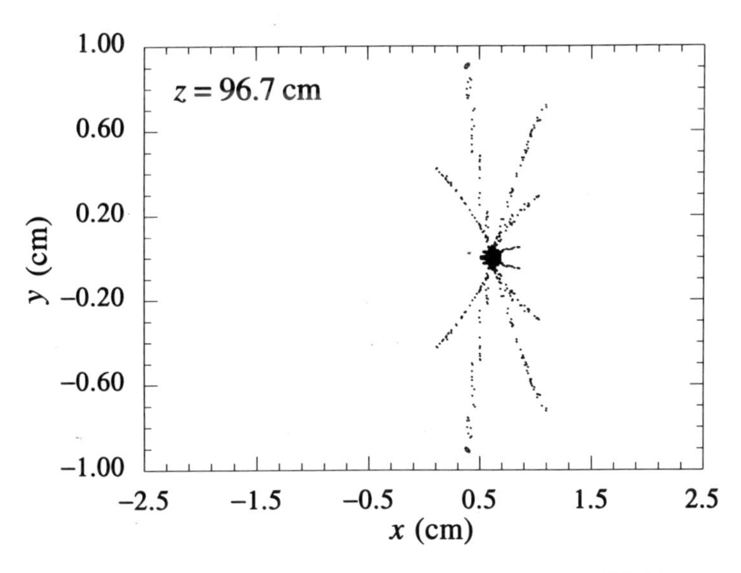

Fig. 14.6 Beam cross-section at $z = 96.7$ cm without self-fields.

show the evolution of the beam cross-section over approximately one wiggler period in the linear stage of the interaction well before saturation, without the inclusion of the self-fields. The inclusion of the self-fields can be expected to alter the beam trajectories to some degree depending upon the magnitude of the fields. One important effect is the spreading of the beam induced by the self-electric field.

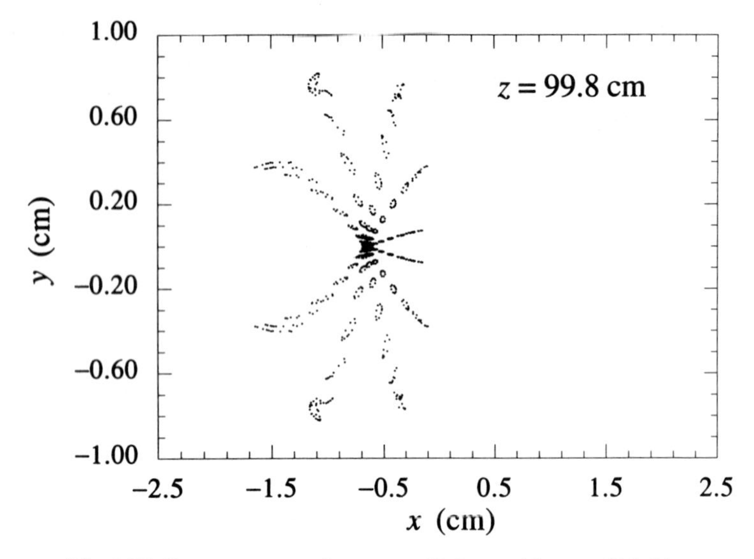

Fig. 14.7 Beam cross-section at $z = 99.8$ cm without self-fields.

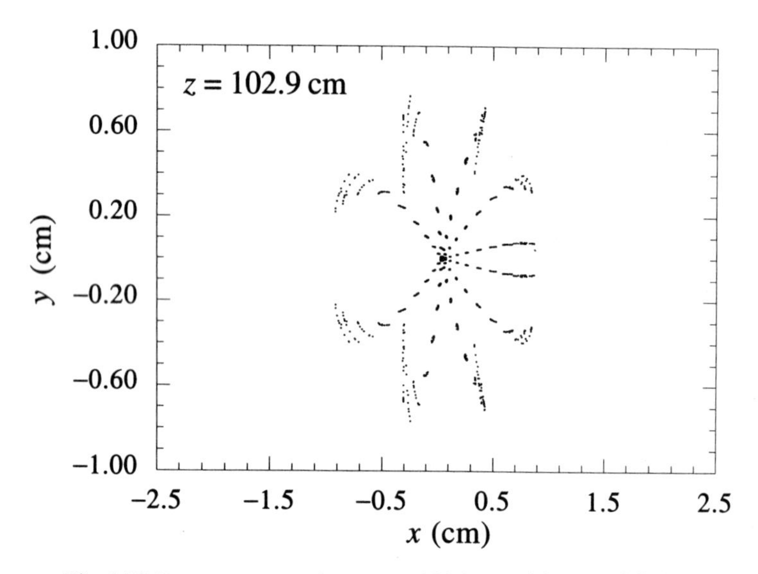

Fig. 14.8 Beam cross-section at $z = 102.9$ cm without self-fields.

Figures 14.9–14.12 show the beam cross-section over the same range with the self-fields included. It is clear that the beam distortion due to the action of the wiggler, radiation and self-fields is complex. However, the principal effect which can alter the interaction with the TE_{21} and TM_{21} modes is the spreading induced in beam cross-section in the x-direction under the action of the self-electric field.

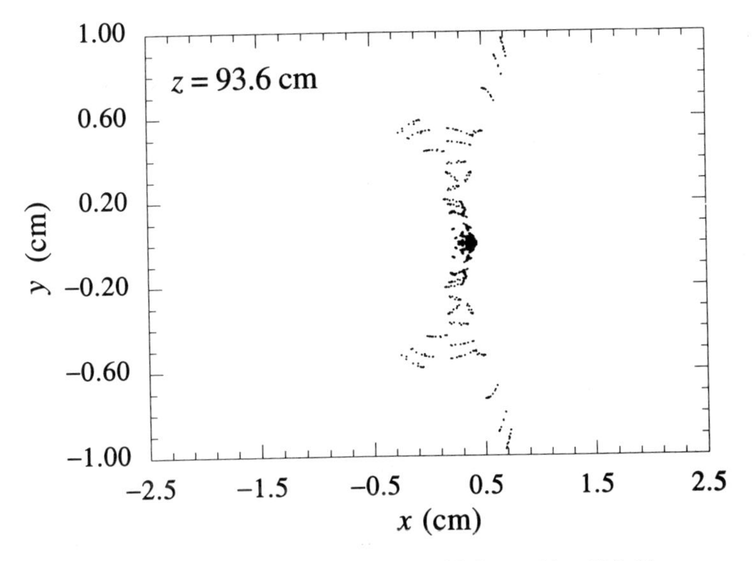

Fig. 14.9 Beam cross-section at $z = 93.6$ cm with self-fields.

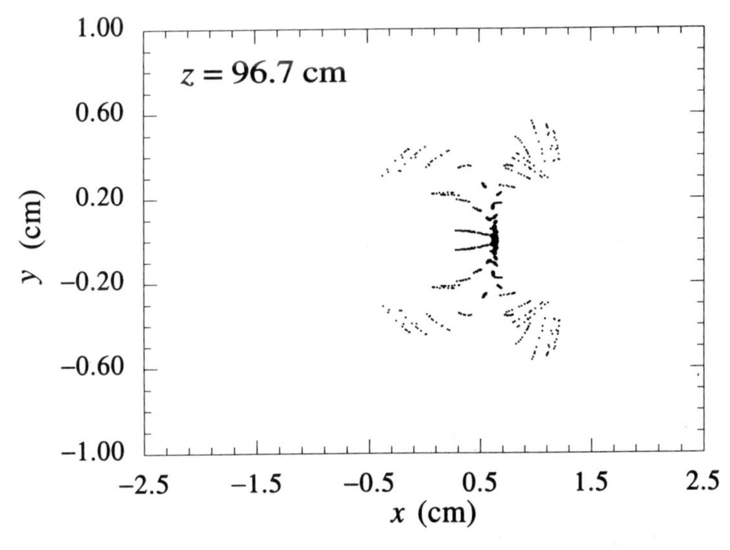

Fig. 14.10 Beam cross-section at $z = 96.7$ cm with self-fields.

The TE_{01} mode will be relatively insensitive to this variation since this field is uniform in the x-direction; hence, relatively little variation in the saturated power in this mode is expected. In contrast, since the TE_{21} and TM_{21} modes vary in x, any beam-spreading in this direction due to the action of the self-fields can be expected to affect the saturated power – in this case to reduce it.

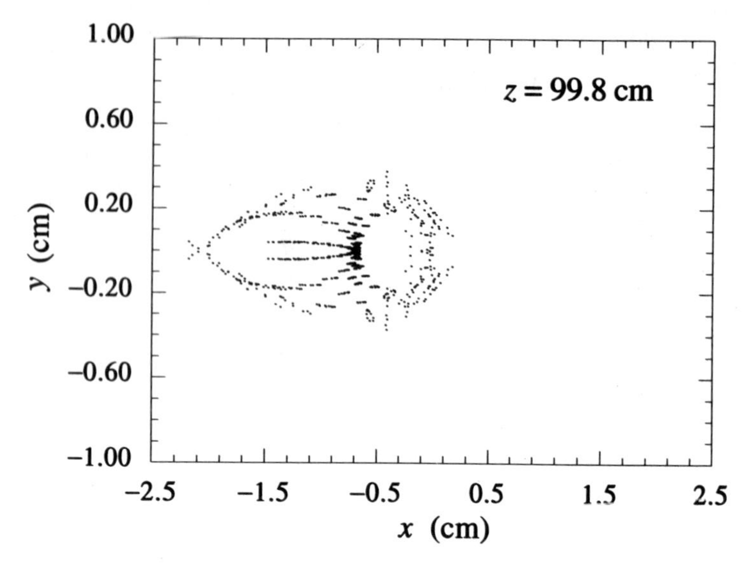

Fig. 14.11 Beam cross-section at $z = 99.8$ cm with self-fields.

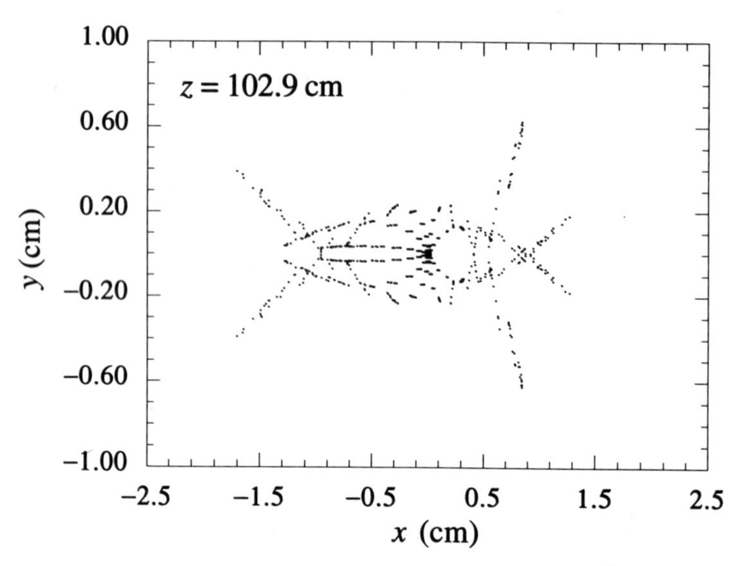

Fig. 14.12 Beam cross-section at $z = 102.9$ cm with self-fields.

The helical wiggler configuration

The Massachusetts Institute of Technology amplifier experiment was discussed in Chapter 13 and we briefly summarize the essential features here. The experiment was driven by a magnetron which produced from 8 to 10 kW at a frequency of 33.39 GHz. The TE_{11} mode was the only wave mode, given the waveguide radius, which could resonantly interact with the beam; however, this experiment operated in the Raman regime and the Gould–Trivelpiece modes must be included. In practice, it was found [14] that only the lowest-order Gould–Trivelpiece mode (for azimuthal mode number $l=0$ and radial mode number $n=1$) was required to obtain reasonable agreement between theory and the experiment in the absence of self-field effects. The experiment operated in three regimes corresponding to an axial magnetic field which was aligned either parallel or anti-parallel (or reversed-field) to the wiggler and the streaming of the electron beam. In the parallel orientation of the magnetic field, two regimes are found which are referred to in the literature as either Group I for which $\Omega_0 < k_w v_\parallel$, or Group II for which $\Omega_0 > k_w v_\parallel$; the experiment was operated for axial fields in both regimes. The third operating regime is the reversed-field case. The maximum operational output power of 61 MW was found to occur in the reversed-field case. Output power of the order of 5 MW was obtained in either the Group I or Group II case. Detailed descriptions of the comparison of the nonlinear theory with this experiment in the absence of self-fields can be found in references [15] and [16].

The first case we consider here is that of a field-reversed configuration in which the nominal experimental magnetic field parameters were an axial field magnitude of 10.92 kG and a wiggler field of 1.47 kG. The transmitted current for these field parameters was 300 A ($\pm$10%), and the axial energy spread of the beam is estimated to be 1.5% as indicated in the experiment. These parameters represent the case of the peak power observed in the experiment of 61 MW.

The comparison of the experiment and theory in the absence of self-fields is shown in Fig. 13.4 in which the power is plotted as a function of axial position, and in which the dots represent the power as measured in the experiment. As shown in the figure, the theory covers two sets of parameters. The first corresponds to the nominal experimental values given above, and the second corresponds to the upper limits on (1) the current, (2) the wiggler field and (3) the input power (due to the experimental uncertainties) of 330 A, 1.55 kG and 10 kW respectively. As is evident in the figure, the agreement between the experimental measurements and theory is good, and virtually all the data points fall between these two curves. The saturated power for these two choices of the current, wiggler field and input power differ only marginally and are close to the 61 MW measured in the experiment. The principal difference is in the saturation length which is due to a small discrepancy in the growth rates for these two cases.

The effect of self-fields on this experiment in the reversed-field regime is relatively small, and the power is plotted as a function of axial position in Fig. 14.13 as computed with the nonlinear theory subject to the inclusion of the self-fields and for the nominal experimental parameters shown in Fig. 14.12. The experimental data are also represented in the figure by the dots. Observe that the self-field effect is marginal and a negligible difference is found between the runs with and without the self-fields for the 300 A parameters.

The negligible effect of the self-fields is interesting since the magnitudes of the space-charge depression here and for the planar wiggler case described above are comparable. The difference between the two cases lies in the orbit dynamics of the

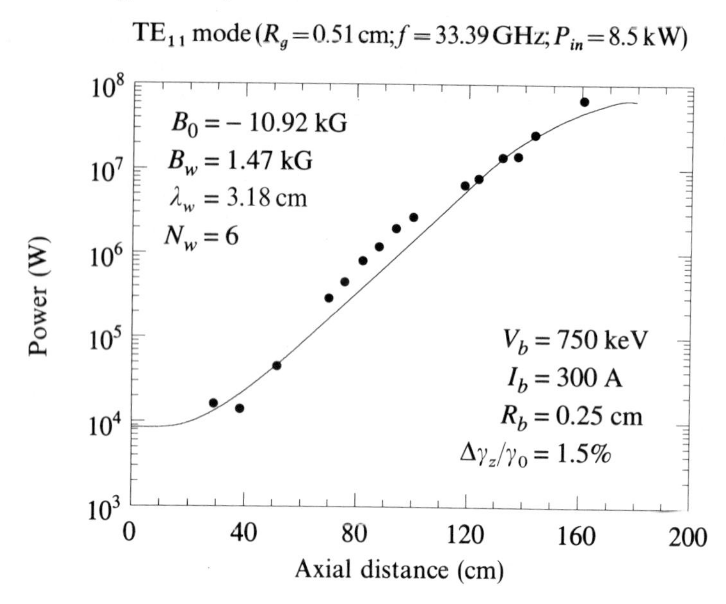

Fig. 14.13 The evolution of the power as determined by theory and from the experiment (dots) for the reversed-field configuration subject to the inclusion of the self-fields.

beam. In the case of the planar wiggler, the wiggler itself provides relatively weak focusing of the beam in the wiggle direction; hence, the beam-spreading due to the self-fields is relatively large. That is not the case for the helical wiggler–axial guide field configuration. Here, the helical wiggler itself provides substantial focusing of the beam which is supplemented by the axial guide field. As a result, the effect of the self-fields on the beam dynamics is relatively unimportant – at least for these parameters. It should also be noted that the beam retains its circular cross-section in these fields; hence, the energy conservation is unaffected by the inclusion of the self-fields and remains good to within approximately 0.1%.

It is also found that self-field effects introduce small modifications to the output power in the Group I and Group II regimes as well. It should be noted here that there is a discrepancy between theory and the experiment in that using the axial energy spread of 1.5% nominally quoted for the experiment results in predicted efficiencies comparable to that found for the reversed-field case. The reason for this discrepancy remains uncertain at this time, but it is speculated that some misalignment exists in the transport system from the cathode to the wiggler that results in the increased axial energy spread for these cases. Be that as it may, we shall employ axial energy spreads of approximately 6.4% in these cases which provide good agreement between the simulation and the measured powers.

The results for the calculated power with and without the self-fields is plotted versus axial position in Fig. 14.14 along with the experimental data (represented by the dots) for an assumed axial energy spread of $\Delta\gamma_z/\gamma_0 = 6.4\%$. It is evident from the figure that the results from theory with and without the self-fields are

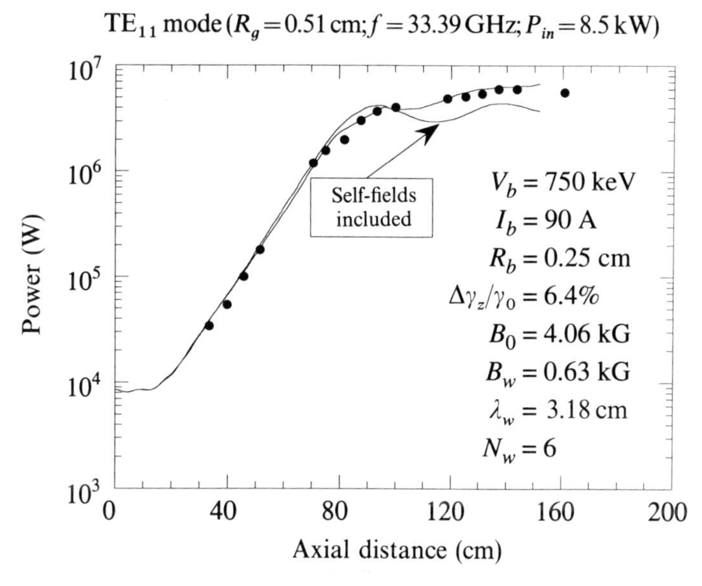

Fig. 14.14 The evolution of the power as determined by theory and from the experiment (dots) for the Group I case with and without the self-fields.

close to the experimental measurements and lie within the experimental uncertainties in the power measurements.

Similar conclusions are found for the Group II case shown in Fig. 14.15, but note that while the simulation is in agreement with the output power measured in the experiment it is not in agreement with the detailed evolution of the signal.

The experiment at the Naval Research Laboratory makes use of a 250 keV modulator capable of producing currents in excess of 100 A, and the electron gun is designed to produce a beam with a radius of 0.4 cm and an axial energy spread of 0.3%. The wiggler field is generated by a bifilar helical coil with a period of 2.54 cm and a total length of 33 wiggler periods including both entrance and exit tapers of five and three wiggler periods respectively in length. The wiggler amplitude can be varied up to approximately 500 G. In addition, the axial guide field can be varied up to 3.2 kG. This ensures operation in the Group I regime. The experiment is configured as an amplifier driven by a coupled cavity TWT which injects up to 5 kW of power over the K_u-band from 12 to 18 GHz. Hence, since the circular waveguide has a radius of 0.815 cm, the interaction is predominantly with the TE_{11} mode. Observe that the space-charge depression for this experiment, $\Delta\gamma_{self}/\gamma_0 \approx 0.53\%$, is comparable to the other two experiments despite the much lower beam current.

In order to simulate this experiment, we consider operation at 16 GHz and use the following model for the wiggler field amplitude in equation (5.80) for the

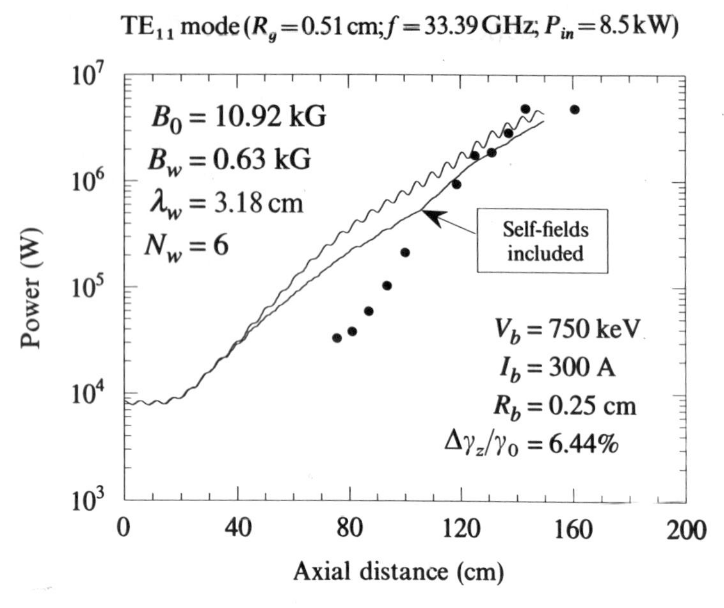

Fig. 14.15 The evolution of the power as determined by theory and from the experiment (dots) for the Group II case with and without the self-fields.

helical wiggler

$$B_w(z) = \begin{cases} B_w \sin^2\left(\dfrac{k_w z}{20}\right); & 0 \leqslant z \leqslant 5\lambda_w, \\ B_w; & 5\lambda_w \leqslant z \leqslant 30\lambda_w, \\ B_w \cos^2\left(\dfrac{k_w(z - 30\lambda_w)}{12}\right); & 30\lambda_w \leqslant z \leqslant 33\lambda_w. \end{cases} \tag{14.28}$$

Preliminary simulations without the inclusion of the self-fields indicated that extremely high efficiencies would be possible. Both the efficiency and beam transmission are plotted as functions of the axial guide field in Fig. 14.16 for the case of a wiggler field amplitude of 300 G. It is clear that the efficiency varies over a wide range from 3% to 33% as the axial guide field increases from 2.2 kG to 2.6 kG. However, the beam transmission falls precipitously with the increase in the efficiency from a value of about 99% at an axial field of 2.2 kG to approximately 5% at an axial field of 2.6 kG. This decline in the beam transmission is due to two factors. The first factor is that the loss of up to 30% of the beam energy to the TE$_{11}$ mode implies that the beam undergoes massive deceleration which is accompanied by an increase in the radius of the wiggler-induced trajectory. The second factor is that the high-power electromagnetic wave acts to kick the beam away from the axis. It should be noted that, as in the case of the Group I and II regimes in the experiment at the Massachusetts Institute of Technology, saturation in this experiment occurs due to beam loss

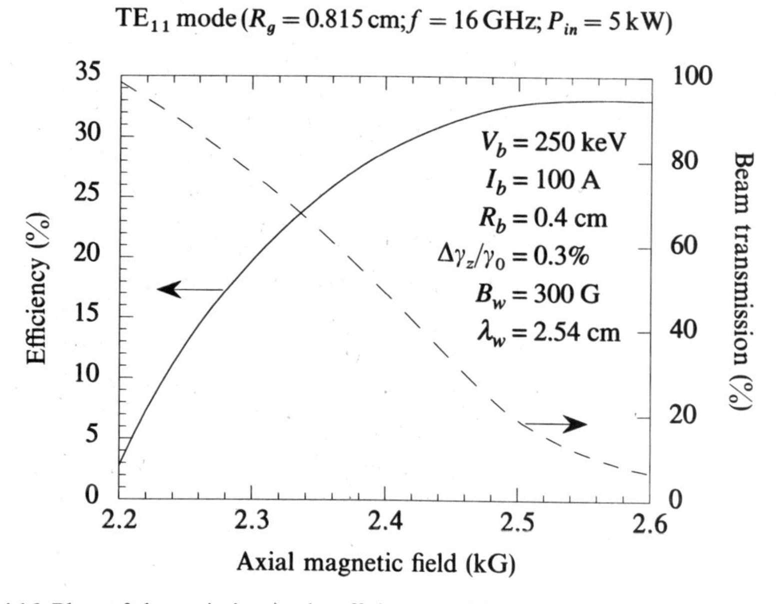

Fig. 14.16 Plots of the variation in the efficiency and beam transmission versus the axial magnetic field in the absence of the self-fields.

rather than the more familiar phase trapping of the beam in the ponderomotive potential formed by the beating of the wiggler and radiation fields. Operation with acceptable levels of beam loss, therefore, was expected to restrict the experiment to efficiencies of approximately 20%.

As might be expected, the effects of the self-fields can act to enhance the beam losses. The effects of the self-fields are more pronounced in this experiment than in the previously analysed experiment at the Massachusetts Institute of Technology (note that the space-charge depressions are comparable for the two experiments) since the beam voltage and axial guide field were lower in the experiment at the Naval Research Laboratory. The efficiency and beam transmission are plotted in Fig. 14.17 as functions of the axial guide field for a wiggler amplitude of 300 G subject to the inclusion of the space-charge fields. It is evident from the figure that for strong axial guide fields in excess of approximately 2.5 kG the efficiency and beam transmission do not differ greatly from those found in the absence of the self-fields. This is because the axial field acts to confine the beam against the spreading induced by the self-fields. In contrast, both the efficiency and beam transmission are substantially less than that found in the absence of the self-fields for weak axial guide fields below about 2.3 kG. In the intermediate regime for axial guide fields in the range of 2.3–2.5 kG, however, the beam transmission is enhanced relative to both the weak and strong guide field cases. This occurs for two reasons. First, the axial guide field is strong enough to provide appreciable confinement of the electron beam. Second, the self-fields are strong enough to

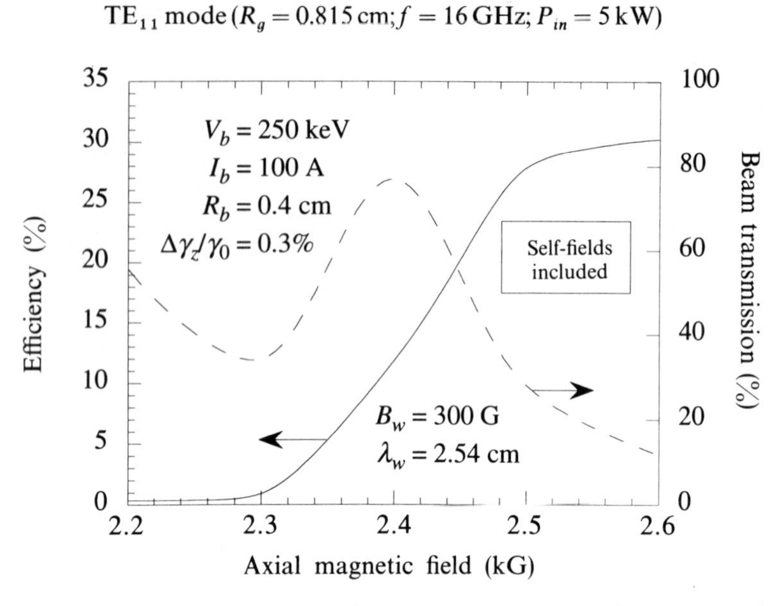

Fig. 14.17 Plots of the variation in the efficiency and beam transmission versus the axial magnetic field in the presence of the self-fields.

cause a reduction in the interaction efficiency; hence, the beam has not lost enough energy and the radiation has not gained enough energy to kick the beam appreciably toward the wall. As a result, we expect that operation with axial guide fields in the neighbourhood of 2.4 kG for a wiggler field of 300 G is preferred, and will result in efficiencies of approximately 10–15%.

With some modification to the original design, this experiment demonstrated high-power and broad-band operation with a maximum output power of 4.2 MW for an efficiency of 18%, a 29 dB gain and a large signal bandwidth (not saturated) greater than 22%. Thus, the experiment met the fundamental design performance goals; specifically, an output power of 1–5 MW, an efficiency greater than 15%, a large-signal gain of 25–30 dB and a large-signal bandwidth greater than 20%. Experimental results are in good agreement with the theoretical predictions and it is important to note that, as predicted, the DC self-fields of the beam played an important role in the interaction.

This is illustrated in Fig. 14.18 in which the variation in beam transmission is plotted versus output power [13]. The solid triangles, circles and diamonds in the figure represent experimental measurements of beam transmission, obtained from a wide range of runs in which the dependence of the output power was studied as a function of wiggler field, guide field and beam voltage. The hollow triangles are from the theory, and the solid line is merely a smooth curve showing the overall variation. It is clear from the figure that good agreement exists between theory and experiment. Note that comparisons of the theoretical predictions without the

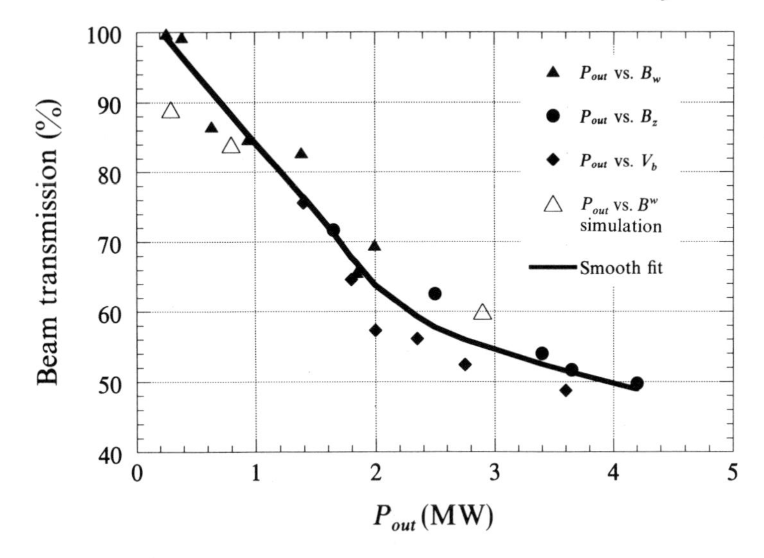

Fig. 14.18 Plot of the variation in beam transmission versus output power from experiment and theory [13].

inclusion of the self-fields exhibit wide discrepancies with the experiment and predict much higher levels of beam transmission.

REFERENCES

1. Freund, H. P. (1993) Space-charge effects in free-electron lasers. *Nucl. Instr. Meth.*, **A331**, 496.
2. Fajans, J., Wurtele, J. S., Bekefi, G., Knowles, D. S. and Xu, K. (1986) Nonlinear power saturation and phase in a collective free-electron laser. *Phys. Rev. Lett.*, **57**, 579.
3. Orzechowski, T. J., Anderson, B. R., Fawley, W. M., Prosnitz, D., Scharlemann, E. T., Yarema, S. M., Sessler, A. M., Hopkins, D. B., Paul, A. C. and Wurtele, J. S. (1986) High gain and high extraction efficiency from a free electron laser amplifier operating in the millimeter wave regime. *Nucl. Instr. Meth.* **A250**, 144.
4. Kirkpatrick, D. A., Bekefi, G., DiRienzo, A. C., Freund, H. P. and Ganguly, A. K. (1989) A millimeter and submillimeter wavelength free-electron laser. *Phys. Fluids B*, **1**, 1511.
5. Conde, M. E. and Bekefi, G. (1991) Experimental study of a 33.3 GHz free-electron laser amplifier with a reversed axial guide magnetic field. *Phys. Rev. Lett.*, **67**, 3082.
6. Fajans, J. and Bekefi, G. (1987) Effect of electron beam temperature on the gain of a collective free-electron laser. *IEEE J. Quantum Electron.*, **QE-23**, 1617.
7. Brau, C. A. (1990) *Free-Electron Lasers*, Academic Press, Boston, p. 63.
8. Orzechowski, T. J., Anderson, B. R., Fawley, W. M., Prosnitz, D., Scharlemann, E. T. and Yarema, S. (1986) High gain and high extraction efficiency from a free-electron laser operating in the millimeter wave regime. *Nucl. Instr. Meth.*, **A250**, 144.
9. Ginzburg, N. S. (1987) Diamagnetic and paramagnetic effects in free-electron lasers. *IEEE Trans. Plasma Sci.*, **PS-15**, 411.
10. Kwan, T. J. T. and Dawson, J. M. (1979) Investigation of the free-electron laser with a guide magnetic field. *Phys. Fluids*, **22**, 1089.

11. Freund, H. P., Jackson, R. H. and Pershing, D. E. (1993) The nonlinear analysis of self-field effects in free-electron lasers. *Phys. Fluids B*, **5**, 2318.
12. Pershing, D. E., Jackson, R. H., Bluem, H. and Freund, H. P. (1991) Improved amplifier performance of the NRL ubitron. *Nucl. Instr. Meth.*, **A304**, 127.
13. Pershing, D. E., Seeley, R. D., Jackson, R. H. and Freund, H. P. (1995) Amplifier performance of the NRL ubitron. *Nucl. Instr. Meth.*, **A358**, 104.
14. Parker, R. K., Jackson, R. H., Gold, S. H., Freund, H. P., Granatstein, V. L., Efthimion, P. C., Herndon, M. and Kinkead, A. K. (1982) Axial magnetic field effects in a collective-interaction free-electron laser at millimeter wavelengths. *Phys. Rev. Lett.*, **48**, 238.
15. Freund, H. P. and Ganguly, A. K. (1992) Nonlinear simulation of a high-power collective free-electron laser. *IEEE Trans. Plasma Sci.*, **20**, 245.
16. Freund, H. P. (1993) Beam transmission in a high-power collective free-electron laser. *Phys. Fluids B*, **5**, 1869.

15

Amplified spontaneous emission and superradiance

Chapters 3 and 4 of this book addressed the issues of incoherent (spontaneous) emission and gain in free-electron laser amplifiers respectively. Incoherent spontaneous emission was treated in Chapter 3 from the single particle point of view. Coherent amplification was treated in Chapter 4 assuming the frequency and amplitude of the signal were given, as would be the case in amplifiers driven by a monochromatic external source. In practice, of course, both emission and amplification occur together. These processes were treated together for the case of oscillators in Chapter 9 where, however, the discussion focused on the case of low-gain oscillators driven by continuous electron beams. In low-gain oscillators, the spontaneously emitted noise circulates through the interaction region many times before it reaches large amplitude. By this time, the radiation can be considered as a superposition of modes of the oscillator structure with amplitudes and phases which evolve in response to the emitted noise. Such a picture was analysed in Section 9.6.3.

A qualitatively different situation exists in a high-gain amplifier in which the gain is sufficient for the spontaneously emitted noise signal to grow to large amplitude before it leaves the interaction region (i.e. in a single pass through the interaction region). The output of such a device is labelled self-amplified spontaneous emission (SASE). The reader should be aware that the terminology for this process and **superradiance** varies within the field. Early work on free-electron lasers made use of intense beam pulse-line accelerators in which the output radiation was derived from the growth of the spontaneously emitted noise in a single pass through the wiggler, and the term **superradiant amplifier** was used for these devices [1]. More recently the term SASE has come into use for this class of device [2, 3], and the term superradiance [4] has been used to describe the spiky nature of radiation which can form at the trailing edge of the electron pulse. However, both uses of superradiance can still be found in the literature.

Generally, the frequency spectrum of SASE is narrower than the incoherent emission based on single particle considerations. This is due to the effect known as gain narrowing. Recall that the frequency width of the emitted radiation for an undulator of length L and period λ_w is given by $\Delta\omega/\omega \approx \lambda_w/L = N_w^{-1}$, where N_w is the number of wiggler periods. When the emitted radiation is amplified in a high-gain system, those frequencies for which the gain is highest are amplified the most. As the gain is exponential with distance, the longer the interaction region, the more narrowly peaked the emitted radiation will be about the frequency of maximum gain. This effect will be calculated for the one-dimensional model of the free-electron laser interaction introduced in previous chapters.

Additional consideration must be given to free-electron lasers driven by pulsed electron beams, such as those produced by r.f. accelerators. In this case the relative slippage between the radiation and electrons discussed for oscillators in Chapter 9 becomes an important effect, and it is necessary to examine the space-time evolution of signals within a pulse [4–8]. An important quantity introduced in this regard is the **cooperation length** [2, 4, 5], the distance at which the beam and radiation slip apart in the time it takes the radiation to propagate one exponentiation length (calculated for a beam of infinite duration). If the beam pulse is of a duration corresponding to a length shorter than the cooperation length, radiation slips out of the pulse modifying the rate of growth. For pulses of a duration corresponding to a length much greater than the cooperation length, the body of the pulse behaves as a continuous beam. However, in the typical case where the radiation travels faster than the beam, the trailing edge of the pulse due to slippage exhibits qualitatively different features from those in the body of the pulse [4–7]. This leads to the interesting effect, mentioned above, known as **superradiance**, whereby a large spike of radiation develops at the trailing edge of the pulse. Again, these effects will be illustrated in this chapter using the one-dimensional model of the free-electron laser interaction introduced previously.

15.1 AMPLIFIED BEAM NOISE

We begin this section by reintroducing the one-dimensional, wiggler-averaged equations describing the space-time evolution of electron trajectories and radiation in a free-electron laser,

$$\left(\frac{\partial}{\partial t}+v_{\|}\frac{\partial}{\partial z}\right)\psi=\omega_{\gamma}\delta\gamma, \tag{9.13}$$

$$\left(\frac{\partial}{\partial t}+v_{\|}\frac{\partial}{\partial z}\right)\delta\gamma=-\frac{\omega}{2\gamma_r}a_w\left[i\delta a(z,t)\exp(i\psi)+\text{c.c.}\right] \tag{9.14}$$

and

$$\left(\frac{\partial}{\partial t}+v_g\frac{\partial}{\partial z}\right)\delta a(z,t)=-\frac{i}{2}\frac{\omega_b^2 a_w}{\omega\gamma_r}\langle\exp(-i\psi)\rangle. \tag{9.25$'$}$$

Here the electron trajectory equations have been transcribed without alteration from Chapter 9, and the radiation evolution equation (9.25′) has been modified by taking eqation (9.25), normalizing the vector potential, $\delta a(z,t)=e\delta A(z,t)/m_ec^2$ and introducing the beam plasma frequency $\omega_b^2\equiv 4\pi e^2 n_b/m_e$, where $J(t_e)=-en_bv_{\|}$ is the beam current density. Recall that the variable t_e represents the time of entrance of a particle which is found at the point z at time t,

$$t_e(z, t) = t - z/v_{\parallel}. \tag{9.22}$$

The characteristics for the particle equations (9.13) and (9.14) are lines of constant entrance time t_e in the (t, z) plane. These are illustrated in Fig. 9.1, and will be important in our subsequent discussion. In terms of the independent pair (z, t_e), equations (9.13), (9.14) and (9.25) can be rewritten

$$v_{\parallel}\frac{d}{dz}\psi(z, t_e) = \omega_{\gamma}\delta\gamma(z, t_e), \tag{9.41}$$

$$v_{\parallel}\frac{d}{dz}\delta\gamma(z, t_e) = -\frac{\omega}{2\gamma_r}a_w[i\delta a(z, t_e)\exp(i\psi) + \text{c.c.}] \tag{9.42}$$

and

$$\left(v_g\frac{\partial}{\partial z} - \varepsilon'\frac{\partial}{\partial t_e}\right)\delta a(z, t_e) = -\frac{i}{2}\frac{\omega_b^2(t_e)a_w}{\omega\gamma_r}\langle\exp(-i\psi)\rangle, \tag{15.1}$$

where

$$\varepsilon' = \left(\frac{v_g}{v_{\parallel}} - 1\right) \tag{15.2}$$

is the rate of slippage similar to the quantity ε defined in (9.44). In particular, over a distance L the radiation slips ahead of the electron beam by a distance $\varepsilon' L$. Note that the coefficient of the time derivative in equation (15.1) is typically negative ($\varepsilon' > 0$); this signifies that the characteristics of the radiation are such that information is carried from large t_e to small t_e values. Physically, this means that radiation entering the interaction region overtakes electrons which entered earlier. This is shown in Fig. 9.1 where the slope of the radiation characteristics is less than that of the electron orbit characteristics.

Our first goal is to determine the spectrum of amplified noise which emerges from the interaction region in the linear regime for the case of a continuous electron beam (i.e. one whose parameters are independent of t_e). Noise is introduced into the system by realizing, as discussed in section 9.6.3, that the variable $Z(z, t_e) = \langle e^{-i\psi}\rangle$ is more properly thought of as representing the average over all electrons entering the interaction region during a time interval T_{ω} which can be taken to be the period of the wave, $T_{\omega} = 2\pi/\omega$. The exact duration of this interval is not important (it could be several multiples of the wave period) so long as the properties of the electron beam and radiation envelope do not vary significantly during this time. We assume these electrons enter at random and uncorrelated times. As the number of electrons is typically large, by the central limit theorem, Z will have a mean $\bar{Z}$ which is calculated by assuming the electrons form a smooth, continuous distribution satisfying the Vlasov equation (9.19), and a small deviation δZ due to the discreteness of the electrons' charge which is a Gaussian random variable. The noise from each electron is uncorrelated with that of other electrons. Thus, we explicitly neglect correlations and assume that the emission comes from a beam of monoenergetic electrons entering the interaction region with a given value of the deviation of the relativistic factor $\delta\gamma_0$. At the end of the calculation we can then sum the noise over the distribution of injected electrons. With these assumptions the noise term can be written

$$\delta Z = \exp\left(-i\omega_\gamma \delta\gamma_0 z/v_\parallel\right)\zeta(t_e), \tag{15.3}$$

where $\zeta(t_e)$ is a Gaussian white noise process satisfying

$$\overline{\zeta(t_e)\zeta(t_e')} = \dot{N}^{-1}\,\delta(t_e - t_e') \tag{15.4}$$

and $\dot{N}$ is the number of electrons entering the interaction region per second. The coefficient of $\zeta(t_e)$ in (15.3) follows from solving (9.41) for the ballistic streaming of electrons, $\psi = \psi_0 + \omega_\gamma\delta\gamma_0 z$, assuming their energies are unperturbed by the radiation. This is appropriate in the linear regime of exponential growth. The random variable ζ then results from consideration of the initial phase, ψ_0, as a random variable.

We now proceed with the calculation by Laplace transforming with respect to the axial coordinate and we introduce the transform variable $i\delta k$. The treatment of the average response of the electrons follows that given by equations (9.56–60), except now the radiation field depends on the entrance time t_e. The result is that the transformed version of (15.1) can be written

$$\left(iD(\delta k, t_e) - \varepsilon'\frac{\partial}{\partial t_e}\right)\delta a(\delta k, t_e) = v_g \delta a(z=0, t_e) - \frac{\omega_b^2(t_e)a_w}{2\omega\gamma_r}\frac{v_\parallel \zeta(t_e)}{\delta k v_\parallel + \omega_\gamma \delta\gamma_0} \tag{15.5}$$

where $D(\delta k, t_e)$ represents the dispersion relation for the complex wave number δk on a beam with parameters evaluated at the entrance time t_e. This quantity is given essentially by equation (9.60),

$$D(\delta k, t_e) = v_g \delta k + \frac{a_w^2\omega_b^2}{4\gamma_r^2}\int d\delta\gamma \frac{\partial f_0/\partial\delta\gamma}{\delta k v_\parallel + \omega_\gamma\delta\gamma}. \tag{15.6}$$

The first term on the right-hand side of (15.5) represents the injected radiation, and the second term represents the spontaneously emitted radiation. For the case of a continuous beam whose parameters do not vary with entrance time, we Fourier transform (15.5) with respect to entrance time, introducing the transform variable $\delta\omega$,

$$\delta a(\delta k, \delta\omega) = -\frac{i}{D(\delta k) + \varepsilon'\delta\omega}\left[v_g\delta a(z=0, \delta\omega) - \frac{\omega_b^2 a_w}{2\omega\gamma_r}\left(\frac{v_\parallel\zeta(\delta\omega)}{\delta k v_\parallel + \omega_\gamma\delta\gamma_0}\right)\right]. \tag{15.7}$$

To determine the level of noise signal at an arbitrary distance from the entrance of the interaction region we must now perform the inverse Laplace transform with respect to the variable δk,

$$\delta a(z, \delta\omega) = \frac{1}{2\pi}\int d\delta k \exp(i\delta k z)\delta a(\delta k, \delta\omega). \tag{15.8}$$

As the noise and injected signal can be considered as independent, we concentrate on the case when only the noise is present. The inverse transform can then be performed easily in two limiting cases. First, if z is small then the contributions to the integration in (15.8) will come from relatively large values of δk. In this case, we can drop the second term in the expression for $D(\delta k)$ which represents the collective response of the beam. This approximation will return the spontaneously emitted noise in the single particle limit, or equivalently, the low-gain limit. The result of the inversion is obtained by picking up contributions at the poles located at $\delta k = -\varepsilon'\delta\omega/v_g$ and $\delta k = -\omega_\gamma\delta\gamma_0/v_\parallel$ respectively, and we obtain that

$$\delta a(z,\delta\omega) = -\frac{z}{2}\frac{\omega_b^2 a_w \zeta(\delta\omega)}{\omega\gamma_r v_g}\left(\frac{\exp(i\theta)-1}{\theta}\right)\exp(-iz\omega_\gamma\delta\gamma_0/v_\parallel), \tag{15.9}$$

where

$$\theta = z\left(\frac{\omega_\gamma\delta\gamma_0}{v_\parallel} - \frac{\varepsilon'\delta\omega}{v_g}\right), \tag{15.10a}$$

or equivalently, using (15.2) for ε',

$$\theta \cong z\left(k + \frac{\delta\omega}{v_g} + k_w - \frac{\omega+\delta\omega}{v_\parallel(\gamma_r+\delta\gamma_0)}\right) \tag{15.10b}$$

is the transit angle ($\theta = 2\Theta$, where Θ is defined following (3.23)) for a particle of energy $\gamma_r + \delta\gamma_0$ and radiation of frequency $\omega + \delta\omega$ and wavenumber $k + \delta\omega/v_g$. Recall that ω and k are some arbitrarily chosen reference frequency and wavenumber, and that the resonant energy γ_r is chosen with respect to this frequency and wavenumber according to $v_\parallel(\gamma_r) = \omega/(k+k_w)$. This part of the calculation is completed by writing the normalized noise power in terms of an integral over the frequency spectrum of the noise,

$$\left|\delta a(z,t_e)\right|^2 = \frac{1}{2\pi\dot{N}}\left(\frac{\omega_b^2 a_w z}{2\omega\gamma_r v_g}\right)^2 \int d\delta\omega d\delta\gamma_0 f_0(\delta\gamma_0)\left(\frac{\sin\Theta}{\Theta}\right)^2, \tag{15.11}$$

where we have made use of the property of the white-noise processes $\zeta(\delta\omega)$,

$$\overline{\zeta(\delta\omega)\zeta^*(\delta\omega')} = \frac{2\pi}{\dot{N}}\delta(\delta\omega - \delta\omega'), \tag{15.12}$$

which follows from (15.4). The actual power per unit area is obtained by restoring the units to the normalized vector potential. This can be compared with the result of (3.29) applied to a beam of length L and cross-sectional area A_{CS}. The two expressions agree within a factor $4c\lambda^2/(v_g A_{CS})$, which accounts for the solid angle into which the power in (15.11) is radiated.

A second limiting case is the one in which z goes to infinity, and the emitted noise is amplified by the beam. In this case the inversion integral (15.8) is dominated by the contribution from the pole at $D(\delta k) + \varepsilon'\delta\omega = 0$. Because of the inherent gain in the system, the pole occurs for complex $\delta k = \delta k_0(\delta\omega) = \delta k_{r0} + i\delta k_{i0}$ and the imaginary part, δk_i, is negative. The resulting expression for the radiation field is then given by

$$\delta a(z, \delta\omega) = -\frac{\omega_b^2 a_w \zeta(\delta\omega)}{2\omega\gamma_r \Omega(\delta\omega, \delta\gamma_0)} \exp\left[i\delta k_0(\delta\omega)z\right] \tag{15.13}$$

where

$$\Omega(\delta\omega, \delta\gamma_0) = \left[(\delta k + \omega_\gamma \delta\gamma_0/v_\parallel)\frac{\partial D(\delta k)}{\partial \delta k}\right]_{\delta k = \delta k_0}. \tag{15.14}$$

The normalized noise power is again expressed as an integral over frequency,

$$\left|\delta a(z, t_e)\right|^2 = \frac{1}{2\pi\dot{N}}\left(\frac{\omega_b^2 a_w}{2\omega\gamma_r}\right)^2 \int d\delta\omega d\delta\gamma_0 f_0(\delta\gamma_0)\frac{\exp\left[2|\delta k_{0i}(\delta\omega)|z\right]}{|\Omega(\delta\omega, \delta\gamma_0)|^2}. \tag{15.15}$$

Due to the exponential factor, $\exp(2|\delta k_{i0}|z)$, the integrand is sharply peaked at the frequency corresponding to maximum gain. The width in frequency of the noise spectrum narrows as z increases, and is given by

$$\Delta\omega = \left(z\left|\frac{\partial^2 \delta k_{0i}}{\partial \delta\omega^2}\right|\right)^{-1/2}. \tag{15.16}$$

This is the effect of gain narrowing discussed in the introduction.

Specific numbers can be given for the case of a cold beam, in which case we can, using (9.63), rewrite $D(\delta k)$

$$D(\delta k) = v_g\left(\delta k + \frac{\kappa_0^3}{\delta k^2}\right), \tag{15.17}$$

where κ_0 is defined according to

$$\kappa_0 = \left(\frac{a_w^2 \omega_b^2}{4\gamma_r^2 v_g v_\parallel^2}\right)^{1/3}, \tag{15.18}$$

and we have taken $\delta\gamma_0 = 0$ without loss of generality. The energy of the beam is taken to be $\gamma_r mc^2$ which defines the reference frequency and wavenumber ω and k. In terms of κ_0, the spatial growth rate δk_{i0} is given approximately by

$$|\delta k_{0i}| \cong \frac{\sqrt{3}}{2}\kappa_0\left[1 - 0.3\left(\frac{\varepsilon'\delta\omega}{v_g \kappa_0}\right)^2\right]. \tag{15.19}$$

Thus, the maximum spatial growth occurs for $\delta\omega = 0$. Using (15.16) the width of the spectrum of amplified noise is written

$$\Delta\omega \cong 1.4\left|\frac{v_g v_\parallel \kappa_0}{v_g - v_\parallel}\right|(z\kappa_0)^{-1/2}. \tag{15.20}$$

As the gain increases, by making the interaction region longer, the spectrum of amplified noise narrows.

15.2 SPACE-TIME EVOLUTION OF DISTURBANCES

The preceding discussion focused on the case of a continuous beam of infinite duration. In this case, while the radiation is continually slipping with respect to the beam electrons, a given time slice of radiation always encounters new electrons which entered at an earlier time, and which have been bunched under the action of other time slices of radiation. This is not the case if the beam pulse is of finite duration. Then electrons at the trailing edge of the beam never encounter radiation, and remain unbunched. Further, a given slice of radiation can leave the electron beam if the interaction length is long enough. (Here we have assumed that the group velocity of the radiation is greater than the beam speed. If the interaction takes place in a waveguide, the opposite may be true, and the leading and trailing edges of the beam pulse then reverse their significance [8,9]). Therefore, it is appropriate to investigate the space and time dependence of growing perturbations which may encounter the temporal boundaries of the beam [4–7]. The problem is analysed qualitatively in Fig. 15.1 where we have drawn characteristics in the (t, z) plane.

A finite duration, noiseless beam pulse enters the interaction region between the times t_{rise} and t_{fall}. A small radiation signal enters at time t_0. The beam

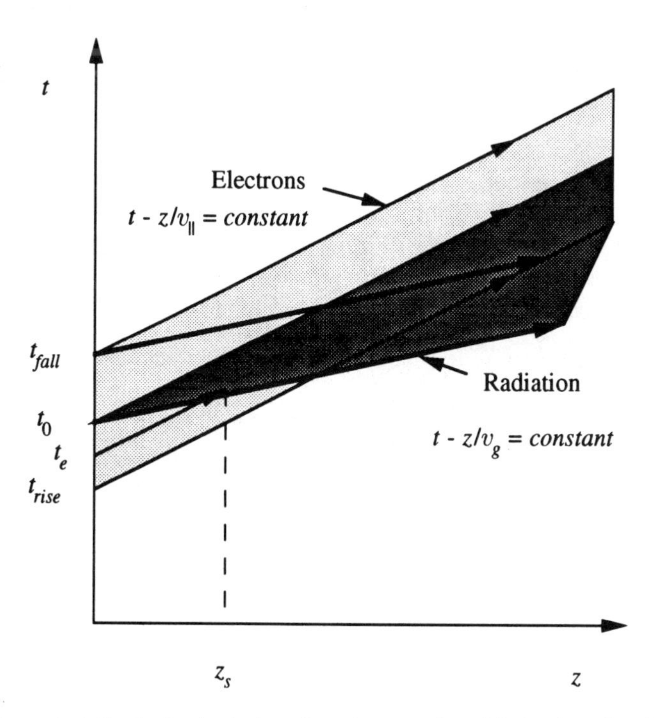

Fig. 15.1 Characteristics in the (t, z) plane for a beam pulse of finite duration. The front of the pulse enters the interaction at t_{rise} and the trailing edge enters at t_{fall}. A small disturbance is initiated at time t_0. The significance of the quantity z_s is discussed in the text.

electrons travel along characteristics $t - z/v_\parallel = t_e = \text{constant}$, while the radiation travels on characteristics $t - z/v_g = \text{constant}$. The location of the beam pulse in the (t, z) plane is indicated by the lightly shaded region of Fig. 15.1. As the radiation and beam pulse propagate, the radiation slips ahead of the electrons entering at t_0, and the disturbance spreads through the beam. The disturbed region is indicated by the darkly shaded region in the figure. Notice that at some distance down the interaction region radiation spills out ahead of the electron pulse, at which time it is no longer amplified by the beam. If the radiation has a characteristic exponentiation rate, κ_0 (defined in (15.18) for a cold beam) the disturbance will have spread in time by an amount $t_c = \varepsilon'/\kappa_0 v_g$ as the disturbance grows by a factor of e. This time is known as the cooperation time, and corresponds to a length

$$L_{co} = v_g t_c = \varepsilon'/\kappa_0, \tag{15.21}$$

known as the **cooperation length** [2, 4, 5]. As a disturbance grows this will be its characteristic extent.

To analyse this situation we return to equation (15.5) and assume that the noise term is small and that the initial radiation signal has the form of a delta function peaked at the entrance time $t_e = t_0$, $\delta a(z = 0, t_e) = \delta(t_e - t_0)$. The solution for $\delta a(\delta k, t_e)$ may then be obtained by introduction of an integrating factor in equation (15.5),

$$\delta a(\delta k, t_e) = \frac{v_g}{\varepsilon'} H(t_0 - t_e) \exp\left[i\varepsilon'^{-1} \int_{t_0}^{t_e} dt'_e D(\delta k, t'_e) \right] \tag{15.22}$$

where $H(t_0 - t_e)$ is the Heaviside function, and we have assumed the radiation speed is greater than the beam speed, $\varepsilon' > 0$. In this case the disturbance is restricted to values of t_e smaller than t_0. This is, it moves forward through the beam affecting electrons entering at times earlier than t_0. To find the evolution of the disturbance with distance down the interaction region we perform the Laplace inversion described by equation (15.8),

$$\delta a(z, t_e) = \frac{v_g}{2\pi\varepsilon'} H(t_0 - t_e) \int dk \exp\left[iF(\delta k, z, t_e)\right], \tag{15.23}$$

where the exponent F is given by

$$F(\delta k, z, t_e) = \delta kz + \frac{1}{\varepsilon'} \int_{t_0}^{t_e} dt'_e D(\delta k, t'_e). \tag{15.24}$$

We specialize now to the case of a cold beam with parameters that are constant within the beam pulse. In this case the exponent F is written

$$F(\delta k, z, t_e) = \delta kz - z_s \left(\delta k + \frac{\kappa_0^3}{\delta k^2} \right), \tag{15.25}$$

where the variable z_s is defined in terms of t_e and t_0,

$$z_s = \frac{v_g}{\varepsilon'}(t_0 - t_e). \tag{15.26}$$

The physical significance of this variable is illustrated in Fig. 15.1. It is the distance the radiation entering at time t_0 must propagate in order to catch up with the electrons which entered at the earlier time t_e. The Laplace inversion may now be carried out using the method of steepest descent. The stationary point of the exponent, δk_s, is determined from $\partial F\ (\partial k)/\partial\delta k = 0$,

$$\delta k_s = \kappa_0\left(\frac{2z_s}{z - z_s}\right)^{1/3} \exp(-i\pi/3). \tag{15.27}$$

Expanding the exponent about this point and evaluating the integral gives [6]

$$\delta a(z,t_e) = \frac{v_g}{\varepsilon'}H(t_0 - t_e)\left(\frac{1}{2\pi i F''}\right)^{1/2} \exp\left[\frac{3i\exp(-i\pi/3)}{2^{2/3}}\kappa_0 z_s^{1/3}\,(z - z_s)^{2/3}\right], \tag{15.28}$$

where the second derivative of the exponent evaluated at the steepest descent point is written

$$F'' = \exp\,(i\pi/3)\frac{3(z - z_s)^{4/3}}{\kappa_0 z_s^{1/3}}. \tag{15.29}$$

The dominant behaviour of the impulse response is determined by the real part of the exponent in (15.28). For a fixed distance down the interaction length, z, the pulse is localized in entrance time corresponding to $z > z_s > 0$. This corresponds, according to (15.26), to the darkly shaded region of Fig. 15.1. A plot of the real part of the exponent versus $\kappa_0 z_s$ for several values of $\kappa_0 z$ is shown in Fig. 15.2.

The radiation pulse moves forward through the beam pulse as it grows. Here positive values of z_s correspond to electron entrance times prior to the disturbance time t_0. The peak of the pulse moves at a velocity intermediate to that of the radiation and electron beam. This velocity can be found by finding the maximum of the exponent as a function of z, $z_{s,peak} = z/3$. The resulting velocity of the pulse is then found by determining z as a function of t [6],

$$v_{pulse} = \frac{3v_g v_\parallel}{2v_g + v_\parallel}. \tag{15.30}$$

As can be seen, the disturbance moves at a speed below the group velocity (assuming it is greater than the beam speed). This effect is essentially the **laser lethargy** discussed in Chapter 9 for oscillators driven by pulsed electron beams. The rate of growth of the disturbance at the moving location of the peak of the pulse is exponential with distance z and is obtained by inserting $z_{s,peak} = z/3$ in (15.28). The result is that the maximum rate of growth is the same as in the continuous beam case (15.18). The width of the disturbance is determined by the second derivative of the exponent of (15.28) with respect to z_s. One finds the width

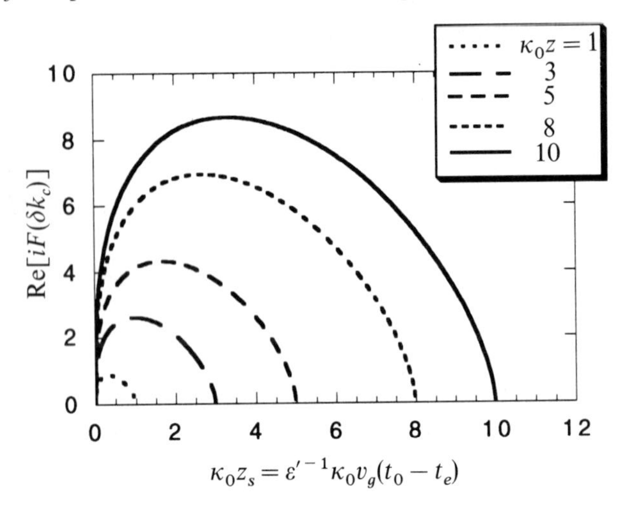

Fig. 15.2 Exponentiation from (15.28) versus entrance time for a disturbance initiated at t_0. Growth is shown for five different distances down the interaction length. The maximum amount of growth occurs for $z_s = z/3$.

measured in entrance times to be

$$\delta t_e = \frac{2\varepsilon'}{3v_g}\left(\frac{z}{\kappa_0}\right)^{1/2}. \tag{15.31}$$

Thus the root-mean-square width of the disturbance grows more slowly than the spatial extent (the darkly shaded region of Fig. 15.1).

Equation (15.28) has important implications for perturbations growing in the trailing edge of a beam pulse. In particular, the growth of a perturbation initiated at $t_0 = t_{fall}$, where t_{fall} is the entrance time of the trailing edge of the pulse, and observed at the location of electrons entering some fixed time earlier corresponds to (15.28) with z_s fixed. In this case the disturbance grows at a rate that is exponential, but proportional to $z^{2/3}$, instead of z. This lower rate of growth is due to the fact that radiation slips ahead of electrons in the trailing edge of the pulse [5], and must be continually regenerated by the bunched electrons.

The preceding discussion can easily by generalized to beams whose parameters vary smoothly with entrance time. The exponent F is given in the general case in (15.24). If we consider a cold beam whose density varies with entrance time then equation (15.25) will be obtained with the replacement

$$z_s\kappa_0^3 \to \frac{v_g}{\varepsilon'}\left|\int_{t_0}^{t_e} dt_e'\,\kappa_0^3(t_e')\right|, \tag{15.32}$$

where $\kappa_0(t')$ is defined in (15.18) with the time-varying beam density.

15.3 SATURATION OF SASE

In the nonlinear regime the saturated radiation signal depends on both the length of the electron bunch, measured in cooperation lengths, as well as the seed signal which grows. A number of studies of the nonlinear problem based on numerical solutions of equations (9.41), (9.42) and (15.1) have been conducted [5–11]. In cases in which the electron phases are uniformly distributed on an interval of 2π at $z = 0$, the spontaneous noise is negligible. Then, if a small monochromatic radiation signal is also injected, this signal grows to saturation in the body of the electron pulse much like the continuous beam studies discussed in Chapter 5. A different behaviour is observed near the trailing edge of the pulse where the signal grows more slowly at first, but ultimately forms a spike which continues to grow as it propagates forward through the beam pulse. The spike grows, as it is able to extract energy from the spent beam which has saturated the growth of the

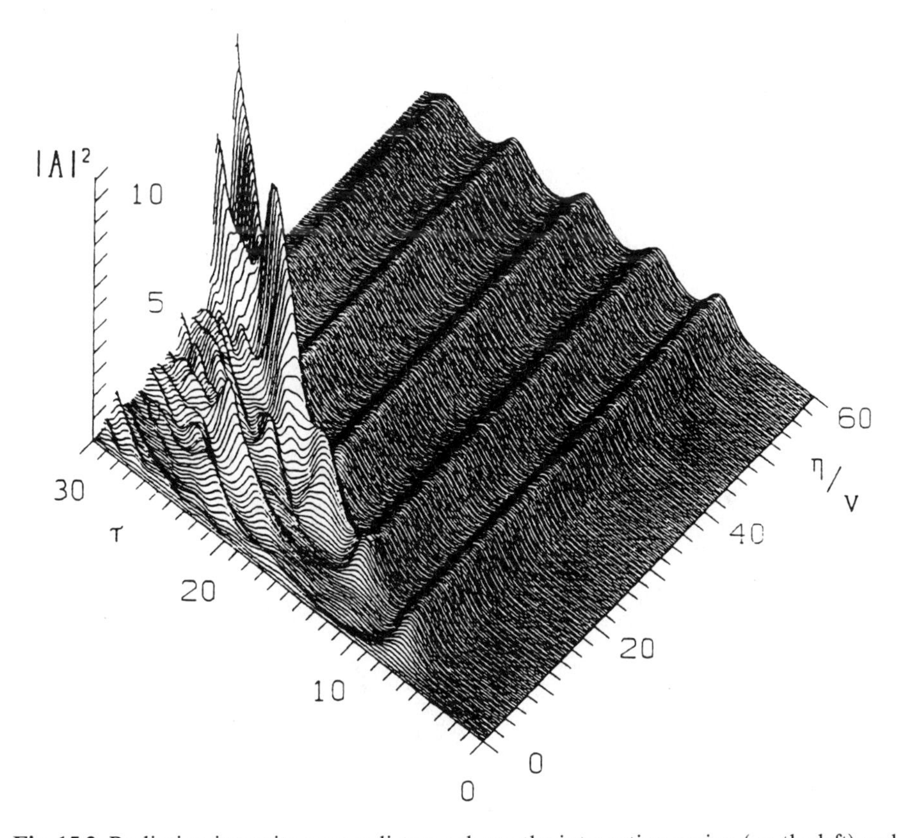

Fig. 15.3 Radiation intensity versus distance down the interaction region (on the left) and entrance time (on the right). Here τ represents $\kappa_0 z$ and η/v represents $\kappa_0 z_s$ with z_s given by (15.26) with t_0 being the trailing edge of the pulse. In this simulation particles are injected uniformly in time so there is no noise. Radiation grows from a monochromatic injected signal. (From reference [10].)

originally injected signal. An example of a simulation illustrating this effect is shown in Fig. 15.3.

If instead of injecting electrons uniformly in phase, one gives the electrons random angles chosen to reproduce the spontaneous noise spectrum, and makes the injected radiation signal negligible, the overall behaviour is qualitatively different [10]. In this case the radiation saturates as in the single-frequency case, but then, due to the presence of spontaneous noise, the sideband instability is excited and the radiation forms spikes throughout the body of the beam pulse. An example of a simulation illustrating this effect is shown in Fig. 15.4.

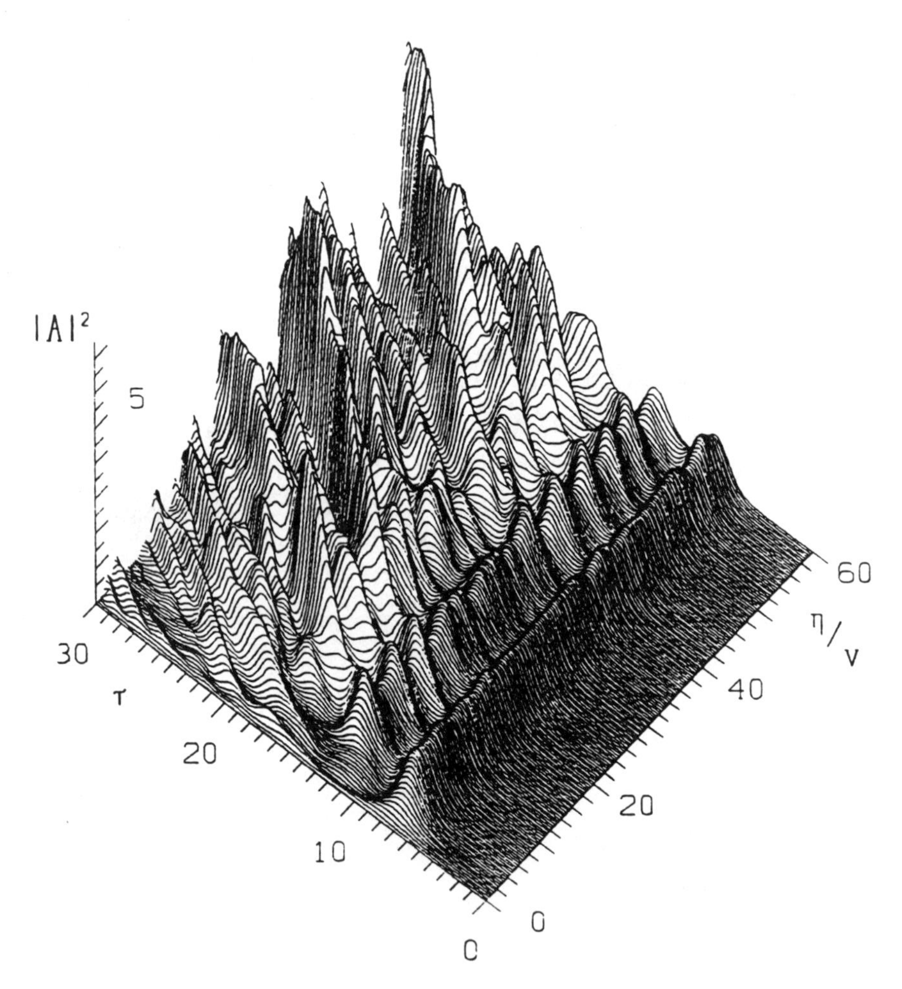

Fig. 15.4 Radiation intensity versus distance down the interaction region (on the left) and entrance time (on the right). Here τ represents $\kappa_0 z$ and η/v represents $\kappa_0 z_s$ with z_s given by (15.26) with t_0 being the trailing edge of the pulse. In this simulation particles are injected nonuniformly in time, generating noise. (From reference [10].)

The difference between these two simulations can be interpreted in the following way [10]. The initial saturation by phase trapping of the radiation only extracts a small part of the available beam energy. Remaining beam energy can still be extracted, but only by a lower frequency wave. This is the cause of the sideband instability. In noiseless simulations, with a monochromatic injected radiation signal, the lower frequency disturbance which will extract additional beam energy is created in the trailing edge of the beam pulse. It must then propagate forward through the pulse to reach electrons in the body of the pulse. This is illustrated in Fig. 15.3. In noisy simulations, the presence of the noise spectrum provides the lower frequency signal which grows to be the sideband instability. In these simulations spikes appear throughout the electron pulse 'spontaneously'. This is illustrated in Fig. 15.4.

It thus seems that, with a sufficiently quite beam, and the ability to inject a monochromatic radiation signal with the beam, one could observe the behaviour of Fig. 15.3. On other hand, if the only signal to be amplified is the spontaneous emission of the electron beam, then the results of Fig. 15.4 are more appropriate.

REFERENCES

1. Parker, R. K., Jackson, R. H., Gold, S. H., Freund, H. P., Granatstein, V. L., Efthimion, P. C., Herndon, M. and Kinkead, A. K. (1982) Axial magnetic field effects in a collective free-electron laser at millimeter wavelengths. *Phys. Rev. Lett.*, **48**, 238.
2. Bonifacio, R., Pellegrini, C. and Narducci, L. (1984) Collective instabilities and high-gain regime in a free electron laser. *Opt. Commun.*, **50**, 373.
3. Kim, K.-J. (1986) Three dimensional analysis of coherent amplification and self-amplified spontaneous emission in free-electron lasers. *Phys. Rev. Lett.*, **57**, 1871.
4. Bonifacio, R. and Casagrande, F. (1985) The superradiant regime of a free electron laser. *Nucl. Instrum. Meth.*, **A239**, 36.
5. Bonifacio, R. and McNeil, B. W. J. (1988) Slippage and superradiance in the high-gain fel. *Nucl. Instrum. Meth.*, **A272**, 280.
6. Bonifacio, R., De Salvo, L., Pierini, P. and Piovella, N. (1990) The superradiant regime of a fel: analytical and numerical results. *Nucl. Instrum. Meth.*, **A296**, 358.
7. Sharp, W. M., Fawley, W. M., Yu, S. S., Sessler, A. M., Bonifacio, R. and De Salvo Souza, L. (1989) Simulation of superradiant free-electron lasers. *Nucl. Instrum. Meth.*, **A285**, 217.
8. Sharp, W. M., Yu, S. S., Pierini, P. and Cerchioni, G. (1990) Waveguide effects in superradiant free-electron lasers. *Nucl. Instrum. Meth.*, **A296**, 535.
9. Sternbach, E. (1991) Computer simulation of superradiant free electron lasers in highly dispersive waveguides. *Nucl. Instrum. Meth.*, **A304**, 663.
10. Graham, R. and Isermann, S. (1991) Nonlinear wave propagation in free-electron lasers. *Phys. Rev. A*, **43**, 3982.
11. Bonifacio, R., De Salvo, L., Pierini, P., Piovella, N. and Pellegrini, C. (1994) A study of linewidth, noise and fluctuations in a FEL operating in SASE. *Nucl. Instrum. Meth.*, **A341**, 181.

Subject index

Author index